BIOSEPARATIONS ENGINEERING

BIOSEPARATIONS ENGINEERING

Principles, Practice, and Economics

MICHAEL R. LADISCH, Ph.D.
Purdue University

A JOHN WILEY & SONS, INC., PUBLICATION

New York • Chichester • Weinheim • Brisbane • Singapore • Toronto

This book is printed on acid-free paper. ♾

Published simultaneously in Canada.

For ordering and customer service, call 1-800-CALL WILEY.

Library of Congress Cataloging-in-Publication Data:

Ladisch, Michael R., 1950-
Bioseparations engineering : principles, practice, and economics / Michael R. Ladisch.
p. cm.
Includes bibliographical references and index.
ISBN 0-471-24476-7 (alk. paper)
1. Biomolecules—Separation. 2. Biochemical engineering. 3. Separation (Technology)
I. Title.

TP248.25.S47 L33 2000
660'.2842—dc21 00-043606

10 9 8 7 6 5 4 3

To Chris, Mark, and Sarah

CONTENTS

Ion Exchange Gradient Chromatography

Size Exclusion (Gel Permeation) Chromatography

Reversed Phase Chromatography

dinated conversion of the manuscript into typeset form. In addition, I wish to acknowledge the patience and encouragement of my friends and training associates during several marathons: Ron Ellis, Wayne Ramsey, and in particular, Professor and Coach, Howie Zelaznik of the Elite Runner's Association.

Most of all, I wish to convey my appreciation and gratitude to my family for their patience, understanding, and support during the long process of writing "the book."

Michael Ladisch

CHAPTER 1

BIOSEPARATIONS

INTRODUCTION

Bioseparations consist of a sequence of recovery and separation steps that maximize the purity of the bioproduct while minimizing the processing time, yield losses, and costs. Solid from liquid separations occur early in a separation sequence since the largest entity encountered in a biotechnology process is the organism that generates and is closely associated with the product. Unicellular organisms typical of cell culture or fermentation processes range from 1 to 20 microns in size and produce macromolecules with molecular weights ranging from 10^3 to 10^6 or small molecules with molecular weights of less than 10^3 Daltons.

The unit operations for the separation of biological molecules differ from their counterparts in the chemical industry. Most biological molecules, and particularly proteins, are destroyed by heat, cannot be evaporated or distilled, are generated at low concentrations, and have a three-dimensional structure that can change during purification, resulting in loss of the molecule's function. Therefore bioproducts are recovered, purified, and concentrated by membranes, adsorption, chromatography, crystallization, and other techniques that do not require heat or extreme pH to achieve fractionation. In comparison, a prevalent technique in the chemical industry is distillation, which exploits differences in the volatility of the molecules being separated and utilizes heat to drive a change in phase to continuously recover a purified and relatively concentrated chemical product through countercurrent contacting.

The purpose of this book is to present the process engineering principles upon which each of the key bioseparation methods are based. These are presented in the order in which they are likely to be first encountered in a purification sequence. The solids/liquid separation by sedimentation, centrifugation, or filtration precede initial product recovery and concentration. Recovery by membrane separation, precipitation, and another round of solid/liquid separations follow. Finally, purification is carried out by crystallization, adsorption, and chromatography. The major unit operations used in bioseparation processes are listed in Table 1.1, together with approximate ranges of fractional yields that can be expected for each operation.

Since over half of the purification costs of biopharmaceuticals lie in the chromatography steps, chromatography is emphasized in this book in Chapters 5 to 9. These chapters discuss

$$\begin{array}{c} \overset{\displaystyle O}{\|} \\ CH \\ | \\ HC-OH \\ | \\ HO-CH \\ | \\ HC-OH \\ | \\ HC-OH \\ | \\ CH_2OH \end{array} \quad \underset{\substack{\text{pH 7 to 8} \\ \text{55 to 65°C}}}{\overset{\substack{\text{Glucose} \\ \text{Isomerase}}}{\rightleftharpoons}} \quad \begin{array}{c} CH_2OH \\ | \\ C=O \\ | \\ HO-CH \\ | \\ HC-OH \\ | \\ HC-OH \\ | \\ CH_2-OH \\ \\ \text{Fructose} \end{array} \tag{1.1}$$

ter and 20 feet high, and are generally operated in a down-flow direction (Boyce, 1986). The pores in the cell membrane are too small to allow the enzyme to diffuse but are large enough to enable the low molecular weight reactant, glucose, to diffuse into the cell, and the product, fructose, together with unreacted glucose, to diffuse out.

An alternate approach involves the rupture of cells and isolation of cell-free glucose isomerase from *Streptomyces*[1] species, followed by the immobilization of the enzyme on a particulate material. The particles are packed into a large column for use as an immobilized enzyme reactor (Antrim, 1995). Antrim (of Genencor) reports four methods for purifying the enzyme on a large scale: chromatography, precipitation, ultrafiltration, and crystallization. Purities, recoveries, and concentration in Table 1.2 illustrate that significant purification of glucose isomerase can be obtained using only a few steps. According to Antrim (1995), the immobilized enzyme should be as pure as possible for this system to function at maximum efficiency. Purified glucose isomerase enzyme from *Bacillus sp.* was shown to have a significant economic benefit over an unpurified enzyme because higher specific activities of immobilized enzyme are

TABLE 1.2 Summary of Large-Scale Purification Techniques for Glucose Isomerase[a]

Method	Purity, %	Recovery, %	Concentration Activity Relative to Starting Preparation
Chromatography (adsorption/desorption)	89	90	5×
Precipitation (quaternary amine)	85–90	92–97	10×
Selective ultrafiltration	78	72	30×
Crystallization	87	96	>10×

Source: From Antrim, 1995, Table 15.3, p. 461. This material is reproduced with permission of Marcel Dekker, Inc. NY.
[a]Initial concentration of cell extract of glucose isomerase was about 30 units/mL.

[1]Glucose isomerase enzyme from *Streptomyces* is a tetrameric protein comprised of four identical subunits and has a molecular weight of 165 kD. The protein is acidic, since its amino acid sequence contains more acidic than basic amino acid residues. In an immobilized form, the enzyme slowly loses activity but is sufficiently stable to be used continuously for 300 days with 1 kg of immobilized enzyme converting 7 tons of dextrose to 42% fructose during this period. This is equivalent to 700 g of product per 100 units of enzyme (Antrim, 1995).

possible, resulting in significantly lower costs of immobilization reagents and carrier to which the enzyme is immobilized (Ladisch et al., 1977).

The partial separation of the products of the reaction, itself, glucose and fructose, is achieved by very large-scale liquid chromatographic systems, as discussed in Chapter 6. This separation enriches the syrup in fructose so that it attains a sweetness that is nearly equivalent to sugar obtained from sugarcane.

1.2 PHYSICAL PROCESSING AND WATER REMOVAL STEPS ARE IMPORTANT SEPARATION METHODS FOR LARGE-VOLUME PRODUCTS

Large industries fractionate renewable materials into starches, protein, oils, and/or fiber and process agricultural commodities (e.g., corn, wheat, and soybeans) and forestry products. For example, wet milling grinds, centrifuges, and screens corn to separate the germ (contains oil) from the starch, and the starch from the glutein (protein) (Hoseney, 1986). In 1996 such processing converted 1.58 billion bushels of corn into more than 50 billion pounds of products including corn starch, modified starch, dextrins, glucose syrups, high-fructose syrups, wet milling by-products, and steep water (Corn Refiners Association, 1997). Dry milling grinds and separates corn into specialty products that provide ingredients for corn flakes, snack foods, and production of beer (Ladisch and Svarczkopf, 1991). The manufacture of pulp and paper utilizes large volumes of water and is principally a physical and chemical process by which the cellulosic components are fractionated from lignin so that the cellulose can be cast into paper (Rydholm, 1965; Wenzel, 1970). Pulp and paper has a U.S. market value of $122 billion/yr (Committee on Bioprocess Engineering, 1992), with some processes utilizing cellulase and hemicellulase enzymes, and fungal cultures as processing aids.

The fermentation of glucose into products ranging from organic acids and amino acids to ethanol presents special challenges in the fractionation of the extracellular fermentation products into substantially water-free and/or extremely pure, high-volume products. Efficient separation processes are important to the continued development of an industry based on the utilization of renewable resources—starch and cellulose—derived through agriculture and silviculture.[2] Separation processes are needed not only to purify products but to efficiently recover them from the large volumes of water in which they are dissolved.

Extracellular fermentation products that have low volatility include antibiotics and organic acids (citric, lactic, and fumeric), and they are recovered and purified by extraction, adsorption, ion exchange and/or crystallization. A well-known volatile product is ethanol, which is recovered from fermentation broth by distillation to obtain a hydrous product. Since ethanol forms an azeotrope with water, its purification to a water-free form requires azeotropic or extractive distillation or adsorption to remove the final 5 to 10% water. Examples of product concentration and water removal are mentioned in Chapters 3 and 5.

[2]The potential for growth in the demand for hexoses derived from corn for industrial uses is significant, but may ultimately find a limit based on economic pressures generated by concurrent demand for corn as food and animal feed. Advances in the fermentation of pentoses to ethanol will improve yields of ethanol from corn, but will also make use of forestry (wood) residues and other cellulosic materials more attractive as a source of fermentable sugars (Ladisch and Svarczkopf, 1991), since about a third of the composition of cellulosic biomass can yield pentoses, and another third is hexoses. Efficient conversion of pentoses to commercially important fermentation products will help wood, agricultural residues, and other cellulose-containing biomass materials to be economically attractive substrates.

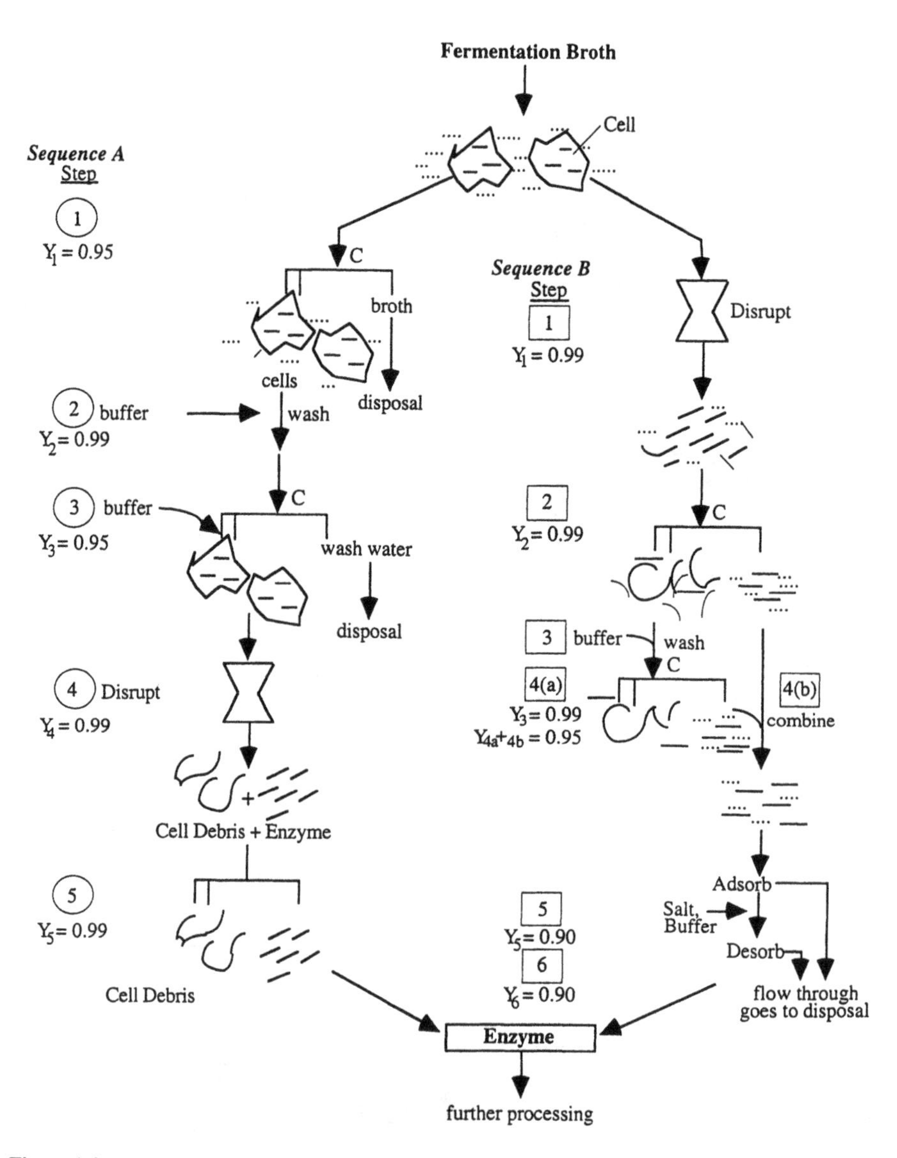

Figure 1.4. Schematic representation of two approaches for recovering a target protein (i.e., an enzyme) from microbial cells. Left side of figure shows washing of cells prior to disruption, and the right side shows the opposite sequence. Contamination of the intracellular product with broth may cause yield losses and extra steps to bring the target molecule to the same extent of enrichment. Yields were arbitrarily assigned as 0.99 (i.e., 99%) for buffer addition or washing, 0.95 for centrifugation or filtration, and 0.90 for adsorption or desorption of protein that is being separated from other constituents in the cytoplasm.

from the liquid containing the enzyme. Yields that might be typically expected for the individual unit operations are indicated below each of the circled step numbers in Figure 1.5. On this basis, the overall yield is

$$Y_{\text{overall}} = \prod_{i=1}^{5} Y_i = (0.95)(0.99)(0.95)(0.99)(0.95) = 0.84 \tag{1.4}$$

If the recovery sequence were to disrupt the cells prior to their separation from the broth, the steps shown on the right-hand side of Figure 1.4 might result. The separation sequence would be more complex, since the enzyme would be both diluted and contaminated with broth components that would need to be removed before the enzyme is further purified. Unlike in sequence *A* which initially captures the protein in concentrated form inside the cells, sequence *B* (step numbers denoted in boxes) is less favorable. It removes compartmentalization of broth from product by prematurely rupturing the cell membrane in step 1 rather than in step 4, as is the case for sequence *A*. The overall yield for sequence *B* would be

$$Y_{\text{overall}} = \prod_{i=1}^{6} Y_i = (0.99)(0.99)(0.99)(0.95)(0.90)(0.90) = 0.747 \tag{1.5}$$

A listing of the bioseparations unit operations is given in Table 1.1 together with some arbitrary limits on expected yields. Regardless of the separation method used, the maximum fractional recovery is set at 0.99 (rather than 1.0), since there is always likely to be a small loss, or loss of product activity, at each step. Optimization of individual recovery and purification steps can lead to yields that exceed 0.95, although the value of the product at a given purity relative to incremental increases in separations costs must also be factored to selection of optimum operating conditions.

1.7 BIOINFORMATICS WILL LEAD TO PRODUCTS THAT BIOSEPARATION METHODS MUST PURIFY

Bioinformatics is an emerging field that seeks to apply computer algorithms to discover patterns in the vast amounts of information on biological organisms that is becoming available from the sequencing of plant, animal, and microbial genomes (Yin, 1999). One of the goals of bioinformatics will be to relate this informational basis of life to its expression in the form of protein biocatalysts upon which the functioning of living organisms is based. This could provide a source of process templates for the design of bioprocesses for manufacture of products via biological systems, to enhance agricultural production using naturally occurring biological phenomena, and to amplify the traits of beneficial organisms for purposes of biologically based remediation of the environment. This focus differs from that of previous approaches in which the protein or organism was often studied principally on a phenomenological basis (i.e., what it did rather than how it worked). Bioinformatics would examine the functioning of biosynthetic pathways at a systems level for the purpose of discovering how a specific pathway is controlled and how it might be amplified or emulated for the purpose of large-scale manufacture of a specific molecule. The principles of biology will be as important to this new field of engineering as chemistry was to the development of chemical engineering. Economic methods of purifying

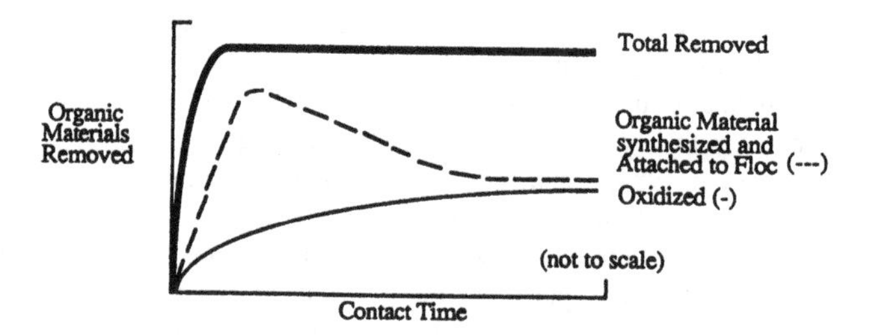

Figure 2.2 Conceptual representation of removal of organic matter in an aerated activated-sluge system (adapted from Bailey and Ollis, 1977, figs. 12.10 and 12.12, pp. 710–711).

ing temperature, and makes a transition from Newtonian to pseudoplastic behavior at concentrations above 10% solids for suspensions of single cells (see reference Atkinson and Mavituna, 1983, pp. 693–694).[1]

For this example an apparent viscosity of 1.8 cp ($0.018 \frac{g}{cm \cdot s}$) is used for a dilute yeast suspension (i.e., at 1.5% solids) that exhibits Newtonian behavior.[2] The settling velocity for yeast is given by Eq. (2.1):

[1]The constitutive equation for a Newtonian fluid is

$$\tau = \mu\gamma \tag{a}$$

where the apparent viscosity, μ_a, and viscosity, μ, calculated from the shear stress, τ, as a function of strain, γ, are the same. The viscosity, μ, of a Newtonian fluid is the slope of the stress–strain curve and is a constant. A pseudoplastic fluid follows the equation:

$$\tau = K\gamma^n, \qquad \text{where } n < 1 \tag{b}$$

and the apparent viscosity decreases with increasing shear rate, $\mu_a = K\gamma^{n-1}$, where K is a proportionality constant used to fit the data, and the viscosity is calculated from the slope of the line connecting the origin of the stress–strain curve to a single point on this curve.

[2]Viscosities of dilute yeast suspensions were reported to follow Einstein's equation by Blanch and Bhavaraju with the viscosity of yeast at 0 to 0.03 volume fraction (i.e., 0 to 3% cells) in water following the equation (Blanch and Bhavaraju, 1976):

$$\mu_S = \mu_L (1 + 2.5\phi) \tag{c}$$

while yeast and spore suspensions at 0.03 to 0.14 volume fraction follows the Vand equation:

$$\mu_S = \mu_L (1 + 2.5\phi + 7.25\phi^2) \tag{d}$$

where μ_S is the viscosity of the suspension, μ_L is the viscosity of the suspending liquid, and ϕ is the volume fraction of cells in the suspension. The viscosity of water at 20°C is 1 cp (= 0.01 g/cm·s). The viscosities of 2% and 10% cell suspensions are 1.05 and 1.325, respectively, by equations (a) and (b), compared to measured values of apparent viscosity of about 2 and 5, at similar cell concentrations (see Atkinson and Mavituna, 1983, pp. 694–695). The discrepancy between the calculated and experimental values may be due to the age of the fermentation broth and morphology of the cell suspension, both of which have a major effect on viscosity (Blanch and Bhavaraju, 1976). This is not directly accounted for in equations (c) and (d). An accurate determination of μ_L is therefore needed to accurately estimate μ_S.

Sedimentation

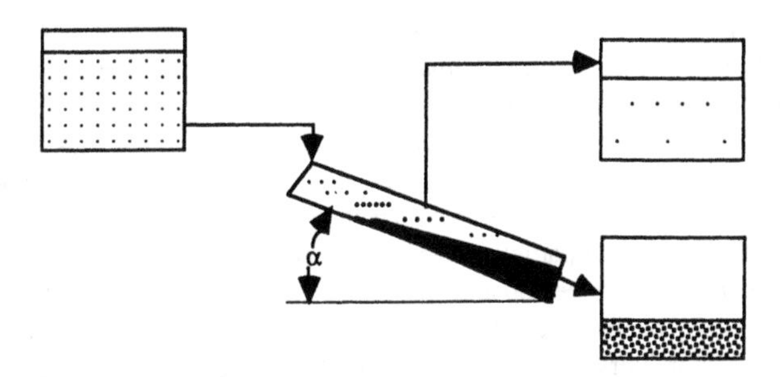

Figure 2.3 Schematic representation of separation of solids by settling.

$$\frac{(7 \times 10^{-4})^2}{(18)(0.018)}(1.03 - 1.00)(980) = 4.45 \times 10^{-5}\,\frac{\text{cm}}{\text{s}}\left(\frac{3600\text{s}}{\text{hr}}\right) = 0.16\,\frac{\text{cm}}{\text{hr}}$$

$$\frac{\text{cm}^2}{\frac{\text{g}}{\text{cm}\cdot\text{s}}} \qquad \frac{\text{g}}{\text{cm}^3} \qquad \frac{\text{cm}}{\text{s}^2}$$

where 980 cm/s^2 is the value of the gravitational constant in consistent dimensions. The terminal velocity for yeast can be compared to a much lower velocity for *E. coli*:

$$\frac{(1 \times 10^{-4})^2}{(18)(0.018)}(1.03 - 1.00)(980) = 9.07 \times 10^{-7}\,\frac{\text{cm}}{\text{s}} = 0.0033\,\frac{\text{cm}}{\text{hr}}$$

Microorganisms will sometimes form flocs that are spherical in shape and have diameters measured in millimeters. Examples include brewer's yeast (flocs with d_p = 0.02 to 1.3 cm), *Aspergillus niger* (*fungi imperfecti*, 0.075 to 0.2 cm), and the bacterium *Zoogloea ramigera* which forms flocs in activated sludges (see Atkinson and Mavituna, 1983, pp. 19–21). The enhanced settling characteristics of these forms of the microorganisms enable facile separation by settling. For example, the settling velocity of floculated brewers yeast with d_p = 0.5 cm, diluted to give a 1.5% suspension, is

$$\frac{(0.5)^2}{(18)(0.018)}(1.03 - 1.00)(980) = 22.7\,\frac{\text{cm}}{\text{s}}$$

if the suspension is assumed to have the same viscosity and density as the single-cell yeast. The yeast pellets would settle almost immediately.

The settling of yeast cells in an ethanol fermentation is complicated by the CO_2 generated by respiring yeast. The CO_2 gas bubbles attach to the cells and cause them to float. This makes recovery of actively growing yeast cells by sedimentation difficult. However, the recovery of the cells after exhaustion of glucose in the medium and heat treatment of the cells at 50°C for 20 minutes surpresses CO_2 production and results in a settling rate of 0.68 cm/hr (Bungay and Millsbaugh, 1984; Walsh and Bungay, 1976). Bungay's clever suggestion that sedimentation of the larger yeast cells from the smaller bacterial cells could moderate bacterial contamination

Centrifugation

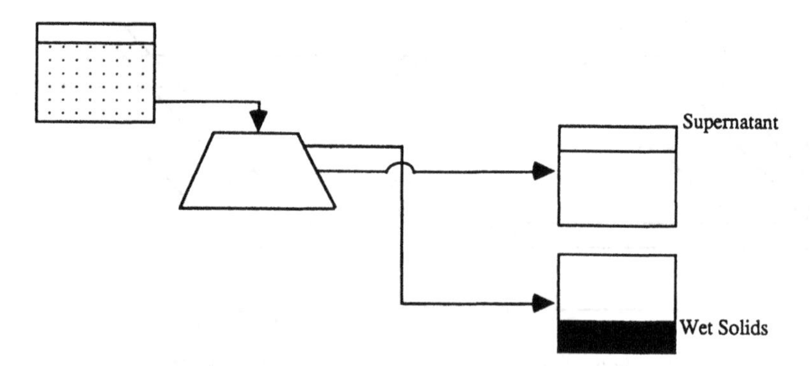

Figure 2.6 Schematic representation of centrifugal separation.

A 20 cm bowl at 10,606 rpm (ω = 1111 radians/s) would experience about 12,588 × **g**. (Note that g denotes gravity, here.)

2.3 THE VOLUMETRIC RATE OF CLARIFIED SUPERNATANT IS MAXIMIZED BY A LARGE-DENSITY DIFFERENCE AND LOW VISCOSITY

The basic relationship between volumetric flowrate for centrifugal sedimentation is (McCabe and Smith, 1967):

$$q_{\text{cut}} = \frac{\pi b \omega^2 (\rho_p - \rho) d_{\text{p,cut}}^2}{18\mu} \; \frac{r_2^2 - r_1^2}{\ln(2r_2/(r_1 + r_2))} \tag{2.3}$$

where

q_{cut} = volumetric flowrate corresponding to the diameter of particle that just reaches one half the distance between r_1 and r_2 (i.e., the cut point)
$d_{\text{p,cut}}$ = particle diameter that reaches the cut point ($d_{\text{p,cut}} > d_p$)
b = width of centrifuge basket
ω = angular velocity
ρ, ρ_p = densities of fluid and particle
r_1, r_2 = inner and outer radii of the layer of solids being collected in the centrifuge, as shown in Figure 2.7
μ = viscosity

The cut point defines the diameter of the particle which reaches half of the distance between r_1 and r_2 (Ladisch and Wankat, 1988; Belfort, 1987) as illustrated in Figure 2.7. If the thickness of the solid layer is small so that $r_1 \cong r_2$, Eq. (2.3) simplifies to

$$q_{\text{cut}} = \frac{2\pi b\omega^2(\rho_h - \rho)d^2_{\text{p,cut}}r_2^2}{9\mu} \tag{2.4}$$

where q is flowrate.

Figure 2.7 illustrates the definitions of the different variables used in Eqs. (2.3) and (2.4) for a sedimentation, tubular bowl, centrifuge.

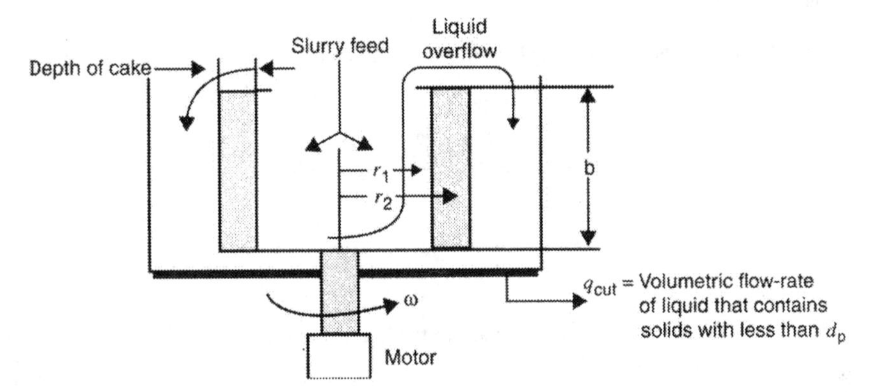

Figure 2.7 Schematic representation of a cut-away view of a tubular bowl centrifuge showing definitions r_1, r_2, q_{cut}.

Smooth operation requires that the load or weight of particles in the centrifuge be balanced, and that the angular velocity or rotational speed not exceed the mechanical stress limits of the bearings, shaft, and materials of construction of the machine. The maximum rotational speeds that can be safely used decrease as the diameter of the centrifuge increases, and must be considered in scaling up a laboratory procedure to a production scale machine. The centrifuge can disintegrate if the limits imposed by mechanical constraints are exceeded. The data of Wang et al. (1979) give a first indication of the effect of the bowl diameter on the maximum rotational speed and the gravitational forces that can be generated for a tubular bowl centrifuge. A tubular bowl with a 4.5 cm diameter has a maximum rpm of 50,000 and a maximum corresponding **g** force of 62,500. At 10.8 cm, the maximum rpm and **g** force are 14,000. At 7.6 cm, the maximum **g** force is 2000.

2.4 CENTRIFUGE SPEED IS LIMITED BY STRESS IN THE BOWL'S WALL AND BY ITS MATERIALS OF CONSTRUCTION

The stress in the bowl wall is generated by its rotation, and is referred to as self-stress, $\mathbf{S_S}$. The total stress $\mathbf{S_T}$, is due to the combination of the self-stress, $\mathbf{S_S}$, and the stress due to the liquid in the bowl, $\mathbf{S_\ell}$ (Smith et al., 1963):

$$\mathbf{S_T} = \mathbf{S_S} + \mathbf{S_\ell} \tag{2.5}$$

The self-stress is given by

$$\mathbf{S_S} = 4.11 \times 10^{-10} n^2 D_b^2 \rho_m \tag{2.6}$$

where

n = speed of rotation in revolutions per minute (rpm)
D_b = bowl diameter in inches
ρ_m = density[4] of the material in the bowl as lb/ft^3

The stress, $\mathbf{S}_\ell$, due to the liquid at the bowl's wall, is given by

$$\mathbf{S}_\ell = 1.03 \times 10^{-10} \frac{n^2 \rho D_b (D_b^2 - D_i^2)}{\delta} \tag{2.7}$$

where

D_i = is the diameter of the inner surface of the liquid, inches
ρ = liquid density in lb/ft^3
δ = wall thickness, inches

The combination of Eqs. (2.5), (2.6), and (2.7) results in

$$\mathbf{S_T} = 4.11 \times 10^{-10} n^2 D_b \left[\rho_m D_b + \frac{\rho (D_b^2 - D_i^2)}{4\delta} \right] \tag{2.8}$$

An example given by Smith et al. (1963) shows that the liquid pressure of a 3 in. layer of water (corresponds to a diameter of 10 in.) at the wall of a 16 in. diameter bowl rotating at 5000 rpm is 500 lb$_f$/square inch (or psi):

$$\mathrm{P} = 2.05 \times 10^{-9} \rho n^2 (D_b^2 - D_i^2)$$

$$= 2.05 \times 10^{-9} (62.4)(5000)^2 (16^2 - 10^2) = 500 \text{ psi} \tag{2.9}$$

For 316 stainless steel, with $\rho_m = 490$ lbs/ft^3, the total stress for a half-inch thick bowl wall is

$$\mathbf{S_T} = (4.11 \times 10^{-10})(5000)^2(16) \left[(409)16 + \frac{62.4(16^2 - 10^2)}{4(0.5)} \right] = 2090 \text{ psi}$$

while the stress due to the liquid, given by Eq. (2.7) is

[4]There are 0.01602 g/mL for 1 lb/ft^3. The density of water at 1 g/mL is equivalent to 62.4 lb/ft^3. There are 6894.76 N/m^2 for 1 psi (psi = lb$_f$/square inch, where lb$_f$ denotes pounds force).

$$\mathbf{S}_{\ell} = 1.03 \times 10^{-10}\left(\frac{(5000)^2(62.4)(16)(16^2 - 10^2)}{0.5}\right) = 802\ psi$$

Hence, for this particular example, only 38% of the stress is from the liquid, while the remainder is from self-stress. While the centrifugal force for a given rotational speed increases in proportion to the diameter of the bowl, D_b, the self-stress increases as D_b^2.

The allowable self-stress sets the maximum centrifugal force (i.e., rotational speed) at which the machine can be safely operated. If the speed is too high, allowable limits of self-stress of the material of construction could be exceeded, and the centrifuge could disintegrate. Equation (2.4) shows that as the difference between the bowl diameter and inner wall diameter decreases, the rotational speed can be higher for the same stress, and illustrates why a larger centrifugal force can be obtained for a smaller diameter centrifuge made of the same material (Smith et al., 1963). Special materials of construction are needed for larger production centrifuges that combine strength with corrosion resistance. According to Erikson, austenetic stainless steel used in the 1950s with yield strengths of 250 N/mm^2 were surpassed by ferritic-austenitic duplex stainless steels with a yield strength of 450 N/mm^2 and a corrosion resistance equal to 316 SS with the exception of chlorides. Proprietary stainless steels that combined both strength (600 N/mm^2) and corrosion resistance similar to 316 SS were introduced in the 1980s (Erikson, 1984).

2.5 THE DISC STACK CENTRIFUGE ENABLES CONTINUOUS AND RAPID PROCESSING OF CELL AND COLLOIDAL SUSPENSIONS

The disc stack centrifuge is used in the biotechnology industry because it is compact, can be contained as a closed system, and can be cleaned in place. The principle of operation of a disc stack centrifuge is fully described by Erikson (1984). This type of centrifuge consists of a rapidly rotating bowl to which a set of conical discs are attached and spaced so that there are passages between the discs or plates (Figure 2.8). The feed of suspended solids is introduced to the

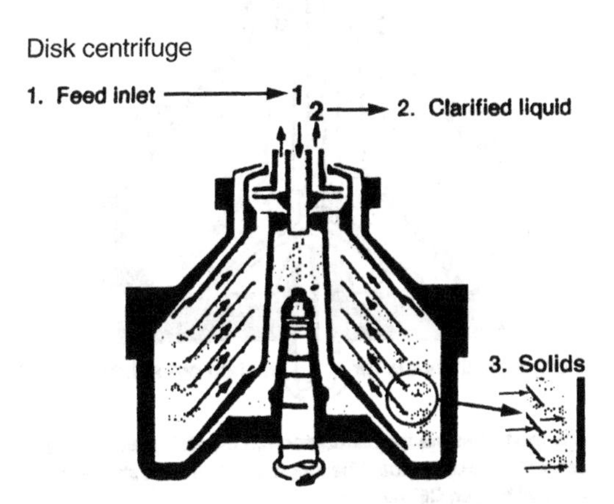

Figure 2.8 Schematic diagram of a cross section of a disc stack centrifuge showing the bowl, discs, rotating shaft, and direction of flows of particles (to the sides) and clarified liquid (out of the top of the centrifuge) (adapted with permission from Erikson, 1984, fig. 1, p. 51).

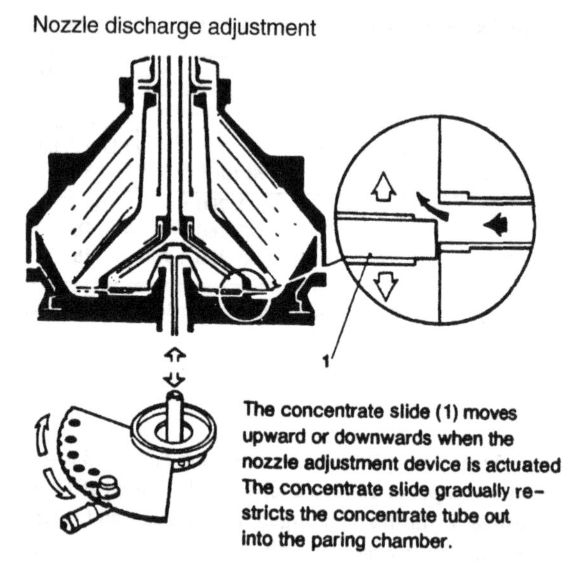

Figure 2.10 Nozzle discharge is adjusted using a collar which forms a partial obstruction to flow through an opening near the bottom of the centrifuge bowl (adapted with permission from Erikson, 1984, fig. 3, p. 52). Reproduced with permission of the American Institute of Chemical Engineers, 1984, AIChE. All rights reserved.

1 Feed
2 Discharge
3 Centripetal pump
4 Discs
5 Sediment holding space
6 Solids ejection ports
7 Piston
8 Bowl valve
9 Opening chamber
10 Opening-water feed
11 Closing-water feed
12 Inner closing chamber
13 Outer closing chamber
14 Sensing-liquid offtake
15 Sensing-liquid clarifying discs
16 Sensing-liquid recycle
17 Sensing-liquid pump
18 Timing unit
19 Magnetic switch
20 Flowmeter

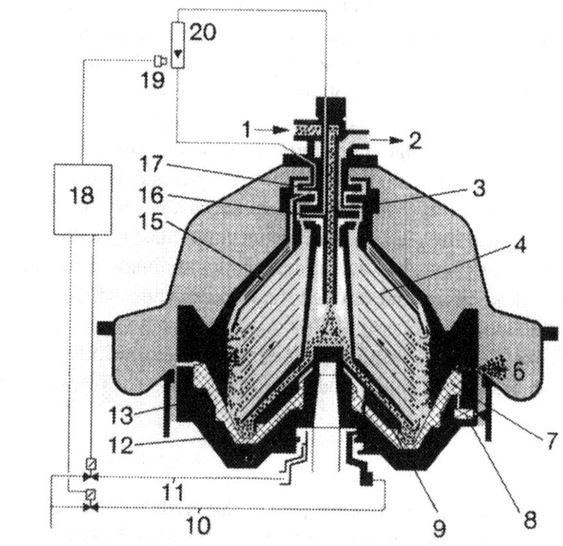

Figure 2.11 Schematic diagram of a self-cleaning clarifier with solids discharge from the side nozzle in a semicontinuous manner (from Hemfort and Kohlstette, 1988, fig. 26, p.23). Reproduced with permission of Westfalia Separator AG.

Figure 2.12 Photograph of a self-cleaning clarifier. Cylindrical housing on the left is the motor, which drives the centrifuge enclosed in the housing on the right (from Hemfort and Kohlstette, 1988, fig. 27, p. 23). Reproduced with permission of Westfalia Separator AG.

An application such as this requires that a solids-ejecting centrifuge or a self-cleaning clarifier be used (Erikson, 1984; Hemfort and Kohlstette, 1988). A solids-ejecting centrifuge has a hydraulically controlled piston that moves the entire centrifuge bowl up and down so that the nozzles (i.e., the openings in the bowl through which the solids discharge) are blocked in the up position. When the piston (and bowl) are lowered, the entire 360° periphery of bowl is in an open position, thus allowing the solids to discharge. The effect of the piston is to intermitantly block passage of the solids concentrate stream and thereby increase the residence time of the slurry. A hydraulic force of 227 Mg (500,000 lb) is needed to keep a 0.6 m bowl closed against the opposing hydraulic force of the fluid in the centrifuge (Erikson, 1984). The solids-processing capacity of the centrifuge is reduced about tenfold compared to an equivalently sized nozzle centrifuge, since its operation throttles the rate of feed slurry entering the centrifuge. Hence the solids-discharging type of centrifuge is used for a low solids containing feed with difficult to separate particles, while a nozzle type of centrifuge is used with larger volumes of high solids feeds of easy to separate particles. Characteristics of an axial solids-ejecting centrifuge are (Erikson, 1984)

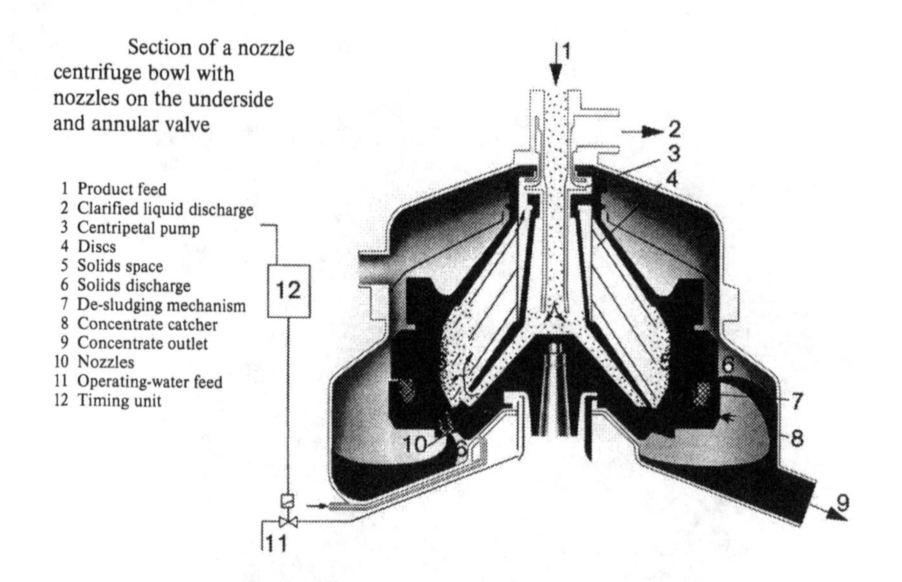

Figure 2.13 Continuously discharging nozzle centrifuge with nozzles on the underside (from Hemfort and Kohlstette, 1988, fig. 30, p.25). Reproduced with permission of Westfalia Separator AG.

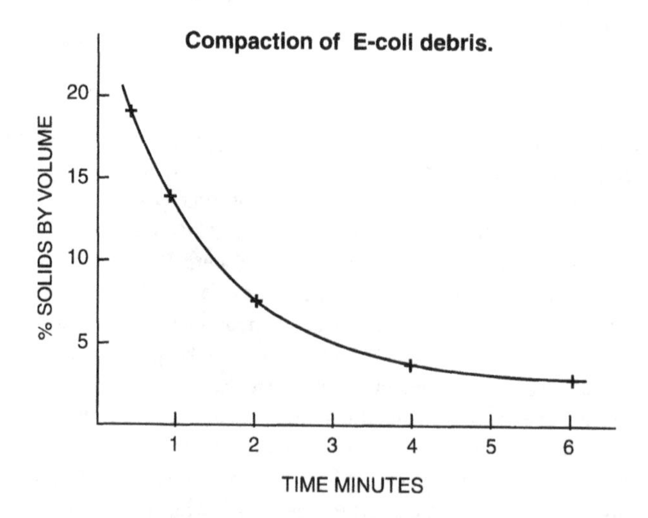

Figure 2.14 Compaction of cell debris from *E. coli* (from Erikson, 1984, fig. 4, p. 52). Reproduced with permission of the American Institute of Chemical Engineers, 1984, AIChE. All rights reserved.

TABLE 2.2 Capacity of Self-cleaning Clarifiers as a Function of Sediment Volume Holding Capacity and Equivalent Clarification Surface

Sediment Volume Holding Capacity, L	Equivalent Clarification Surface (Σ), m^2	Volumetric Capacity of *E. coli* Feed, L/hr
0.25	1,900	50 to 70
1.5	10,600	300 to 400
3.9	31,000	900 to 1200
16.	51,600	1400 to 2000
17.	68,400	1900 to 2600

Source: Adapted from Figure 43, p. 34, of Hemfort and Kohlstette, 1998. The material is reproduced with permission of Westfalia Separator AG Company.

Diameter	60 cm
Bowl speed	6800 rpm
Maximum force	$15{,}000 \times \mathbf{g}$
Σ	240,000 m^3
Solids space	15 L
Maximum liquid capacity	23 m^3/hr
Maximum solids discharge	1.8 m^3/min

A different design that gives similar results is a self-cleaning clarifier or axial solids-ejecting centrifuge (Figures 2.11 and 2.12) that discharges the solids in a semicontinuous manner (Hemfort and Kohlstette, 1988). However, unlike the solids ejector separator design, the solids concentrate stream leaves the centrifuge from side ports during both the operation of the centrifuge and clean-in-place procedures. A hydraulically actuated valve or plate controls the size of the opening at the periphery of the bowl (Hemfort and Kohlstette, 1988). A centrifuge design with discharge from the underside of the annular valve is shown in Figure 2.13.

The capacities for handling of *E. coli* streams for one manufacturer's self-cleaning clarifiers is summarized in Table 2.2. The sediment holding space for these industrial centrifuges (see Figure 2.11) ranges from 250 mL to 17 L. The centrifuges process between 50 and 2600 L of *E. coli* broths/hr (Hemfort and Kohlstette, 1988). Larger centrifuges are now probably avail-

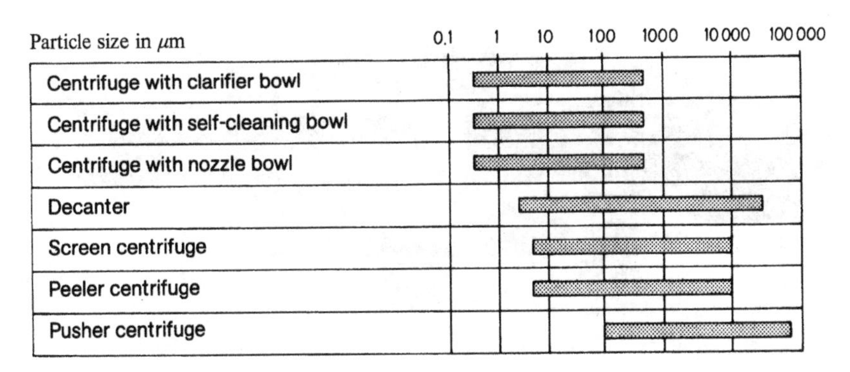

Figure 2.15 Centrifuge type as a function of particle size (from Hemfort and Kohlstette, 1988, fig. 25). Reproduced with permission of Westfalia Separator AG.

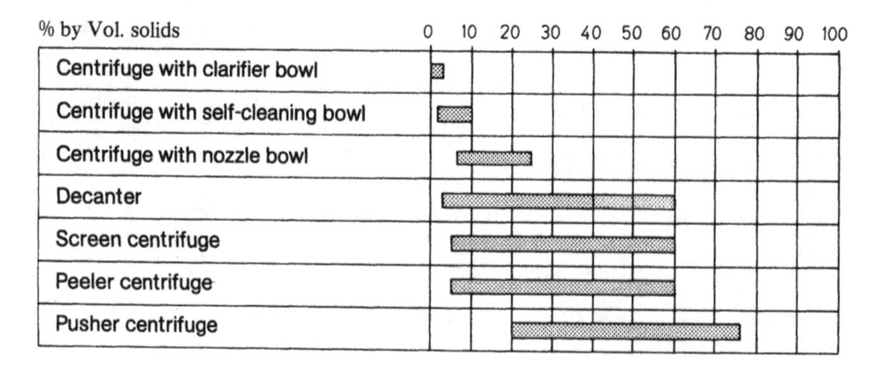

Figure 2.16 Percentage of solids concentration (by volume) in feed (from Hemfort and Kohlstette, 1988, fig. 26). Reproduced with permission of Westfalia Separator AG.

able. The data of Table 2.2 are useful for illustrating the capacity as a function of equivalent clarification surface (Σ).

The selection criteria for different types of centrifuges are summarized in Figure 2.15 as a function of particle size and the solids content of the feed (expressed on a % volume basis) (Hemfort and Kohlstette, 1988). The self-cleaning clarifier bowl operates semicontinuously, while the nozzle bowl and solids-ejecting centrifuges operate continuously. Both the particle size range (1 to 10 micron) and volume concentrations (0.5 to 20%) of fermentation and cell culture broths are compatible with the first three types of centrifuges listed in Figure 2.15 and Figure 2.16.

2.6 A DECANTER CENTRIFUGE IS LESS EFFICIENT FOR RECOVERY OF MICROORGANISMS THAN A DISC CENTRIFUGE

The decanter is a horizontal rotating centrifuge where the screw (scroll) rotates at a faster speed than the bowl in order to continuously convey the separated solids to the outlet (Figures 2.17

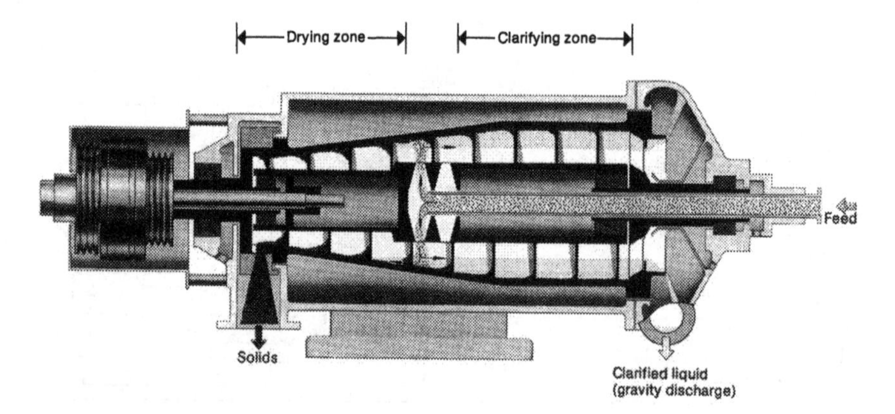

Figure 2.17 Cross sectional diagram of a counter-current decanter type centrifuge (from Hemfort and Kohlstette, 1988, fig. 44, p. 35). Reproduced with permission of Westfalia Separator AG.

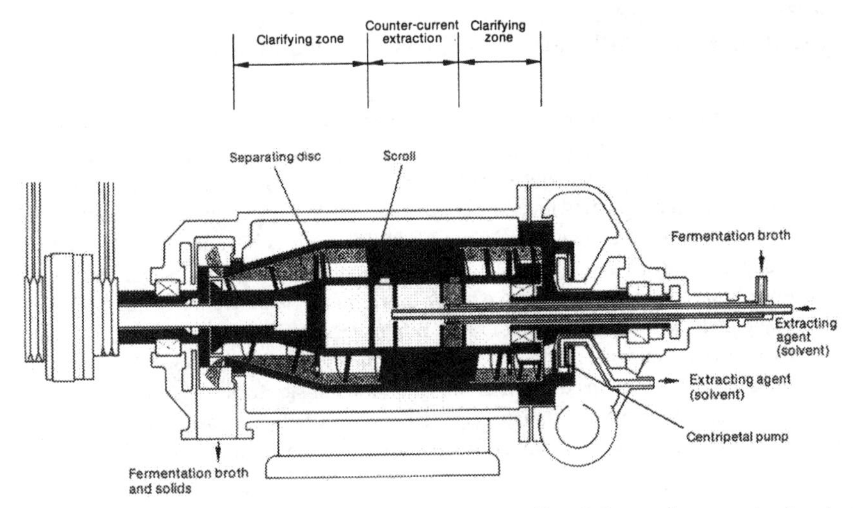

Figure 2.18 Cross-sectional diagram of a decanter type centrifuge being used as an extraction device (from Hemfort and Kohlstette, 1988, fig. 46, p. 36). Reproduced with permission of Westfalia Separator AG.

and 2.18). The decanter is typically not effective for recovering microorganisms unless a floculant is added, or the cells are already concentrated. An example is the recovery of a single-cell protein in which the microbial biomass has already been preconcentrated in a disc centrifuge. A possible application of the decanter is for the direct extraction of an antibiotic from the mycelium. The extract containing the antibiotic leaves as the light phase, while the mycelium from which the antibiotic has been extracted leaves the decanter as a slurry of solids in the fermentation broth. The manufacturer of the machine reports that the fermentation broth does not need to be filtered, and that there are 1.7 theoretical stages in this type of extractor (Hemfort and Kohlstette, 1988).

2.7 STERILITY, CONTAINMENT, AND HEAT FROM MECHANICAL WORK AFFECT DESIGN OF PROCESS CENTRIFUGES

The scale-up part of a centrifugation step must account for the heat sensitivity of the product, since the mechanical work due to the rapidly rotating bowl can cause a temperature rise of 3 to 20°C in the solid and liquid discharges. If the product is deactivated at a higher temperature, cooling will need to be incorporated into the centrifuge.

The machine itself must be designed and installed to allow for its sterilization prior to dissembly for maintenance purposes. Another challenge is containment. The machine must be placed in a work area designed to prevent or contain release of aerosols that may be formed as the material passes through the centrifuge. Aerosols have the potential to escape into the environment around the centrifuge and to carry pathogenic substances to which the operator should not be exposed.

2.8 CENTRIFUGE DESIGN FOR BIOTECHNOLOGY PROCESSES INCORPORATES CLEANING-IN-PLACE AND STERILIZATION CAPABILITIES

Cleaning-in-place (CIP) has the objective of cleaning process lines and equipment and avoiding bacterial growth without opening and disassembling the apparatus. Cleaning-in-place was first developed for the dairy industry many years ago (Hemfort and Kohlstette, 1988). Cleaning fluids are passed through the process lines and equipment and they may include hot water and 2% lye at 85 to 90°C to dissolve organic sediment, and 0.5% nitric acid solution at 85 to 90°C to dissolve inorganic deposits and repassivate the stainless steel surfaces against corrosion. Once the centrifuge bowl and lines are cleared out, the bowl, hood, chamber, and solids catcher can be filled completely with disinfecting solution. While cleaning-in-place may be sufficient for some types of food and enzyme products, the production of vaccines and protein biotherapeutic products require that the system be steam sterilizable. A steam-sterilizable system requires extra steam lines, which complicates installation and increases installation cost of a process centrifuge. A schematic diagram of the layout of such a system is shown in Figure 2.19.

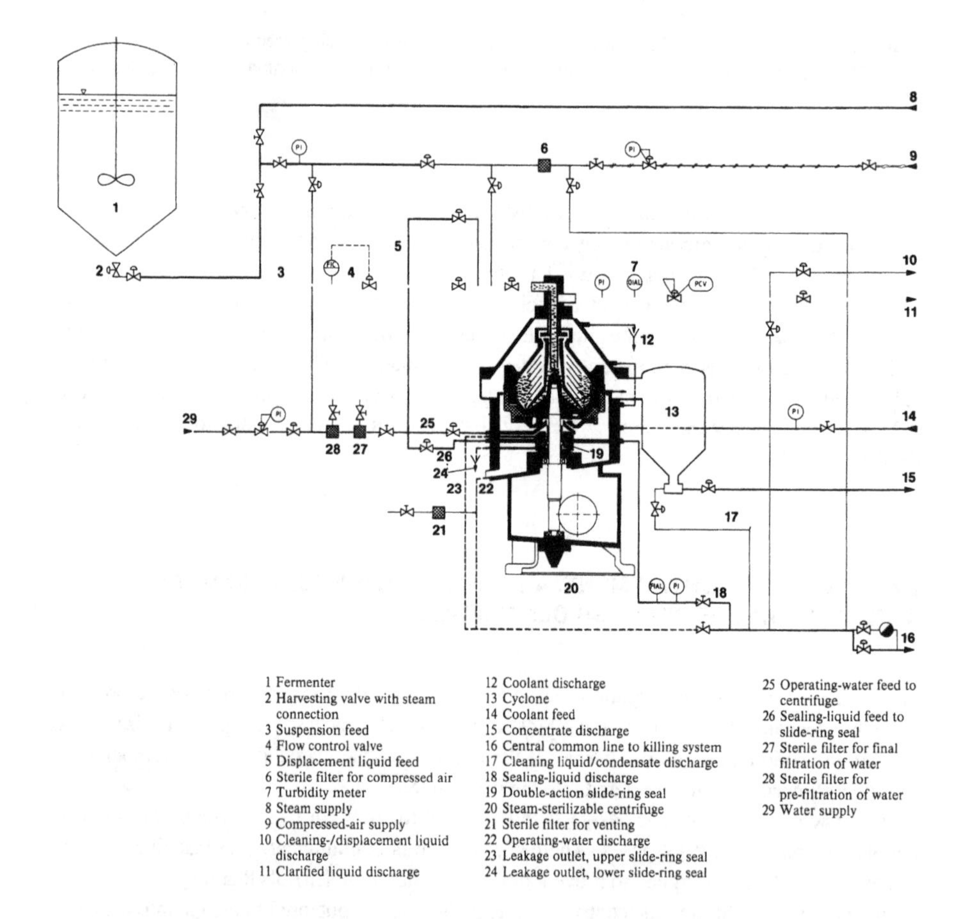

Figure 2.19 Diagram of a steam sterilizable centrifuge system (from Hemfort and Kohlstette, 1988, fig. 42, p. 33). Reproduced with permission of Westfalia Separator AG.

Sterilization is carried out with saturated steam at between 120°C (2 bar) to 127°C (2.5 bar) between 90 to 120 min at 120°C or 60 to 90 min at 127°C. When the system is cooled, sterile air is admitted into the system in order to avoid a vacuum from developing and promoting the entry of foreign bacteria when the system is cooled. Sterile filters are used for both the water (items 27 and 28 in Figure 2.19) and air (item 21) that contact internal parts of the centrifuge. A steam-sterilizable system is designed so that all sterilizable parts are made of stainless steel, with seals made of special elastomers and valves specified for sterile operation. According to one manufacturer, diaphragm valves are used almost exclusively in practice (Hemfort and Kohlstette, 1988). The gear chamber, which is nonsterile, must be completely isolated from the bowl and other centrifuge parts that are in direct contact with the product and therefore must be sterile. The piping must be laid out and installed so that there are no deadends, so that the system can be completely drained.

Figure 2.19 also shows coolant water inlet (item 14) and outlet (item 12) lines (Hemfort and Kohlstette, 1988). Cooling during centrifuge operation is necessitated by the heat generated by the mechanical work of the centrifuge. The processing of the clarified liquid and solids concentrate streams through the centrifuge can result in temperature rises of 3 to 6°C for the liquid and 15 to 20° for the solid. Circulation of a coolant through the jacket of the centrifuge is therefore needed to avoid a temperature rise that could promote loss of heat sensitive bioproducts (Erikson, 1984).

2.9 CENTRIFUGE CONTAINMENT IS NECESSARY FOR PROCESSING OF SOME TYPES OF BIOTECHNOLOGY PRODUCTS

The three levels of containment defined by the US National Institutes of Health (NIH) for processes containing 10 L or more of culture are BL1-LS, BL2-LS, and BL3-LS.* As summarized by Alford in the context of automatic control systems (Alford, 1991), all BL-LS levels require that culture fluids not be removed from a closed system or other primary containment equipment unless the viable organisms have been inactivated by a validated procedure. In addition a validated sterilization procedure must be used before containment equipment is opened up for maintenance or other purposes. For BL1-LS processes, sample collection, addition of materials, and the transfer of culture fluids from one closed system to another must be done in a manner that minimizes the release of aerosols or contamination of exposed surfaces. These operations are to be carried out in a manner that prevents the release of aerosols or contamination of exposed surfaces for BL2-LS and BL3-LS processes. BL2-LS and BL3-LS operations also require that mechanical seals associated with the closed system must be designed to prevent leakage, or be fully enclosed in ventilated housings with use of appropriate filters. Sensing devices that monitor the integrity or containment during operations are also required.

Large-scale production of approved drugs produced by *E. coli* K12, *S. cerevisiae*, and *B. subtilis* was exempted from the BL1 level of containment in a revision of the NIH guidelines in 1987, although it is not clear how this affects current biotechnology plant design. However, cell culture bioreactors and recovery systems that are used to produce pathogenic virus vaccines, tumor antigens, and recombinant proteins must be designed for biocontainment. These designs may be much more stringent than the minimum containment required to maintain as-

*The reader should be aware that the guideliness are subject to change, and direct reference should be made to the FDA publications for current information. This section is only intended to introduce the topic and does not serve as a source document.

ceptic operation at all times. According to NIH guidelines all cell cultures require a minimum of BL-1 containment or B1-1LS for volumes exceeding 10 L (Nelson and Geyor, 1991). The NIH guidelines are relevant to industrial processes given the expectation of the FDA that "The production of new drugs and biologics by recombinant DNA technology should generally follow the NIH guidelines for Research Involving Recombinant DNA Molecules (Nelson and Geyor, 1991)." Engineering for containment is complex, and in the case of centrifuges, the containment of aerosols is important and must also be factored into the overall design of the recovery system.

2.10 FILTRATION

Filtration is effective when the solids are large enough to be stopped by a filter cloth when pumped (filter press, Figure 2.20) or pulled (vacuum filtration, Figure 2.21) through the filter cloth. The particle size of the solids will generally be larger than several hundred microns (Figure 1.3 in Chapter 1) and amenable to recovery by backflushing the filter element (in the filter press) or cutting off the accumulated solids (in vacuum filtration). While vacuum filtration (Figure 2.21) can be operated continuously, a filter press (Figure 2.20) is a cyclical or batch operation that alternates between filtration and backflushing to recover the accumulated solids.

2.11 A FLUID'S SUPERFICIAL VELOCITY, OR FLUX, THROUGH FILTER CAKE IS PROPORTIONAL TO A PERMEABILITY COEFFICIENT EXPRESSED IN UNITS OF DARCIES

The equation for pressure drop as a function of flowrate of an incompressible fluid through porous media reflects a gradual transition from laminar to turbulent flow. Consequently a general pressure drop equation contains both viscous and inertial terms:

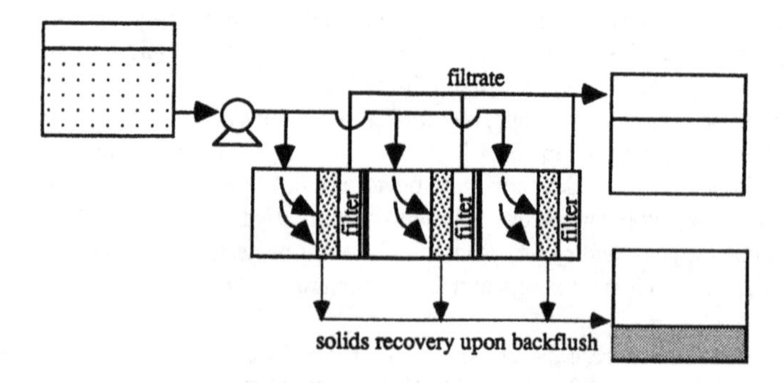

Figure 2.20 Schematic representation of a filter press.

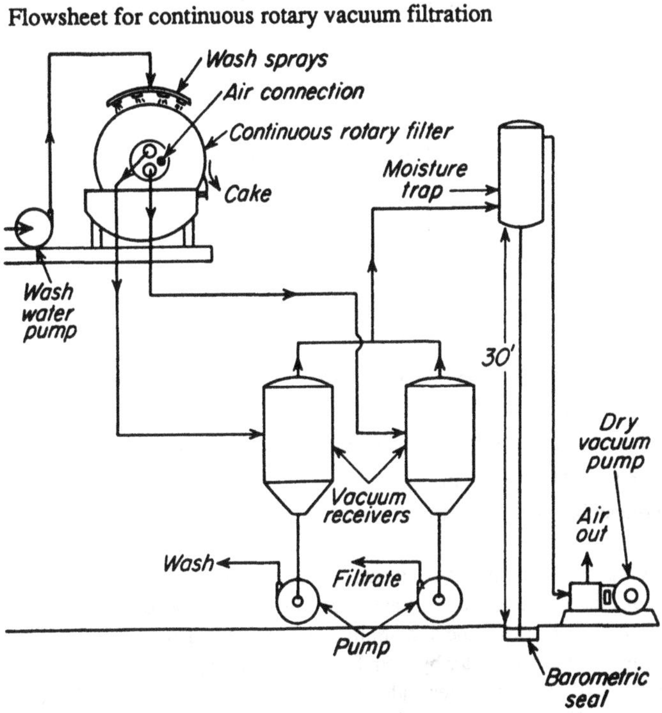

Flowsheet for continuous rotary vacuum filtration

Figure 2.21 Flowsheet for continuous rotary vacuum filtration. (Reprinted from McCabe and Smith, Figure 2B-11, page 882, with permission, McGraw-Hill, N.Y.)

$$\frac{\Delta p}{L_c} = a\mu\upsilon + b\mu\upsilon^2 \tag{2.10}$$

where a and b represent viscous and inertial resistance coefficients, respectively (Boucher and Alves, 1963). The inertial term ($b\mu\upsilon^2$) is small relative to the viscous term ($a\mu\upsilon$) for the slow flowrates encountered during filtration and flow through consolidated porous media. Hence Eq. (2.10) simplifies to an equation that includes only the viscous resistance term:

$$\frac{\Delta p}{L_c} = a\mu\upsilon \tag{2.11}$$

Equation (2.11) is a form of Darcy's law and is the basis of equations that describe filtration.

Filtration requires the fluid to flow through a porous bed of solids that increases in depth as the solids accumulate. The flux of the liquid, υ, is proportional to the pressure drop across the

The distinction between υ and u is made here, since constants for characterizing pressure drops and cake resistances reported in the literature are likely to be based on total filter area with the rate of fluid moving through the filter being defined simply in terms of "linear velocity" without specification of whether a superficial or interstitial velocity is being used. Usually fluid velocities associated with filtration denote the superficial velocity or flux. Attention to this detail will help ensure that the values of constants used reflect cake or membrane resistances in a manner that is consistant with the definition of the velocity on which the measurement of these constants was based.

Combination of Eqs. (2.12) and (2.13) results in the basic equation for filtration at a constant pressure drop (Belter et al., 1988):

$$\boxed{\frac{1}{A}\frac{dV}{dt} = \frac{\Delta p}{\mu(R_m + R_c)} = \upsilon} \tag{2.15}$$

The medium's resistance, R_m, is constant (and usually small[8] unless small solids pass into the media and plug it), while the cake's resistance, R_c, is the dominating resistance which increases as the total volume of filtered solids and cake depth, L_c, increases. Equations (2.12) and (2.15) show that under a constant pressure, the flux of fluid through the cake will decrease as the solids retained on the filter cloth accumulate and result in an increase in the thickness of the filter cake, L_c; see Eq. (2.12). The compressible nature of biological materials leads to a decrease in void fraction and therefore a decrease in the permeability k and increase in resistance R_c, Eq. (2.15), as the cake compresses. The permeability of some types of cakes may be so poor that a solid that adds structure to the cake must be added to the slurry, prior to filtration, in order to achieve reasonable filtration rates. This type of additive is called a *filter aid*.

2.12 DIATOMACEOUS EARTH AND PERLITES (VOLCANIC ROCK) SERVE TO ENHANCE PERMEABILITY OF FILTER CAKES AND AID FILTRATION OF FERMENTATION BROTHS

Bacterial and fungal fermentation broths will contain slimy or very fine solids that will compress when filtered and form a dense, impermeable cake. The fine solids in the cake can also plug the filter media in which the weave or pores are small enough to retain them. Solid, incompressible materials, added to the solids being filtered, increase the porosity of the cake and facilitate passage of the liquid through it. These additives improve filtration and are appropriately named filter aids (McCabe and Smith, 1967).

The two types of filter aids that are most effective for fermentation-derived bioproducts are diatomaceous earths and perlites. Diatomaceous earths are skeletal remains of tiny aquatic plants. Perlites are volcanic rocks processed into a coarse powder. Their pH's range from 7.0 to 10, and they can enhance the flowrate by a factor of 200 to 7500× over filtrations where the filter aids are not used (Table 2.3) (Belter et al., 1988). A wide range of bioproducts are processed using filter aids as shown in Table 2.4. The commercial filter aids are listed in the order of increasing particle size, with the largest size material giving the fastest relative flowrate but also the lowest clarity in the filtrate (Belter et al., 1988).

[8] The pressure difference across the filter's septum, filter cloth, or membrane is typically between 10 and 20 mm Hg (McCabe and Smith, 1967).

TABLE 2.3 Typical Properties of Filter Aids

	Density, kg/m^3				
Grade	Dry	Wet	pH	Water Adsorption, %	Relative Flowrate
Diatomaceous earths					
Standard					
Super Cel	130	280	7.0	260	200
512 Hyflo					
Super Cel	140	280	10.0	250	500
535	190	280	10.0	250	1400
560	310	320	10.0	220	7500
Perlites					
Terracel 300	110	260	7.5	—	300
Terracel 500	130	240	7.5	—	900

Source: Reproduced with permission from Belter, Cussler, and Hu (1988), table 2.2-1, p. 20. Reprinted by permission of John Wiley & Sons, Inc.

Note: Data are taken from Johns-Manville brochures.

2.13 FILTRATION FOR STREPTOMYCIN RECOVERY REQUIRES COAGULATION OF THE MYCELIA AND ADDITION OF A FILTER AID

Filtration has been used in the recovery of small molecules and extracellular products where larger volumes of fermentation broth are to be processed and the microbial biomass is not the product of the filtration step. Filtration requires preprocessing or pretreatment of the broth in order to enhance the otherwise poor flow properties of microbial solids (Hacking, 1986; Aiba et al., 1973). For example, heat pretreatment, adjustment of the pH of the broth to promote coagulation of the mycelial biomass, and addition of diatomaceous earth as a filter aid has enabled the industrial filtration of *Streptomyces griseus* (Aiba et al., 1973). The fermentation broth was

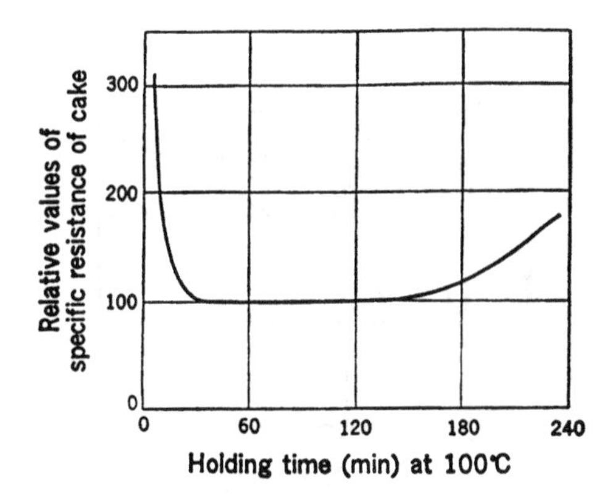

Figure 2.22 Filtration and antibiotic recovery characteristics of mycelium from *Streptomyces griseus*. Resistance of the cake to filtration is a minimum for mycelia treated at between 30 and 150 minutes at 100°C (with permission from Aiba et al., 1973, fig. 13.4b, pp. 356–357, Academic Press, copyright).

Laboratory apparatus for simulating rotary filter

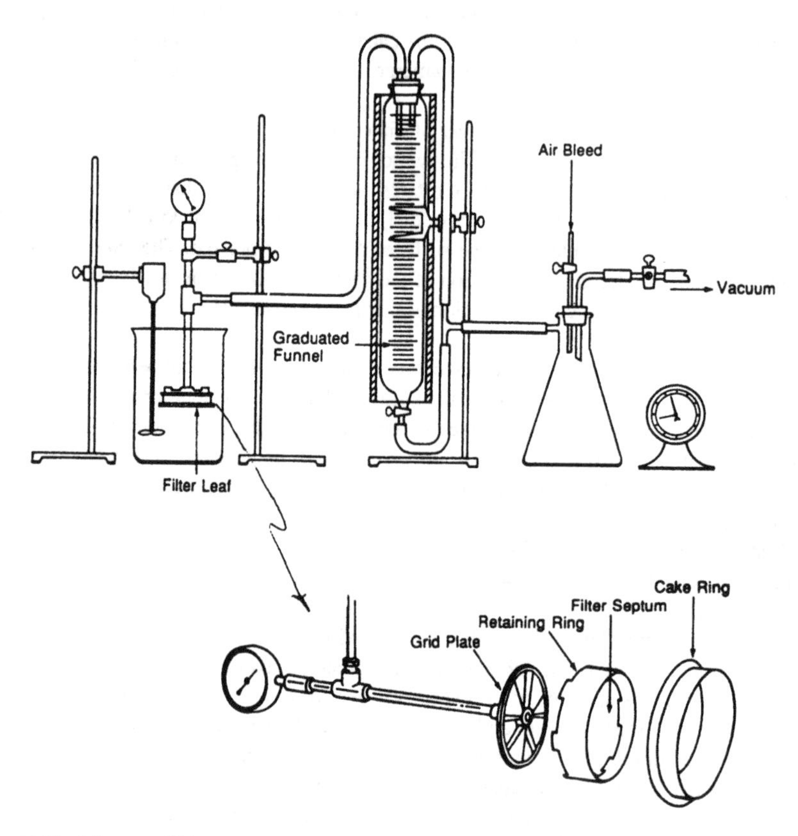

Figure 2.27 Diagram of laboratory apparatus for simulating rotary vacuum filtration rates (from Belter et al., 1988, figs. 2.5-2 and 2.5-3, p. 38, with permission from John Wiley & Sons, Inc.).

2.15 PENICILLIN G FROM *PENICILLIUM CHRYSOGENUM* IS RECOVERED BY ROTARY FILTERS PRIOR TO ITS HYDROLYSIS BY IMMOBILIZED PENICILLIN ACYLASE

Penicillin G is the precursor for the production of 6-aminopenicillanic acid (abbreviated 6-APA) and is obtained from large scale fermentation of *Penicillium chrysogenum* in fermentation vessels having volumes of several hundred thousand liters. The 6-APA product contains the basic beta lactam ring structure and is a building block for production of semisynthetic penicillins including amoxicillin, carbenicillin, hetacillin, bacanpicillin, and cyclacillin (Harrison and Gibson, 1984). The fermentation requires about 10 days and attains a final penicillin concentration of 24 to 31 g/L. The broth is filtered through a rotary precoat filter where the solids and mycelia are removed and the clarified broth containing the penicillin collected for further processing. The organic material and microbial biomass retained on the filter are properly disposed of, while the penicillin in the clarified broth is extracted, subjected to a series of crystallizations, and with the crystallized Penicillin G again recovered using a filter (Harrison and Gibson, 1984). Once the filtration steps are completed, the Penicillin G is passed through a reactor packed with beads to which the enzyme penicillin acylase is immobilized, and the Penicillin G is hydrolyzed to 6-APA as shown in Figure 2.28:

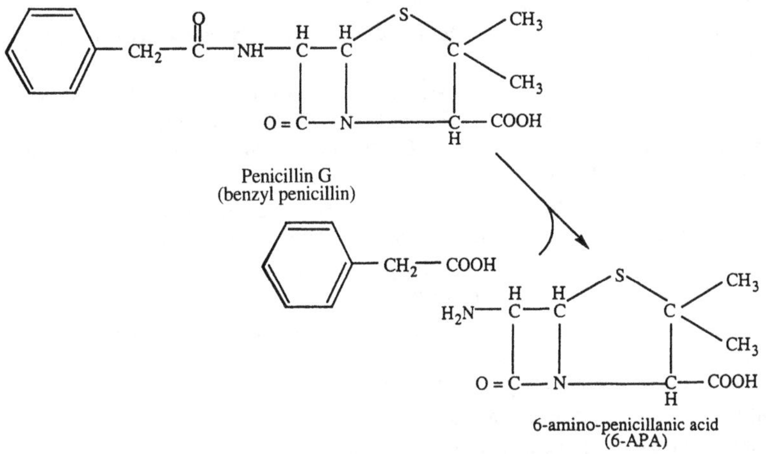

Penicillin G
(benzyl penicillin)

6-amino-penicillanic acid
(6-APA)

Figure 2.28 Hydrolysis of penicillin G (benzyl penicillin) gives 6-APA as a product.

REFERENCES

Aiba, S., A. E. Humphrey, and N. F. Millis, 1973, *Biochemical Engineering*, Academic Press, New York, 355–370.

Alford, J. S., 1991, Automatic control systems, in *Drug Biotechnology Regulation*, Y-Y. H. Chin and J. L. Gueriguian, eds., Marcel Dekker, New York, 144–153.

Ambler, C. M., 1952, The evaluation of centrifuge performance, *Chem. Eng. Progr.*, 48(3), 150–158.

Atkinson, B., and F. Mavituna, 1983, *Biochemical Engineering and Biotechnology Handbook*, Nature Press, Macmillan, New York, 17–22, 389–395, 675, 693–712.

Bailey, J. E., and D. F. Ollis, 1977, *Biochemical Engineering Fundamentals*, McGraw-Hill, New York, 709–713.

Belfort, G., 1987, in *Membrane Separation Technology: An Overview, in Advanced Biochemical Engineering*, H. R. Bungay, G. Belfort, eds., Wiley, New York, 239–297.

Belter, P. A., E. L. Cussler, and W-S. Hu, 1988, *Bioseparations: Downstream Processing for Biotechnology*, Wiley-Interscience, New York, 1–9, 15–18, 22–30, 238–239, 242–247.

Birnie, G. D., and D. Rickwood, 1978, *Centrifugal Separations in Molecular and Cell Biology*, Butterworths, London, 1–6.

Blanch, H. W., and S. M. Bhavaraju, 1976, Bioengineering report: Non-Newtonian fermentation broths; rheology and mass transfer, *Biotechnol. Bioeng.*, 18, 745–790.

Boucher, D. F., and G. E. Alves, 1963, Fluid and particle mechanics, in *Chemical Engineers' Handbook*, R. H. Perry, C. H. Chilton, and S. D. Kirkpatrick, eds., 5–52 to 5–53.

Bungay, H. R., and M. P. Millsbaugh, 1984, Cross-flow lamellar settlers for microbial cells, *Biotechnol. Bioeng.*, 26, 640–641.

Erikson, R. A., 1984, Disk stack centrifuges in biotechnology, *Chemical Eng. Progress*, 80(12), 51–54.

Greenkorn, R. A., and D. P. Kessler, 1972, *Transfer Operations*, McGraw-Hill, New York, 254–263.

Hacking, A. J., 1986, Economic Aspects of Biotechnology, Cambridge University Press, Cambridge, 127, 131–133.

Harrison, F. G., and E. D. Gibson, 1984, Approaches for reducing the manufacturing costs of 6-aminopenicillinic acid, *Process Biochem.*, 19(1), 33–36.

Hemfort, H., and W. Kohlstette, 1988, *Centrifugal Clarifiers and Decanters for Biotechnology*, Technical Scientific Documentation 5, Westfalia Separator AG, Oelde, Germany, 22–26.

Herbolzheimer, E., and A. Acrivos, 1981, Enhanced sedimentation in narrow tilted channels, *J. Fluid. Mech.*, 108, 485–499.

Ladisch, M. R., and P. C. Wankat, 1988, Scale-up of bioseparations for microbial and biochemical technology, in *The Impact of Chemistry on Biotechnology: Multidisciplinary Discussions*, ACS Symp., Ser. 362, M. Phillips, S. P. Shoemaker, R. D. Middlekauff, and R. M. Ottenbrite, eds., American Chemical Society, Washington, DC, 72 to 101.

Lehmann, H. R., and K-H. Zettier, 1985, *Separations for the Dairy Industry*, Technical Scientific Documentation 7, 2nd rev. ed., Westfalia Separator AG, Oelde, Germany, 5–8.

McCabe, W. L., and J. C. Smith, 1967, *Unit Operations of Chemical Engineering*, McGraw-Hill, New York, 162–164, 751–787, 879–904, 928–929.

Nelson, K. L., and S. Geyor, 1991, Bioreactor and process design for large scale mammalian cell culture manufacturing, in *Drug Biotechnology Regulation*, Scientific Basis and Practices, Y-Y. H. Chin and J. L. Gueriguian, eds., Marcel Dekker, New York, 112–143.

Pepplar, J. E., 1978, Yeasts, *Ann. Rep. Fermentation Processes*, 2, 191–202.

Smith, J. C., C. M. Ambler, H. L. Bullock, D. A. Dahlstrom, L. A. Dale, R. C. Emmett, C. F. Gurnham, R. F. McNamara, A. W. Michalson, S. A. Miller, J. Nardi, J. Y. Oldshue, J. T. Roberts, and E. R. Vrablik, 1963, Liquid-liquid systems, in *Chemical Engineers Handbook*, R. H. Perry, C. H. Chilton, and S. D. Kirkpatrick, eds., McGraw-Hill, New York, 19–86 to 19–100.

Walsh, T. J., and H. R. Bungay, 1979, Shallow-depth sedimentation of yeast cells, *Biotechnol. Bioeng.*, 21, 1081–1084.

Wang, D. I. C., C. L. Cooney, A. L. Demain, P. Dunnill, A. E. Humphrey, and M. D. Lilly, 1979, *Fermentation and Enzyme Technology*, Wiley-Interscience, New York, 261–267.

PROBLEMS

2.1 **a.** Develop a possible flow diagram for the preparation of an immobilized glucose isomerase from *Bacillus coagulans* based on manufacture of pellets consisting of whole, inactivated cells.

b. Calculate amount of enzyme that is to be used in a bioreactor (immobilized enzyme) in order to produce 3162 kg of fructose/year in the form of a 42% fructose syrup. Assume that there are 355 operating days/year, and that the enzyme has an operational life of 300 days. Express your answer in terms of kg *and* enzyme units.

2.2 The flocculation of brewer's yeast is being considered as a preparation step for recovering the yeast from fermentation broth. Because of the special properties of the broth, it is not possible to obtain flocculated particles that are greater than 5 mm. A previous laboratory study had shown that a vessel with a 100 cm straight side was sufficient to recover 99% of flocculated cells. Flocs with 50 mm size settled within a one minute. However, the yeast flocs that had been formed in the laboratory were much larger than those obtained in production where the size was 5 mm. What is the settling velocity for 5 mm yeast floc particles? Will the 100 cm, straight-sided vessel be sufficient to achieve 99% recovery within one minute? If not, suggest an alternate approach.

2.3 Cell mass (at 20 g/l) from a streptomycin fermentation is to be filtered through a vacuum drum filter. Conditions are as follows:

pH = 3.8

Temperature = 90°C

Δ_p = 12 psig

ρ_p = 1.12 (particle)

ρ = 1.09 (fluid)

Filter aid is to be added to the broth at a level of 150 g/L. Tests on a filter having 1.5 m^2 area give the filtration times tabulated below. Viscosity of the filtrate is 4×10^{-3} g · cm^{-1} · sec^{-1}. Cake porosity (ε) is 0.4 R_c = (c · α · V)/A where c = mass of solid deposited on the filter per unit volume filtrate and α is the average specific cake resistance.

Filtration Time, min/L	0.6	1.4	4.5	5.0	6.0
Volume of filtrate collected (L)	5.0	20.0	65.0	75.0	85.0

a. What is the value of α and R_m? Do the values of α and R_m suggest that a different filter aid should be used? Why or why not?

b. An attempt is made to decrease the cost of filter aid. A decrease of 50 g/L decreases cake porosity by 5%. Calculate the value of overall Δ_p in filter and concentration. Would you recommend that further evaluation of cost savings (by reducing filter aid) be carried out? Why or why not? Note: α = constant · $(1 - \varepsilon)/\varepsilon^3$.

c. Another approach being considered for this separation is the use of a centrifuge. A sharples 6P tubular bowl centrifuge with a 10.8 cm diameter and 10.8 cm height run at 15,000 rpm is available in the pilot plant. If the properties of the broth and q_c are as listed, determine whether w is centrifuge could be used if:

q_c = 600 liter/hr

$\rho_p - \rho$ = 0.03 g/mL

$\mu = 4 \times 10^{-3}$ g · cm^{-1} · s^{-1}

$d_{p,cut} = 10^{-6}$ m (i.e., 1 micron)

2.4 a. A different fermentation (from that in Problem 2.3) is carried out, and the resulting broth has different properties although the cell mass in this streptomycin fermentation is still ~20 g/L at the end of the fermentation. Pilot facility runs in which this broth is filtered using a system having an area of 1.5 m^2 give the results and data below. Filter aid is added to give a concentration of 10 g/L. However, for this case, the pressure drop is limited to 5 psig. The production scale facility is to process 5000 L batches with a filtration time not to exceed 60 minutes per batch. Calculate the area of the filter which would be required for this pilot plant. State your assumptions and show your work. Use the data provided below.

μ	2×10^{-3} g · cm^{-1} s^{-1}
Temp	85°C
Δp	5 psig
ρ	1.07 (for broth)
ρ_p	1.10 (for particle)
pH	3.9
ε	0.4 (cake porosity)

Working equations are:

$$\int_0^t dt = \frac{\mu}{A\Delta p} \int_0^V (R_c dv + R_m dv)$$

and

$$\int_0^t dt = \frac{\mu}{A g_c \Delta p} \int_0^V \left(\frac{\alpha c v}{A} + R_m \right) dv$$

Filtration Time, min/L	Volume of Filtrate Collected
0.5	5.0
1.2	20.0
3.8	65.0
4.2	72.0
5.0	85.0

b. Pectinase enzyme is to be isolated from about 20,000 l broth of *Aspergillis ochraceus*. The enzyme is extracellular, and is precipitated when 12,000 kg ammonium sulfate is added to a cell free broth. Give a process flowsheet for the recovery of this enzyme. Clearly identify potential problem areas.

2.5 The production of a recombinant protein involves fermentation of *E. coli*, with the production of protein being intracellular. The protein forms precipitates in the form of submicron size inclusion bodies inside the cell. The protein in the inclusion bodies is not active. It is desired to recover inclusion bodies so that the protein can later be recovered in an active form.

a. Give a conceptual process diagram that indicates the key steps in recovery of the protein in an active form (10 pts.)

b. As part of this process, a centrifuge is designed to recover the *E. coli* cells (size = 1 micron, $\rho = 1.03$) where the thickness of the liquid layer is small compared to the bowl radius. Later it is found that addition of a flocculating agent causes the cells to clump together into spheres of 14 micron size ($\rho = 1.07$). How much increase in volumetric throughput could be expected for the second case, relative to the first? (10 pts.)

Complete recovery of the cells is needed. State your basis and show your work.

2.6 The steady-state operation of a centrifuge for recovering *E. coli* results in a 1–inch layer of cells next to the wall of the centrifuge. The centrifuge is operated at a bowl rpm of 6000. The diameter of the centrifuge is 14 inches, with a wall thickness of 0.20 inches and the density of both the feed slurry and water are the same.

a. Estimate the self-stress for this centrifuge if the density of the material in the bowl is 6.5 g/mL.

b. Estimate the fraction of the total stress due to the liquid/solids that form the cake next to the wall.

c. Does the estimated self-stress exceed the design limit of ferritic austenitic stainless steel used to fabricate the centrifuge bowl? If not, what is the maximum rpm that the centrifuge can be operated at? If the self-stress limit is exceeded, how slow must the bowl's rpm be so that the self-stress limit is not exceeded.

2.7 Filtration theory applies to other forms of packed beds, including liquid chromatography columns. The pressure drop across an LC column packed with a dextran gel is given by

$$\Delta p = \frac{150(LV_o\mu(1-\varepsilon)^2)}{g_c D_p \varepsilon^3}$$

where V_0 is the average interstitial velocity.

The dextran gel packing material consists of spheres of crosslinked dextran having an average diameter of 120 microns. When packed in a bed with a 60 cm depth, the pressure drop, Δp, is 100 psig. At identical conditions, a second column of 100 cm length gives a Δp, which exceeds 800 psig and flow essentially stops after a short period of operation. Inspection of the 100 cm column shows a compacted region near the column outlet, and a change in shape of the particles near the outlet from spherical to a shape that approximates a cylinder. The average void fraction, ε, of the 100 cm column is 0.33, while that of the 60 cm column is 0.36.

Propose an explanation of why the pressure drop is so much higher in the 100 cm long column. Be specific. Does the change in ε account for the increase? Is it the change in shape, or is it something else? (50 pts.)

Note: $S_p/v_p = 6/D_p$ for spheres, cubes, and cylinders.

CHAPTER 3

MEMBRANE SEPARATIONS

INTRODUCTION

Membranes enable filtration to be extended to separations of colloids, cells, and molecules by microfiltration, ultrafiltration, or reverse osmosis. The different separation processes, summarized in Figure 3.1, show that membrane and other types of filtration processes overlap with each other (Wheelwright, 1991), although microfiltration is commonly associated with particles, while ultrafiltration and reverse osmosis separate at a macromolecular or molecular level. Membrane filters sterilize clarified solutions by removing microorganisms with a 0.45 or 0.22 μm cut-off membrane, while pore sizes of ultrafilters are in the range of about 5 to 5000 nm, and correspond to molecular weight cutoffs of 5 to 500 KD. Reverse osmosis membranes are semipermeable with an effective pore size of 0.5 to 5 nm. The basic equipment for membrane separations consists of a pump and the filtration module, as well as the associated hold tanks (Figure 3.2).

3.1 MICROFILTRATION MEMBRANES REMOVE PARTICLES WHOSE SIZES RANGE FROM 0.1 TO 10 MICRONS

Microfiltration utilizes inorganic or organic membranes having pores of 0.1 to 10 microns, although the narrow range of 0.2 to 2 microns is more typical (Ladisch and Wankat, 1988; Moto et al., 1987; Olson, 1983; Patel et al., 1987). The membranes have flat, pleated sheet, or hollow fiber configurations. The individual hollow fibers are 1 to 2 mm in diameter and are arranged in a bundled configuration which resembles a heat exchanger. Pressure from the pump that pushes the cell slurry through the hollow fibers also provides the driving force for the liquid to permeate the membrane. The rate at which the liquid permeates the membrane is on the order of 100 $L/(hr \cdot m^2)$ for a tubular microfilter at a yeast cell concentration of 100 g/L compared to

*The author acknowledges the major contribution of materials from Professor Munir Cheryan (University of Illinois) and thanks him for his suggestions. The author thanks Phil Hall and Nathan Mosier for their reading and helpful suggestions to this chapter, while they were graduate students at Purdue University.

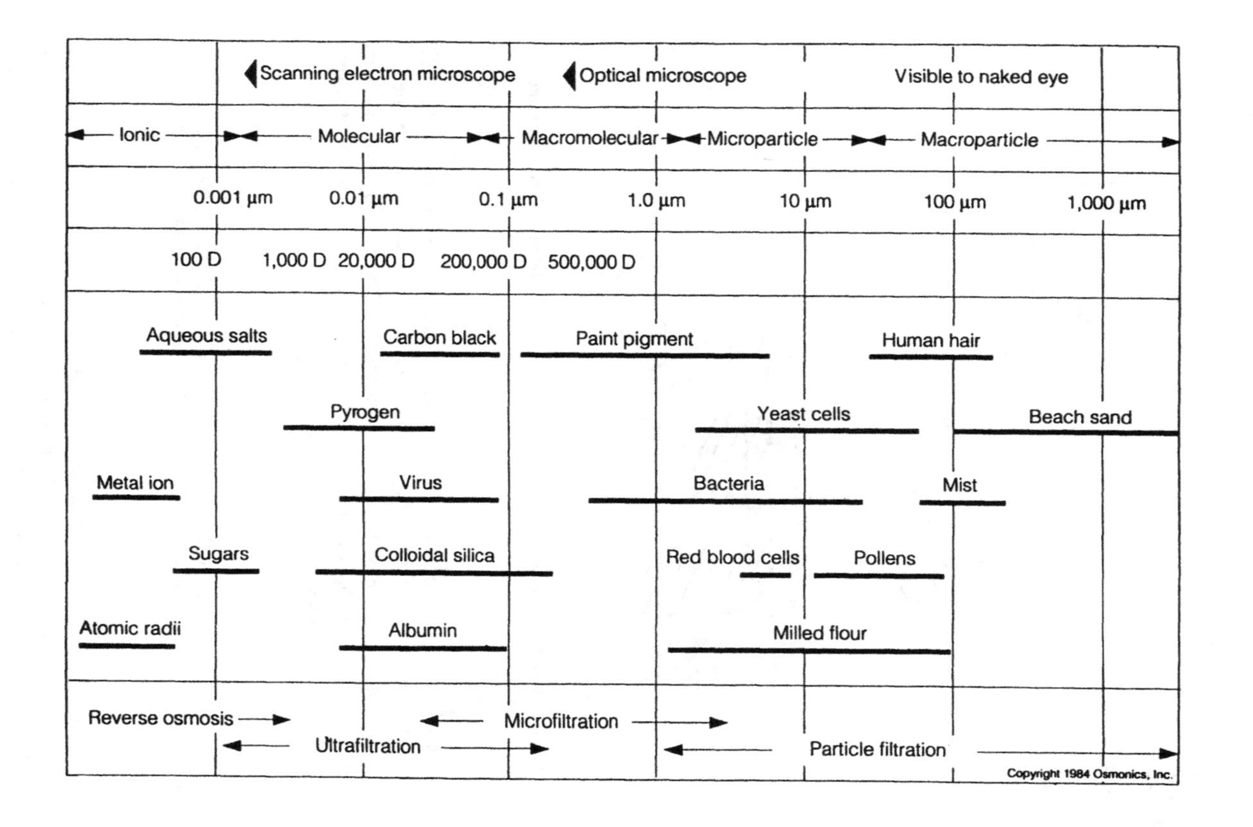

Figure 3.1 Classification of particles by size and applicable filtration separation process (courtesy of Osmonics, Inc., Minnetonka, MN, reprinted with permission, copyright 1984, Osmonics).

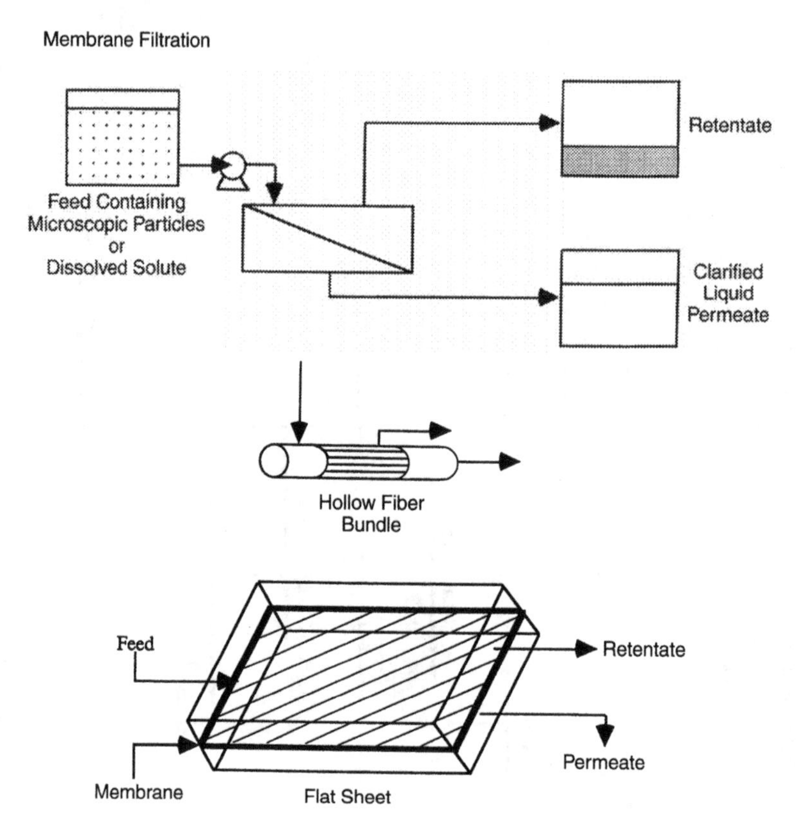

Figure 3.2 Schematic representation of membrane filtration systems showing feed, permeate, and rententate streams.

40 L/(hr · m^2) for a hollow fiber for a yeast cell concentration of 250 g/L (Patel et al., 1987). The actual pore size can vary from the rated pore diameter and hence allow some particles to leak through. For example, quantitative cell retention occurs for tubular polypropylene (0.2 micron pore size) and polysulfone hollow fibers (0.007 micron pore) but not for pleated sheet acrylic based polymer membranes (0.2 micron pore size).

3.2 MOLECULAR FILTRATION BY ULTRAFILTRATION AND REVERSE OSMOSIS UTILIZES SUPPORTED MEMBRANES WITH NANOMETER-SIZE PORES

Molecular ultrafiltration is a form of filtration carried out on a molecular level using membranes whose pores will prevent passage of solutes having molecular weights ranging from 10^2 to greater than 10^5. The membrane allows water and molecules smaller than the product to pass while rejecting the product itself. The rejected portion is the retentate, and the fluid passing through the membrane is the permeate. Since the flux increases with increasing pore size, the molecular weight cutoff is selected to be as large as possible while still resulting in retention of the solute. Laboratory scale ultrafiltration is typically carried out using flat sheet membranes, while industrial systems will use bundles of hollow fibers, or flat sheet membranes configured in stacks that enable fluid flow to pass across the membranes. Relatively high fluid tangential velocities are used in crossflow configurations to minimize buildup of the solute molecules next to the membrane's surface that would otherwise restrict flux. Ultrafiltration membranes consist of a supporting matrix upon which a film or membrane is attached. Otherwise, the equipment configuration is similar to that for microfiltration. The back pressures in ultrafiltration are as high as 600 psig, with the driving force provided by the pumping pressure. This type of filtration is used to concentrate proteins (McCabe and Smith, 1967).

Reverse osmosis membranes are characterized by pore sizes and molecular weight cutoffs are smaller than for ultrafiltration. Reverse osmosis pressures can be as high as 1000 psig, with a major application of this type of membrane being the removal of salts from water. The permeate, water, is the product, and the retentate is discarded (McCabe and Smith, 1967).

Both reverse osmosis and ultrafiltration are subject to concentration polarization in which the buildup of the rejected solute against the membrane causes a decrease in flux. A decrease in effective membrane porosity is also caused by fouling in which components adsorb on the membrane (Michaels, 1981). While induced turbulance or rapid fluid movement across the membrane surface reduces concentration polarization, avoidance of fouling requires that the feed stream be prefiltered so that it is devoid of particulates, cells, and colloidal matter.

3.3 FLUX, j_v, THROUGH A MEMBRANE FOLLOWS DARCY'S LAW WHEN THE OSMOTIC PRESSURE DIFFERENCE ACROSS THE MEMBRANE IS SMALL

The superficial velocity or flux through a membrane is proportional to the membrane's permeability as well as the difference between hydraulic and osmotic pressure driving forces (Belter et al., 1988; Belfort, 1987):

$$\boxed{j_v = K_p(\Delta P - \sigma\Delta\Pi)} = \frac{\Delta P}{\mu}\cdot\frac{1}{(L_c/k)} - \frac{\sigma\Delta\Pi}{\mu}\cdot\frac{1}{(L_c/k)} \tag{3.1a}$$

where

j_v = volumetric flux, $\frac{\text{L}}{\text{m}^2\cdot\text{day}}$ or $\frac{\text{L}}{\text{m}^2\cdot\text{hr}}$ (a rate of 1 $\frac{\text{L}}{\text{m}^2\cdot\text{day}}$ is equivalent to a superficial velocity expressed as 0.1 cm/day)

K_p = permeability constant = $\frac{k}{L_c\mu}$ (where k is a permeability constant expressed in darcies, while K_p has dimensions of $\frac{\text{cm}}{\text{s}\cdot\text{atm}}$)

μ = viscosity, cp

ΔP = transmembrane pressure drop $P_{in} - P_{out}$ (where $P_{in} > P_{out}$ and ΔP is positive)

σ = Staverman reflection coefficient; represents interaction between solute and solvent in the membrane according to Belfort (1987)

$\Delta\Pi = \Pi_{in} - \Pi_{out}$ = osmotic pressure difference

The osmotic pressure difference, $\Delta\Pi$, of the fluid is adjusted by the parameter σ to account for the effect of solutes which leak through the membrane. If the membrane were to reject all solutes so that none would pass through the membrane, then $\sigma = 1$. If both solute and solvent were to freely pass through the membrane, $\sigma = 0$. When the passage of solute through the membrane lies between these extremes, $0 < \sigma < 1$. The solute which is too large to pass through the membrane is called the retained or rejected solute.

The symbol Π_{in} denotes the osmotic pressure[1] of the concentrated solution (or retentate) on the inlet side of the membrane (upstream side of Figure 3.3), while Π_{out} is the osmotic pressure on the side of the membrane where the unretained fluid (i.e., permeate) passes (downstream side of Figure 3.3):

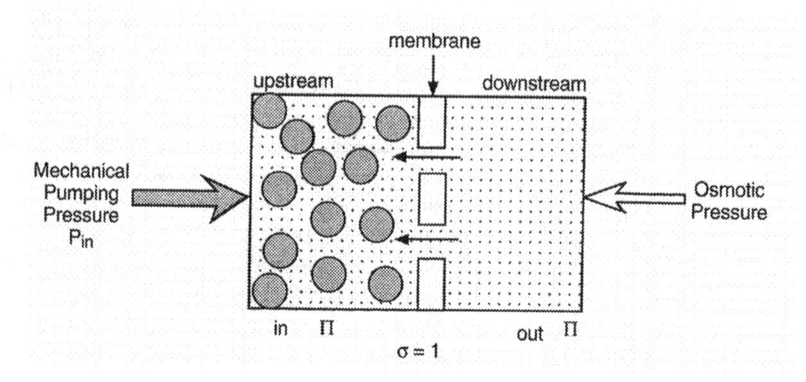

Figure 3.3 Schematic representation of osmotic pressure.

Since the pumping pressure in a membrane separation has the goal to increase the solute concentration on the upstream side, the osmotic pressure Π_{in} will also increase, resulting in an increase in the pressure difference. Hence the net flux or movement of solvent through the membrane will eventually approach zero for a given pumping pressure. The pumping pressure can be increased, but a limit is reached where the pressure is too large to be practical or exceeds the mechanical stability of the membrane. The maximum tolerable membrane pressures range from about 7 to 70 bar, depending on the type of membrane and the materials used in its synthesis. However, as described later in this chapter, osmotic pressure is not always the dominant factor that determines flux at high concentrations and/or for macromolecular solutes.

3.4 THE GIBBS AND VAN'T HOFF EQUATIONS PROVIDE A BASIS FOR CALCULATING ESTIMATES OF OSMOTIC PRESSURES

Cheryan (1986) gives an excellent synopsis of the derivation of the Gibbs equation, as well as the van't Hoff equation as a special case of the Gibbs equation, for osmotic pressure. For an open system in which matter and energy may leave the system, terms that represent the mass of the system are part of the Gibbs equation:

[1]The osmotic pressure is the pressure which results when a solvent (e.g., water) flows across a membrane to dilute a solution of a macromolecule dissolved in the same solvent on the other side of the membrane (large circles on upstream side of Figure 3.3). The macromolecule cannot pass through the membrane, since it is too large. The solvent is small enough to diffuse through to the solute side and dilute the solute. The dilution of a fixed amount of solute molecules with the solvent increases the total volume of the solute solution, and therefore causes the back pressure of the solute side (left-hand side of Figure 3.3) to increase until further passage of the solvent to the solute side stops. The pressure on the upstream side must then be increased to force solvent to flow to the downstream side (see Belter et al., 1988).

$$dG = -SdT + VdP + \mu_1 dn_1 + \mu_2 dn_2 + \mu_3 dn_3 + \ldots = -SdT + VdP + \Sigma\mu_i dn_i \qquad (3.1b)$$

where

μ_i = chemical potential of component i
n_i = number of moles of component i

The chemical potential is defined as a driving force expressed as a change in the free energy of the system as a result of the change in the composition of the system:

$$\mu_i = \left(\frac{\partial G}{\partial n_i}\right)_{T,P,n_j} \qquad (3.2)$$

where j denotes all the other components besides i. This nomenclature defines component 1 as denoting water.

The osmotic pressure for a system in which the solvent vapor behaves as an ideal gas and the liquid is incompressible is derived from the Gibbs equation as shown in Eqs. (3.3) and (3.4), below. The Gibbs equation is rewritten as

$$\left(\frac{\partial G}{\partial P}\right)_{T,n_i} = V \qquad (3.3)$$

The second derivative of Eq. (3.3) is equivalent to the derivative of the chemical potential from Eq. (3.4):

$$\left(\frac{\partial \mu_i}{\partial P}\right)_{T,n_i,n_j} = \left(\frac{\partial^2 G}{\partial n_i \partial P}\right)_{T,n_j} = \left(\frac{\partial V}{\partial n_i}\right)_{T,P,n_j} = V_i \qquad (3.4)$$

where V_i = partial molar volume of component i. Equation (3.4) results in the definition of the partial molar volume in terms of the chemical potential:

$$V_i = \left(\frac{\partial \mu_i}{\partial P}\right)_{T,n_i,n_j} \qquad (3.5)$$

and rearrangement of Eq. (3.5) results in the form

$$d\mu_i = V_i dP \qquad (3.6)$$

Upon substitution of the ideal gas law for V_i, Eq. (3.6) becomes

$$d\mu_i = RT\frac{dP_i}{P_i} \qquad (3.7)$$

A change in the chemical potential will result from a change in vapor pressure due to a change in the solute concentration. Integration of Eq. (3.7) results in an expression for the chemical potential:

$$\mu_i - \mu_i^o = RT \ln \frac{P}{P^o} \tag{3.8}$$

where $P = x_1 P^o$ (Raoult's law), P is the vapor pressure of the solution, and P^o is the vapor pressure of the pure solvent at that temperature. Rearrangement of Eq. (3.8) gives

$$\mu_1^o - \mu_1 = -RT \ln x_1 \tag{3.9}$$

where μ_1^o is the chemical potential of pure water and x_1 is the mole fraction of water. The fraction of water x_1 is always less than one, so $\ln x_1$ is negative and the right-hand side of Eq. 3.9 is positive. Since $\mu_1^o > \mu_1$, the water will move from the side that has the higher chemical potential (i.e., pure water) to the lower chemical potential, which is the side that contains the solute dissolved in water. The flow of water in the opposite direction therefore requires application of an external, hydraulic pressure, P^*, which is greater than the pressure, P_o, in order to raise the value of the chemical potential, μ_1. This is expressed in Eq. (3.10) which combines Eq. (3.9) with a pressure term:

$$\mu_1 = \mu_1^o + RT \ln x_1 + V_1 \int_{P_o}^{P^*} dP \tag{3.10}$$

where V_1 is the molar volume of an incompressible liquid. The pressure P_{in} in Figure 3.3 is equivalent to P^*.

The osmotic pressure is defined by the applied pressure, P^*, which causes the chemical potentials to be balanced, $\mu_1 = \mu_1^o$, and is represented by the integrated form of Eq. (3.10):

$$\mu_1 - \mu_1^o = RT \ln x_1 + V_1 (P^* - P_o) = RT \ln x_1 + V_1 \pi = 0 \tag{3.11}$$

or

$$\boxed{\Pi = \frac{-RT}{V_1} \ln x_1} \tag{3.12}$$

Equation (3.12) is a form of the Gibbs equation.[2]

[2]The general form of Gibbs's free energy equation is (Cheryan, 1986)

$$G = H - TS = E + PV - TS \tag{e}$$

since

$$H = E + PV \tag{f}$$

where G = Gibbs's free energy, H = enthalpy, T = absolute temperature, S = entropy, P = pressure, and V = volume. The first law of thermodynamics is

(footnote continued)

Cheryan (1986) also shows that the van't Hoff equation is a special case of Eq. (3.12) for a binary system consisting of water and a solute. If the mole fraction of water, x_1, is large relative to the mole fraction at the solute, x_2, then

$$\ln x_1 = \ln (1 - x_2) \cong -x_2 \tag{3.13}$$

Since $x_1 + x_2 = 1$, and if $n_1 >> n_2$ the mole fraction of the solute can be represented by

$$x_2 = \frac{n_2}{n_1 + n_2} \cong \frac{n_2}{n_1} \tag{3.14}$$

where n_1 and n_2 are moles of water and solute, respectively. Therefore Eq. (3.12) combined with Eqs. (3.13) and (3.14) gives the van't Hoff equation:

$$\boxed{\Pi = C_2RT = \frac{n_2}{V_1 n_1} RT} \tag{3.15}$$

where V_1 is the molar volume of the solvent (volume of solvent/moles of solvent), n_1 is the moles of solvent, and $V_1 n_1$ is the volume of solvent. Therefore C_2 is the molar concentration of the solute. Since calculation of osmotic pressure for a membrane separation system is usually based on the bulk phase concentration of a single dominating solute, this concentration will be denoted by C_B, that is to say, $C_2 = C_B$. Note that the concentrations used in Eqs. (3.12) and (3.15) are different. The osmotic pressure in the Gibbs's model, Eq. (3.12), is calculated from the mole fraction of water, while the van't Hoff model is based on the concentration of the solute. The formalism of the van't Hoff equation is analogous to the ideal gas law, although its physical significance is obviously different. Note that the dissociation of salts in aqueous solutions increases osmotic pressure, and this must be accounted for in Eq. 3.15, as described in section 3.5.

$$dE = dq + dw \tag{g}$$

where q is the heat produced and w is work. A reversible change in the system is given by

$$dq - TdS = 0 \tag{h}$$

which is the second law if only PV work is done and there is no change in the composition:

$$dw + PdV = 0 \tag{i}$$

The change in the Gibbs free energy is the derivative of combined equations (e) and (f):

$$dG = dE + PdV + VdP - TdS - SdT \tag{j}$$

Equation (j), combined with Eqs. (g) to (i), gives

$$dG = VdP - SdT \tag{k}$$

This synopsis is from Cheryan (1986).

Table 3.1 compares the osmotic pressures calculated from the two equations and shows that both underpredict osmotic pressure, although the Gibbs model is closer to the measured values. The van't Hoff equation should only be used at low molar concentrations where osmotic pressure is a linear function of solute concentration, since it is based on the approximation $x_2 \cong n_2/n_1$. At higher concentrations, the van't Hoff equation significantly underpredicts the osmotic pressure as is illustrated in Table 3.1. Despite this limitation, the van't Hoff equation can be readily used for many types of biotechnology products, particularly macromolecules, because their maximum concentrations are often much less than 1 molar. The equivalence between mole and weight concentration in Table 3.2 illustrates that a large concentration of a high molecular weight component expressed on a weight basis corresponds to a low mole concentration. This helps explain the widespread use of the van't Hoff equation, for it utilizes the familiar formalism of expressing solution concentrations in terms of the dissolved solute rather than the concentration of water. Table 3.1 also illustrates that the osmotic pressure of the low molecular weight solute, sucrose, can readily approach several hundred atmospheres for solutions that are thick syrups.

TABLE 3.1 Osmotic Pressure of Aqueous Sucrose Solutions at 30°C

	Osmotic Pressure, atm[b]		
Molality[a] of Sucrose Solution	van't Hoff's Model Equation (3.15)	Gibbs's Model Equation (3.12)	Experimental
0.991	20.3	26.8	27.2
1.646	30.3	47.3	47.5
2.366	39.0	72.6	72.5
3.263	47.8	107.6	105.9
4.108	54.2	143.3	144.0
5.332	61.5	199.0	204.3

Source: Reproduced with permission from Cheryan (1986), Table 1.3, p. 21. Reprinted from Ultrafiltration Handbook, with permission from Technomic Publishing Co., Inc., 1986.

[a]Value of the gas law constant, R, is

$$82.506 \frac{\text{cm}^3 \cdot \text{atm}}{\text{mole} \cdot \text{K}} \frac{\text{L}}{1000 \cdot \text{cm}^3} = 0.082506 \frac{\text{L} \cdot \text{atm}}{\text{mole} \cdot \text{K}}$$

[b]Molality is defined as the number of moles of solute per 1000 g of solvent.

TABLE 3.2 Equivalence of Mole and Weight Concentrations for Different Molecular Weight Components

	Concentrations	
Molecular Weight	moles/L	g/L
180	0.05	9
	0.1	18
	1.0	180
12,000	0.0005	6
	0.001	12
	0.01	120
66,000	0.0005	33
	0.001	66
	0.01	660

Osmotic pressure for bovine serum albumin

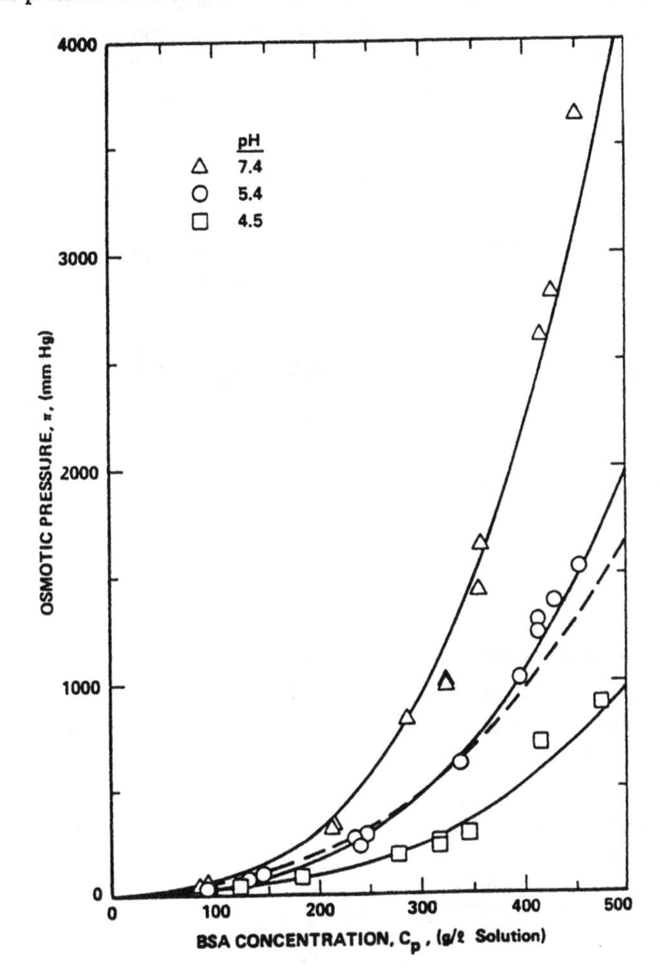

Figure 3.4 Osmotic pressure as a function of BSA concentration and 25°C in 150 mM NaCl at pH 7.4, 5.4, and 4.5. Solid lines represent best fit of a polynomial to the data. Dashed line is from the three term virial equation fig. 1, p. 70. Reprinted from Journal of Membrane Science, Vol. 20, 1984, Vilker et al., Figure 1, page 70, with permission from Elsevier Science.

An example of the deviation of the van't Hoff equation from linearity is given by measured osmotic pressures for proteins above 100 g/L is shown by Vilker et al. (1984; see Figure 3.4.). For example, the globular protein, bovine serum albumin, dissolved in 150 mM NaCl at pH 4.5, 5.4 and 7.4, respectively, exhibits osmotic pressures which follow a parabolic function of protein concentrations. At low concentrations (below about 50 g/L and perhaps up to 100 g/L) the osmotic pressures are the same at all three pH values and are consistant with values from the van't Hoff equation (3.15).[3] An osmotic pressure coefficient of 0.27 $\frac{\text{mm} \cdot \text{Hg}}{\text{g} \cdot \text{L}}$ $(= \Pi/c_B)$ was ob-

[3]Values over which the van't Hoff correlation is applicable were estimated from Figure (3.4), where the three curves converge at a BSA concentration below 100 g/L.

served as the concentration of the bovine serum albumin approached zero. Significant deviations from linearity occurred at BSA (bovine serum albumin) concentrations above 100 g/L, with an osmotic pressure of 3690 mm Hg (4.86 atm) being observed at pH 7.4 for a BSA concentration of about 440 g/L. A virial form of the van't Hoff equation, proposed by Vilker et al. (1984) to represent the osmotic pressure at large solute concentrations, is

$$\Pi = \frac{RT}{M_{\mathrm{BSA}}}\left(C_{\mathrm{B,wgt}} + B_2 C_{\mathrm{B,wgt}}^2 + B_3 C_{\mathrm{B,wgt}}^3\right) \tag{3.16}$$

where the concentration of the solute in the bulk phase (weight per volume basis) is denoted by $C_{\mathrm{B,wgt}}$ (g BSA/L) and is equivalent to $C_2 \cdot M_{\mathrm{BSA}}$. The virial coefficients are given by

$$B_2 = (V_{\mathrm{BSA}} + R_1 S_1)\left(\frac{N_{\mathrm{A}}}{1000\, M_{\mathrm{BSA}}}\right) \tag{3.17}$$

$$B_3 = \left(V_{\mathrm{BSA}}^2 + 2R_1 S_1 V_{\mathrm{BSA}} + \frac{(R_1 S_1)^2}{3}\right)\left(\frac{N_{\mathrm{A}}}{1000\, M_{\mathrm{BSA}}}\right)^2 \tag{3.18}$$

where

N_{A} = Avogadro's number = 6.0236×10^{23} molecules/mole
M_{BSA} = 69,000 = M_{B} = molecular weight of BSA[4]
V_{BSA} = molecular volume of BSA = $\frac{4}{3}\pi ab^2$ (assumes prolate ellipsoid)
= 1.5×10^{-19} cm^3 (for BSA)

$$R_1 = \frac{a}{2}\left\{1 + \frac{(1-c^2)}{2c}\ln\frac{(1+c)}{(1-c)}\right\}$$

$$S_1 = 2\pi b^2\left\{1 + \frac{\sin^{-1} c}{c(1-c^2)^{1/2}}\right\}$$

where

$\pi = 3.14159$
$c^2 = 1 - (1/p^2)$
p = axial ratio = a/b = 3.2
a, b = long and short axes, respectively, of an ellipsoid

Polynomial fits of the data gave the solid lines in Figure 3.4, while the dotted line is from the virial expansion shown in Eq. (3.16) with the parameter values defined as indicated.[5] The virial equation (3.16) only fits the data for pH 5.4. Data taken at pH 7.4 shows a higher osmotic

[4]The nominal molecular weight and isoelectric point were reported to be 69 kD and pH 4.72, respectively. The apparent molecular weight of the bovine serum albumin appears to be larger than the molecular weight of 66,100 derived from the amino acid sequence due to the presence of 5% dimers or oligomers of the protein and/or other high molecular weight protein contaminants (Vilker et al., 1984).
[5]Note: The virial equation is readily handled using a spreadsheet.

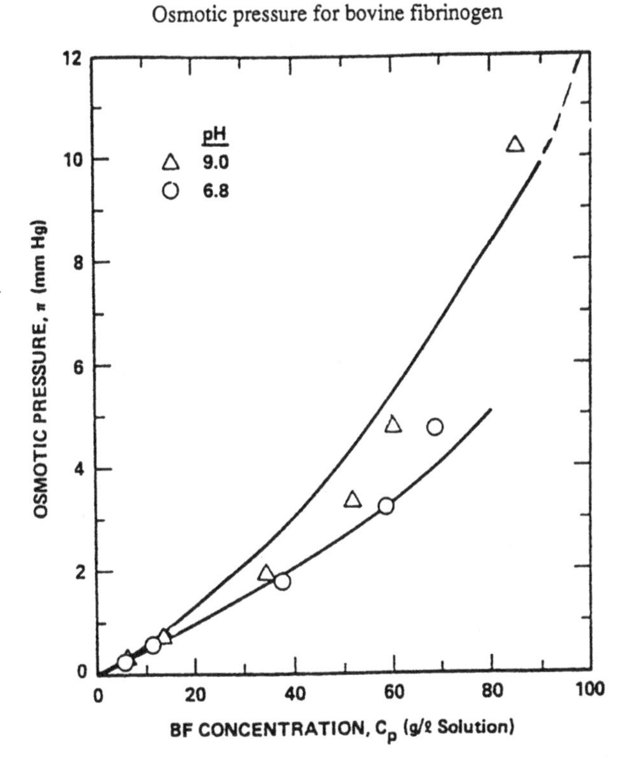

Figure 3.5 Osmotic pressure of bovine fibrinogen (MW = 390 kD) as a function of fibrinogen concentration in 0.15 M NaCl at pH 6.8 and 9.0. Solid curves were calculated by fitting data at each pH to separate cubic equations for which the fibrinogen molecular weight was the same. The dashed curve is from the three-term virial equation, Eq. (3.16), for which the molecule was modeled as a cylinder. (Reprinted from Journal of Membrane Science, Vol. 20, 1984, Vilker et al., Figure 1, page 71, with permission from Elsevier Science.)

pressure; and pH 4.5 a lower pressure (curves not shown in Figure 3.4). The pH presumably affects both size and shape of the protein, and hence would require that the parameters in Eqs. (3.17) and (3.18) be adjusted. Low-density lipoprotein is another prolate ellipsoid with $a = 21.4$ nm and $b = 9.2$ nm. As would be expected, the osmotic pressure for a concentration of 68 g/L of low-density lipoprotein is lower than for a similar concentration on a weight/volume basis for the BSA (MW = 69,000), since the molecular weight of low-density lipoprotein (MW = 3,000,000) is much larger. A water-soluble polymer, polyethylene oxide (MW = 609,000) gave similar results as well, although there was no pH dependence because it is not a charged molecule (Vilker et al., 1984).

Bovine fibrinogen is a rod-shaped protein with a molecular weight in the range of 340,000 to 400,000 and an isoelectric point of pH 5.5. A molecular weight of 390 kD was used in this work. Its osmotic pressure deviates from linearity at 40 g/L and is higher at pH 9.0 than at pH 6.8 (Figure 3.5). The dotted line that is superimposed on the polynomial fit between 85 and 100 g/L for the data taken at pH 9.0 shows that the osmotic pressures calculated from Eqs. (3.16) to (3.18), using parameters for a cylindrical molecule,[5] fit with the data:

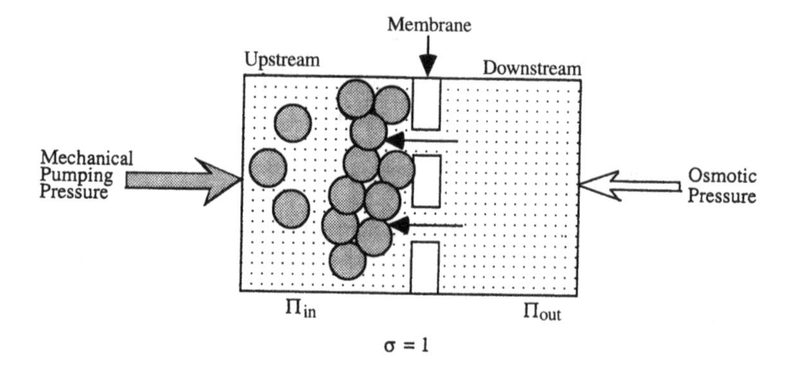

Figure 3.6 Schematic representation showing concentration of a solute or concentration polarization.

romolecules or small particulates to the surface of the membrane. A gel forms at the membrane's surface and eventually blocks flow through the membrane. This is called *fouling*, since the gel does not spontaneously dissipate when the hydrostatic pressure is removed. The avoidance of fouling requires costly pretreatment and cleaning procedures (Belfort, 1987).

Wheelwright (1991) proposes an equation for flux through a membrane, j_V, that is analogous to Eq. (2.15) for filtration, but where the resistances are defined in terms of membrane properties and solution viscosity, μ:

$$j_V = \frac{\Delta P}{\mu(R_{m,mem} + R_{m,gel})} \tag{3.21}$$

where

$R_{m,mem}$ = resistance due to the membrane $\left(\frac{\text{atm} \cdot \text{s}}{\text{cp} \cdot \text{cm}}\right)$

$R_{m,gel}$ = resistance due to concentration polarization or fouling that blocks pores in the membrane $\left(\frac{\text{atm} \cdot \text{s}}{\text{cp} \cdot \text{cm}}\right)$

The resistances $R_{m,mem}$ and $R_{m,gel}$ are a function of membrane characteristics and filtration operational conditions. The velocity of the flow across the membrane surface, the type of feed, and temperature affect the value of the resistance coefficient, $R_{m,gel}$, which reflects the formation of a concentrated layer of solute molecules next to the membrane.

3.7 FLUX INCREASES WITH INCREASING TEMPERATURE AND FLUID VELOCITY ACROSS THE MEMBRANE'S SURFACE

The viscosity of a liquid is inversely proportional to temperature and proportional to feed concentration, and hence flux should increase with increasing temperature and decreasing feed concentration given that viscosity is in the denominator of Eq. (3.1a). This is illustrated by the data of Cheryan (1986) for ultrafiltration of defatted soy flour extract (Figure 3.7) across hollow fibers. An increase in velocity across the membrane surface also increases flux as is illustrated

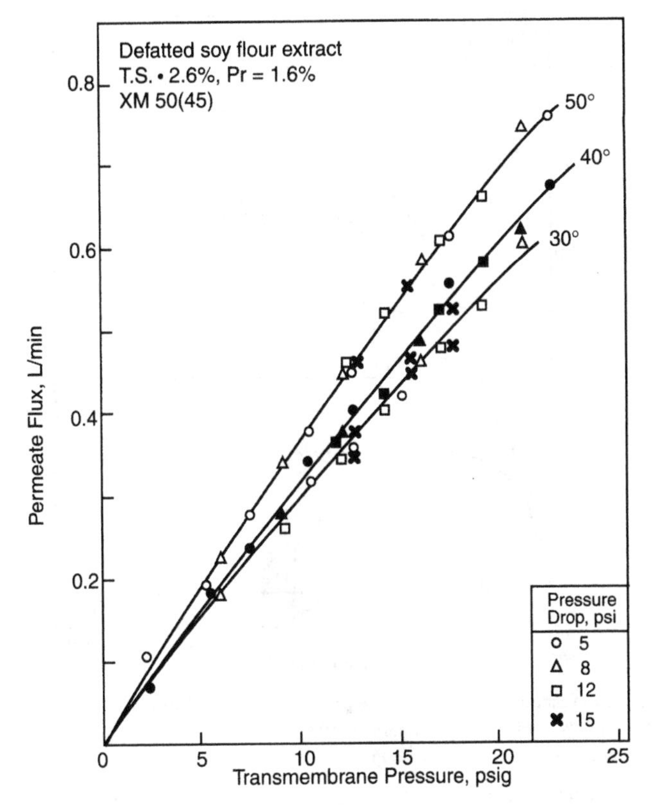

Figure 3.7 Hollow-fiber ultrafiltration. Flux as a function of transmembrane pressure and temperature for defatted, soy flour extracts (from Cheryan, 1986, fig. 4.5, p. 78, with permission from Ultrafiltration Handbook, Technomic Publishing Co., Inc.).

for skim milk in Figure 3.8. These phenomena are described by a momentum balance for a hollow fiber membrane system *when concentration polarization is not a dominating factor* and flow through the channel is laminar ($Re < 1800$) (Cheryan, 1986):

$$j_v = \frac{\varepsilon_{\text{surface}} r^2 \Delta P}{8\mu\Delta x} \tag{3.22}$$

where

j_v = volumetric flux

$\varepsilon_{\text{surface}}$ = surface porosity of the membrane (ratio of open area of the pores to the total area of the membrane)

r = mean pore radius in the membrane

ΔP = transmembrane hydraulic pressure (assume that $\Delta P << \sigma\Delta\Pi$)

μ = viscosity of fluid which permeates the membrane

Δx = thickness of the membrane's skin

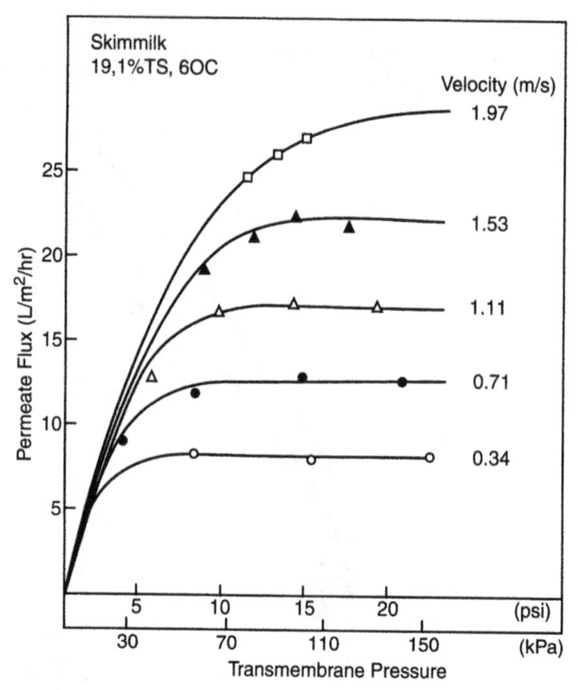

Figure 3.8 Flux as a function of transmembrane pressure for feed velocities of 0.34 to 1.97 m/s for skim milk (19.5% total solids at 60°C) (from Cheryan, 1986, fig. 4.5, p. 78). Reprinted from Ultrafiltration Handbook, with permission, from Technomic Publishing Co., Inc., 1986.

When the feed concentration or pressure is large, the flux becomes independent of pressure and mass transfer becomes the controlling factor. Therefore Eq. (3.22) is not valid in this region. The concept of transition between pressure and mass transfer controlling regimes is concisely summarized in Figure 3.9 (from Cheryan, 1986).

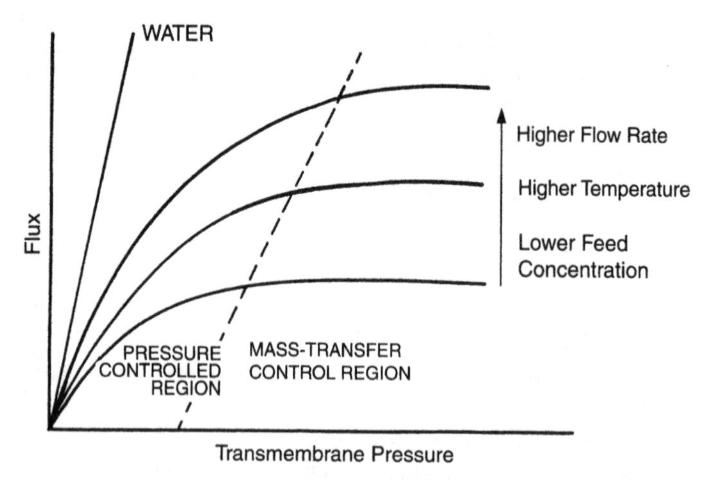

Figure 3.9 Conceptual representation of transition from pressure controlled to membrane controlled regions for flux as a function of pressure (from Cheryan, 1986, fig. 4.12, p. 82). Reprinted from Ultrafiltration Handbook, with permission, from Technomic Publishing Co., Inc., 1986.

3.8 PORE OCCLUSION AND CONCENTRATION POLARIZATION CAN BE MORE IMPORTANT THAN OSMOTIC PRESSURE IN DETERMINING FLUX FOR MEMBRANE FILTRATION OF PROTEINS

The nominal molecular weight cutoff of membranes used to study concentration of several proteins in Vilker's experiments were about 30 kD (i.e., an Amicon PM 30 membrane) (Vilker et al., 1984). These membranes were 99% impermeable to the macrosolutes. The saline reflection coefficient was 2×10^{-4}, (by Eq. 3.1a), since the Na^+ and Cl^- freely passed through the membrane with NaCl showing a linear dependence on pressure (Figures 3.10, 3.11) as would be expected from Eq. (3.20), since NaCl does not contribute a significant osmotic pressure difference. The measurements of flux through the membrane showed that the highest fluxes for a given applied pressure occurred for protein solutions exhibiting the highest, not the lowest osmotic pressures, namely, bovine serum albumin at pH 7.4 (Figure 3.10) and bovine fibrinogen at pH 9.0 (Figure 3.11). As Eq. 3.1a indicates, the opposite should have been observed.

Another reason that the leveling off of flux at relatively low transmembrane pressures is unexpected is that the membrane has large pores. Since the osmotic pressure of the high molecular weight protein is relatively small at the conditions of the experiment, an increase in flux should occur when the transmembrane pressure increases. Instead the flux leveled off at a relatively

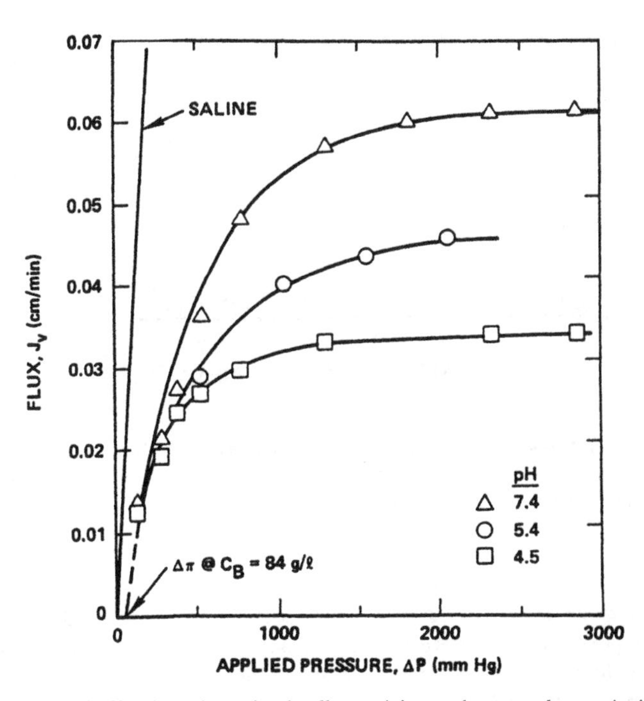

Figure 3.10 Flux of buffer through a stirred cell containing a sheet membrane, Amicon HFA-180 (180,000 molecular weight cutoff) (from Vilker et al., 1984, fig. 2, p. 70, with permission of Elsevier Science).

1

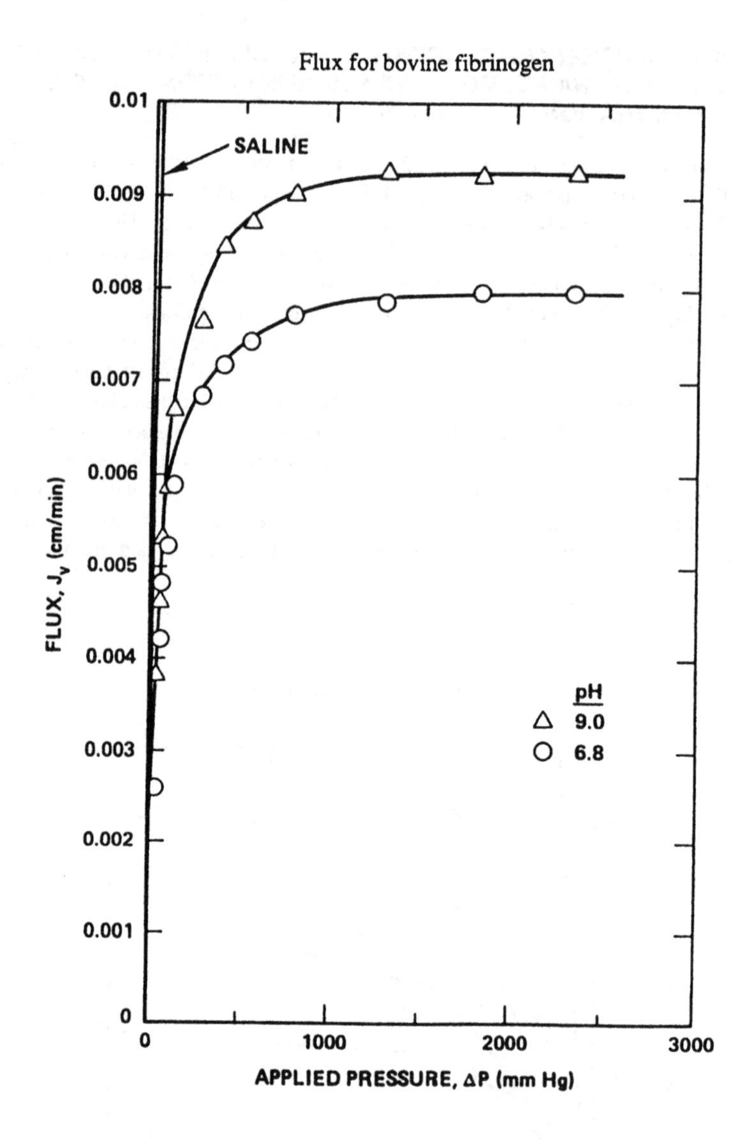

Figure 3.11 Bovine fibrinogen ultrafiltration flux from a stirred cell as a function of pressure at bulk fibrinogen concentration of 53 g/L and in 0.15 M NaCl at pH 6.8 and 9.0. Curves are visual data fits. (From Vilker et al., 1984, Figure 4, pg. 72, with permission of Elsevier Science.)

low 1000 to 2000 mm Hg (1.3 to 2.6 atm). Vilker et al. (1984) concluded that polarization of highly charged proteins at the surface of the membrane limits transmembrane flux at pressures commonly used for ultrafiltration of these solutions (Vilker et al., 1984). Hence osmotic pressure has only a secondary effect on flux when proteins are concentrated using membranes of this type. The flow-through the pores are restricted due to protein molecules that partially block or occlude the openings of the pores in a manner which is illustrated in Figure 3.6.

3.9 FLUX EQUATIONS ARE CLASSIFIED INTO OSMOTIC PRESSURE DEPENDENT, HYDRAULIC PRESSURE DEPENDENT, AND PRESSURE INDEPENDENT (CONCENTRATION POLARIZATION) REGIMES

Flux as a function of transmembrane pressure can be divided into three distinct regimes, and a transition region as illustrated in Figure 3.12. Osmotic pressure plays a role at low pressures and low solute concentrations (regime *i* in Figure 3.12). Hydraulic pressure is the dominant driver of flux as transmembrane pressure increases (regime *ii*). At the highest pressures, formation of a viscous or gel-type layer of the rejected solutes on the membrane restricts permeate flow through the membrane. At this point, mass transfer controls and pressure has little or no effect on flux. The flux is therefore constant and independent of pressure (regime *iii*). The cause of this effect is referred to as *concentration polarization* (Cheryan, 1986).

Concentration polarization is envisioned as a balance between the rate at which solute molecules are transported to and deposited on a highly concentrated zone of solute next to the membrane by fluid convection, and diffusion of the solute from the concentrated layer back into the bulk phase. This is represented by

$$j_s = j_v C_B \tag{3.23}$$

where

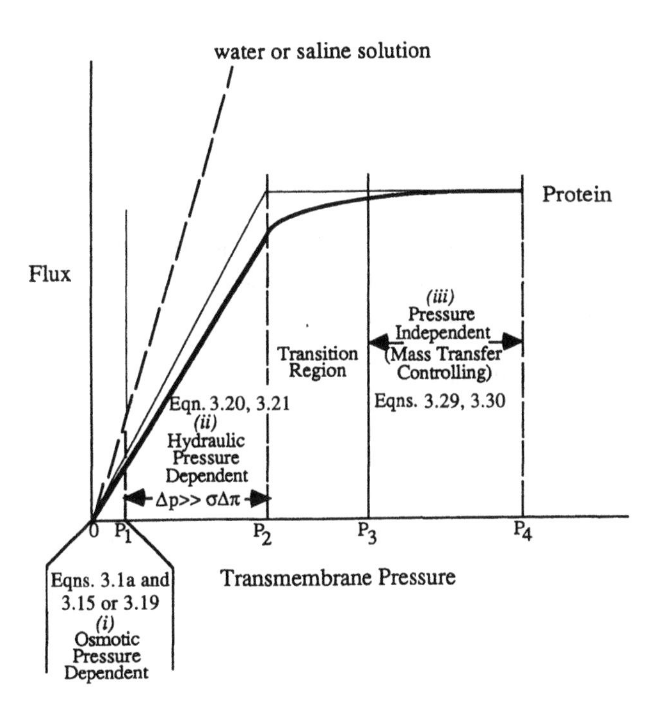

Figure 3.12 Conceptual representation of (*i*) osmotic pressure dependent, (*ii*) hydraulic pressure dependent, and (*iii*) pressure independent regions of membrane flux.

TABLE 3.6 Properties of Hollow Fiber and Tubular Membrane Units

Specification		Type of Membrane		
		Hollow Fiber	Tubular	Dimensions
d_h	Diameter	0.11	1.25	cm
L	Length	63.5	240.0	cm
n_c	Number of channels (fibers or tubes)	660.0	18.0	—
v	Superficial fluid velocity	100.0	200.0	cm/s
ΔP	Pressure drop over length if tube	0.9	2.0	kg/cm^2
γ_w	Shear rate at wall	7272.0	—	1/s

Source: From Cheryan (1986), p. 87. Reprinted from Ultrafiltration Handbook, with permission from Technomic Publishing Co., Inc., 1986.

for $0.029 \leq B \leq 0.05$, from page 74, and

$$L_c = \frac{0.1\gamma_w d_h^3}{D} = \frac{(0.1)(\gamma_m)(0.11)^3}{7.0 \times 10^{-7}} = 190\gamma_m = 1.38 \times 10^6 \text{ cm}$$

$$\frac{\left|\frac{1}{\text{s}}\right| \text{cm}^3 \Big|}{\frac{\text{cm}^2}{\text{s}}}$$

where

$$\gamma_w = \frac{(8)(100)}{0.11} \frac{\frac{\text{cm}}{\text{s}}}{\text{cm}} = 7272 \frac{1}{\text{s}}$$

Based on this calculation, the length of the hollow fiber, which is 63.5 cm according to Table 3.6, fits the criteria:

$$L < L_c \quad \text{and} \quad L > L_v$$

resulting in selection of model ii from Table 3.4.

3. Select the appropriate exponents from Table 3.4, based on results from ii, combine with Eq. (3.31), and calculate the value of the Sherwood number, *Sh*. Model ii in Table 3.4 ($Re < 1800, L < L_c, L > L_v$), also known as the *Leveque solution* (Cheryan, 1986), is used:

$$Sh = A' (\text{Re})^{\alpha} (\text{Sc})^{\beta} \left(\frac{d_h}{L}\right)^{\omega} \tag{3.34}$$

$$= (1.86)\,(1416)^{0.33}\,(11{,}095)^{0.33}\left(\frac{0.11}{63.5}\right)^{0.33} = 54$$

The expression for the Schmidt number, Sc, from Table 3.3 is calculated using the properties given in Table 3.5:

$$\mathrm{Sc} = \frac{\mu}{\rho D} = \frac{0.008}{(1.03)(7.0\times 10^{-7})} = 11{,}095$$

$$\frac{\left|\dfrac{\mathrm{g}}{\mathrm{cm}\cdot\mathrm{s}}\right|}{\left|\dfrac{\mathrm{g}}{\mathrm{cm}^3}\right|\ \dfrac{\mathrm{cm}^2}{\mathrm{s}}\left|\right.}$$

4. Calculate the mass transfer coefficient, k_{mt}, from definition of Sh in Table 3.3 and the value of the Sherwood number from step 3:

$$\frac{k_{mt} d_h}{D} = Sh = 54$$

$$k_{mt} = \frac{D\,Sh}{d_h} = \frac{(7.0\cdot 10^{-7})(54)}{0.11}\ \frac{|\mathrm{cm}^2/\mathrm{s}|}{|\mathrm{cm}|}$$

$$= 3.44\cdot 10^{-4}\,\frac{\mathrm{cm}^3}{\mathrm{cm}^2\cdot\mathrm{s}}\left(\frac{1\ \mathrm{L}}{1000\ \mathrm{cm}^3}\right)\left(10^4\,\frac{\mathrm{cm}^2}{\mathrm{m}^2}\right)\left(\frac{3600\ \mathrm{s}}{\mathrm{hr}}\right)$$

$$= 12.4\,\frac{\mathrm{L}}{\mathrm{m}^2\cdot\mathrm{hr}}$$

5. Use Eq. (3.29) together with the value of the mass transfer coefficient from step (4) and estimated values of bulk and gel phase concentrations of protein from Table 3.5, to calculate flux, j_v:

$$j_v = k_{mt}\ln\left(\frac{C_{B,gel}}{C_{B,bulk}}\right) = 12.4\ln\left(\frac{22}{3.1}\right) = 24.3\,\frac{\mathrm{L}}{\mathrm{m}^2\cdot\mathrm{hr}}$$

An analogous sequence of calculations results in a flux of 80.7 L/($\mathrm{m}^2\cdot$ hr) for the tubular membrane configuration. The results of the calculations for both cases are summarized in Table 3.7. The higher flux of the tubular unit reflects its operation in the turbulent regime, compared to the laminar regime for the hollow fibers. However, the turbulent flow regime consumes more energy, which could be an important cost consideration in the processing of a high-volume product such as milk.

This example, taken from Cheryan's *Ultrafiltration Handbook*, is one of many calculation procedures which he clearly presents. Complete discussions of membrane chemistry, structure, and function; applications; and process design of ultrafiltration systems are offered in this

TABLE 3.7 Summary of Values of Dimensionless Groups Used to Calculate Flux for the Ultrafiltration of Milk

	Parameter	Value		
		Hollow Fiber[a]	Tubular Membrane[b]	Dimensions
j_v	Flux	24.3	80.7	$L/(m^2 \cdot hr)$
k_{mt}	Mass transfer coefficients	12.4	40.5	$L/(m^2 \cdot hr)$
Sh	Sherwood number	54.0	2008.0	Dimensionless
Re	Reynolds number	1416.0	32,188.0	Dimensionless
Sc	Schmidt number	11,096.0	11,096.0	Dimensionless
d_h/L	Diameter/length ratio	0.00173	NA[b]	Dimensionless

[a]Flow regime for hollow fiber is laminar.
[b]Flow regime for tubular membrane is turbulent; hence $\omega = 0$ and d_h/L is not applicable.

Handbook, which is an excellent source of constants, properties, and suppliers for different types of membranes (Cheryan, 1986).

Foods contain significant quantities of low molecular weight solutes and exhibit relatively high osmotic pressures (Table 3.8), but their effect on the flux that occurs through ultrafiltration and microfiltration membranes is small. Hence mass transfer resistances are chosen as the basis for estimating fluxes. Ultrafiltration and microfiltration membranes have larger pores than reverse osmosis membranes (Figure 3.1) and the dominating resistances are likely to reflect concentration polarization at relatively modest solute concentrations. When flux is to be estimated for a reverse osmosis membrane, the dominating resistance may be more difficult to identify because both osmotic pressure and mass transfer resistance in the boundary layer next to the membrane's surface are important.

The concentration range over which osmotic pressure provides a major resistance to flux is also the range over which osmotic pressure is a linear function of solute concentration. Hence the linear form of the van't Hoff equation, Eqs. (3.15) or (3.19), is sufficient. Above the linear range, concentration polarization and fouling dominate resistance to flux to such a large extent

TABLE 3.8 Osmotic Pressure of Foods

Food	Total Solids, %	Experimentally Measured Osmotic Pressure, atm
Milk	11	6.8
Whey	6	6.8
Orange juice	11	15.6
Apple juice	14	20.4
Grape juice	16	20.4
Coffee extract	18	34.0

Source: Adapted from Cheryan (1986), table 1.4, p. 22. Reprinted from Ultrafiltration Handbook, with permission from Technomic Publishing Co., Inc., 1986.

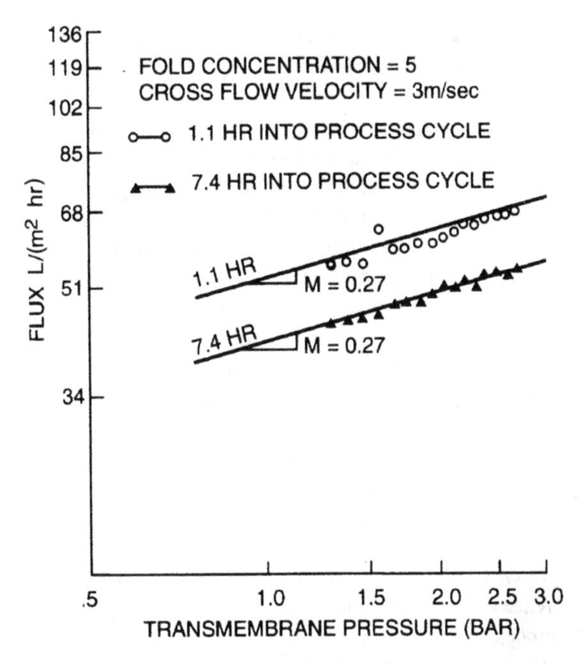

Figure 3.13 Flux in L/m^{2}hr as a function of transmembrane pressure for a flat sheet ultrafiltration membrane (from O'Sullivan et al., 1984, fig. 4, p. 70). Reproduced with permission of the American Institute of Chemical Engineers, 1984, AIChE. All rights reserved.

that the osmotic pressure contribution is often ignored. This may help explain why nonlinear osmotic pressure equations are only infrequently used.

The effect of the pressure difference, ΔP, across the membrane has also been represented by O'Sullivan et al. (1984):

$$j_v = k_f(\Delta P)^M \tag{3.35}$$

where k_f is a proportionality constant and M is the value of the slope of the line of flux as a function of transmembrane pressure (Figure 3.13). The proportionality constant, which decreases with increasing time on stream, probably represents fouling. The slope of the lines in Figure 3.13 are the same at different times, indicating that M represents a reversible compaction[11] or concentration polarization effect that increases with increasing pressure and flux. The use of Eq. (3.35) requires that experimental data be available for determining values of k_f and M. Hence the obtained first estimate of flux is less general than the dimensional analysis of Cheryan.

[11]The flux of clean, deionized water through a membrane can decrease with increasing pressure due to the compaction of the membrane. This effect is generally small and partially reversible except when a new membrane is first started up (i.e., broken in) and the rate of increase in flow with pressure is initially larger than when membrane has been in use for some time.

3.11 SOLUTE FLUX IN DIALYSIS IS BASED ON A CONCENTRATION GRADIENT, NOT A PRESSURE GRADIENT

Dialysis achieves separation through a concentration gradient in which the solute diffuses through the membrane, with some backdiffusion of water. This differs from the other membrane separations discussed so far in which a hydraulic (pumping) pressure is applied to the feed in order to force water through the membrane and concentrate the solute or particles. Unlike reverse osmosis, ultrafiltration, or microfiltration, the operation of a dialysis system attempts to balance the hydrostatic pressure on both sides of the membrane in order to avoid stress on the membrane. Industrial scale dialysis was practical over 35 years ago, using large membrane areas. One example cites a system with an area of 3278 ft^2 or 300 m^2; (see Tuwiner et al., 1962).

The general equation for mass transfer due to dialysis, j_d, across the membrane takes the form of the familiar Darcy equation (Tuwiner et al., 1962):

$$\boxed{j_d = K_{P,\text{dialysis}} \Delta C} \tag{3.36}$$

where $K_{P,\text{dialysis}}$ represents a permeability coefficient for dialysis and ΔC is the transmembrane concentration difference (rather than the pressure difference). The permeability coefficient reflects two parallel processes of diffusion. One is for diffusion of the solute into the membrane (represented by $2\lambda/D_{sm}$), and the other is for diffusion within the membrane ($L_{mem}/(D_{mem}K)$:

$$\frac{1}{K_{P,\text{dialysis}}} = \frac{2\lambda}{D_{sm}} + \frac{L_{mem}}{D_{mem}K} \tag{3.37}$$

where

λ = molecular displacement parameter = L_{mem}/m (where m is the number of steps that the molecule takes when it moves across the membrane)
D_{sm} = diffusion coefficient for solute diffusing into the membrane
D_{mem} = diffusion coefficient of solute within the membrane
L_{mem} = thickness of membrane
K = partition coefficient for the solute between the two sides of the membrane

The thickness of the membrane is significantly greater than the average step size for the molecular displacement; hence $L_{mem} >> \lambda$ and Eq. (3.37) reduces to

$$K_{P,\text{dialysis}} \cong \frac{D_{mem}K}{L_{mem}} \tag{3.38}$$

The utility of Eq. (3.38) is illustrated by the experimentally measured diffusion of methyl and ethyl ureas through the plant cell membranes of *Chara ceratophilla*. The data are plotted according to Eq. (3.37), and they give an intercept (= $2\lambda/D_{sm}$) of close to zero and a slope of the line equivalent to L_{mem}/D_{mem} (Figure 3.14).

Membrane performance is characterized by a performance factor, U_o (sometimes also referred to as a coefficient of performance). This factor is expressed in terms of the amount of

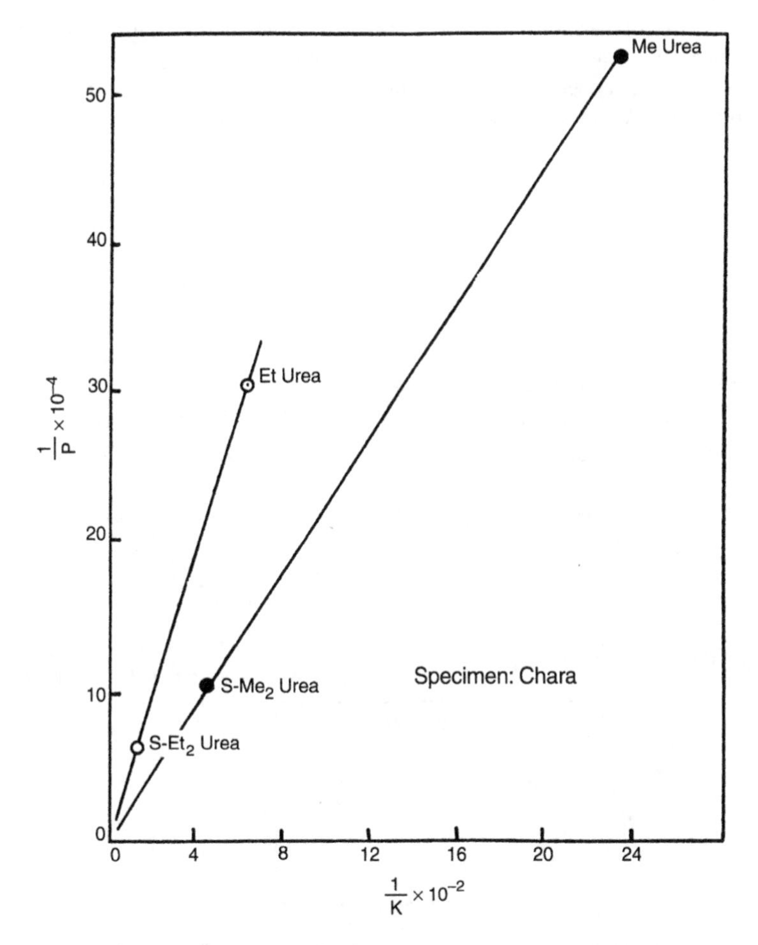

Figure 3.14 Graph showing linear relationship between inverse permeability coefficient ($1/P$) versus inverse of the partition coefficient ($1/K$) for diffusion of ethyl and methyl ureas through a plant membrane (from Tuwiner et al., 1962, fig. 2.4, p. 51, with permission).

solute recovered (dry weight basis) per unit time and membrane area, A, for a given concentration difference between the feed and dialysate as (Tuwiner et al., 1962)

$$\boxed{U_o = \frac{W}{A C_{lm}}} \tag{3.39}$$

where

W = rate of solute recovery in the dialysate, lb/hr

A = membrane area, ft^2

C_{lm} = average concentration (g/mL)

U_o = performance factor, $\frac{\text{lb} \cdot \text{mL}}{\text{hr} \cdot \text{ft}^2 \cdot \text{g}}$

The dimensions of the performance coefficient reflect the measurements used to determine its value, and hence are somewhat unusual, with the form of Eq. (3.39) being analogous to the definition of a heat transfer coefficient. The parameter C_{lm} represents the log mean concentration driving force for diffusion across the membrane, and it is given by

$$C_{\mathrm{lm}} = \frac{C_{\mathrm{f}} - C_{\mathrm{d}} - C_{\mathrm{o}}}{\ln\left(\dfrac{C_{\mathrm{f}} - C_{\mathrm{d}}}{C_{\mathrm{o}}}\right)} \tag{3.40}$$

where

C_{f} = concentration of feed to the dialysis unit
C_{d} = concentration of the diffusate (solute that has diffused across the membrane)
C_{o} = outlet concentration of the dialysate (liquid from which the solute has been removed)

These quantities are represented in Figure 3.15. The pressures on the diffusate (P_{d}) and dialysate sides (P_{o}) are assumed to be the same.

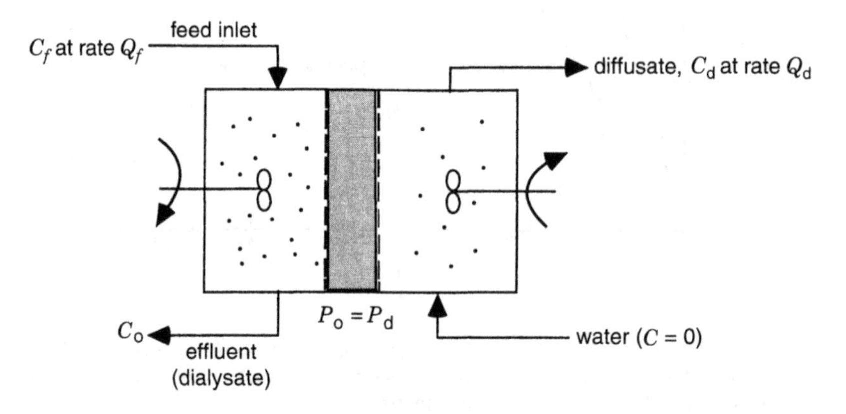

Figure 3.15 Schematic representation of principles of dialysis.

The maximum rate of solute transfer per unit area is given by :

$$\frac{W_{\max}}{A} = U_{\mathrm{o}} C_{\mathrm{f}} \tag{3.41}$$

where C_{f} is the concentration of the solute in the feed. For example, if

$$C_{\mathrm{f}} = 0.2 \frac{\mathrm{g}}{\mathrm{mL}}$$

and

$$U_o = 2.0 \frac{\text{lb} \cdot \text{mL}}{\text{hr} \cdot \text{ft}^2 \cdot \text{g}}$$

then

$$\frac{W_{max}}{A} = U_o C_f = (2.0)\,(0.2) = 0.4 \frac{\text{lb}}{\text{hr} \cdot \text{ft}^2} \tag{3.42}$$

The concentration of the solute in the diffusate (right-hand side of Figure 3.15) will be significant relative to the inlet or feed concentration, and hence the actual rate of solute transferred, W, is a fraction, F, of the maximum rate:

$$\frac{W}{A} = F \frac{W_{max}}{A} \quad \text{and} \quad C_{lm} = F C_f \tag{3.43}$$

The ratio, M, of the rate of diffusate, Q_d, to feedrate, Q_f ($M = Q_d/Q_f$), combined with the % recovery, R, of the solute in the diffusate enables the split of the solute, and therefore the concentration of the solute in the diffusate and effluent streams, to be calculated. A material balance around a dialysate unit shown in Figure 3.15 is

$$Q_d C_d = Q_f\,[C_f - C_o] = Q_f \left[C_f - \left(1 - \frac{R}{100} \right) C_f \right] = Q_f C_f \left(\frac{R}{100} \right) \tag{3.44}$$

where

R = % recovery of solute in diffusate $\frac{C_f - C_o}{C_f} \cdot 100$
C_f = concentration of solute in feed
Q_f = feedrate
C_d = concentration of solute in diffusate
Q_d = flowrate of diffusate

If the ratio of the flowrates of diffusate to feed is given by $M = Q_d/Q_f = 1.1$ and half of the total weight of the solute diffuses through the membrane and is recovered in the diffusate stream ($R = 50\%$), the concentration of the solute in the diffusate is

$$C_d = C_f \frac{Q_f}{Q_d} \frac{R}{100} = C_f \frac{1}{M} \left(\frac{R}{100} \right) = (0.2) \left(\frac{1}{1.1} \right) \left(\frac{50}{100} \right) = 0.0909 \tag{3.45}$$

then the log mean concentration can be calculated by Eq. (3.40):

$$C_{lm} = \frac{0.2 - 0.0909 - 0.1}{\ln \left(\frac{0.2 - 0.0909}{0.1} \right)} = 0.1044$$

The fraction of the maximum rate at which the solute is transferred is given by Eq. (3.43):

$$F = \frac{C_{lm}}{C_f} = \frac{0.1045}{0.2} = 0.522$$

The transfer capacity in the absence of back diffusion of water into the feed is given by Eqs. (3.36) and (3.43) where the transmembrane concentration difference is expressed as a log mean concentration:

$$j_d = U_o C_{lm} = \frac{W}{A} = F\frac{W_{max}}{A} = (0.522)\,(0.4) = 0.2089\,\frac{\text{lb}}{\text{ft}^2 \cdot \text{hr}} \tag{3.46}$$

If the ratio of feed to diffusate is extremely large ($M = 1000$), then $C_d = 0.0001$ and $C_{lm} = 0.144$. The value of F is therefore 0.72. If $M = 0.6$, $C_d = 0.167$, then $C_{lm} = 0.0606$ and F is 0.303. These calculations may be summarized in a graphical form that specification of the % recovery, R, and the ratio of diffusate to feed, M, enables quick determination of the fraction F. Once F is known, Eq. (3.42) can be used to calculate the mass transfer per unit area. The amount of dialysis membrane area needed to carry out dialysis of 100 lb/hr of a solute is obtained by:

$$A = \frac{100 \cdot \text{lb/hr}}{0.2089\ \text{lb/ft}^2 \cdot \text{hr}} = 479\ \text{ft}^2 \left| \frac{1}{10.76}\,\frac{\text{m}^2}{\text{ft}^2} \right| = 44.5\ \text{m}^2$$

A measure of relative rates of dialysis is given by the 50% escape time, namely, the time required for half of the solute to dialyze through a membrane. The rate for batch dialysis is

$$k_{dialysis} t = -M \log\left(\frac{C_f - C_o}{C_f}\right) = -M \log\left(\frac{R}{100}\right) \tag{3.47}$$

where

$k_{dialysis}$ = dialysis rate constant
t = time
M = ratio of volume of feed to total volume of product (i.e., diffusate)
C_f = concentration of solute in feed
C_o = concentration of solute in dialysate

When $C = 0.5C_o$, Eq. (3.47) becomes

$$t_{1/2} = \frac{-M}{k_{dialysis}} \log\left(\frac{C_o - 0.5C_o}{C_o}\right) = 0.301\,\frac{M}{k_{dialysis}} \tag{3.48}$$

The values of $t_{1/2}$ for various solutes aqueous acetic acid are given in Table 3.9. The type of membrane for which these measurements were made is not given, although it is likely that the membrane is cellulose based. The data presented in Table 3.9 illustrate that the dialysis rates for small molecules relative to proteins (i.e., insulin and ribnuclease) are significantly different so that a clean separation of salts from proteins can be achieved. The values of the escape times are in the same range as modern microfiltration or diafiltration systems. This suggests that di-

TABLE 3.9 Comparative Dialysis Rates of a Number of Solutes

		$t_{1/2}$, min	
Substance	Molecular Weight	In 0.1N Acetic Acid	In 50% Acetic Acid
Sodium chloride	58	10	37
Urea	60	10	
Phenylalanine	165	33	85
Aspartic acid	133	40	150
Ornithine monohydrochloride	132	26	58
Glucose	180	35	60
Sucrose	342	60	192
Raffinose	504	190	240
DNP-glycine	241	—	96
DNP-Gly-Leu	354	—	240
Polymyxin B hydrochloride	1220	60	—
Polypeptin hydrochloride	1150	66	—
Tyrosidine A hydrochloride	1270	—	1740
Bacitracin A	1500	162	522
Gramicidin A	3800	—	4560
Insulin	6000	No detectable escape	—
Ribonuclease	13,900	No detectable escape	—

Source: Adapted from Cheryan (1986), Table 14.3, p. 32. Reprinted from Ultrafiltration Handbook, with permission from Technical Publishing Co., Inc., 1986.

alysis is a viable option for industrial-scale membrane processes for salt removal. However, while dialysis is a commonly used laboratory technique, it is rarely found on an industrial scale for the desalting of proteins.

ENGINEERING CONCEPTS OF MEMBRANE APPLICATIONS

Membranes used in the industry and in the laboratory are available in flat sheet, spiral wound, and hollow fiber configurations. Their application to concentration, filtration, and protein recovery in the food industry, and particularly, the dairy industry is widespread. Membranes are also widely used in the biotechnology products industry, where flat sheet and hollow fiber membranes are used for cell recovery, washing, and concentration. The concepts that govern selection and design of membrane filtration systems are illustrated by applications that range from the diafiltration of bacterial and yeast cells to the recovery of products from mammalian cell culture. Practical issues that must be considered include the operational lifetimes of different types of membranes, testing of membranes for leaks and porosities, probability of bacterial propagation during sterile filtering designed to remove microorganisms from a bioproduct, estimating physical property data, and anticipating the impact of concentration polarization effects.

3.12 MEMBRANE SEPARATIONS OF SMALL PARTICLES UTILIZE FIBROUS OR PARTICULATE DEPTH FILTERS AND ISOTROPIC (SYMMETRIC) SCREEN FILTERS

Microfilters remove cells, cell debris, and colloids by physical retention of the particles or molecules on one side of the filter media in the manner illustrated in Figure 3.16 (Wheelwright,

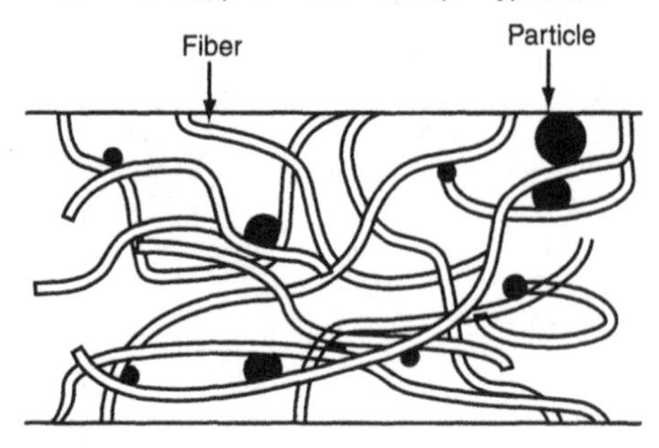

Figure 3.16 Depth filters. Figure representation of fibrous depth filter (from Wheelwright, 1991, fig. 9.2, p. 102, with permission).

1991). *Depth filters* trap the particles as they are pushed into the tortuous paths between the fibers that make up the matrix of the filter (Figure 3.16). Particles that are retained by the filter make the pathways even more tortuous, and then further promote the capture of more particles. This type of filter is manufactured from fibrous materials or fibers, including cellulose in the form of paper, fabric, or compacted bundles (Figure 3.17). Other types of depth filters are based on fiberglass or nylon (Figure 3.18). Particle-based depth filters are made from sintered steel, alumina, silver (Figure 3.19), and ceramics. An example of a stainless steel particle based depth

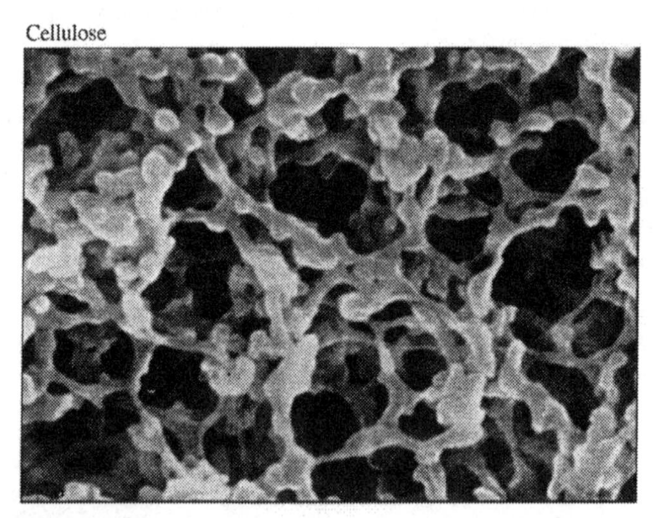

Scanning electron micrograph of surface of MCE membrane filter

Figure 3.17 Scanning electron micrograph (*top view*) of mixed cellulose ester membrane filter (courtesy of Poretics, 1995, *Microfiltration and Laboratory Products Catalog*, p. 40): Membranes are available in different average pore size cutoffs that range from 0.10 to 8.0 microns with membrane thickness of 150 microns.

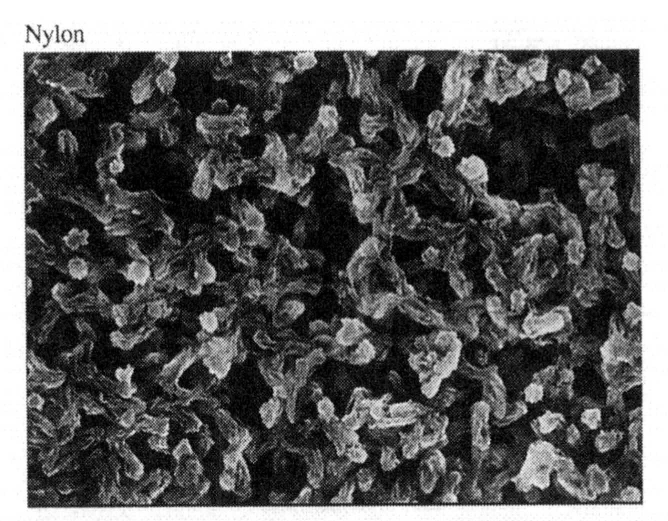

Photomicrograph of the surface of nylon membrane

Figure 3.18 Scanning electron micrograph (*top view*) of nylon membrane (courtesy of Poretics, p. 44, *1995 Catalog*). Pore sizes of different membranes range from 0.22 to 5 microns.

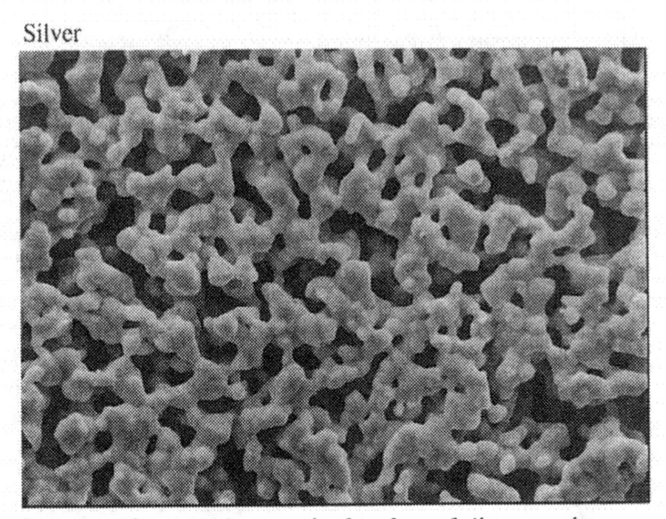

Scanning electron micrograph of surface of silver membrane filter, magnified 1,500X.

Figure 3.19 Scanning electron micrograph of silver membrane at 1500 × magnification, top view (courtesy of Poretics, p. 37, *1995 Catalog*). Pores sizes of different silver membranes range from 0.2 to 5.0 microns.

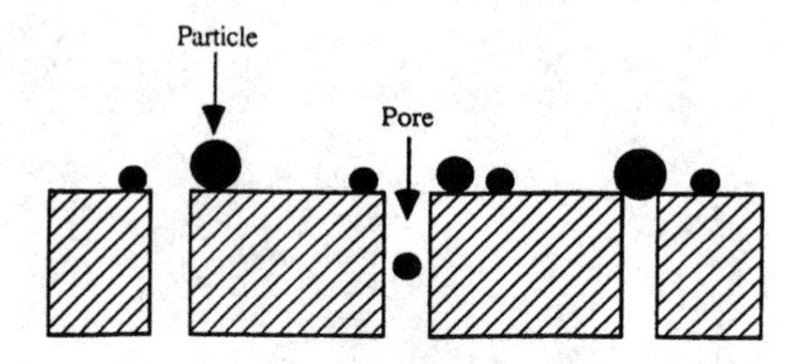

Figure 3.20 Screen filters: Schematic representation (from Wheelwright, 1991, p. 102, with permission).

filter is the 2 micron cutoff, sintered steel end fitting that is commonly used as an end fitting on analytical chromatography columns. Depth filters can be relatively inexpensive, but some types have the disadvantage of shedding fibers into the filtrate or entrapping the particles being filtered in a manner that makes it difficult to recover them. A depth filter is operated in a deadend mode[12] and will often be followed by a screen in order to filter out the small particles that may have been released from the depth filter (Wheelwright, 1991).

Screen filters (Figure 3.20) have a controlled pore size, and they consist of a microporous membrane designed to reject particles that are larger than the membrane's rated pore size cutoff. These filters are further divided into two categories:

1. Isotropic, symmetric membranes
2. Anisotropic, assymmetric membranes

According to Wheelwright, microfiltration membranes are typically isotropic, while ultrafiltration membranes are anisotropic (Wheelwright, 1991).

Isotropic membranes have defined pores that typically follow a tortuous path. An exception is illustrated in Figure 3.21 where the pores are relatively straight. The pore diameters in such a membrane will be slightly larger than the particle that they retain (Figure 3.22). Isotropic membranes with a relatively rough surfaces may capture (and become plugged by) deformable materials such as cells or cell debris that are swept across its surface. Hence isotropic membranes are typically used in systems where filtration is carried out in a deadend mode.

Anisotropic membranes with a pore-limiting skin are smooth and allow particles or molecules that are larger than the pores to freely move across the surface without getting caught (Wheelwright, 1991). An example of this type of membrane is given by the smooth surface on the inside of a hollow fiber (Figure 3.23). An anisotropic membrane has only a small capacity when used in a deadend mode, since particles quickly plug the pores in the filter structure.

[12]When filters are operated in a deadend mode, the process stream flows directly at the membrane, that is the flow is in a direction normal (at a right angle) to the filter, and nearly all of the liquid addressed to the filter passes through it (Olson, 1983). Crossflow filtration refers to systems where the solution passes over the plane of the filter in a parallel or tangential direction. A significant amount of liquid passes out of the filter together with the retained particles.

Screen Filter

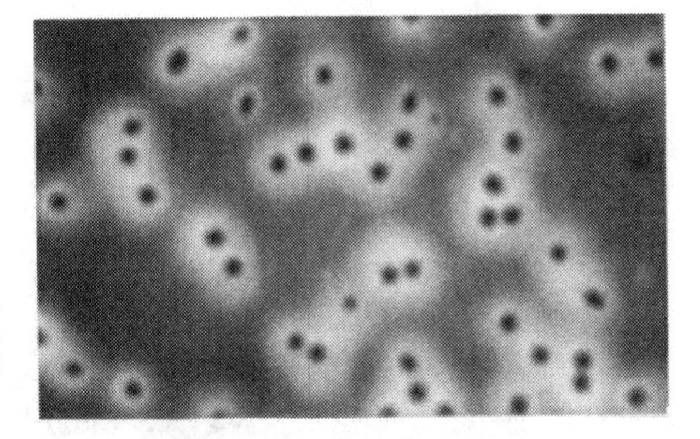

Scanning electron micrograph of surface of Poretics PCTE membrane filter.

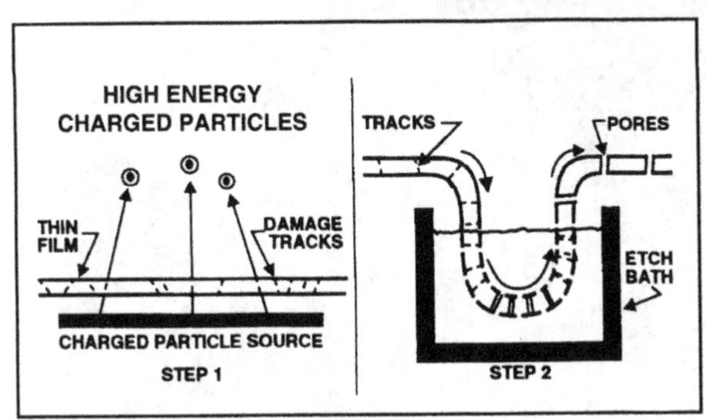

How polycarbonate screen membrane filters are made. In the first step, thin plastic film is exposed to ionizing radiation, forming damage tracks. In the second step, the tracks are preferentially etched out into pores by a strong alkaline solution.

Figure 3.21 Scanning electron micrograph of surface of a polycarbonate track-etch membrane (courtesy of Poretics, 1995 catalog).

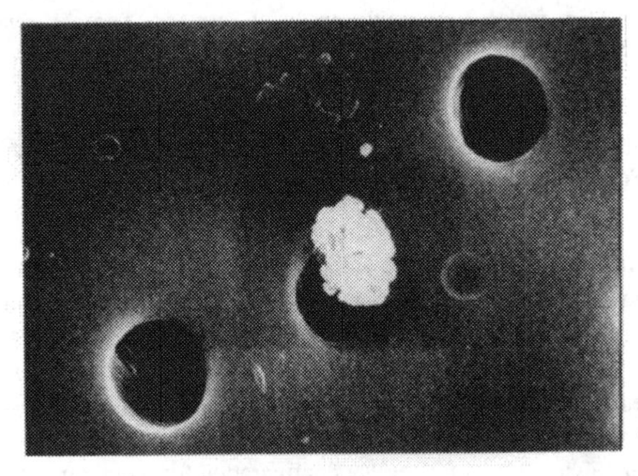

Dust particle on surface of PCTE screen membrane filter.

Figure 3.22 Dust particle covering a pore on surface of membrane (courtesy of Poretics, pp. 14, 15, 1995

Hollow-Fiber Membrane

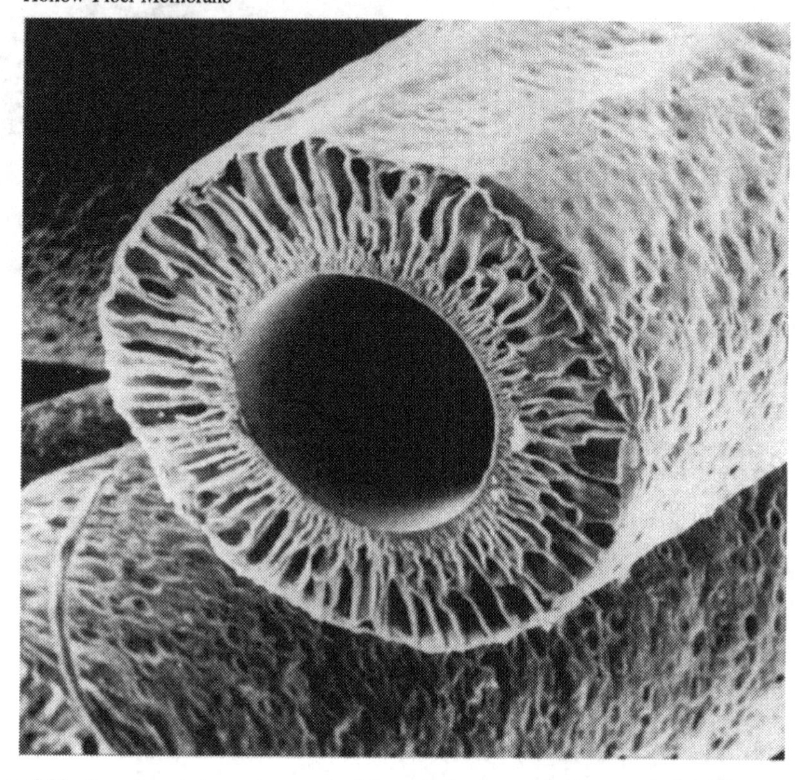

Figure 3.23 Scanning electron micrograph of the end of an assymmetric support membrane structure, with a pore-limiting skin (inner part of fiber). The thick wall of the hollow fiber imparts strength, while the skin or active membrane (0.1 micron thick) gives the controlled pore size. Flow occurs on the inside of the hollow fiber (diameter of 0.5 mm) in a direction parallel to the axis (i.e., along the length of the tube) and generates a sheer force that minimizes concentration polarization of the solute retained on the inside of the fiber. The support structure enables intermittant backflushing to be used (see Figure 3.25) to periodically force removal of foreign materials (i.e., foulants) and the flux inhibiting layer away from the membrane (skin) surface. Courtesy of Rhomicon.

3.13 MEMBRANES ARE PACKAGED IN FLAT-SHEET OR HOLLOW-FIBER CARTRIDGE CONFIGURATIONS

Membranes are configured in process equipment in order to give the desired flow pattern, as well as to maximize the amount of filter surface area per unit volume. Flat sheets are utilized in a plate and frame configuration that is similar in concept to the schematic representation of a filter shown in Figure 2.20, except that a membrane would be placed on top of the filter, with the filter providing the necessary mechanical support (O'Sullivan et al., 1984). The channels are narrow, and fluid flow occurs across the membrane surface, with the retentate leaving the filter in a continuous manner (Figure 3.2). An alternate configuration for membrane sheets is in the form of a spiral wound module (Figure 3.24). This configuration consists of a membrane that is laid on top of a mesh carrier, and wound into a spiral (sometimes called a *jellyroll*). The

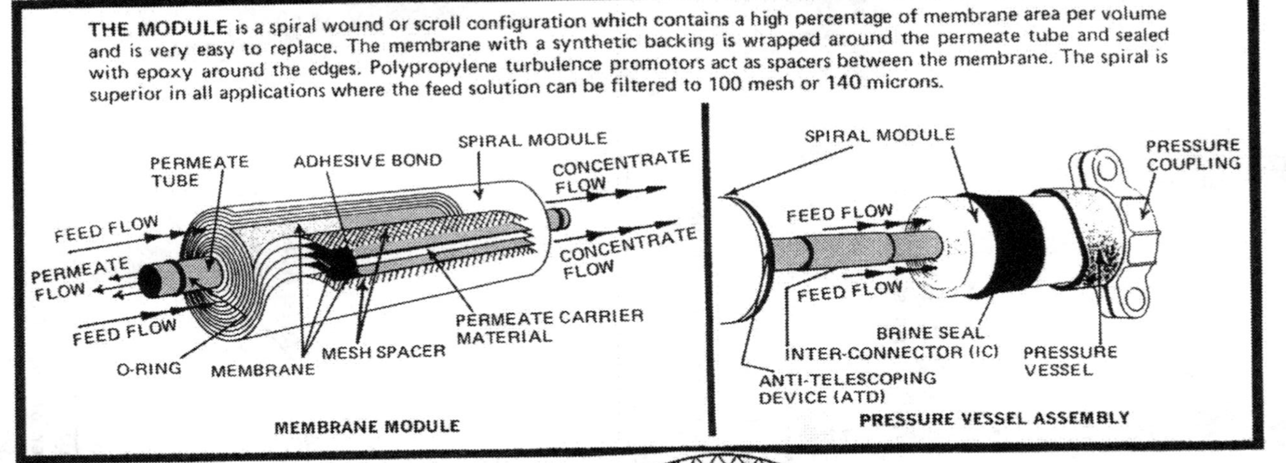

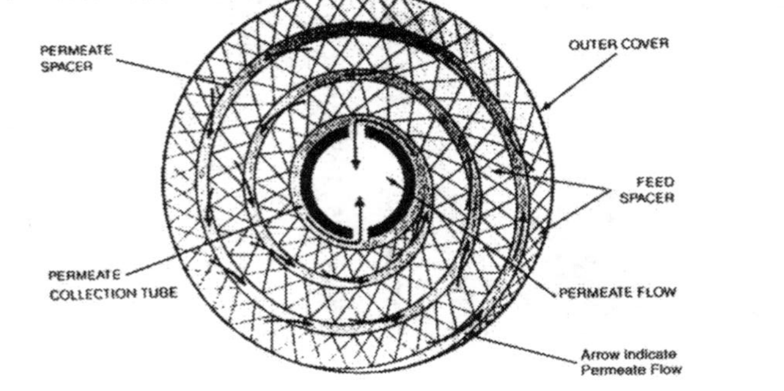

Figure 3.24 Diagrams of flat sheet, spiral wound membrane configuration (courtesy of Osmonics, Inc.). Cross section of a spiral wound element, showing a spiral path taken by the permeate to the central collection tube. Feed is pumped into the feed channel in a direction perpendicular to the plane of this cross-sectional drawing (from Cheryan, 1986, p. 162, with permission) with the permeate taking the path to the central collection tube as indicated in the cross-sectional schematic below the figure (from [illegible] 162). Reprinted from Ultrafiltration Handbook, with permission from Technomic Publishing Co., Inc., 1986.

Hollow Fiber Configuration

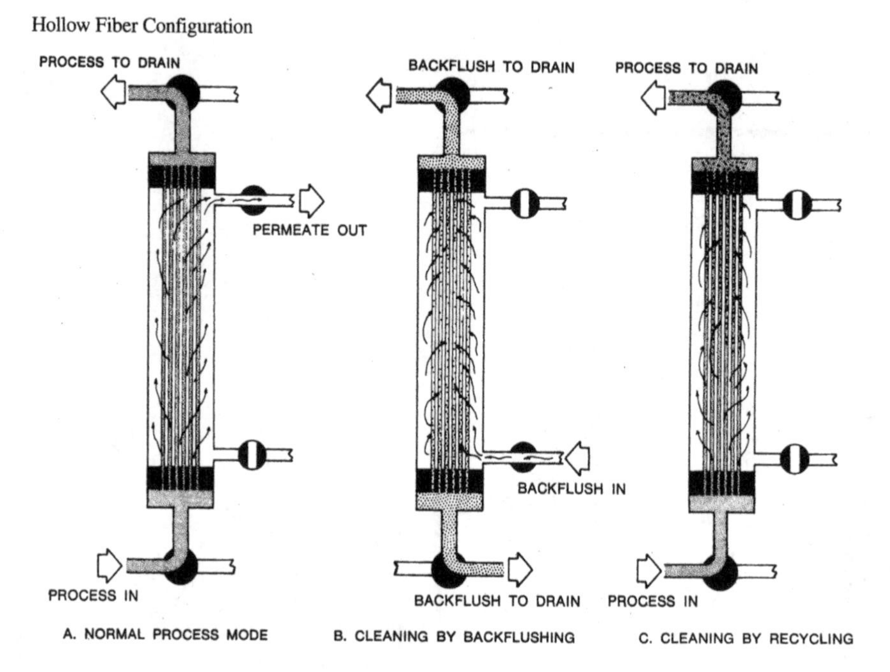

Figure 3.25 Hollow-fiber separation devices (courtesy of Amicon).

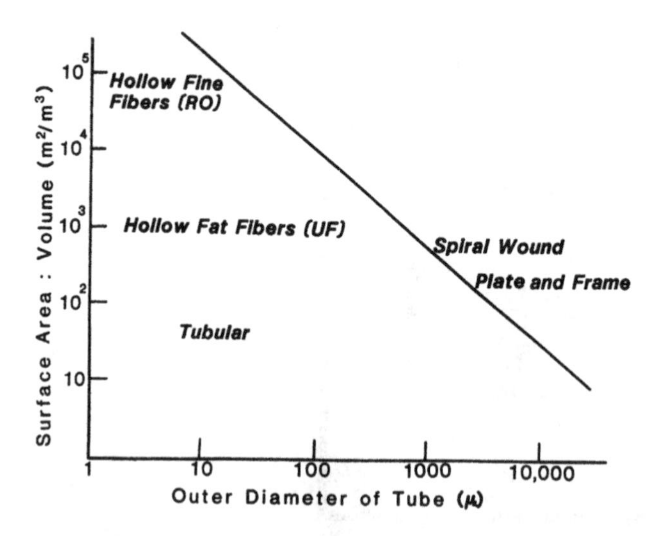

Figure 3.26 Approximate correlation of tube diameter and surface area; volume ratios showing trend for different types of membrane systems (from Cheryan, 1986, fig. 5.18, p 143). Reprinted from Ultrafiltration Handbook, with permission from Technomic Publishing Co., Inc., 1986.

feed flows in an axial direction between the sheets, while the permeate moves in a spiral or radial direction toward the center of the module (see inset at the bottom of Figure 3.24).

The hollow fiber configuration consists of a bundle of hollow fibers that are several millimeters in diameter. During the normal process mode, the feed flows inside the fibers and the permeate passes to the outside of the fiber (Figure 3.25). Intermittent cleaning of the membrane is achieved by flowing fluid in the opposite direction, from outside to inside, or by passing a clean solvent (usually water) through the system as shown in Figure 3.25. The benefit of a hollow fiber membrane systems is their large surface area per unit volume, which is on the order of 10^5 m^2/m^3 compared to 10^2 to 10^3 for a spiral wound membrane. The surface area that can be achieved is inversely proportional to the outer diameter of the tube or membrane (Figure 3.26).

3.14 THE PRESSURE AT WHICH GAS FLOWS THROUGH A WETTED MEMBRANE GIVES A MEASURE OF ITS PORE SIZE: BUBBLE-POINT TEST

The bubble-point pressure is defined as the minimum pressure required to force air through a filter that has been prewetted with water or alcohol (Poretics Company, 1995). This test (ASTM Procedure F316-70) is based on the capillary action of water in the pores of the membrane, where the gas pressure required to displace water from the pores is inversely proportional to the diameter of the pores (Wheelwright, 1991):

$$P = \frac{4\gamma(\cos \Psi)}{d_{\text{pore}}} \tag{3.49}$$

where

P = bubble-point pressure
d_{pore} = pore diameter, microns
γ = surface tension at the interface between air and water, dynes/cm
Ψ = liquid–solid contact angle[13]

The contact angles of three types of surfaces are shown in Figure 3.27. The bubble-point test was proposed as a way of detecting holes or tears in the membrane that could cause the membrane to leak as well as pore size distribution when the surface area is less than 1000 cm^2 (2,252). The membrane's integrity is confirmed if the pressure applied to one side of the membrane reaches a minimum level before a steady stream of air bubbles is observed passing from its surface (Wheelwright, 1991). Cheryan shows that Eq. 3.49 significantly overpredicts the bubble-point pressure. For example, for water with $\gamma = 73$ dynes/cm at 20°C for a contact angle of zero, Eq. 3.49, is

$$P = \frac{41.8}{d_p} \tag{3.50}$$

[13]The contact angle θ is the angle of a line drawn tangent to a drop of liquid placed on a solid surface as illustrated in Figure 3.27. Surface (a) is resistant to wetting, (b) is partially wetted, and (c) wetted. The contact angle is $\theta = 180$ for a surface that is unwetted and $\theta = 0$ for a completely wetted surface (Yang et al., 1990).

where P is pressure measured in psi. The bubble-point pressures calculated from Eq. 3.50 are significantly different than the measured bubble point (Table 3.10). The differences are due to the difficulty in estimating the contact angles and accounting for the tortuosity of the pores. However, the bubble-point pressure is still a useful, experimentally measured parameter that indicates the diameter of the largest pore. Therefore it gives a measure of the smallest particle that can leak through a membrane.

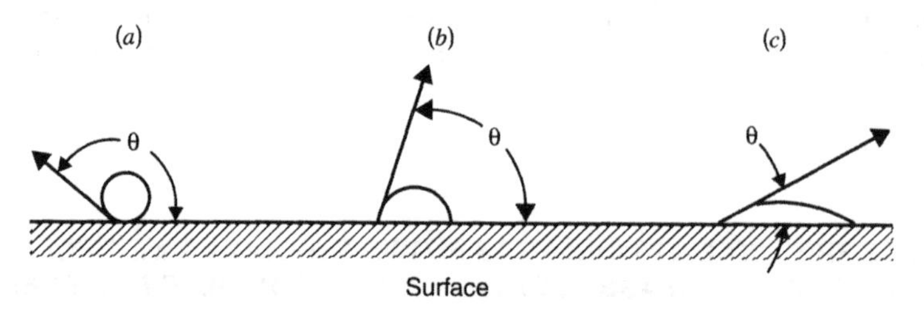

Figure 3.27 Schematic representation of contract angle.

An example is given by Olson for gas flow through a microfilter. Point *a* in Figure 3.28 represents the air diffusing through the thin layer of water in the pores, *b* the onset of gas flow due to expelling of water from the largest pores, and *c* the flow due to the smaller pores in the membrane. The slope of the line reflects the sharpness of the pore size distribution, with the vertical line, starting at point *c*, indicating a narrow pore size distribution. A line with a shallow slope would indicate a broader pore size distribution. The pressures required for testing a microfiltration membrane are on the order of 10 to 100 psig, while the pressures for ultrafiltration membranes are much higher. Hence alternate procedures are recommended for testing ultrafiltration membranes, and these may include use of solvents with a lower surface tension, or scanning electron microscopy to detect pores for membranes in which the pores are larger than 50 nm (Cheryan, 1986).

TABLE 3.10 Comparison of Calculated and Measured Bubble-Point Pressures for Microfiltration Membranes (in psi)

Pore Size (micron)	Bubble Point from Eq. (3.50)	Measured Bubble Point[a]
0.10	418	250
0.22	190	55
0.80	52	**16**

Source: Adapted from Cheryan, (1986), table 3.2, p. 55, based on data of Brock. Reprinted from Ultrafiltration Handbook, with permission from Technomic Publishing Co., Inc., 1986.

[a]Measurement made with a commercial membrane.

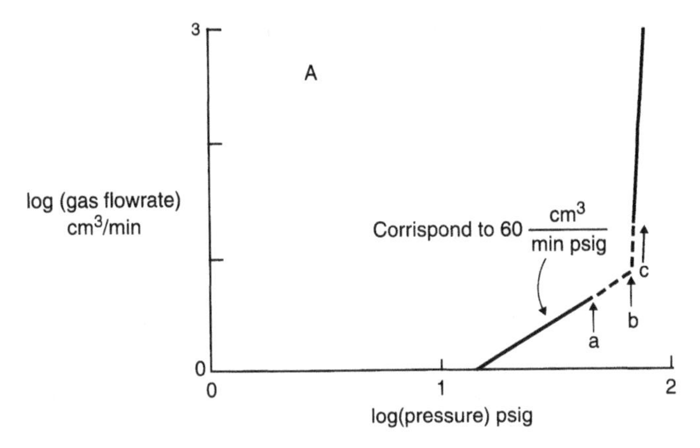

Figure 3.28 Log-log plot showing the onset of gas flow through a wetted microfiltration membrane as a function of differential air pressure. Reprinted from Process Biochemistry, Vol. 18, 1983, Page 32, Olson, Figure 1, with permission from Elsevier Science.

3.15 STERILIZATION OF HUMAN PLASMA PROTEINS, HARVESTING OF RECOMBINANT MICROBIAL CELLS, AND RECOVERY OF CELL CULTURE PRODUCTS ARE APPLICATIONS OF MICROFILTRATION

The utility of microfiltration in the processing of biotechnology products is illustrated in several applications. Removal of bacteria and fungi from 5% dextrose, 5% albumin, or other plasma proteins is an important final step before the sterile filling of the product into vials. The purpose of the microfiltration through a 0.2 μm membrane is to achieve the complete capture and removal of bacteria and bacterial debris that could be present at trace concentrations. The efficiency and rapidity of filtration are both important, since processing of the aqueous solution through a validated batch filter must have a probability of contamination that is less than 10^{-6} (Olson, 1983) and a processing time that is less than two hours for 1000 L of solution. The rationale for a relatively fast filtration is based on the premise that a short residence time reduces the risk that bacterial cells, if initially present in the solution, would multiply and generate bacterial contaminants (endotoxins[14]) during the filtration process. Olson gives the example that 16 daughter cells could propagate from each parent bacteria during a two hour filtration if a doubling time of 30 minutes is assumed. However, if the filtration required eight hours, Olson reports that 3096 cells could result (Olson, 1983).[15] The more rapidly that the bacteria are removed, the less time they have to grow, and the lower the risk of propagation of contaminants.

Another example of use of microfiltration is in the recovery and washing of recombinant *Saccharomyces cerevisiae* and *E. coli* (Bailey et al., 1990) using crossflow filtration. The combination of filtration with washing is named diafiltration, since it combines the functions of di-

[14]Endotoxins are complexes of lipopolysaccharides and proteins from the outer membrane of gram-negative bacteria. Toxicity is due to the lipopolysaccharide, which consists of lipid, oligosaccharide, and polysaccharide regions and has a molecular weight in the range of 10 kD (Held et al., 1997).

[15]A doubling time of 30 minutes would result in 2 bacteria from each daughter cell in the first 30 minutes, 4 in an hour, 8 in 1.5 hours, 16 in 2 hours, and $2^{(n_{hrs}/0.5)}N_o$ for the indicated number of hours (n_{hrs}) and the initial number of cells (N_o). Hence, for 8 hours, the total number of bacteria would be $2^{(8/0.5)} = 65{,}536$. Olson gives a value of 3096, but this probably assumes a given removal rate of bacteria from the filter so that the average residence time of the bacteria is less than 8 hours.

TABLE 3.12 Summary of Estimated Permeabilities and Resistances for Membrane Filters at Δp = 0.68 atm

Membrane				$R_{m,tot}$ Calculated Value	
				If μ = 200 cp,	If μ = 500 cp,
Type (Nominal Cutoff)	Area, m^2	j_v Flux,[a] $L/m^2 \cdot hr$	K_p cm/s · atm	$\frac{atm \cdot s}{cp \cdot cm}$	$\frac{atm \cdot s}{cp \cdot cm}$
Hollowfiber					
0.1 μm	4.18	66	2.70×10^{-3}	1.86	0.74
0.1 μm	0.46	65	2.65×10^{-3}	1.86	0.74
500 kD	6.97	23	0.94×10^{-3}	5.32	2.13
Flat sheet					
0.45 μm	2.79	94	3.84×10^{-3}	1.30	0.52

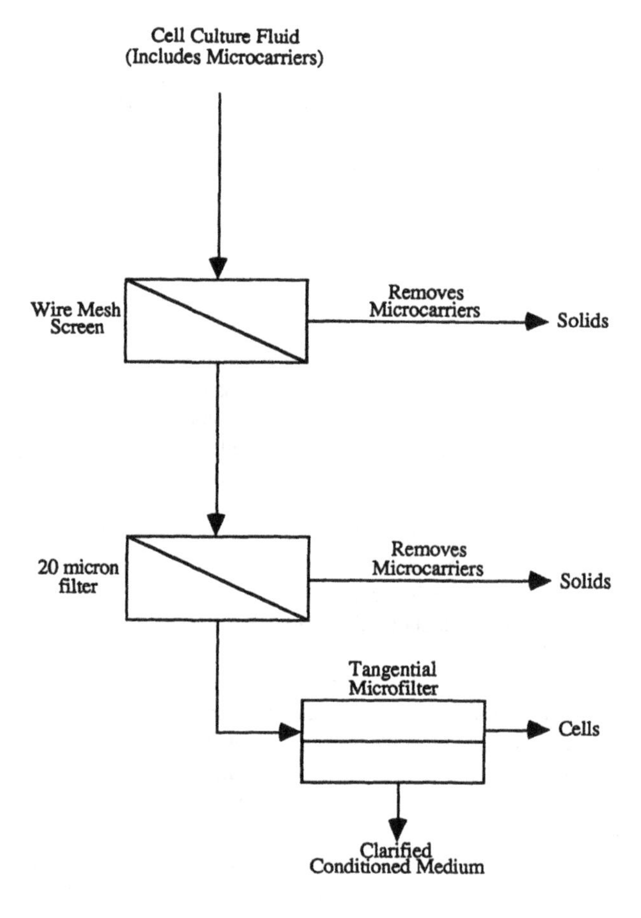

Figure 3.29 Summary of filtration steps for harvesting product from cell culture fluid (based on description of Nelson and Geyor, 1991).

TABLE 3.13 Industrial Products Derived from Mammalian Cell Culture

Product	Culture Volume	Culture System
Vaccines		
Foot-and-mouth	3000 L	Suspension
Polio	1000 L	Microcarrier
Rabies	325 L	Microcarrier
Hormones		
Human growth hormone (hGH)		Roller bottle
Erythropoietin (EPO)		Roller bottle
Enzymes		
Urokinase	200 L	Multiplate-propagator
Tissue plasminogen activator (t-PA)	10,000 L	Suspension
Cytokines and lymphokines		
Lymphoblastoid interferon (natural)	8000 L	Suspension
Monoclonal antibodies (MAb)	2,000 L	Suspension

Source: Adapted from Nelson and Geyor (1991), table 1, p. 113. The material is reproduced with permission by Marcel Dekker, Inc., NY.

conditioned medium was harvested every 24 or 48 hours using a sequence of filtration steps (Figure 3.29) in which the medium was passed through a screen and then a 20 micron filter to remove microcarriers.[18] The resulting filtrate was passed through a tangential microfilter system with submicron size pores for final clarification and removal of cells (Figure 3.29). The largest particles are removed first, followed by the capture of the cells or cell debris on filters with successively smaller pore size cutoffs. The rationale is that removal of the larger particles as a first step will prevent particles from plugging succeeding membranes, which have smaller pores. Selection of pore size cutoffs for ultrafiltration of proteins by the same principles would seem to be a logical approach to reduce blockage by proteins that are larger than the pore size cutoff of membranes located downstream. However, this is not always true as is illustrated by the experiments of Prouty (1991) and Kelly and Zydney (1997). The selection of membrane filters, and specification of the sequence in which they are used to avoid plugging, is not intuitive, as discussed in the next section.

A first estimate of the membrane areas for cell culture applications can be generated if an average flux and maximum processing time are specified. Various examples from the literature indicate that the maximum processing time should be no more than two hours. If the average flux is assumed to be 10 L/(m$^2 \cdot$ hr), then the area required for processing 10^4 L of culture fluid ($= V_{culture}$) in two hours is

$$A_{membrane} = \frac{V_{culture}}{j_v t} = \frac{10{,}000}{(10)(2)} = 500\ \mathrm{m}^2 \tag{3.53}$$

The value of a flux of 10 is used as a reference point only. Actual values of flux need to be determined from experimental data and rate equations associated with microfiltration of a particu-

[18]Some cell lines for vaccine production must be attached to a solid substrate or surface for growth, and are denoted as being anchorage-dependent. While these cells will not grow in free cell suspension, they will attach to small particles, known as microcarriers, and grow. The particles have a density that facilitates their being kept in suspension by gentle stirring or fluidization.

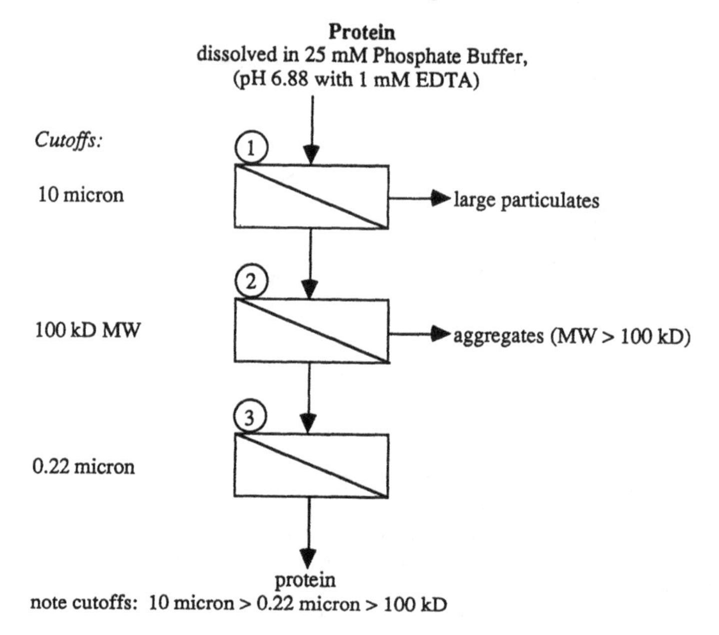

Figure 3.31 Schematic representation of microfiltration experiment of Kelly and Zydney, 1997, demonstrating impact of protein aggregation on reducing flux.

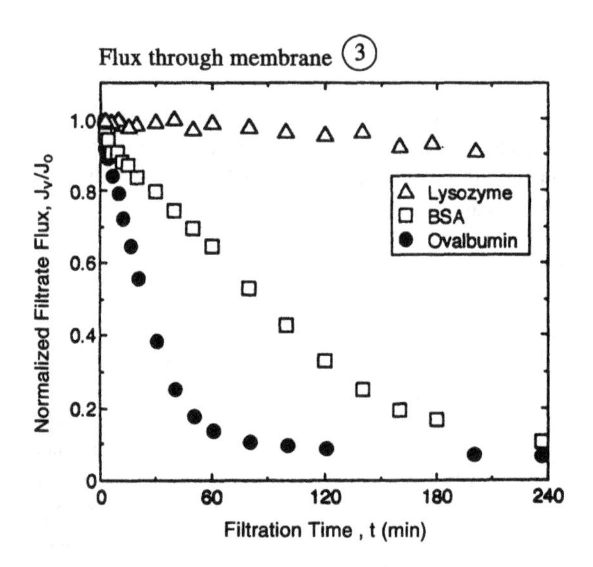

Figure 3.32 Normalized filtrate flux during the sequential filtration of 100*K* prefiltered solutions of lysozyme, BSA, and ovalbumin at 14 kPa and 600 rpm through 0.22 μm PVDF membranes that had first been used to filter a small volume of the unfiltered protein (from Kelly and Zydney, 1997, fig. 5, p. 97, with permission of Wiley-Liss, Inc., a subsidiary of John Wiley & Sons, Inc.).

TABLE 3.14 Summary of Protein Physical Properties Used in Experiments of Kelley and Zydney

Protein	Source	MW (kD)	Number of Internal Disulfide Linkages, –S–S–	Number of Free Thiol Groups, –SH	pI[1]
BSA	Bovine serum	67	17	1	4.7
Cys-BSA	Bovine serum	67	18	0	4.7
Ovalbumin	Chicken egg	40	1	4	4.6
Lysozyme	Chicken egg	14	4	0	11.0
β-Lactoglobulin	Cow's milk	18	2	1	5.3
α-Lactalbumin	Cow's milk	15	4	0	5.1
Pepsin	Pig's stomach	36	3	0	1.0
Myoglobin	Horse muscle	17	0	0	7.0

Source: Kelly and Zydney (1997). The material is reproduced with permission of Wiley-Liss, Inc., a subsidiary of John Wiley & Sons, Inc.

Proteins that do not have exposed sulfuydryl groups can also cause a decline in flux but by a different mechanism. For example, cysteinylated-bovine serum albumin, lysozyme, α-lactalbumin, pepsin, and myoglobin do not have free –SH groups (Table 3.14). These groups, even when they are microfiltered, will also cause a decline in flux passing through a 0.22 μm pore size polyvinylidene fluoride membrane. This was attributed to protein adsorption and direct deposition of large protein aggregates on the membrane Figure 3.34.

Coefficients for the rate of flux decline shown in Figure 3.35 and Figure 3.38 were determined from the slope of the data plotted in Figures 3.34 and 3.37 using the definition

$$K = \frac{1}{j_V}\frac{dj_V}{dt} \tag{3.54}$$

The graphs of K as a function of filtration time give further insights into the mechanisms that help to explain decreases in flux. As shown in Figure 3.35, the values of K decrease sharply and

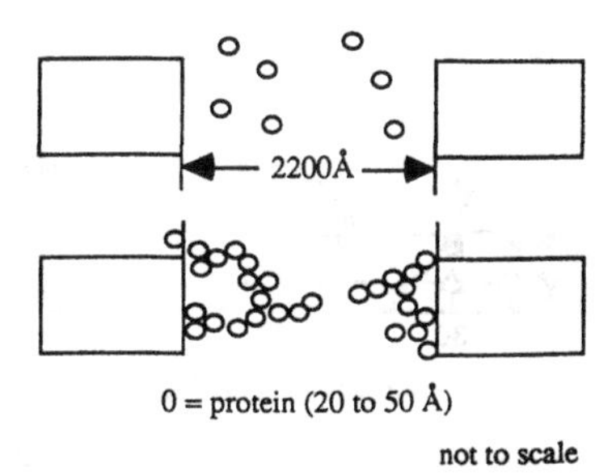

Figure 3.33 Conceptual illustration of how protein aggregates might cause pores to be blocked (the small ovals represent individual protein molecules).

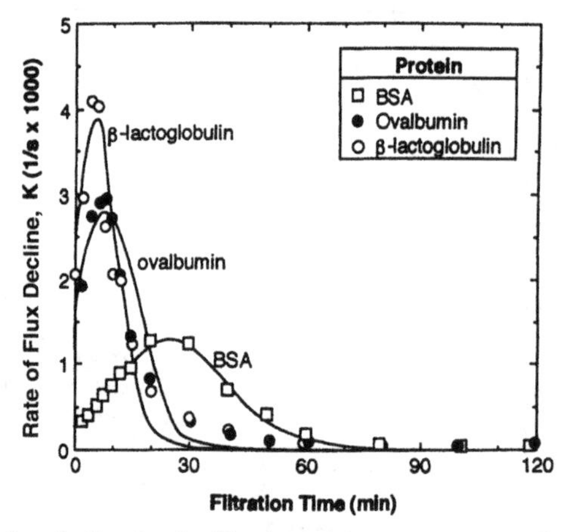

Figure 3.38 Rate of flux decline for the filtration of 2 g/L solutions of β-lactoglubulin, BSA, and ovalbumin through 0.22 μm PVDF membranes at 14 kPa and 600 rpm. Solid curves are model calculations (from Kelly and Zydney, 1997, fig. 3, p. 94, with permission of Wiley-Liss, Inc., a subsidiary of John Wiley & Sons, Inc.).

β-lactoglobulin) shows a decline in flux (Figures 3.37 and 3.38) but for the different reason discussed earlier in this section. The fouling times[20] for these proteins range from 8 minutes for β-lactoglobulin and 9 minutes for ovalbumin to 25 minutes for BSA. The rates of flux declines show maxima at 8, 10, and 25 minutes for β-lactoglobulin, ovalbumin, and BSA, respectively (Figure 3.38), with the normalized filtrate rate decreasing more rapidly for the smaller protein (ovalbumin) than the larger protein (BSA) as shown in Figure 3.39.

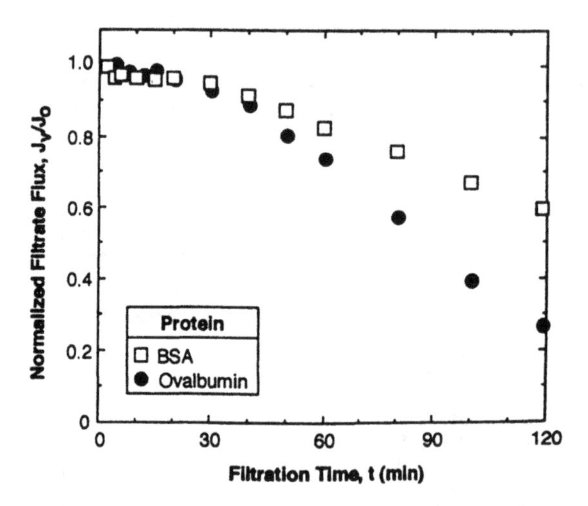

Figure 3.39 Normalized filtrate flux during the filtration of 100*K* prefiltered solutions of BSA and ovalbumin through 0.22 μm PVDF membranes at 14 kPa and 600 rpm (from Kelly and Zydney, 1997, fig. 4, p. 94, with permission of Wiley-Liss, Inc., a subsidiary of John Wiley & Sons, Inc.).

[20]Fouling time is defined as the time required for $j_v = 0.20\,j_{v,o}$, where $j_{v,o}$ is the flux at the start of the filtration.

3.17 DIALYSIS AND EVAPORATION OR REVERSE OSMOSIS PROCESSES REMOVE ETHANOL FROM BEER TO YIELD BEVERAGES WITH ALCOHOL CONTENT REDUCED

The market for beers with reduced alcohol content[21] is increasing due to social, legal, and health issues. However, it is difficult to directly ferment a beer that is low in alcohol because partial fermentation processes leave residual, unfermented sugars, and hence give an undesirable sweet taste. Distillation or evaporation at atmospheric pressure can remove the ethanol, but the flavor is changed due to loss of volatile flavor components, or development of off-flavors caused by reactions promoted by the temperature of the distillation process. Hence lower temperature (40 to 65°C), vacuum distillation, or evaporation is preferred if the alcohol is to be removed through a vapor/liquid separation process.

TABLE 3.15 Comparison of Composition of a Normal Beer to Reduced Alcohol Beers Prepared Using Thin Film Evaporation at 30°C, 0.04 Bar and about 1 Second Contact Time

Flavor Components	Compositions in mg/mL		
	Reduced Alcohol Beers[a]		
	At 2.02% Ethanol	At 0.84% Ethanol	Normal Beer (Control) at 4.02% Ethanol
	Fermentation by-products		
Acetaldehyde	31.0	16.6	28.4
Methyl-formate/acetate	10.4	12.8	16.2
Ethyl acetate	8.8	1.2	28.3
n-Propanol	4.2	2.0	9.5
Isobutanol	3.4	1.0	10.7
Isoamylacetate	2.5	1.3	4.1
2-Methyl, - and 3-Methyl, 1-butanol	31.6	7.9	72.1
Diacetyl	0.08	0.07	0.05
2,3-Pentanediene	0.03	0.002	0.02
	Other properties		
pH	4.57	4.63	4.52
CO_2, %	0.46	0.42	0.44
Fullbodiness	0/10	Not reported	9/10
Purity of taste	8/10	Not reported	3/10
Purity of smell	8/10	Not reported	1/10

Source: Adapted from Kruger, et al. (1990), tables 2, 3 and 4.
[a]% is given on a weight basis.

[21]The alcohol contents of normal beers are as high as 6.6% (weight/volume basis). Non- and low-alcohol beers are defined differently in different countries. In the United Kingdom, a nonalcohol beer is a product that contains less than 0.1% v/v alcohol. A low-alcohol beer has an alcohol content of 0.5 to 1.5%. Light beers contain 1.6 to 2.5% v/v. In the United States, beverages containing less than 0.5% alcohol cannot be called beer (Attenborough, 1988), while in Germany the upper limit for the alcohol content of low-alcohol beers is 1.5%, by weight. Light or Lite beers in Australia are typically in the range of 2.5% (by volume) with the alcohol content of a reduced alcohol beer being at 0.9% (Attenborough, 1990).

The low temperatures of reverse osmosis, dialysis, and vacuum distillation processes are attractive, since these minimize thermal degradation of flavor components. Reverse osmosis can be utilized in a manner that decreases change in flavor, although some low molecular weight flavor components are still lost from the retentate, that is, the low alcohol beer. These molecules are able to permeate the membrane together with the ethanol (Attenborough, 1988). Other factors that are reported to affect the taste and body of beer include alcohol content, colloidal structure, and the composition of fermentation by-products, some of which are listed in Table 3.15 (Kruger et al., 1990). Reduction of the alcohol content of the beer to below 2% reduces the body and flavor of the beer, with the remaining components no longer balancing out the flavoring effects lost when the ethanol is removed. A comparison of the flavor components, pH, dissolved CO_2, and sensory characteristics (as measured by a 10 person taste panel) shows measureable differences in concentrations of flavor components and sensory evaluation, while pH and dissolved CO_2 are about the same (Table 3.15). The 2% beer has lower levels of acetates as well as decreased levels of the higher molecular weight alcohols, and a higher concentration of diacetyl. While the "purity" of taste and smell was ranked higher for the 2% beer than the normal 4% beer by 8 out of 10 test panelists, the body of the beer was ranked lower, if not nonexistant. As noted by Kruger et al. (1990), "it is understandable that the values of purity of smell are better for the reduced alcohol beer, as the concentration of fermentation by-products is lower than in the control beer. The lower body of the reduced alcohol beer is reported to be consistent with a disruption of the colloidal characteristics of the beer, and therefore its flavor or body." Some of the sensory losses due to a reduced ethanol content are counteracted by other flavor components, but this counterbalancing effect is also lost when the ethanol content approaches 2% (Kruger et al., 1990).

These interesting observations on the flavor of a low-alcohol beverage are from an industrial study on rapid alcohol removal using a low-residence time thin film evaporator.[22] The alcohol was evaporated from the beer as an aqueous ethanol vapor, while leaving behind a significant fraction of nonethanol flavor components. The separation was achieved in about 1 second at temperatures of 30 to 40°C under a vacuum of 0.04 bar. Changes in the colloidal structure of the beer were decreased by first diluting the beer by 8 to 15% so that evaporation returns it back to its original volume (Kruger et al., 1990).

The removal of alcohol from beer may also be achieved in a series of dialysis modules based on cellulose (Cuprophan®) membranes where the feed to the dialysis unit contains 5% (by volume) alcohol and the module is operated at cellar temperatures (0 to 4°C). Drinking quality water is passed in a countercurrent direction to the feed on the other side of the membrane, with the dialyzed product leaving the unit at 3.5% ethanol, and the dialysate (water) stream, containing 1%. The dialysis water stream contains dissolved CO_2 to avoid CO_2 loss from the beer, and also contains small amounts of flavor components to minimize the quantities that permeate through the membrane from the beer. The resulting beer is called "Diat" beer, but still contains a substantial alcohol content. It would appear that a primary function of the dialysis step is to obtain the proper flavor.

[22]Distillation and evaporation have been used for over 100 years to produce beers with a reduced alcohol content. Short residence times and low temperatures are desirable in order to minimize detrimental effects on flavor compounds (see Table 3.15). A thin film vacuum evaporator achieves this objective and consists of a heated rotating core over which beer, pumped into the device, is spread out over the surface of the core in a very thin, turbulent film due to the centrifugal force. The alcohol is evaporated from the beer at about 30 to 40°C and a vacuum of 0.04 bar. The reduced alcohol content product leaves the device and is cooled. The contact time is less than 1 second, hence minimizing chemical reactions. An apparatus of this type made by Alfa-Laval (i.e., the Centri-Therm) has been in use by the Guiness and Kirin breweries. Thin film evaporation has also found application for concentrating other bioproducts including antibiotics, enzyme solutions, yeast extracts, blood plasma, fruit juices, insulin, amino acids, coffee and tea extracts, and vitamin solutions (Kruger et al., 1990).

Fouling of the membrane can be an issue, and consequently the cellulosic dialysis membranes are cleaned weekly using dilute alkali at pH 11, followed by hot water and then ambient temperature rinse water. The membranes are replaced about every 6 months.

A hybrid approach that facilitates alcohol reduction combines dialysis across a cellulose membrane at 0°C with vacuum distillation. The alcohol content is decreased from 4.2% to about 1.5% in a process that integrates a membrane separation with a low-temperature vacuum distillation to process 1000 L/hr of beer. The staging of the two unit operations is schematically illustrated in Figure 3.40 with the dialysis unit on the left and distillation on the right. The dialysis process operates at 2.1 bar (= 2.07 atm) pressure, with a transmembrane pressure of 0.1 bar (= 0.09869 atm) and cellar temperature so that there is no thermal damage to the product retained within the module. Coupled with vacuum distillation of the permeate at a temperature that is at 50 to 65°C, the dialysis process gives a product with the appropriate beer taste and an alcohol content which is inversely proportional to the residence time of the beer in the dialysis step (Attenborough, 1988, 1990). The molecules that impart flavor to the beer are kept in the system due to their recycle in a concentrated form from the bottom of the distillation tower. When this stream is countercurrently contacted with the beer, which is on the other side of the membrane, backdiffusion of some of the flavor molecules may occur. While a cocktail of flavor components can be added to a low alcohol beer, to make up for losses, the dialysis process of Figure 3.40 is preferable since the net transfer of flavor components can be minimized and the beer flavor maintained.

A combined dialysis/vacuum distillation plant was reported to be a source of a beer containing 3 to 3.5% alcohol for 1988 Olympics in Seoul, Korea (Attenborough, 1988). The feed to the dialysis process consisted of 1000 L/hr beer with 4.2% (by volume) alcohol. The ethanol content was reduced to 1.5% by dialysis, and then blended with a portion of the feed beer con-

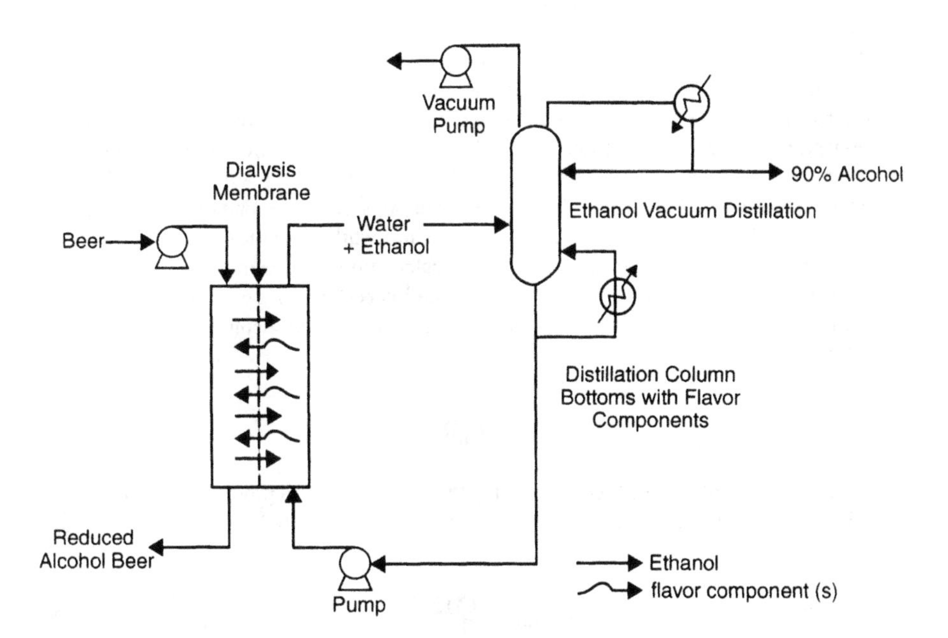

Figure 3.40 Schematic represented of dialysis process for removing alcohol to obtain a reduced alcohol product (adapted from Attenborough, 1988, fig. 9, p. 44).

taining 4.2% ethanol, to obtain a final product with 3 to 3.5% alcohol. A similar process is apparently used to obtain a 2 to 2.5% Pils beer starting with a feed containing 5% alcohol.

Reverse osmosis is another potential method, which could be used to reduce the alcohol content of beer. In this case a cellulose acetate or nylon membrane is selected that allows alcohol and water to pass through it, while most of the flavor components are retained since they have a higher molecular weight than either the water or ethanol. The operating pressures of a reverse osmosis system range from 30 to 80 bar. As is the case for dialysis, the beer is initially diluted with water to minimize damage to its colloidal structure, as when the ethanol and water is pushed through the membrane, and the flavor components concentrated. Leakage of salt through the membrane, and membrane fouling, affect the selection of the types of membranes and operating rates (Kruger et al., 1990). Some leakage occurs, and therefore flavor components must still be added back to the retentate, that is, the reduced alcohol beer (Attenborough, 1988). The removal of alcohol from beer through reverse osmosis illustrates a condition where the osmotic pressure of the permeate, namely ethanol, must be accounted for in calculating flux through the membrane.

A reverse osmosis system utilizes a feed that has been diluted with water to obtain the desired alcohol content of the final product. Benefits of the dilution may include preserving the colloidal structure and body of the beer and reducing the plugging of the membrane (Attenborough, 1988; Kruger et al. 1990). The diluted feed is passed through the reverse osmosis unit at 30 bar (= 2.96 atm) pressure in order to concentrate the diluted beer back to its original volume while passing ethanol through the membrane. Some flavor losses occur because flavor components that have a molecular weight or size similar to ethanol also pass through the membrane. While reverse osmosis has the benefit that it can be operated at cellar temperatures, the membrane costs are apparently significant, with annual replacement costs being up to 7% of the original capital cost (Attenborough, 1990). The minimum alcohol content achievable by reverse osmosis is apparently 0.5%. It is apparently not economic to obtain lower concentrations by this method (Attenborough, 1988).

The calculation of the pressure required to process beer into a reduced alcohol product by ultrafiltration is illustrated by a hypothetical example in which beer initially at 9.5 proof is processed through a reverse osmosis membrane to obtain a reduced alcohol retentate with flavor components of the type listed in Table 3.16, and at a final alcohol content of 4.78 proof or 2.39% by volume (equivalent to 1.9% weight/volume). A schematic diagram of the material flows for this example is given in Figure 3.41. For this example, we assume that part of the acetaldehyde and part of the ethyl acetate also permeate the membrane, while all of the other components are retained. This behavior is assumed to illustrate the calculation of the osmotic pressures of multicomponent mixtures, so it may not reflect the separation performance of an actual membrane.

The osmotic pressure of an aqueous solution of a 4% by volume alcohol may be calculated using Eq. (3.55):

$$\pi = C_B RT \tag{3.55}$$

where the weight % of alcohol[23] corresponding to 4% by volume (8 proof) is read from Table 3.17. The bulk phase (feed) concentration of alcohol in the diluted beer is

$$C_B = \frac{19.0\text{g}}{\text{L}} \cdot \frac{1 \text{ mole ethanol}}{46.07 \text{ g ethanol}} = 0.412 \frac{\text{moles}}{\text{L}} = 4.12 \times 10^{-4} \frac{\text{moles}}{\text{cm}^3}$$

[23]The correlation between proof (a measure of the % of alcohol by volume) and weight % alcohol is not linear. The data in Table 3.17 gives a convenient comparison between the different measures of alcohol concentration.

TABLE 3.16 Compositions of Upstream (Retentate) and Downstream (Permeate) Flows in a Reverse Osmosis Process for Removing Alcohol from Beer

Component		Upstream (In)		Downstream (Out)	
Name of Component	MW	g/L	moles/L	g/L	moles/L
Ethanol	46.07	19.0	0.412	19.0	0.412
Acetaldehyde	44.05	22.0	0.449	6.0	0.136
Methyl formate	61.06	5.0	0.0819	3.0	0.0491
Methyl acetate	74.08	5.0	0.0674	3.0	0.0849
Ethyl acetate	88.11	20.0	0.227	2.0	0.0227
n-Propanol	60.10	9.0	0.150	0.0	0.0
Isobutanol	74.12	11.0	0.148	0.0	0.0
Isoamylacetate	130.14	4.0	0.0307	0.0	0.0
2-Methyl-and 3-Methyl-1-butanol	88.15	7.2	0.0817	0.0	0.0
Diacetyl	86.09	0.1	0.00116	0.0	0.0
2,3 Pentadiene	68.12	0.05	0.00073	0.0	0.0
Total (moles/L)	—	—	1.65	—	0.648
π (atm)	—	—	48.84	—	15.94

Source: Adapted from Attenborough (1988); Kruger et al. (1990).

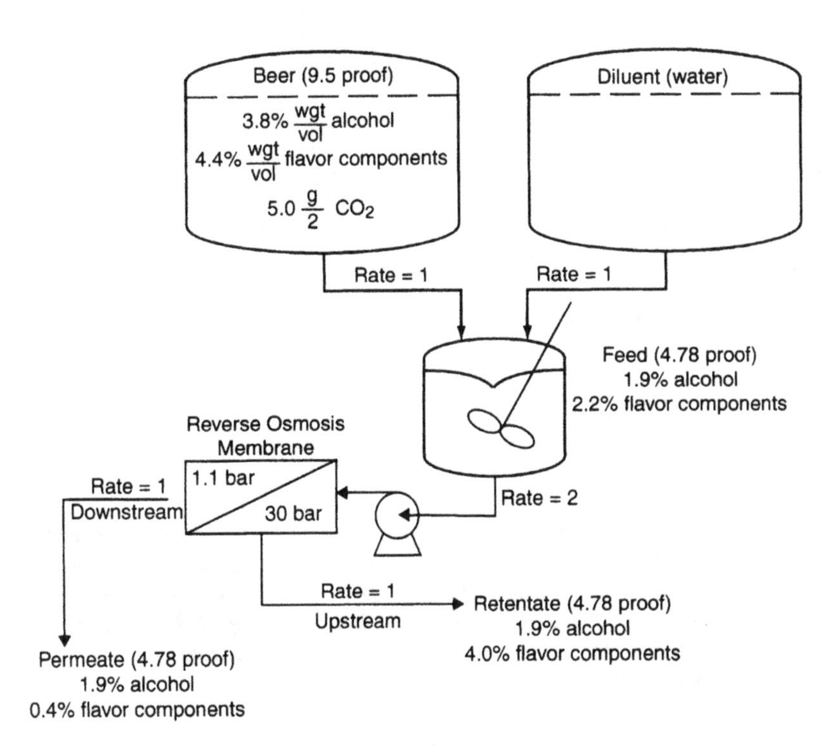

Figure 3.41 Process diagram for production of low alcohol beer using reverse osmosis (adapted from Attenborough, 1988, fig. 7, p. 43). Note that the extent of loss of flavor components in the permeate is an arbitary value, but it is indicated here for illustration purposes.

TABLE 3.17 Correlations of Proof, Volume %, Weight %, and Specific Gravity for Ethanol/Water Solutions

Volume % Ethyl Alcohol	US Proof (Degrees at 60°F)	Weight % Ethyl Alcohol	Specific Gravity at 60°/60°F
0.0	0	0.00	1.0000
0.5	1	0.39	0.9993
1.0	2	0.80	85
1.5	3	1.19	78
2.0	4	1.59	70
2.5	5	1.99	63
3.0	6	2.39	56
3.5	7	2.79	49
4.0	8	3.19	42
4.5	9	3.60	35
5.0	10	4.00	28
5.5	11	4.40	21
6.0	12	4.80	15
6.5	13	5.21	08
7.0	14	5.61	02
7.5	15	6.02	0.9896
8.0	16	6.42	90
8.5	17	6.83	78
9.0	18	7.23	78
9.5	19	7.64	72
10.0	20	8.05	66
10.5	21	8.45	60
11.0	22	8.86	54
11.5	23	9.27	49
12.0	24	9.68	43

Source: Adapted from table 22, p. 138, Anonymous (1962).

At 20°C, the osmotic pressure, π, of the ethanol is

$$\pi = (4.12 \times 10^{-4})(82.06)(273 + 20) = 9.9 \text{ atm}$$

$$\left|\frac{\text{moles}}{\text{cm}^3}\right|\frac{\text{atm}\cdot\text{cm}^3}{\text{mole}\cdot\text{K}}\left|\text{K}\right|$$

However, the alcohol concentrations on both sides of the membrane are the same, and hence they do not effect the flux. Rather, the higher-molecular weight components in Table 3.16 affect the backpressure. Assuming that there is no leakage of the ethanol across the membrane (i.e., $\sigma = 1$) the applied pressure would need to be greater than 16.6 atm (or 244 psia) in order to obtain flux of ethanol across the membrane. If the permeability coefficient is assumed to be $K_p = 5$, then the flux is given by Eq. (3.1a):

$$j_v = K_p[\Delta P - \sigma\Delta\pi] = K_p[(P_{in} - P_{out}) - (1)(\pi_{in} - \pi_{out})] \quad (3.1a)$$

or

$$j_v = (5)\,[(P - 1.0) - (1)\,(0 - 16.6)] \tag{3.56}$$

where the outlet pressure P_{out} is 1 atm, and the osmotic pressure on the outlet or permeate side is assumed to be 0.1 atm due to leakage of a low concentration of flavor components. The flux equation becomes

$$j_v = (5)\,[(P_{in} - 1) - (1)\,(16.6 - 0.1)] = 5\,[P_{in} - 17.5] \tag{3.57}$$

A hydraulic pressure of 30 atm would result in a flux of

$$j_v = 5\,(30 - 17.5) = 62.5\ \mathrm{L/(m^2 \cdot hr)}$$

REFERENCES

Alberts, B., D. Bray, J. Lewis, M. Raff, K. Roberts, and J. D. Watson, 1989, *Molecular Biology of the Cell*, 2nd ed., Garland Publishing, New York, 54 to 55, 1203 to 1212.

Anonymous, Poretics—An Osmonics Company, 1995, *Microfiltration and Laboratory Products Catalog*.

Anonymous, Miscellaneous Applications, in Ethyl Alcohol, Enjay Co., Division of Humble Oil and Refining, New York, 117–171 (1962).

Atkinson, B., and F. Mavituna, 1983, *Biochemical Engineering and Biotechnology Handbook*, Nature Press, Macmillan, 17–22, 389–395, 675, 693–712, New York.

Attenborough, W. M., 1988, Evaluation of processes for the production of low- and non-alcohol beer, *Ferment.*, (April) 40–44.

Attenborough, W. M., 1989, Processing systems for the production of reduced alcohol and near beers, pp. 295–299. From private communication, George Duncomb, Miller Brewing Company, November, 1990.

Bailey, F. J., R. T. Warf, and R. Z. Maigetter, 1990, Harvesting recombinant microbial cells using crossflow filtration, *Enz. Microb. Technol.*, 12, 647–652.

Belfort, G., 1987, Membrane separation technology: An overview, in *Advanced Biochemical Engineering.*, H. R. Bungay and G. Belfort, eds., J. Wiley, New York, 239–297.

Belter, P. A., E. L. Cussler, and W.-S. Hu, 1988, *Bioseparations: Downstream Processing for Biotechnology*, Wiley-Interscience, New York, 1–9, 15–18, 22–30, 238–239, 242–247.

Cheryan, M., 1986, *Ultrafiltration Handbook*, Technomic Publishing, Lancaster, PA, 1–26, 53–62, 76–103, 127–169, 235–242.

Cheryan, M., 1998, *Ultrafiltration and Microfiltration Handbook*, Technomic Publishing, Lancaster, PA.

Dullien, F. A. L., 1992, *Porous Media: Fluid Transport and Pore Structure*, 2nd ed., Academic Press, New York, 325–329.

Dutton, G., 1996, Transgenic animal-based protein products move toward clinical trials, *GEN*, 16(9), 37.

Held, D. D., R. J. Mehigh, C. M. Woogie, S. P. Crump, and W. K. Kappel, 1997, Endotoxin reduction in macromolecular solutions: Two case studies, *Pharm. Technol.*, 21(4), 32–38.

Kelly, S. T., and A. L. Zydney, 1997, Protein fouling during microfiltration: Comparative behavior of different model proteins, *Biotechnol. Bioeng.*, 55(1), 91–100.

Kruger, E., B. Oliver-Danman, G. Sommer, M. Metscher, and H. Berger, 1990, The production of reduced alcohol beers, Technical document from Alfa-Laval, Provided by G. Duncombe, Miller Brewing Company, November, 1990.

Ladisch, M. R., and P. C. Wankat, 1988, Scale-up of bioseparations for microbial and biochemical technology, in *The Impact of Chemistry on Biotechnology: Multidisciplinary Discussions*, ACS Symp., Ser. 362, M. Phillips, S. P. Shoemaker, R. D. Middlekauff, and R. M. Ottenbrite, eds., American Chemical Society, Washington, DC, 72–101.

Liley, P. E., Y. S. Tohloukian, and W. R. Gambill, 1963, Physical and chemical data, *Chemical Engineers Handbook*, R. H. Perry, C. H. Chilton, and S. D. Kirkpatrick, eds., 3–201.

McCabe, W. L., and J. C. Smith, 1967, *Unit Operations of Chemical Engineering*, McGraw-Hill, New York, 751–787, 879–904.

Michaels, A. S., 1981, Ultrafiltration: an adolescent technology, *Chem. Tech.*, 36–43, Vol. 11.

Mota, M., C. Lafforgue, P. Strehaiano, and G. Goma, 1987, Fermentation coupled with microfiltration: kinetics of ethanol fermentation with cell recycle, *Bioprocess Eng.*, 2, 65–68.

Nelson, K. L., and S. Geyor, 1991, Bioreactor and process design for large scale mammalian cell culture manufacturing, in *Drug Biotechnology Regulation, Scientific Basis and Practices*, Y-Y. H. Chin and J. L. Gueriguian, eds., Marcel Dekker, New York, 112–143.

Olson, W. P., 1983, Process microfiltration of plasma proteins, *Process Biochem.*, 18, 29–33.

O'Sullivan, T. J., A. C. Epstein, S. R. Korchin, and N. C. Beaton, 1984, Applications of ultrafiltration in biotechnology, *CEP*, (January), 68–75.

Patel, P. N., M. A. Mehaia, and M. Cheryan, *J. Biotechnol.*, 5, 1–16.

Prouty, W. F., 1991, Production-scale purification processes, in *Drug Biotechnology Regulation*, 13, Y-Y. H. Chien and J. L. Gueriguian, eds. Marcel Dekker, New York, 221–262.

Rosenbaum, E. J., 1970, *Physical Chemistry*, Appleton-Century-Crofts, Meredith, New York, 550–555, 611, 652.

Tuwiner, S. B., L. P. Miller, and W. E. Brown, 1962, *Diffusion and Membrane Technology*, Am. Chem. Soc. Monograph Series, Reinhold Publishing, New York, 48–52, 290, 296–303.

Van Reis, R., S. Gadan, L. N. Frautschy, S. Orlando, E. M. Goodrich, S. Saksena, R. Kuriyel, C. M Simpson, S. Pearl, and A. Zydney, 1997, High Performance Tangential Filtration, *Biotechnol. Bioeng.*, 56(1), 71–82.

Vilker, V. L., C. K. Colton, K. A. Smith, and D. L. Green, 1984, The osmotic pressure of concentrated protein and lipoprotein solutions and its significance to ultrafiltration, *J. Membrane Sci.*, 20, 63–77.

Wheelwright, S. M., 1991, *Protein Purification: Design and Scale up of Downstream Processing*, Hanser Publishers, Munich, 1–9, 61, 213–217.

PROBLEMS

3.1 Chinese hamster ovary cells (ATCC No. CCL61) transfected to express recombinant human rt-PA (MW of about 65 kD) were grown at 37°C in stainless steel fermentors to obtain densities of (0.5 to 4) $\cdot$ 10^6 cells/mL, in a serum-free cell culture media. The cells are to be separated from the culture media using Microdyn® hollow fiber modules from AKZ0. These modules each contain 3000 fibers of 0.2 micrometer pore cut-off polypropylene membranes, with an inner fiber diameter of 0.6 mm, and a total area of 2.5 m^2. The stainless steel housings that contain the fibers are connected to a 1200 L fermentor, from which cell suspensions are pumped using a positive displacement feed pump, at rates of either 920 or 1840 L/h using a Waukesha pump. The membranes are regenerated and sanitized after each use. (This description is taken from R. van Reis, L. C. Leonard, C. C. Hsu, and S. E. Builder, *Biotechnol. Bioeng.*, 38, 413–422, 1991).

The objective of carrying out the microfiltration is to recover kg quantities of rt-PA while maintaining cell viability so that the cells do not rupture and add additional contaminating protein, endotoxin, or nucleic acids to the secreted protein product.

The filtration of a fluid is to be carried out at a shear rate that is not to exceed 4000 s^{-1}, and it is to be completed in 3 hours or less. Estimate the minimum membrane area that will be required, and maximum amount of fluid that can be processed for a 3000 fiber module.

Given: Hollow fiber length = 0.435 m; I.D. = 6×10^{-4} m; cell culture density = 1030 kg/m^3; cell culture dynamic viscosity = 1×10^{-3} Ns/m^2. j_v/TMP = 5.975 L/(m^2 · hr · kPa) where J_v = flux and TMP = transmembrane pressure.

$$N_{Re} = \frac{2\rho Q}{\mu n \pi r} = \text{Reynolds number}$$

$$\gamma_W = \frac{4Q}{n\pi r^3} = \text{wall shear rate } (\text{s}^{-1})$$

$$\Delta p = \frac{2\gamma_w L \mu}{r} = \text{pressure drop (Pa)}$$

$$\text{TMP} = \frac{P_{in} + P_{out}}{2} - P_f = \text{transmembrane pressure}$$

where

L = length of hollow fiber, m
r = inside radius of hollow fiber, m
n = number of hollow fibers
Q = cell suspension feed rate, m^3/s
ρ = cell suspension density, kg/m^3
P_{in} = inlet pressure = 25,000 Pa
P_{out} = outlet pressure = 100 Pa
P_f = filtrate pressure = 100 Pa

3.2 The production of a low-alcohol beer involves a series of steps. One of these steps is the concentration of the nonalcoholic components that give flavor to a beer. The composition of the retentate and filtrate from such a process is shown in Table 3.16.

a. Assume that if the permeability coefficient is 5 for the membrane and the transmembrane pressure at which flux begins to occur is 19 bar. Estimate the value of σ that would result in a positive flux across the membrane for beer that has the composition given in Table 3.16.

b. Estimate the area of the membrane that would be required to process 1000 L/hr of beer if the maximum pressure is 40 bar instead of the 30 bar shown in Figure 3.41.

c. Estimate the cost of the power requirement for processing 1000 L/hr of beer. Assume that the power required is electrical and the cost of electricity is $0.05/kwh.

3.3 A fermentation broth component is to be sterile filtered in order to remove bacteria that have a doubling time of 15 minutes, and an initial concentration of 10 cells/mL. The membrane filtration step involves recirculation of fluid across the membrane, and return of the retentate to a large hold tank until the filtration is completed. The volume of feed is initially 10,000 L and the retentate at the completion of this batch filtration is 1000 L. The overall hold time from the beginning to the end of the filtration is 2 hours, with the final permeate volume being 90% of the initial volume. What is the bacterial concentration of the retentate at the end of the sterile filtration process? What is a possible effect of bacterial growth in the retentate? State your assumptions.

CHAPTER 4

PRECIPITATION, CRYSTALLIZATION, AND EXTRACTION

INTRODUCTION

Precipitation may be induced by addition of neutral salts, organic solvents, or an acid or base to change the pH. These methods are based on altering the solvent's dissolution capacity for a given solute by decreasing the amount of water associated with a given solute, decreasing the dielectric constant of the solvent, or decreasing the repulsion of the solutes for each other. Flocculation is a form of precipitation in which a change in pH or addition of a hydrolytic enzyme destabilizes the colloid of protein micelles by decreasing the charge on the micelle's surfaces. This reduces their electrostatic repulsion or steric hindrance so that their interaction will lead to irreversible agglomeration and formation of a floc or gel (Reuttiman and Ladisch, 1991).

4.1 THE ADDITION OF NEUTRAL SALTS, OR AN ACID OR BASE TO AQUEOUS SOLUTIONS, INDUCES THE SOLUTE TO PRECIPITATE

Precipitation occurs due to a salting-out (hydrophobic) effect or when a salting-in (electrostatic) effect combined with a salting-out (hydrophobic) effect results in a net decrease in solubility. Salting-in causes solute solubility to increase with increasing salt concentration, while salting-out causes a decrease in solubility with increasing salt. An example of salting-out and salting-in effects is given by hemoglobin whose solubility either increases or decreases with increasing salt concentration depending on the salt (Figure 4.1). The overall solubility of the hemoglobin reflects a combination of salting in and salting out effects as represented in Figure 4.2.

Precipitation may also be induced by adjusting the pH of the solution so that it is close to the protein's isoelectric pH, which is referred to as the pI or isolectric point. Above the pI, the surface of a protein is negatively charged, and below the pI it is positively charged. The number

The author thanks Nathan Mosier, Phil Hall, and Cheng-Hong Li for their reading and helpful suggestions regarding this chapter while they were graduate students at Purdue University.

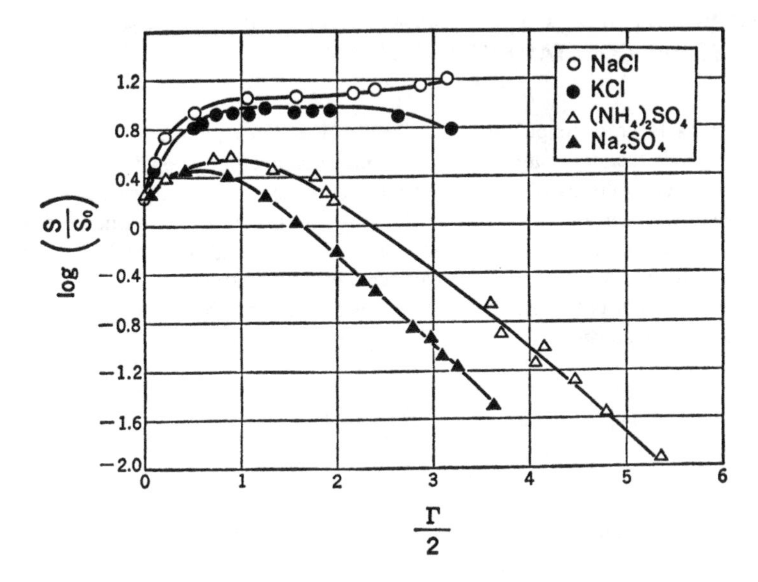

Figure 4.1 Normalized solubility of hemoglobin as a function of ionic strength for different salts (from Cohn and Ferry, 1943, fig. 13.12, p. 368, with permission).

of positive and negative charges are equal, and the net charge is zero at a pH that is close to the pI. At this pH, precipitation may result since the electrostatic repulsion between molecules is at a minimum. Precipitation at the pI is enhanced by decreasing ionic strength of the solution or by increasing its temperature. Denaturation and inactivation of the precipitated protein often accompanies isoelectric precipitation. Therefore precipitation by adjusting pH is often used to remove unwanted proteins by precipitating them while leaving the product protein in solution.

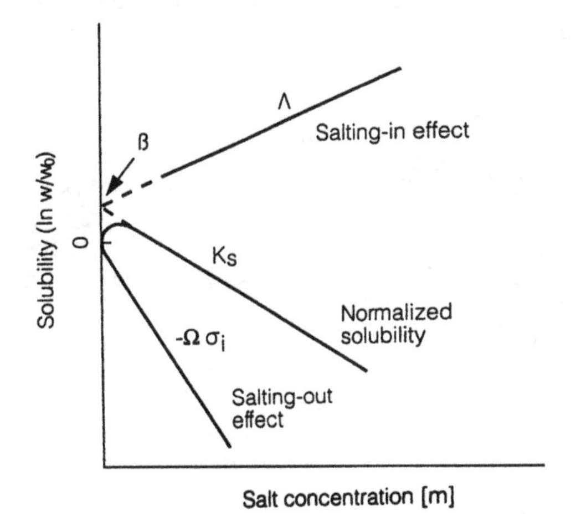

Figure 4.2 Schematic representation of combined salting-in (electrostatic) and salting-out (hydrophobic) effects on solubility of a protein in aqueous solution resulting in a normalized solubility curve (from Wheelwright, 1991, fig. 3.2, p. 33, with permission).

4.2 ALCOHOLS DECREASE SOLVATING POWER OF WATER BY LOWERING THE DIELECTRIC CONSTANT OF THE SOLUTION

Water is a poor solvent for nonpolar solutes. The dissolution of a nonpolar substance such as a protein or a gas is thermodynamically unfavorable. The hydrogen bonding between water molecules is altered by the dissolution of the solute, since the solute forces the water molecules to form a cavity into which it fits (Figure 4.3). New hydrogen bonds are formed between the water molecules surrounding the cavity now occupied by the solute. This results in a small decrease in the enthalpy, ΔH, as well as a significant decrease in the entropy, ΔS, due to the increase in the degree of order of solvent molecules surrounding the solute. The overall free energy change of the process, ΔG (cal/gmol) is represented by

$$\Delta G = \Delta H - T\Delta S \tag{4.1}$$

where

$\Delta H =$ change in enthalpy, cal/gmole

$\Delta S =$ change in entropy, $\dfrac{\text{cal}}{\text{gmol} \cdot \text{K}}$

$T =$ temperature, K

When the water that surrounds the protein is drawn into the bulk phase by an added solvent or second solute (Figure 4.4), a significant increase in entropy occurs accompanied by a relatively small change in enthalpy. Hence ΔG, as represented in Eq. (4.1), can become negative, and precipitation will occur. Denaturation of the protein in the precipitate is minimized by adding precooled solvents to a precooled solution, and keeping the temperature at or below 0°C. The beneficial effect of cooling to preserve protein activity is conceptually consistent with Eq. (4.1), since a lower temperature will give a smaller value of $T\Delta S$, and therefore a smaller negative change in the value of ΔG.

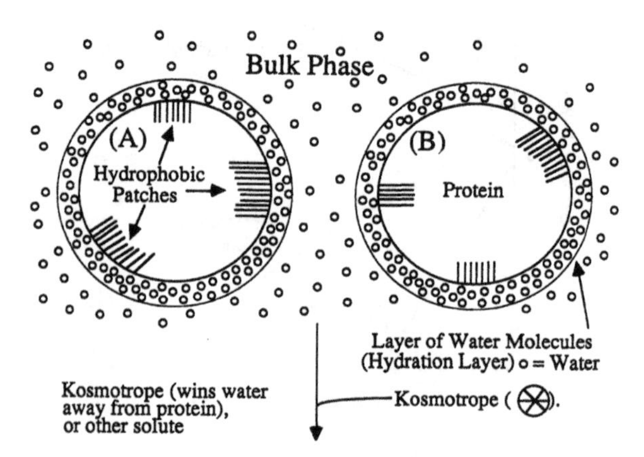

Figure 4.3 Schematic representation of hydrophobic interaction caused by disruption of hydration layer surrounding a protein molecule or an inert surface (*A*) and a second protein (*B*) (adapted from Roe, 1989, fig. 31, p. 223, using the description of Eriksson, 1989).

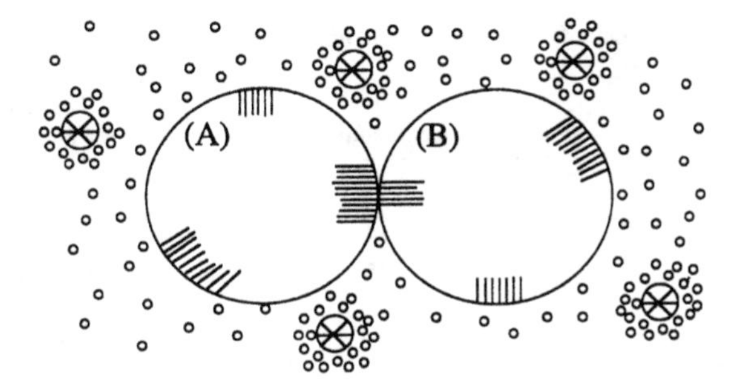

Figure 4.4 Interaction of A and B as follows from Figure 4.3.

The addition of alcohols such as methanol, ethanol, and isopropanol or a ketone such as acetone lowers the dielectric constant of a solution and hence reduces the solvating power of the solvent (water) causing a precipitate to form. A conceptual representation of precipitation due to decrease in the dielectric constant is given by the Born charging equation for a particle (Washington, 1969):

$$\Delta G = \frac{q^2}{2R_{\text{sphere}}}\left(1 - \frac{1}{\varepsilon}\right);\ \text{while per g} - \text{ion},\ \Delta G = -\frac{N_A(Ze)^2}{2R_{\text{sphere}}}\left(1 - \frac{1}{\varepsilon}\right) \tag{4.2}$$

where

$\Delta G =$ Gibbs free energy
$q =$ charge
$R_{\text{sphere}} =$ radius of a sphere (assumed to be an ion)
$\varepsilon =$ dielectric constant
$N_A =$ Avogadro's constant $= 6.02 \times 10^{23}$ molecules/mole
$Ze =$ charge on the ion
$e =$ electronic charge
$Z =$ valence

Equation (4.2) shows that a change in free energy occurs when a sphere of radius R and charge q is moved from vacuum to a solvent with a dielectric constant ε. The corresponding change in enthalpy is

$$\Delta H = -\frac{N_A(Ze)^2}{2R_{\text{sphere}}}\left(1 - \frac{1}{\varepsilon} - \frac{T}{\varepsilon^2}\frac{\partial \varepsilon}{\partial T}\right) \tag{4.3}$$

ΔG, ΔH, and solubility decrease as the dielectric constant ε of the solvent decreases. The dielectric constants of the different solvents in Table 4.1 suggest that acetone would give the greatest decrease in the dielectric constant of water, and hence the largest decrease in solubility of a third solute. While qualitatively useful, the theory is "clearly a gross approximation," according to

TABLE 4.1 Dielectric Constants of Selected Ionizing Solvents

Solvent	Temperature, °C	Dielectric Constant, ε (dimensionless)
Water	25	78.3
HF (anhydrous liquid)	0	84.0
Methanol	25	32.6
Ethanol	25	24.3
Acetone	25	20.7

Source: From Waddington (1969), table 1-2, p. 3.

Waddington (1969), since "the solvent near the ion cannot be treated as a continuous dielectric, and in the ion's immediate neighborhood the electric fields are so intense, that the solvent molecules are likely to be coordinated."

Cold, *anhydrous* liquid hydrogen fluoride is mentioned here given its ability to dissolve polysaccharides and proteins. Cellulose dissolves in HF to form conducting solutions, while water-soluble proteins, silk fibroin, and collagen also dissolve freely. Insulin, ribonuclease, and lysozyme have been recovered from HF with full retention of activity after their dissolution at low temperature. Vitamin B_{12} survives dissolution in cold, anhydrous HF and can also be recovered with full biological activity (Waddington, 1969).

4.3 NEUTRAL SALTS ADDED TO SOLUTIONS OF AMINO ACIDS OR PROTEINS CAUSE PRECIPITATION BY HYDROPHOBIC INTERACTIONS

A protein molecule in water holds a film of water in an ordered structure at its surface (Figure 4.3). This film, also known as a hydration layer, covers patches of hydrophobic, nonpolar amino acids such as alanine, methionine, tryptophan, and phenylalanine that are interspersed with hydrophilic areas consisting of polar amino acids. The removal of water molecules from this layer into the less ordered bulk phase results in an increase in entropy ($\Delta S > 0$) with a small change in the enthalpy. The free-energy change of the process, represented in Eq. 4.1, is negative and hence can occur spontaneously, resulting in hydrophobic interaction and precipitation (Figure 4.4). Salts that promote this type of change in phase are kosmotropes (i.e., "water structure maker"). Kosmotropes win water away from the protein, so the water is excluded from the protein's hydration layer, and is trapped in the bulk solution by hydration of the salt ions (Figure 4.4). Hydrophobic patches are then exposed on the protein's surface, and they interact with hydrophobic patches on other protein molecules or on a solid surface, such as a chromatography stationary phase (Roettger et al., 1989, 1990). The proteins will aggregate with one another or adsorb on a solid surface. In other words, "the salt ions attract around themselves the more polarizable molecules of the medium (in this case water), thereby squeezing out other components such as proteins" from solution (Cohn and Ferry, 1943, p. 609).

A protein precipitate formed in the presence of a kosmotrope, such as ammonium sulfate, retains its activity. This activity is recovered once the precipitate is separated from the salt and redissolved in an appropriate aqueous buffer. Isoelectric precipitation, which depends on minimizing electrostatic repulsion, is more likely to denature the protein. Protein activity is not recovered, and the precipitate may fail to redissolve in buffer.

The ions that are most effective in salting-out and precipitating proteins from aqueous solutions, namely, *kosmotropes*, create the greatest structuring of water (Roe, 1989). *Chaotropes*, or water structure breakers, have the opposite effect. Chaotropes promote dissolution, and in some cases, cause the unfolding and denaturation of the protein. Salting-out ions increase the surface tension of water, while salting-in ions have the opposite effect. The effectiveness of salts in precipitating or salting out proteins follows the order of the lyotropic series:

$$\textit{Most precipitation} \quad \text{phosphate} > \text{sulfate} > \text{acetate} > \text{chloride} \quad \textit{least} \tag{4.4}$$

Monovalent cations associated with these anions are more effective in promoting precipitation than divalent cations. These cations follow the order

$$\textit{Most precipitation} \quad NH_4^+ > K^+ > Na^+ \quad \textit{least} \tag{4.5}$$

The kosmotrope NH_4^+ is the most effective in promoting precipitation and Na^+ is the least effective. The ordering of anions and cations in Eq. (4.4) and (4.5) can be presented in an expanded form by the Hofmeister or lyptropic series.[1] Shih et al. (1992) show that the

[1] The ranking of the salts with respect to their action as chaotropes or kosmotropes is called the *lyotropic series*, or the Hoffmeister series after the scientist who identified it. There is variation in the ranking of ions within this series, with different references using different conventions or giving different sequences (Bailey and Ollis, 1977; Eriksson, 1989; Roe, 1989; Wheelwright, 1991). An attempt is made here to present these rankings in an internally consistent manner. Wheelwright (1991) lists the chaotropic series for anions for protein solubilization (from Knuth and Burgess) as follows:

Most dissolution of protein (chaotropes)

$$CBr_3COO^- > CCl_3COO^- > SCN^- > I^- > Br^- > CF_3COO^- >> CH_3COO^- > \text{citrate} > \text{phosphate} > \text{sulfate} \tag{n}$$

least dissolution (kosmotropes)

Roe (1989) lists the anions in a similar order:

$$SCN^- > I^- > ClO_4^- > NO_3^- > Br^- > Cl^- >> CH_3COO^- > \text{sulfate} > \text{phosphate}$$

where SCN^- has the least salting-out effect (i.e., highest dissolution) and phosphate the largest salting-out effect. Ericksson (1989) classifies ClO_4^-, I^-, and SCN^- as chaotropes, and gives the chaotropic series as:

$$SCN^- > I^- > ClO_4^- > NO_3^- > Br^- > Cl^- > \text{sulfate}$$

The order in which these salts promote hydrophobic interaction (i.e., protein precipitation from solution) follows the opposite sequence (i.e., sulfate $> Cl^- > Br^- > NO_3^- > ClO_4^- > I^- > SCN^-$).

Bailey and Ollis (1977) give the Hofmeister series of anions in the order of diminishing effectiveness of salting out:

$$\text{citrate}^{3-}, \text{tartrate}^{2-}, SO_4^{2-}, F^-, IO_3^-, H_2PO_4^-, CH_3COO^-, B_2O_3^-, Cl^-, ClO_3^-, NO_3^-, ClO_4^-, I^-, SCN^-$$

This represents the reverse order of the chaotropic series in equation (n). For cations the chaotropic series is (Wheelwright, 1991):

$$\textit{Most dissolution of protein} \quad \text{Guanidine} > Ca^{2+} > Li^+ > Na^+ > K^+ \quad \textit{least dissolution}$$

Roe (1989) gives a more complete listing:

$$Ba^{2+} > Ca^{2+} > Mg^{2+} > Li^+ > Cs^+ > Na^+ > K^+ > Rb^+ > NH_4^+$$

The chaotropic series proceeds from left to right while the kosmotropic series proceeds from right to left in equations (n) and (o) above (Bailey and Ollis, 1977; Wheelwright, 1991).

K'_S = constant obtained from slope of solubility curve (dimensionless since solubility of solute expressed in terms of concentration). The symbol (′) denotes the constant for solubility data expressed on a weight per volume basis.

K'_S is usually referred to as a salting out constant, although it actually represents a combination of salting-in and salting-out effects, as represented in Figure 4.2. The constant β' is obtained by extrapolating the solubility data at high-salt concentration to a hypothetical condition of zero ionic strength or salt concentration. The ionic strength Γ is defined by (Rosenbaum, 1970)

$$\frac{\Gamma}{2} = \frac{1}{2} \sum Z_1^2 C_1 \tag{4.7}$$

where

Z_1 = valences of cationic and anionic species
C_1 = molar concentration of cationic and anionic species, moles/L

For example, 1 M Na_2SO_4 gives the value for ionic strength of 3, since the valence of Na is 1 and there are two moles of Na per mole of salt. The valence of SO_4 is (−2), and there is one mole of it per mole of salt. Equation 4.7 therefore results in a value of 3:

$$\frac{\Gamma}{2} = \frac{1}{2}[(1)^2(2) + (-2)^2(1)] = \frac{1}{2}[2+4] = 3$$

The Cohn equation, Eq. (4.6), is broadly applicable to gases and low molecular weight solutes as well as to proteins. It applies when the salt concentrations are expressed on a mole fraction, molar (moles/kg solution), or molal (moles/L solvent) concentration or weight basis.[3] The range of salt concentrations over which this equation applies will depend on both the salt and the solute as is illustrated for an amino acid (cystine) in Figure 4.5 or a protein (hemoglobin) in Figure 4.1. The chaotropes, NaCl, KCl, and $CaCl_2$, increase the solubility of the solute with increasing concentration, while the kosmotropes $(NH_4)_2SO_4$ and Na_2SO_4 decrease solubility

[3]The logarithmic relation of Eq. (4.6) is also referred to as the Cohn and Edsall equation. In general, solubility correlations are expressed as logarithmic equations. The activity coefficient, γ_R, for an uncharged molecule, R, dissolved in water at low ionic strength is (Cohn and Edsall, 1943)

$$\log \alpha_R = \log \alpha_E + K_S\left(\frac{\Gamma}{2}\right) = -K_R + K_S\left(\frac{\Gamma}{2}\right) \tag{p}$$

where K_S is a salting-out constant with a positive value (i.e., solubility of the uncharged molecule increases with increasing ionic strength). Solubility of a dipolar ion, R^*, at low ionic strength is

$$\log \gamma_{R^*} = \log \gamma_E + K_S\left(\frac{\Gamma}{2}\right) = (-K_R + K_S)\left(\frac{\Gamma}{2}\right) \tag{q}$$

where $\log \gamma_E$, and $K_R(\Gamma/2)$ represents electrostatic effects which are a function of the molecule's dipole moment and distribution of charge. The linear relation between the decrease in the log of the protein solubility with increasing ionic strength or salt concentration was demonstrated in 1935, according to Cohn and Ferry (1943). Modern chromatography literature uses logarithmic relations of this type to describe retention behavior of solutes in reversed phase, ion exchange, and hydrophobic interaction chromatography.

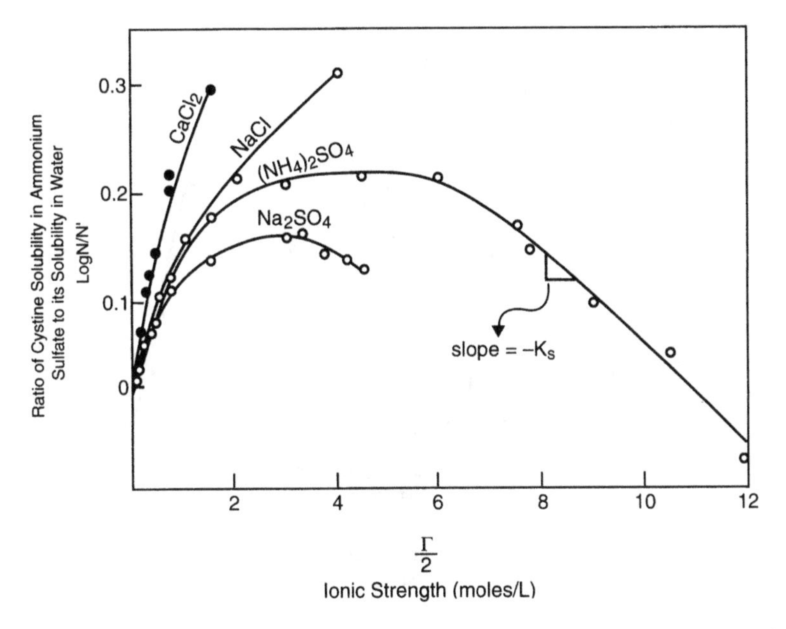

Figure 4.5 Solubility of cystine in several salts illustrates salting-in ($CaCl_2$ and NaCl) and salting-out ($(NH_4)_2SO_4$ and Na_2SO_4) effects as a function of ionic strength. Ammonium sulfate curve follows Cohn's Eq. (4.10) (adapted from Cohn, and Ferry, 1943, fig. 4, p. 243).

with increasing concentration. The ionic strength required to achieve an onset of decreasing solubility is lower for hemoglobin ($\Gamma/2 > 1.5$) than for cystine ($\Gamma/2 > 4$).

The Cohn equation captures the key ranges of interest when the objective is to carry out amino acid or protein precipitation (Figures 4.5 and 4.6). The range of salt concentrations over which the equation applies is small, but it gives an excellent fit of the decreasing solubility of amino acids and proteins with increasing salt concentration. The use of weight fraction (S/S_0); molar concentration (N) as molality (moles/1000 g of solvent) or molarity (moles/L solvent); mole fraction (represented as N/N_0); the ratio of moles of solute in a salt solution, N, relative to its solubility in water, N', in place of S; or the weight or mole concentration of the salt in place of $\Gamma/2$ works equally well. However, the values and dimensions of K_s and β will differ for each of these formalisms. An example is given by the equation that represents solubility of cystine in aqueous ammonium sulfate. The data for cystine solubility follow the correlation (Cohen, 1943)

$$\log N = \beta'' - K_s''\left(\frac{\Gamma}{2}\right) = \beta''' - K_s'''\left(\frac{I}{2}\right) \tag{4.8}$$

or

$$\log N = (-4.61) - (0.044)\left(\frac{\Gamma}{2}\right) = (-4.69) - (0.027)\left(\frac{I}{2}\right) \tag{4.9}$$

where

$N =$ mole fraction of the amino acid

fective in promoting precipitation than the Na^+ cation, according to the sequence in Eq. (4.5), but this is not the case for cystine, which is more soluble in $(NH_4)_2SO_4$ than Na_2SO_4 at equivalent ionic strengths (see Figure 4.5). However, the divalent Ca^{++} cation associated with chloride is more effective than the monovalent Na^+ in promoting solubility (see top two curves in Figure 4.5). This is consistent with the chaotropic series (refer to equation (o) in footnote 1).

The presence of another amino acid, glycine, together with cystine increased the solubility of the cystine in aqueous NaCl (Cohen, 1943). This is attributed to the large dielectric constant of glycine, which increases ΔG and ΔH of the solution, (Eqs. (4.2) and (4.3)) and therefore solubility of the cystine. While a protein would not typically be found in solution with pure amino acids, other components are likely to be present. These could significantly affect the solubility of the protein or other target molecule. (The target molecule refers to the solute which is to be separated from the other components). It is difficult to predict exactly the solubility of the target molecule in a complex biological solution, and hence it is necessary to make experimental measurements for determining K_s and β. The equations presented here are meant to be used together with heuristic rules to select appropriate ranges of salt concentration for protein precipitation.

4.5 THE SEPARATION FACTOR FOR TWO PROTEINS, LYSOZYME AND α-CHYMOTRYPSIN, IS CALCULATED FROM THEIR DISTRIBUTION COEFFICIENTS: AN EXAMPLE

The separation factor is a ratio of distribution coefficients. A distribution coefficient specifies the ratio of concentration of a single solute distributed between two phases. The distribution coefficient, $K_{D,i}$, for a protein, i, which is at equilibrium and distributed between the supernate and precipitate phases is

$$K_{D,i} = \frac{\Delta C_{\text{supernate},i}}{\Delta C_{\text{precipitate},i}} = \text{slope} \tag{4.12}$$

where

$C_{\text{supernate},i}$ = protein concentration in the supernatant for component i, moles/L
$\Delta C_{\text{supernate},i}$ = difference between initial and final concentrations of solute in supernate
$C_{\text{precipitate},i}$ = protein concentration in the precipitate phase for component i, moles/L

The value of $K_{D,i}$, is less than one since the decrease in concentration of the protein in the supernate phase is less than the increase in concentration in the precipitate phase. The slope of $K_{D,i}$ is negative, as illustrated in Figure 4.7, since the concentration decreases in one phase and increases in the other. The distribution coefficient could also be expressed as a ratio of solute concentration in the precipitate phase to the solute concentration in the liquid phase (see the right-hand side of Figure 4.7), although this convention is apparently not used.

Figure 4.7 represents the lines that would be formed by plotting concentrations of the protein in the supernate against concentrations in the precipitate. The ratio of the protein in the two phases is not necessarily constant. For example, the elliptical curve, on the left side of Figure 4.7 illustrates the shape of the curve for a solute where precipitation is favored at high solute concentrations in the supernate as indicated by a steep slope of the tangent line (i). Another possibility is that precipitation occurs at a low supernate concentration as indicated by the tangent line (*ii*) in Figure 4.7. The calculation of a mass balance from the distribution coefficients requires that the volumes of the supernate and precipitate phases be known so that total mass in the two phases can be computed.

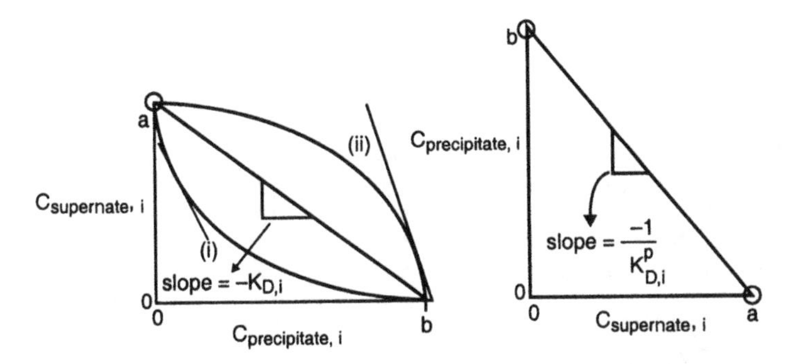

Figure 4.7 Schematic representation of curves that give separation factors.

Lysozyme and chymotrypsin exhibit a linear distribution coefficient as illustrated in Figure 4.8. The distribution coefficient for lysozyme in the presence of α-chymotrypsin ($K_{D,lysozyme}$) ranges from about 0.03 to 0.13 at pH 5.0 in sodium phosphate and sodium sulfate solutions as shown in Table 4.4 (Shih et al., 1992). The slope is steeper at the higher sodium phosphate buffer concentration as represented in Figure 4.8. Hence $K_{D,lysozyme}$ is larger at 2.53 than at 2.09 M sodium phosphate (Table 4.4) for α-chymotrypsin in solution with lysozyme, where $K_{D,\alpha\text{-chymo}}$ ranges from 0.055 to 0.080. $K_{D,\alpha\text{-chymo}}$ is higher at 3.05 than at 2.82 M sodium phosphate as illustrated in Figure 4.8.

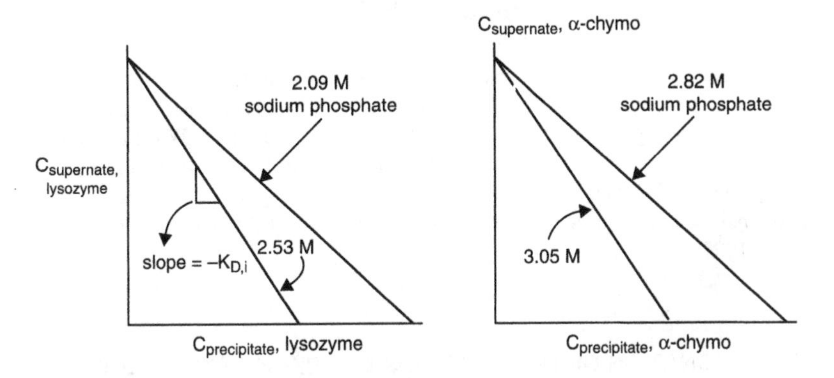

Figure 4.8 Schematic representation of linear distribution coefficients for lysozyme and α-chymotrypsin.

The separation factor, α, for lysozyme (component $i = 1$) from α-chymotrypsin ($i = 2$) is

$$\alpha = \frac{K_{D,1}}{K_{D,2}} = \frac{K_{D,\text{lysozyme}}}{K_{D,\alpha\text{-chymo}}} \tag{4.13}$$

In Eq. (4.13), it is assumed that the presence of one protein does not affect the other. Thus for Na_2SO_4, where the measurements of single components for both lysozyme and α-chymotrypsin

TABLE 4.4 Experimentally Measured Distribution Coefficients for Precipitation of Pure Component Lysozyme and α-Chymotrypsin from Sodium Phosphate and Sodium Sulfate Buffers at pH 5.0

	Salt			Distribution Coefficient[b] (Supernatant/Precipitate given by Eq. (4.4)	
Protein	Type	pH	Concentration, moles/L	$K_{D,lysozyme}$	$K_{D,\alpha\text{-chymo}}$
Lysozyme	Sodium Phosphate[a]	5.0	2.09	0.05	—
		5.0	2.53	0.13	—
	Na_2SO_4	5.0	1.56	0.03	—
α-Chymotrypsin	Sodium phosphate[a]	5.0	2.82	—	0.055
		5.0	3.05	—	0.080
	Na_2SO_4	5.0	1.56	—	0.055

Source: Data from Shih et al. (1992). Reprinted by permission of Wiley-Liss, Inc., a subsidiary of John Wiley & Sons, Inc.

[a]The pH of the phosphate solutions was adjusted by adjusting the ratio of Na_2HPO_4 to H_3PO_4 (Shih et al., 1992).
[b]Values of distribution coefficients are approximate, since they were read from figures 10 to 13 in Shih et al. (1992).

were carried out at a pH of 5 and salt concentration of 1.56 M (Table 4.4), the separation factor is

$$\alpha = \frac{K_{D,lysozyme}}{K_{D,\alpha\text{-chymo}}} = \frac{0.030}{0.055} = 0.545$$

The small values of both of the distribution coefficients indicate that almost all of the protein is precipitated. The value of 0.545 for the separation factor means that the change of lysozyme concentration in the supernate is less than the change for chymotrypsin. While the supernate is slightly enriched in lysozyme relative to the chymotrypsin, the fractionation of lysozyme from chymotrypsin using Na_2SO_4 is not effective because both proteins precipitate at these conditions. An analogous conclusion applies to chymotrypsin and BSA (data not shown here, see Shih et al., 1992).

4.6 FRACTIONATION OF TWO PROTEINS BY PRECIPITATION REQUIRES THAT THEIR SOLUBILITIES ARE SIGNIFICANTLY DIFFERENT FROM EACH OTHER

Fibrinogen, hemoglobin, and myoglobin precipitate at widely different ionic strengths, thus suggesting selective precipitation is possible in some cases (Figure 4.6). The precipitation of fibrinogen would be nearly complete at an ionic strength of 3 moles/L, while myoglobin is almost completely soluble at up to 8 M ammonium sulfate. For example, at an ionic strength of 2 M ammonium sulfate, the solubility of fibrinogen is 2.5 g/L, while at an ionic strength of 2.5, the solubility of fibrinogen is 0.1 g/L. The concentrations of fibrinogen in the precipitate phase, where the precipitate phase is assumed to be 10% of the volume of the supernate, are close to zero at an ionic strength of 2, and 24 g/L at an ionic strength of 2.5. The distribution coefficient corresponding to these concentrations from Eq. (4.12) is

$$K_{\text{D,fibrinogen}} = -\frac{\Delta C_{\text{supernate},i}}{\Delta C_{\text{precipitate},i}} = -\frac{2.5 - 0.1}{0 - 24.0} = 0.1$$

Myoglobin would remain completely soluble at this ionic strength (Figure 4.6). If the initial solubility of the myoglobin is 3.16 g/L, then from Eq. (4.12),

$$K_{\text{D},i} = -\frac{3.16 - 3.16}{0 - 3.16} = 0$$

Hence the distribution coefficient for myoglobin approaches zero. The separation factor, for this example, would be

$$\alpha = \frac{K_{\text{D,fibrinogen}}}{K_{\text{D,myoglobin}}} = \frac{0.1}{K_{\text{D,myoglobin}} \to 0} = \text{infinite}$$

A complete fractionation should be possible. Practical considerations, however, preclude a complete separation of one protein from the other. The form of the logarithmic solubility Eq. (4.6) can give a concentration that would approach zero, but there would still be a very small amount of fibrinogen in solution, by definition. Furthermore, since precipitation occurs with the solid phase in contact with, and partially including the liquid phase, there will be a small amount of the soluble protein included in the precipitated phase. Hence the precipitate should be viewed as a solid phase that is rich in fibrinogen but also includes some soluble myoglobin. The liquid phase that is rich in myoglobin contains a small amount of the fibrinogen. The presence of other proteins will also affect the solution behavior of the target protein itself, much like glycine affects solubility of cystine.

4.7 pH, TEMPERATURE, AND INITIAL CONCENTRATION ALSO AFFECT PROTEIN SOLUBILITY

Hemoglobin attains a constant or increasing solubility at ion concentrations of about 0 to 3 M NaCl and KCl (Figure 4.1), while the solubility decreases with respect to increasing concentrations of ammonium sulfate and sodium sulfate. Cohn and Ferry's analysis and clear representation of solubility data for hemoglobin shows that β, the hypothetical solubility at an ionic strength of zero, is a function of both pH and temperature, since the intercept changes with respect to pH and temperature. K_S for hemoglobin in concentrated solutions of ammonium sulfate is independent of temperature and pH as indicated by the parallel lines in Figures 4.9 and 4.10. The slopes of the solubility curves are the same regardless of pH (Figure 4.10), and temperature (Figure 4.9).

Solubility of hemoglobin is a function of temperature as shown in Figure 4.9. The intercept is larger for the lines that represent data at 0°C, compared to results obtained at 25°C. The heat of solution of the protein affects its solubility and hence the intercept β should change with temperature. The intercept also increases with an increase in pH (Figure 4.10). A change in pH changes the ionization of the protein and therefore its solubility, even when the protein is dissolved in concentrated salt solutions. Consider the comparison of egg albumin to hemoglobin

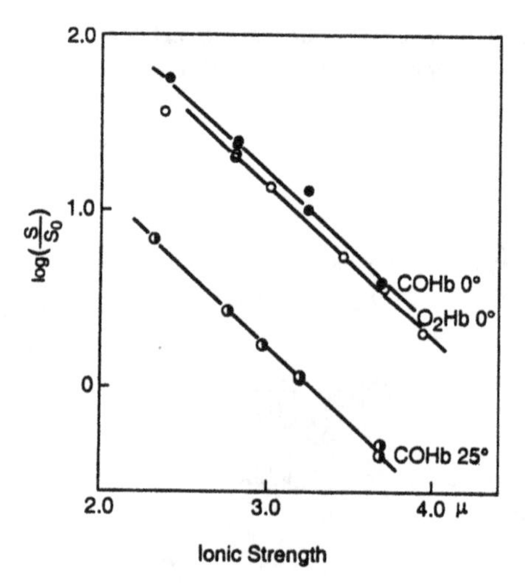

Figure 4.9 Effect of temperature (data of A. A. Green in J. Biol. Chem., 93, 507, 1931 as plotted in Cohn and Ferry, 1943, fig. 6, p. 605).

in Figure 4.11 which super-imposes curves for the two proteins. The difference between the β that was measured at the indicated pH's and β_n, the hypothetical protein solubility at its isoelectric point, approaches zero at a pH of about 4.5 for egg albumin, and 6.6 for hemoglobin. These pH values coincide with the respective isoelectric points of the two proteins.

The reason that β refers to a hypothetical solubility is demonstrated in Table 4.5. The values of S_0 that correspond to β′ far exceed limits where the solute would dissolve in a liquid, and the

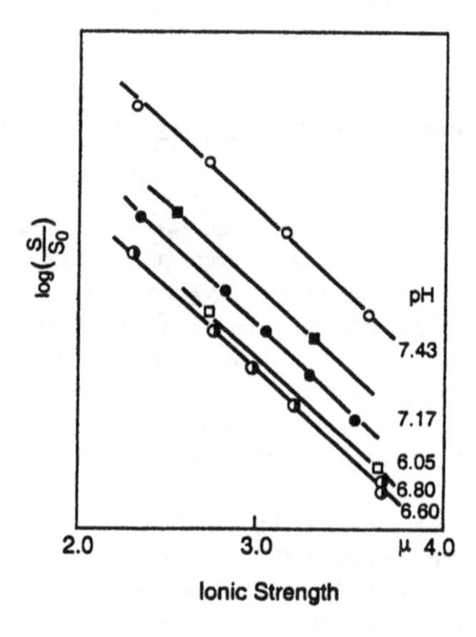

Figure 4.10 Effect of pH on the solubility of hemoglobin phosphate buffer (data of A. A. Green in *J. Biol. Chem.*, *93*, 507, 1931 as plotted in Cohn and Ferry, 1943, fig. 6, p. 605).

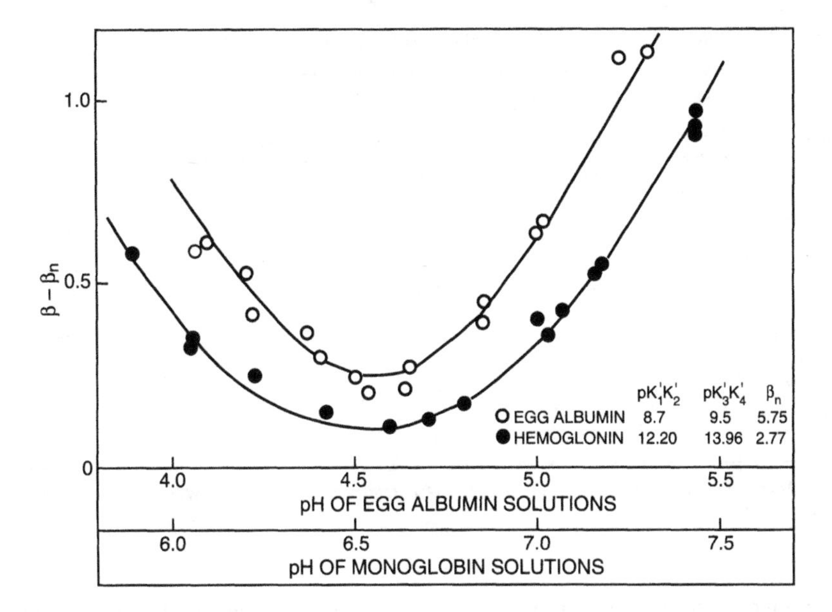

Figure 4.11 Solubility of hemoglobin and egg albumin as function of pH (data of A. A. Green, *J. Biol. Chem.*, *93*, 517, 1934 (as presented in Cohn and Ferry, 1943, fig. 7, p. 606).

TABLE 4.5 Values of Parameters Determined by Fitting Eq. (4.6) to Lysozyme Solubility Data

Salt	pH	β'	S_o, mg/mL	K_S' (dimensionless) Based on Salt Concentration	K_S' (dimensionless) Based on Ionic Strength
Sodium sulfate	4	2.93	851.1	1.949	0.683
	6	3.47	2951.2	1.946	0.682
	7	3.51	3235.9	2.085	0.695
	8	3.55	3448.1	2.136	0.712
	9	3.71	5128.6	2.380	0.780
	11	3.45	2818.4	2.295	0.765
Sodium phosphate	5	3.70	5011.9	1.303	0.914
	6	3.91	8128.3	1.659	0.870
	7	3.84	6918.3	1.890	0.799
	8	3.77	5888.4	2.095	0.720
Sodium chloride	4	1.48	30.2	0.526	0.526
	5	1.70	50.1	0.426	0.426
	6	1.47	29.5	0.481	0.481
	8	1.18	15.1	0.480	0.480
	9	0.903	8.0	0.447	0.447
	10	0.848	7.0	0.453	0.453
	11	0.839	6.9	0.297	0.297

Source: Adopted from Shih et al. (1992), table I, p. 1158. Reprinted by permission of Wiley-Liss, Inc., a subsidiary of John Wiley & Sons, Inc.

Note: The protein solubility data were expressed in terms of concentration; hence β has dimensions of concentration while K_S' is dimensionless. S_o is calculated from β using Eq. (4.6), where $\Gamma/2 = 0$, so $S_o = 10^{\beta'}$.

TABLE 4.6 Values of Salting-out Constants for Amino Acids and Proteins

	K_S, L/mole			
Solute	NaCl	$(NH_4)_2SO_4$	Na_2SO_4	Phosphate
Cystine	—	0.05	—	—
Leucine	0.09	—	—	—
Tyrosine	0.31	—	—	—
Lactoglobulin	—	—	0.63	—
Hemoglobin (horse)	—	0.71	0.76	1.00
Hemoglobin (human)	—	—	—	2.00
Myoglobin	—	0.94	—	—
Egg albumin	—	1.22	—	—
Fibrinogen	1.07	1.46	—	2.16

Source: Adapted from Cohn and Ferry (1943), table 6, p. 604. Reprinted by permission of Wiley-Liss, Inc., a subsidiary of John Wiley & Sons, Inc.

liquid is the dominating phase. Consider, for example, lysozyme dissolved in sodium sulfate. At pH 4.0 its hypothetical, extrapolated solubility is 851 mg/mL, while at pH 7 the solubility is 3235 (Table 4.5). These hypothetical solubilities exceed a reasonable level as do many of the other values of β reported in Table 4.5.

The data in Table 4.6 illustrate that proteins have higher K_S values (0.6 to 2.16 L/mole) than amino acids ($K_S = 0.05$ to 0.3 L/mole) and that they are therefore more sensitive to salting-out effects than amino acids. For example, the solubility of leucine decreases by about 11% when NaCl is added to give a 0.5 M solution, while fibrinogen solubility decreases by about 1000% over the same range of salt concentration (Cohn and Ferry, 1943, p. 604).

Recent studies by Shih et al. (1992) confirm that the variation of K_S as a function of pH is small except at extreme pH values (Table 4.5). The effect of pH on K_S' is noticeable at high pH values for lysozyme, since the isoelectric point of lysozyme is at pH = 10.5. The protein has a minimum solubility at its isoelectric point, since the net charge on the macromolecule approaches zero. K_S' also changes when pH < 5. At an acidic pH (< 5), the anions of salts will bind with the protein and thereby reduce its net charge and solubility. The solubilities of some proteins are a function of their initial concentration. This type of protein has been labeled as a Type II protein. The precipitation behavior of a Type I protein is independent of initial protein concentration. Lysozyme has been proposed to fit the definition of a Type I protein while chymotrypsin and BSA would represent a Type II protein system.

4.8 HEAT OF SOLUTION EFFECTS CAN BE SIGNIFICANT FOR PROTEINS

Heats of solution vary significantly with the type and concentration of the salt, as well as with respect to the protein itself. If the heat of solution of a protein is negative, dissolution is exothermic and a decrease in temperature increases solubility. For example, horse carboxyhemoglobin is about 10-fold more soluble at 0°C than at 25°. Its heat of solution is −14,011 cal/g mole when dissolved in phosphate buffer with an ionic strength of between 2.2 and 4 (Cohn and Ferry, 1943). When the heat of solution is positive, dissolution is endothermic and an increase in temperature increases solubility. For example, lactoglobulin has ΔH of +4200 cal/g mole in water. Its solubility in water increases from 0.35 g/L at 5°C to 0.58 g/L at 25°C (Cohn and Ferry, 1943). An addition of a second component may alter the temperature dependence and therefore the apparent heat of solution. Lactoglobulin has a ΔH of +2400 cal/g mole in 0.5 M glycine compared to $\Delta H = 4200$ in water. While an increase in temperature results in an increase in solubility of this protein in both water and glycine solution, the increase is less for lactoglobulin in 0.5 M glycine than in water.

The heat of solution, ΔH (cal/gmole), is related to the solubility of the protein at two temperatures by the thermodynamic relation:

$$\Delta H = \frac{R \ln (S_2/S_1)}{(1/T_1 - 1/T_2)} \tag{4.14}$$

S_1 and S_2 are the solubilities at temperatures T_1 and T_2, respectively. Temperatures are in °K. R is the gas law constant = 1.98 cal/(gmole · K). Equation (4.14) enables calculation of heat of solution for carboxyhemoglobin in pH 6.6 phosphate buffer, for example, where the protein has a solubility of 22.7 g/L at 0°C and 2.58 g/L at 25°C. The heat of solution is therefore indicated to be negative:

$$\Delta H = \frac{1.98 \ln (2.58/22.7)}{\dfrac{1}{273+0} - \dfrac{1}{273+25}} = -14{,}011 \; \frac{\text{cal}}{\text{gmol}}$$

This approach is only useful in obtaining a magnitude of order estimate of heats of solution, and does not apply at higher protein concentrations (Cohn and Ferry, 1943).

4.9 THE SALTING-OUT CONSTANT, K_S, COMBINES SALTING-OUT AND SALTING-IN EFFECTS THAT CHARACTERIZE HYDROPHOBIC INTERACTIONS

The combination of salting-in and salting-out effects by the salting-out constant K_S can be represented as (Melander and Horvath, 1977; Wheelwright, 1991)

$$\boxed{K_S = \Omega\sigma - \Lambda} \tag{4.15}$$

where

K_S = slope of the normalized solubility curve defined by the Cohn Eq. (4.6), and originally referred to as a "salting-out" constant by Cohn and his peers.

$\Omega\sigma$ = an intrinsic salting-out coefficient based on hydrophobic interactions

Ω = nonpolar contact area parameter

σ = molal surface tension increment

Λ = salting-in coefficient proportional to dipole moment of protein at high salt concentration.

The molal surface tension increment, σ(erg/cm^2), in Eq. (4.15) is defined by the surface tension of an inorganic salt solution:

$$\sigma = \frac{\gamma - \gamma_o}{m} \left| \frac{\text{dyn} \cdot \text{g}}{\text{cm} \cdot \text{mol}} \right| \tag{4.16}$$

where

$\gamma =$ surface tension of the salt dissolved in the solution
$\gamma_o =$ surface tension of pure water = 72 dyne/cm at 25°C
$m =$ molal concentration of the salt (moles/L)

The surface tension increment, σ, has dimensions of surface tension of the solution (dyne/cm) divided by the molal concentration of the salt dissolved in the solvent that has a surface tension of γ_o (dyne/cm).[5] According to Eq. (4.15) the experimental data should give a straight line when K_S is plotted against σ as illustrated in Figure 4.12.

One interpretation of the constant slopes, and therefore the pH independent nature of K_S for proteins is that a protein molecule behaves as a charged species at low salt concentration and as a neutral dipole at high salt concentrations. As explained by Melander and Horvath (1977), the charged protein, even though it is a macromolecule, may be viewed as an ion when in solution. The Debye-Hückel[6] theory can then be used as a first approximation to describe the protein's behavior at low ionic strength. As the salt concentration increases the average distance between charged species becomes smaller, and ionic binding and clustering of salt at the macromolecule's surfaces is believed to occur. As the salt concentration increases even more, the proteins are shielded by a layer or atmosphere of ions, and they behave as if they were molecules with a neutral dipole. Consequently at high salt concentrations, the electrostatic-free energy of the protein becomes proportional to the effective dipole moment of the protein. This leads to the definition of the salting-in coefficient, Λ (the upper line in Figure 4.2):

$$\boxed{\Lambda = \left(\frac{D}{RT}\right)\mu} \tag{4.17}$$

where

$D =$ constant
$R =$ gas law constant
$\mu =$ dipole moment of the protein
$T =$ temperature, K

[5]The dimensions for surface tension measurements may be expressed in a number of different ways, with different dimensions being associated with the same values in different publications. For example, the x-axis in Figure 4.14 gives the dimensions of the surface tension increment of

$$10^3 \left(\frac{\text{dyne} \cdot \text{g}}{\text{cm} \cdot \text{gmol}} \right) + \frac{\text{dyne}}{\text{cm}} \cdot \frac{1}{\left. \frac{\text{gmol}}{\text{L}} \right| \frac{\text{L}}{1000\ \text{g}}} = \frac{\text{dyne}}{\text{cm M}}$$

This is equivalent to dyne/cm divided by the molal concentration as shown.

[6]The Debye-Hückel theory addresses the calculation of deviations from ideality of solutions of strong electrolytes. It is based on the statistical calculation of the average distributions of ions around any particular ion in solution. Deviations from ideality of the ionic chemical potential due to electrical charge on the ion is expressed in terms of an activity coefficient (Rosenbaum, 1970).

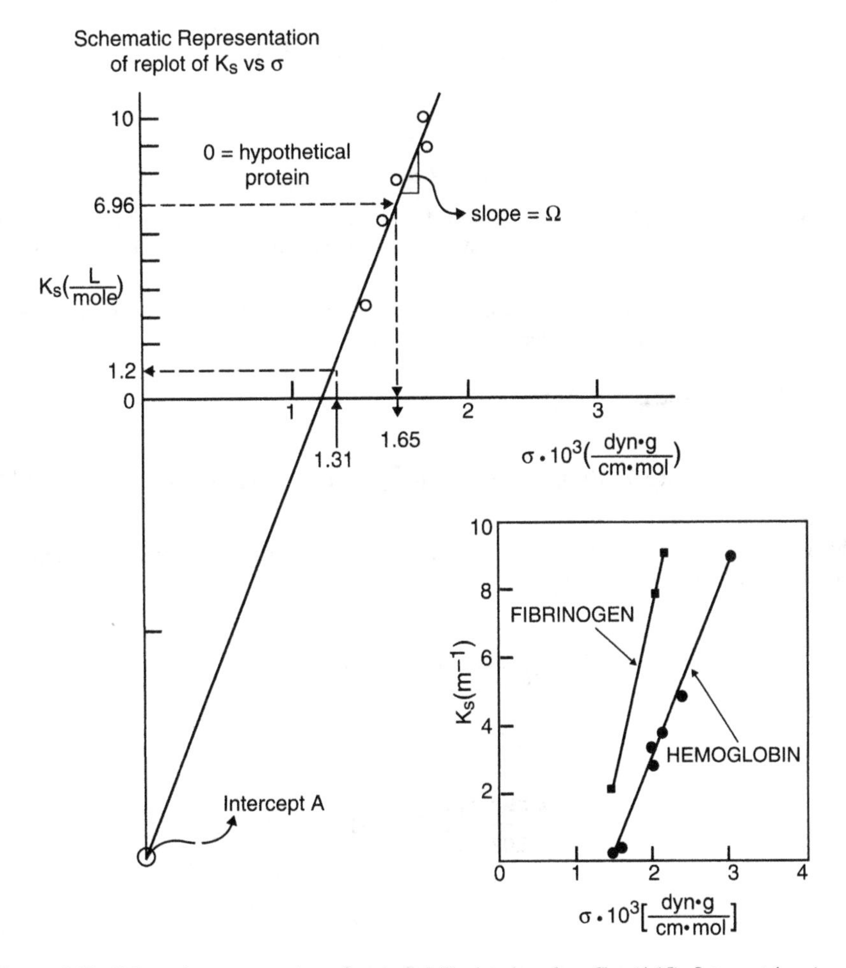

Figure 4.12 Schematic representation of replot of K_S data based on Eq. (4.15). Intercept is $-\Lambda$ and represents the salting-in coefficient which is proportional to the dipole moment μ. The slope gives the value of Ω, which is proportional to the magnitude of the nonpolar contact area, Φ, between protein molecules. Inset shows data for fibrinogen and hemoglobin (from Melander and Horvath, 1977, fig. 3, p. 207 with permission from Academic Press.).

The parameter, Λ, does not change significantly when pH and temperature changes.

The salting-out coefficient, $\Omega\sigma$, reflects the hydrophobic or nonpolar contact area,[7] Φ, between two precipitating molecules, and the molal surface tension increment of the salt, σ. The parameter Ω is a function of the contact area, Φ (Melander and Horvath, 1977):

[7]The contact area, denoted by Φ, represents the surface area of the protein molecule which is dehydrated when the protein precipitates. Alternately, Φ represents the surface area of the protein which must be wetted when dissolution occurs. The contact area of the protein is less than the total surface area of a protein molecule, since a significant portion of the total area of crystalline (precipitated) proteins is covered with water molecules (Melander and Horvath, 1977).

$$\Omega = \frac{1}{RT}[N\Phi + 4.8N^{1/3}(\kappa^e - 1)V^{2/3}] \tag{4.18}$$

where

$N =$ Avogadro's number = $6.0225 \cdot 10^{23}$ molecules/mole

$\kappa^e =$ 1.277 = proportionality constant for correcting the macroscopic surface tension of the solvent to a molecular dimension (dimensionless) (Melander and Horvath, 1977, p. 207).

$V =$ 18 cm^3/mol = molar volume of the solvent (in this case the solvent is water where 1 g = 1 cm^3 = $18 \cdot (10^8$ Å/cm$)^3$ cm^3/gmole = $18 \cdot 10^{24}$ $Å^3$/gmole

If the solvent is water, Eq. (4.17) becomes

$$\Omega = \frac{1}{RT}[6.02 \cdot 10^{23}\Phi + 4.8(6.02 \cdot 10^{23})^{1/3}(1.277 - 1)(18 \cdot 10^{24})^{2/3}] \tag{4.19}$$

The magnitude of the nonpolar contact area, in square angstroms, for a protein in an aqueous solvent is obtained by rearranging Eq. (4.19):

$$\Omega = \frac{1}{RT}[6.02 \cdot 10^{23}\Phi + 7.71 \cdot 10^{24}]$$

to

$$\Phi = \frac{\Omega RT - 7.71 \cdot 10^{24}}{6.02 \cdot 10^{23}} \cong 411\,\Omega - 12.8 \tag{4.20}$$

where

$$R = 8.904 \cdot 10^{23} \quad \text{and } T = 273 + 5 = 278 \text{ K}$$

Neither R nor the temperature were specified by Melander and Horvath (1977). If the temperature is 5°C, and the value of R is as indicated, the coefficient of 411 in Eq. (4.20) results.

4.10 HYDROPHOBIC CONTACT AREAS Φ OF SELECTED PROTEINS, RANGING FROM 20 TO 42% OF TOTAL SURFACE AREA, MAY BE DETERMINED FROM K_S

The definition of Ω given by Eq. (4.18) should allow determination of hydrophobic contact area by Eq. (4.20). For example, the plot of K_S as a function of σ is linear for fibrinogen and hemoglobin, as illustrated by the inset in Figure 4.12 (Melander and Horvath, 1977). The intercepts are reported to give a value for the constant term in Eq. (4.17) of

$$\frac{D}{RT} = 1.73 \times 10^{-3} \tag{4.21}$$

An implicit assumption is made that D has the same value for different globular proteins (Melander and Horvath, 1977), thereby enabling the contact parameter, Ω, to be calculated using Eq. (4.22), which is a rearranged form of Eq. (4.15):

$$\Omega = \frac{K_S + \Lambda}{\sigma} = \frac{K_S + (D/RT)\mu}{\sigma} = \frac{K_S + (1.73 \cdot 10^{-3})\mu}{\sigma} \tag{4.22}$$

Use of Eq. (4.22) requires that a value of the salting-out coefficient be already determined and that the ratio of D/T be a constant. Measurement of protein solubility as a function of the salt concentration and plotting the data according to Eq. (4.6) gives a line whose slope is K_S. When Eq. (4.22) is combined with Eq. (4.20), an expression for Φ results, as reported by Melander and Horvath (1977):

TABLE 4.7 Molal Surface Tension Increments of Various Salts

Salt	Surface Tension Increment (σ), $\times 10^3 \frac{dyn \cdot g}{cm \cdot mol}$	Salt	Surface Tension Increment (σ), $\times 10^3 \frac{dyn \cdot g}{cm \cdot mol}$
KSCN	0.45[a]	K_2-taratrate	1.96
$NaClO_3$	0.55	$Ba(NO_3)_2$	2.00
NH_4I	0.74	LiF	2.00
LiI	0.79	Na_2HPO_4	2.02
KI	0.84	$NiSO_4$	2.10
NH_4NO_3	0.85	$MgSO_4$	2.10
$KClO_3$	0.86	$MnSO_4$	2.10
NaI	1.02	$CuSO_4$	2.15
$NaNO_3$	1.06	$(NH_4)_2SO_4$	2.16
NH_4Br	1.14	$ZnSO_4$	2.27[c]
$LiNO_3$	1.16	Na_2-tartarate	2.35
LiBr	1.26	K_2SO_4	2.58
KBr	1.31	Na_3PO_4	2.66
NaBr	1.32	Na_2SO_4	2.73
CsI	1.39	Li_2SO_4	2.78
NH_4Cl	1.39	$FeCl_3$	2.78
$KClO_4$	1.40	$BaCl_2$	2.93
$FeSO_4$	1.55	K_3-citrate	3.12
LiCl	1.63[b]	$MgCl_2$	3.16[d]
NaCl	1.64	$CaCl_2$	3.66[e]
$CsNO_3$	1.57	$K_4Fe(CN)_6$	3.9[f]
$CuSO_4$	1.82	$K_3Fe(CN)_6$	4.34

Source: Reproduced from Melander and Horvath (1977), table IV, p. 212, with permission from Academic Press.

Note: The deviation from Eq. (4.18) is less than 1.5% except in the cases indicated.

[a]Concentration of 0 to 0.5 m only.

[b]Concentration of 0 to 2.0 m only.

[c]Maximum deviation, 8%.

[d]Maximum deviation, 18%.

[e]Maximum deviation, 28%.

[f]Average value of two sets of data.

TABLE 4.8 Nonpolar (Contact) Surface Areas for Selected Proteins Based on Salting-out Data from $(NH_4)_2SO_4$

Protein	Dipole Moment, μ (Debye)	Contact Area as % of Total Surface Area	Φ Nonpolar Contact Area, Eq. (4.23), $Å^2$	Total Surface Area,[a] ($Å^2$)	Measured Salting-out Constant K_S, L/mole
Myoglobin	170	42	1500	3700	5.50
Ovalbumin	250	28	1925	6950	5.84
Hemoglobin	480	25	2300	9200	3.86
Albumin	380	20	1930	9500	3.61

Source: Adapted from Melander and Horvath (1977), table II, p. 208. The material is reproduced with permission of Academic Press.

[a]Assumes spherical molecules.

$$\Phi = 411\Omega - 12 = 411\left(\frac{K_S + (1.73 \cdot 10^{-3})\mu}{\sigma}\right) - 12 \tag{4.23}$$

where

μ = dielectric constant of the protein
σ = molal surface tension increment

For ammonium sulfate, the molal surface tension increment is $\sigma = 2.17$ (Table 4.7). The contact areas of four proteins, together with corresponding values of K_S of these proteins in ammonium sulfate solutions appear in Table 4.8. The total surface areas in this table were calculated by Melander and Horvath (1977) from the molecular weight of the proteins and a specific volume of 0.73 cm^3/g that was assumed to be the same for all four proteins. The contact areas are shown to range from 20 to 42% of the total surface area of the proteins.

4.11 K_S MAY BE CALCULATED FROM THE PROTEIN'S DIPOLE MOMENT (μ), CONTACT AREA (Φ), AND SURFACE TENSION INCREMENT (σ): OVALBUMIN EXAMPLE

The tabulation of molal surface tension increment, σ, for 44 salts (Table 4.7) enables K_S to be calculated. Therefore the change in protein solubility at a high salt concentration can be estimated when one salt is replaced by another, if the contact area Φ and the dipole moment μ are known for the protein.

The value of K_S can be calculated from Eq. (4.15), since Λ can be obtained from Eq. (4.17) for $D/RT = 1.73 \cdot 10^{-3}$. Ω is from a rearranged form of Eq. (4.23):

$$\Omega = \frac{\Phi + 12}{411} \tag{4.24}$$

where Φ is the hydrophobic or non-polar surface area (in Å^2) characteristic of that protein. Equations (4.25) to (4.28) illustrate an example calculation for ovalbumin in aqueous ammonium sulfate. Ovalbumin has a dipole moment of $\mu = 250$ Debyes and a nonpolar contact area of $\Phi = 1925$ Å^2 as given in Table 4.8. The use of Eq. (4.16) gives the numerical value of the salting-in coefficient, Λ:

$$\Lambda = \left(\frac{D}{RT}\right)\mu = (1.73 \cdot 10^{-3})(250) = 0.4325 \tag{4.25}$$

Equation (4.24) gives the value of the nonpolar contact area parameter, Ω, at a temperature of 5°C for a contact area of 1925 Å^2 as in

$$\Omega = \frac{\Phi + 12}{411} = \frac{1925 + 12}{411} = 4.71 \tag{4.26}$$

The molal surface tension increment for ammonium sulfate in water at 25°C is $\sigma = 2.16$ as given in Table 4.7. The values of Λ (from Eq. (4.25)), Ω (from Eq. (4.26)), and σ enable calculation of K_S using Eq. (4.15):

$$K_S = \Omega\sigma - \Lambda = (4.71)(2.16) - 0.4325 = 9.74 \tag{4.27}$$

If NaCl ($\sigma = 1.64$ from Table 4.7) replaces ammonium sulfate, the slope of K_S becomes

$$K_S = \Omega\sigma - \Lambda = (4.71)(1.64) - 0.4325 = 7.29 \tag{4.28}$$

The literature values of μ, σ, and Φ are likely to be limited to a few proteins for the immediate future. Hence experimental measurements will continue to be necessary to generate the solubility curves from which K_S and β would be estimated. The experiments for a single protein would require at least three different salts, at concentrations that are high enough so that the data fall in the range of decreasing solubility. The K_S values obtained from the slope of the normalized solubility curves would then be replotted against the respective values of σ for each salt from Table 4.7 in order to give a plot of the type illustrated in Figure 4.14. The values of Ω and $-\Lambda$ are obtained from the slope and intercept of the plot of K_S in terms of σ. Note that if the target molecule is not a globular protein, the parameters Ω and $-\Lambda$ may represent different interactions than developed above. Other types of molecules that are subject to precipitation include polypeptides, polysaccharides, high molecular weight dyes, DNA, RNA, and long-chain organic acids. While the dominating mechanisms that cause precipitation may be different for these molecules than for globular proteins, the methods of presenting the data and obtaining constants is still valid and useful for estimating the effects of changing types and concentrations of salt.

The plotting of the data and determination of constants, as in Figure 4.12, could prove to be useful for selecting different salts for optimizing a precipitation process. Alternately, these data could also be used for selecting salts to be used as mobile phase components for hydrophobic interaction chromatography (see Chapter 8). In this context the molal surface tension increments in Table 4.7 provide a numerical representation of the Hofmeister or lyotropic series, where the chaotropic salts have the lowest values of σ and kosmotropes the highest.

4.12 GRAPHING OF K_S AGAINST MOLAL SURFACE TENSION INCREMENT (σ) ENABLES CALCULATION OF PROTEIN SOLUBILITY IN DIFFERENT SALT SOLUTIONS: AN EXAMPLE

An example of how a replot of K_S in terms of σ (Figure 4.12) might be used is given here. This example shows how a salt might be identified that will give the same extent of precipitation as KBr but at a concentration that is five times lower. The approach is to obtain a solubility relation, determine K_S for the new condition from the K_S versus σ data, and then use this value to obtain σ. Once σ is known, the salts that correspond to the molal surface tension increment can be obtained from Table 4.7.

For this example, assume that $\sigma = 1.31$, which corresponds to the salt KBr (Table 4.7); K_S is 1.2. Also assume that at a given KBr concentration, the protein concentration in the supernate is reduced 1000-fold from its solubility at a low ionic strength. This protein is represented in the graph of Figure 4.12. An experiment shows that the protein's solubility begins to decline as the KBr concentration approaches 0.2 M.

First, the solubility equation is adapted to properly reflect the question that was posed: What is the ratio of solubilities of the same protein in two different salt solutions? Equation (4.6) is expressed on a weight fraction basis to give the form

$$\ln\left(\frac{W}{W_o}\right) = \ln(f_1) = \beta - K_S m_1 \tag{4.29}$$

Equation (4.29) is equivalent to

$$f_1 = \exp(\beta - K_s m_1) \tag{4.30}$$

where m_1 = molal concentration of salt "1." The ratio of the weight fraction f_1 of the target molecule in solution with salt "1" at concentration m_1, to that of f_2 for salt "2" at concentration m_2, is

$$\frac{f_1}{f_2} = \frac{\exp(\beta - K_S m_1)}{\exp(\beta - K_S m_2)} = \exp[K_S(m_2 - m_1)] \tag{4.31}$$

where

$$f_1 = \frac{W_1}{W_o}$$

$$f_2 = \frac{W_2}{W_o}$$

The subscripts 1 and 2 denote different salts, while W_o, W_1, W_2 are the weights of the same protein dissolved in buffer (no salt), or in buffer containing salts 1 and 2, respectively. β is a function of temperature and pH but is eliminated from Eq. (4.31), since the temperature and pH are the same for both salts. The salting-out parameter, K_S, is defined by Eq. (4.15), and when substituted into Eq. (4.31), gives the expression

$$\frac{f_1}{f_2} = \exp[(\Omega\sigma - \Lambda)(m_2 - m_1)] = \exp[K_S(m_2 - m_1)] \quad (4.32)$$

Ω and Λ are properties of the protein determined from the slope and intercept of the line in Figure 4.12, while σ is determined by the type of salt (Table 4.7). The salt concentration required to achieve a 1000-fold reduction in dissolved protein can then be calculated using Eq. (4.32), where $f_1 = 0.9$, $m_1 = 0.2$ moles/L, and $K_S = \Omega\sigma - \Lambda = 1.2$:

$$\frac{f_1}{f_2} = \frac{0.9}{0.0009} = 1000 = \exp[1.2(m_2 - 0.2)] \quad (4.33)$$

This assumes that protein solubility decreases according to the Cohn equation when the concentration of KBr is above 0.2 M. Recall that $K_S = 1.2$ is read from Figure 4.12 for $\sigma = 1.31 \times 10^{-3}$. Rearrangement of Eq. (4.33) gives the value of m_2:

$$m_2 = \frac{\ln(f_1/f_2)}{K_S} + m_1 = \frac{\ln 1000}{1.2} + 0.2 = 5.956\ \frac{\text{moles}}{\text{L}} \quad (4.34)$$

The molal concentration of a salt that would achieve the same reduction in dissolved protein, but at a five times lower concentration, can now be estimated. Rearrangement of Eq. (4.32) to solve for K_S gives Eq. (4.35), where $m_2 = 5.96/5 = 1.19$ and $m_1 = 0.2$ moles/liter:

$$K_S = \frac{\ln(f_1/f_2)}{(m_2 - m_1)} = \frac{\ln(1000)}{(1.19 - 0.2)} = 6.96 \quad (4.35)$$

Reference to Figure 4.14 shows that a value of σ of about 1.65×10^{-3} corresponds to $K_S = 6.96$. The listing of molal surface tension increments in Table 4.7 indicates that NaCl ($\sigma = 1.64 \times 10^{-3}$) is a good candidate for further experimentation. The same result could be obtained numerically by fitting the data with Eq. (4.15) in order to obtain values of Ω and $-\Lambda$, and then using the resulting equation to calculate K_S as a function of σ.

4.13 THERMODYNAMICS OFFER AN EXPLANATION FOR BOTH SALTING-IN AND SALTING-OUT EFFECTS

The molal free energy, ΔG_{solute}, of a solute in a dilute solution is given by

$$\Delta G_{\text{solute}} = \Delta G^{\circ} + RT \ln x \quad (4.36)$$

where

x = mole fraction of the protein

ΔG° = free energy associated with transferring a solute from a hypothetical gas phase into a liquid phase

A protein or other nonvolatile solute is obviously not transferred from a gas to a liquid phase. However, this general treatment enables definition of the energetics of the dissolution process in an internally consistent manner while also applying to the case of a gas dissolving in a liquid.

The condition for equilibrium between the precipitate and the pure solute dissolved in the fluid phase is represented by the free energy, ΔG_{ppt}:

$$\Delta G_{ppt} = \Delta G^{o} + RT \ln x_{sat} \tag{4.37}$$

where x_{sat} is the mole fraction of the solute dissolved in the fluid to give a saturated solution:

$$\ln x_{sat} = \frac{\Delta G_{ppt} - \Delta G^{o}}{RT} \tag{4.38}$$

The mole fraction of proteins in saturated solution is very low (i.e., $x_{sat} << 1$), and hence the saturation concentration can be represented by

$$x_{sat} = \frac{W/M_r}{55.5} \tag{4.39}$$

where

W = concentration of solute, g/L

M_r = molecular weight of the solute

and 55.5 is the molarity of water (55.5 moles/L). Thus, for example, if 100 g/L of BSA (M_r = 66,000) is saturating, this is equivalent to 1.52×10^{-3} moles/L or 1.52 mmolar or $x_{sat} = 2.73 \times 10^{-5}$. The combination of Eqs. (4.38) and (4.39) results in

$$\ln\left(\frac{W/M_r}{55.5}\right) = \frac{\Delta G_{ppt} - \Delta G^{o}}{RT} \tag{4.40}$$

which upon rearrangement gives

$$\boxed{\ln W = \left(\frac{\Delta G_{ppt} - \Delta G^{o}}{RT}\right) + \ln(55.5 M_r)} \tag{4.41}$$

Equation (4.41) is the thermodynamic basis from which the Cohn equation is derived as described by Melander and Horvath (1977), starting from the definitions of Sinanoğlu and Abdulnur. This derivation is summarized by Eqs. (4.42) to (4.55) below.

The free energy associated with transferring a species from a hypothetical gas phase into solution is (Melander and Horvath, 1977)

$$\Delta G^{o} = \Delta G_{cav} + \Delta G_{e,s} + \Delta G_{vdw} + RT \ln\left(\frac{RT}{P_o V}\right) \tag{4.42}$$

$$= \Delta G_{cav} + \Delta G_{e,s} + \text{constant} \tag{4.43}$$

where

ΔG_{cav} = free-energy change for formation of a cavity in the solvent to accommodate the solute molecule

$\Delta G_{e,s}$ = free-energy change due to electrostatic interactions between solvent and solute

ΔG_{vdw} = free-energy change due to van der Waals interactions between the solute and solvent

$RT \ln (RT/P_o V)$ = change in free volume

P_o = pressure

T = absolute temperature

V = volume

The free-energy change of cavity formation, ΔG_{cav}, and the free-energy change of electrostatic interactions between solvent and solute (= protein), $\Delta G_{e,s}$, are assumed to dominate the unitary free-energy change, $\Delta G°$. The *change* in free energy due to van der Waals interactions (ΔG_{vdw}) between solute (i.e., protein) and solvent, and *changes* in free volume [$= RT \ln(\frac{RT}{PV})$] are assumed to be small. Hence these two terms in Eq. (4.42) are treated as constants.[8]

The free energy for formation of a cavity by the solute (protein), was defined by Sinanoğlu and then modified by Melander and Horvath to include a parameter for the molal surface tension increment, Φ, as defined in Eq. (4.20). This parameter replaced the total surface area, A, as in

$$\Delta G_{cav} = [NA + 4.8N^{1/3}(\kappa^e - 1)V^{2/3}]\gamma^o + [NA + 4.8N^{1/3}(\kappa^e - 1)V^{2/3}]\sigma m$$

$$= \text{constant} + [N\Phi + 4.8N^{1/3}(\kappa^e - 1)V^{2/3}]\sigma m \tag{4.44}$$

Hence the change in free energy associated with the cavity formation is based on the contact or hydrophobic area on the surface of the protein (or other macromolecule) rather than total surface area.

The electrostatic term, $\Delta G_{e,s}$ is represented by

$$\Delta G_{e,s} = A - \frac{B(m^{1/2})}{1 + C(m^{1/2})} - D\mu m \tag{4.45}$$

where

A = constant that is inversely proportional to the size of the protein, and directly proportional to the net charge on the protein at low ionic strength

[8] The reason for adding a neutral salt to a protein solution is to cause it to precipitate, so there is obviously a change. However, the molar concentration of a protein in an aqueous solution is initially small due to the high molecular weights of proteins (M_r = 10,000 to 150,000), hence making this assumption reasonable. The $RT \ln\left(\frac{RT}{P_o V}\right)$ term was assumed to be constant when "only the salt concentration in the aqueous solution changes and there are no specific interactions between the proteinaceous solute and salt." While selective interactions between a protein and some types of salts and other ligands can occur (see the discussion of immobilized metal affinity chromatography in Chapter 9), such interactions in the context of salting-out are likely to be small, given the large amount of salt required to drive the precipitation process. Hence the two terms on the right-hand side of Eq. (4.42) are considered to be independent of salt concentration.

$B =$ constant that is proportional to the net charge on the protein at low ionic strength
$C, D =$ fitted constants
$m =$ molal concentration of protein

Equation (4.5) can be simplified for limiting conditions. For example, at the protein's isoelectric point, the constants A and B approach zero, and Eq. (4.45) becomes

$$\Delta G_{e,s} \cong -D\mu m \tag{4.46}$$

Another condition would be represented by a high salt concentration. In this case m is very large. Equation (4.45) simplifies to

$$\Delta G_{e,s} = A - \frac{B}{C} - D\mu m \tag{4.47}$$

The combination of Eqs. (4.41) and (4.42) results in the expression

$$\ln W = \left[\frac{\Delta G_{ppt} - (\Delta G_{cav} + \Delta G_{e,s} + \Delta G_{vdw})}{RT}\right] - \ln\left(\frac{RT}{P_o V}\right) + \ln(55.5M_r) \tag{4.48}$$

where ΔG_{cav} and ΔG_{es} are defined by Eqs. (4.44) and (4.45), and $(\Delta G_{ppt} - \Delta G_{vdw})/RT$ and $\ln (RT/P_o V)$ are treated as constant terms. The constant terms in Eq. (4.48) are lumped together to give

$$\ln W = \text{constant} - \frac{1}{RT}\Delta G_{cav} - \frac{1}{RT}(\Delta G_{es} - A)$$

$$= \ln W_o - \frac{1}{RT}\Delta G_{cav} - \frac{1}{RT}(\Delta G_{es} - A) \tag{4.49}$$

where $G_{es} - A$ is defined by Eq. (4.45). A key step in this derivation is the combination of all of the salt independent terms from Eqs. (4.41), (4.42) and (4.44) into a constant and then recognizing that a constant term defines the initial concentration of the solute as $\ln W_o$ (Melander and Horvath, 1977). This clever simplification by Melander and Horvath (1977) gave the form

$$\boxed{\ln\left(\frac{W}{W_o}\right) = -\frac{1}{RT}(\Delta G_{es} - A) - \frac{1}{RT}(\Delta G_{cav})} \tag{4.50}$$

Equation (4.50) is further modified by the insertion of terms for $\Delta G_{es} - A$ from Eq. (4.45) and ΔG_{cav} from Eq. (4.44) to give

$$\ln\left(\frac{W}{W_o}\right) = \frac{1}{RT}\left[\frac{B(m^{1/2})}{1 + C(m^{1/2})} + D\mu m\right] - \frac{1}{RT}[N\Phi + 4.8N^{1/3}(\kappa^e - 1)V^{2/3}]\sigma m \tag{4.51}$$

where $\ln W_o$ is the solubility of the protein in pure water and (W/W_o) represents a fractional solubility. Note that the constant term in Eq. (4.44) is also assumed to be absorbed into $\ln W_o$. The parameter Φ represents the magnitude of the non-polar contact area between protein molecules. Equation (4.51) can be rewritten as

$$\ln\left(\frac{W}{W_o}\right) = \frac{1}{RT}\left[\frac{B(m^{1/2})}{1 + C(m^{1/2})}\right] + \frac{D\mu}{RT}m - \frac{1}{RT}[N\Phi + 4.8N^{1/3}(\kappa^e - 1)V^{2/3}]\sigma m \quad (4.52)$$

Equation (4.52) is restricted to the region of solubility in Figure 4.1 where salting-out effects dominate and protein concentration decreases with increasing salt concentration. The first term on the right-hand side of Eq. (4.52) simplifies to a constant when $Cm^{1/2} >> 1$:

$$\frac{1}{RT}\left[\frac{Bm^{1/2}}{1 + Cm^{1/2}}\right] \rightarrow \frac{1}{RT}\left[\frac{Bm^{1/2}}{Cm^{1/2}}\right] = \frac{B/C}{RT} = \beta \quad (4.53)$$

The constant β represents the salting-out region (Figure 4.2) where the Debye term approaches a constant value. The other coefficients in Eq. (4.52) represent the salting-in and salting-out coefficients, respectively:

$$\text{salting-in} = \Lambda = \left(\frac{D}{RT}\right)\mu \quad (4.17)$$

$$\text{salting-out} = \Omega = \frac{1}{RT}[N\Phi + 4.8N^{1/3}(\kappa^e - 1)V^{2/3}] \quad (4.18)$$

Equation (4.52) thus simplifies to Eq. (4.54) when the definitions given in Eqs. (4.17), (4.18), and (4.53) are used. There is a limitation, however. Equation (4.54) *is restricted to high salt concentrations* where the normalized solubility (as shown in Figure 4.2) is dominated by salting-out effects:

$$\boxed{\ln\frac{W}{W_o} = \beta + \Lambda m - \Omega\sigma m = \beta - (\Omega\sigma - \Lambda)m} \quad (4.54)$$

where m is the molar concentration of salt. The definition of K_S is obtained from Eq. (4.54), based on the slope of the normalized solubility curve in Figure 4.2, where the slope of the line is

$$K_S \equiv -\frac{d\ln W}{dm} = \frac{-d}{dm}[-\ln W_o + \beta + \Lambda m - \Omega\sigma m] \quad (4.55)$$

The derivative from Eq. (4.55) gives Eq. (4.15):

$$\boxed{K_S = \Omega\sigma - \Lambda} \quad (4.15)$$

The series of σ values in Table 4.7 together with Eq. (4.15) demonstrate both the qualitative and quantitative utility of the framework that was developed by Melander and Horvath for es-

TABLE 4.9 Molal Salting-out Constants for Hemoglobin and Fibrinogen for Various Salts

		K_S, m^{-1}	
Salt	σ	Hemoglobin[a]	Fibrinogen[b]
KCl	—	0.28	2.09
NaCl	1.64	0.62	—
$MgSO_4$	2.10	2.85	—
K-phosphate (pH 6.6)	—	3.38	7.94
$(NH_4)_2SO_4$	2.16	3.86	9.18
K_2SO_4	2.58	4.97	—
Na_2SO_4	2.73	4.97	—
Na-citrate (pH 6.6)	—	8.97	—

Source: From Melander and Horvath (1977), table I, p. 206. The material is reproduced with permission of Academic Press.
[a]Recalculated from the data of Green, as cited by Melander and Horvath (1977).
[b]Recalculated from the data of Florkin, as cited by Melander and Horvath (1977).

timating Ω, σ, Λ, and K_S. Their theory is consistent with experimentally measured salting-out constants for hemoglobin and fibrinogen, which show that K_S increases as the molal surface tension increment, σ, increases (Table 4.9). This is consistent with the Hofmeister series where citrates and sulfate are more effective than chlorides in causing precipitation (see footnote 1). A large value of K_S indicates that precipitation of the protein will occur over a relatively narrow range of salt concentrations. This is the case, for example, when sulfates are added to a solution containing hemoglobin where the hemoglobin has a K_S of 3.86 L/mole or fibrinogen where K_S is 9.18 L/mole.

4.14 PROTEIN MICELLES IN MILK PRECIPITATE BY ENZYME-INDUCED COAGULATION

The clotting or coagulation of milk is the basis of the cheese industry. It is mentioned here because clotting may play a role in the initial stages of purifying biopharmaceuticals from the milk of transgenic mice, sheep, pigs, and cows.[9] Milk is a complex biological fluid consisting of water, fat, proteins, lactose, and inorganic constituents as shown in Table 4.10 (Ruettimann and Ladisch, 1987). The caseins are the predominant proteins and are designated as α_{s1}-, α_{s2}-, β-, and κ-casins. They consist of between 169 and 209 amino acid residues with molecular weights of 19 to 25.2 kD, and they exist as a colloidal dispersion in micelles which include calcium, magnesium, inorganic phosphate, and citrate. The salts found in milk are only partially soluble (Table 4.11) with insoluble, colloidal calcium phosphate participating in the stabilization of the micelle structures. These particles have a broad size range of 20 to 600 nm and exist as a stable

[9]Transgenic mice are obtained when 2 picoliters (2×10^{-6} mL) containing several hundred copies of a cloned gene are injected into fertilized eggs that contain two pronuclei—one from the male and the other from the female. These later form the nucleus of the single-cell embryo. The eggs are transplanted into the mother, where between 10% and 30% of the eggs develop, and where a small fraction of those eggs that develop contain the foreign DNA integrated into their chromosomes. The mice that carry the foreign gene are referred to as transgenic animals (Watson et al., 1992). A transgenic sheep is obtained by injecting a gene under control of the β-lactoglobulin promoter, which is active only in mammary tissue, into the pronucleus of ova. The ova are transplanted into foster mothers. Offspring that express the transgene would do so only into the milk where the protein can be recovered.

TABLE 4.10 Average Composition of Milk

Component	Weight, %
Water	87.2
Fat	3.7
Protein	3.5
Lactose	4.9
Ash	0.7

Source: From Ruettimann and Ladisch (1987), table 2a, p. 579. Reprinted from Eng. Microb. Technol., Vol. 9, 1987, with permission from Elsevier Science.

TABLE 4.11 Salt Distribution in Milk

Component	Total, mg/100 mL	As Soluble Salt, mg/100 mL
Calcium	132.1	51.8
Magnesium	10.8	7.9
Phosphorous	95.8	36.3
Citric acid	156.6	141.6

Source: From Ruettimann and Ladisch (1987), table 2(b), p. 579; data from Johnson (1974). Reprinted from Eng. Microb. Technol., Vol. 9, with permission from Elsevier Science.

colloidal dispersion. The κ-casein is located at the outer layer or coat of the micelle, and is believed to protect the calcium sensitive core of the micelle, which contains the α- and β-caseins, from precipitation. While the casein micelles are approximately spherical, their exact structure is not completely known. In addition to casein proteins, whey proteins are also found in milk. These include β-lactoglobulin, α-lactalbumin, bovine serum albumin, immunoglobulins, and other minor proteins (Table 4.12).

TABLE 4.12 Whey Proteins in Skim Milk

Component	Concentration, g/L
β-Lactoglobulin	2.0 to 4.0
α-Lactalbumin	0.6 to 1.7
Serum albumin	0.4
Immunoglobulin G_1	0.3 to 0.6
Immunoglobulin G_2	0.05 to 0.1
Immunoglobulin A	0.05 to 0.15
Immunoglobulin M	0.05 to 0.1
Secretory component	0.02 to 0.1

Source: From Ruettimann and Ladisch (1987), table 3, p. 579. Reprinted from Eng. Microb. Technol., Vol. 9, 1987, with permission from Elsevier Science.

TABLE 4.13 Characteristics of Casein Micelles

Characteristic	Average Value
Diameter	130–160 nm
Surface	8×10^{-10} cm^2
Volume	2.1×10^{-15} cm^3
Density (hydrated)	1.0632 g/cm^3
Mass	2.2×10^{-15} g
Water content	63%
Hydration	3.7 g H_2O/g protein
Voluminosity	4.4 cm^3/g protein
Molecular weight (hydrated)	1.3×10^9 daltons
Molecular weight (dehydrated)	5×10^9 daltons
Number of peptide chains (MW 30 000)	10^4
Number particles per ml milk	10^4–10^{16}
Whole surface of particle	5×10^4 cm^2/ml milk
Mean free distance between micelles	240 nm

Source: From Ruettimann and Ladisch (1987), table 1, p. 578. Reprinted from Eng. Microb. Technol., Vol. 9, 1987, with permission from Elsevier Science.

The coagulation of the casein proteins from whey is the basis of cheese manufacture where a bacterial culture is first added to the milk followed by rennet. Rennet is a proteolytic enzyme that causes milk to clot or coagulate. The coagulation is initiated when the rennet hydrolyzes the bond between the phenylalanine 105 and methionine 106 residue of κ-casein to release a water soluble macropeptide while destabilizing the micelle. This leaves a hydrophobic surface while at the same time causing a decrease in the electrostatic repulsion between casein micelles. Once a critical extent of hydrolysis has occurred, coagulation begins and quickly proceeds to form flocs. The resulting flocs rapidly crosslink within minutes to hours to form a gel matrix that entraps the milk fat and whey proteins (Ruettimann and Ladisch, 1987, 1991). The gel is cut, resulting in the curd and the whey. The curd undergoes further processing into the final cheese product (Ruettimann and Ladisch, 1987). Precipitation processes are potentially important to the recovery of complex bioproducts from transgenic animals, as well as in the manufacture of cheese. Concentration and recovery of biotherapeutic protein could be achieved by coagulation if the target protein is associated with the micelle.

Centrifugation of milk or filtration of micelles could serve as an initial separation step, if the protein is a soluble protein in the serum. The micelles have a density greater than water and hence can be removed by centrifugation. Their size is large enough that membrane filtration could also be considered (Table 4.13). The membrane could retain the micelles while allowing the serum (i.e., whey) to pass as the permeate. Whether recovered by centrifugation or by membrane filtration, the protein product would be in the liquid, namely, the supernate from the centrifuge or the permeate from the membrane separation.

Genetically engineered animals have been available since 1980, but it is only since 1995 that new products, which are difficult to manufacture in any other way, have been obtained from transgenic animals for the purpose of clinical trials. A transgenic goat has been shown to carry the gene to produce BR-96[10], a monoclonal antibody that could prove effective in delivering conjugated anticancer drugs. Factor IX, which is needed by Hemophilia B patients, has been produced at levels of 0.6 g/L in mice. Fibrinogen has been expressed at concentrations in excess of 5 g/L, and protein C[10] in excess of 0.4 g/L in the milk of transgenic sheep. Other examples include the production of human lactoferrin[10] and human serum albumin from transgenic dairy cattle. These products have potential utility in treating gastrointestinal disorders, and the treatment of trauma after severe blood loss (Thayer, 1996; Dutton, 1996). A cow is an attractive subject, since it can produce 10,000 L/year of milk, with each liter containing about 30 to 35 g of the product protein (Johnson, 1974).

CRYSTALLIZATION

Crystallization is a nucleation process that usually occurs from a concentrated solution, where additives, if any, are at low concentrations and serve as nucleation sites for crystal growth. Nucleation is the generation of crystals from a solution of molecules or ions that are in constant motion. Nucleation occurs when the concentration of the solute exceeds saturation *and* clusters of the solute molecules are present that are sufficiently large to provide the nuclei for crystal growth. While it is often difficult to make a clear distinction between crystallization and precipitation processes, there are differences in the nature of the products. Precipitates often have a poorly defined morphology and are characterized by a small particle size. Crystals have a well-defined morphology and a large particle size and are usually formed in a secondary nucleation process induced by the introduction of crystals to a supersaturated solution (Mullin, 1972; Tavare, 1995). The largest cluster that exists before spontaneous crystallization or precipitation may occur from a homogeneous solution is the critical cluster or nucleus (Walton, 1967). Subcritical, critical, and supercritical clusters or particles are defined by the number of molecules or ions of solute, n, associated with each type of particle:

$$n < n^* \qquad \text{(subcritical)}$$

$$n = n^* \qquad \text{(critical)}$$

$$n > n^* \qquad \text{(supercritical)}$$

Particles that have $n \leq n^*$ molecules are referred to as embryos. When $n = n^*$, the embryo is said to have attained a critical size, and formed a critical nucleus for inducing spontaneous precipitation or crystallization. The term nucleus is used for all particles which initiate growth, and these can range from 10 nm to several microns in size. Primary nucleation can occur by either a homogeneous and heterogeneous mechanism. Homogeneous nucleation refers to crystallization when a solid interface is absent. Heterogeneous nucleation occurs in the presence of a foreign interface. Secondary nucleation occurs in the presence of a crystal interface of the same solute (Tavare, 1995).

Crystallization plays an important role in the recovery and purification of the protease inhibitor Crixivan® which is used to treat HIV infected people. This protease inhibitor is manufactured using organic synthesis steps, and it can be crystallized from the organic solvents used in its manufacture. A discussion of the development of protease inhibitors is given in Chapter 9.

[10]BR-96 is a monoclonal antibody designed to deliver the conjugated anticancer drug BR-96/doxorubicin which is currently in Phase II clinical trials as of May, 1996. Protein C is a complex, highly processed human plasma protein involved in blood clotting. It acts locally, leaving the rest of the coagulation system intact. However, protein C is not yet commercially available. Lactoferrin is a natural antibiotic which is undergoing clinical trials for the treatment or prevention of gastrointestinal infections in premature infants, patients undergoing chemotherapy and radiotherapy, AIDS patients, and other people whose immune systems are compromised (Thayer, 1996; Dutton, 1996).

Crystals of zinc insulin derived from recombinant human insulin have been obtained (Chance et al., 1981a, 1981b). Although crystallization is apparently not a means of purifying insulin on a large scale, recovery and concentration of the insulin product by precipitation may be employed in the later stages of commercial recombinant insulin production (Kroeff et al., 1989). Recombinant insulin production is discussed in Chapter 8.

4.15 A NUCLEUS OR CRITICAL CLUSTER KNOWN AS AN EMBRYO IS REQUIRED FOR CRYSTALLIZATION

Walton (1967) represents the formation of a critical cluster in terms of molecular aggregation of a solute B:

$$B + B \Leftrightarrow B_2 + B \Leftrightarrow B_3 + B + \ldots \tag{4.56}$$

Eventually a critical cluster or embryo forms that consists of x molecules

$$B_{x-1} + B \Leftrightarrow B_x \tag{4.57}$$

which then undergoes nucleation:

$$B_x + B \rightarrow B_{x+1} \downarrow \tag{4.58}$$

and crystal growth:

$$B_{x+1} + B \rightarrow \text{crystal growth} \tag{4.59}$$

The steps in the evolution of a crystal follow the sequence of cluster → embryo → nucleus → crystal. A crystal represents the most highly organized type of nonliving matter. The atoms, molecules, or ions that make up a crystal are arranged in orderly three-dimensional arrays called *space lattices* (McCabe and Smith, 1967).

Supersaturation is required for crystal formation and growth. Supersaturation may be achieved by cooling a solution if the solubility of the salt decreases with decreasing temperature. If the solubility is not a strong function of temperature, evaporation of a portion of the solvent may be necessary. If cooling is not effective, and evaporation is not feasible, supersaturation may be achieved by adding a third component. The third component forms a mixed solvent in which the solubility of the solute is significantly reduced (McCabe and Smith, 1967). This third approach is referred to as salting-out, as has been discussed in sections 4.9 to 4.14.

4.16 THE ANALYSIS OF A CRYSTALLIZATION PROCESS IS BASED ON DIFFERENCES IN CHEMICAL POTENTIAL AND THE SATURATION RATIO

The equilibrium state of the solution is defined by the chemical potential of the crystal and solution phases. If they are the same, that is, $\mu_1 = \mu_2$, the free energy is zero ($\Delta G = 0$), and the system is at equilibrium. At a supersaturated condition, the chemical potential of the solution will be greater than that of the crystal or solid phase:

$$\mu_1 > \mu_2 \quad \text{or} \quad \mu_{\text{solution}} > \mu_{\text{crystal}}$$

The difference in the chemical potential between the crystal and solution phases ($\mu_2 - \mu_1$) is negative and therefore, the Gibbs free-energy change, ΔG, between the two phases is also negative (Nielsen, 1964). This difference is expressed so that ΔG is assigned a negative sign:

$$-\Delta G = (\mu_1 - \mu_2)\Delta n = (\mu_{\text{solution}} - \mu_{\text{crystal}})\Delta n \tag{4.60}$$

where

μ_1 = chemical potential of the solution phase = μ_{solution}, ergs/molecule
μ_2 = chemical potential of the crystal phase = μ_{crystal}, ergs/molecule
Δn = number of molecules that are crystallized

The condition $\mu_1 - \mu_2 > 0$ denotes that spontaneous precipitation from a homogeneous solution is probable (Nielsen, 1964). Equation (4.60) may also be expressed in terms of a driving force for a change in phase on a specific or "per molecule" basis. The free energy on a specific basis is referred to by the symbol ϕ as in

$$\phi = -\frac{\Delta G}{\Delta n} = \mu_1 - \mu_2 = \mu_{\text{solution}} - \mu_{\text{crystal}} \tag{4.61}$$

The driving force for phase change can be calculated from the temperature T_o at which the solution of a given concentration would be at its solubility limit. This is done by writing the precipitation parameter, ϕ, in terms of the temperature, and the specific latent heat of crystallization (i.e., latent heat per molecule or ion):

$$\phi = \frac{-\Delta G}{\Delta n} = \int_{T_o}^{T} \Delta S' dT \cong \Delta H'\left(\frac{T - T_o}{T_o}\right) \tag{4.62}$$

where

T = temperature of the solution, K
T_o = temperature at which solubility is equal to the actual concentration of the solute
$\Delta H'$ = specific latent heat of crystallization, ergs/molecule
$\Delta S'$ = entropy, erg/molecule

This precipitation parameter ϕ can be expressed in terms of the ratio of activity a of the dissolved solute and activity a_o of the pure solute molecules in the crystal:

$$\phi = KT \ln\left(\frac{a}{a_o}\right) = KT \ln\left(\frac{\gamma_c C}{\gamma_{\text{sat}} C_{\text{sat}}}\right) \tag{4.63}$$

where

ϕ = precipitation parameter, ergs

K = Boltzmann's constant[11] = 1.38×10^{-16} erg/K = $\frac{R_o}{N_A}$

T = temperature, K

C = concentration of solute in solution

C_{sat} = saturation concentration of solute in solution at temperature, T

γ_c, γ_{sat} = activity coefficients at concentration, C, and saturation concentration, C_{sat}

For an ideal solution, each of the activity coefficients and their ratio will be 1. Alternately, even if the solution is not ideal, the activity coefficients may have similar values, and their ratio will be close to 1. In both cases the supersaturation ratio, S, is defined by (McCabe and Smith, 1967; Walton, 1967):

$$S = \frac{\gamma_c C}{\gamma_{sat} C_{sat}} \cong \frac{C}{C_{sat}} \tag{4.64}$$

The difference in chemical potentials on a *mole basis* is analogous except that the gas law constant replaces the Boltzmann constant (McCabe and Smith, 1967):

$$\varnothing = \hat{\mu}_1 - \hat{\mu}_2 = \hat{\mu}_{solution} - \hat{\mu}_{crystal} = nR_oT \ln S \tag{4.65}$$

where

R_o = gas constant = 8.3143×10^7 ergs/(gmole · K) the subscript "o" denotes a gas constant that is expressed in dimensions associated with chemical potential.

n = number of ions for a non-ionic crystal, the number of ions per neutral molecule is one (i.e., $n = 1$).

$\hat{\mu}_1 = \hat{\mu}_{solution}$ = chemical potential of solution phase, ergs/gmole

$\hat{\mu}_2 = \hat{\mu}_{crystal}$ = chemical potential of crystal phase, ergs/gmole

Equations (4.62) to (4.65) provide the basis for developing work functions that represent the energetics of crystallization. These correlations also form the starting point for development of equations that represent solute concentration and solution temperature at which crystallization occurs and the rate at which nucleation proceeds.

[11]The gas law and Boltzmann constants are related to each other through Avogadro's number (McCabe and Smith, 1967) with the Boltzmann constant representing a specific basis and the gas law constant a mole basis:

$$K = \frac{R_o}{N_A} = \frac{8.3143 \cdot 10^7 \ \frac{\text{ergs}}{\text{gmole} \cdot \text{K}}}{6.0225 \cdot 10^{23} \ \frac{\text{molecule}}{\text{gmole}}} = 1.3805 \cdot 10^{-16} \frac{\text{ergs}}{\text{molecule K}}$$

The dimensions for Boltzmann's constant, K, are usually expressed as ergs/K rather than the more explicit ergs/(molecule ·K) (Mullin, 1972, p. 142). Note that cal = ergs.

4.17 THE WORK FOR FORMING AN EMBRYO IS ASSOCIATED WITH BUILDING THE SURFACE OF A CRYSTAL AND INCREASING ITS VOLUME

Embryos are particles consisting of no more than n^* molecules where n^* represents the number of molecules in a cluster of critical size. Embryo formation results in generation of crystal faces or surfaces, and the increase in the volume of the solid particles that make up the embryo. A work function that is analogous to a total energy of activation may be used to represent this process. The combination of the two types of work is represented as (McCabe and Smith, 1967)

$$W_{crystal} = \gamma A_p - \emptyset N_{moles} \tag{4.66}$$

where

γA_p = surface energy of the particle (i.e., work required to form surface)
$\emptyset N_{moles}$ = work to form volume
$W_{crystal}$ = work function, ergs
γ = surface tension, ergs/cm^2
A_p = surface area of particle, cm^2
$\emptyset$ = $\hat{\mu}_1 - \hat{\mu}_2$, difference in chemical potential given by Eq. (4.65)
N_{moles} = mass of solid particles expressed in gmoles

Embryonic crystals are unlikely to consist of spheres. Nonetheless, the hydraulic radius and surface area of multi-faced particles can be approximated by a sphere. The use of spherical dimensions enables a generalized work function to be formulated. The surface area of a sphere is

$$A_p = \pi D_p^2 \tag{4.67}$$

while the volume is

$$V_{particle} = \frac{\pi}{6} D_p^3 \tag{4.68}$$

The mass of the individual embryo particles on a mole basis, N_{moles}, can then be defined by

$$N_{moles} = \frac{V_{particle}}{V_{molal}} = \frac{\pi D_p^3/6}{V_{molal}} \tag{4.69}$$

where

D_p = diameter of the spherical embryo, cm
$V_{particle}$ = volume of the spherical embryo, cm^3
V_{molal} = molal volume of the solid, cm^3/gmole

The substitution of terms for surface area, Eq. (4.67), and the mass of crystals, Eq. (4.69), into Eq. (4.66), enables the work function to be related to the diameter and density of the crystal:

$$W_{\text{crystal}} = \gamma[\pi D_p^2] - (\hat{\mu}_1 - \hat{\mu}_2)\left[\frac{\pi D_p^3}{6V_{\text{molal}}}\right]$$

$$= \pi D_p^2\left(\gamma - (\hat{\mu}_1 - \hat{\mu}_2)\frac{D_p}{6V_{\text{molal}}}\right) \tag{4.70}$$

4.18 THE DIAMETER OF AN EMBRYO AT EQUILIBRIUM REPRESENTS THE CRITICAL DIAMETER AT WHICH CRYSTALLIZATION MAY BE INDUCED

The formalism of Eq. (4.70) enables definition of the critical diameter of an embryo, $D_{p,eq}$. This represents the size at which crystallization from a supersaturated solution may be induced. At the critical diameter or size, the change in work with particle diameter, D_p, is zero:

$$\frac{d(W_{\text{crystal}})}{dD_p} = 0 = \pi D_{p,eq}\left(2\gamma - (\hat{\mu}_1 - \hat{\mu}_2)_{eq}\frac{3D_{p,eq}}{6V_{\text{molal}}}\right) \tag{4.71}$$

The difference in the chemical potentials (mole basis) at this equilibrium condition is given by

$$(\hat{\mu}_1 - \hat{\mu}_2)_{eq} = \frac{4\gamma V_{\text{molal}}}{D_{p,eq}} \tag{4.72}$$

The combination of Eq. (4.65) and Eq. (4.72) results in the Kelvin equation which relates solubility of a solute to its particle size (McCabe and Smith, 1967; Nielsen, 1964):

$$\ln S = \frac{4\gamma V_{\text{molal}}}{nR_oTD_{p,eq}} \tag{4.73}$$

Upon rearrangement, an expression for the size of the critical nucleus is obtained:

$$\boxed{D_{p,eq} = \frac{4\gamma V_{\text{molal}}}{nR_oT\ln S}} \tag{4.74}$$

where

$D_{p,eq}$ = particle diameter (as a sphere) at equilibrium, cm; that is, the size of critical nucleus where $n = n^*$ and n is the number of molecules.

n = number of ions per neutral molecule

R_o = gas law constant = $8.3143 \cdot 10^7$ ergs/(gmole · K)

T = temperature, K

$S =$ saturation ratio as defined in Eq. (4.64), dimensionless

$\gamma =$ surface tension, ergs/cm^2

$V_{molal} =$ molal volume, cm^3/gmole

4.19 THE WORK FUNCTION IS A THERMODYNAMIC EXPRESSION THAT REPRESENTS A SURFACE TENSION AND TEMPERATURE DEPENDENT ACTIVATION ENERGY FOR HOMOGENEOUS NUCLEATION

The combination of Eqs. (4.65), (4.70), and (4.74) results in the expression for a work function (i.e., an activation energy) for crystal formation from a critical nucleus.[12] Since this work function is defined based on an equilibrium diameter, it is denoted by $W_{crystal,eq}$ (McCabe and Smith, 1967):

$$W_{crystal,eq} = \frac{16\pi\gamma^3 V_{molal}^2}{3(nR_oT \ln S)^2} \tag{4.75}$$

[12]An alternate derivation is given by Mullin (1972) where the quantity of work to form a crystal, W, is the sum of the work required to form the surface, W_S (a positive quantity) and work to form the bulk of the particle (a negative quantity):

$$W = W_S + W_V = a\gamma - V\Delta p = 4\pi r^2\gamma - \frac{4}{3}\pi r^3\left(\frac{2\gamma}{r}\right) = \frac{4}{3}\pi r^2\gamma \tag{u}$$

where a and V are surface area and volume of a spherical particle, respectively, and Δp is the pressure difference between the vapor phase and interior of a liquid droplet. The Gibbs-Thomson (Kelvin) equation gives the increase in vapor pressure of a liquid droplet as its size increases:

$$\ln\frac{p_r}{p^*} = \frac{2M_r\gamma}{RT\rho r} \cong \ln\frac{C_r}{C^*} = \ln S \tag{v}$$

where p_r is the vapor pressure over a liquid droplet of radius r and p^* is the equilibrium saturation vapor pressure over a flat surface. ρ is the density of the droplet. Equation (v), solved for r, gives

$$r = \frac{2M_r\gamma}{RT\rho \ln S} \tag{w}$$

when the expression for r is substituted into equation (u), equation (4.74) is obtained, since $V_m = M_r/\rho$, and $D_{p,eq} = 2r$. According to Mullin (1972), equation (v) is similar in form to an equation due to Ostwald (1900), as corrected by Freundich (1909):

$$\ln\frac{C_1}{C_2} = \frac{2M_r\gamma}{RT\rho}\left(\frac{1}{r_1} - \frac{1}{r_2}\right) \tag{x}$$

where C_1 and C_2 = solubilities of spherical particles of radius r_1 and r_2, γ = surface energy (tension) of solid particle in contact with the solution, and ρ = density of the solid. Equation (x), while inexact, gives a rationale to the observation that particles of a solute, if small enough and suspended in a solvent, can dissolve to give a concentration which is temporarily greater than the normal solubility of that substance. See Mullin (1972) for a discussion.

A *homogeneous* supersaturated solution can be stable indefinitely when $n \leq n^*$ until the start of crystallization. Crystallization occurs when there are statistical fluctuations in the free energy in small parts of the system (Nielsen, 1964, p. 3). If a seed crystal with $n > n^*$ is introduced into the solution, crystallization is spontaneous. Once a favorable condition for crystallization is attained, the rate of nucleation, rather than equilibrium phenomena, dominates the crystallization process. The rate equation for a primary homogeneous nucleation mechanism is given by (McCabe and Smith, 1967; Tavare, 1995; Walton, 1967):

$$J = \frac{dN}{dt} = A \exp\left(\frac{-W_{\text{crystal,eq}}}{KT}\right) = A \exp\left(\frac{-W_{\text{crystal,eq}}}{R_o T/N_A}\right) \tag{4.76}$$

where

dN/dt = rate of appearance of crystal particles
$W_{\text{crystal,eq}}$ = work function (energy of activation) for crystal formation defined at the critical or equilibrium particle size, ergs
A = preexponential factor[13] $\approx 10^{25}$ = rate at which individual molecules or particles strike the surface of a cluster
K = Boltzmann's constant[11] = $1.3805 \cdot 10^{-16}$ ergs/K
R_o = gas law constant = $8.3143 \cdot 10^7$ ergs/gmole · K
N_A = Avogadro's number
T = temperature, K

Combination of Eqs. (4.75) and (4.76) gives the rate of nucleation[14] as (McCabe and Smith, 1967; Tavare, 1995):

$$J = \frac{dN}{dt} = A \exp\left[\frac{-16\pi\gamma^3 V_{\text{molar}}^2 N_A}{3n^2 (R_o T)^3 (\ln S)^2}\right] \tag{4.77}$$

[13] The value of the preexponential factor is not known. It is taken to be analogous to the constant that represents nucleation of water drops from supersaturated water vapor. For water this constant is on the order of 10^{25} (McCabe and Smith, 1967). This value has been used over the years, since it appears to give equations that correlate with the data.

[14] An approximation of Eq. (4.77) is given by a power-law expression based on concentration difference (Tavare, 1995):

$$J = \frac{dN}{dt} = k_n (C - C_{\text{sat}})^m = k_n \Delta C^m \tag{y}$$

where

k_n = rate constant for crystallization
m = exponent, generally in range of 3 to 4; if process is primary nucleation, $m \cong 10$
C = solute concentration in solution
C_{sat} = maximum solute solubility at a given temperature

This empirical expression for the crystallization equation (y) has the same form as Eq. (3.36) for dialysis, which is another unit operation driven by a concentration difference. Equation (y) alone is of limited usefulness, since nucleation, once initiated, occurs so rapidly that calculation of its initial rate is not as important as being able to estimate the concentration, temperature, and surface tension at which a rapid rate of crystallization is induced. For this, the form of Eq. (4.77) is needed.

The symbol J is widely used in the literature on crystallization to denote the rate of crystallization and should not be confused with flux. Critical supersaturation, S^*, is defined here as the concentration at which one nucleus is formed per second per unit volume:

$$\frac{dN}{dt} = 1 = A \exp\left(\frac{-\hat{C}(\gamma, T)}{(\ln S^*)^2}\right) = A \exp\left(\frac{-\hat{B}\gamma^3}{(\ln S^*)^2 T^3}\right) \tag{4.78}$$

where

$$\hat{C}(\gamma, T) = \text{lumped parameter} = \frac{16\pi\gamma^3 V_{\text{molal}}^2 N_A}{3n^2 (R_o T)^3} = \hat{B}\left(\frac{\gamma}{T}\right)^3$$

$$\hat{B} = \text{constant} = \frac{16\pi V_{\text{molal}}^2 N_A}{3n^2 R_o^3}$$

This gives a definition of a solute concentration, below which nucleation is very slow and above which it is almost instantaneous, as represented by the curve for homogeneous crystallization in Figure 4.13.

Numerical values of the interfacial surface tension, γ, between precipitate and the surrounding solution have been estimated to be on the order of $\gamma = 80$ to 100 ergs/cm^2 for ordinary salts (McCabe and Smith, 1967) at room temperature, although exact values are difficult to deter-

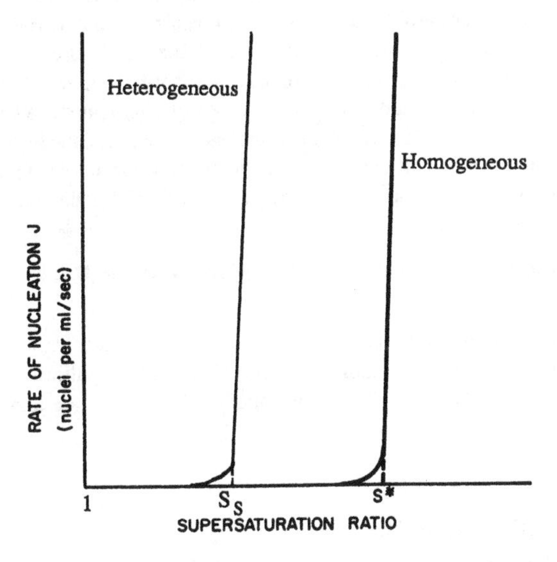

Figure 4.13 Homogeneous nucleation rate, J (nuclei per mL/s), as a function of the degree of supersaturation (S = actual concentration/solubility) (based on Walton, 1967, fig. 1–3, p. 6, and Tavare, 1995, fig. 30, p. 59 with permission from Kluwer Academic/Plenum Publishers).

mine (Tavare, 1995). When a nucleation rate of one nucleus per second per cm^3 ($dN/dt = 1$) is defined to represent the initiation of an observable crystallization process, the supersaturation ratio, S, given in Eq. (4.64), may be calculated as a function of interfacial tension from a rearranged form of Eq. (4.78). For example, the critical supersaturation ratio, S^*, for an interfacial tension of 100 ergs/cm^2, a molal volume of 50 cm^3/gmole, and a temperature of 277 K, is given by

$$\ln S^* = \left[\frac{-1}{\ln 10^{-25}} \cdot \frac{16\pi(100)^3(50)^2(6.02 \cdot 10^{23})}{3 \cdot (8.3143 \cdot 10^7 \cdot 277)^3} \right]^{1/2} = 5.99 \tag{4.79}$$

which is equivalent to a critical supersaturation ratio of $S^* = \exp(5.99) = 399$ ($= C/C_{sat}$). McCabe and Smith (1967) point out that the value of S obtained in this manner is much larger than that of any supersaturation ratio for which crystallization is induced at an industrial scale. Crystallization is often observed to be initiated at values of about 1 to 200 (Tavare, 1995). Nucleation from a homogeneous solution is therefore unlikely to accurately reflect an industrial process. Most crystallization processes are likely to represent a heterogeneous phenomenon in which the presence of other solutes or a solid particle lowers the energy barrier to nucleation and catalyzes the nucleation process (Walton, 1967).

True Examples of Spontaneous Nucleation Are Rare

Careful examination of reported cases of spontaneous nucleation has often shown that crystallization is often induced in some other way (Mullin, 1972; Tavare, 1995). According to Mullin (1972), aqueous solutions carefully prepared in the laboratory in the presence of atmospheric dust may contain 10^6 to 10^8 solid particles/cm^3. Even if carefully filtered, these solutions still contain 10^3 particles/cm^3. Atmospheric dust can often contain particles of the product itself. This is especially true in facilities where samples of the crystalline material have been prepared or handled. Seed crystals or particles do not have to consist of the material being crystallized in order to be effective as nucleation sites, although crystallization is often intentionally induced by innoculating or seeding a supersaturated solution with small particles of the same substance to be crystallized. For example, addition of a few drops of colloidal silica causes prolific precipitation of glycine crystals. Heterogeneous nucleation also occurs in biologic materials, where chemical bonding as well as other physical factors may play a role in processes including calcification (bone and kidney stone formation) and formation of gall stones (Walton, 1967).

The Onset of Rapid Nucleation Is Extremely Sensitive to the Supersaturation Ratio

The experimentally measured interfacial tension of a crystal in water may be used to illustrate why supersaturation is required to obtain an observable rate of crystallization. This example is from McCabe and Smith (1967, p. 762). The following properties are known for potassium chloride (KCl):

$$M_r = 74.56$$

$$n = 2\ (KCl = K^+ \text{ and } Cl^-)$$

$$\rho_{crystal} = 1.988$$

$$V_m = \frac{M_r}{\rho_{crystal}} = \frac{74.56\ |g/gmole|}{1.988\ \ |g/cm^3|} = 37.5\,\frac{cm^3}{gmole}$$

The apparent interfacial energy obtained for heterogeneous nucleation at 27°C is given by

$$\gamma = \text{apparent interfacial energy} = 2.5\ \frac{\text{ergs}}{\text{cm}^2}$$

where

$$T = 27 + 273 = 300\ \text{K}$$

The apparent interfacial energy of 2.5 was selected because it is at the midpoint between 2.0 and 3.0 measured for heterogeneous nucleation of KCl solutions, which is close to the interfacial energy of 2.8 ergs/cm^2 measured for supersaturated KCl solutions seeded with microscopic KCl crystals. Use of these parameter values in Eq. (4.77) gives

$$\frac{dN}{dt} = 10^{25} \exp\left[\frac{-16\pi(2.5)^3(37.5)^2(6.0225 \cdot 10^{23})}{3(2)^2(8.3143 \cdot 10^7 \cdot 300)^3(\ln S)^2}\right]$$

$$\frac{\left|\frac{\text{ergs}^3}{\text{cm}^6}\right|\frac{\text{cm}^6}{\text{gmole}^2}\left|\frac{\text{molecules}}{\text{gmole}}\right|}{\left|\frac{\text{ergs}^3}{\text{gmole}^3 \cdot \text{K}^3}\right|\ \text{K}^3\ \Big|}$$

The expression that makes up the exponential term is dimensionless. If the critical nucleation rate is set at one as in Eq. (4.78):

$$\frac{dN}{dt} = 1 = 10^{25} \exp\left[\frac{-0.00357}{(\ln S)^2}\right] \tag{4.80}$$

or

$$\ln 10^{-25} = \frac{-0.00357}{(\ln S)^2} \tag{4.81}$$

TABLE 4.14 Nucleation (Crystallization) Rates of KCl as a Function of Supersaturation at 300 K

Supersaturation Ratio: $S = C/C_{sat}$ (dimensionless)		Nucleation Rate at 300 K:
S	$\ln S$	dN/dt (particles/cm^3 · s)
1.005	0.00498	$4.70 \cdot 10^{-38}$
1.006	0.00598	$4.72 \cdot 10^{-19}$
1.007	0.00698	$1.37 \cdot 10^{-7}$
1.007906	0.007875	1.0
1.009	0.00896	$4.86 \cdot 10^{5}$

Source: Values from Eq. 4.80.

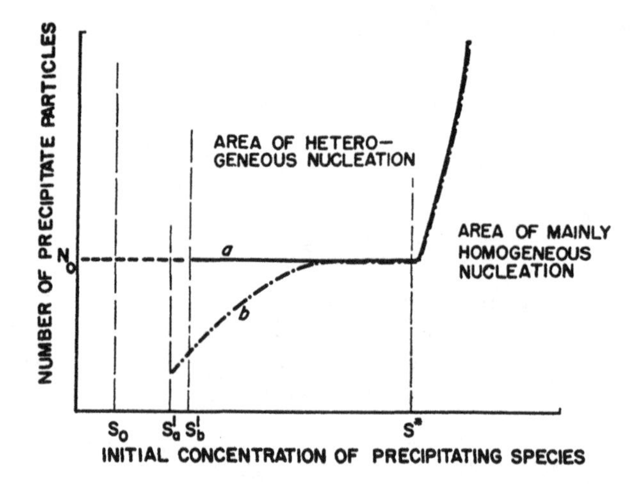

Figure 4.14 Typical curves for the number of precipitate particles produced as a function of the initial precipitant concentration. Curve a is for a system of N_o impurity particles, each equally efficient in catalyzing heterogeneous nucleation. Curve b corresponds to the precipitation onto N_o impurities of varying nucleation efficiency. Heterogeneous nucleation occurs at a supersaturation s', and homogeneous nucleation occurs at a much higher (critical) supersaturation s^*; s_o is the solubility of the precipitant (from Walton, 1967, fig. 1–12, p. 25).

Upon rearrangement, Eq. (4.81) results in an expression for calculating S:

$$S = \exp\left[\left(\frac{-0.00357}{\ln(10^{-25})}\right)^{1/2}\right] = 1.007906 \tag{4.82a}$$

The supersaturation ratio, in Eqs. (4.80) to (4.82) is not represented as S^*, since this value which is close to one indicates a heterogeneous phenomenon, while the critical supersaturation ratio denotes a concentration for the onset of rapid nucleation under homogeneous conditions (Figure 4.13). Homogeneous nucleation is associated with much higher supersaturation values (Table 4.14), and these values are denoted with an S^* (Figures 4.13 and 4.14).

The effect of the supersaturation ratio S on the rate of nucleation (number of particles/(cm^3 · s)) is illustrated in Table 4.14. The onset of heterogeneous nucleation is very sensitive to slight variations in the critical supersaturation ratio, S. This is consistent with the metastable behavior of a supersaturated solution, where the attainment of supersaturation does not ensure that crystallization will occur.

4.20 THE INDUCTION PERIOD FOR CRYSTALLIZATION IS PROPORTIONAL TO THE CUBE OF SURFACE TENSION (γ^3) AND INVERSELY PROPORTIONAL TO (ln S)^2

The induction period is the sum of time required to form a critical nucleus, time required for the nucleus to grow to a detectable size, and the system's relaxation time. The induction period may be measured by visual observation, particle size analysis, solution turbidity, concentration, or conductivity, as long as the induction period exceeds 10 seconds. Shorter time periods require a technique analogous to stopped flow kinetics for their quantitative determination

(Tavare, 1995). An empirical observation that the induction period and rate of nucleation, J, are inversely related enables an apparent surface tension to be estimated from experimental data. The hypothesized relation is (Tavare, 1995):

$$t_{ind} \alpha J^{-1} = A \exp\left[\frac{\hat{B}\gamma^3}{T^3(\ln S)^2}\right] \qquad (4.82b)$$

where the "α" symbol in Eq. (4.82b) denotes "proportional to." If this approximation can be made, rearrangement of Eq. (4.82b):

$$\ln t_{ind} \alpha \ln A + \hat{B}\frac{\gamma^3}{T^3(\ln S)^2} \qquad (4.83a)$$

where

$\hat{B} = \frac{16\pi V_{molal}^2 N_A}{3_n^2 R_o^3} = \text{constant}$

$\gamma =$ surface tension

$T =$ temperature, °K

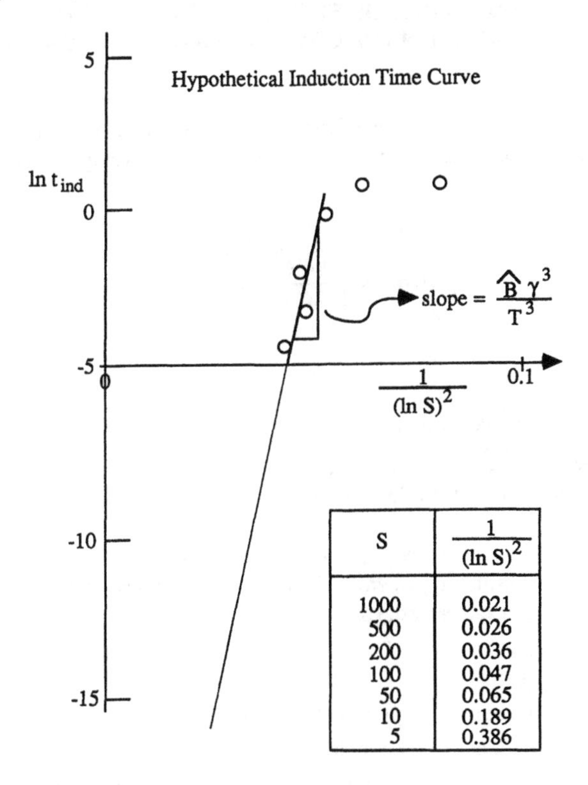

S	$\frac{1}{(\ln S)^2}$
1000	0.021
500	0.026
200	0.036
100	0.047
50	0.065
10	0.189
5	0.386

Figure 4.15 Schematic illustration of an induction time curve for *homogeneous* crystallization as a function of the saturation ratio, *S* (adapted from Tavare, 1995, fig. 31, p. 62 with permission from Kluwer Academic/Plenum Publishers).

The other parameters are the same as defined for Eq. (4.74). Tavare emphasizes that "the induction period can neither be regarded as a fundamental property of the crystallization system nor can be relied on to yield basic information of a nucleation process (1995, p. 61)." It is nonetheless useful in providing a numerical measure of a crystallizing system.

A plot of the type illustrated in Figure 4.88b can be used to estimate the apparent interfacial tension for *homogeneous* crystallization if the induction period can be accurately measured as a function of the supersaturation ratio, S. A hypothetical curve for induction time as a function of the supersaturation ratio (S) is given in Figure 4.15. The slope is determined from the linear portion of the data that occurs before the induction time begins to level off. The intercept term $\ln A$ should have a value on the order of 46 to 70 if the values of A are between 10^{20} to 10^{30}, respectively. The slope of the data is steep. Slight variations in the data will give significant change in the slope. This would result in large variations of γ. According to Tavare (Tavare, 1995), the plotting and interpretation of data from stopped flow kinetic experiments in this manner has shown that the induction periods range from milliseconds to seconds when the supersaturation ratio is between 1 to 200. The corresponding interfacial tensions are from 10 to 200 mJ/m^2. The more soluble species show a lower interfacial tension than less soluble species. While this approach enables a first estimate of γ, the difficulty in obtaining suitably precise data may limit its utility.

4.21 SURFACE TENSION MAY BE ESTIMATED FROM THE RATIO OF SOLUTE CONCENTRATION IN THE CRYSTAL TO ITS CONCENTRATION IN SOLUTION

An alternate approach to determining interfacial tension is provided by an equation that was obtained by Mersmann on the basis of regular solution theory, according to Tavare (1995). This equation is

$$\gamma = 0.414KT\left[\frac{\rho_c N_A}{M_r}\right]^{2/3} \ln\left[\frac{C_{\text{crystal}}}{C}\right] \tag{4.83b}$$

where

$\gamma =$ interfacial surface tension, mJ/m^2
$\rho_c =$ density, kg/m^3
$K =$ Boltzmann constant $= 1.38 \cdot 10^{-23}$ J/K

Note that the ratio of concentrations, C_{crystal}/C, is not the same as the supersaturation ratio, S. An example calculation using Eq. (4.82a) is given by Tavare (1995) for barium sulfate in aqueous solution at 18°C. The values of the parameters are

$T =$ 273 + 18 = 291 K
$\rho_c =$ 4500 kg/m^3 = 4.5 g/cm^3 = density of crystal
$M_r =$ 233 kg/kmol = 233 g/mole
$C =$ $0.93 \cdot 10^{-5}$ $kmol/m^3$ = 9.3 μmoles/L = concentration of solute in a solution from which crystals have formed
$C_{\text{crystal}} =$ 19.3 $kmol/m^3$ = 19,300 μmoles/L = concentration of solute in crystal

Substitution of these values into Eq. (4.83b) gives

$$\gamma = (0.414)(1.38 \cdot 10^{-23})(291)\left[\frac{4500 \cdot 6.02 \cdot 10^{23}}{233}\right]^{2/3} \ln\left(\frac{19.3}{0.93 \cdot 10^{-5}}\right)$$

$$\gamma = 0.00127 \frac{\text{J}}{\text{m}^2}$$

This example, for barium sulfate, may be applicable to other types of molecules, although this author is not aware of calculations for intermediate molecular weight solutes (M_r = 500 to 3000), or for proteins, that have been systematically validated with experimental data and published in the literature.

Protein crystals contain large channels and holes filled with solvent molecules. Consequently the density of the crystal will not be as large as for a salt or small molecule. Since proteins also have a large molecular weight, the surface tension as given by Eq. (4.83b), will be lower for proteins than for salts, where a protein might have $\gamma = 0.0038$ J/m^2, $\rho_c = 1100$ kg/m^3, M_r = 10,000, and the other values are the same as for the barium sulfate example. This estimate of surface tension suggests that crystallization of proteins can be readily achieved. This appears to be inconsistent with experiments where up to several months have been required to obtain protein crystals of 0.5 mm in size, Eq. (104). The low supersaturation ratio of the protein may help to explain this effect. While a low surface tension favors a shorter induction time, a small value of S has the opposite effect. Crystallization of proteins is a key method for generating samples from which protein structure can be deduced using X-ray crystallography. The large-scale production of protein biopharmaceuticals and other bioproducts is more likely to be associated with conditions that represent precipitation processes rather than crystallization.

4.22 PATTERNS OF PARTICLE ACCUMULATION AS A FUNCTION OF SOLUTE CONCENTRATION SOMETIMES DISTINGUISH HETEROGENEOUS FROM HOMOGENEOUS NUCLEATION: CHOLESTEROL AND CITRIC ACID EXAMPLES

Homogeneous nucleation is negligible if $S < S^*$, but once S^* is attained, the rate is so rapid that the formation of crystals would appear to be instantaneous as illustrated in Figure 4.13. However, nucleation usually occurs at supersaturation ratios that are much lower than that estimated from Eq. (4.78) (McCabe and Smith, 1967; Mullin, 1972; Tavare, 1995). Since Eq. (4.78) is based on the assumption of homogeneous nucleation, heterogeneous nucleation mechanisms must also be considered when this equation fails to anticipate the supersaturation ratio at which crystallization occurs.

The expected characteristics for the transition from heterogeneous to homogeneous nucleation are schematically presented in Figure 4.14 and represent the general trends observed in many crystallization experiments. If there are N_0 foreign particles initially present, and each particle is equally efficient in catalyzing heterogeneous nucleation, the number of particles stays constant, as represented by line a in Figure 4.14. Under heterogeneous conditions with all of the initial particles having the same nucleation efficiently, the particles grow in size, rather than increasing their numbers. Conversely, since the number of particles is constant up to S^*, the size of the particles must increase as represented in line *a* in Figure 4.17. If the initial number of N_0 foreign particles have varying nucleation efficiency, the particle count will increase with increasing solute concentration as represented by line *b* in Figure 4.14. In this case the average particle size may decrease at lower values of *S*, as shown in Figure 4.17, because some seed particles that are not efficient as nucleation sites will redissolve and then become part of the solute pool that precipitates. These particles are initially small, and thereby decrease the average size. At a high enough supersaturation ratio, however, all of the particles will grow, and the average size will increase.

The transition from heterogeneous to homogeneous nucleation occurs at the critical supersaturation ratio S^*, and should result in a dramatic increase in the number of particles formed, accompanied in a sharp decrease in the average size. The decrease in size, represented by Figure 4.17, occurs because the particles that provide the nucleation sites are propagated from the solute molecules themselves, and they might be measured initially as clusters consisting of 10 to 100 molecules. These newly formed nucleation sites will be small in size and large in number compared to the N_0 particles introduced to achieve heterogeneous nucleation. Hence the average particle size decreases. The sharp changes in these curves at a measurable value of *S* allow the critical supersaturation ratio, S^*, to be identified and thereby provide a handle by which interfacial energies can be estimated using a rearranged form of Eq. (4.78).

Above $S = 15$ (which corresponds to a cholesterol concentration of 20×10^{-4}, the number of particles increases suddenly, but the size drops below the threshold of the counter. The onset of homogeneous nucleation at $S = 15$ leads to an estimate of interfacial energy for cholesterol in a solution of 15 ergs/cm^2.

Consider, for example, the crystallization of cholesterol from 63.3% aqueous ethanol, where the transition from a homogeneous to a heterogeneous process gives measurable changes in particle size and particle count. Crystallization is initiated by addition of ethanol, so cholesterol particles form and provide the source of foreign nuclei. These nuclei are sometimes referred to

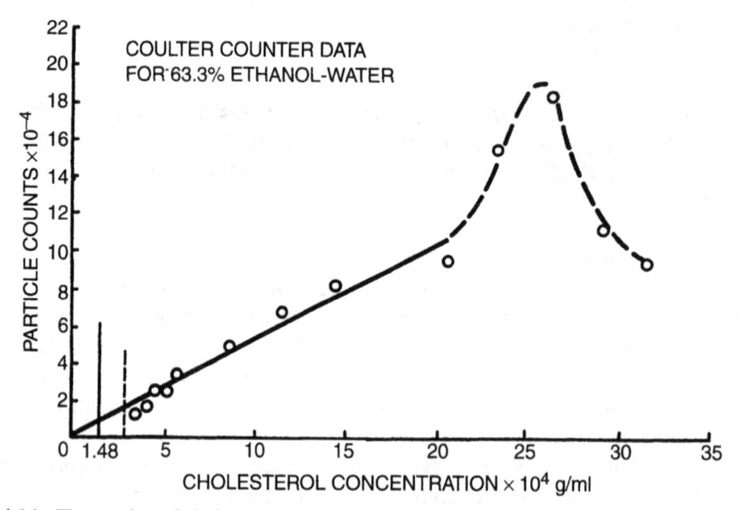

Figure 4.16 The number of cholesterol particles precipitated from aqueous ethanolic solution. Particle numbers were established by dilution × 100 after precipitation and then using a Coulter counter. Nucleation is heterogeneous up to a supersaturation ratio of approximately 15, the source of impurity nuclei being the cholesterol itself (from Walton, 1967, fig. 1–12, p. 25). Above $S = 15$ (which corresponds to a cholesterol concentration of 20×10^{-4}, the number of particles increases suddenly, but the size drops below the threshold of the counter. The onset of homogeneous nucleation at $S = 15$ leads to an estimate of interfacial energy for cholesterol in a solution of 15 ergs/cm^2.

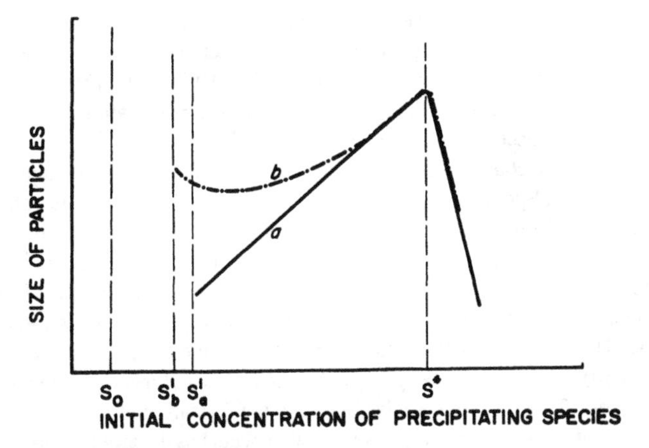

Figure 4.17 Average precipitate particle size as a function of initial precipitant concentration. A maximum particle size is usually found at or about the concentration corresponding to that causing homogeneous nucleation (from Walton, 1967, figs. 1–15, page 39).

as impurities, since they would normally not be present in the solution. Figure 4.16 shows that the number of particles increases linearly from about 10,000 to about 100,000 as the initial cholesterol concentration increases from about $3 \cdot 10^{-4}$ g/mL to $20 \cdot 10^{-4}$ g/mL (0.3 to 2 mg/mL). A saturated solution of cholesterol corresponds to $1.48 \cdot 10^{-4}$ mg/mL. The number of particles increases sharply when the cholesterol concentration is between $20 \cdot 10^{-4}$ and $25 \cdot 10^{-4}$ g/mL.

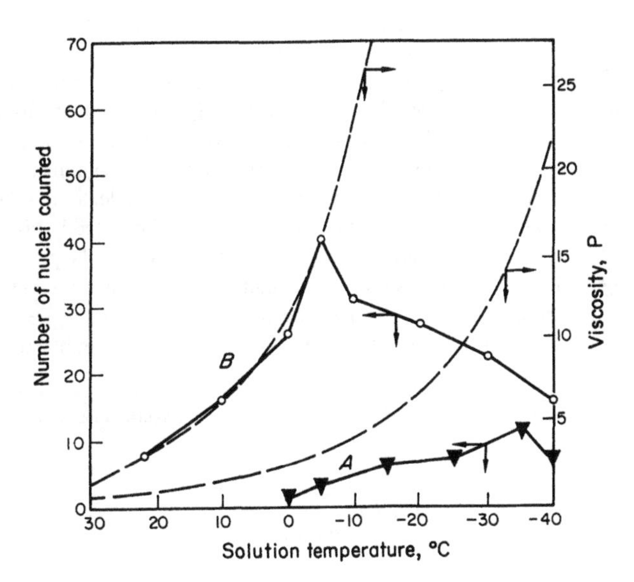

Figure 4.18 Spontaneous nucleation in supercooled citric acid solution: Curve *A*, 4.6 kg of citric acid monohydrate per kg of "free" water (initially saturated at a temperature of 62°C). Curve B, 7.0 kg/kg (initially saturated at a temperature of 85°C) (from Mullin, 1972, fig. 5.5, p. 145). Reproduced with permission from CRC Press, LLC.

At about $25 \cdot 10^{-4}$ g/mL the size of the individual particles decreases dramatically, with many of the particles falling to a size that is too small to be detectable by the particle measuring instrument when the cholesterol concentration exceeds $25 \cdot 10^{-4}$ g/mL. This causes the number of particles to appear to decrease at the highest cholesterol concentrations as shown in Figure 4.16 (Walton, 1967) at concentrations above $25 \cdot 10^{-4}$ g/mL.

A decrease in particle count is not always due to a transition from homogeneous to heterogeneous crystallization. Spontaneous nucleation of a supercooled citric acid solution leads to an increase and then a decrease in the number of particles as the crystallizations are carried out at progressively lower temperatures (Figure 4.18). Figure 4.18 presents the data for citric acid by plotting the number of nuclei counted at each temperature as a function of *decreasing* temperature. (Note that the x-axis shows the temperatures in descending value rather than in ascending order). The viscosity of the solution increases as the temperature decreases (dotted lines in Figure 4.18), while the number of particles formed over a specified time interval passes through a maximum. This phenomenon is not due to the formation of an undetectable particle size. Rather it is related to an increase in viscosity that occurs as the temperature of the solution is decreased. The increase in viscosity restricts molecular movement and inhibits crystal formation (Mullin, 1972). The number of particles is also lower at the higher temperatures (to the left of Figure 4.18), but this is probably due to the increased solubility at the higher temperatures.

4.23 THE TRANSITION FROM HETEROGENEOUS NUCLEATION AT LOW SUPERSATURATION TO HOMOGENEOUS NUCLEATION AT HIGH SUPERSATURATION FACILITATES ESTIMATES OF INTERFACIAL SURFACE TENSION

The influence of an impurity is significant, since it lowers the energy barrier to nucleation and catalyzes the nucleation process. Crystallization that is catalyzed by an impurity is referred to as *heterogeneous nucleation*, with the number of particles formed being constant, or increasing to a constant level, as illustrated in Figure 4.14. A sharp increase in the number of particles occurs when there is a transition from catalyzed, or heterogeneous nucleation, to homogeneous nucleation. Homogeneous nucleation dominates when $S \geq S^*$, as indicated in Figure 4.14. This enables the critical supersaturation ratio to be directly determined from the particle counts and particle size (Figure 4.16). S^* is assumed to correspond to the sharp increase in the number of particles and decrease in size. Once S^* is known, the interfacial energy at the precipitant/solvent interface can be deduced for a given temperature using a rearranged form of Eq. (4.78)[15]:

[15] It should be noted that Walton (1967), from which the data of Table 4.15 is taken, gives a different expression even though it is obtained from an equation that is analogous to Eq. (4.77) except that it is expressed on a specific or per molecule basis. Walton uses

$$J = A \exp\left[\frac{+16\pi\gamma^2 V_{\text{molal}}^2}{3(KT)^3(n \ln S^*)^2}\right] \tag{aa}$$

to obtain

$$\ln S^* = \left[\frac{32\gamma^3 V_{\text{molal}}^2}{(kT)^3 n^2 \ln A}\right]^{1/2} \tag{ab}$$

(footnote continued)

TABLE 4.15 Interfacial Energies and Critical Cluster Sizes Calculated from Homogeneous Nucleation Data

Precipitate	Critical Supersaturation Ratio, S^*	Surface Tension (Interfacial Energy), $ergs/cm^2$	Critical Diameter, Å
$BaSO_4$	500 to 1000	116	11
$PbSO_4$	28 to 40	74	13
$SrSO_4^b$	39	81	12
$PbCO_3$	106	105	11
$SrCO_3$	30	86	12
CaF_2	80	140	9
MgF_2	30	129	9
AgCl	5.5[b]	72	15
AgBr	3.7	56	15
Ag_2SO_4	19	62	14
$Ca(C_2O_4)$	31	67	13
$CH_2(NH_2)COOH$[a]	2.1	40	30
Cholesterol[a]	13	17	28

Source: From Walton (1967), table 1–5, p. 30. The material is reproduced with permission of John Wiley & Sons, Inc.

Note: The interfacial energies apply to aqueous solutions. The nucleation data are for compact spherical nuclei. This table was compiled from particle size data listed in literature references using equation (ab) in footnote 15. Correlations have not been made for ionic activities.

[a]These data apply to aqueous ethanol solutions.

[b]The critical supersaturation for silver chloride changes with ionic environment; the maximum value is quoted.

$$\gamma^3 = \frac{T^3(\ln S^*)^2 \ln A}{\hat{B}}$$

Interfacial energies were determined by Walton for a number of solutes by reading S^* from experimentally observed step changes and calculating γ. These range from 17 to 140 $ergs/cm^2$ (Table 4.15). According to these measurements the critical diameters for inducing homogeneous nucleation range from 9 to 30 Å, with the largest interfacial energies coinciding with the smallest critical diameters.

4.24 HETEROGENEOUS NUCLEATION CAN GIVE RISE TO ANOMALIES UPON SCALE-UP

The first sign of nucleation often appears in one region of a crystallization vessel and it may coincide with a local, high degree of supersaturation, or a surface containing imperfections that acts as a catalyst for nucleation. On a laboratory scale the surface of a test tube or beaker can

when $J = 1$. This is equivalent to

$$\hat{B} = \frac{32V_{\text{molal}}^2}{n^2K^3} \qquad \text{(ac)}$$

The form of equation (ab) does not follow directly from (aa), and Walton does not give an explanation. Nonetheless, the interfacial energies (i.e., surface tensions) reported in Table 4.15 are given without further explanation.

serve as a nucleation site. Consequently results can be different for an experiment carried out in a vessel with a small volume and large surface to volume ratio, compared to a vessel with a large volume and a small area to volume ratio. In the case of a solution cooled below its saturation point, the presence of a foreign particle or a catalytic surface can appear to induce nucleation at degrees of supercooling lower than those required for spontaneous nucleation.

Mullin presents a useful analysis (1972) and graphical representation (Figure 4.19) based on the overall free-energy change, $\Delta G'_{\text{crit,hetero}}$ associated with formation of a critical nucleus for heterogeneous nucleation relative to the free energy change for homogeneous nucleation:

$$\Delta G'_{\text{crit,hetero}} = \phi_n \Delta G_{\text{crit,homo}} \tag{4.84}$$

where

ϕ_n = function of contact angle between crystalline deposit and solid surface

$\Delta G_{\text{crit,homo}}$ = free energy change associated with homogeneous nucleation

The relation between heterogeneous and homogeneous nucleation can be envisioned in terms of a phase diagram (Figure 4.19) that represents surface energies at the boundaries of two solid

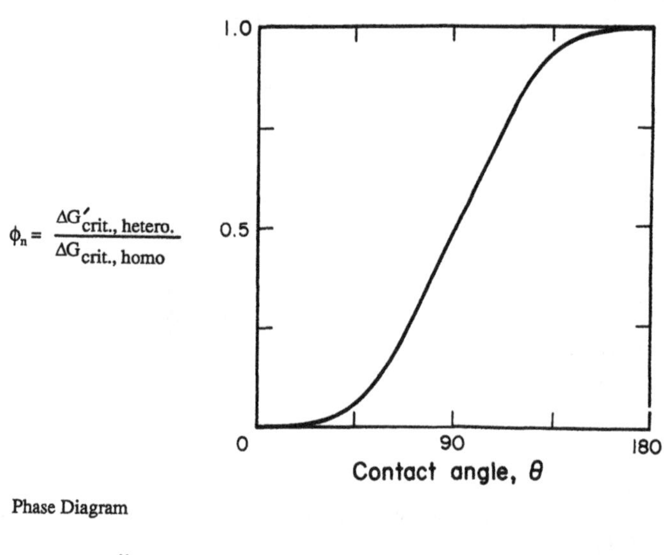

Phase Diagram

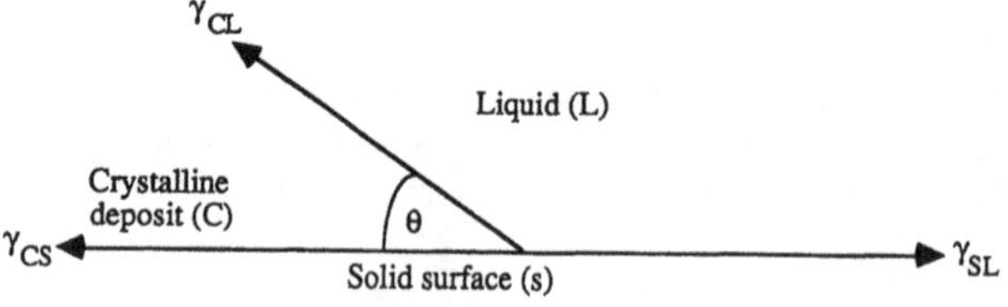

Figure 4.19 Ratio of free energy of heterogeneous nucleation to homogeneous nucleation as a function of contact angle. $\Delta G'_{\text{crit,hetero}}$ refers to the overall free energy change (i.e., an activation energy) associated with the formation of a critical nucleus under heterogeneous conditions. $\Delta G'_{\text{crit,homo}}$ denotes homogeneous conditions (from Mullin, 1972, figs. 5.7 and 5.6, pp. 148–149). Reproduced with permission from CRC Press, LLC.

phases and one liquid phase. The angle, θ, of contact between the crystalline deposit and the solid surface is

$$\cos\theta = \frac{\gamma_{SL} - \gamma_{CS}}{\gamma_{CL}} \tag{4.85}$$

where

γ_{SL} = interfacial energy between solid and liquid
γ_{CS} = interfacial energy between the solid crystalline phase and foreign solid surface
γ_{CL} = interfacial energy between the crystalline phase and liquid phase

The angle θ corresponds to the angle of wetting in liquid–solid systems. The factor ϕ_n is a function of the contact angle θ (Mullin, 1972):

$$\phi_n = \frac{(2 + \cos\theta)(1 - \cos\theta)^2}{4} \tag{4.86}$$

Therefore, when $\theta = 180$, the system behaves as in homogeneous nucleation in the absence of a foreign solid surface and the solid crystalline material, and $\phi = 1$. When $0 < \theta < 180°$, $\phi_n < 1$, and nucleation occurs more readily at heterogeneous conditions than for homogeneous conditions because the overall excess free energy is less for heterogeneous nucleation. If $\theta = 0$, $\phi_n = 0$, and the free energy of nucleation is zero, seeding of a supersaturated solution with crystals of the required solute would be needed since no nuclei need be formed in solution to achieve onset of crystallization. The ratio of free energies of heterogeneous to homogeneous nucleation as a function of contact angle is represented graphically in Figure 4.19.

The representation of the effect of contact angle by Mullen (1972) clearly illustrates the major effect that contaminants and surfaces can have on the interpretation and application of laboratory data to large-scale crystallization processes. Careful control of experimental conditions may help to reduce the presence of foreign particles, but the sensitivity of crystallization to external factors strongly suggests the need for large-scale experiments for specification of industrial processes.

4.25 MIERS PLOTS REPRESENT CRYSTALLIZATION PATHS FOR SOLUTIONS BROUGHT TO SUPERSATURATION BY COOLING AND THROUGH SOLVENT REMOVAL BY EVAPORATION

The Miers plot represents the different states of solute concentration as a function of temperature and helps to summarize the concept of supersaturation in crystallization (Figure 4.20). A solution at point *G* in Figure 4.20 may follow several paths. Labile products, whose solubilities are directly and strongly proportional to temperature, are amenable to crystallization by cooling, while those that are heat stable can be brought to a supersaturated state by heating to evaporate solvent. Crystallization by cooling follows line *GHP*, until point *P*, or in some cases, point *D*, is reached. An alternate path is represented by *GH′P′*, in which supersaturation is attained by evaporating solvent at a constant temperature. In practice, combinations of evaporation, followed by cooling may be carried out, resulting in a path located in the region somewhere between *GHP* and *GH′P′*. An example is given by line *GH″P″*. In all cases the map of the

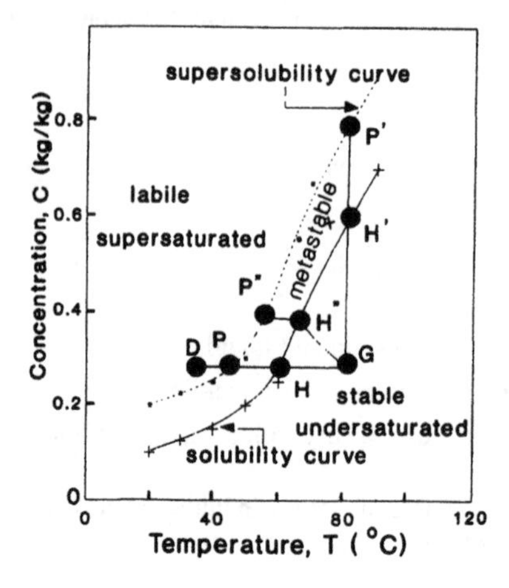

Figure 4.20 Concentration/temperature map of solubility curves for a typical salt—Mier's plot (from Tavare, 1995, fig. 4, p. 17 with permission from Kluwer Academic/Plenum Publishers).

crystallization process shows how a solute will pass through the metastable region between curves that represent the solubility (solid line) and supersaturation (dotted line).

4.26 NUMERICAL SOLUTIONS OF MATERIAL BALANCES GIVE CURVES THAT REPRESENT THE DECREASE IN SOLUTE AS A FUNCTION OF TIME FOR HETEROGENEOUS CRYSTALLIZATION

The change in the concentration of the solute in solution during a crystallization process can be based on either its concentration in solution or the amount of solid crystals formed. The driving force for crystallization is given by ΔC:

$$\Delta C = C - C_{\text{sat}} \tag{4.87}$$

where

$C =$ solute concentration in solution, kg solute/kg solvent

$C_{\text{sat}} =$ solute solubility at saturation at a given temperature

The decrease of the solute in solution, and therefore the decrease in the driving force, occurs due to solute deposition for the nucleation process and during crystal growth (Tavare, 1995):

$$-\frac{d\Delta C}{dt} = J_{\text{mass}} + A_{\text{T}}R \tag{4.88}$$

where

$J_{\text{mass}} =$ nucleation rate on a mass basis, kg solute/(s · kg solvent)

A_T = total crystal surface area, m^2/kg solvent
R = overall growth rate based on mass deposition, kg solute/$m^2 \cdot s$

Note that the nucleation rate is expressed in terms of mass rather than the number of particles, and hence is denoted as J_{mass} rather than J. The nucleation rate, J_{mass}, used here is in the form of equation (y) (in footnote 44) multiplied by the coefficient M_T^j which represents the density of the suspension of crystals:

$$J_{mass} = k_{mass} M_T^j \Delta C^n \tag{4.89}$$

where

k_{mass} = nucleation rate constant
M_T = M_{TO} (W/W_o) = suspension density, kg crystal/kg solvent
j = exponent of magma density (= 1, in this case)
M_{TO} = initial suspension density due to seed crystals
W = weight of crystals during any time t of the crystallization process, kg
W_o = initial weight of seed crystals, kg

The surface area of the crystals upon which solute is deposited is given by a similar equation:

$$A_T = A_{TO}\left(\frac{W}{W_o}\right)^{2/3} \tag{4.90}$$

where

A_{TO} = initial total crystal surface area, m^2/kg solvent

Similarly the characteristic average size of the crystals, denoted by L, is given by

$$\bar{L} = \bar{L}_o\left(\frac{W}{W_o}\right)^{1/3} \tag{4.91}$$

where

$\bar{L}$ = average initial size of the crystals, m

The void fraction of the slurry, ε, is represented by

$$\varepsilon = 1 - \left[\frac{W}{\rho_c V}\right] \tag{4.92}$$

where

W = weight of crystals, kg
ρ_c = density of crystals, kg crystals/m^3 (= g/cm^3)

V = volume of crystal suspension or crystal bed, m^3

The initial weight of seed crystals charged, as well as their initial size, can be used to calculate the initial suspension density, M_{TO}:

$$M_{TO} = \frac{W_o}{W_s} \tag{4.93}$$

and the initial surface area,

$$A_{TO} = \frac{FW_o}{\rho_c L_o W_s} \tag{4.94}$$

where

W_s = weight of solvent, kg
F = ratio of surface to volume, shape factor

The expression for the overall growth rate is based on mass deposition, R kg/($m^2 \cdot$ s),

$$R = A_T k_{crystals} \Delta C^m \tag{4.95}$$

where

$k_{crystals}$ = overall growth rate constant for crystallization $(\text{kg/m}^2 \cdot \text{s} \left(\frac{\text{kg}}{\text{kg}}\right)^m)$
m = dimensionless exponent for concentration driving force term, ΔC

The combination of Eqs. (4.89) and (4.95) with Eq. (4.88), together with the definitions in Eqs. (4.90) to (4.94) gives an expression that may be used to simulate the change in solute concentration in the solvent phase during crystallizer operation:

$$-\frac{d(\Delta C)}{dt} = M_T^j k_{mass} \Delta C^n + A_T k_{crystals} \Delta C^m \tag{4.96}$$

The utility of the approach of Tavare (1995) as outlined in Eqs. (4.88) to (4.96) is given by two examples, also taken from Tavare (1995, pp. 96–99).

The first example is for a fluidized bed crystallization process that is used to grow crystals of a uniform size under conditions where nuclei are not formed (J_{mass} = 0). This is a heterogeneous process where seed crystals are added and then grown. The time required to double the weight of seeds charged is to be calculated using the following data:

J_{mass} = 0 (no nucleation)
ΔC_o = $1.5 \cdot 10^{-2}$ kg solute/kg solvent initially present
$k_{crystals}$ = 0.1 $(\text{kg/m}^2 \cdot \text{s} \left(\frac{\text{kg}}{\text{kg}}\right)^m)$
m = 2 (dimensionless)

$$W_s = 8000 \text{ kg}$$
$$W_o = 80 \text{ kg}$$
$$L_o = 550 \times 10^{-6} \text{ m}$$
$$F = 6 \text{(dimensionless)}$$
$$\rho_c = 2000 \frac{\text{kg}}{\text{m}^3}$$

Since $J_{\text{mass}} = 0$, the first term in Eq. (4.96) is zero, and the equation simplifies to

$$-\frac{d(\Delta C)}{dt} = A_T k_{\text{crystals}} \Delta C^m \tag{4.97}$$

The initial surface area of the seed crystals, A_{TO}, is calculated from Eq. (4.94):

$$A_{TO} = \frac{FW_o}{\rho_c L W_s} = \frac{(6)(80)}{2000(550 \cdot 10^{-6})8000} = 0.0545 \frac{\text{m}^2}{\text{kg}}$$

$$\frac{\text{kg}}{\left|\frac{\text{kg}}{\text{m}^3}\right|\text{m}\left|\text{kg}\right|}$$

Once the value of A_{TO} is known, the final surface area, $A_{T,f}$ is calculated from Eq. (4.90) for spherical particles:

$$A_{T,f} = A_{TO}\left(\frac{W}{W_o}\right)^{2/3} + 0.0545\left(\frac{2W_o}{W_o}\right)^{2/3} = 0.0865 \frac{\text{m}^2}{\text{kg}}$$

where the condition that $W = 2W_o$ was specified previously. Two approaches can be used to calculate the time required to double crystal size. The first approach is to calculate the mean specific surface area of the crystals:

$$\bar{A}_T = \frac{A_{T,f} - A_{T,O}}{\ln\left(\frac{A_{T,f}}{A_{T,O}}\right)} = \frac{0.0865 - 0.0545}{\ln\left(\frac{0.0865}{0.0545}\right)} = 0.0693 \frac{\text{m}^2}{\text{kg}}$$

Then we insert this as a constant average value into Eq. (4.97):

$$-\frac{d(\Delta C)}{dt} = (0.0693)(0.1)\Delta C^2$$

and integrate the equation:

$$-\int_{\Delta C_o}^{\Delta C} \frac{d(\Delta C)}{\Delta C^2} = 0.00692 \int_o^t dt \tag{4.98}$$

where ΔC_o is specified and ΔC is obtained from a material balance for the weight of the crystals in the crystallizer:

$$W_f = W_o + (\Delta C_o - \Delta C)W_s \tag{4.99}$$

Equation (4.99) solved for ΔC gives

$$\Delta C = \frac{W_o - W_f}{W_S} + \Delta C_o = \frac{80 - 160}{8000} + 0.015 = 0.005 \frac{\text{kg solute}}{\text{kg solvent}}$$

since

W_f = final weight of crystals = 2 W_o or double the initial weight, as specified

The solution of Eq. (4.98) gives

$$\frac{1}{\Delta C} - \frac{1}{\Delta C_o} = 0.00692t$$

which, when solved for t, gives

$$t = \frac{1}{0.00692}\left(\frac{1}{0.005} - \frac{1}{0.015}\right) = 19{,}267 \text{ s} = 5.35 \text{ hr}$$

The growth of the seed crystals requires a significant time period.

The effect of using a constant, mean crystal size to carry out the crystallization is small but significant. This is shown by a second method for calculating the required time by numerically integrating Eq. (4.97). This is done by substituting Eqs. (4.90) and (4.99) into (4.97) to obtain

$$-\frac{d\Delta C}{dt} = A_{TO}\left(\frac{W}{W_o}\right)^{2/3} R = A_{TO}\left(\frac{W_o + (\Delta C_o - \Delta C)W_s}{W_o}\right)^{2/3} R \tag{4.100}$$

where

$R =$ $0.1\, \Delta C^2 = k_{crystals} A_T \Delta C^2$ kg solute/kg solvent
$A_{TO} =$ 0.0545 m^2/kg solvent
$W_o =$ 80 kg
$\Delta C_o =$ 0.015 kg solute/kg solvent
$W_s =$ 8000 kg

Equation (4.100) then gives

$$-\frac{d\Delta C}{dt} = 0.00545\,(2.5 - 100\Delta C)^{2/3}(\Delta C)^2 \tag{4.101}$$

Numerical integration of Eq. (4.101) using a Runge-Kutta integration technique (Forsythe et al., 1977), adapted for an Excel spreadsheet by Billo (1997) gives a time of 17600 s (= 4.89 hr) to attain a value of $\Delta C = 0.005$.[16] There is an obvious difference between the calculation assuming a constant average surface area and one that changes with the time course of the crystallization process.

The material balance on the solid crystals, sometimes referred to as a *solid-side representation*, must reflect the two types of crystals in a seeded crystallizer. These are the seed crystals S and the nuclei N that are subsequently generated. The population density function based on total solvent capacity, as well as the weight of crystals formed, are measures of the results of the crystallization process. The composition of the crystals is also an important measure of a crystallization process, since the goal of the process is to obtain a pure product. The derivations of equations for the material balance usually assume that the crystals are pure, and hence composition is not an explicit parameter.

The size distribution of crystals depends not only on the rates of nucleation and growth but also on the generation and destruction of particles by breakage and agglomeration. Since there are a large number of particles in a crystallization process, a continuous variable can be used to represent a discrete distribution where (Mullin, 1972)

$$\lim_{\Delta L \to 0} \frac{\Delta N}{\Delta L} = \frac{dN}{dL} = n \tag{4.102}$$

where

ΔN = number of particles in the size range ΔL per unit volume of the system

ΔL = size range between L_1 and L_2

n = concentration of particles, number/length · volume

[16]This footnote sets up and demonstrates the use of a Runge-Kutta integration of Eq. (4.101) within an Excel spreadsheet. The setup of the first calculation is summarized below. Subsequent iterations would require Microsoft Excel to be run. For the first integration step, the initial value is $\Delta C = 0.015$ at t = 0 (cell G22). The first iteration is calculated in row 23 of the spreadsheet. The step size is given by $dx = 255.08$. The equations in each cell are

$$\text{TA1 : cell C23 -Area0} \cdot (\text{a - b} \cdot \text{G22})^{2/3} \cdot 0.1 \cdot (\text{G22})^2 \cdot \text{dx}$$

$$\text{TA2 : cell D23 -Area0} \cdot (\text{a - b} \cdot (\text{G22} + \tfrac{\text{TA1}}{2}))^{2/3} \cdot 0.1 \cdot (\text{G22} + \tfrac{\text{TA1}}{2})^2 \cdot \text{dx}$$

$$\text{TA3 : cell E23 -Area0} \cdot (\text{a - b} \cdot (\text{G22} + \tfrac{\text{TA2}}{2}))^{2/3} \cdot 0.1 \cdot (\text{G22} + \tfrac{\text{TA2}}{2})^2 \cdot \text{dx}$$

$$\text{TA4 : cell F23 -Area0} \cdot (\text{a - b} \cdot (\text{G22 + TA3}))^{2/3} \cdot 0.1 \cdot (\text{G22 + TA3})^2 \cdot \text{dx}$$

$$\Delta C \text{ : cell G23 G22} + \tfrac{1}{6}(\text{TA1} + 2 \cdot \text{TA2} + 2 \cdot \text{TA3} + \text{TA4})$$

The result in cell G23 becomes the initial value for the next iteration that is carried out in the next row (row 24). Note that Area0 is equivalent to $A_{T,0}$ and that TA1, TA2, TA3, and TA4 are assigned to their respective columns using the INSERT NAME, DEFINE commands in Excel (Billo, 1997, pp. 57–61). The equations in each set of four cells follow the classical fourth-order Runge-Kutta method (Forsythe et al., 1977):

$$y_{n+1} = y_n + \frac{1}{6}(k_0 + 2k_1 + 2k_2 + k_3)$$

where, for the function $y = f(y, t)$,

$$k_0 = hf(y_n, t_n)$$
$$k_1 = hf(y_n + \tfrac{1}{2}k_0, t_n + \tfrac{h}{2})$$
$$k_2 = hf(y_n + \tfrac{1}{2}k_1, t_n + \tfrac{h}{2})$$
$$k_3 = hf(y_n + k_2, t_n + h)$$

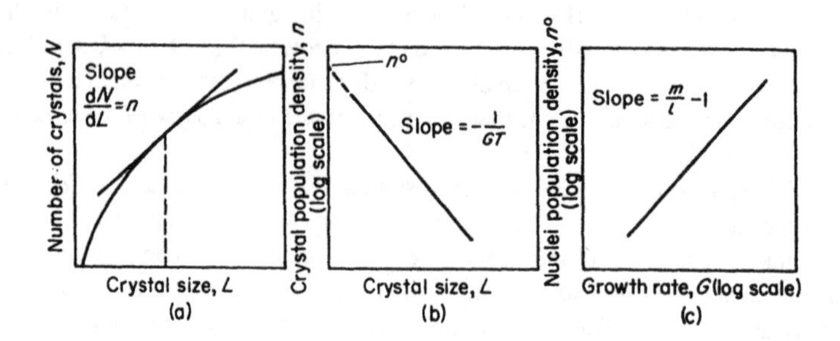

Figure 4.21 Plots illustrating the concepts of (*a*) crystal size distribution, (*b*) population density as a function of crystal size, and (*c*) growth kinetics (from Mullin, 1972, fig. 10.2, p. 340). Reproduced with permission from CRC Press, LLC.

The differential dN/dL represents the slope of the curve illustrated in Figure 4.21. The number of crystals in the size range of L_1 and L_2 is therefore given by

$$\Delta N = \int_{L_1}^{L_2} n dL \tag{4.103}$$

The concepts and utility for population balances for a continuous mixed suspension, mixed product removal crystallizer are presented by Mullin (1972) and discussed here. The balance across this type of crystallizer is presented in Figure 4.22.

The balance at steady state across the crystallizer in Figure 4.22 is

$$\begin{bmatrix}\text{crystals growing}\\ \text{into the arbitrary}\\ \text{size range, } \Delta L\end{bmatrix} = \begin{bmatrix}\text{crystals growing}\\ \text{out of size range,}\\ \Delta L\end{bmatrix} + \begin{bmatrix}\text{crystals in size range}\\ \Delta L\text{, that are removed}\\ \text{from crystallization}\end{bmatrix}$$

$$[n_1 G_1 V \Delta t] = [n_2 G_2 V \Delta t] + [Q\bar{n}\Delta L \Delta t] \tag{4.104}$$

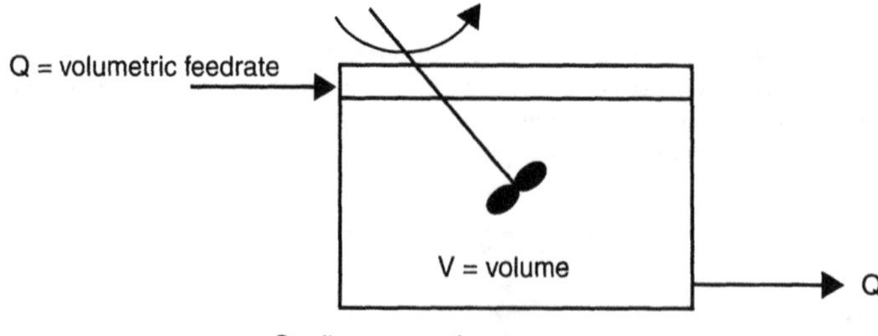

Figure 4.22 Schematic representation of mixed, product removal crystallizer.

Rearrangement of Eq. (4.104) enables a differential form to be written:

$$\frac{V(n_2G_2 - n_1G_1)}{\Delta L} + Q\bar{n} = 0 \tag{4.105}$$

In the limit, $\Delta L \rightarrow 0$,

$$V\frac{d(nG)}{dL} + Qn = 0 \tag{4.106}$$

Since the mean residence time is defined by $V/Q = t$. We can assume that growth is independent of size[17] so that $dG/dL = 0$, and Eq. (4.106) simplifies to

$$\frac{dn}{dL} + \frac{n}{G\bar{t}} = 0 \tag{4.107}$$

Rearrangement of Eq. (4.107), followed by integration between the initial population of seed crystals, n_o, at a relative size defined as the zero point (i.e., $L = 0$) as shown in Figure 4.21, and the population leaving the crystallizer at size L gives

$$\int_{n_o}^{n} \frac{dn}{n} = \int_{0}^{L} -\frac{dL}{G\bar{t}}$$

so

$$n = n_o \exp\left(-\frac{L}{G\bar{t}}\right) \tag{4.108}$$

Equation (4.108) represents a number-size distribution. A plot of ln n as a function of L should give a straight line as is schematically illustrated in Figure 4.21. Since the slope is $-1/G\bar{t}$, if the average residence time is known, the crystal growth rate can be calculated.

A more general form of the material balance requires an accumulation term, since both crystal breakage and spurious nucleation occurs. The generalized material balance has the form (Mullin, 1972)

[17] This simplifying assumption is known as the ΔL law of crystal growth (McCabe and Smith, 1967): i.e., that all crystals have the same geometric shape; shape does not change during growth; concentration difference, $C - C_{sat}$, is constant and the same for all crystals; and that the overall growth coefficients are the same for each face of the crystal. The equation for the linear growth rate is:

$$D_p^Y = \frac{dD_p}{dt} = \frac{2(C - C_{sat})KM_r\lambda}{L_{crystal}} \tag{z}$$

where λ is a shape factor and K is an average mass transfer coefficient for the entire crystal. The total growth is given by $\Delta D_p = D_p^Y \Delta t$. This is called the ΔL law because the symbol L was commonly used to denote crystal size at the time that this relationship was derived (by McCabe in 1929).

$$\frac{Q_i n_i}{V} + B = \frac{Q_e n_e}{V} + D + \left[\frac{\partial n}{\partial t} + \frac{\partial (nG)}{dL} + n\frac{\partial (\ln V)}{\partial t}\right] \tag{4.109}$$

where

$B =$ birth function
$D =$ death function
$Q_i, n_i =$ inlet flowrate and population density at inlet, respectively
$Q_e, n_e =$ outlet flowrate and population density, respectively

The total number of crystals up to a size L is

$$N = \int_0^L n dL = \int_0^L n_o \exp\left(-\frac{L}{GT}\right) dL = n_o GT\left[1 - \exp\left(-\frac{L}{GT}\right)\right] \tag{4.110}$$

Equation (4.110) also represents the *zeroth moment* of the distribution. For large values of L, N is the same as the total number of crystals in the system, and Eq. (4.110) simplifies to

$$N_T = n_o G\overline{t} \tag{4.111}$$

While the zeroth moment represents the number of crystals, the *first moment* of the distribution represents the system's cumulative length (i.e., the length if all of the crystals were laid side by side):

$$L = \int_0^L nLdL = \int_0^L n_o L \exp\left(\frac{-L}{G\overline{t}}\right) dL$$

which gives

$$L = n_o GT\left[GT\left(1 - \exp\left(\frac{-L}{G\overline{t}}\right) - L\exp\left(\frac{-L}{G\overline{t}}\right)\right)\right] \tag{4.112}$$

If L is very large so that all of the particle sizes are included, then the total length is

$$L_T = n_o (G\overline{t})^2 \tag{4.113}$$

when L is very large.

The *second moment* gives the surface area:

$$A = \beta \int_0^L nL^2\, dL \tag{4.114}$$

where β is a surface shape factor. For $L \rightarrow \infty$,

$$A_T = 2\beta n_o(G\bar{t})^3 \tag{4.115}$$

The *third moment* gives the mass, W,

$$W = \alpha\rho\int_o^L nL^3 dL \tag{4.116}$$

where α is the volume shape factor, and ρ is the crystal density. The determination of particle size is still commonly based on sieve analysis, and as pointed out by Mullin (1972, p. 397), the equivalent diameter obtained from sieve analysis is likely to be the second largest dimension, since crystals are never true spheres. This is illustrated in Figure 4.23 (from Mullin, 1972, fig. 11.14, p. 397), which shows the various particle shapes that would all pass through the same sieve-aperature diameter. When L is very large, the total weight is

$$W_T = 6\alpha\rho n_o(GT)^4 \tag{4.117}$$

The shape factors are defined by a length or diameter parameter, d_p, and volume or shape factors denoted by α or β, respectively. The volume, mass, and surface area are given by

$$V = \alpha d_p^3 \tag{4.118}$$

$$W = \alpha\rho d_p^3 \tag{4.119}$$

$$A = \beta d_p^2 \tag{4.120}$$

Shape factors for several types of geometries are summarized in Table 4.16.

An overall, surface to volume factor, or specific surface shape factor F, may be defined using Eqs. (4.118) to (4.120):

$$F = \frac{\beta}{\alpha} \tag{4.121}$$

where

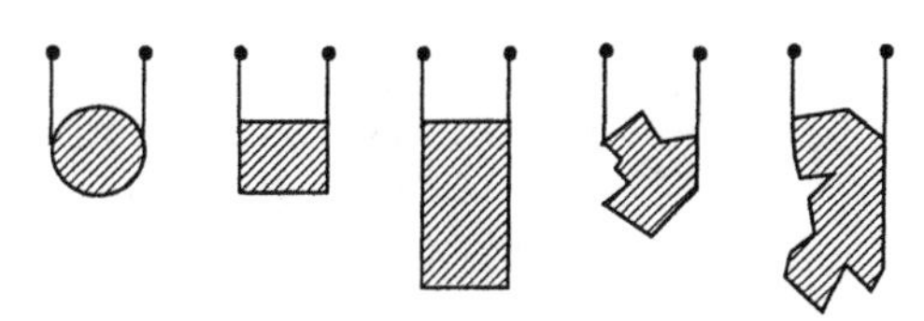

Figure 4.23 Schematic representation of irregularly shaped particles passing through a sieve opening.

TABLE 4.16 Volume and Surface Area Shape Factors for Selected Geometries

Geometry	Volume Shape Factor, α	Surface Area Shape Factor, b
Sphere	$\pi/6$	π
Cube	1	6
Octahedron (d_p = length of an edge)	$\sqrt{2}/3$	$2\sqrt{3}$
Block (5:2:1)	1.25	8.6

Source: Tabulated from data of Mullin (1972), p. 397. The material is reproduced with permission of CRC Press LLC.

$$\beta = \frac{A}{d_p^2}$$

$$\alpha = \frac{V}{d_p^3}$$

The individual surface to volume ratio is given by

$$\frac{A}{V} = \frac{\beta d_p^2}{\alpha d_p^2} = \frac{F}{d_p} \tag{4.122}$$

while the surface to mass ratio is

$$\frac{A}{W} = \frac{\beta d_p^2}{\alpha \rho d_p^3} = \frac{F}{\rho d_p} \tag{4.123}$$

Dimensionless forms of Eqs. (4.110), (4.112), (4.114), and (4.116) are obtained by defining the ratio

$$X = \frac{L}{G\bar{t}} = \frac{\text{crystal}}{\left[\begin{array}{l}\text{size of crystal that}\\ \text{has grown over the}\\ \text{drawdown time}\end{array}\right]}$$

Substitution of the expression for X into these equations gives

$$\frac{N}{N_T} = N(X) = 1 - \exp(-X) \tag{4.124}$$

$$\frac{L}{L_T} = L(X) = 1 - (1 + X)\exp(-X) \tag{4.125}$$

$$\frac{A}{A_T} = A(X) = 1 - \left(1 + X + \frac{X^2}{2}\right)\exp(-X) \tag{4.126}$$

$$\frac{W}{W_T} = W(X) = 1 - \left(1 + X + \frac{X^2}{2} + \frac{X^3}{6}\right)\exp(-X) \tag{4.127}$$

The different mass distribution is represented by (Mullin, pg 343, 1972):

$$\frac{dW(X)}{dX} = \frac{1}{6}X^3 \exp(-X) \tag{4.128}$$

A plot of Eqs. (4.127) and (4.128) is given in Figure 4.24, where it is shown that a mixed crystallizer will have a dominant site fraction at $X = 3$ with the median size fraction (at $W(X) = 0.5$) of 3.67. This dimensionless correlation makes it convenient to calculate the crystal size distribution without knowing the dependence of growth rate on supersaturation (Mullin, 1972).

EXTRACTION AND LEACHING

Extraction utilizes differences in solubilities of biological molecules between two different liquid phases, or between a solid and a liquid. Leaching is a form of extraction where a product is recovered from a solid such as microbial cell mass or a solid biological material. Examples of extraction processes are the recovery of oil from soybeans using hexane, and the partitioning

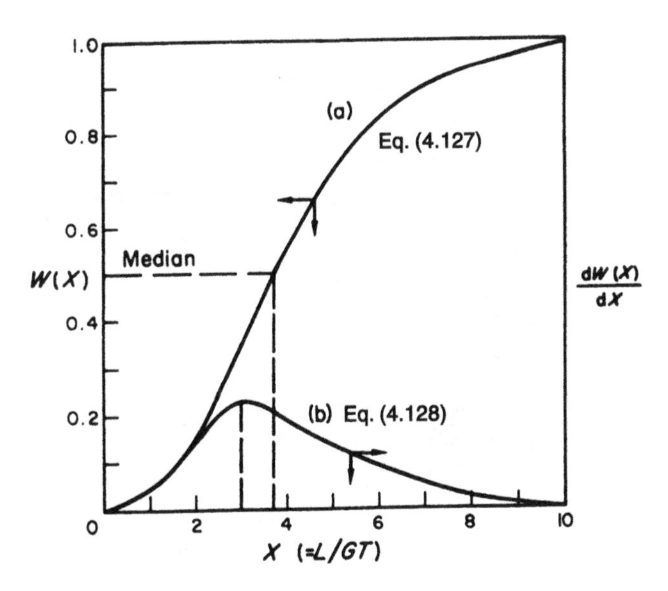

Figure 4.24 (*a*) Cumulative and (*b*) differential mass distributions represented by Eqs. (4.121) and (4.122) (from Mullin, 1972, fig. 10.3, p. 343). Reproduced with permission from CRC Press, LLC.

of a fermentation bioproduct between two aqueous phases. Leaching of sugars from hydrolyzed biomass is an example of leaching where water is used to remove sugar from a solid material.

4.27 "LEACHING IS THE PREFERENTIAL SOLUTION OF ONE OR MORE CONSTITUENTS OF A SOLID MIXTURE BY CONTACT WITH A LIQUID SOLVENT" (TREYBAL, 1968)

Leaching is one of the oldest unit operations in the chemical industries (Treybal, 1968). While often associated with the metallurgical industry, this method has many applications in the recovery of biobased and agricultural products. The material to be leached is placed in a vessel, and solvent is passed through it. The solvent dissolves the extractables and removes them. Completion of this unit operation leaves solvent in the bottom layer of the bed as illustrated in Figure 4.25. Consequently a countercurrent operation as shown in Figure 4.21 may be used to improve recovery. Examples include sugar from sugarbeet or hydrolyzed biomass, oil from soybeans and cottonseed, tannin from the bark, and pharmaceutical compounds from plant roots and leaves. This unit operation is based on the diffusion of the solvent into the solid, dissolution of the solute into the solvent, and diffusion of the solution out of the solid. These steps are usually rate limiting, with the transfer of the solvent into and out of the contacting device being achieved rapidly relative to the rates of diffusion and dissolution. The contacting device may consist of a conveyer belt as illustrated in Figure 4.26 or a series of vessels that hold the solids.

Efficient leaching of metal ones is facilitated by particles that are crushed into the smallest possible size which still allows solvent to pass through a bed of the solids. Treybal (1968) gives the example of a copper ore, where leaching requires 4 to 8 hours for 0.25 mm particles, 5 days for 6 mm particles, and 4 to 6 years for 15 cm lumps. The leaching of natural products (oils, sugars, or pharmaceuticals) is carried out using solids that are small enough to facilitate contact with the solvent while minimizing the diffusion path length, and large enough to facilitate flow

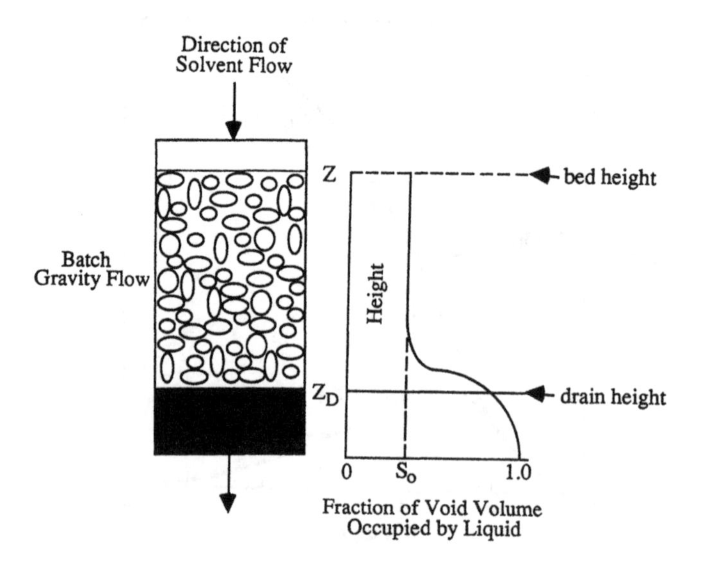

Figure 4.25 Schematic representation of illustrating drain height for leaching. Note that not all of the fluid drains freely.

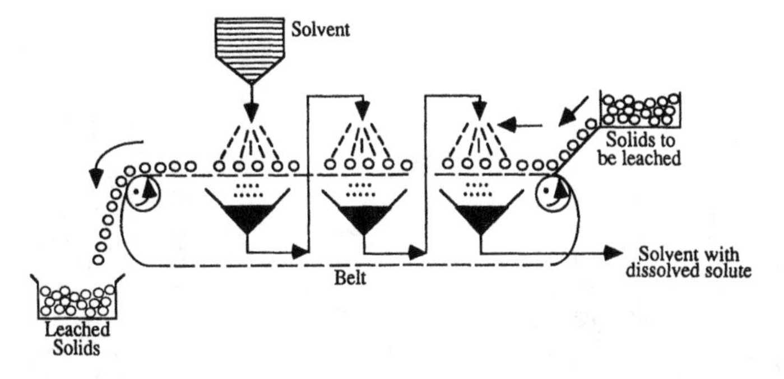

Figure 4.26 Schematic representation of leaching process: countercurrent leaching with solids on a conveyer belt. Reprinted with permission from Academic Press Ltd., London, UK.

through a compressible bed. Hence smaller is not necessarily better for some types of biomaterials. This guideline also reflects the cellular structure of plant and mammalian cells, and the benefit of using a cell membrane to contain some of the cells components that cannot diffuse through the cell wall while the lower molecular weight product can diffuse out. The cell wall thus serves as a natural membrane that fractionates the target molecule from other cell constituents such as proteins or colloidal substances. Biomaterials are often flaked or cut into thin slices, known as *cosettes*, in a manner that preserves the cell wall structure. This approach is used for leaching sugar from sugar beets where the cells are deliberately left intact (Treybal, 1968).

Pharmaceutical products from plant roots, stems, and leaves are typically recovered from dried materials where the drying process causes the cell wall to be ruptured, thus potentially releasing all of the cell's contacts and providing better contact between solvent and solute. Soybeans, which are rolled and flaked before oil is extracted, are an example where the preprocessing largely ruptures the cell wall so that enhanced contact of the solvent (hexane) with the soybean oil occurs. The theory and design equations for these potentially important unit operations are given by Treybal (1968).

4.28 SUPERCRITICAL CARBON DIOXIDE IS AN EFFECTIVE EXTRACTOR FOR SOLID BIOPRODUCTS

The critical point of a pure component is defined by a temperature and pressure above which it behaves neither as a gas or a liquid. A fluid compressed to a pressure above its critical pressure at a temperature that is above its critical temperature is referred to as a *supercritical fluid.* The isotherms near the critical point in the pressure temperature diagrams of Figure 4.27 undergo only small changes in slope when the reduced temperature ($T_r = T/T_c$) is between 1.05 and 1.2, and this illustrates that small changes in temperature and pressure cause a supercritical fluid to undergo large changes in volume (and therefore density). Since the isotherms near the critical point are nearly flat. For example, CO_2, with a critical temperature and pressure of 31.06°C and 73.83 bar, exhibits isotherms with shallow slopes. Therefore relatively small changes in reduced densities occur when the reduced temperature ($T_r = T/T_c$) is between 1.05 and 1.2. The ability of the supercritical fluid to make a transition from low to high density provides its unique solvating power. Other beneficial properties of supercritical CO_2 are (Singh and Rizvi, 1995) as follows:

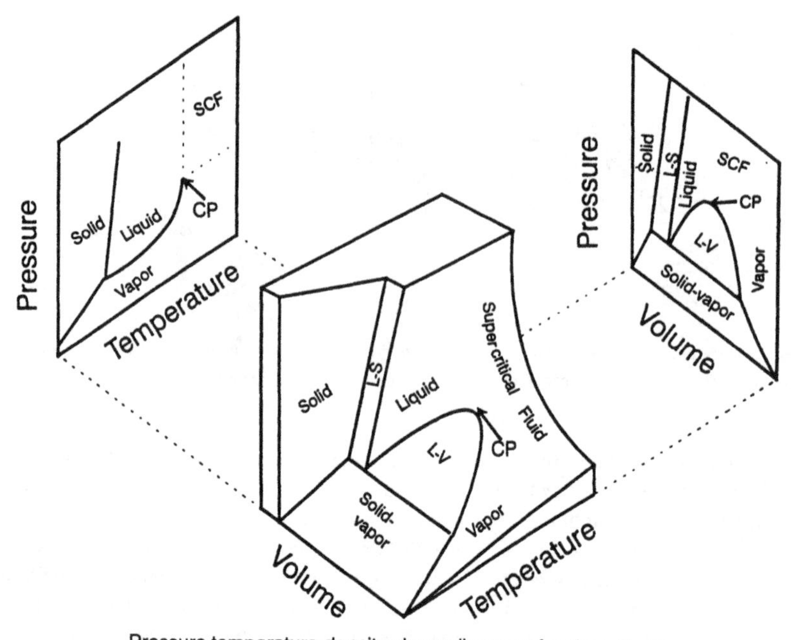

Pressure temperature-density phase diagram of carbon dioxide.

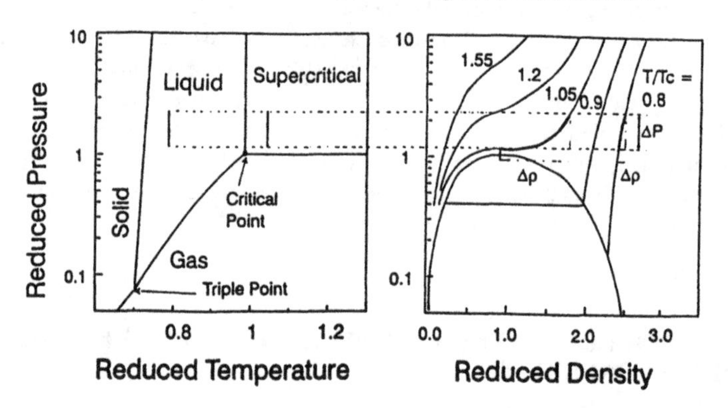

Figure 4.27 (*a*) Pressure-density phase diagram of CO_2; (*b*) density variations (represented as Δp) for liquid and supercritical CO_2 near the critical point (from Singh and Rizvi, 1995, figs. 3.1 and 3.2, p. 61 with permission from Marcel Dekker, Inc.).

1. Enhanced mass transfer of the solute in a supercritical fluid compared to a liquid due to the supercritical fluid's lower velocity and higher diffusion coefficient.
2. Rapid penetration into porous solid, since the fluid has no surface tension.
3. Nonflammable, nontoxic, inexpensive, generally recognized as safe (GRAS), and a low critical temperature.

Supercritical CO_2 was first introduced to the food industry in 1978, with the start-up of a plant in Germany for decaffination of green coffee beans. The conceptual basis of a supercritical fluid extraction process, superimposed on a phase diagram, is illustrated in Figure 4.28

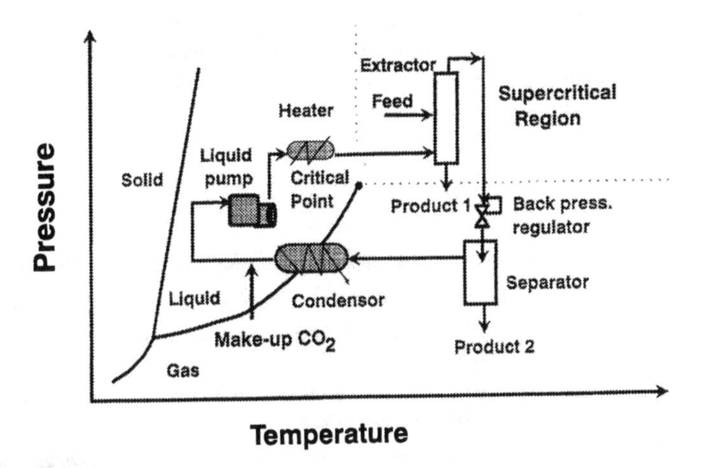

Figure 4.28 Main components of a supercritical extraction process superimposed on a phase diagram for CO_2; (from Singh and Rizvi, 1995, fig. 3.7, p. 80 with permission from Marcel Dekker, Inc.).

(Singh and Rizvi, 1995). The extractor is operated in the supercritical region while the other parts of the process are run at a temperature and pressure below the critical point. Other applications include decaffination of tea and extraction of flavor from hops (used in beer brewing) with plants in Germany, Australia and the United States. The extraction and refining of edible oils is an area of research, since current processes of 800 atm and 70°C are needed and since the higher temperature and pressure is where soybean triglycerides and CO_2 become misible (Singh and Rizvi, 1995). An excellent overview of theory, design, and applications of supercritical fluids for bioseparations is given by Singh and Rizvi (1995).

SUMMARY AND PERSPECTIVES

The unit operations associated with the recovery and purification of biological molecules are principly liquid based, and they are carried out at close to isothermal conditions. A solution phase, modified to obtain a solid by precipitation, coagulation, or crystallization, will be subject to separations by centrifugation, filtration, and microfiltration. Their design and placement in a separation sequence consisting of multiple unit operations must consider the separations steps that follow or precede them. The thermodynamics and rate processes that define the results of the individual processing steps assist in this selection process. The analysis presented in this chapter has attempted to describe the fundamental principles of each of the unit operations in terms of thermodynamics and rate processes that are generally applicable. For example, the development of flux equations for membrane separation were later shown to be applicable to dialysis, while the theory associated with salting-in and salting-out of proteins was shown to be applicable to principles of hydrophobic chromatography. Leaching and supercritical extraction were only briefly mentioned, since books containing chapters on their theory and applications to bioproducts are available.

Overall, this chapter has attempted to explain the principles of the unit operations based on mechanical force, filtration, membranes, or precipitation. These unit operations, together with chromatographic separations discussed in the next chapter, will then be revisited in Chapter 8 in the context of case studies which show how they fit together for purifying biomolecules. The case studies of Chapter 8 show how sequences of different separation steps are used for the manufacture of biotechnology products, and illustrate the principles, practice, and economics involved in multiple-step purification processes.

REFERENCES

Bailey, J. E., and D. F. Ollis, 1977, *Biochemical engineering fundamentals*, McGraw-Hill, NY, 166–168.

Billo, E. J., Excel for Chemists, A Comprehensive Guide, Wiley-VCH, 37–60, 330–348 (1997).

Chance, R. E., E. P. Kroeff, and J. A. Hoffmann, 1981, *Insulins, growth hormone, and recombinant DNA technology*, Raven Press, NY, 71–85.

Chance, R. E., E. P. Kroeff, J. A. Hoffman, and B. H. Frank, 1981, Chemical, physical, and biologic properties of biosynthetic human insulin, *Diabetes Care*, 4(2), 147–154.

Cohen, E. J., and J. T. Edsall, 1943, *Proteins Amino Acids and Peptides as Ions and Dipolar Acids*, Am. Chem. Soc. Monograph Series, Reinhold Publishing, New York, 477, 496–497, 500–505.

Cohen, E. J., and J. D. Ferry, 1943, Interactions of proteins with ions and dipolar ions in *Proteins, Amino Acids, and Peptides As Ions and Dipolar Ions*, E. J. Cohn and J. T. Edsall, eds., *Am. Chem. Soc.* Monograph Series No. 90, Reinhold Publishing Corporation, NY, 587, 597, 602–622.

Dutton, G., 1996, Transgenic animal-based protein products move toward clinical trials, *GEN*, 16(9), 37.

Edsall, J. T. 1943, Dipolar ions and acid-base equilibria, *Proteins, Amino Acids, and Peptides as Ions and Dipolar Ions*, E. J. Cohn and J. T. Edsall, eds., *Am. Chem. Soc. Monograph Series*, NY, 111–115, 242–250.

Eriksson, K-O., 1989, Hydrophobic interaction chromatography, in *Protein Purification: Principles, High Resolution Methods, and Application*, J-C. Janson and L. Ryden, eds., VCH, NY, 207–226.

Forsythe, G. E., M. A. Malcom, and C. B. Moler, 1977, *Computer Methods for Mathematical Computations*, Prentice-Hall, Englewood Cliffs, NJ, 110–148.

Harris, E. L. V., 1992, Concentration of the extract, in protein purification methods: A practical approach, E. L. V. Harris and S. Angal, eds., IRL Press, Oxford, 149–164.

Johnson, A. H., 1974, in Fundamentals of Dairy Chemistry (B. H. Webb, A. H. Johnson, J. A. Alford, eds.), AVI Publishing Co., Westport, Conn., 1–57.

Kroeff, E. P., R. A. Owens, E. L. Campbell, R. D. Johnson, and H. I. Marks, 1989, Production scale purification of biosynthetic human insulin by reversed-phase high-performance liquid chromatography, *J. Chromatogr.*, 461, 45–61.

McCabe, W. L., and J. C. Smith, 1967, *Unit operations of chemical engineering*, McGraw-Hill, NY, 751–787, 879–904.

Melander, W., and C. Horvath, 1977, Salt effects on hydrophobic interactions in precipitation and chromatography of proteins: An interpretation of the lyotropic series, *Arch. Biochem. Biophys.*, 183, 200–215.

Mullin, J. W., 1972, *Crystallization*, CRC Press, Cleveland, OH, 43–46, 139–150, 335–363, 394–400.

Nielson, A. F., 1964, *Kinetics of precipitation*, Pergamon Press, NY, 1–39.

Roe, S., 1989, Separation based on structure, in *Protein Purification Methods—A Practical Approach*, B. I. V. Harris and S. Angal, eds., IRL Press, Oxford, 200–216, 221–230.

Roettger, B. F., J. A. Myers, M. R. Ladisch, and F. E. Regnier, 1990, Mechanisms of Protein Retention in Hydrophobic Interaction Chromatography, in *Protein Purification from Molecular Mechanisms to Large-Scale Processes*, ACS Symposium Ser. 427, M. R. Ladisch, R. C. Willson, C-d. Painton, and S. E. Builder, (eds.), 80–92.

Roettger, B. F., J. A. Myers, M. R. Ladisch, and F. E. Regnier, 1989, Adsorption Phenomena in Hydrophobic Interaction Chromatography, *Biotechnol. Prog.*, 5(3), 79–88.

Rosenbaum, E. J., 1970, *Physical Chemistry*, Appleton-Century-Crofts Educational Division, Meredith Corp., NY, 550–555, 611, 652.

Ruettimann, K. W., and M. R. Ladisch, 1991, *In situ* observation of Casein Micelle Coagulation, *J. Colloid. Interface Sci.*, 146(1), 276–287.

Ruettimann, K. W., and M. R. Ladisch, 1987, Casein Michelles: Structure, properties and enzymatic coagulation, *Enz. Microb. Technol.*, 9, 578–589.

Shih, Y-C., J. M. Prausnitz, and H. W. Blanch, 1992, Some characteristics of protein precipitation by salts, *Biotechnol. Bioeng.*, 40, 1155–1164.

Singh, B., and S. H. Rizvi, 1995, Bioseparations with critical fluids, in *Bioseparation Processes in Foods*, R. K. Singh and S. H. Rizvi, eds., Marcel Dekker, NY, 59–111.

Tavare, N. S., 1995, *Industrial Crystallization, Process Simulation Analysis and Design*, Plenum Press, NY, 1–20, 59–66, 79–102.

Thayer, A., 1996, Firms boost prospects for transgenic drugs, *Chemical and Engineering News*, 74(35), 23–24.

Treybal, R. E., 1968, *Mass-transfer operations,* 2nd edition, McGraw-Hill Book Company, NY, 628–675.

Waddington, T. C., 1969, *Non-Aqueous Solvents*, Studies on modern chemistry, Appleton-Century-Crofts, NY, 1–10, 28.

Walton, A. G., 1967, The formation and properties of precipitates, Interscience Publishers, NY, 1–43.

Watson, J. D., M. Gilman, J. Witowski, M. Zoller, 1992, *Recombinant DNA*, 2nd edition, Scientific American Books, W. H. Freehold, NY, 16–17, 214, 224, 225, 256–257, 466, 467, 480.

Wheelwright, S. M., 1991, *Protein Purification Design and Scale up of Downstream Processing*, Hanser, Munich, 25–29.

SUGGESTED READING

Alberts, B., D. Bray, J. Lewis, M. Raff, K. Roberts, and J. D. Watson, 1989, *Molecular Biology of the Cell*, 2nd Ed., Garland Publishing, New York, 54, 58, 65, 116, 172–173, 209–210, 212–216, 445.

Baldwin, M. A., F. E. Cohen, and S. B. Pruisner, 1995, Prion protein isoforms, a convergence of biological and structural investigations, *J. Biol. Chem.*, 270(33), 19197–19200.

Borman, S., 1992, Anticancer drug: Boost to Taxol supply planned, *Chem. Eng. News*, 70(10), 4.

Borman, S., 1996, Epothilone epiphany: Total synthesis; compound that shares Taxol's mechanism of action and is more potent in in vitro tests, *Chem. Eng. News*, 74(52), 24–26.

Borman, S., 1997a, (-Peptides: Nature improved?, Peptide analogs from (-amino acids fold into defined three dimensional stuctures similar to those of natural peptides, *Chem. Eng. News*, 75(24), 32–35.

Borman, S., 1997b, Proteins to order: Program determines amino acid sequence from three dimensional structure data, *Chem. Eng. News*, 75(40), 9–10.

Borman, S., 1997c, Nobel prize for prions: Pruisner wins medicine award for infectious protein concept, *Chem. Eng. News*, 75(41), 4.

Borman, S., 1998a, Peptoids eyed for gene therapy applications: Despite lack of intramolecular hydrogen bonds, peptide analogs fold into helices, *Chem. Eng. News*, 76, 56.

Borman, S., 1998b, Prion research accelerates: Spurred by fears that Mad Cow Disease is spreading to humans, researchers strive to understand abnormal form of brain protein, *Chem. Eng. News*, 76(6), 22.

Branden, C., and J. Tooze, 1991, *Introduction to Protein Structure*, Garland Publishing, New York, 4–5, 11–15.

Carey, F. A., 1996, *Organic Chemistry*, McGraw Hill, New York, 630–631.

Chesebro, B., 1998, BSE and prions: Uncertainty about the agent, *Science*, 279(5347), 42–43.

Creighton, T. E., 1993, *Proteins: Structures and Molecular Properties*, W. H. Freeman, New York, 25–28, 38–39, 54, 79–81, 100, 165–167, 176–177, 189, 217–221, 223–235, 253–256, 264–270, 464–465.

Dale, B. E., and J. P. McBennett, 1992, Stability of high temperature enzymes: Development and testing of new predictive model, in *Biocatalysis at Extreme Temperatures: Enzyme Systems near and above 100°C*, M. W. W. Adams and R. M. Kelly, eds., Am. Chem. Soc., Washington, DC, 136–152.

Degussa, A. G., 1985, *Amino Acids for Animal Nutrition*, Degussa A. G., Frankfurt, 6–14.

Dennis, E. A., 1994, Diversity of group types, regulation, and function of phospholipase A2, *J. Biol. Chem.*, 269(18), 13057–13061.

Evans, R. M., 1965, *The Chemistry of the Antibiotics Used in Medicine*, Pergamon Press, Oxford, 48–50.

Gross, N., H. Dawley, J. Carey, M. Trinephi, and K. L. Miller, 1997, Mad cows—and humans: A Nobel prize hasn't quelled the controvery over the disease's cause, *Business Week*, 3558, 80–82.

Hegde, R. S., J. A. Mastrianni, M. R. Scott, K. A. Defea, P. Trembley, M. Torchia, S. A. DeArmond, S. B. Pruisner, and V. R. Lingappa, 1998, A transmembrane form of the prion protein in neurodegenerative disease, *Science*, 279(5352), 827–834.

Henry, M. P., 1989, Ion-exchange chromatography of proteins and peptides, in *HPLC of Macromolecules: A Practical Approach*, IRL Press, Oxford, 99, 105.

Hevehan, D. L., and E. De Bernardez Clark, 1997, Oxidative renaturation of lysozyme at high concentrations, *Biotechnol. Bioeng*. 54(3), 221–230.

Horton, H. R., et al., 1993, *Principles of Biochemistry*, Neil Patterson Publishers/Prentice-Hall, Englewood Cliffs, NJ, 3–6, 3–7.

Jacobs, M., 1998, Editor's Page: Where's the beef?, *Chem. Eng. News*, 76(6), 3.

Jaenicke, R., and R. Rudolph, 1990, Folding proteins, in *Protein Structure: A Practical Approach*, T. E. Creighton, ed., IRL Press, Oxford.

Kaiser, E. T., 1988, Taming the chemistry of proteins, in *Biotechnology and Materials Science: Chemistry for the Future*, Mary L. Good, ed., Am. Chem. Soc., Washington, DC, 45–60.

Kilman, S., and S. Warren, 1998, Old rivals fight for new turf-biotech crops, B1, B6 (May 27).

Krinski, T. L., 1992, Emerging polymeric materials based on soy protein, in *Emerging Technologies for Materials and Chemicals from Biomass*, R. M. Rowell, T. P. Schultz, and R. Narayan, ACS Symp. Ser. 476, Am. Chem. Soc., 299–312.

Kucera, E., 1965, Contribution to the theory of chromatography: Linear non-equilibrium elution chromatography, *J. Chromatogr.*, 19, 237–248.

Kuchel, P. W., and G. B. Ralston, 1988, *Schaum's Outline of Theory and Problems of Biochemistry*, McGraw-Hill, New York, 57–59.

Lewin, B. 1997, *Genes VI*, Oxford University Press, Oxford.

Maachupalli-Reddy, J., Brian D. Kelley, and E. DeBernardez Clark, 1997, Effect of inclusion body contaminants on oxidative renaturation of hen egg white lysozyme, *Biotechnol. Progr.*, 13(2), 144–150.

Meister, A., 1988, Glutathione metabolism and its selective modification, *J. Biol. Chem.*, 263(33), 17205–17208.

Melcher, R., and A. Barrett, 1998, Grains that taste like meat?, *Business Week*, 3579, 44.

Morrison, R. T., and R. N. Boyd, 1973, *Organic Chemistry*, 3rd ed., Allyn and Bacon, Boston.

Murray, R. K., P. A. Mayes, D. K. Granner, and V. W. Rodwell, 1990, *Harper's Biochemistry*, 22nd ed., Appleton and Lange, Norwalk, CT, 21–57, 124–127, 530–532.

Nass, K. K., 1988, Representation of the solubility behavior of amino acids in water, *AIChE J.*, 34(8), 1257–1266.

News Item, 1998, Paclitaxel production, marketing heats up, *Chem. Eng. News*, 76(24), 11–12.

News Item, 1992, Metabolism of anticancer drug Taxol to be studied, *Chem. Eng. News*, 70(6), 24.

Nozaki, Y., and C. Tanford, 1971, The solubility of amino acids and two glycine peptides in aqueous ethanol and dioxane solutions, *J. Biol. Chem.*, 246(7), 2211–2217.

Orella, C. J., and D. J. Kirwin, 1989, The solubility of amino acids in mixtures of water and alophatic alcohols, *Biotechnol. Progr.*, 5(3), 89–91.

Pearlman, R., and T. H. Nguyen, 1989, Formulation strategies for recombinant proteins: Human growth hormone and tissue plasminogen activator, in *Current Communications in Molecular Biology—Therapeutic Peptides and Proteins: Formulation, Delivery, and Targeting*, D. Marshak and D. Liu, eds., Marcel Dekker, NY, 23–30.

Perry, J. H., R. H. Perry, C. H. Chilton, and S. D. Kirkpatrick, 1968, *Chemical Engineers' Handbook*, 4th ed., McGraw-Hill, New York, 3–201.

Petersen, A., 1998, Merck posts disappointing profit and stock sinks 7.1%, *Wall Street J.*, B8 (July 22).

Pruisner, S. B., 1977, Prion diseases and the BSE crisis, *Science*, 278(5336), 245–251.

Righetti, P. G., E. Ginazza, C. Gelfi, and M. Chiari, 1990, Isoelectric focussing, in *Gel Electrophoresis of Proteins: A Practical Approach*, 2nd ed., B. D. Hanes and D. Rickwood, eds., IRL Press, Oxford, 149–216.

Shuler, M. L., and F. Kargi, 1992, *Bioprocess Engineering: Basic Concepts*, Prentice Hall, Englewood Cliffs, NJ, 432.

Tokatlidis, K., S. Salamitou, T. Fujino, P. Beguin, P. Dhurjati, J. Millet, and J.-P. Aubert, 1993, Duplicated segment of Clostridium thermocellum cellulases, *Protein Folding, in vivo and in vitro*, ACS Symp. Ser. 526, J. L. Cleland, ed., Am. Chem. Soc., Washington, DC, 38–45.

Tyn, M. T., and T. W. Gusek, 1990, Prediction of diffusion coefficients of proteins, *Biotechnol. Bioeng.*, 35, 327–338.

Vogel, G., 1997, Pruisner recognized for once-heretical prion theory, *Science*, 278(5336), 214.

Walsh, C. T., L. D. Zydowsky, and F. D. McKeon, 1992, Cyclosporin A, the cyclophilin class of peptidyprolyl isomerases, and blockade of T-cell signal transduction, *J. Biol. Chem.*, 267(19), 13115–13118.

PROBLEMS

4.1 Show how the expression for the "work function" or energy of activation of crystallization given in Eq. (4.75) is obtained.

4.2 Figure 4.29 (from Roettger et al.) shows that the retention of protein when eluted through a hydrophobic interaction chromatography column using eluents of ammonium salts of iodine, chloride, acetate, or sulfate. The retention time (i.e., the time at which the protein peak elutes) increases, respectively, for small amounts of protein injected into a chromatography column packed with Butyl-Toyopearl 650S, and then eluted with 1.0 M concentrations of the salts indicated in Figure 4.29. Explain a probable mechanism that would be consistent with the observed retention behavior. Answer this question by considering salting-in or salting-out effects, chaotropes versus kosmotropes, and hydrophobic interactions.

4.3 Figure 4.30 from Roettger et al., 1990 shows the retention behavior of the protein, myoglobin from a liquid chromatography column, as a function of increasing ammonium salt concentration, for ammonium salts of sulfate, phosphate, acetate, chloride, and iodide. Construct a probable precipitation curves that show solubility of the myoglobin between 0.4 and 1.4 M salt. Assume that solubility is inversely proportional to the retention time of the myoglobin. Give the precipitation curve as a function of ionic strength.

4.4 Which salts are expected to be more effective in promoting solubility of a protein:

a. Na_2SO_4 or NaCl?

b. $CaCl_2$ or NaCl?

Explain your reasoning.

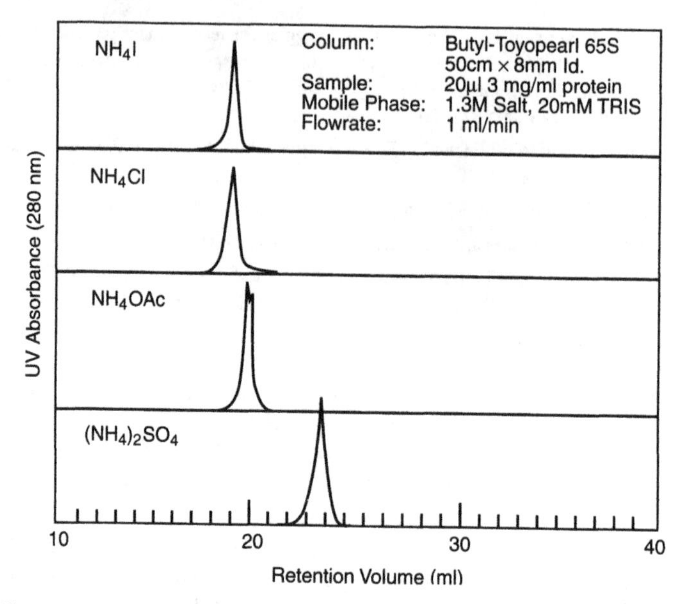

Figure 4.29 Figure for homework problem 4.2 (Roettger et al., with permission).

4.5 Lysozyme is being used as a model protein to test for separation characteristics of a precipitation/adsorption procedure. The lysozyme is first dissolved in sodium sulfate at pH = 8. The solution is then passed over an anion exchange column where the sulfate ion exchanges for the chloride ion. After a brief period of operation, the column plugs up, and flow is no longer possible. Subsequent analysis of the run shows that a concentrated salt solution was used to condition the column, prior to the introduction of the lysozyme, and significant salt had been left in the column, so the ionic strengths of the Na_2SO_4 and NaCl are about the same during the early and middle parts of the run. Explain.

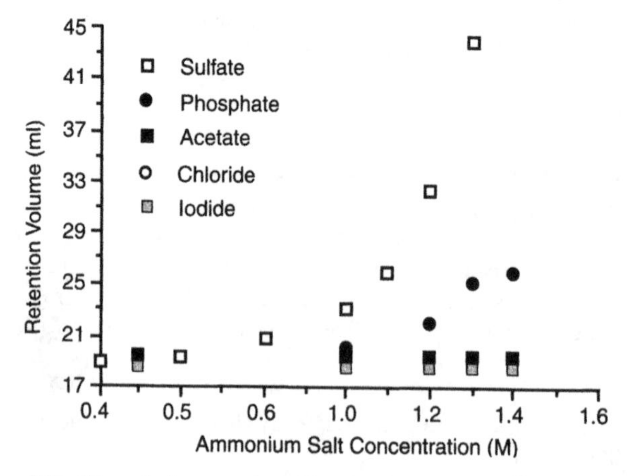

Figure 4.30 Figure for homework problem 4.3 (Roettger et al., with permission).

4.6 The precipitation of a protein is carried out using either methanol or acetone.

a. In the first experiment the addition of methanol to a solution of protein at room temperature causes the protein to precipitate. The slurry is then centrifuged to recover the solid (protein). The liquid is decanted, and the solid plug remains. Buffer is added. However, the protein does not redissolve when buffer is added. The protein was denatured.

A second experiment is carried out in which the same amount of methanol is added, but in this case both the protein solution and methanol are precooled to 4°C, and the precipitate is removed by centrifugation that is carried out at 4°C. The protein redissolves when buffer is added to the precipitate. Give an explanation in terms of the free energy change that occurs upon addition of methanol.

b. What is the relative change in free energy when equal mole amounts of acetone or methanol are added to a protein solution? Which solvent would be expected to be more effective in precipitating protein?

4.7 The complex solution behavior of proteins is addressed by Cohn and Edsall (1943, p. 587) found that egg albumin solubility in 2.14 M ammonium sulfate decreases from 3.18 and 2.09 g/L between 0° and 12°C, and increases from 1.81 to 2.24 g/L between 20 and 29°C. Estimate the accompanying heat of solutions. How would heat of solution effects impact recovery of therapeutically useful protein generated in egg (yolks)?

4.8 The curve of K_S in terms of σ for a hypothetical protein is given in Figure 4.12. What is the ratio of solubility of this protein in NaCl to its solubility in $CaCl_2$ solution if the concentrations are 1 M for each of the salts. If the salt concentration were reduced to 0.8 M and the objective were to achieve the same protein solubility as observed for NaCl, which salt would you recommend to be tried to obtain this solubility?

4.9 A crystallizer is seeded with crystals under conditions at which nucleation from the supersaturated solution itself is negligible. The objective is to grow larger crystals having a uniform size. The properties for this crystallization are given below. How long is required to increase the weight of the seed crystals by 50%?

$A_{TO} = 0.040 \text{ m}^2/\text{kg solvent}$

$W_O = 80.0$

$\Delta C_O = 0.012 \dfrac{\text{kg solute}}{\text{kg solvent}}$

$W_S = 8000 \text{ kg}$

$a = 2.5$ (lumped constant)

$b = 100$ (lumped constant)

CHAPTER 5

PRINCIPLES OF LIQUID CHROMATOGRAPHY

INTRODUCTION

This chapter introduces the different modes of chromatography and the governing equations for chromatography carried out with solutes that exhibit linear equilibria. The practical aspects of packing, operating, and monitoring liquid chromatography separations, and the theory for analyzing elution profiles for different types of chromatographic separations under linear conditions are described here.

5.1 LIQUID CHROMATOGRAPHY SYSTEMS ARE CLASSIFIED BY PRESSURES THAT CHARACTERIZE THEIR OPERATION: HPLC, LPLC, AND MPLC

Chromatography columns are packed beds of small particles that cause the mobile phase pumped through the column to experience a pressure drop. In the case where the pressure drop is 200 to several thousand psig, the column is labeled as belonging to a high-pressure liquid chromatography system (abbreviated HPLC). Pressures below 50 psig are considered to be representative of low-pressure liquid chromatography (or LPLC). Liquid chromatography carried out at pressures between these two values are referred to as moderate or medium pressure LC (MPLC). HPLC is also used to denote high-performance liquid chromatography. High performance designates a chromatography system that gives rapid and complete resolution of multiple components. High pressure is one characteristic that is usually associated with high performance, and hence has resulted in the use of the term HPLC to denote both high performance and high pressure. However, not all high-performance systems are high pressure, nor is high-pressure operation synonymous with high performance.

The author thanks Craig Keim and Nathan Mosier for their reading and helpful suggestions regarding this chapter while they were graduate students at Purdue University.

Pressure is a major issue in chromatography development and scale-up, since the stationary phases that often give the best results are ones that consist of particles having the smallest size but that give the highest pressure drop. Pressure drops that can be tolerated at a process scale are often significantly lower than those typical of laboratory scale systems. Hence packing characteristics of the column, and the effect of particle size on pressure drop, must be considered in chromatography design, with the material that has the smallest average particle size at a tolerable pressure drop often being viewed as the most attractive technical choice.

5.2 THIS CHAPTER PRESENTS THE PRINCIPLES AND PRACTICES OF ANALYZING AND SCALING-UP CHROMATOGRAPHY COLUMN PERFORMANCE FROM EXPERIMENTAL MEASUREMENTS

Methods and equations for extrapolating experimentally measured chromatographic separations obtained at one set of conditions to different flowrates, column dimensions, sample volumes, and concentrations and different particle sizes of stationary phases form the basis of scaling-up chromatographic separations. The purpose of this chapter is to present the process engineering principles upon which unit operations for chromatography and adsorption are based. A brief comparison of chromatography to liquid and vapor phase adsorption is also provided, since engineering students are likely to be familiar with process-scale adsorptive separation methods and analytical-scale chromatographic systems. Such a comparison is intended to provide the student with a reference point that relates the new material covered in this chapter to their existing knowledge of adsorptive separations.

Steps that precede adsorption and chromatography are briefly mentioned in order to define how the streams obtained that enter a chromatography or adsorption column. However, since 50% of the purification costs of many bioproducts lie in the chromatography steps (Knight, 1990), and since the unit operations of distillation and extraction are given in other books (King, 1971; Treybal, 1968; Wankat, 1990), the focus of this chapter is on chromatography. The manner in which various types of stationary phases might be selected for chromatography are discussed in Chapter 8, while the impacts of biochemistry and molecular biology on development of new affinity chromatography separations are discussed in Chapter 9. The methods for selecting the sequence of separation steps are introduced in this chapter. Further discussion on the impact of regulatory considerations, and the need for, and identification of orthogonal separation sequences is given in Chapter 8.

This chapter presents the theory for analysis of scale-up (or scale-down) of given characteristics of adsorption and chromatography columns when the appropriate stationary phase with the desired chemistry and selectivity has been selected, tested, and shown to give the desired purification. The modeling of thermodynamic and rate processes of chromatographic and adsorptive separations are discussed at a macroscopic rather than molecular level. The goal is to determine the impact of column dimensions, flowrates, peak resolution, and column plate counts for a given separation. Correlations are presented that enable calculation of changes in resolution as a function of particle size of the stationary phase, volume or concentration of the sample, and flowrate. These are related to resolution (i.e., extent of purification), throughput, and cost. The models developed in this chapter, combined with knowledge of the molecular basis of the separation discussed in Chapters 8 and 9, are intended to provide guidance for the process of purification development, and the application of principles of chromatographic separations to practical purification processes.

Definitions

Chromatographic separations are classified as being either isocratic or gradient. An isocratic separation is carried out at constant conditions: flowrate, eluent composition, and temperature. Gradient chromatography is carried out at changing conditions. The most common parameter that is changed in gradient chromatography is the composition of the eluent as a function of time. Definitions and nomenclature are discussed in the following sections so that the student can become familiar with the language of chromatography.

5.3 LIQUID CHROMATOGRAPHY SYSTEMS CONSIST OF COLUMNS, INJECTORS, DETECTORS, PUMPS, FRACTION COLLECTORS, AND STATIONARY AND MOBILE PHASES

Liquid chromatography is based on the reversible partitioning of a molecule between the liquid phase in which it is dissolved and a bed of solid particles packed in a column. Differences in partitioning between different molecules, as the liquid phase moves a pulse of sample containing these molecules through the column, results in their separation. The particles are held in a stationary configuration by a frit or filter that is placed at each end of the column. Hence the packed bed of adsorbent particles is referred to as the stationary phase or a fixed bed. Since the stationary phase forms the medium over which chromatographic separations occur, it is also known as the separations or chromatographic medium, or chromatographic stationary phase. The instrument consists of an eluent reservoir, pump, injection valve for introducing the sample, a column, and a detector at the outlet of the column (Figures 5.1 to 5.3). The injection valve (Figure 5.2) is replaced by a feed tank and switching valve in process systems (Figure 5.4). Low dead volume fittings (Figure 5.3) minimize mixing.

Detectors commonly used in chromatography are ultraviolet spectrophotometers, conductivity meters, and differential refractometers. Ultraviolet (UV) spectrophotometers and refrac-

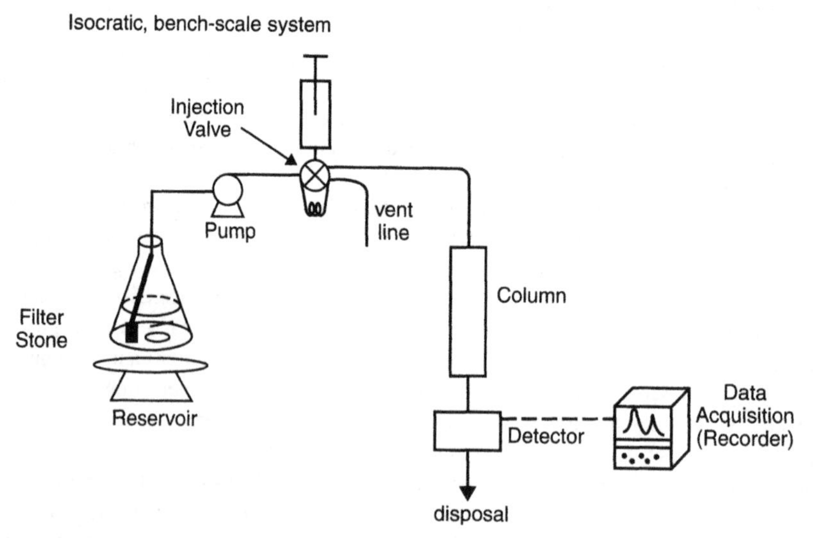

Figure 5.1 Schematic representation of a single-pump isocratic system with injector, column, and detector.

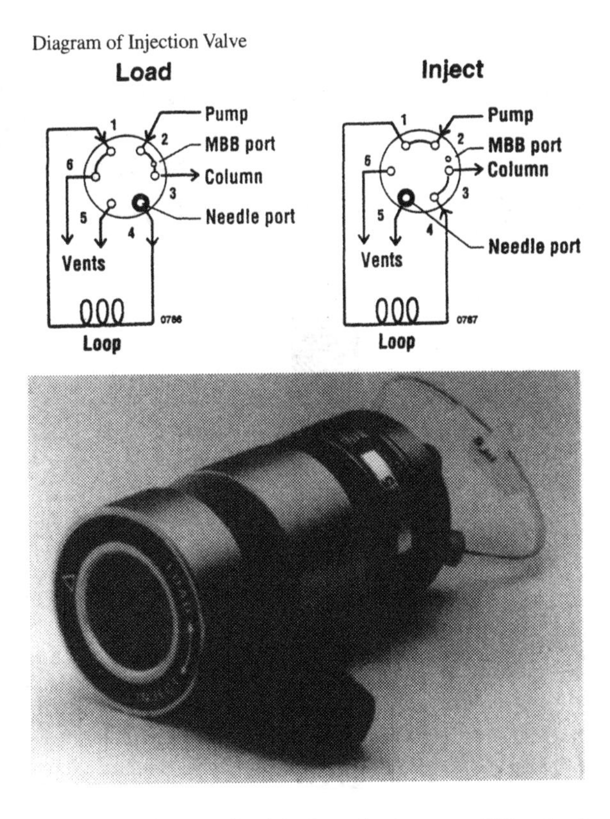

Figure 5.2 Diagram of an injection valve (courtesy of Rheodyne).

Low dead volume fittings

Zero Dead-Volume (ZDV)
In theory, a fitting which adds no extra-column volume to the system when making a connection. Commercially manufactured ZDV fittings actually have a finite, but insignificant, dead-volume in terms of practical chromatography.

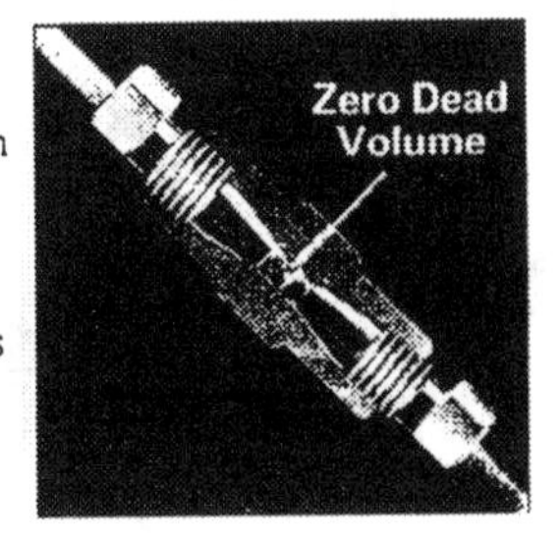

Figure 5.3 Zero dead volume fittings minimize extra column dispersion (courtesy of UpChurch Scientific, 1998).

V_{col} = volume of an empty column[2]

The extraparticulate void fraction is determined by measuring the time that is required for a pulse of a soluble, high molecular weight, nonadsorbed solute to pass through the column at a constant eluent flowrate. Blue dextran, a water-soluble dextran of molecular weight of 2,000,000 D, which is derivatized with a blue dye, is often used for this purpose. Since this dextran has a high molecular weight, it is excluded from stationary phase particles whose pore diameters are less than 200 Å. The elution volume, V_o, of an excluded component, such as blue dextran, is given by

$$V_o = q_{flow} t_o \tag{5.2}$$

where q_{flow} is the flowrate in mls/min and t_o is the time that has elapsed between when a small pulse of blue dextran is injected into the column, and when its peak center elutes from the column. The subscript "o" denotes quantities associated with the extraparticulate void volume. This is illustrated in Figure 5.6

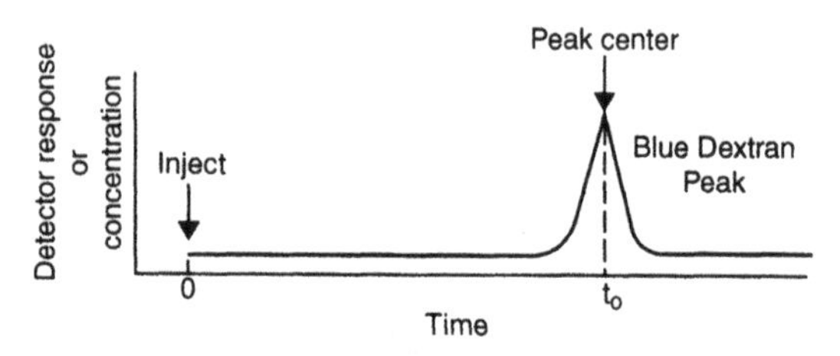

Figure 5.6 Schematic representation of excluded peak eluting at void volume of column.

5.4 THE TARGET MOLECULE IS THE MOLECULE THAT IS TO BE RECOVERED IN A PURIFIED FORM

The partitioning effects of chromatographic and adsorptive separations are based on differential binding or molecular sieving of the target molecule relative to other molecules with respect to the stationary phase. Ion exchange, reversed and normal phase, hydrophobic interaction, and affinity chromatography achieve separations based on differences in adsorption or binding of different molecules on the stationary phase. Size exclusion (also known as gel permeation) chromatography achieves partitioning based on a differential sieving effect in which small molecules penetrate the adsorbent or stationary phase particles, to the exclusion of the larger molecules that are too large to penetrate at least some of the pores in the stationary phase. The smaller molecules spend more time in a column packed with the stationary phase than the larger ones. Hence the smaller molecules elute later and thereby separate from the larger molecules, which elute first.

The selection of stationary phases is carried out so that the target molecule is purified based on different principles in different purification steps. If the first step is ion exchange chroma-

[2]The nominal column volume is given by the diameter and length of the column. Slight variations in the column diameter along the length of the column can result in a column volume that is different than that calculated from its diameter measured at the inlet and length. Hence a preferred method for measuring column volume is to fill the column with water, empty it, and measure the volume of water that is collected. While inconvenient or impractical for process scale columns or prepacked columns, this approach is useful for columns packed and used in the laboratory, since it gives a more accurate measure of column columes from which the void fractions are calculated. Void fractions that are widely used in various types of chromatography equations and models.

tography, which separates based on charge, the next step may be hydrophobic interaction, which separates based on hydrophobic interactions. Consequently the target molecule may adsorb on the stationary phase in one step, and then appear in the flow-through in the next. Flow-through refers to the eluent that flows through the column and carries unretained components from the injected sample with it.

5.5 THE NOMENCLATURE OF CHROMATOGRAPHY IS SUMMARIZED IN SCHEMATIC DIAGRAMS

Figures 5.7 and 5.8 show the terminology associated with the column. Figures 5.9 to 5.11 illustrate the concepts of gradient chromatography and adsorptive separations.

The frits in Figure 5.7 are sintered metal or polymeric filters with particle size cut-offs ranging from 2 to 20 microns. These are placed at both ends of the column in order to retain the stationary phase, which is indicated by the shaded region. The average particle sizes of stationary phases that consist of spherical particles will range from 5 to 300 microns. The smaller particle sizes are used in small diameter analytical chromatography columns, while the larger particles are used for larger diameter preparative or process-scale chromatography columns.

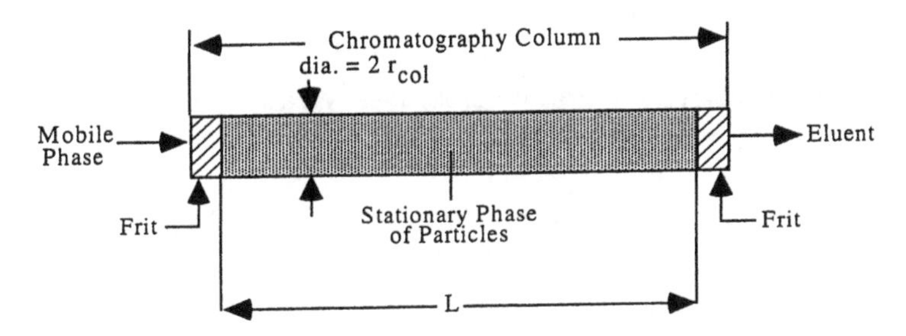

Figure 5.7 Schematic diagram showing terminology for liquid chromatography column.

The graphical notation in Figures 5.7 and 5.8 represents a cylindrical chromatography column as a horizontal rectangle. The concentration profiles of individual bands are given as a function of distance from the inlet of the column or as a function of time as indicated in Figure 5.8. Analytical scale columns are often operated in a horizontal or inclined mode, while most preparative or process scale columns are operated in a vertical configuration.[3] The distance from the column inlet is denoted by z, while the column's bed length is L. While one solute is more strongly retained than the other, both elute at isocratic conditions of constant flowrate and constant mobile phase composition. The less strongly retained component (speckled band in Figure 5.8) elutes before the more strongly retained one. The shading that represents particles that make up the stationary phase is not indicated in Figures 5.8 to 5.11 so that solute bands can be clearly visualized for explanation purposes.

Chromatographic media are often selected to strongly retain one solute only and to weakly retain the others. A change in the composition of the mobile phase is then required in order to

[3]Analytical scale columns are defined here as chromatography columns that have stationary phase bed volumes of 1 to 20 mL. Stationary phase bed volumes of 20 to 1000 mL are associated with preparative columns. Columns with bed volumes that are greater than 1 L are process-scale. The stationary phase bed volume is given by the volume calculated from the bed dimensions ($= \pi r^2 L$), and it does not include the frits. The bed volume (corresponds to empty column volume, excluding the frits) and is measured as described in footnote 2.

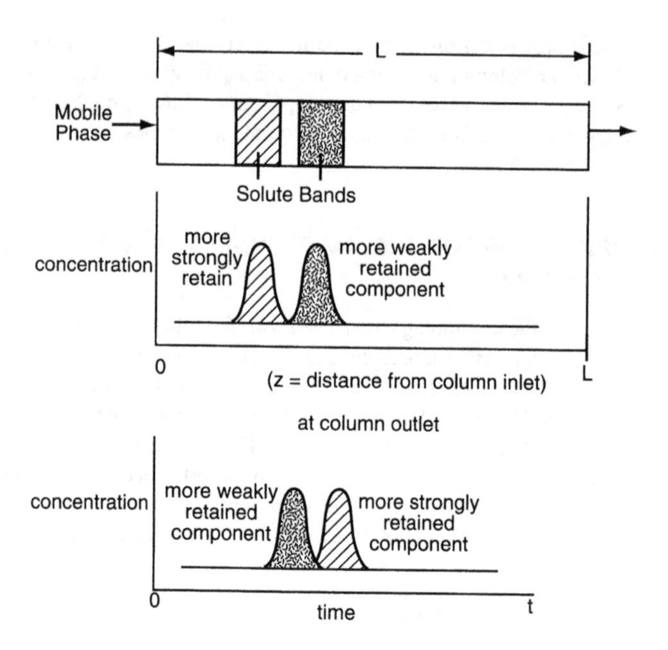

Figure 5.8 Schematic diagram showing relation of eluting peaks to the position of the peaks within the column.

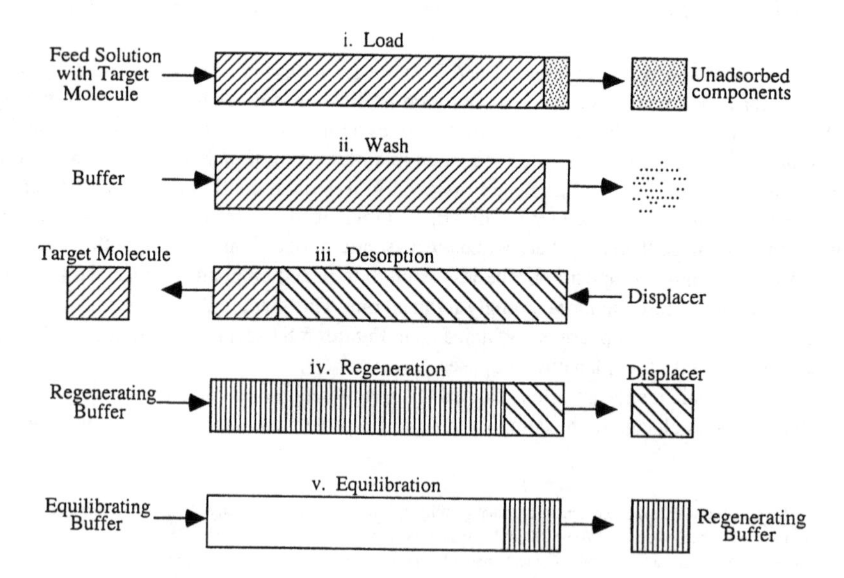

Figure 5.9 Schematic diagram showing steps that make up an adsorption/desorption process.

desorb the solute that has partitioned into the stationary phase and is strongly held there. For example, in the case of ion exchange chromatography a concentration of 0.1 to 2 M NaCl may be used to displace an adsorbed protein. The Cl^- ion exchanges for a negatively charged protein in anion exchange chromatography, while Na^+ exchanges for a positively charged protein in cation exchange chromatography. Multiple charged sites on the surface of the protein can be involved in the ion exchange process.

Chromatography separates components in samples that are injected as pulses of feed. When the feed is pushed continuously through the column until breakthrough of the strongly adsorbed component occurs, the mode of operation is referred to as adsorption instead of chromatography. The steps in carrying out a single run in adsorption are illustrated in Figure 5.9. These consist of (i) loading, (ii) washing to remove unretained (unadsorbed) components, (iii) desorbing the retained target, molecule using a displacer or modifier, (iv) regenerating to clean trace amounts of solute and other adsorbed components from the stationary phase, and then (v) re-equilibrating with the same buffer that is used to dissolve the solute in the feed solution for the loading step. The equilibrating buffer washes the regenerating buffer out of the column. The dissolved ion, chemical species, or organic solvent that is used to desorb the target molecule is referred to as the displacer or modifier. Examples of displacers or modifiers are NaCl or KCl in ion exchange chromatography, methyl mannoside in affinity chromatography, and acetonitrile in reversed phase chromatography.

Affinity or on–off chromatography differs from adsorption, since the feed is introduced into the column as a pulse and the displacer is passed through the stationary phase in the same direction as is used for loading the column with the retained solute. When displacer of a constant concentration is pumped through the column, desorption is said to occur by a step gradient, since there is a sharp change in concentration of the displacer upon initiation of the desorption (Figure 5.10).

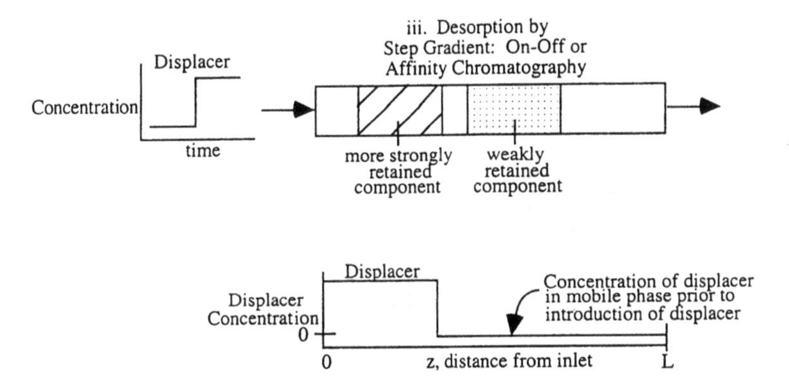

Figure 5.10 Schematic diagram illustrating concept of on-off chromatography.

5.6 GRADIENT CHROMATOGRAPHY IS A FORM OF ADSORPTION

Gradient chromatography differs from adsorption by the manner in which the sample is applied to the column, and from on–off chromatography by the sample size and the manner in which the concentration of the displacer varies during the course of a run (compare Figures 5.9, 5.10, and 5.11). In gradient chromatography the sample is applied as a small pulse, and is followed immediately by the introduction of displacer. There is usually no wash step between introduc-

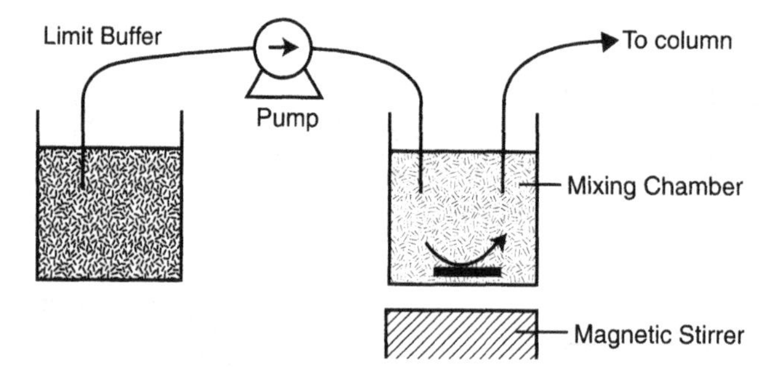

Figure 5.15 Gradient-forming system consisting of two connected beakers and a pump between the beakers that is used to control rate of addition of limit buffer.

nol/water azeotrope. The vapor is passed through a bed of previously dried desiccant,[4] which retains the water while allowing the ethanol to pass (Figure 5.14). The water-free ethanol water vapors are then condensed. The appearance of water in the product coincides with a temperature rise near the outlet of the bed. The feed is stopped and the adsorbed water is removed by passing a noncondensable gas such as low humidity, CO_2 or N_2 over the bed (Ladisch et al., 1984; Lee et al., 1991; Westgate et al., 1992; Beery et al., 1998).

5.7 GRADIENTS ARE FORMED BY COMBINING TWO OR MORE LIQUID BUFFERS TO GIVE A TIME-VARYING CHANGE IN DISPLACER CONCENTRATION

A gradient system has several eluent tanks or containers whose contents are mixed in controlled proportions to obtain a displacer concentration that changes with time. Several methods can be used to obtain the desired gradient. The simplest laboratory apparatus consists of two beakers connected as shown in Figures 5.15 or 5.16. The limit buffer (left side of Figure 5.15) flows into the mixing chamber at the same time that the mobile phase is pumped from the mixing chamber into the column. The adjustment of flowrate R_1 (denotes flowrate of limit buffer to mixing chamber) relative to R_2 (denotes flowrate of liquid from mixing chamber to column) will give different gradients of displacer concentration as shown in Figure 5.16.

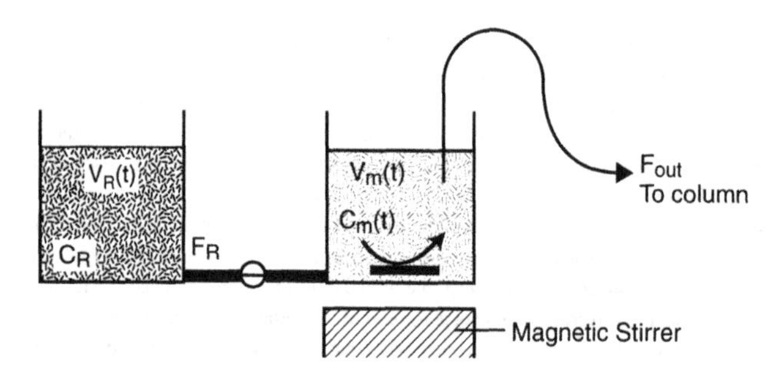

Figure 5.16 Gravity-fed mixing chamber gives linear gradient.

The volume of eluent for gradient elution typically corresponds to about three to five bed volumes of mobile phase per run. A convex gradient will cause weakly retained peaks that elute close together early in the chromatogram. For example, Figure 5.17 shows gradients for columns that have been equilibrated with the starting composition of buffer from the mixing chamber. Peaks that are more retained and elute later in the chromatogram will elute further apart as shown in Figure 5.17. The concave gradient of Figure 5.18 will cause the opposite effect as il-

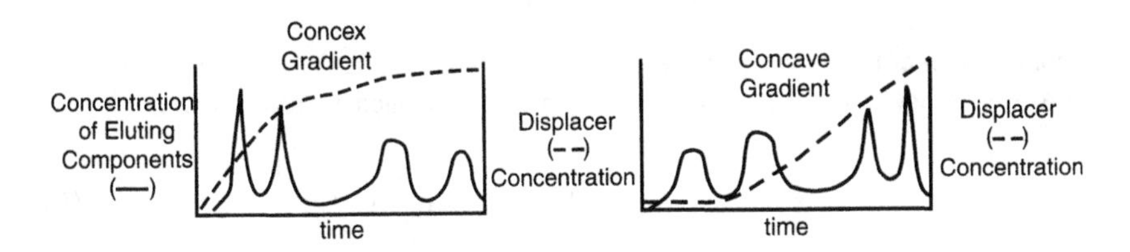

Figure 5.17 Schematic diagram showing convex and concave gradients.

lustrated on the right side of Figure 5.17, while the linear gradient will give an elution pattern that is between the two. The limit buffer is given its name because it represents the ending (or limiting) concentration of the displacer of the mobile phase at the end of the run (Roe, 1989).

Velayudhan (1989) reviewed the derivation of an equation that gives the gradient as a function of concentrations, volumes, and flowrates of the system in Figure 5.16. When the reservoir and the mixing chamber are open, and at the same (atmospheric) pressure, the levels in the two containers remain equal. Hence each unit of volume of mobile phase that is pumped from the mixing chamber can be assumed to consist of equal parts of solution from the reservoir and the mixing chamber, respectively. A material balance around the reservoir and mixing is given by

$$F_R C_R - F_{out} C_M(t) = \frac{d}{dt}[V_M(t) C_M(t)] \tag{5.3}$$

where

F_R = flow from reservoir, mL/min (a constant)
C_R = concentration of modifier in reservoir (a constant), mg/mL
F_{out} = flowrate of mobile phase from mixing chamber, mL/min
$C_M(t)$ = concentration of modifier in mobile phase leaving the mixing chamber, mg/mL
$V_M(t)$ = volume of mobile phase in mixing chamber

Since the levels between the two reservoirs remain constant, the volumes in the two reservoirs are the same, although they both change with time:

$$V_M(t) = V_R(t) \tag{5.4}$$

where V_R is the volume of liquid in the reservoir. An integral material balance can be used to specify the overall volume of the system at any point in time:

$$V_R(t) + V_m(t) + \int_0^t F_{mixer,out} dt = V_{R,o} + V_{M,o} \tag{5.5}$$

where

$V_{R,o}$ = initial volume of limit buffer in reservoir, mL
$V_{M,o}$ = initial volume of mobile phase in mixing chamber, mL

The combination of the mobile phase volumes in the two chambers is the total volume of the system. The change in volume as a function of time, t, is obtained by differentiating Eq. (5.5):

$$\frac{dV_R(t)}{dt} + \frac{dV_m(t)}{dt} + F_{out} = 0 \tag{5.6}$$

When subject to Eq. (5.4), the material balance for the system is

$$\frac{dV_R(t)}{dt} = \frac{dV_m(t)}{dt} = \frac{-F_{out}}{2} \tag{5.7}$$

The volume of the mobile phase in the mixing chamber starting at $t = 0$ follows from Eq. (5.7):

$$\int_{V_{m,o}}^{V_m(t)} \frac{dV_m(t)}{dt} = -\frac{F_{out}}{2} \quad \text{or} \quad V_m(t) = V_{m,o} - \frac{F_{out}}{2} t \tag{5.8}$$

The change in the volume of mobile phase in the limit buffer reservoir is given by

$$\frac{dV_R}{dt} = -F_R \tag{5.9}$$

and the combination of Eqs. (5.9) and (5.7) gives

$$F_R = \frac{F_{out}}{2} \tag{5.10}$$

The flowrate from the reservoir with the limit buffer is half of the flowrate from the mixing chamber. Substitution of expressions for Eqs. (5.8) and (5.10) into Eq. (5.3) gives a differential equation that describes the rate of change of the mobile phase:

$$\frac{F_{out}}{2} C_R - F_{out} C_M(t) = \frac{d}{dt}\left[\left(V_{M,o} - \frac{F_{out}}{2} t\right) C_M(t)\right] \tag{5.11}$$

Upon rearrangement, Eq. (5.11) becomes a linear, first-order ordinary differential equation:

$$\left(V_{M,o} - \frac{F_{out}}{2} t\right) \frac{dC_M(t)}{dt} + F_{out} C_M(t) = \frac{F_{out}}{2} C_R \tag{5.12}$$

Solution of Eq. (5.12) gives

$$C_M(t) = \left[\frac{V_{M,o} - \frac{F_{out}}{2}t}{V_{M,o}}\right](C_{M,o} - C_R) + C_R = C_{M,o} + (C_R - C_{M,o})\frac{F_{out}}{2V_{M,o}}t \quad (5.13)$$

Equation (5.13) shows that the gradient forming system of Figure 5.16 will give a linear change in concentration with the slope of the gradient being determined by the difference in concentrations of the limit buffer reservoir and the ratio of the outlet flowrate to the initial volume of the mobile phase. Velayudhan (1989) points out that the gradient will deviate from linear behavior as the volume of the mobile phase in the mixer decreases, and the vortex formed by the mixing action becomes a large fraction of the total volume. The formation of convex or concave gradients requires that the volumes of the chambers be allowed to change at different rates, as represented in Figure 5.18.

Example 2. Generating a Linear Gradient

Two buffers containing acetonitrile are to be combined to give a linear gradient starting at 10% acetonitrile, upon injection of the sample, and to attain 55% acetonitrile after 60 minutes. If the mobile phase flowrate to the column is 2 mL/min, and the initial volumes of buffer in each of the two reservoirs are 200 mL, (1) can this gradient be achieved? (2) If not, what is the flowrate required if the limit buffer is set at 90% acetonitrile? (3) For this gradient, estimate the times when the four components of Example 1 would elute from the column. (4) Briefly describe how a linear gradient can be achieved if two pumps were used to pump out the reservoirs separately, with the two streams being mixed as shown in Figure 5.19, for a constant combined flowrate of 3.75 mL/min?

Solutions

1. *Can the gradient be achieved?* The initial acetonitrile concentration of the mixing chamber buffer must be 10% acetonitrile, as specified by the initial inlet concentration. Use Eq. (5.13) to calculate the initial concentration of the limit buffer. From the information given,

   $C_{M,o} = 100$ g/L

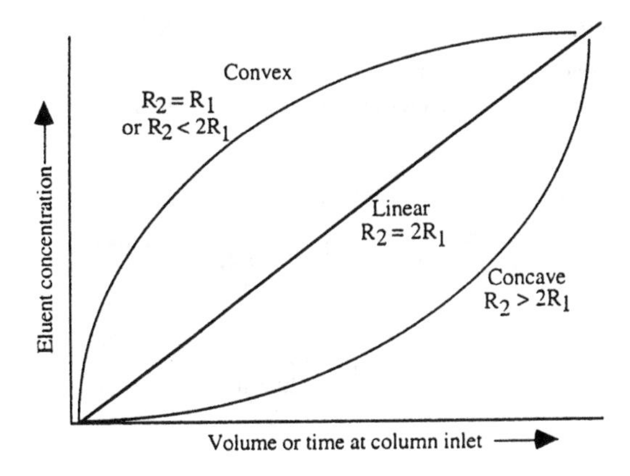

Figure 5.18 Gradients generated by different rates of flow: R_1 = flowrate of limit buffer to mixing chamber and R_2 = flowrate of mixed buffer from mixing chamber to column (from Roe, 1989, figs. 15 and 16, and table 8, p. 194–195, Harris, with the permission of Oxford University Press).

$C_M = 550$ g/L

$V_{M,o} = 200$ mL

$F_{out} = 2$ mL/min

$t = 60$

From Eq. (5.13),

$$C_R = C_{M,o} + \frac{C_M(t) - C_{M,o}}{\left(\dfrac{F_{out}t}{2V_{M,o}}\right)}$$

$$C_R = 100 + \frac{550 - 100}{(2 \cdot 60)/(2 \cdot 200)} = \boxed{1600\,\frac{g}{L}} \quad \text{(not reasonable)} \tag{5.14}$$

No, the gradient cannot be achieved (a limit buffer concentration of 1600 g/L that is greater than 100% acetonitrile is not physically reasonable). Either a faster flowrate or longer time is needed.

2. *Calculate the flowrate to the column.* From the information given

$$C_R = 900 \text{ g/L}$$

Then from Eq. (5.13),

$$F_{out} = \frac{(C_M(t) - C_{M,o})}{(t/2V_{M,o})(C_R - C_{M,o})} = \frac{550 - 100}{(60/2 \cdot 200)(900 - 100)} = \boxed{3.75\,\frac{mL}{min}}$$

3. *Calculate expected elution times.* From Example 1, the first peak elutes at the void volume and three peaks elute at acetonitrile concentrations of:

Peak 2 250 g/L

Peak 3 300 g/L

Peak 4 520 g/L

Since the column extraparticulate void fraction is not given, the time at which the first peak elutes is not known. The elution time for peaks 2 to 4 is

$$t_2 = \frac{C_m(t) - C_{m,o}}{\dfrac{F_{out}}{2V_{m,o}}(C_R - C_{m,o})} = \frac{250 - 100}{\dfrac{3 \cdot 75}{2 \cdot 200}(900 - 100)} \tag{5.15}$$

$$t_2 = \frac{150}{7.5} = \boxed{20 \text{ min}}$$

Similarly

$$t_3 = \frac{300 - 100}{7.5} = \boxed{26.7 \text{ min}}$$

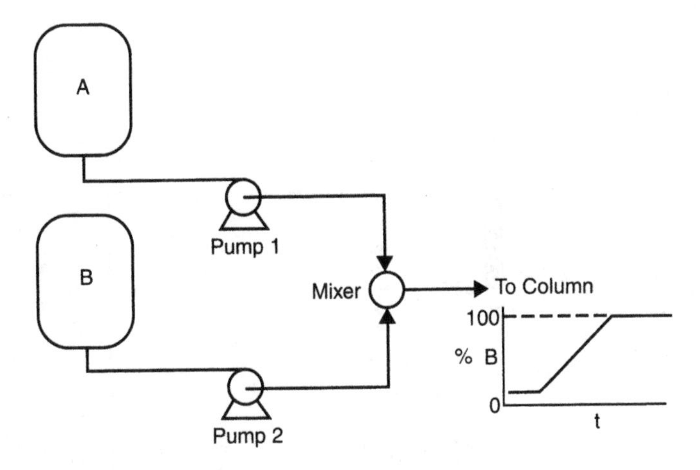

Figure 5.19 Gradient formed by mixing streams from two different reservoirs.

$$t_4 = \frac{520 - 100}{7.5} = \boxed{56 \text{ min}}$$

4. At a constant flowrate to the column, the flowrate of pump 1 (initial acetonitrile concentration) would need to decrease, while the flowrate of pump 2 (limit acetonitrile

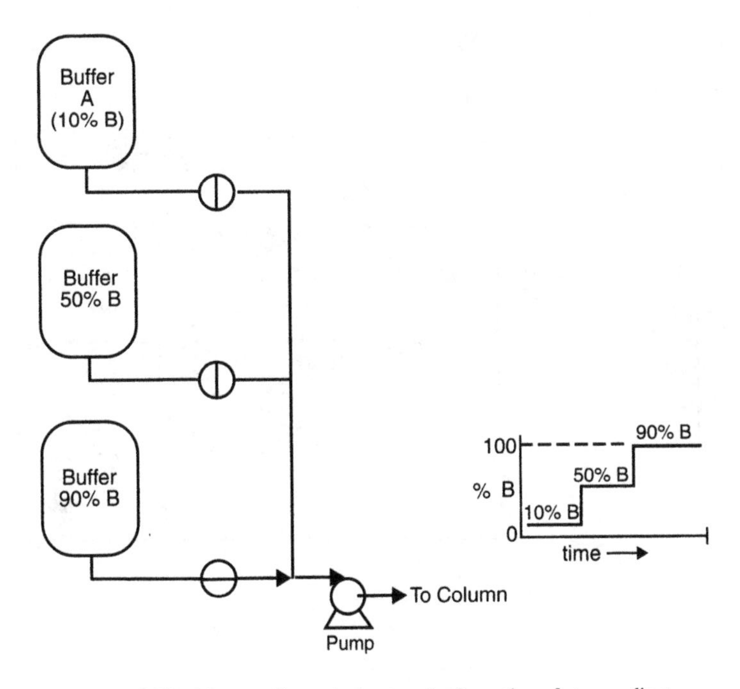

Figure 5.20 Diagram illustrating system for formation of step gradient.

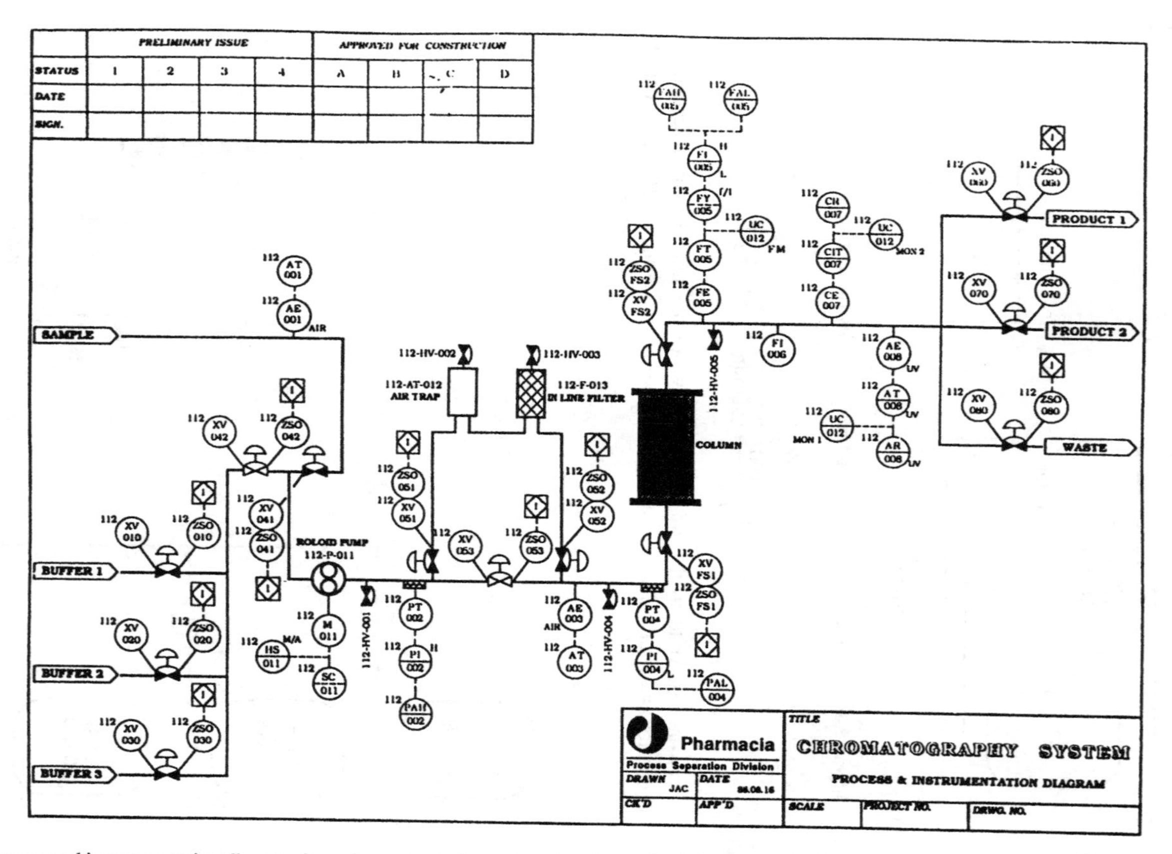

Figure 5.21 Process and instrumentation diagram for a chromatography system showing multiple buffers, a single pump, an in-line air trap and filter, column, and product lines (from Sofer and Nystrom, 1989, fig. 28, p. 71, with permission by Academic Press).

concentration) would need to increase, with both pumps changing their rate in a continuous manner so that a constant overall flowrate is maintained.

An alternate approach to gradient formation involves multiple pumps and reservoirs, with the flowrates of the two pumps being changed in a controlled manner to achieve the desired composition, as shown in Figure 5.19. This type of system is commonly used at the analytical and preparative scale, since various types of gradients can be readily attained using computer controlled pumps.

Gradient chromatography on a process-scale frequently utilizes step gradients (see Figure 5.20). Step gradients are better defined and less subject to variability from one run to the next compared to continuously changing gradients since only one pump is required to generate a step gradient. Automated valves switch from one buffer to the next on a predetermined, timed sequence, in order to achieve rapid changes in displacer concentration as shown in Figure 5.20. Since these increases approximate steps, this mode of gradient formation is referred to as giving a step gradient. The process and instrumentation diagram in Figure 5.21 illustrates a piping layout of this type of system (Sofer and Nystrom, 1989).

Packing Columns

The method by which the stationary phase is packed into the column is critical to achieving a uniform bed and good separations. The basic methods of slurry-packing liquid chromatography columns, and determining the column's efficiency based on the plate height concept, are discussed in this section.

5.8 LIQUID CHROMATOGRAPHY COLUMNS ARE PACKED USING LIQUID SLURRIES

The major method by which analytical, preparative, and process-scale columns are packed involves slurrying the resin and then pumping the slurry into the column using a packer bulb or other type of design that provides extra volume above the bed (Lin et al., 1988). The apparatus for packing an analytical scale columns is illustrated in Figure 5.22. The stationary phase is slurried in buffer or water, loaded into the packer bulb, and pumped into the column under a constant pressure and not constant flowrate (Lin et al., 1988). As the column is packed, the flowrate will decrease, since the void volume between particles decreases, and the bed itself compacts. The pressure is set at a limit that is below the crushing strength of the particles. Packing is complete when a constant flowrate of mobile phase is attained at the packing pressure. A similar method can be used for preparative columns, although the free volume above the bed is provided by the column itself. In the case process-scale column such as shown in Figure 5.23, once the column is packed, a plunger to which the inlet distributor plate is attached will be lowered onto the bed in order to eliminate the dead space above it.

Large columns may also be packed by pumping a slurry of the stationary phase from a separate tank into the column rather than load it through an empty column section or dead space. Alternately, large particle size polymeric stationary phases (100 to 300 micron size) whose moisture is at 40% to 50% may be packed as free-flowing particles. After the stationary phase is poured into the column, the mobile phase is pumped into the column in an upflow direction to expand and backwash the bed as well as to bring the particles to an equilibrium moisture. The bed is then allowed to settle, and the column is operated in a downflow direction to compact the bed. Alternately, some columns are packed by simply pumping stationary phase slurry, and then compacting it without backwashing. Once the packing of the bed has been completed, the

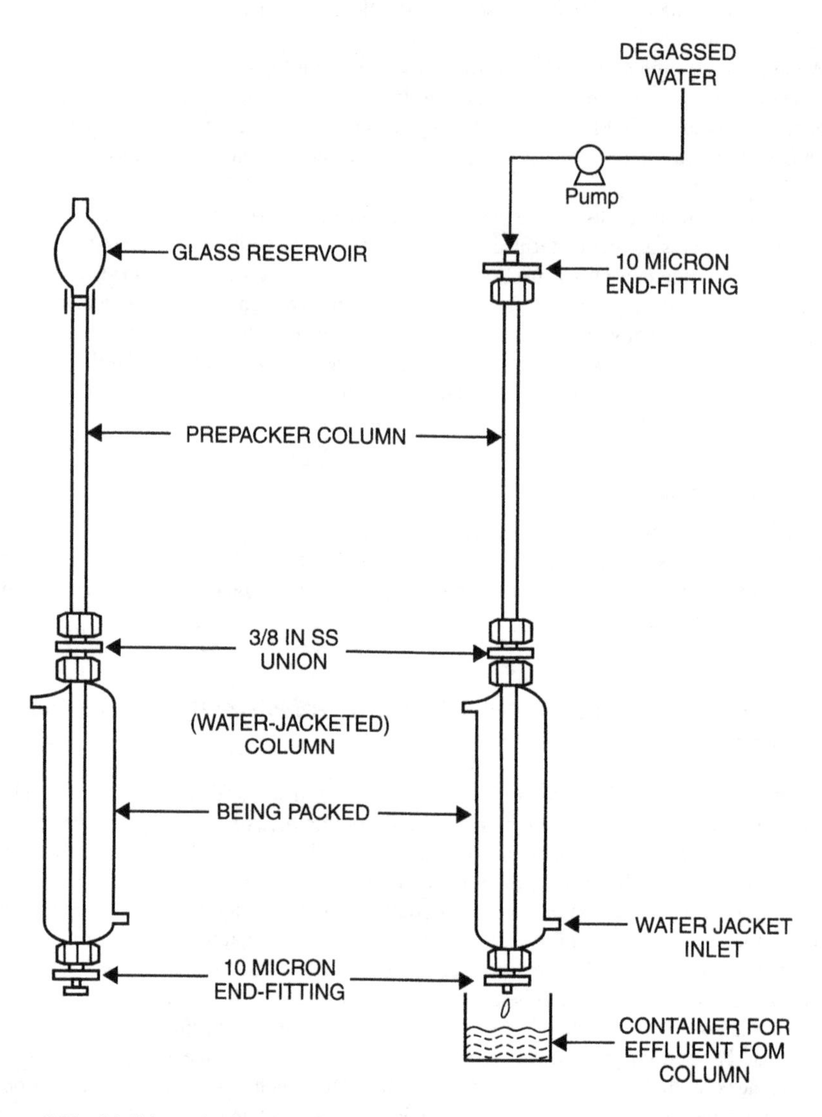

Figure 5.22 (*a*) Schematic diagram of vertical slurry packing system for ion exchange resins and polymeric materials used for carbohydrate chromatography. The steps in packing an analytical column are fill column with the solvent (i.e. the mobile phase), connect the packer bulb, fill the packer bulb with a slurry containing the resin (stationary phase), fill the packer bulb with solvent, and then connect the apparatus to the pump (shown in (*b*)). (*b*) For the packing of resin used in carbohydrate separations, the temperature is raised from ambient (20°C) to 80°C over a two-hour period. These columns are usually operated at 80 to 85°C. Once the column is packed, it is removed from the packer bulb, and capped with a column end fitting consisting of frit that holds the resin within the column.

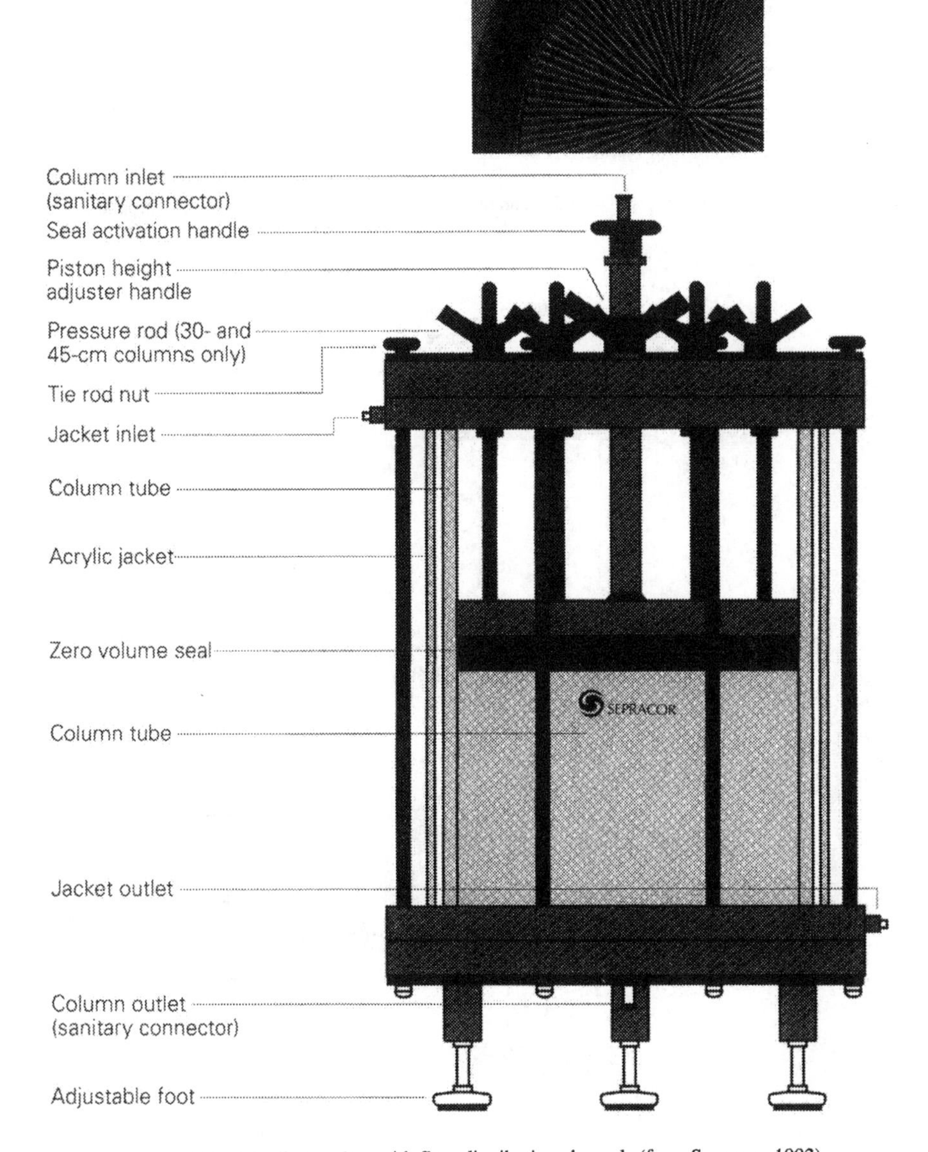

Figure 5.23 Distributor plate with flow distribution channels (from Sepracor, 1992).

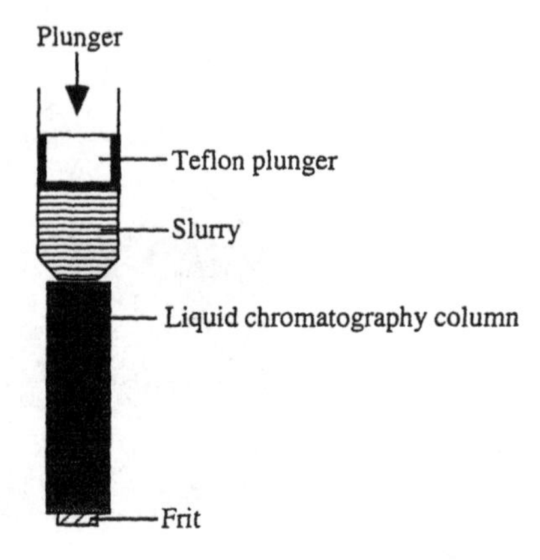

Figure 5.24 Schematic diagram of axial compression system for packing a liquid chromatography column.

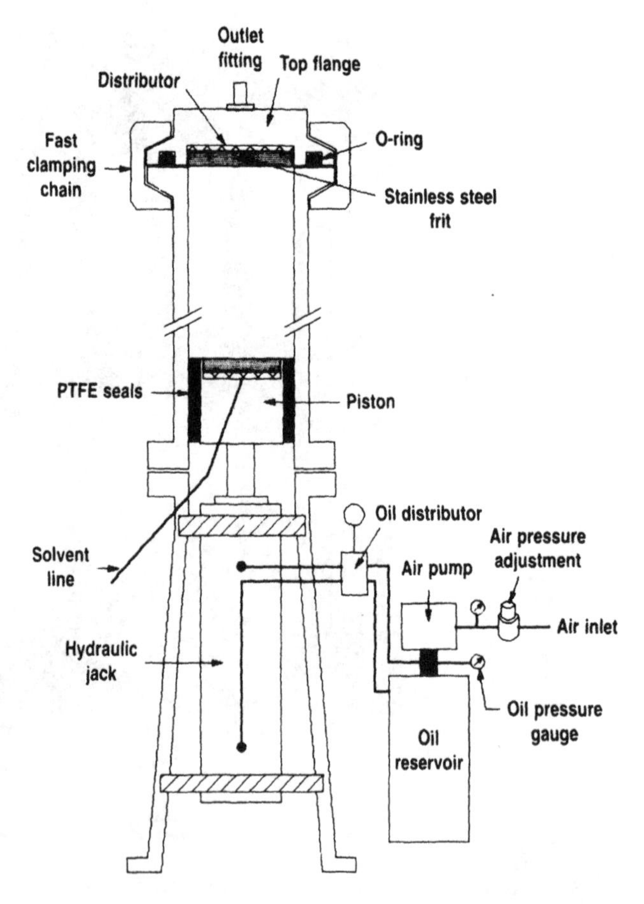

Figure 5.25 Schematic diagram showing mechanical elements of an axial compression system (from Prochrom, 1993).

distributor plate is lowered and positioned on top of the stationary phase. An industrial column that is associated with this type of operation is shown in Figure 5.23.

Piston based column-packing procedures for slurries of silica particles make it possible to efficiently carry out silica-based chromatography on a large scale. A slurry is used because a dry material is difficult to pack and handling of large volumes of a small particle medium that resembles a dust (particle sizes of 20 to 40 microns) can be challenging. Slurry packing by axial compression is carried out in several ways (Colin et al., 1990; Martin and Shalon, 1985). For silica particles, one approach utilizes a large-diameter column section as a slurry reservoir, with a teflon piston pushing the slurry into a smaller diameter column (i.e., the chromatography column) as illustrated in Figure 5.24 (Martin and Shalon, 1985). Once the column is packed, the column inlet is closed with a frit, and the column is ready for use. This type of packing procedure is referred to as axial compression. Since the piston is removed from the column after packing, and the pressure due to compression by the piston is absent during operation of the column.

An alternate piston packing method uses the column as both the slurry receiver and the packed column, and the piston is left in place during operation of the column as illustrated in Figure 5.25 (Colin et al., 1990; Prochrom, 1993). The slurry is made up in a separate tank and then pumped into the column against a sealed piston located at the bottom of the column. The head plate, which has a frit to retain the stationary phase, is then secured onto the top of the column, and the piston is pushed up. As the slurry is pushed against the frit, solvent is expressed through the solid and the frit and then out of the top of the column. The bed volume and extraparticulate void volume of the stationary phase decreases as the piston moves up (Figure 5.26). Once the column is packed the piston is kept in place to maintain the packing pressure. When the pressure is maintained during the operation of the column, as well as during its packing, the method is referred to as dynamic axial compression.

Once the stationary phase has been compressed to a predetermined pressure drop, the column is ready for use. Feed sample and mobile phase are pumped through the column to carry out liquid chromatographic separations based on the principles described earlier in this chapter.

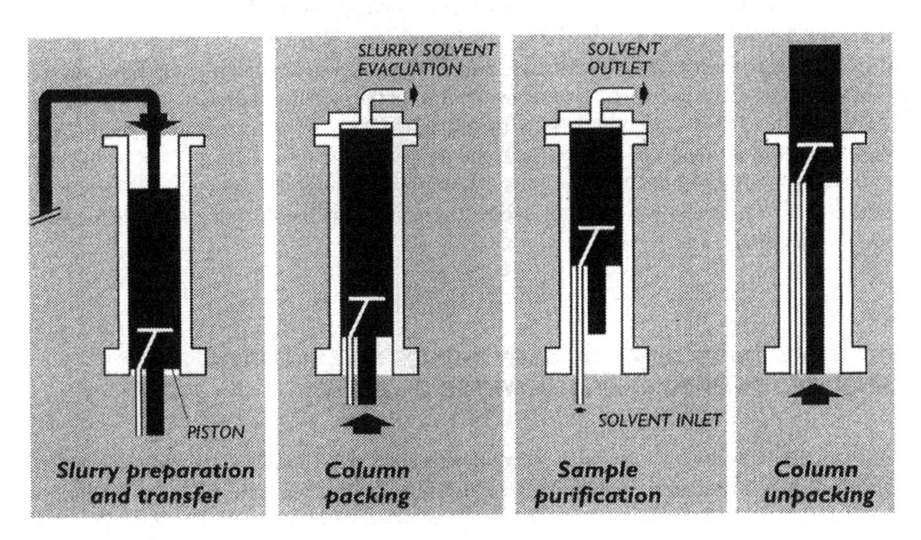

Figure 5.26 Schematic representation of slurry packing, operation, and unpacking of an axial compression chromatography column (from Colin et al., 1990, figs. 1 and 3, p. 302).

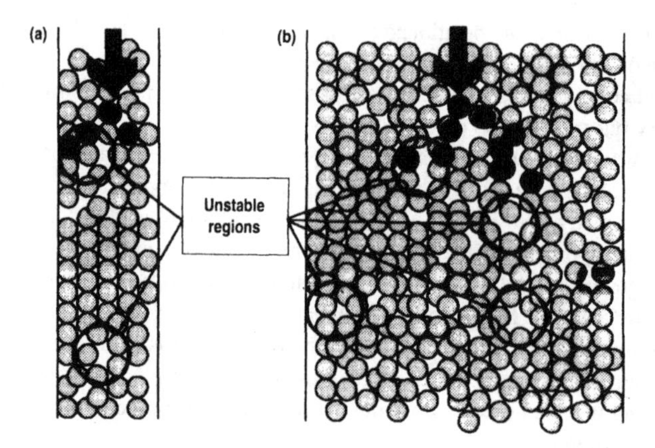

Figure 5.27 Illustration of packed bed irregularities that can lead to column void formation during its operation.

When the stationary phase is at the end of its operational lifetime, an axial compression column is unpacked by removing the headplate, and pushing the piston up to extrude the solid material, as illustrated on the right of Figure 5.26.

The compression of the bed of silica particles using a hydraulically driven piston minimizes the effect of intrabed voids by effectively pressing the particles together at pressures of up to 100 bar. Axial compression methods are associated with silica based supports, since these materials resist crushing. Silica-based materials are commonly used for reversed phase or normal phase chromatography (discussed in Chapter 8). The gels or supports used for size-exclusion chromatography, or the polymeric materials that are typical of ion exchange or hydrophobic interaction chromatography, are generally too soft and lack the compressive strength needed for packing by axial compression systems.

As the column diameter increases, the wall no longer supports the packed bed, and the bed itself can be unstable (Colin et al., 1990; Ladisch and Tsao, 1978). One consequence is that a dense plug may form at the outlet of the column. This plug can effectively stop flow, since it causes the pressure drop to exceed operational limits of the system's design (Ladisch and Tsao, 1978). Another effect of bed instability is the collapse of bridged particle regions (Figure 5.27) that form when a column is slurry packed, and the generation of a void volume near the inlet of the bed, or formation of channels in the interior of the bed as the particles in these regions settle. Intracolumn voids cause mixing and/or channeling which result in band broadening and loss of resolution.

5.9 SOME TYPES OF STATIONARY PHASES UNDERGO SIGNIFICANT SWELLING WHEN HYDRATED IN WATER OR BUFFER

An example of swelling of a dry stationary phase is given by Sephadex G-75 particles that will almost double their diameter within 90 seconds of being placed in water or aqueous buffer of swelling (Figure 5.28) (Monke et al., 1990). The characteristic relaxation time is on the order of 15 seconds, where relaxation time, τ, as a function of particle diameter, d_p, can be estimated from

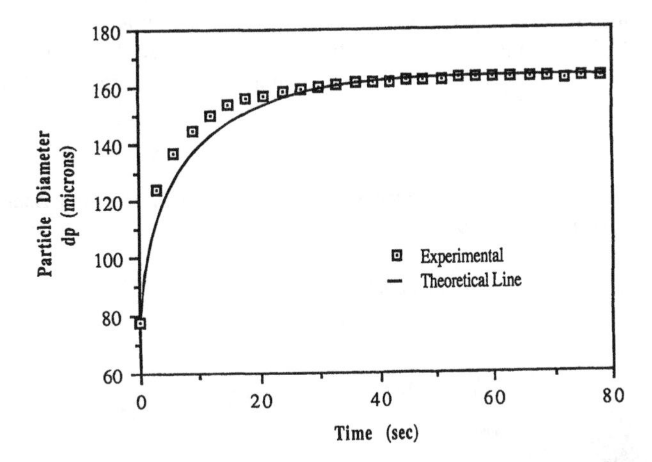

Figure 5.28 Comparison of the Tanaka-Fillmore theory (solid line—from Eq. (5.17)) to experimental data for swelling of Sephadex G-75 in water for an initial particle diameter (dry) of 78.1 μm (from Monke et al., 1990, fig. 6, p. 380 with permission by American Chemical Society and American Institute of Chemical Engineers).

$$\tau = \frac{d_p^2}{(4\pi^2 D)} \tag{5.16}$$

where D is the diffusion coefficient of water in the gel (= $6.3 \cdot 10^{-7}$ cm^2/s) as determined from experimental measurements. The diameter of a gel particle, d_p, as a function of time, t, is given by

$$\boxed{d_p = d_p^{\infty} - (d_p^{\infty} - d_p^{o})\left(\frac{6}{\pi^2}\exp\frac{-t}{\tau}\right)} \tag{5.17}$$

where

d_p^{∞} = final particle diameter at a long time
d_p^{o} = particle diameter at $t = 0$
τ = relaxation time, s

Equation (5.17) has the same form as the leading term in the solution to the analogous scalar diffusion equation (Crank, 1975; Monke et al., 1990). The observation that the theory of Tanaka and Fillmore, Eq. (5.17), is consistent with experimental data for swelling of Sephadex G-75 in water (compare calculated line against the data points in Figure 5.28 suggests that rapid packing can be achieved by mixing the Sephadex gel with water or buffer, holding for about two minutes and then pumping the slurry into a column (Monke et al., 1990).

Example 3. Stationary Phase Swelling

A dextran gel crosslinked with epichlorohydran is supplied so that the dry particles have diameters ranging from 78 to 107 microns, and swollen sizes of 163 to 232 microns respectively. How much time is required for the smallest and largest particles to swell to 99.9% of the final size? 99.99% of their final size? What is the recommended hold time if the particles are to attain 99.99% of their size before being pumped into the chromatography column? Assume that $D = 5 \cdot 10^{-7}$ cm^2/s.

Solution

Use Eqs. (5.16) and (5.17). First calculate τ using Eq. (5.16). The numerical results are shown in Eqs. (5.18) and (5.19):

For $d_p = 78$ μm (use size of dry beads),

$$\tau_{78} = \frac{d_p^2}{4\pi^2 D} = \frac{\left(78\ \mu\text{m} \cdot 10^{-4}\ \text{cm}/\mu\text{m}\right)^2}{4\pi^2(5 \cdot 10^{-7})} = 3.08\ \text{s} \tag{5.18}$$

For $d_p = 107$ μm,

$$\tau_{107} = \frac{(107 \cdot 10^{-4})^2}{4\pi^2(5 \cdot 10^{-7})} = 5.80\ \text{s} \tag{5.19}$$

Calculate t using Eq. (5.17):

$$t = \tau_{78}\left(-\ln \frac{\pi^2(d_p^\infty - d_p)}{6(d_p^\infty - d_p^0)}\right) = \tau_{78} \ln\left[\frac{6(d_p^\infty - d_p^0)}{\pi^2(d_p^\infty - d_p)}\right]$$

The numerical results are shown in Eqs. (5.18) and (5.19) for the different particle sizes and extents of swelling:

For 78 μm particle size ($d_p^0 = 78$) for $d_p = 0.999 d_p^\infty$,

$$d_p = 0.999 d_p^\infty = 0.999(163)$$

$$t = 3.08 \ln\left[\frac{6(163 - 78)}{\pi^2(163 - 0.999 \cdot 163)}\right] = \boxed{17.7\ \text{s}} \tag{5.20}$$

For 78 μm particle size at $d_p = 0.9999 d_p^\infty$,

$$t = 3.08 \ln\left[\frac{6(163 - 78)}{\pi^2(163 - 0.9999 \cdot 163)}\right] = \boxed{24.8\ \text{s}} \tag{5.21}$$

For $d_p^0 = 107$ microns,

$$t = 5.80 \ln\left[\frac{6(232 - 107)}{\pi^2 232(1 - 0.999)}\right] = \boxed{33.6\ \text{s}} \tag{5.22}$$

$$t = 5.80 \ln\left[\frac{6(232 - 107)}{\pi^2 232(1 - 0.9999)}\right] = \boxed{46.9 \text{ s}} \tag{5.23}$$

These calculations indicate that a hold time of 47 seconds would be sufficient if the longest time is used as the basis. Other parameters, including settling of a packed bed and equilibration of particles with the running buffer, require hours and hence extend the period of time needed to pack a column.

Particles cause pore diffusional resistances that result in band broadening and overlap of separated peaks. Since the molecules in the moving fluid are transported more quickly than molecules that explore the stagnant fluid in the pores, some molecules are left behind and take longer to pass through the column. This is a mechanism by which small molecules can be separated from larger ones. However, diffusion inside pores also contributes to dispersion and band broadening of the injected pulse. One of the goals of liquid chromatography is to obtain rapid mass transfer in order to achieve sharp peaks. Small particles are desired because the path length, over which a molecule diffuses when it penetrates and then leaves the particle, is shorter, and dispersion (peak broadening) is less than for larger particles. While porous spherical, par-

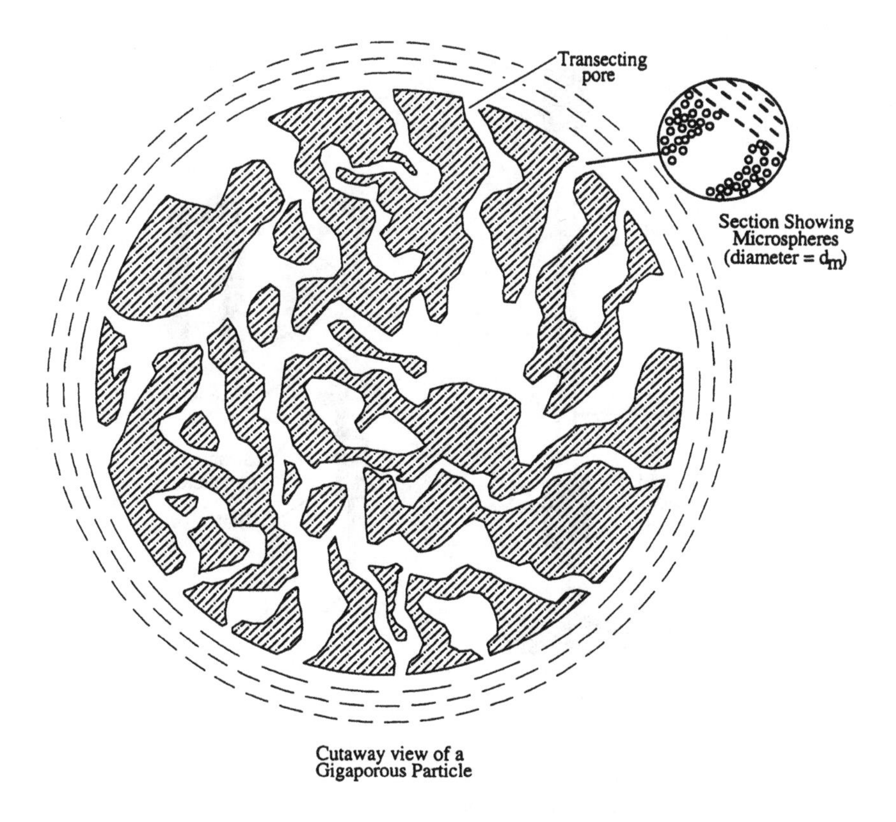

Figure 5.29 Schematic representation of a gigaporous particle, showing the concept of flow through a particle due to transecting pores (from Pfieffer et al., 1996, fig. 1, p. 934). Reproduced with permission of the AIChE. All rights reserved.

ticulate stationary phases are the dominant stationary phases used for analytical, preparative, and process chromatography, stationary phases with reduced porosities are under development.

The opposite approach has also been examined, where particles with transecting pores, namely gigaporous particles (Figure 5.29) have been studied for convective flow through them. If convective flow occurs, large particles consisting of agglomerates of smaller particles would, in effect, increase the ratio of void fraction in which convective flow occurs, to the void fraction in which diffusion is the dominant mechanism. This offers the possibility of more efficient columns through enhanced mass transfer. The concept is presented at this point of the chapter so that the student can be cognizant of a major difference between the gel type size exclusion phase described in Example 3, and a rigid gigaporous stationary phase, discussed below.

5.10 CONVECTIVE FLOW THROUGH GIGAPOROUS PARTICLES WITH TRANSECTING PORES MAY OCCUR AT HIGH PRESSURES

The flow of fluid through individual gigaporous particles of a stationary phase with large transecting pores has been proposed to enable convective transport of the solute through the interior of individual particles if the pressure is sufficiently large. This concept of a gigaporous particle is represented in Figure 5.29. The possibility of flow through was experi-

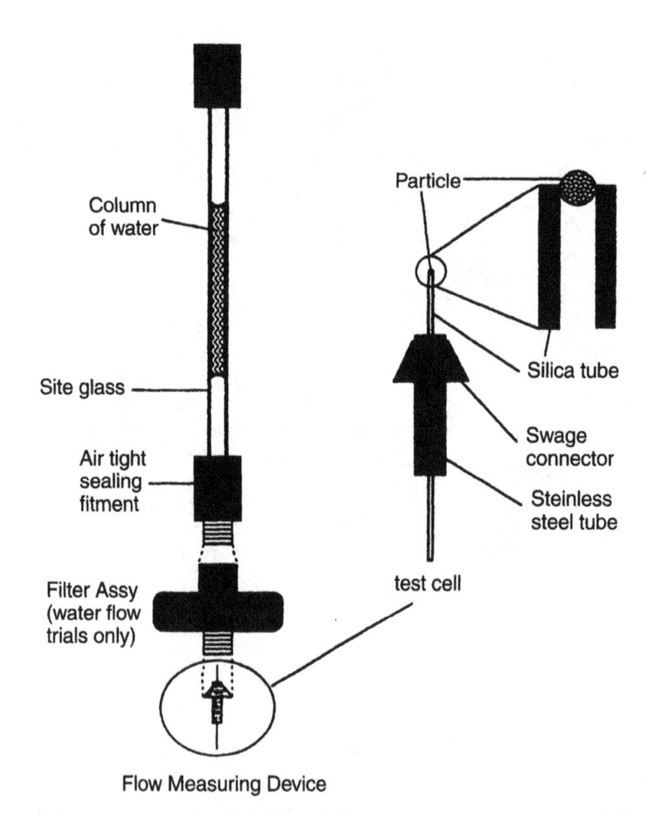

Figure 5.30 Experimental system used to measure flow through individual gigaporous particles (from Pfieffer et al., 1996, fig. 5, p. 936). Reproduced with permission of the AIChE. All rights reserved.

mentally tested by "potting" a particle in a small glass (silica) tube and measuring the flowrates as a function of pressure drop. Figure 5.30 shows the apparatus. Figure 5.31 plots the flowrate through the particle which is in the order of 10^{-3} μL/s at pressure drops of 25 to 35 bar (equivalent to about 360 to 507 psi). Permeability inside of the particle was estimated by the Carman-Kozenzy equation:

$$K = \frac{\varepsilon_p^3 d_m^2}{150(1-\varepsilon_p)^2} \tag{5.24}$$

where

ε_p = particle void fraction inside of particle, as defined by Eq. (5.60) as discussed later in this chapter

d_m = diameter of microspheres that make up the particle, m

K = permeability, m^2

The particle void fraction, inside of particle, is sometimes called in the intraparticle void fraction, and is a function of the size of solute used to measure it. A complete explanation is given in section 5.15. An average measured permeability of 8×10^{-15} m^2 was obtained from the slope of the line of the flowrate vs. pressure drop curves such as the one given in Figure 5.31. The pressure gradient across the particle was related to fluid velocity by

$$v_p = \frac{K}{\mu} \nabla P \tag{5.25}$$

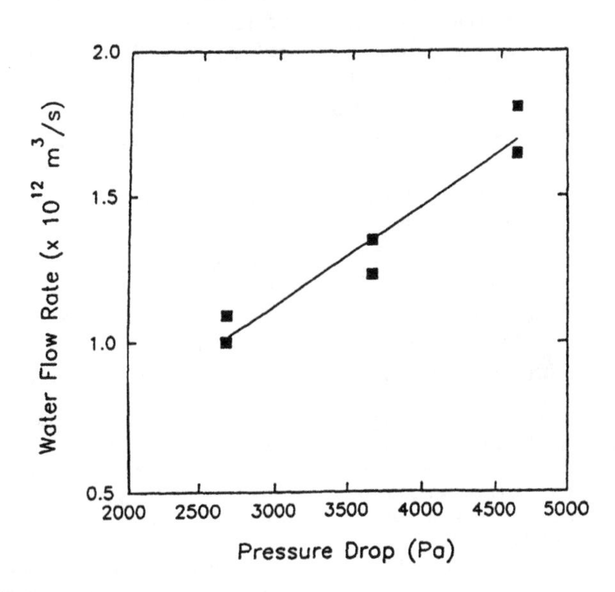

Figure 5.31 Fluid velocity versus pressure curve for a gigaporous particle. Slope of line is K/μ as given by Eq. (5.24) (from Pfieffer et al., 1996).

where

v_p = fluid velocity within the particle
∇P = pressure gradient within the particle
μ = viscosity of fluid

The data showed that flow can be forced through the particles.[5] However, the permeability calculated from experimentally measured $v_p\mu/\nabla p$ using Eq. (5.25) was larger than the permeability calculated from Eq. (5.24), which gave a value on the order of 0.5 to $1 \cdot 10^{-15}$ m^2. Although the permeability of particles can be estimated, precise values of ε_p and d_m, as are required by Eq. (5.24), are unlikely to be available. Therefore experimentally determined values of K based on flow experiments are preferred (Pfeiffer et al., 1996). The data suggest that the intraparticle structure of individual gigaporous particles behave as an inhomogeneous assembly of microparticles.

5.11 PELLICULAR PARTICLES, POLYMER MONOLITHS, ROLLED STATIONARY PHASES, AND BUNDLES OF HOLLOW FIBERS REPRESENT OTHER FORMS OF STATIONARY PHASES

This section on packing columns has described packing and flow characteristics of columns packed with billions of particles. The column domains are defined by both interstitial and intraparticulate pores and void spaces, which contribute to band broadening. A goal of chromatography column development has been to reduce band broadening and achieve sharper peaks by optimizing packing characteristics and improving mass transfer effects. This has resulted in the development of columns in which the flow paths between particles or the intraparticle voids within particles have been reduced. For example, one approach in analytical chromatography uses pellicular particles where the porosity of each particle is limited to an outer shell to reduce the intraparticle diffusion path lengths. Unlike a porous stationary phase, the interior of a pellicular particle has no pores as illustrated in Figure 5.32.

Other forms of stationary phases[6] that give packed columns with reduced interstitial porosities have been reported, and these are briefly summarized below. Stationary phases have not always resulted in sharp peaks, but they are potentially useful because they illustrate concepts that may enable higher flowrates and enhanced mass transfer to be achieved for process-scale chromatography.

In one alternate approach, the chromatography column resembles a single large particle where all the pores are connected, fluid moves by convection in all of the pores, and solute movement is dominated by convection. The column is prepared by filling a tube with a polymerization mixture containing porogens. After the polymerization has been completed a solvent is pumped into the column that dissolves the porogens. The removal of the porogens leave pores in the polymeric matrix. The resulting column consists of a continuous matrix of polymer with channels (i.e., pores) that are generated by the dissolution of the porogens by the solvent (Figure

[5]Similar observations have been reported for hydrodynamic properties of packed textile beds. While most of the fluid will pass through interyarn zones of a fabric, a small portion will flow between fibers and through intrayarn zones. The spaces between fibers would be analogous to the transecting pores of gigoporous particles. Experimental values of K derived from pressure drop versus flow measurements were larger than values of K estimated from the void fractions and physical dimensions of the fabrics. The value of K was in the range of 0.4 to $10 \cdot 10^{-11}$ m^2 for flow of fluid through flat sheets of fabric (van der Breckel and de Jong, 1989).

[6]Five basic techniques are used to prepare the majority of synthetic polymers, according to Svec and Fréchet (1996). These are bulk, solution, dispersion, emulsion, and suspension polymerization. These methods give small particles that contain small channels or pores within the particles into which molecules from the moving fluid can diffuse. However, the pores are too small to allow convective fluid flow.

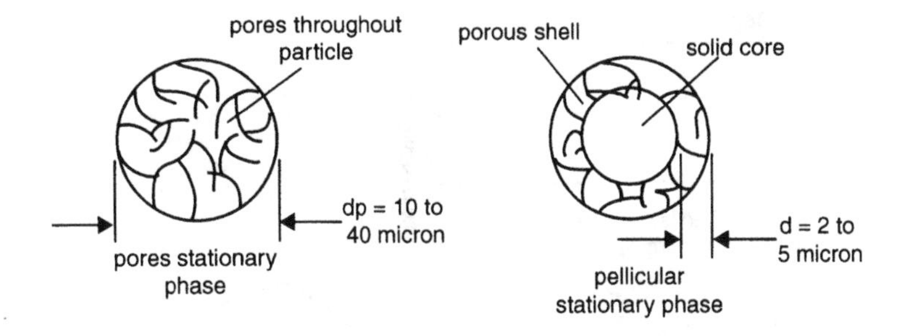

Figure 5.32 Schematic diagram comparing porous to pellicular stationary phases.

5.33). The solvent is washed from the column and displaced with the mobile phase to be used for the chromatographic separations (Hjertén et al., 1991; Svec and Fréchet, 1996).

Under operational conditions, all of the pores in such a column are swept by mobile phase. Diffusional effects are reduced compared to a column packed with porous particulates, since intraparticulate void volumes are absent. This enables solute molecules to be rapidly transported to the surfaces of the stationary phase and rapid separation with less dispersion than would occur for a packed column of particles that contains both stagnant and mobile phase regions. A column of this type with dimensions of 8 × 50 mm separated a three component mixture in less than 60 seconds (Svec and Fréchet, 1996). This column was a single large particle with a bed volume of 2500 μL. This is much larger than an individual particle in a packed bed, where the volume of the particle is $3.35 \cdot 10^{-5}$ μL for a characteristic diameter of 40 microns. Scale-up is an issue, however, since the logistics of making homogeneous, molded cylinders of polymers on 1 to 1000 L scales still need to be proved.

Another type of continuous stationary phase is obtained by rolling a woven matrix into a cylindrical shape and pulling it into a column as was first reported by Yang et al. (1992) for Nomex-Kevlar fabric. Development of the packing technique resulted in the procedure illustrated in Figure 5.34 (Hamaker et al., 1998). In this case the stationary phase was a textile consisting of a 60 : 40 cotton (cellulose): a polyester blend that was prepared as described by Hamaker et al. (1998). The material was then rolled into a cylinder and packed into a glass column as indicated in Figures 5.34 and 5.35. The ends of the fabric were cut to size, and plungers on each end of the column were adjusted so that the distributor was situated on top of the rolled

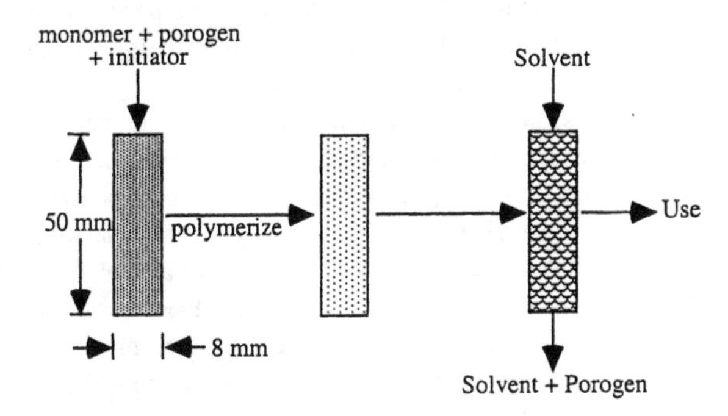

Figure 5.33 Schematic diagram summarizing formation of a polymeric (continuous) stationary phase.

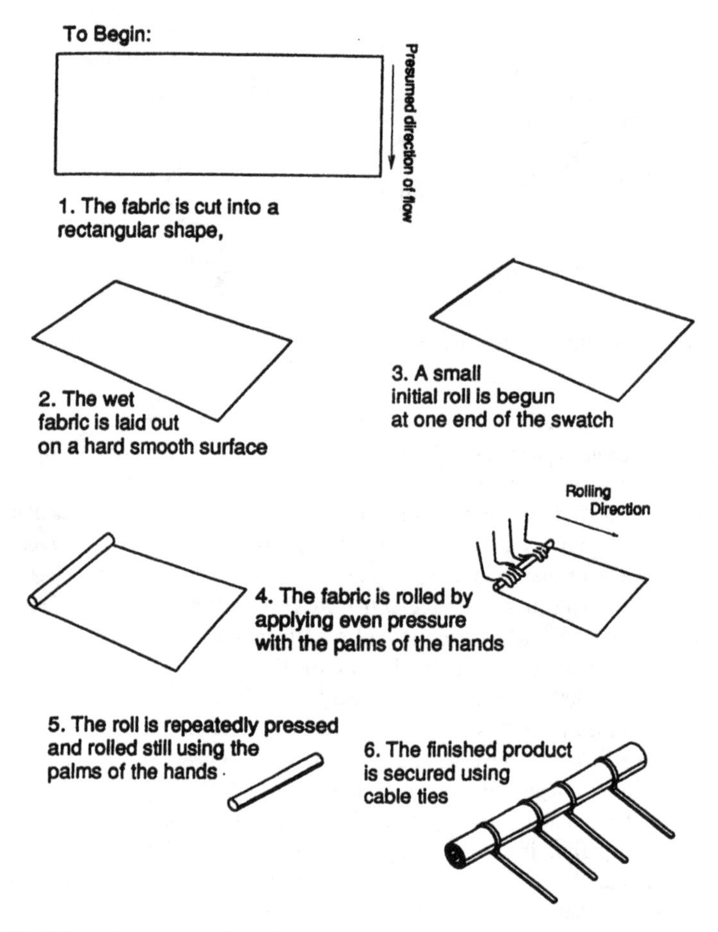

Figure 5.34 Schematic representation of a rolled stationary phase. Preparation and rolling of stationary phase (from Hamaker et al., 1998, fig. 4, p. 23, with permission by American Chemical Society and American Institute of Chemical Engineers).

stationary phase (Figure 5.36). After packing, the column was operated in the same manner as a standard chromatography column, and gave separation of two peaks in 60 seconds (Figure 5.37). This type of column can be scaled up, with the limitation being the length and width of the fabrics that are available or could be made on a large scale.

Forms of stationary phases that have reduced discontinuities, and therefore the potential for enhanced intraparticle mass transfer also include stacked thin membranes, compressed polyacrylamide gels, rolled cellulose sheets and matrices, macroporous beads embedded into a web of polymer (Svec and Fréchet, 1996), or fused silica capillaries (0.53 mm i.d. by 1070 mm) packed with four fibers of cellulose acetate and hollow fibers (Kearny, 1997; Kiso et al., 1986). These stationary phases, like the cast polymer monolith, are characterized by a reduced interstitial porosity; that is to say, there are less voids between the structures that make up the stationary phase. A challenge in design of this type of stationary phase is the optimization of the stationary phase porosity so that the benefits of enhanced mass transfer are not negated by loss of capacity due to loss of accessible surface area as the intraparticulate pores become larger.

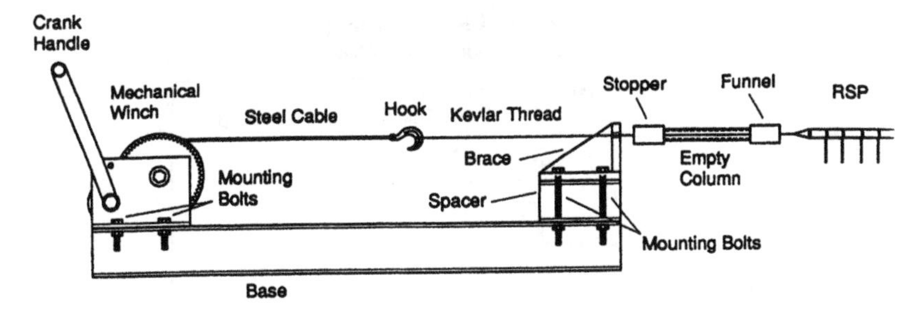

Figure 5.35 Packing (from Hamaker et al., 1998, fig. 5, p. 24, with permission by American Chemical Society and American Institute of Chemical Engineers).

Reduced discontinuities can also be achieved by a system of parallel channels. An example is given by modules consisting of 120 to 27,000 parallel hollow fibers, with each fiber having a diameter of 120 microns. When coated with a stationary phase, a bundle of parallel hollow fibers approximates a separation media with uniform channels (±1% variation) and minimal dispersion. The dispersion is comparable to that predicted by the Aris-Taylor theory for spread of a nonadsorbent peak in a cylindrical tube and is given by

$$\sigma^2 = \frac{d^2 L}{96 D v} \tag{5.26}$$

where

σ^2 = variance of the peak (4σ = width of a symmetrical chromatography peak at its base)

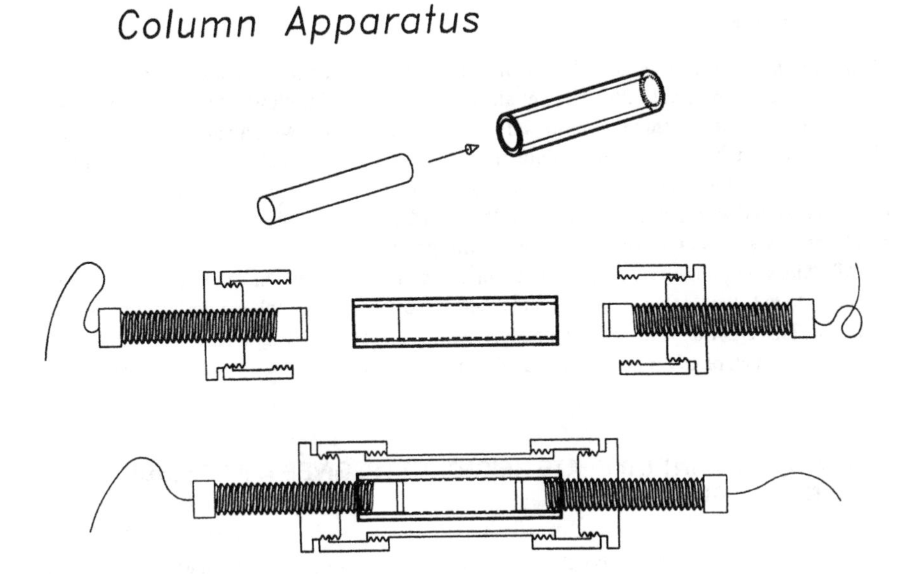

Figure 5.36 Adjusting plungers (from Hamaker et al., 1998, fig. 3, p. 22, with permission by American Chemical Society and American Institute of Chemical Engineers).

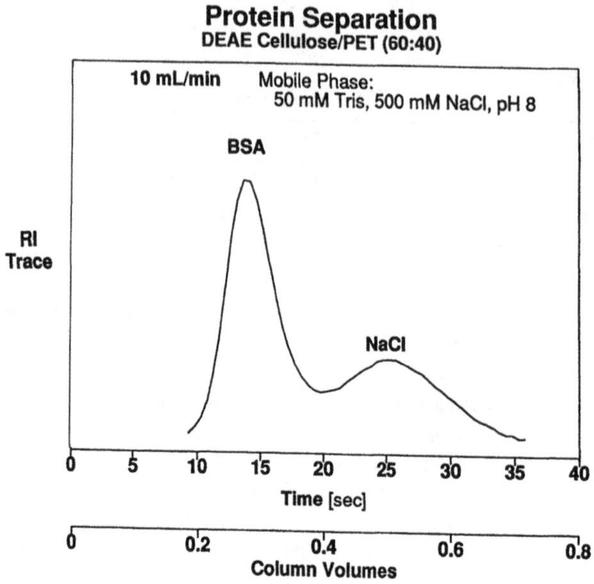

Figure 5.37 Separation of a large molecule (BSA, $MW = 66$ kD) from a small molecule (NaCl) (from Hamaker et al., 1998, fig. 15, p. 29, with permission by American Chemical Society and American Institute of Chemical Engineers).

L = length of a tube or hollow fiber, cm

d = inside diameter of the tube or hollow fiber, cm

D = diffusion coefficient of solute in mobile phase, cm^2/s

v = mobile phase velocity, cm/s

Dispersion is minimized by a high velocity, short channel, length, and a small-diameter flow path, as indicated by Eq. (5.26) (Ding et al., 1989). Individual hollow fibers have the potential to achieve chromatographic separations if their walls are coated with an appropriate stationary phase into which differential partitioning of different molecular species can occur. A challenge, however, is to obtain exactly equivalent pressure drops across a bundle of the hollow fibers so that dispersion due to slight variations in flowrates between the fibers and variations in solute residence times among the different fibers is minimized.

All of these columns require a consistent measure of their dispersion characteristics that allows one column to be compared to another and a numerical criterion for column performance to be established. This criterion is the plate count, N, from which the plate height, H, can be obtained. The methods by which plate counts are measured and calculated are discussed in the next section.

5.12 PLATE COUNT OR PLATE HEIGHT (HETP) GIVES A FIRST INDICATION OF PACKING EFFICIENCY

The dispersion of a pulse of nonadsorbed solute, injected onto the chromatography column, gives a measure of the efficiency of the packed bed or stationary phase. The less the extent of dispersion, the better the efficiency. Dispersion is proportional to the width of the peak that elutes from the outlet of the column, and is measured by plate count, N, based on the peak width of a Gaussian peak, as illustrated in Figure 5.38.

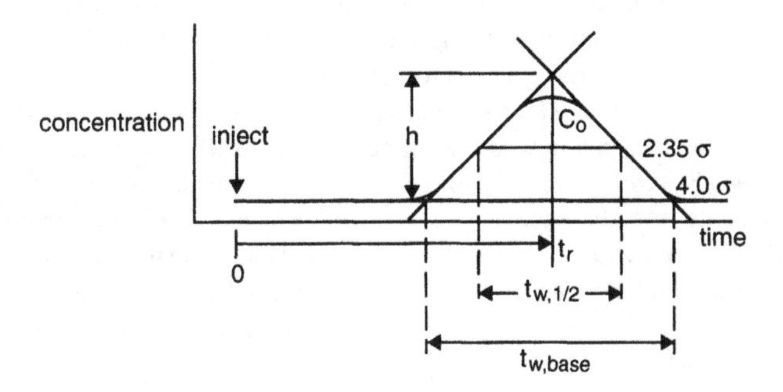

Figure 5.38 Gaussian peak.

The number of plates for the peak represented in Figure 5.38 is given by

$$N = 16\left(\frac{t_r}{t_{w,\text{base}}}\right)^2 = 5.54\left(\frac{t_r}{t_{w,1/2}}\right)^2 \tag{5.27}$$

where

t_r = retention time; the time that has elapsed between when the pulse was injected into the column and the center of the peak elutes from the column

$t_{w,1/2}$ = peak width at half-height, measured in units of time

$t_{w,\text{base}}$ = peak width at the base of the peak, measured in units of time

h = peak height; the distance between the base of the peak and the intersection of tangent lines drawn to the two sides of the peak

The number of plates is proportional to the column efficiency (Grushka et al., 1975). Each plate or height equivalent to one theoretical plate (HETP) for chromatography (Martin and Synge, 1941) is defined as "the thickness of the layer" (the width of a column cross section) "such that the solution issuing from it is in equilibrium with the mean concentration of the solute in the non-mobile phase throughout the layer." Unlike a distillation column, the chromatography column does not physically contain plates. Rather, the plate is a concept used to provide a measure of the peak's spreading relative to the distance that it has migrated (Grushka et al., 1975).

The column is envisioned as being divided into equal volume elements or plates having equal heights, H, as illustrated in Figure 5.39 for a column section showing six plates.

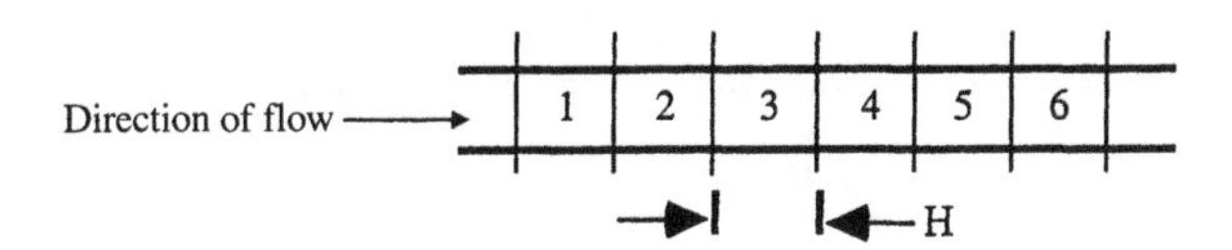

Figure 5.39 Concept of plates in a chromatography column.

The partitioning of the solute between the mobile and stationary phases at each plate is assumed to be so fast that the distribution of the solute between liquid and solid phases is at equilibrium before the mobile phase reaches the next plate (Grushka et al., 1975). The extent of partitioning at each plate is assumed to be constant. In the case of an excluded nonadsorbing solute, the equilibrium assumption is satisfied by definition, since the solute does not partition into the stationary phase and therefore equilibrium at each plate is satisfied.

Other assumptions of the HETP or plate model are that diffusion does not occur between plates in the axial direction, that flow is discontinuous, and that the sample is injected onto the column as a sharp pulse (Dirac delta function). Flow is envisioned to proceed in increments that have volumes ranging from that volume contained by one plate, to a volume that is infinitesimally small (Grushka et al., 1975; Martin and Synge, 1941). Despite these approximations, the plate theory gives a useful measure of column efficiency and forms the basis of experimental interpretation of column performance. A small plate height, or conversely, a large plate count, indicates an efficient column.

5.13 POISSON AND GAUSSIAN DISTRIBUTION EQUATIONS MAY BE USED TO CALCULATE ELUTION PROFILES OF SINGLE CHROMATOGRAPHY PEAKS

The plate theory indicates the peak shape to be that of the Poisson distribution,[7] which at large plate numbers becomes a Gaussian or bell-shaped curve with the distribution of solute given by (Grushka et al., 1975):

[7]The Poisson distribution function is given by the probability function

$$Y_n(k) = \frac{a^k \exp(-a)}{k!} \tag{a}$$

where $a = \sigma^2$ and k is the probability of k occurrences in n events. The Gaussian distribution is

$$p(x) = \frac{1}{a\sqrt{2\pi}} \exp\left[\frac{-(x-b)^2}{2a^2}\right] \tag{b}$$

where $\sigma^2 = a^2$, for the variance, and a and b are constants. A complete explanation is given in Gelbert et al. (1975). For a linear equilibrium, the volume associated with a single stage or plate, $V_{\text{tot,plate}}$ (as described by Figure 5.39), was defined by van Deemter et al. (1956) as

$$V_{\text{tot,plate}} = V_{\text{o,plate}} + K_D V_{\text{s,plate}} = HA_o(\varepsilon_b + \varepsilon_p) \tag{c}$$

where

$V_{\text{o,plate}}$ = volume of the moving phase (i.e., the volume of mobile phase between particles in the stationary phase) for a single plate

$V_{\text{s,plate}}$ = volume of the immobile phase (i.e., the volume of the fluid inside the particles that make up the stationary phase) for a single plate

$K_D = C_{\text{s,plate}}/C_{\text{o,plate}}$ = coefficient that gives the distribution of solute between fluid in the stationary phase and the moving (mobile) phase in cases where linear equilibria apply

These definitions were used (van Deemter et al., 1956) to suggest the approximate criterion of

$$V > 100 \cdot V_{\text{tot,plate}} \tag{d}$$

for when the Poisson distribution function may be approximated by a Gaussian distribution function for calculating or fitting a chromatography elution profile. In equation (d), V denotes total volume of mobile phase required to elute the peak.

$$C = C_o \exp\left[\frac{-(V - V_R)^2}{2\sigma_V^2}\right] \tag{5.28}$$

where

C = concentration of solute

C_o = concentration of solute at the peak maximum (see Figure 5.38); this is *not* the same as the concentration of solute at the inlet of the column

$V_R = q_{\text{flowrate}}\, t_R$ = retention volume

t_R = retention time

$V = q_{\text{flowrate}}\, t$ = volume of mobile phase passed through column

t = elapsed time between injection of sample (at $t = 0$), and time at which concentration is measured

σ_v^2 = peak variance measured in units of volume, where the peak width at the base is 4σ

Since the plate count is given by (Grushka et al., 1975),

$$N = \frac{V_R^2}{\sigma_v^2} = \frac{q_{\text{flowrate}}^2 (t_R)^2}{q_{\text{flowrate}}^2\, \sigma_t^2} = \frac{16 t_R^2}{W^2} = \frac{L^2}{\sigma^2} \tag{5.29}$$

where

σ_v^2 = variance measured in terms of volume

σ_t^2 = variance measured in terms of time

σ^2 = variance measured in terms of distance

W = peak width = 4σ; expressed in dimensions of volume, time, or distance

$\sigma^2 = \dfrac{W^2}{(4)^2} = \dfrac{W^2}{16}$

L = column length

then the expression for the Gaussian distribution in Eq. (5.28) can be combined with Eq. (5.29) and give the expression (Schwartz and Smith, 1953)

$$\boxed{C = C_o \exp\left[\frac{-N(V - V_R)^2}{2V_R^2}\right]} \tag{5.30}$$

The plate height is the rate of change in the variance of the peak relative to the distance that it has migrated. This can be expressed by the derivative

$$H = \frac{d\sigma^2}{dL} \tag{5.31}$$

which for a uniform column becomes σ^2/L, since the change for each plate is the same. On the basis of the definition in Eq. (5.31) and the assumptions of the plate theory,

$$H = \frac{\sigma^2}{L} = \frac{L\sigma^2}{L^2} \tag{5.32}$$

Substitution of Eq. (5.29) into Eq. (5.32) gives the relationship between plate height, H, and plate count, N (Grushka et al., 1975):

$$H = L\left(\frac{\sigma^2}{L^2}\right) = \frac{L}{N} \tag{5.33}$$

The number of plates that define a useful column is a function of the intended application. Some types of industrial columns will have plate counts of 50 to 200 per meter. A good column, according to Grushka et al. (1975) would have 1000 plates, and an analytical column 10,000 plates/meter.

The plate height is a function of factors that control diffusion as well as the flowrate of the liquid. As clearly described by Martin and Synge, p. 1363 (1941): "There is an optimum rate of flow in any given case, since diffusion from plate to plate becomes relatively more important the slower the flow of liquid and tends of course to increase the H.E.T.P. Apart from this, the H.E.T.P. is proportional to the rate of the flow of the liquid and to the square of the particle diameter. Thus the smallest H.E.T.P. should be obtainable by using very small particles and a high pressure difference across the length of the column. The H.E.T.P. depends also on the diffusibility of the solute in the solvent employed, and in the case of large molecules, such as proteins, this will result in serious decrease in efficiency as compared with solutes of molecular weights of the order of hundreds." The diffusibility of a macromolecule such as a protein is 10 to 100 times less than the diffusibility of a small molecule, such as water or salt. The analysis of Martin and Synge summarizes some of the challenges of developing, testing, scaling-up, and optimizing industrial chromatography columns, which are discussed later in this book.

Example 4. Determining Column Efficiency: Plate Count

A spherical, particulate stationary phase is packed into a column that measures 10 mm i.d. by 20 cm long. The average particle size of this material (wet) is 100 microns. The first step in determining the dispersion characteristics of this column, and therefore the goodness of packing, is to inject a 50 μL sample of the water soluble polysaccharide, blue dextran (MW = $2 \cdot 10^6$ D) at an initial concentration of 2 mg/mL and to measure the plate count at a flowrate of 1 mL/min. The blue dextran is a large molecule that does not penetrate nor adsorb to the stationary phase particles that make up the packed bed. The resulting chromatogram is shown in Figure 5.40a. The elution profile measured at the outlet of the column has a maximum concentration of 0.1 mg/mL, while the peak maximum at the intersection of the two tangents is 0.103. A second column, of the same dimensions, is packed with a stationary phase that has an average particle diameter of 40 microns and is evaluated in the same manner. Calculate the (1) plate counts, and (2) generate elution profiles using the data obtained from the first chromatogram. (3) Also calculate the extraparticulate void volumes for the two columns. Do the plate counts and void fractions indicate well packed column?

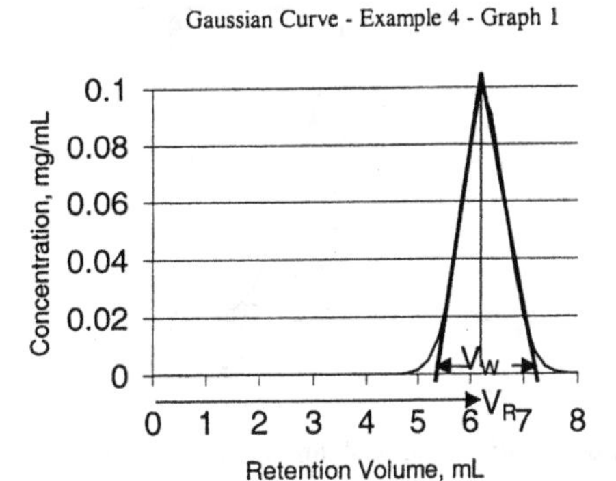

Figure 5.40a Schematic representation of a peak with a retention volume of 6.3 mL, example 4.

Solution

1. Use Eq. (5.29) to first estimate plate count, based on the definitions given in Figure 5.40a. The construction of the tangents is illustrated in Figure 5.40a. On this basis the retention volume (V_R = 6.3 mL) and peak width (V_w = 2 mL) are determined and are used to calculate the plate count, N.

$$N = 16\left(\frac{V_R}{V_W}\right)^2 = 16\left(\frac{6.3}{2.0}\right)^2 = 159 \tag{5.34}$$

We use Eq. (5.30) together with the graphically determined plate count to generate the profile in Figure 5.41. There is only a small difference between the curve with $N = 159$ (dotted line) and the curve for $N = 200$. (The value of $N = 200$ was used with Eq. (5.30) to generate the peak for this example.) Hence the graphical interpretation is only useful as a first estimate for calculating the plate count.

The second column packed with the 40 micron particle size should have a sharper peak according to the hypothesis of Martin and Synge (1941). The square of the ratio of particle sizes is

$$\left(\frac{100 \text{ microns}}{40 \text{ microns}}\right)^2 = 6.25 \tag{5.35}$$

and the plate count would be increased to 1250 for the column packed with 40 micron particles from 200 for the column packed with 100 micron particle:

$$N_{40\text{ micron}} = N_{100\text{ micron}}(6.25) = (200)(6.25) = 1250 \tag{5.36}$$

The total mass under the peak eluting from the column packed with the 100 micron particles is proportional to the area under the peak. This area is approximated by two triangles as indicated in Figure 5.40b below.

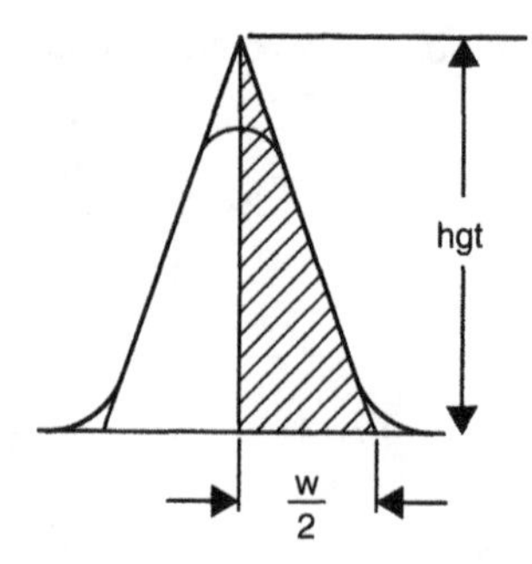

where $w/2$ is half the width of the peak and the area of the shaded triangle is given by

$$\text{Area of triangle} = \frac{(\text{hgt})(w/2)}{2} \tag{5.37}$$

The area under the peak is double this value. The area of one triangle for the Gaussian peak from the column packed with 100 micron particles is the product of the concentration at height (hgt) and the peak width, estimated from Figure 5.40a. The area under the peak is twice this value, or 0.101 as in

$$\text{Area of triangle} = \frac{(0.101)(2/2)}{2} \cdot 2 = 0.101 \text{ mg} \tag{5.38}$$

$$\left(\frac{\text{mg}}{\cancel{\text{mL}}}\right)(\cancel{\text{mL}})$$

Based on a plate count of 1250, the width of the peak follows from Eq. (5.29) since $V_R = qt_R$ (where q denotes flowrate):

$$V_w = \frac{4V_R}{\sqrt{N}} = \frac{(4)(6.3)}{\sqrt{1250}} = 0.713 \text{ mL}$$

Since the total mass of solute (and therefore, the area) under the peak is the same in both cases, the peak height for the column packed with 40 micron particles can be estimated from the area of the triangle:

$$\text{hgt} = \frac{2(\text{area of triangle})}{V_w} = \frac{0.101 \text{ mg}}{0.713 \text{ mL}} = 0.142 \frac{\text{mg}}{\text{mL}} \tag{5.39}$$

Equation (5.30) can then be used to generate the expected elution profile for the column packed with 40 micron particles, where

$C_o = 0.142$ mg/mL

$N = 1250$

$V_R = 6.3$ mL

The result is the narrow peak in Figure 5.41.

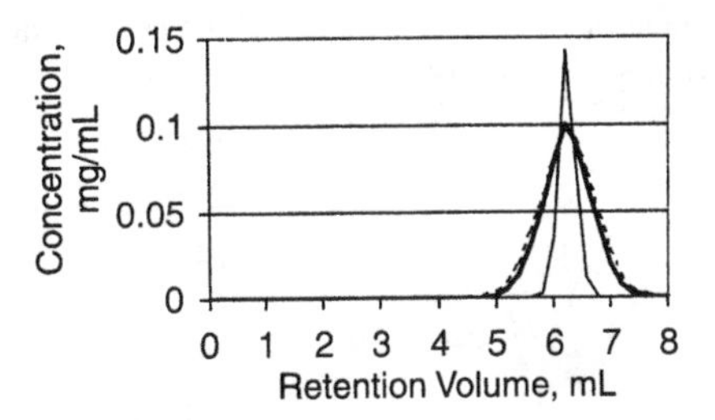

Figure 5.41 Figure illustrating curve generated by equation 5.30, example 4.

The simple Gaussian correlation is useful in comparing the efficiency of different columns. It does not reflect changes due to sample size, sample concentration, or flowrate. Hence a more complete expression is needed (derived in Section 5.15 of this chapter).

2. The extraparticulate void volume can be estimated from the retention time of the blue Dextran, assuming that it is excluded from the stationary phase and is not adsorbed. In this case, the definition for retention volume in Eq. (5.2) is used, and the retention volume is read directly from the graph as 6.3 mL, which is the volume at which the peak center elutes for both columns. The nominal column volume may be calculated from the dimensions of the column and is 15.7 mL. Consequently the extraparticulate void fraction is 6.3/15.7 or about 0.40. This is in the upper range of the expected void fractions for columns of spherical columns that typically have expected extraparticulate fractions ranging from 0.35 to 0.40. The possibility that the actual column volume is larger than the nominal volume calculated from column dimensions should be considered (this would decrease the void fraction). Alternately, the presence of a void at the inlet, or poor packing, could result in a smaller plate count, and this may also result in a void fraction that is at the high end of the expected range.

5.14 MANY PEAKS THAT ELUTE FROM CHROMATOGRAPHY COLUMNS ARE SKEWED DUE TO INTRACOLUMN AND EXTRA COLUMN DISPERSION EFFECTS: EXPONENTIALLY MODIFIED GAUSSIAN PEAKS

An exponentially modified Gaussian peak model was proposed for estimating plate counts from skewed peaks of the type illustrated in Figure 5.42:

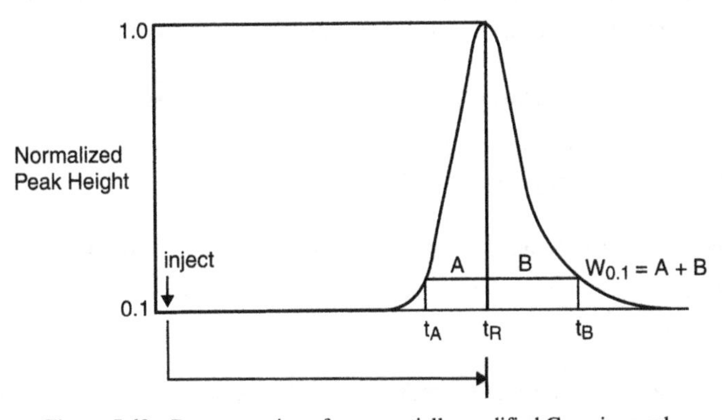

Figure 5.42 Representation of exponentially modified Gaussian peak.

The equation to determine the plate count for a skewed peak, with the variables as represented in Figure 5.42, is (Foley and Dorsey, 1983)

$$N_{EMG} = \frac{41.7\left(\frac{t_R}{W_{0.1}}\right)^2}{\frac{B}{A} + 1.25} \tag{5.40}$$

where

N_{EMG} = plate count for an exponentially modified Gaussian peak
t_R = elapsed time between injection of sample and elution of peak maximum
$W_{0.1}$ = peak width (in units of time) at 10% height as indicated in Figure 5.42
A = peak width from peak center to left-hand side (leading edge) of peak in units of time
B = peak width from peak center to right-hand side (tailing edge), in units of time
B/A = assymetry factor, dimensionless

Note that Foley and Dorsey (1983) define the peak maximum to be the same as the top of the peak, rather than the intersection of two tangents as was the case for a Gaussian peak. Equation (5.40) was found to give values of plate count that were accurate to within ±1.5% for peak assymetries (= A/B) ranging from 1 to 2.76 (Foley and Dorsey, 1983). In the case where the peak is Gaussian and $A = B$, the ratios of peak widths at half height ($W_{0.5}$); 0.3 height ($W_{0.3}$) and 0.1 height ($W_{0.1}$) should be 0.5487 : 0.7231 : 1 (Foley and Dorsey, 1983). The equation that represents an exponentially modified Gaussian peak is reported to give a more accurate calculation of plate count than the equation for a Gaussian peak. While true Gaussian peaks are rarely observed, Eq. (5.27) is still commonly used to estimate N from graphical representations of the chromatographic peaks, probably due to convenience.

The plate heights that are calculated from Gaussian or skewed peak equations give a first indication of the efficiency of the column. However, a change in flowrate, or particle size of the stationary phase, as well as a change in the concentration and/or volume of the injected sample, can lead to different elution profiles and therefore different apparent plate counts. A material balance enables a more general approach to be used in interpreting elution profiles from a chromatography column and extrapolating the results to different conditions. This approach is introduced in the next section and developed further throughout this chapter.

Example 5. Determining Plate Count for Gaussian and Skewed Peaks

Tabulated data of a chromatographic peak is obtained through digital data acquisition. The data for the two peaks are listed in Table 5.1.

1. Calculate the plate count for peaks 1 and 2, assuming that these peaks are Gaussian.
2. Calculate the asymmetry factors and plate counts for peaks 1 and 2, using the equation for exponentially modified Gaussian peaks.

Solution

1. A rearranged form of Eq. (5.30) represents the concentration profile for a Gaussian peak where $N/2$ is the slope of a rearranged form.

Table 5.1 List of Concentration as a Function of Retention Time for Example 5

Peak 1		Peak 2	
t_R, s	Concentration, mg/mL	t_R, s	Concentration, mg/mL
120	0	120	0
240	0	240	0
300	0.0014	300	0
324	0.0130	324	0.004
348	0.0533	348	0.036
360	0.0797	360	0.069
378	0.10	378	0.100
390	0.0904	390	0.090
402	0.0668	402	0.070
414	0.0404	414	0.045
426	0.0199	426	0.023
438	0.00805	438	0.010
450	0.0027	450	0.005
452	0.00072	462	0.003
486	0	486	0.002
500	0	500	0.001

$$-\ln \frac{C}{C_o} = \frac{N}{2} \frac{(V - V_R)^2}{V_R^2} = \frac{N q_{flow}^2 (t - t_R)^2}{2 q_{flow}^2 t_R^2}$$

$$\text{A plot of } -\ln (C/C_o) \text{ versus } \left(\frac{t - t_R}{t_R} \right)^2 \tag{5.41}$$

should give a straight line that goes through zero, with its slope = $N/2$, if the peak is Gaussian. Both leading and tailing edges of the peak should be plotted.

The plot of the leading edge (i.e., the left side of the peak obtained from the tabulated data of this example) is obtained from Eq. (5.41). Both peaks 1 and 2 give straight lines with the slopes being equal to $N/2$ as shown in Figure 5.43. Even though the peaks appear similar, the plate count associated with peak 2 would seem to be 1/3 greater than peak 1 for the following reason:

Peak 1: $N/2$ = slope = 100 or N = 200.

Peak 2: $N/2$ = slope = 157 or N = 314.

The plot of the tailing edge (the right side of the peak) from the tabulated data, shown in Figure 5.44, gives the same result for peak 1 as the plot in Figure 5.43. For the tailing edge, the slope is 99.8 and N is again close to 200. However, the result for peak 2 is different, with a variation occurring between the straight line, expected from Eq. (5.30), and the data points. The data for the tailing edge of peak 2 tapers off relative to its leading edge, Figure 5.45, thus indicating a skewed peak.

2. The plot of the tabulated data in the form of concentration against elution volume, on an expanded scale, further indicates that there is a difference between peaks 1 and 2,

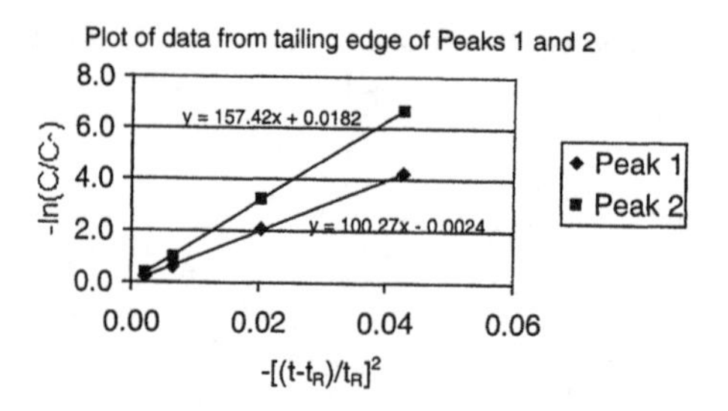

Figure 5.43 Replot of data from Table 5.1, example 5, tailing edge of peaks 1 and 2.

although the difference in the tailing edge of peak 2 relative to peak 1 would be difficult to detect from the elution profile Figure 5.45. The numerical values of Table 5.1 are needed. The use of Eq. (5.40a), which assumes that the peak follows an exponential modified Gaussian form, gives the plate counts for the two peaks as shown in (3) of this example.

3. Calculate asymmetry factors and plate counts using equations for exponentially modified peaks:

Peak 1. A and B, as defined in Figure 5.42, have the values as indicated, and they give the ratio

$$\frac{B}{A} = \frac{436 - 378}{378 - 321} = 1.02 \tag{5.42}$$

where 321 and 436 s are the retention times at which the leading and tailing edges of the peak are 10% of the given peak maximum of 0.1, as defined by Foley and Dorsey (1983) and calculated by linear interpretation of the tabulated data given in Table 5.1. Note that the maximum peak height is defined here by the top of the actual peak, rather than by the peak maximum determined from the intersections of the two tangents as given in Figure 5.38 and used in Example 4, for a Gaussian peak.

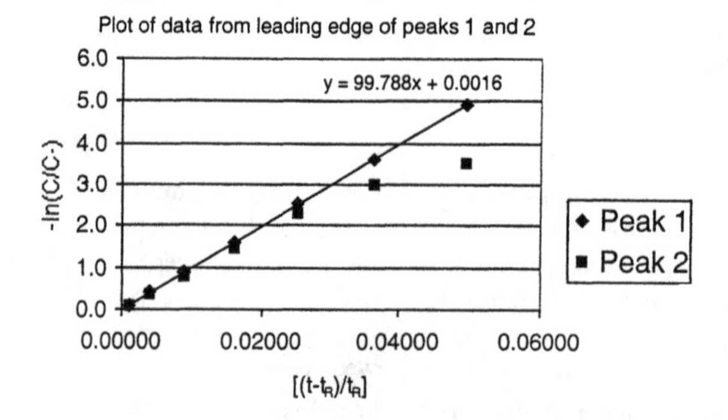

Figure 5.44 Plot generated from leading edge of peaks 1 and 2 (from Table 5.1, example 5).

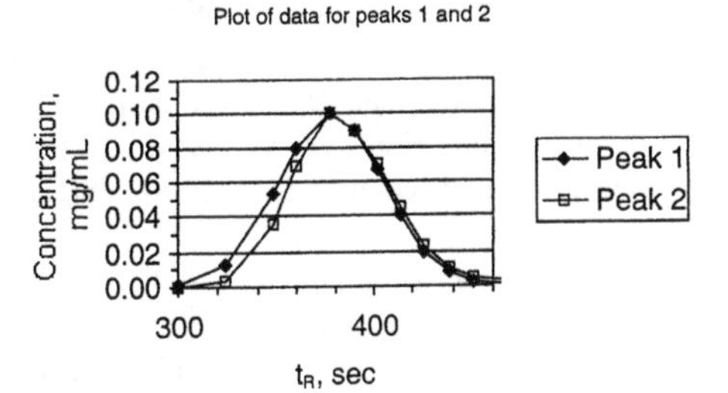

Figure 5.45 Plot of data in form of peaks (from Table 5.1, example 5).

The retention time for peak 1 is taken as 378, with width of the peak at 10% of the peak height given by

$$W_{0.1} = 436 - 321 = 115 \text{ s} \tag{5.43}$$

These values result in a plate count for the exponentially modified Gaussian peak as

$$N_{\text{EMG}} = \frac{41.7\left(\dfrac{t_R}{W_{0.1}}\right)^2}{(B/A) + 1.25} = \frac{41.7\left(\dfrac{378}{115}\right)^2}{1.02 + 1.25} = 199 \tag{5.44}$$

Peak 2. A and B have values of

$$\frac{B}{A} = \frac{438 - 378}{378 - 332} = 1.30 \tag{5.45}$$

where 332 and 438 s are the retention times at which the leading and tailing edges of the peak are at 10% of the peak's maximum. These values were obtained from the tabulated values for peak 2 from Table 5.1. The width of the peak at 10% of the maximum peak height is given by

$$W_{0.1} = 438 - 332 = 106 \tag{5.46}$$

The values result in the plate count calculated in Eq. (5.47) for an exponentially modified Gaussian peak of

$$N_{\text{EMG}} = \frac{41.7\left(\dfrac{378}{106}\right)^2}{1.30 + 1.25} = 208 \tag{5.47}$$

The peaks in this example have a similar appearance. However, small changes in their shape, or a change in the method used to estimate their plate count, can result in differences in the values of the plate count. Both Gaussian and skewed peaks can be evaluated using the formula for an exponentially modified peak. The variability of a single measurement leads to the suggestion that multiple measurements must be taken to obtain a plate count for purposes of comparing the performance of different columns. This example also illustrates the importance of using a consistent measure of plate count if different columns are to be compared.

The definition of the plate count is based on a Gaussian distribution equation. It can also be derived from a material balance and related to the dimensionless groups used in the chemical engineering literature to represent momentum and mass transfer phenomena. This second approach sets the stage for simulating elution profiles from chromatography columns, and is developed in the following section based on the derivation of Athalye et al. (1992).

5.15 ONE-DIMENSIONAL MODEL OF DIFFERENTIAL CHROMATOGRAPHY ENABLES SIMULATION OF ELUTION PROFILES

The movement of a solute, dissolved in the mobile phase, through a packed bed of stationary phase is given by the one-dimensional material balance. The derivation is taken from a review by Athalye, Lightfoot et al. (1992). It starts with the material balance

$$\varepsilon_b\left(\frac{\partial C_m}{\partial t} + v\frac{\partial C_m}{\partial z} - D\frac{\partial^2 C_m}{\partial z^2}\right) = -(1-\varepsilon_b)\frac{\partial C_s}{\partial t} \tag{5.48}$$

where

C_m = concentration of a solute in the mobile phase

C_s = concentration of a solute in the stationary phase where the stationary phase includes the solid and the fluid within the pores that are found within the individual stationary phase particles

v = interstitial velocity, cm/s

ε_b = interstitial void fraction, namely the void fraction between the particles that make up the stationary phase, dimensionless

D = convective, axial dispersion coefficient that accounts for axial molecular diffusion, hydrodynamic band broadening, and mixing effects due to complicated flow paths around the particles that make up the stationary phase, cm^2/s

z = axial coordinate, distance from the inlet of the bed, cm

t = time, s

Solute transport inside of a spherical, porous particle is given by

$$\frac{\partial C_p}{\partial t} = \frac{D_p}{r^2}\cdot\frac{\partial}{\partial r}\left(r^2\frac{\partial C_p}{\partial r}\right) - f(C_p, C_{surf}) \qquad \text{for } o \le r \le R \tag{5.49}$$

where

C_p = local concentration of solute present in the liquid within the pore, mg/mL

D_p = effective diffusivity of the solute in the liquid within the pores, cm^2/s

C_{surf} = local concentration of solute adsorbed on pore surfaces that are internal to the stationary phase where the stationary phase consists of the solid matrix and pore space, mg/mL

$f(C_p, C_{surf})$ = function that represents the local rate of sorption within the stationary phase particle

The local concentrations of solute (C_p and C_{surf}) are based on a unit volume of stationary phase that includes both the solid matrix and the pore space, rather than the individual respective phase volumes (liquid and solid) within a stationary phase particle. Linear adsorption kinetics are assumed so that the local sorption rate is given by

$$f(C_p, C_{surf}) = k_a \left(C_p - \frac{C_{surf}}{K_D} \right) \tag{5.50}$$

where

k_a = forward rate constant for adsorption, s^{-1}

K_D = distribution coefficient; ratio of mass of adsorbed solute to unadsorbed solute in the pores of a stationary phase, dimensionless

Hence the overall stationary phase concentration of solute, is given by

$$C_s = \frac{3}{4\pi R^3} \int_0^R (C_p + C_{surf}) 4\pi r^2 dr \tag{5.51}$$

The link between the concentration of the solute in the mobile phase outside of the particle near the entrance of the pore and the solute inside the particle is

$$-D_p \frac{\partial C_p}{\partial r} = k_c \left(\frac{1}{\varepsilon_p^*} C_p - C_m \right) \quad \text{at } r = R \tag{5.52}$$

where

k_c = fluid phase mass transfer coefficient, cm/s

D_p = effective diffusivity of solute in the liquid filled pores, cm^2/s

ε_p^* = intraparticle pore fraction that is accessible to the solute

ε_p = intraparticle pore fraction based on pore fraction that is accessible to the smallest sized molecule (this is usually the mobile phase)

The initial conditions for the concentrations of the solute and fluid in the stationary phases can be written as

$$C_m(z, t = 0) = \frac{m_o}{A_o \varepsilon_b} \delta(z) \tag{5.53}$$

where

m_o = mass of solute initially injected as a pulse to the column, mg
A_o = cross-sectional area of the column, cm^2
$A_o\varepsilon_b$ = cross-sectional area of the column that corresponds to interstitial void fraction

and the initial condition is

$$C_s\,(z, t = 0) = 0 \tag{5.54}$$

These conditions represent the case of differential chromatography where a short pulse of solute in the feed Eq. (5.53) is applied to a column that is devoid of the solute, Eq. (5.54). The pulse is represented by the dirac delta[8] function $\delta(z)$. The boundary condition for the column is given by

$$\lim_{z \to \text{large}} C_m = \lim_{z \to \text{large}} C_s = 0 \tag{5.55}$$

Condition (5.55) represents the long column assumption[1] in which the boundary conditions have little effect on peak shape, and the concentration field decays rapidly. At isocratic (i.e., constant flowrate) conditions, and long times, the solution of Eqs. (5.48), (5.49), and (5.52) gives a Gaussian function of the concentration of the solute in the mobile phase axial coordinate z with respect to time, t:

$$\boxed{C_m(z, t) = \frac{m_o x}{A_o \varepsilon_b \sqrt{2\pi H z_o}} \exp\left[\frac{-(z - z_o)^2}{2Hz_o}\right]} \tag{5.56}$$

where

z = axial coordinate of the packed bed, cm
$z_o = xvt$ = mean position of a solute peak at time t, cm
v = interstitial fluid velocity, cm/s
x = fraction of initial solute in the moving liquid phase at long times (i.e., ratio of solute in mobile phase to the amount of solute initially injected), dimensionless
H = plate height (HETP), cm

[8]Dirac defined a delta function $\delta(t)$ to be zero if $t \neq 0$ and to be infinite at $t = 0$ in such a way that

$$\int_{-\infty}^{\infty} \delta dt = 1 \tag{d}$$

This is an impulsive function which is very large in a vanishingly small region and very small outside of it, such that its integral over the region is finite (Roe, 1989).

The fraction of the solute in the moving fluid phase is defined in terms of the extraparticulate void fraction ε_b, the intraparticulate void fraction ε_p^*, and the distribution coefficient, K_D:

$$x = \frac{\varepsilon_b}{\varepsilon_b + (1 - \varepsilon_b)\varepsilon_p^*(1 + K_D)} \tag{5.57}$$

where

$$\varepsilon_p^* = \frac{V_{R,i} - V_{R,o}}{V_{col} - V_{R,o}} \quad \text{for component } i \tag{5.58}$$

V_{col} is the volume of the empty column, $V_{R,i}$ is the retention volume of component i, and $V_{R,o}$ is the retention volume of an excluded component (e.g., blue dextran). K_D is defined as the equilibrium constant for adsorption. K_D is a ratio of the mass of solute that is adsorbed on the stationary phase to the solute that is dissolved in the liquid that fills the pores, inside the particle, at equilibrium. Since ε_p^* is the species-dependent porosity that represents the fractional volume of the stationary phase particle that is accessible to the solute, the size exclusion characteristic is incorporated into the value for ε_p^*, while adsorption or binding of the solute is handled separately by K_D. When adsorption does not occur (e.g., in size exclusion chromatography) and $K_D = 0$, Eq. (5.57) simplifies to

$$x = \frac{\varepsilon_b}{\varepsilon_b + (1 - \varepsilon_b)\varepsilon_p^*}$$

$$= \frac{\dfrac{V_{R,D}}{V_{col}}}{\dfrac{V_{R,D}}{V_{col}} + \left(\dfrac{V_{col} - V_{R,D}}{V_{col}}\right)\left(\dfrac{V_{R,i} - V_{R,D}}{V_{col} - V_{R,D}}\right)} = \frac{\dfrac{V_{R,D}}{V_{col}}}{\dfrac{V_{R,D}}{V_{col}} + \dfrac{V_{R,i}}{V_{col}} - \dfrac{V_{R,D}}{V_{col}}} = \frac{V_{R,D}}{V_{R,i}} \tag{5.59}$$

where $V_{R,D}$ is the retention volume of the excluded component, blue dextran.

The porosity of the stationary phase as a fraction of the total column volume is based on the definition given in Eq. (5.60):

$$\boxed{\varepsilon_p = \frac{V_{R,i} - V_{R,o}}{V_{col}}} \tag{5.60}$$

This definition is sometimes used if the separation is based solely on differences in size of the molecules being fractionated. In this definition the fraction of initial solute in the moving liquid phase would be given by

$$x = \frac{\varepsilon_b}{\varepsilon_b + \varepsilon_p} = \frac{\dfrac{V_{R,D}}{V_{col}}}{\dfrac{V_{R,o}}{V_{col}} + \dfrac{V_{R,i} - V_{R,D}}{V_{col}}} = \frac{V_{R,D}}{V_{R,i}} \tag{5.61}$$

which is the same result as obtained from Eq. (5.59).

The limiting case of a solid stationary phase that has no pores ($\varepsilon_p^* = 0$) results in $x = 1$ for Eq. (5.57). If $K_D = 0$, then Eq. (5.57) becomes Eq. (5.59) because all of the solute is in the fluid phase that is located within the particle or makes up the flowing fluid outside the particle. Another limiting case for $K_D = 0$ is given by a stationary phase that is so porous that the mobile phase in which the solute is dissolved penetrates almost all of the volume occupied by the stationary phase, and ε_p^* in Eq. (5.58) approaches 1.0 or ε_p approaches $(1 - V_{R,o}/V_{col})$. Then x, as given by Eq. (5.57) approaches $\varepsilon_b/(\varepsilon_b + 1 - \varepsilon_b) = \varepsilon_b$. In the case of Eq. (5.61), $\varepsilon_b + \varepsilon_p$ approaches 1 and again $x = \varepsilon_b$.

The phase ratio is used in some of the chromatographic literature as a convenient measure of mass transfer area per unit column volume. A larger phase ratio would imply greater surface for adsorption compared to a smaller one. The volume occupied by a stationary phase and any stagnant fluid that is inside the stationary phase, to the volume occupied by the mobile (moving fluid) phase, is given by $(1 - \varepsilon_b)/\varepsilon_b$, namely by the phase ratio, ϕ. Other definitions of the phase ratio are based on ε_p or ε_p^*, where ϕ is given by $(1 - \varepsilon_p - \varepsilon_b)/(\varepsilon_p + \varepsilon_b)$ or $(1 - \varepsilon_p^* - \varepsilon_b)/(\varepsilon_p^* + \varepsilon_b)$. Careful reading of the literature is needed, when these definitions occur, so that the basis is clearly understood and an internally consistent basis of comparison for different stationary phases is selected.

At long times the distribution of a nonadsorbed solute is proportional to the ratio of the fluid phase that moves through the column (i.e., the mobile phase) relative to the fluid phase that is sequestered or held within the stationary phase and is not moving. This situation is achieved for size exclusion chromatography (abbreviated SEC) using chromatography columns packed with lightly crosslinked dextran gels to form spherical particles that can take up 5 to 50 times their weight in water, as shown in Figure 5.46.[9] The value of x will usually fall between the extremes of $1 < x < \varepsilon_b$ for size exclusion chromatography.

Comparison of Eq. (5.30) to Eq. (5.56) suggests the analogies of

$$C_o = \frac{m_o x}{A_o \varepsilon_b \sqrt{2\pi H z_o}} = \frac{\text{mass of solute in mobile phase}}{\text{volume of mobile phase}} \tag{5.62}$$

Now, recalling that $H = L/N$, we have

$$\frac{-N(V - V_R)^2}{2V_R^2} = \frac{-N(z - z_o)^2}{2(Lz_o)} = \begin{bmatrix}\text{plate count; a} \\ \text{measure of} \\ \text{dispersion}\end{bmatrix} \cdot \begin{bmatrix}\text{fractional distance that solute} \\ \text{has migrated relative to a fixed} \\ \text{position in the column}\end{bmatrix} \tag{5.63}$$

[9]Dextran is a polymer of α, 1-6 linked glucose and is produced by the bacterium *Leuconostoc mesenteroides*. Dextran matrices are formed by crosslinking the polymer with epichlorohydrin. This material is sold under the name Sephadex® by Amersham-Pharmacia-Biotech. The crosslink, represented schematically in Figure 5.46 stabilizes the matrix and prevents dissolution of the dextran polymers in water. The matrix is chemically stable over a pH range of 2 to 12. While autoclavable, such matrices are also biodegradable, making sterile operation important (Roe, 1989) in order to prevent microbial contamination and degradation of chromatography columns packed with this material.

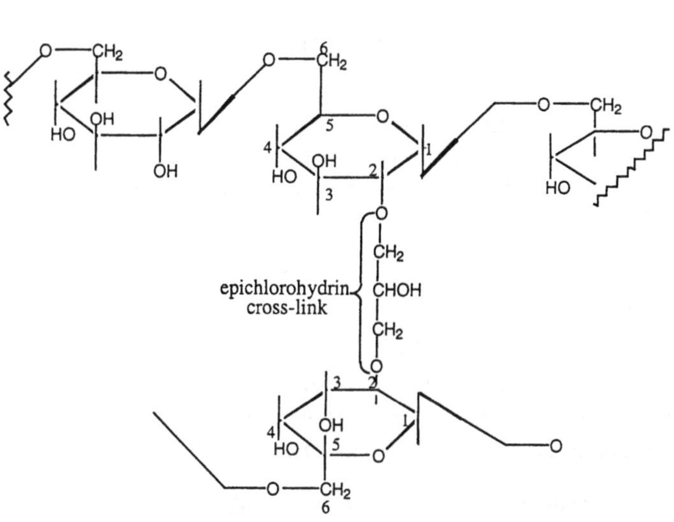

Figure 5.46 Structure of Sephadex®.

Equations (5.62) and (5.63) connect the retention time of chromatographic elution profiles with the void fractions within the column through x, the effect of mass transfer resistances on peak width through H, N, and the effect of the peak's residence time on its width through Z and Z_o.

Example 6. Simulating Elution Profiles for the Desalting of a Protein

The separation of a salt from protein is referred to as desalting. This separation is often carried out using size-exclusion chromatography (SEC), which is used in many industrial processes that involve protein purification. The molecules are separated on the basis of differences in size, with no adsorption occurring. The first step in developing appropriate conditions is to carry out a bench-scale separation. The following data are available for a bench-scale system, for the separation of bovine serum albumin ($MW = 66{,}000$) from NaCl ($MW = 58.44$):

Column diameter	5 cm
Column length	30 cm
Void fraction, ε_b	0.38
Retention volume, blue dextran	224. mL
Retention volume, BSA	250. mL
Retention volume, NaCl	442. mL
Flowrate of mobile phase	4.2 mL/min
HETP	0.1 cm
Maximum mass of protein loaded	25. mg
Maximum mass of NaCl loaded	100. mg

As a first approximation, assume that the sample mass is loaded in a manner that it approaches a dirac delta function. In the first case, the mobile phase consists of a buffer that suppresses adsorption of the BSA so that $K_D = 0$. In a second run, the mobile phase composition is changed, and the BSA is retained due to adsorption and gives $K_D = 1.5$. K_D is the distribution coefficient as defined for Eq. (5.50).

Calculate the expected elution profiles. Will the column give a separation of the two components for both cases?

Solution

1. Calculate interstitial velocity:

Column area, $A_o = \pi r^2 = \pi (5/2)^2 = 19.6 \text{ cm}^2$

Column volume, $V_{col} = \pi r^2 h = \pi (5/2)^2 \, 30 = 589$

Extraparticulate void fraction, $\varepsilon_b = 224/589 = 0.38$

Interstitial area $= A_o \varepsilon_b = (19.6)(0.38) = 7.47 \text{ cm}^2$

$$\text{Interstitial velocity} = \frac{q}{A_o \varepsilon_b} = \frac{4.2}{7.47} = 0.563 \text{ cm/min}$$

For protein (BSA) (first case), $K_D = 0$ by definition, no adsorption,

$$\varepsilon_p^* = \frac{V_{R,i} - V_{R,o}}{V_{col} - V_{R,o}} = \frac{250 - 224}{589 - 224} = 0.071 \tag{5.64}$$

$$x_{BSA} = \frac{\varepsilon_b}{\varepsilon_b + (1 - \varepsilon_b)\varepsilon_p^*(1 + K_D)}$$

$$= \frac{0.38}{0.38 + (1 - 0.38)(0.071)(1)} = 0.896 \tag{5.65}$$

For salt (NaCl), $K_D = 0$ by definition, no adsorption,

$$\varepsilon_p^* = \frac{442 - 224}{589 - 224} = 0.597 \tag{5.66}$$

$$x_{NaCl} = \frac{0.38}{0.38 + (1 - 0.38)(0.597)(1)} = 0.507 \tag{5.67}$$

2. These parameters and the calculated results from Eqs. (5.64) to (5.67) are then substituted into Eq. (5.56) where the masses, m_o, for the two components are 25 and 100 mg, respectively, and Z is the length of the column, while Z_o represents the distance that the peaks have traveled. The results are summarized in Table 5.2.

A portion of a spreadsheet used to carry out the calculations, and the resulting elution profiles, are reproduced in Figure 5.47 and Table 5.3. The calculations for $K_D = 1.5$ are analogous for the BSA except that x_{BSA} is smaller, since the BSA reversibly adsorbs:

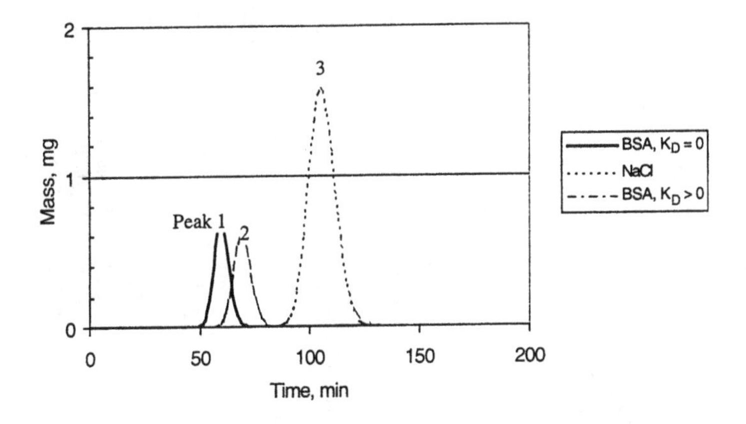

Figure 5.47 Plot of result from example 6.

$$x_{\text{BSA}} = \frac{0.38}{0.38 + (1 - 0.38)(0.071)(1 + 1.5)} = 0.775 \qquad (5.68)$$

This causes the BSA peak to elute later as indicated by the peak at 69 minutes in Figure 5.47. Note that there is significant dilution of both the salt and BSA peaks due to

TABLE 5.2 Tabulated Results of Calculations

Type of Chromatography		Size Exclusion	Size Exclusion for NaCl Adsorption ($KD > 0$) for BSA
Column diameter		5 cm	5 cm
Column length		30 cm	30 cm
Retention volume, blue dextran		224 mL	224 mL
Retention volume, BSA		250 mL	? mL
Retention volume, NaCl		442 mL	442 mL
Column area		19.6 cm^2	19.6 cm2
Column volume		589 mL	589 mL
Interstitial area (= $A_o \varepsilon_b$)		7.47 cm^2	7.47 cm2
Void fraction, ε_p		0.38 dimensionless	0.38 dimensionless
Flowrate		4.2 mL/min	4.2 mL/min
HETP (H)		0.1 cm	0.1 cm
m_{protein}		25 mg	25 mg
m_{NaCl}		100 mg	100 mg
Interstitial velocity		0.563 cm/min	0.563 cm/min
K_D	no adsorption	0 by definition	1.5 by definition
ε_p^*	BSA	0.071 by Eq. (5.58)	0.071 by Eq. (5.58)
x_{protein}	Eq. (5.57)	0.896 for BSA	0.775 for BSA
K_D	no adsorption	0 by definition	0 by definition
ε_p^*	NaCl	0.597 by Eq. (5.58)	0.597 by Eq. (5.58)
x_{NaCl}	Eq. (5.57)	0.507 for NaCl	0.507 for NaCl

TABLE 5.3 Results from Spreadsheet Calculation using Eq. 5.56

Time, min	$Z_{o,BSA} = x_{BSA}vt$, cm (Distance Traveled by BSA Peak)	$Z_{o,NaCl} = x_{NaCl}vt$, cm (Distance Traveled by NaCl Peak)	Z, Length of Column (Confusing Notation)[1]	Pre-exp Term, Eq. (5.56) BSA, $K_{D,ads} = 0$	Pre-exp Term, Eq. (5.66) NaCl, $K_{D,ads} = 0$	$C_m(z, t)$, Eq. (5.56) BSA, $K_{D,ads} = 0$	$C_m(z, t)$, Eq. (5.56) NaCl, $K_{D,ads} = 0$	$Z_{o,BSA} = x_{BSA}vt$, cm (Distance Traveled by BSA Peak, $K_D > 0$)	Pre-exp Term, Eq. (5.56) BSA $K_{D,ads} > 0$	$C_m(z, t)$, Eq. (5.56) BSA $K_{D,ads} > 0$
						Peak 1	Peak 3			Peak 2
0	0.00	0.00	30	#DIV/0!	#DIV/0!	#DIV/0!	#DIV/0!	0	#DIV/0!	#DIV/0!
1	0.50	0.29	30	5.33	16.093	0.000	0.000	0.436	4.974	0.000
2	1.01	0.57	30	3.77	11.379	0.000	0.000	0.872	3.517	0.000
3	1.51	0.86	30	3.08	9.291	0.000	0.000	1.308	2.872	0.000
47	23.69	13.40	30	0.78	2.347	0.000	0.000	20.491	0.726	0.000
48	24.19	13.69	30	0.77	2.323	0.001	0.000	20.927	0.718	0.000
49	24.70	13.98	30	0.76	2.299	0.003	0.000	21.363	0.711	0.000
50	25.20	14.26	30	0.75	2.276	0.008	0.000	21.799	0.703	0.000
51	25.70	14.55	30	0.75	2.253	0.021	0.000	22.235	0.697	0.000
52	26.21	14.83	30	0.74	2.232	0.048	0.000	22.671	0.690	0.000
53	26.71	15.12	30	0.73	2.210	0.097	0.000	23.107	0.683	0.000
54	27.22	15.40	30	0.73	2.190	0.175	0.000	23.543	0.677	0.000
55	27.72	15.69	30	0.72	2.170	0.281	0.000	23.979	0.671	0.000
56	28.22	15.97	30	0.71	2.150	0.407	0.000	24.415	0.665	0.001
57	28.73	16.26	30	0.71	2.132	0.533	0.000	24.851	0.659	0.003
58	29.23	16.54	30	0.70	2.113	0.633	0.000	25.287	0.653	0.008
59	29.74	16.83	30	0.69	2.095	0.686	0.000	25.723	0.648	0.019
63	31.75	17.97	30	0.67	2.027	0.414	0.000	27.467	0.627	0.195
64	32.26	18.25	30	0.67	2.012	0.303	0.000	27.903	0.622	0.283
65	32.76	18.54	30	0.66	1.996	0.207	0.000	28.339	0.617	0.379
66	33.26	18.82	30	0.66	1.981	0.132	0.000	28.775	0.612	0.472
67	33.77	19.11	30	0.65	1.966	0.080	0.000	29.211	0.608	0.546
68	34.27	19.39	30	0.65	1.952	0.045	0.000	29.647	0.603	0.591
69	34.78	19.68	30	0.64	1.937	0.024	0.000	30.083	0.599	0.598
70	35.28	19.96	30	0.64	1.923	0.012	0.000	30.519	0.595	0.569

TABLE 5.3 Continued

Time, min	$Z_{o,BSA} = x_{BSA}vt$, cm (Distance Traveled by BSA Peak)	$Z_{o,NaCl} = x_{NaCl}vt$, cm (Distance Traveled by NaCl Peak)	Z, Length of Column (Confusing Notation)[1]	Pre-exp Term, Eq. (5.56) BSA, $K_{D,ads} = 0$	Pre-exp Term, Eq. (5.66) NaCl, $K_{D,ads} = 0$	$C_m(z, t)$, Eq. (5.56) BSA, $K_{D,ads} = 0$	$C_m(z, t)$, Eq. (5.56) NaCl, $K_{D,ads} = 0$	$Z_{o,BSA} = x_{BSA}vt$, cm (Distance Traveled by BSA Peak, $K_D > 0$)	Pre-exp Term, Eq. (5.56) BSA $K_{D,ads} > 0$	$C_m(z, t)$, Eq. (5.56) BSA $K_{D,ads} > 0$
71	35.78	20.25	30	0.63	1.910	0.006	0.000	30.955	0.590	0.509
72	36.29	20.53	30	0.63	1.897	0.003	0.000	31.391	0.586	0.431
73	36.79	20.82	30	0.62	1.883	0.001	0.000	31.827	0.582	0.345
74	37.30	21.11	30	0.62	1.871	0.000	0.000	32.263	0.578	0.261
75	37.80	21.39	30	0.62	1.858	0.000	0.000	32.699	0.574	0.189
76	38.30	21.68	30	0.61	1.846	0.000	0.000	33.135	0.571	0.129
77	38.81	21.96	30	0.61	1.834	0.000	0.000	33.571	0.567	0.085
78	39.31	22.25	30	0.60	1.822	0.000	0.000	34.007	0.563	0.053
79	39.82	22.53	30	0.60	1.811	0.000	0.000	34.443	0.560	0.032
80	40.32	22.82	30	0.60	1.799	0.000	0.000	34.879	0.556	0.018
81	40.82	23.10	30	0.59	1.788	0.000	0.000	35.315	0.553	0.010
82	41.33	23.39	30	0.59	1.777	0.000	0.000	35.751	0.549	0.005
83	41.83	23.67	30	0.59	1.766	0.000	0.000	36.187	0.546	0.003
84	42.34	23.96	30	0.58	1.756	0.000	0.001	36.623	0.543	0.001
85	42.84	24.24	30	0.58	1.745	0.000	0.002	37.059	0.540	0.001
86	43.34	24.53	30	0.57	1.735	0.000	0.004	37.495	0.536	0.000
87	43.85	24.81	30	0.57	1.725	0.000	0.008	37.931	0.533	0.000
88	44.35	25.10	30	0.57	1.715	0.000	0.014	38.367	0.530	0.000
89	44.86	25.38	30	0.57	1.706	0.000	0.026	38.803	0.527	0.000
90	45.36	25.67	30	0.56	1.696	0.000	0.044	39.239	0.524	0.000

[1]Confusing since Z is also used to denote variable distance from inlet of column in same equations.

dispersion within the column. The initial mass injected for the BSA and NaCl was 25 and 100 mg, respectively, in a very small volume (i.e., dirac delta function assumption). By the time the two peaks elute from the column, their maximum concentrations are significantly reduced, since the two components elute over large volumes.

For example, in the case of $K_D = 0$ (BSA), with times and concentrations read from the tabulated spreadsheet calculations,

$$\text{Volumes of BSA peak: } (73 - 48) \text{ min } (4.2 \frac{\text{mL}}{\text{min}}) = 105 \text{ mL} \tag{5.69}$$

A summary of the column and stationary phase characteristics, and the calculated values used together with Eq. (5.56) to generate the graph, is given in Table 5.2. The mass at the top of this peak is about 0.686 mg. Consequently the sample was diluted by a factor of 36.4. Unlike Eq. (5.30), where an experimental determination of C_0 is needed, the concentration term in Eq. (5.56) is based on column characteristics and the initial mass of solute loaded onto the column as a dirac delta function.

The elution profiles in Figure 5.47 were generated using Eq. (5.56) together with the mass (m_{protein} or m_{NaCl}) of solute ($= m_o$); appropriate values of x from Eqs. (5.65), (5.67), or (5.68); the interstitial area ($= A_o\varepsilon_b$) from the column dimensions and void fraction (given in problem statement at the beginning of the example); the plate height, H (given in the example problem statement as 0.1 cm); the mean position of the solute, z_o, calculated from x, v, and t; and the axial coordinate of the bed, z, which in this case is given as 30 cm (i.e., the column length). The calculation was set up in an Excel spreadsheet to show the concentration at the column outlet. A portion of the spreadsheet calculations is summarized in Table 5.3. The columns marked peak 1, peak 2, and peak 3 in Table 5.3 coincide with the peaks 1, 2, and 3 in Figure 5.47. The calculation for peak 2 (NaCl) is only partially shown in Table 5.3.

5.16 SAMPLE (FEED) VOLUMES AFFECT THE CALCULATION OF PLATE COUNT DUE TO CONTRIBUTIONS OF THE FEED VOLUME TO PEAK WIDTH

The effect of feed volume on calculation of plate count was analyzed by van Deemter, Zuiderweg, and Klinkenberg (1956). They showed that the width of the elution curve, V_w, at a small feed volume is constant when the sample (feed) volume, V_{spl}, satisfies the condition

$$V_{spl} < 0.5 V_{\text{tot,plate}} \sqrt{N} \tag{5.70}$$

where

$$V_{\text{tot,plate}} = V_{\text{o,plate}} + \frac{V_{\text{s,plate}}}{K^*} = V_{\text{o,plate}} + K V_{\text{s,plate}}$$

$$= H(A_o\varepsilon_b) + KHA_o(\varepsilon_p) = HA_o(\varepsilon_b + K\varepsilon_p) \tag{5.71}$$

and where

$V_{\text{tot,plate}}$ = effective plate volume; volume of mobile and immobile (stationary) phase for one theoretical plate, mL

$V_{o,plate}$ = volume of mobile phase (i.e., fluid between particles) in one theoretical plate, mL
$V_{s,plate}$ = volume of immobile (stationary) phase in one theoretical plate, mL
K^* = ratio of concentration of solute in mobile phase to concentration of solute in immobile (stationary) phase[10] = 1/K
K = distribution coefficient
H = height of one theoretical plate, cm
ε_b = extraparticulate void fraction as defined by Eq. (5.1)
ε_p = intraparticulate void fraction as defined by Eq. (5.60) as discussed in Section 5.15
A_o = cross-sectional area of chromatography column, cm = πR^2_{col}
N = plate count = HL
L = column length, cm

The plate volume is defined here as the sum of the fluid volume associated with the moving fluid or mobile phase external to the particles, and the fluid that is held stationary or immobile inside the stationary phase particles. If Eq. (5.70) is satisfied, the width of the eluting peak will be constant and will be given by

$$V_{w,o} = 4V_{tot,plate}\sqrt{N} = \text{constant} \tag{5.72}$$

for a given flowrate. In this case, the elution curve is Gaussian.

If a large sample volume is used where

$$V_{spl} > 3V_{tot,plate}\sqrt{N} \tag{5.73}$$

The relation between plate count, N, and peak width, V_w, is given by

$$V_w = V_{spl} + V_{tot,plate}\sqrt{2\pi N} \tag{5.74}$$

The rationale presented by van Deemter is based on their derivation of the expected elution profile for a Gaussian peak near the outlet of the chromatography column, given by

$$\frac{C}{C_o} = \frac{1}{V_{tot,plate}} \int_{V-V_{spl}}^{V} \frac{1}{(N-1)!} \left(\exp\left(\frac{-V'}{V_{tot,plate}} \right) \right) \left(\frac{V'}{V_{tot,plate}} \right)^{N-1} dV'$$

$$\approx \frac{1}{V_{tot,plate}\sqrt{2\pi N}} \int_{V-V_{spl}}^{V} \exp\left[\frac{-\left(\dfrac{V'}{V_{tot,plate}} - N \right)^2}{2N} \right] dV' \tag{5.75}$$

[10]Van Deemter et al. (1956) present the linear distribution of solute in terms of its ratio between mobile and immobile phases (K^*) rather than using the current convention that expresses this distribution as the ratio between immobile and mobile phases (i.e., $q = KC$).

where the second term in Eq. (5.75) is an approximation when $N - 1 \approx N$, and $V > V_{spl}$, and where

V' = parameter (denotes volume) to be integrated

V = volume of mobile phase that has passed through the column, mL

V_{spl} = volume of feed, mL

Another approximation made by van Deemter et al. (1956) in Eq. (5.75) is the replacement of the Poisson distribution in the integrand by an approximation for the Gaussian distribution.

The Poisson distribution function is:

$$\frac{1}{(N-1)!}\exp\left(-\frac{V'}{V}\right)\left(\frac{V'}{V_{tot,plate}}\right)^{N-1}$$

while the Gaussian distribution function is given by:

$$\cong \frac{1}{\sqrt{2\pi V'/V_{tot,plate}}}\exp\left\{\frac{-\left(\frac{V'}{V_{tot,plate}} - N\right)^2}{2N}\right\}$$

Hence, an approximation to Gaussian distribution function results when $V >> V_{tot,plate}$,

$$\cong \frac{1}{V_{tot,plate}\sqrt{2\pi N}}\exp\left\{\frac{-\left(\frac{V'}{V_{tot,plate}} - N\right)^2}{2N}\right\}$$

Since $V' = \pi R_{col}^2 L\varepsilon_b$ and $V_{tot,plate} = \pi R_{col}^2 H\varepsilon_b$, the ratio of $V'/V_{tot,plate} = L/H = N$ when $V'/V_{tot,plate}$ is very large. The expression (5.75) can be further simplified using the dimensionless groups:

$$\bar{V} = \frac{V}{V_{tot,plate}\sqrt{N}} = \text{dimensionless elution volume}$$

$$\bar{V}_{spl} = \frac{V_{spl}}{V_{tot,plate}\sqrt{N}} = \text{dimensionless feed volume}$$

to obtain Eq. (5.76) from Eq. (5.75):

$$\frac{C}{C_o} = \frac{1}{\sqrt{2\pi}}\int_{\bar{V}-\bar{V}_{spl}}^{\bar{V}} \exp\left(-\frac{1}{2}(\bar{V}' - \sqrt{N})^2\right)d\bar{V}' \tag{5.76}$$

where V' denotes the variable to be integrated. The integration of Eq. (5.76) gives the expression for calculating the concentration of the peak at its maximum:

$$\left(\frac{C}{C_o}\right)_{max} = \text{erf}\left(\frac{V_{spl}}{2V_{tot,plate}\sqrt{2N}}\right) = \text{erf}\left(\frac{a}{2\sqrt{2}}\right) \tag{5.77}$$

V_{spl} denotes the sample volume in Eq. (5.77). The width of the elution curve as a function of sample volume is given by

$$\frac{V_w}{V_{tot,plate}\sqrt{N}} = a + 2\delta + \sqrt{2\pi}\left(\frac{\bar{V}_{spl} + \delta}{\delta}\right)$$

$$\cdot \exp\left(\frac{1}{2}\delta^2\right)\left[\text{erf}\left(\frac{\bar{V}_{spl} + \delta}{\sqrt{2}}\right) - \text{erf}\left(\frac{\delta}{\sqrt{2}}\right)\right] \tag{5.78}$$

where

$$a = \frac{V_{spl}}{V_{tot,plate}\sqrt{N}} = \bar{V}_{spl} \tag{5.79}$$

$$\delta = (a + \delta)\exp\left(\frac{1}{2}\delta^2\right)\exp\left[-\frac{1}{2}(a + \delta)^2\right] \tag{5.80}$$

Equations (5.78) through (5.80) are useful for calculating the peak width when the sample size is no longer small (i.e., $V_{spl} > 0.5\ V_{tot,plate}\sqrt{N}$). The definitions of V_w, V_{spl} (the sample size in units of volume), and $V_{R,f}$, are given in Figure 5.48. The parameter δ in Eq. (5.80) is *not* the dirac delta function.

The volume of mobile phase that corresponds to the center of the peak and at which the maximum concentration is achieved, is a function of sample volume and column volume as represented in Figure 5.48 and given by

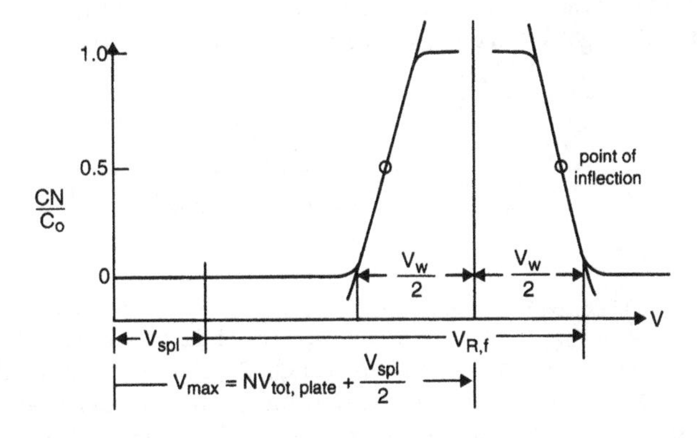

Figure 5.48 Representation of peak resulting from large sample volume using notation of van Deemter et al., 1956).

$$V_{max} = NV_{tot,plate} + \frac{V_{spl}}{2} \tag{5.81}$$

where

$NV_{tot,plate}$ = column volume that is accessible to the solute, mL, with $V_{tot,plate}$ as given by Eq. (5.71)

V_{spl} = sample (feed) volume, mL

The amount of mobile phase that must be passed through the column to remove all of the solute from the column is the final retention volume, or $V_{R,f}$ as indicated in Figure 5.48 and given by

$$V_{R,f} = NV_{tot,plate} - \frac{V_{spl}}{2} + \frac{V_w}{2} \tag{5.82}$$

If the dimensionless feed volume a is large enough,

$$a = \frac{V_{spl}}{(V_{tot,plate})(\sqrt{N})} > 3 \tag{5.83}$$

then

$$V_{R,f} = NV_{tot,plate} + 0.31 V_{w,o} \tag{5.84}$$

where $V_{w,o}$ is defined by Eq. (5.72). Equation (5.84) shows that the final retention volume is independent of feed volume if the condition in Eq. (5.83) is satisfied. The final retention volume and height and width of the peak is measurable by experiment. Hence Eq. (5.84) or Eq. (5.82) can be used to calculate $NV_{tot,plate}$ if $V_{R,f}$ and $V_{w,o}$ are measured. Equation (5.77) can be rearranged to

$$\left(\frac{C}{C_o}\right)_{max} = \text{erf}\left(\frac{V_{spl}}{2\sqrt{V_{tot,plate}}(\sqrt{2NV_{tot,plate}})}\right) \tag{5.85}$$

Hence the value of $NV_{tot,plate}$ from Eq. (5.82) or Eq. (5.83), when substituted into Eq. (5.85), can be used to calculate $V_{tot,plate}$ from the measured value of the peak height. N is thereby determined by simultaneous solution of Eq. (5.85) and Eq. (5.82) or Eqs. (5.83) and (5.84).

The elution volume at the center of the peak (i.e., peak maximum) as denoted by V_{max}, together with the sample volume is used by Barford et al. (1978) to calculate the ratio of the plate height at small sample size (= H) to the plate height at a large sample volume (= H'). This is obtained as follows:

$$\frac{H}{H'} = \frac{\dfrac{\cancel{L}(V_w - V_{spl})^2}{1\cancel{6}\left(V_{max} - \dfrac{V_{spl}}{2}\right)^2}}{\dfrac{\cancel{L}V_w^2}{1\cancel{6}V_{max}^2}} = \frac{\dfrac{(V_w - V_{spl})^2}{V_w^2}}{\dfrac{\left(V_{max} - \dfrac{V_{spl}}{2}\right)^2}{V_{max}^2}} \tag{5.86}$$

Equation (5.86) shows that the plate heights will be the same if V_{spl} has small values. Otherwise, Eq. (5.86) gives a convenient means to calculate a plate height that is normalized to a given sample volume.

Example 7. Elution Characteristics of an Injected Sample

Calculate:

1. The height of the peak at its maximum (this is not the same as the plate height)
2. The peak width
3. The retention volume of the peak
4. The total retention volume as measured from the end of the sample injection for 2.5 mL injections of D2O and bovine serum albumin (BSA) for the liquid chromatography column having the dimensions and characteristics shown below

Column diameter			3.8 cm			
Column length			20 cm			
Column volume			227 mL			
ε_b			0.4 dimensionless			
ε_p			0.3 dimensionless			
V_{spl}			2.5 mL			
D_2O	K	1	H	0.15 cm	N_{D2O}	133
BSA	K	0	H	0.10 cm	N_{BSA}	200

Solution

The following equations as derived by van Deemter et al. (1956) are used:

Calculate $V_{tot,plate} = HA_o(\varepsilon_b + K\varepsilon_p) = 0.15\left[(3.8/2)^2\pi\right](0.4 + (1)(0.3)) = 1.19$ from Eq. (5.71).

$$\text{Since } V_{tot,plate} = \begin{cases} D_2O & 1.19 \text{ mL} \\ \text{BSA} & 0.45 \text{ mL} \end{cases}$$

$$\frac{V_{spl}}{(V_{tot,plate})(N^{0.5})} = \begin{cases} D_2O & 0.18 \text{ dimensionless feed volume} \\ \text{BSA} & 0.39 \text{ dimensionless feed volume} \end{cases}$$

Calculate peak max as

$$\frac{C}{C_o} = \text{erf}\left(\frac{V_{spl}}{2V_{tot,plate}\sqrt{2N}}\right) = \text{erf}\left(\frac{2.5}{2(1.19)\sqrt{2 \cdot 133}}\right) \quad \text{from Eq. (5.77)}$$

D_2O	0.0724	dimensionless
BSA	0.1545	

Calculate peak width by solving for delta in Eq. (5.80) using solver in Excel spreadsheet. The equation to be solved is a rearranged form of Eq. (5.79):

$$0 = \delta - (0.18 + \delta)\exp\left(\frac{1}{2}\delta^2\right)\exp\left[-\frac{1}{2}(0.18 + \delta)^2\right]$$

The expression is considered to be solved when the right hand side of the equation above is about 10^{-7} (≅) for the values indicated using the solver the values are:

for	δ	gives a value close to zero:
D_2O	0.910	$8.246 \cdot 10^{-7}$
BSA	0.812	$7.931 \cdot 10^{-7}$

Using δ to calculate dimensionless peak width from Eq. (5.78),

$$d = \frac{V_w}{V_{\text{tot,plate}}\sqrt{N}} = 0.18 + 2(\delta) + \sqrt{2\pi}\left(\frac{0.18 + \delta}{\delta}\right)\exp\left(\frac{1}{2}\delta^2\right)\left[\text{erf}\left(\frac{0.18 + \delta}{\sqrt{2}}\right) - \text{erf}\left(\frac{\delta}{\sqrt{2}}\right)\right]$$

where $\delta = 0.910$ for D_2O. Values for d are from the equation above as

D_2O	2.40	dimensionless
BSA	3.06	

Calculate peak width from result of Eq. (5.78) for D_2O is:

$$V_w = (d)(1.19)\sqrt{133} = (2.40)(1.191)\sqrt{133} = 22.2$$

The peak widths (V_w) for the two components are

D_2O	33.0 mL
BSA	19.6 mL

Calculate peak retention volume from Eq. (5.81):

$$V_{\max} = NV_{\text{tot,plate}} + \frac{V_{\text{spl}}}{2} = (133)(1.19) + \frac{2.5}{2} = 159.7$$

This gives retention volumes for D_2O and BSA:

D_2O	160.0
BSA	92.0

Calculate final peak retention volume (when all the sample has eluted) using Eq. (5.82). This volume is measured from the end of the feed pulse, as shown in Figure 5.48:

$$V_{r,f} = NV_{tot,plate} - \frac{V_{spl}}{2} + \frac{V_w}{2} = (133)(1.19) - \frac{2.5}{2} + \frac{33.0}{2} = 174$$

The final retention volumes are:

D_2O	174 mL (not including volume of feed pulse)
BSA	99.3

Equation (5.84) applies to the limiting case of $a = V_{spl}/((V_{tot,plate})(N^{0.5})) > 3$, where $V_{w,o}$ is obtained from Eq. (5.72). A sample size of 41.3 mL (= V_{spl}) is needed for $a > 3$. While this condition is not satisfied, Eq. (5.84) still gives an answer that is similar to that from Eq. (5.82) even though $V_{spl} = 2.5$. The values for D_2O and BSA from Eq. (5.82) are:

	$V_{r,f}$	$V_{w,o}$
D_2O	175.8	55.0
BSA	98.7	25.7

5.17 CONTRIBUTIONS TO PEAK BROADENING DUE TO PARTICLE SIZE AND FLOWRATE EFFECTS ARE GIVEN BY THE VAN DEEMTER EQUATION

The plate height expression as a function of interstitial velocity, v, and other factors is given by the general form (Grushka et al., 1975):

$$H = A + \frac{B}{v} + C_m v + C_s v \tag{5.87}$$

where

H = HETP, as defined in Figure 5.39
A, B = constants
C_m = resistance to mass transfer of the solute in the mobile phase
C_s = resistance to mass transfer of the solute in the stationary phase

Equation (5.87) assumes that the contributions to zone spreading are independent of one another and can be linearly added. This is a reasonable approximation for many types of biomolecules, since they are present at relatively low concentrations and hence the presence of one solute has only a small effect on the elution profile of the other. Equation (5.87) may give a parabolic shape, as shown by the solid line in Figure 5.49, where the velocity at which the minimum in the curve follows from the derivative of Eq. (5.87):

$$\frac{dH}{dv} = 0 = \frac{B}{v_{min}^2} + C_m + C_s \tag{5.88}$$

which upon rearrangement gives

$$v_{min} = \sqrt{\frac{B}{C}} \tag{5.89}$$

where $C = C_m + C_s$.

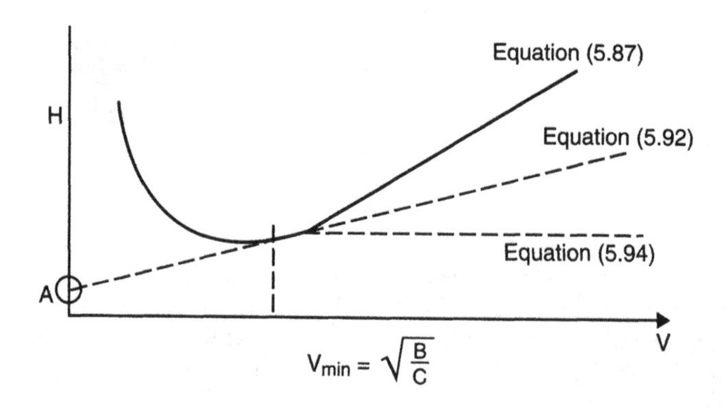

Figure 5.49 Schematic representation of limiting forms of van Deemter equation.

Eddy diffusion and mass transfer effects are coupled, with the overall contribution to the plate height, H, of the two terms being smaller than their sum. The equation for H in which the contribution of these effects is coupled is given by

$$H = \frac{B}{v} + C_s v + \frac{1}{\frac{1}{A} + \frac{1}{C_m v}} \tag{5.90}$$

Hence, when the velocity becomes large, the third term in Eq. (5.90) becomes

$$\frac{1}{\frac{1}{A} + \frac{1}{C_m v}} \rightarrow \frac{1}{\frac{1}{A}} = A \tag{5.91}$$

while the first term in Eq. (5.90) vanishes. Equation (5.90) therefore results in

$$H = A + C_s v \tag{5.92}$$

with the characteristic curve showing a lower slope than that associated with Eq. (5.87), as indicated in Figure 5.49. A different type of coupling expression was proposed by Yang et al. (1992), which resulted in a plate height equation of the form

$$H = A + \frac{B}{v} + \frac{DCv}{D + Cv} \tag{5.93}$$

that at high velocity reduces to

$$H = A + D \tag{5.94}$$

This would give a flat or velocity independent plate height at high velocity, as illustrated in Figure 5.49. Other forms of plate height equations are given in a comprehensive review by Hamaker and Ladisch (1996).

Particle size effects are incorporated into a reduced form of Eq. (5.87), and they are given by (Grushka et al., 1975; van Deemter et al., 1956):

$$h = \frac{H}{2R} = 2\lambda + 2\gamma \frac{D}{v2R} + C\frac{v2R}{D} \tag{5.95}$$

$$= A' + \frac{B'}{v} + C'v \tag{5.96}$$

where D is the axial dispersion coefficient as defined as given for Eq. (5.87) and

$$v = \text{reduced velocity} = \frac{vd_p}{D}$$

$$h = \frac{H}{2R} = \text{reduced plate height} \tag{5.97}$$

$$R = \text{radius of average particle in the packed bed}$$

Values for A', B', and C' are lumped constants from Eq. 5.95 and can be found by fitting a plot of h against v using Eq. (5.96), where h is obtained for columns packed with different particle size stationary phases and run over a range of flowrates (i.e., interstitial velocities). The pressure drop will limit the maximum flowrate at which a column can be run. While the comparison of packing efficiencies of different columns can be achieved by comparing elution profiles of a given solute that is run at the same interstitial velocities, and the same composition of mobile phase, and the same particle sizes, at least one of these parameters is likely to change as the separation is scaled up. This will change the width of the eluting peak. The variation due to changes in either or both are handled by writing a material balance across the column, and defining particle size and velocity effects in terms of the dimensionless Reynolds (Re), Schmidt (Sc), Peclet (Pe), Nusselt (Nu), and Dankohler (Da) numbers, as is described in the next section.

5.18 THE EFFECT OF PARTICLE SIZE, FLOWRATE, SOLUTE, AND TEMPERATURE ON PLATE HEIGHT IS MODELED USING DIMENSIONLESS NUMBERS RE, SC, PE, NU, AND DA (DERIVATION OF ATHALYE, LIGHTFOOT ET AL.)

The height equivalent to a theoretical plate, H, in Eq. (5.33) gives a measure of column characteristics at a given flowrate and for a given stationary phase. If a stationary phase is used that has a different average particle size, or if it is run at a different flowrate and/or temperature, the value of H must be determined for it. Van Deemter et al. (1956) proposed that an equation of the type shown in Eq. (5.98) be used to relate H to both flowrate and particle size, through the concept of a reduced plate height, where the reduced plate height is the number of particles per plate as given by

$$h = \frac{H}{2R} \tag{5.98}$$

where R is the average radius of the stationary phase particles used to pack the column, cm.

The effect of velocity, mass transfer, and convection on the reduced plate height, as explained by Athalye et al. (1992) and Van Deemter et al. (1956), is

$$h = \frac{2}{\text{Pe}} + \frac{(1-x)^2}{3(1-\varepsilon_b)} \text{ReSc}\left[\frac{1}{\text{Nu}} + \frac{m}{10} + \frac{m^*}{\text{Da}}\right] \tag{5.99}$$

where definitions and significance of the dimensionless groups are (Athalye et al., 1992; Cussler, 1997):

$$\text{Pe} = \frac{2Rv}{D} = \text{particle Peclet number} = \frac{\text{convective mass transfer}}{\text{diffusive mass transfer}} \tag{5.100}$$

$$= \frac{\text{flow velocity}}{\text{diffusion velocity}}$$

$$\text{Nu} = \frac{2Rk_c}{D_m} = \text{fluid-phase Nusselt number}$$

$$= f(\text{Re}, \text{Pr}) = \text{analogy to heat transfer} \tag{5.101}$$

where Pr denotes the Prandtl number (ratio of momentum diffusivity to thermal diffusivity).[11]

$$\text{Da} = \frac{R^2 k_a}{D_p} = \text{Damköhler number} = \frac{\text{adsorptive uptake}}{\text{intraparticle pore diffusion}} \tag{5.102}$$

$$\text{Re} = \frac{2Rv\varepsilon_b\rho_b}{\mu} = \text{Reynolds number} = \frac{\text{inertial force}}{\text{viscous force}} \tag{5.103}$$

The particle number, Re, is based on the radius of the particle and superficial velocity (i.e., velocity of approach = $\upsilon = v\varepsilon_b$) and therefore uses $\upsilon\varepsilon_b$ for the velocity term. Since the interstitial velocity, v, is based on the flow of the fluid through the open cross-sectional area of the packed column, the superficial velocity will be slower by the factor ε_b (i.e., $\upsilon = v\varepsilon_b$):

$$\text{Sc} = \frac{\mu}{\rho_b D_m} = \text{Schmidt number} = \frac{\text{diffusivity of momentum}}{\text{mass diffusivity}} \tag{5.104}$$

$$\text{ReSc} = \frac{2Rv\varepsilon_b}{D_m} = \text{reduced velocity} \tag{5.105}$$

[11]The Nusselt number is actually developed for heat transfer. In the case of laminar fluid flow in a pipe, heat transfer curves developed using the Nusselt number can be employed for convective fluid transport by substituting the Sherwood number for the Nusselt number, and Schmidt number for the Prandtl number.

$$m = \frac{D_m}{\varepsilon_p^* D_p} = \frac{D_m}{\varepsilon_p^* D_m/2.5} = \frac{2.5}{\varepsilon_p^*} = \text{intraparticle diffusion resistance} \tag{5.106}$$

$$m^* = \frac{3}{2} m \left(\frac{K_D}{1 + K_D} \right)^2 = \text{sorption kinetic term} \tag{5.107}$$

where

k_a = forward rate constant of adsorption, 1/s

K_D = equilibrium adsorption constant for ratio of mass of solute adsorbed to its concentration in the pore liquid

k_c = concentration-based fluid mass transfer coefficient, cm/s

v = interstitial fluid velocity, cm/s

x = fraction of solute in the mobile (moving fluid) phase at long times, as defined in Eq. (5.57)

ρ_b = density of mobile phase, g/mL

μ = viscosity of mobile phase g/(cm·s)

ε_p^* = intraparticulate void fraction as defined by Eq. (5.58).

The diffusivity of the solute in the fluid within a pore is less than its diffusivity in a free unbounded solution. However, the factor of 2.5 in Eq. (5.106) is arbitrary and presumably selected by Athalye, Lightfoot et al. (1992), since it gave a good fit of the equation to the data. The diffusion coefficients and diffusivities are defined as

D = convective axial dispersion coefficient, cm^2/s

D_m = effective solute diffusivity in unbounded solution (mobile phase), cm^2/s as given for various types of molecules in Eqs. (5.110), (5.111), and (5.113) below.

D_p = effective solute diffusivity in pore fluid, $cm^2/s = D_m/2.5$

The intraparticle diffusivity D_p is proportional to the diffusivity of the solute in free solution which in turn is affected by the ratio of solute size to pore size. The choice of 2.5 for calculating D_p is heuristic or arbitrary, according to Athalye et al. (1992) who chose the value of D_p as being $D_m/2.5$.

For size exclusion chromatography $K_D = 0$ (no adsorption) and the sorption kinetic term is $m^* = 0$. Equation (5.99) simplifies to

$$\boxed{h = \frac{2}{\text{Pe}} + \frac{(1-x)^2}{3(1-\varepsilon_b)} \text{ReSc} \left[\frac{1}{\text{Nu}} + \frac{m}{10} \right]} \tag{5.108}$$

While the effect of particle size is not explicitly shown in Eqs. (5.99) and (5.108), particle size is incorporated into the expression for h and through the dimensionless numbers Pe, Nu, Re, and Da.

The correlations by which values of Pe, Nu, and m may be calculated are a function of the values of ReSc that fall into the ranges ReSc << 1; 1 < ReSc < 100; and 100 < ReSc < 1000. However, the diffusivities are independent of ReSc and are calculated as outlined in Eqs. (5.109) to (5.115). Dispersion is related to the effective solute diffusivity in an unbounded solution by

$$D = \frac{D_{\mathrm{m}}}{\tau} = \frac{D_{\mathrm{m}}}{1.4} \tag{5.109}$$

where D and D_{m} are as defined above, and

τ = tortuosity factor for interstitial spaces between spherical particles in a packed bed = 1.4

The values of D_{m} will be a function of the solute molecule's size and shape, as well as the temperature and viscosity of the solution. Tyn and Gusek (1990) correlated the diffusion coefficients for dilute solutions of 86 proteins that have a random coil conformation using the correlation:

$$D_{\mathrm{m}} = 5.78 \cdot 10^{-8} \frac{T}{\mu_o R_g} \tag{5.110}$$

where

T = temperature, K

μ_o = viscosity of water in which protein is dissolved, at 20°C, 1.006 cP (note that 1 cP = 0.01 g/(cm·s)

R_g = radius of sphere that is equivalent to the volume occupied by the solute, cm (note that 1 angstrom = 10^{-8} cm; a 10 angstrom diameter sphere would have a volume of 524 Å^3 or $5.24 \cdot 10^{-22}$ cm^3)

This correlation was able to predict the diffusion constant to within ±20% of measured values for most of the proteins considered by Tyn and Gusek and is recommended for DNA, as well. Proteins that are known to have a rodlike shape gave measured constants that were 60% to 80% higher than the values calculated from Eq. (5.110). For this type of macromolecule at a temperature of 20°C, Tyn and Gusek recommend:

$$D_{\mathrm{m}} = \frac{1.69 \cdot 10^{-5}}{0.1\mathrm{L}} \tag{5.111}$$

$$L = \sqrt{12} R_g \tag{5.112}$$

where

L = length of the protein that is calculated from the molecule's radius of gyration, R_g, as given in Eq. (5.112).

Polyethylene glycol (PEG) is another molecular probe that is often used to characterize the porosity of chromatography columns. The free solution diffusion constant at 25°C is given by Eq. (5.113):

$$D_{m,PEG} = C_2 (MW)^{-C_1} = 5.16 \cdot 10^{-5}(MW)^{-0.43} \tag{5.113}$$

The temperature dependent correlation for D_m in Eq. (5.110) also requires a temperature dependent correlation for the viscosity of the fluid in which the solute is dissolved. The viscosity of water ranges from 1.792 cP at 0°C, 1.000 at 20°C to 0.2838 at 100°C, and follows the correlation

$$\mu = \exp\left[-4.5318 - \frac{220.57}{149.39 - T}\right] \tag{5.114}$$

where

T = temperature, K

Viscosities of other solutions can be obtained from tables in handbooks (Dean, 1973; Perry et al., 1963).

It should be noted that the diffusion coefficient, D_m, for small molecules, and particularly ions are larger. Furthermore the diffusion of small ions is affected by the presence of others. While the diffusion coefficient of species A at infinite dilution in solvent B (denoted D_{AB}) is given by the Nernst-Haskel equation, the calculated diffusion coefficient may change with salt concentration (Sherwood et al., 1975). As the salt concentration increases, D_{AB} decreases. As the salt concentration increases further, D_{AB} increases again after passing through a minimum. This behavior has been observed for aqueous organic solutions, as well as salts. Table 5.4 illustrates this effect for $CaCl_2$, NaCl, and KCl in water (Sherwood et al., 1975).

The value of the Peclet number for ReSc $<< 1$ is given by

$$\text{Pe} = \frac{2Rv}{D} = \frac{2Rv}{D_m/1.4} \tag{5.115}$$

where v = interstitial velocity, cm/s, and the axial dispersion (diffusion) coefficient is a factor 1.4 lower than the diffusion coefficient for the solute in an unbounded mobile phase, D_m. A factor of 1.4 is used here (instead of 2.5 in the case of intraparticulate diffusion), since the tortuosity of the void spaces between particles and their associated mass transfer resistance is less than that offered by void spaces within the particles. The Nusselt number, Nu, for ReSc < 1 is given by

$$\text{Nu} = \left[\beta^3 + \gamma^3 \,\text{ReSc}\right]^{1/3} \tag{5.116}$$

where

$$\gamma = \frac{1.09}{\varepsilon_b} \quad \text{and} \quad \beta = 4.0 \tag{5.117}$$

TABLE 5.4 Diffusion Coefficients of Selected Salts in Water

	Temperature, °C	Salt Concentration	D_{AB}, cm²/s
NaCl	18	0.4 M	$1.17 \cdot 10^{-5}$
		0.8	$1.19 \cdot 10^{-5}$
		2.0	$1.23 \cdot 10^{-5}$
$CaCl_2$	25	Dilute	$1.35 \cdot 10^{-5}$
		0.16 N	$1.11 \cdot 10^{-5}$
		2.25	$1.32 \cdot 10^{-5}$
KCl	18	0.4	$1.46 \cdot 10^{-5}$
		0.8	$1.49 \cdot 10^{-5}$
		2.0	$1.58 \cdot 10^{-5}$

Source: Adapted from Sherwood et al. (1975), table 2.11, p. 37. The material is reproduced with permission of The McGraw-Hill Companies.

At a higher flowrate, but still within the creep flow range, where 1 < ReSc < 100, Gunn's equation may be used to calculate the Pe number:

$$\text{Pe} = \left(\frac{\text{ReSc}}{1-\varepsilon_b}\right)^{-1/6} \tag{5.118}$$

where ε_b is the extraparticulate void fraction as defined by Eq. (5.1).

$$\varepsilon_b = \frac{V_o}{V_{col}} \tag{5.1}$$

and where

V_o = retention volume of an unretained (large) molecule, mL
V_{col} = column volume, mL

Equation (5.118), combined with Eq. (5.100), gives a correlation for estimating the axial dispersion coefficient, D, for 1 < ReSc < 100:

$$D = 2Rv\left(\frac{\text{ReSc}}{1-\varepsilon_b}\right)^{1/6} \tag{5.119}$$

Note that Eq. (5.119) gives a velocity dependent form of D, while Eq. (5.115) uses $D = D_m/1.4$, since the flowrate is so slow that the diffusional effect dominates and is constant when ReSc << 1. The corresponding Nusselt number for ReSc > 1 is given by

$$\text{Nu} = \gamma(\text{ReSc})^{1/3} \tag{5.120}$$

where γ is defined by Eq. (5.117).

At higher flowrates, where 100 < ReSc < 1000, the Peclet number, as obtained from Eq. (5.118), appears to continue to be applicable even though it is defined to be limited to 1 < ReSc < 100. This observation is based on comparison of the correlation of Eq. (5.118) to a limited number of data points by Athalye et al. (1992). A more complex, and possibly more general correlation for dispersion as a function of ReSc and characteristics of the packed bed, is mentioned by Athalye et al. (1992) and developed more fully by Dullien (1992) based on correlations originally reported by Gunn and his coworkers. These complex expressions, while judged to be less empirical, will not be developed here. Rather the reader is referred to the detailed discussion of Dullien, 1992. The correlations presented in Eqs. (5.109) to (5.120) are likely to coincide with the upper bounds of axial dispersion, and hence they are presented here as framework by which a first estimate can be obtained for quantifying the impact of solute diffu-

TABLE 5.5 Parameters for the Chromatography Column of Example 8

	Stationary phase	
R (radius)	0.002 cm = 20 microns	0.02 cm
Temperature	20 C	20 c
Column diameter	5 cm	5 cm
Column length	30 cm	30 cm
Retention volume, blue dextran	224 mL	224 mL
Retention volume, BSA	250 mL	250 mL
Retention volume, NaCl	442 mL	442 mL
Column area	19.6 cm^2	19.6 cm^2
Column volume	589 mL	589 mL
Interstitial area (= $A_o\varepsilon_b$)	7.47 cm^2	7.47 cm^2
Void fraction, ε_b	0.38 dimensionless	0.38 dimensionless
Flowrate (middle range of interest)	4.2 mL/min	4.2 mL/min
$m_{protein}$ (mass of protein)	25 mg	25 mg
m_{NaCl} (mass of salt)	100 mg	100 mg
Interstitial velocity	0.563 cm/min	0.563 cm/min
	Properties	
BSA		
K_D no adsorption	0 by definition	0
ε_p^* BSA	0.071 by Eq. (5.58)	0.071 by Eq. (5.58)
$x_{protein}$ Eq. (5.57)	0.896 for BSA	0.896 for BSA
radius of gyration	30.6 angstrom (Tyn and Gusek, 1990)	30.6
K_D no adsorption	0 by definition	0 by definition
NaCl		
ε_p^* NaCl	0.597 by Eq. (5.58)	0.597 by Eq. (5.58)
x_{NaCl} Eq. (5.57)	0.507 for NaCl	0.507 for NaCl
	Mobile phase buffer (water)	
Viscosity	1 cp (= 10^{-2} g/(cm/s))	1
Density	1 g/mL	1
Mass transfer parameters		
$m\ D_m/D_p^*\varepsilon_p$	35.10 assume $D_p = D_m/2.5$	35.10 Athalye (p. 80)
Gamma Nu at large ReSc	2.87 per Athalye	2.87
Beta Nu at lower ReSc	4	4
m^* Athalye (Eq. 7)	0 for BSA	0

TABLE 5.6 Summary Values from Parts of the Solution for Example 8: Case 1 (Base Case) for $R = 0.002$ cm

Interstitial Velocity (v), cm/s	Re = $2Rv$ dens/visc (visc in g/cm·s dimensionless)	Solution Diffusion Infinite Dilution (D_m), cm^2/s	Convective Axial Dispersion Coefficient, 20°C (D), cm^2/s	Equation Form for Calculating Axial Dispersion Coefficient
9.375E-07	0.00000	5.53E-07	3.95E-07	Eq. (5.58)
9.375E-06	0.00000	5.53E-07	3.95E-07	
0.00009	0.00004	5.53E-07	3.81E-07	Eq. (5.113)
0.00094	0.00038	5.53E-07	5.59E-06	
0.00469	0.00188	5.53E-07	3.65E-05	
0.00938	0.00375	5.53E-07	8.20E-05	***BASE flowrate***
0.01406	0.00563	5.53E-07	1.32E-04	
0.01875	0.00750	5.53E-07	1.84E-04	
0.04688	0.01875	5.53E-07	5.36E-04	
0.06563	0.02625	5.53E-07	7.94E-04	
0.09375	0.03750	5.53E-07	1.20E-03	
0.11250	0.04500	5.53E-07	1.49E-03	
0.14063	0.05625	5.53E-07	1.93E-03	
0.18750	0.07500	5.53E-07	2.70E-03	
0.37500	0.15000	5.53E-07	6.07E-03	
0.56250	0.22500	5.53E-07	9.73E-03	
0.75000	0.30000	5.53E-07	1.36E-02	

Sc = vis/(dens D_m) dimensionless	*ReSc*	*Peclet Number*	*Nusselt Number*	*Damköhler Number*	*m BSA Eq. (5.57)*	m^*	h
1.81E+04	0.007	0.00949	4.003		3.51E+01	0	210.8
1.81E+04	0.068	0.09486	4.033		3.51E+01	0	21.1
1.81E+04	0.678	0.98524	4.308		3.51E+01	0	2.0
1.81E+04	6.776	0.67123	5.424		3.51E+01	0	3.1
1.81E+04	33.879	0.51331	9.275		3.51E+01	0	4.6
1.81E+04	67.757	0.45731	11.686		3.51E+01	0	5.8
1.81E+04	101.6	0.42742	13.377		3.51E+01	0	6.8
1.81E+04	135.5	0.40741	14.723		3.51E+01	0	7.7
1.81E+04	338.8	0.34971	19.982		3.51E+01	0	12.7
1.81E+04	474.3	0.33064	22.354		3.51E+01	0	15.9
1.81E+04	677.6	0.31156	25.176		3.51E+01	0	20.4
1.81E+04	813.1	0.30223	26.753		3.51E+01	0	23.4
1.81E+04	1016.4	0.29120	28.819		3.51E+01	0	27.8
1.81E+04	1355.1	0.27757	31.720		3.51E+01	0	35.1
1.81E+04	2710.3	0.24728	39.964		3.51E+01	0	63.8
1.81E+04	4065.4	0.23113	45.747		3.51E+01	0	92.2
1.81E+04	5420.6	0.22031	50.352		3.51E+01	0	120.4

sional properties, temperature, packed bed characteristics, and flowrate on peak spreading in liquid chromatography columns.

Example 8. Estimating Reduced Plate Counts for a Size Exclusion Column

A column for carrying out size exclusion chromatography has the properties given in Table 5.5

1. For $R = 0.002$ cm, calculate and graph the predicted reduced plate height curve as a function of ReSc (from ReSc = 0.007 to ReSc = 678).
2. Construct a second curve that shows the reduced plate height as a function of the corresponding interstitial velocities that range from about $9.375 \cdot 10^{-7}$ to 0.09375 cm/s.
3. How does the plot of h against ReSc change if the average particle size is increased from $R = 0.002$ cm (= 20 microns) to $R = 0.02$ cm (= 200 microns)? Is there a difference in the slopes of the lines?
4. Does the plot of h against interstitial velocity show any difference in h at equivalent interstitial velocities?
5. What is the HETP, H, at the flowrate of 4.2 mL/min for the column packed with 40 micron particles ($R = 0.002$ cm)? For a column packed with 400 micron particle ($R = 0.02$ cm)?

Solutions

1. Adsorption is not a factor for this particular separation. Consequently Eq. (5.108) will be used to generate h as a function of ReSc. This requires that D_m and μ be calculated at 20°C, that Pe and Nu be calculated using Eqs. (5.109), (5.115), (5.116), and (5.117) at ReSc < 1; or Eqs. (5.118), (5.119), and (5.120) at $1 <$ ReSc < 100. The solution is illustrated in tabular form (and calculated using an Excel spreadsheet). The columns in Table 5.6 with the Damköhler number and m^* are blank because there is no adsorption, and hence these parameters do not enter into calculation of the values of h against ReSc. The results are graphed in Figure 5.50.
2. The second case is analogous to the first, except $R = 0.02$. The calculations are summarized as in Table 5.7. The results are also graphed in Figure 5.50, and the data fall on the same curve as the results from part 1.

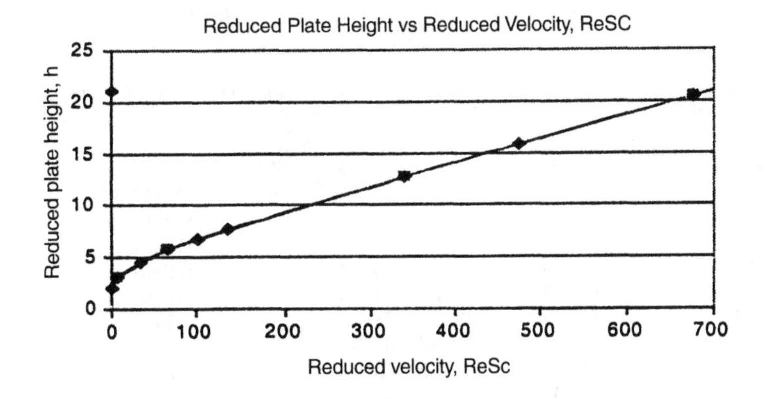

Figure 5.50 Plot of reduced plate height vs. reduced velocity, ReSc, example 8.

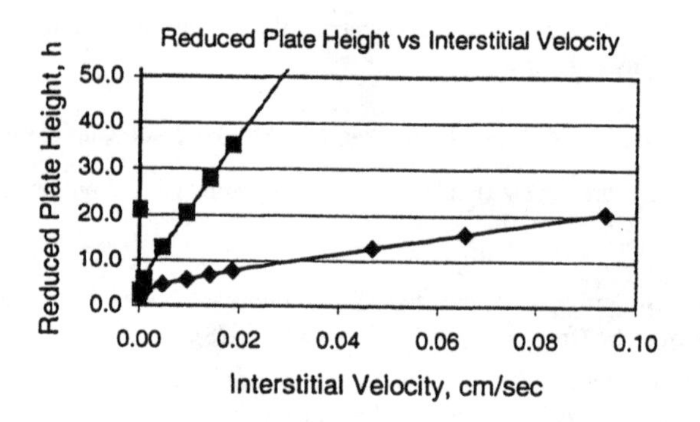

Figure 5.51 Plot of reduced plate height vs. interstitial velocity, example 8.

TABLE 5.7 Case 2 (Experimental Case) for Case II: R = 0.02 cm

Interstitial Velocity (v), cm/s	Re = $2Rv$ dens/visc (visc in g/cm · s dimensionless)	Solution Diffusion Infinite Dilution BSA (D_m), cm^2/s	Convective Axial Dispersion Coefficient, 20°C (D), cm^2/s	Equation Form for Calculating Axial Dispersion Coefficient
9.375E-07	0.00000	5.53E-07	3.95E-07	Eq. (5.109)
9.375E-06	0.00004	5.53E-07	3.95E-07	
0.00009	0.00038	5.53E-07	5.59E-06	Eq. (5.113)
0.00094	0.00375	5.53E-07	8.20E-05	
0.00469	0.01875	5.53E-07	5.36E-04	
0.00938	0.03750	5.53E-07	1.20E-03	
0.01406	0.05625	5.53E-07	1.93E-03	
0.01875	0.07500	5.53E-07	2.70E-03	
0.04688	0.18750	5.53E-07	7.87E-03	
0.06563	0.26250	5.53E-07	1.17E-02	
0.09375	0.37500	5.53E-07	1.77E-02	
0.11250	0.45000	5.53E-07	2.19E-02	
0.14063	0.56250	5.53E-07	2.84E-02	
0.18750	0.75000	5.53E-07	3.97E-02	
0.37500	1.50000	5.53E-07	8.90E-02	
0.56250	2.25000	5.53E-07	1.43E-01	
0.75000	3.00000	5.53E-07	2.00E-01	

Sc =$\mu/(\rho D_m)$ *dimensionless*	*ReSc*	*Peclet Number*	*Nusselt Number*	*Damkohler Number*	*m*	m^* *BSA Eq. (5.108)*	*h*
1.81E+04	0.07	0.095	4.033	—	3.51E+01	0	21.085
1.81E+04	0.68	0.949	4.308	—	3.51E+01	0	2.123
1.81E+04	6.78	0.671	6.069	—	3.51E+01	0	3.124
1.81E+04	67.76	0.457	11.840	—	3.51E+01	0	5.790
1.81E+04	338.79	0.350	20.035	—	3.51E+01	0	12.736
1.81E+04	677.57	0.312	25.209	—	3.51E+01	0	20.412
1.81E+04	1016.36	0.291	28.845	—	3.51E+01	0	27.828
1.81E+04	1355.15	0.278	31.741	—	3.51E+01	0	35.127
1.81E+04	3387.87	0.238	43.061	—	3.51E+01	0	78.034
1.81E+04	4743.02	0.225	48.169	—	3.51E+01	0	106.306

TABLE 5.7 Continued

Sc $=\mu/(\rho D_m)$ *dimensionless*	*ReSc*	*Peclet Number*	*Nusselt Number*	*Damköhler Number*	*m*	m^*	*h BSA Eq. (5.108)*
1.81E+04	6775.75	0.212	54.247	-	3.51E+01	0	148.512
1.81E+04	8130.90	0.206	57.645	-	3.51E+01	0	176.570
1.81E+04	10163.62	0.198	62.094	-	3.51E+01	0	218.578
1.81E+04	13551.50	0.189	68.342	-	3.51E+01	0	288.457
1.81E+04	27102.99	0.168	86.103	-	3.51E+01	0	567.156
1.81E+04	40654.49	0.157	98.562	-	3.51E+01	0	845.282
1.81E+04	54205.98	0.150	108.481	-	3.51E+01	0	1123.140

3. The plot of reduced plate height against reduced flowrate results in a single characteristic curve, where the different values of the parameter (here the size of the stationary phase particles) is incorporated into the dimensionless parameters used in the correlation. The corresponding graphs for parts 1 and 2 are given in Figures 5.50 and 5.51. The graph of h as a function of ReSc (Figure 5.50) and graph of h as a function of interstitial velocity (Figure 5.51—lower curve for $R = 0.002$, upper curve for $R = 0.020$) illustrate the differences between the two types of plots.
4. The graph of h against ReSc (i.e., reduced velocity) in Figure 5.50 gives a characteristic curve in which the particle size effect is incorporated into the Re number. The plot of h against interstitial velocity shows a major difference in Figure 5.51, however, since the different flowrates will give the same value of h. The larger particle size causes greater extent of dispersion, and therefore a larger value of h at a given flowrate.
5. The value of the HETP, H, at the base flowrate of 4.2 mL/min is calculated from Eq. (5.97). Results are in Tables 5.6 (for $R = 0.002$ cm) and 5.7 (for $R = 0.02$ cm):

$$h = H/2R$$

For small particles $\quad H = 2Rh = 2\,(0.002)\,(5.8) = 0.232$ cm

For large particles $\quad H = 2\,(0.02)\,(20.4) = 0.816$ cm

5.19 MASS TRANSFER AND ADSORPTION KINETICS ALSO IMPACT PLATE HEIGHT

The general form of the plate height correlation, Eq. (5.99), is applicable to ion exchange, reversed or normal phase, and hydrophobic interaction chromatography. The mass transfer coefficient (k_c), the adsorption rate constant (k_a), and the equilibrium distribution coefficient (K_D) enter into the plate height equation through the Nusselt number Eq. (5.101), Damköhler number Eq. (5.102), and the sorption kinetic term Eq. (5.107). The expression for mass transfer and adsorption rate constants will change for different ranges of flowrate (i.e., different values of ReSc), while K_D is a velocity-independent equilibrium constant that depends on the type of solute, the mobile phase composition, and the type of stationary phase.

For ReSc < 1, the Nusselt number is given by Eq. (5.116), and for ReSc > 1 it is given by Eq. (5.120). Alternately, the Nusselt number Eq. (5.101) can be calculated from the characteristic dimension or diameter (= $2R$) of the particles that make up the stationary phase, the effective diffusivity of the solute in an unbounded fluid, D_m, and the overall mass transfer coefficient, k_c.

The mass transfer coefficients between phases for ion exchange and sorption[12] give a measure of mass transfer resistances in the fluid outside particles, and as well as mass transfer inside the particles due to diffusion resistance within the pores and/or within the solid itself. The rate at which solute adsorbs or becomes fixed to the surface of the adsorbent is incorporated into the mass transfer coefficients (Sherwood et al., 1975). As summarized by Levenspiel (1972), the resistances for a catalyst particle are as follows:

1. *Fluid film resistance.* Solute diffuses from main body of fluid to the exterior surface of the particle at a rate of

$$\frac{\partial q}{\partial t} = \frac{k_f a}{\rho_B}(C - C_i) \tag{5.121}$$

2. *Pore diffusion resistance.* For porous particle, most of the mass transfer surface area is within the particle, so the solute moves through the fluid-filled spaces within each particle, in spaces where the solute is smaller than the characteristic dimension of these spaces:

$$\frac{\partial q}{\partial t} = k_p a(q_i - q) \tag{5.122}$$

3. *Surface resistance.* The solute will ultimately interact or become reversibly retained at the surface of the particle.

The parameters in Eqs. (5.121) and (5.122) are defined as

q = solid phase concentration of adsorbed solute, moles/g = C_s
q_i = concentration of solute in stationary phase at the interface with the fluid phase
t = time, s
ρ_B = density of stationary phase: mass of stationary phase/volume of bed, g/mL
k_f = mass transfer coefficient for solute transport from flowing fluid phase to the surface of the stationary phase, cm/s
k_p = mass transfer coefficient for solute transport through the stationary phase, cm/s
a = interfacial area of stationary phase, area per volume of column, cm^2/mL
C = concentration of solute in fluid phase, moles/L
C_i = concentration of solute in fluid phase at the interface between fluid and stationary phase, moles/L

[12] The adsorption of molecules from a gas mixture or liquid solution onto the internal surface of a porous solid, or the adsorption of ions by an ion exchange resin or material, with these ions displacing other ions originally in the resin, is referred to as sorption (Sherwood et al., 1975). In sorption, the concentrations in the fluid and solid phases inside the bed depend on both position and time. Also, according to Sherwood et al., "a steady state is never reached in practical conditions because a bed loses its sorptive ability, and permits feed to pass through unchanged in its composition" when equilibrium is approached (Sherwood et al., 1975, p. 548).

Unlike catalysis, adsorption does not involve a reaction in which the adsorbed species is converted into a different molecule. The concentrations of the solutes in liquid chromatography are generally low enough so that the heat of adsorption effect is small. For a single idealized pore, the resistances can be presented in the conceptual framework shown in Figure 5.52. Not all of the resistances occur in series or in parallel, and hence their effects are not readily added together using a simple expression.

The determination of resistances can be simplified by identifying the dominating resistance and then using a single equation to obtain an approximation for the overall resistance. In the case of sorption or chromatography, the contribution of resistance 3 due to adsorption at the surface is likely to be small, namely the actual sorption or ion exchange rates are likely to be rapid compared to the rate at which the solute is transported into (or out of) the pore, although for monovalent ion exchange such as Na^+ exchanging for H^+ this may not be the case (Sherwood et al., 1975). The parameter q_m in Eq. (5.124) corresponds to an ion exchange resin in which all ion exchange groups on the surface are associated with H ions:

$$Na^+ + RH \rightarrow H^+ + RNa \tag{5.123}$$

The concentrations terms that correspond to Eq. (5.123) are

$$(C, q_m - C) \quad \text{and} \quad (C_o - C, q) \tag{5.124}$$

where C_o is the concentration of H^+ at equilibrium with the solid, and q is the concentration of Na^+ that has ion exchanged for H^+ on the surface.

The kinetic driving force for reaction (i.e., ion exchange), Eq. (5.123), is defined by Thomas as

$$\rho_B \frac{\partial q}{\partial t} = \kappa a \left[C\left(1 - \frac{q}{q_m}\right) - \frac{1}{K}(C_o - C)\frac{q}{q_m} \right] \tag{5.125}$$

where

a = interfacial area

$b = C\left(1 - \frac{q}{q_m}\right) - \frac{1}{K}(C_o - C)\frac{q}{q_m}$ = kinetic driving force

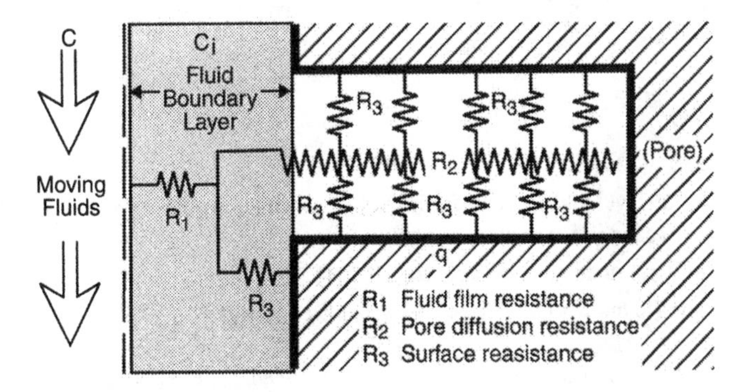

Figure 5.52 Schematic representation of resistances to mass transfer as summarized by Levenspiel, 1972.

κ = kinetic coefficient
K = equilibrium constant

The interfacial contact area of a sphere in a packed bed is given by

$$a = \frac{\pi(2R)^2(1-\varepsilon_b)}{\left(\frac{\pi(2R)^3}{6}\right)} = \frac{6(1-\varepsilon_b)}{d_p}\frac{\text{cm}^2}{\text{cm}^3}, \qquad \text{where } d_p = 2R \tag{5.126}$$

The correlation of solid-state diffusion in the stationary phase, based on empirical fit of measured overall mass transfer resistances for ion-exchange processes is given by

$$k_p = \frac{10D_p}{d_p(1-\varepsilon_b)} \tag{5.127}$$

where

D_p = particle-phase diffusivity, cm^2/s, estimated from D_m as discussed in Section 5.18.
d_p = particle diameter = $2R$

Multiplying Eqs. (5.126) and (5.127) together gives

$$k_p a = \frac{60D_p}{d_p^2} \tag{5.128}$$

where

$k_p a$ = solid-phase mass transfer coefficient per unit volume of the stationary phase, cm^3/s, where a is the interfacial contact area given in Eq. (5.126)

The calculation of k_f can be attempted from the correlation in Eq. (5.129) which is based on fitting data for a fixed bed of spherical particles with a void fraction $\varepsilon_b = 0.4$ to 0.44:

$$k_f = j_D \varepsilon_b v \text{Sc}^{-2/3} \qquad \text{for } 10 < \text{Re}_p < 2500 \tag{5.129}$$

where

j_D = flux
v = interstitial velocity, cm/s
$\varepsilon_b v$ = superficial velocity, which is the velocity which would exist without stationary phase packed in the column, cm/s
Sc = Schmidt number, as defined in Eq. (5.104)
Re_p = particle Reynold's number as defined in Eq. (5.103)

The values of v and Sc in Eq. (5.129) are readily calculated or are known. The interstitial velocity, v, is defined by the flowrate, column diameter, and column void fraction, while the

Schmidt number is obtained from the viscosity and density of the mobile phase. If the mobile phase is water, μ is given by Eq. (5.114). The diffusivity, D_m, of the solute is estimated by Eqs. (5.110), (5.111), or (5.113) if the solute is a protein or PEG. Otherwise, the data can be obtained from the literature. The flux, j_D, may be determined from a graphical correlation of the type illustrated in Figure 5.53 in which previously measured data for the flux, j_D, as a function of the Reynolds's number, Re, gives a characteristic curve. The flux, j_D, can be obtained from this type of graph or its equivalent numerical representation if the Reynolds's number is known. The value of k_f is then obtained from Eq. (5.129). The plot of the type illustrated in Figure 5.53 is based on data obtained for both gases and liquids. The liquid systems consisted of water with isobutanol, methyl ethyl ketone, benzoic acid, or 2-napthol, benzene with salicylic acid, and *n*-succinic acid in *n*-butanol or acetone. The particle sizes in the packed beds ranged from 0.056 to 1.61 cm, with Sc of 159 to 12,300.

Another correlation that incorporates the effect of ε_b has the form (Sherwood et al., 1975)

$$\varepsilon_b j_D = 1.10\left(\frac{\mathrm{Re}}{1-\varepsilon_b}\right)^{-1/2} \tag{5.130}$$

for $0.25 < \varepsilon_b < 0.50$ and $40 < \mathrm{Re}/(1-\varepsilon_b) < 4000$.

Once the values of k_p and k_f are measured, κ will still need to be estimated. The process of mass transfer requires that the flux $(\partial q/\partial t)$ of solute to the stationary phase due to the kinetic driving force given by Eq. (5.125) is equivalent to the flux through the fluid film given by the rate expression of solute transport from the fluid in Eq. (5.121) for fluid film resistance, which equals the flux in the pore as given by Eq. (5.122) for pore diffusion resistance:

$$\underbrace{\frac{\kappa a}{\rho_b}\left[c\left(1-\frac{q}{q_m}\right)-\frac{1}{K}\frac{q}{q_m}(C_o-C)\right]}_{\text{kinetic term}} = \underbrace{\frac{k_f a}{\rho_B}(C-C_i)}_{\text{film diffusion}} = \underbrace{k_p a(q_i-q)}_{\text{pore diffusion}} \tag{5.131}$$

where

q_i = equilibrium concentration of ion adsorbed at the interface between the stationary and fluid phases (i.e., on the surface of the stationary phase)

The expression for q_i can be specified by a Langmuir type isotherm, for example,

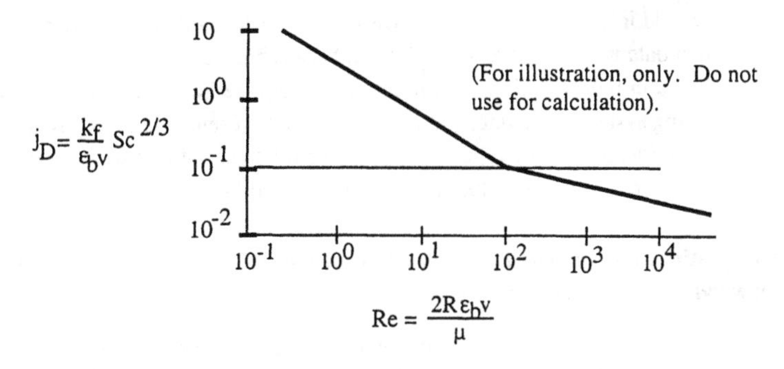

Figure 5.53 Schematic representation of a plot that is used to estimate flux, j_D, from the particle Reynolds number.

$$q_i = q_m \frac{K\left(\frac{C_i}{C_o}\right)}{1+(K-1)\left(\frac{C_i}{C_o}\right)} = \frac{q_m K C_i}{(C_o - C_i) + KC_i} \tag{5.132}$$

In the case of a constant pattern condition (Figure 5.54) when $C = \frac{1}{2}C_o$ (i.e., concentration of the solute at the inflection point of a break-through curve for constant pattern conditions so that $C/C_o = q/q_m = \frac{1}{2}$), Eq. (5.131) becomes

$$\underbrace{\kappa\left[\frac{C_o}{2}\left(1-\frac{1}{2}\right)-\frac{1}{K}\frac{1}{2}\left(C_o-\frac{C_o}{2}\right)\right]}_{\text{kinetic term}} = \underbrace{k_f\left(\frac{C_o}{2}-C_i\right)}_{\text{film diffusion}} = \underbrace{k_p\rho_{Bed}\left(q_i-\frac{q_m}{2}\right)}_{\text{pore diffusion}} \tag{5.133}$$

where ρ_{Bed} represents the bulk density of the stationary phase as g dry stationary phase/mL bed.

A constant pattern condition refers to the constant pattern or shape of breakthrough patterns as a function of column length, L, as shown in Figure 5.54.

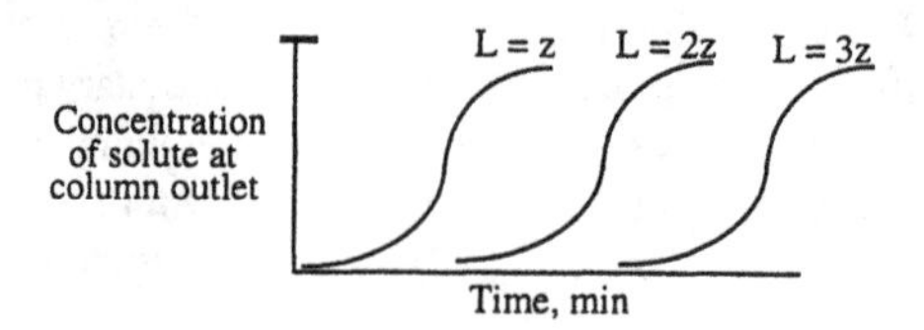

Figure 5.54 Schematic representation of breakthrough patterns for a constant pattern condition.

Other approaches that have been used by chemical engineers to measure and correlate mass transfer and these are summarized by Sherwood et al. (1975). Additional selected correlations for mass transfer at liquid/solid interfaces are tabulated by Cussler (1997). Despite the availability of a calculation method, the results that are obtained must be viewed carefully, since the Reynolds numbers encountered in chromatography will be much lower (Re < 1), and the composition of fluids are different (proteins in aqueous salt solutions) than those used to develop these correlations.

The data upon which the characteristic lines are based as reported in handbooks or texts do not include measurements for biomolecules. Data that confirm that biomolecules fall on the same characteristic curves would increase the confidence with which such a correlation may be used. The availability of such data would help to close the gap between the current practice of analyzing mass transfer effects in bioseparations and the more generalized approaches associated with chemical engineering systems. The data and correlations are presented here to provide a possible starting point for identifying the type of data that would be needed, as well as give an indication of the magnitude of order of values of mass transfer coefficients.

Example 9. Estimating the Kinetic Mass Transfer Coefficient for an Ion Exchanger (from Sherwood et al., 1975, ex. 10.5, p. 583)

The use of Eq. (5.125) to model an ion-exchange process requires that the kinetic driving force, κ, be known. Determination of κ is illustrated by an example given by Sherwood et al. (1975) for a cation exchange resin with a capacity of 4.9 milliequivalents/g resin (dry basis) (= q_m) for a 10 mM aqueous NaCl (= C_o) being passed through the bed at 25°C at a superficial fluid velocity of 1.0 cm/s. The properties of the stationary phase are as follows:

ρ_B = 0.7 g resin (dry basis)/cm^3
R = 1 mm (d_p = cm), average radius of ion exchange particles
ε_b = 0.4
$\varepsilon_b v$ = 1.0 cm/s (this is a large velocity compared to most chromatographic systems)
D_{m,Na^+} = 1.2 · 10^{-5} cm^2/s (in free solution)
D_{p,Na^+} = 0.094 · 10^{-5} cm^2/s (inside the resin particles)
K = 1.2 (= equilibrium constant between sodium ions in solution and hydrogen ions associated with the cation exchanger)
C_o = 10 mM = 0.01 M NaCl

Use the condition of $C = C_o/2$ as an average appropriate concentration to obtain a first estimate of the kinetic driving force, κ.

Solution

The equivalence of kinetic, film diffusion, and pore diffusion processes given in Eq. (5.133) is used to determine κ. In order to calculate κ, k_f or k_p must be determined. Values of q, C, C_i, and k_p must first be determined using information given in the problem statement in order to determine κ from the first equality given in Eq. (5.133):

$$\kappa\left[C\left(1-\frac{q}{q_m}\right)-\frac{1}{K}\frac{q}{q_m}(C_o-C)\right]=k_f(C-C_i)$$

Obtain k_f from a plot of the type shown in Figure 5.53. Note that Figure 5.53 is for explanation purposes only, and is not suitable for this calculation. Rather refer to a plot of data, or a correlation in the chemical engineering literature, such as Sherwood et al. (1975, fig. 6.18, p. 244).

1. Calculate the particle's Reynolds number (d_p = ZR):

$$\mathrm{Re}=\frac{d_p\varepsilon_b \upsilon\rho_B}{\mu}=\frac{(0.2)(0.4)(1.0)}{(0.00913)}=8.76$$

$$\frac{\mathrm{cm(cm/s)\,(g/cm^3)}}{\mathrm{g\cdot cm/s}}$$

2. Use correlation for mass transfer between flowing liquids and particles in a packed bed, such as given by Sherwood et al. (1975, fig. 6.18).

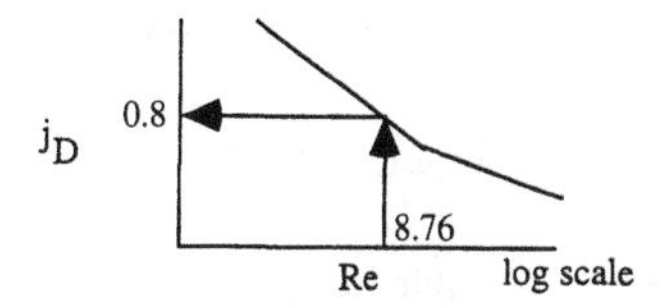

3. Calculate k_f from Eq. (5.129) and with Sc from Eq. (5.104):

$$k_f = j_D \varepsilon_b \upsilon Sc^{-2/3} = j_D \varepsilon_b \upsilon \left(\frac{\mu}{\rho_b D_{m,Na}} \right)^{-2/3}$$

$$= (0.8)(1.0) \left(\frac{0.00913}{(1)(1.2 \cdot 10^{-5})} \right)^{2/3}$$

$$\frac{cm}{s} \cdot \frac{\left| \frac{g}{cm \cdot s} \right|}{\left| \frac{g}{cm^3} \right| \frac{cm^2}{s} \right|}.$$

$$k_f = 0.8(761)^{-2/3} = \boxed{9.5 \cdot 10^{-3} \frac{cm}{s}}$$

4. Calculate k_p from Eq. (5.127):

$$k_p = \frac{10(D_{p,Na^+})}{d_p(1 - \varepsilon_b)} = \frac{(10)(0.094 \cdot 10^{-5})}{(0.2)(1 - 0.4)} = \boxed{7.83 \cdot 10^{-5} \frac{cm}{s}}$$

5. Calculate equilibrium expression using Eq. (5.132):

$$q_i = \frac{q_m K C_i}{(C_o - C_i) + K C_i} = \frac{(4.9)(1.2) C_i}{(0.01 - C_i) + 1.2 C_i} = \frac{5.88 C_i}{0.01 + 0.2 C_i} \tag{5.134}$$

6. Use Eq. (5.133) together with calculated values of k_f (from step 3), k_p (from step 4), and q_i (from step 5) to solve for C_i at the midpoint of the breakthrough curve ($q = q_m/2$; $C = C_o/2$):

$$k_f \left(\frac{C_o}{2} - C_i \right) = k_p \rho_{Bed} \left(q_i - \frac{q_m}{2} \right) \tag{5.135}$$

$$(9.5 \cdot 10^{-3}) \left(\frac{0.01}{2} - C_i \right) 1000 \frac{\text{milli-equivalents}}{\text{mole}} = (7.83 \cdot 10^{-5})(0.7) \left(\frac{5.88 C_i}{0.01 + 0.2 C_i} - \frac{4.9}{2} \right)$$

Solve for C_i using the Excel spreadsheet solver tool.

$$C_i = 0.004667$$

Use Eq. (5.134) from step 5 and value of C_i calculated from Eq. (5.135) to calculate q_i:

$$q_i = \frac{5.88C_i}{0.01 + 0.2C_i} = \frac{5.88(4.997 \cdot 10^{-3})}{0.01 + (0.2)(4.997 \cdot 10^{-3})} = 2.570 \frac{\text{meq}}{\text{g}}$$

7. The results from step 6, combined with the expression for κ in Eq. (5.133), give an estimate of the kinetic mass transfer coefficient:

$$\kappa\left[C\left(1 - \frac{q}{q_m}\right) - \frac{1}{K}\frac{q}{q_m}(C_o - C)\right] = K_f(C - C_i)$$

$$\kappa\left[0.005(1 - 0.5) - \frac{1}{1.2}(0.5)\left(0.01 - \frac{0.01}{2}\right)\right] = (9.5 \cdot 10^{-3})(0.005 - 0.004667)$$

$$\kappa = \frac{3.163 \cdot 10^{-6}}{4.17 \cdot 10^{-4}} = 0.00758 \frac{\text{cm}}{\text{s}}$$

Note that 1 mMole = 1 milli-equivalent of Na^+. The concentrations C and q at the midpoint of constant pattern behavior are

$$C = \frac{C_o}{2} = \frac{0.01}{2} = 0.005 \frac{\text{mequiv}}{\text{cm}^3}$$

$$q_m = \frac{4.9}{2} = 2.45 \frac{\text{mequiv}}{\text{g}}$$

The concentrations of the sodium at the liquid and solid interface that correspond to C and q_m are given by C_i and q_i, respectively, as calculated above.

The solution of Sherwood et al. (1975) gives $C_i = 4.61 \cdot 10^{-3}$ mequiv/cm^3 with $q_i =$ 2.48 mequiv/gram. Therefore, from Eq. (5.131),

$$k_f(C - C_i) = 9.5 \cdot 10^{-3}\,(0.005 - 0.00461) = 3.125 \cdot 10^{-6} \frac{\text{cm}}{\text{s}}$$

$$\kappa = \frac{3.125 \cdot 10^{-6}}{C\left(1 - \frac{q}{q_m}\right) - \frac{1}{K}\frac{q}{q_m}(C_o - C)}$$

$$= \frac{3.125 \cdot 10^{-6}}{0.005(1 - 0.5) - \frac{1}{1.2}(0.5)(0.01 - 0.005)} = 0.00750 \frac{\text{cm}}{\text{s}}$$

A number of other methods can also be used to estimate κ. As shown by Sherwood et al., the calculated values may range from 10^{-5} to 10^{-2} cm/s, depending on the assumptions and correlations used to carry out the calculations. The value calculated here differs slightly from that reported by Sherwood et al., since the value for C_i, obtained by solving Eq. (5.135) is 0.004667, rather than the 0.0046065 obtained by Sherwood.

The kinetic mass transfer coefficient, which is analogous to k_c in Eq. (5.101), can be used to estimate the Nusselt number which in turn is used to calculate plate height using Eq. (5.99). This assumes that an estimate or experimentally determined value of k_a is also available. Furthermore, this approach requires that mass transfer is a determinant of peak spreading and that the capacity factor of the solute reflects its equilibria with the stationary behavior. This is a key measure of separation, as discussed in the next section.

5.20 CHROMATOGRAPHIC CAPACITY FACTORS ARE DETERMINED FROM PEAK RETENTION

The types of chromatographic separations shown in Figures 5.8 through 5.11 define the expected characteristics for column chromatography. If dispersion did not occur, the inlet and outlet concentrations of a pulse containing a single component would be the same, as indicated in Figure 5.55. A pulse injected onto the column would remain a pulse, with no peak broadening or dilution.

In this ideal case, a binary mixture, when injected into the column would separate into two bands and elute in the same volume in which the solute was injected, as represented in Figure 5.56. For a two component mixture, where each component is present in equal amounts, the concentration in the eluting bands would be half of the inlet concentration, as represented in Figure 5.56.

The chromatogram in Figure 5.56 represents the case of ideal chromatography. However, since mass transfer resistances result in peak dispersion, the ideal peaks of Figure 5.56 will broaden, thus spreading their mass over a larger volume as shown in Figure 5.57. The center of the peak mass, however, represented by t_1 and t_2, is the same in both Figures 5.56 and 5.57.

The discussion of peak broadening, and the simulation of peaks eluting from a chromatography column, given earlier in this chapter, outlines numerical procedures by which elution profiles for individual peaks can be calculated. If the presence of one component does not affect the retention behavior of the other, a combination of simulations for individual peaks may be used to represent the eluting peaks, as illustrated in Example 6, in order to construct a chromatogram. While Gaussion peak behavior, and lack of interference between components, may be found in many cases, the data required to simulate even this type of simple behavior may not be available for a newly developed separation. Therefore process scale-up and simulation often involves obtaining elution profiles for a separation at the bench scale, analyzing chromatograms obtained over a limited range of conditions, and then scaling-up the result to a larger column.

The first step is to define the retention behavior of the individual peaks in terms of a retention factor or capacity factor k' for cases where linear equilibrium is approximated. The capacity factor is a measure of how strongly a solute interacts with the stationary phase and is directly pro-

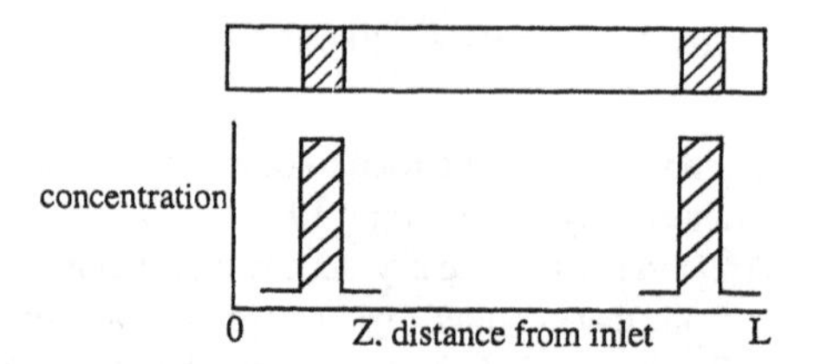

Figure 5.55 Schematic of pulse shape for ideal chromatography (no dispersion).

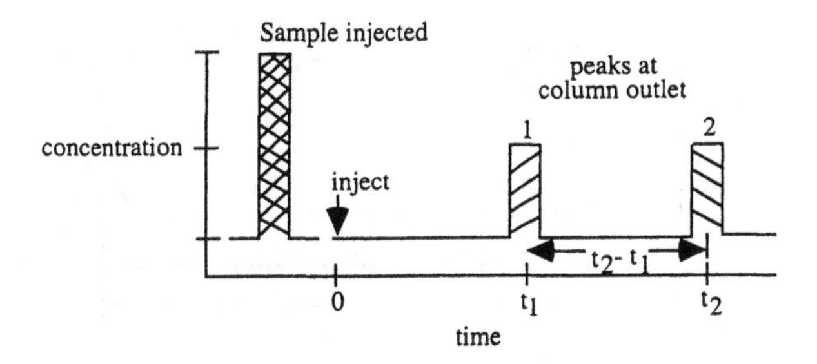

Figure 5.56 Schematic representation of separation of two components in a single pulse under ideal chromatography conditions.

portional to the distribution coefficient, K_D, which is given by Eq. (5.136) in the case of linear equilibrium. The expression for a linear isotherm is simply

$$q_{eq} = K_D C_{eq} \tag{5.136}$$

where

q_{eq} = solute concentration associated with the stationary phase at equilibrium, $\dfrac{\text{moles solute}}{\text{volume of stationary phase}} = C_{s,eq}$

K_D = distribution coefficient or equilibrium constant

C_{eq} = solute concentration in mobile phase which is in equilibrium with solute concentration associated with the stationary phase

For linear equilibrium, the capacity factor, k', is

$$k' = \frac{q_{eq}}{C_{eq}}\frac{V_s}{V_o} = \frac{q_{eq}}{C_{eq}}\phi = K_D\phi \tag{5.137}$$

where

ϕ = ratio of stationary to mobile phase (i.e., phase ratio)

$$= \frac{1-\varepsilon_b}{\varepsilon_b} \text{ or } \frac{1-(\varepsilon_b+\varepsilon_p)}{\varepsilon_b} \tag{5.138}$$

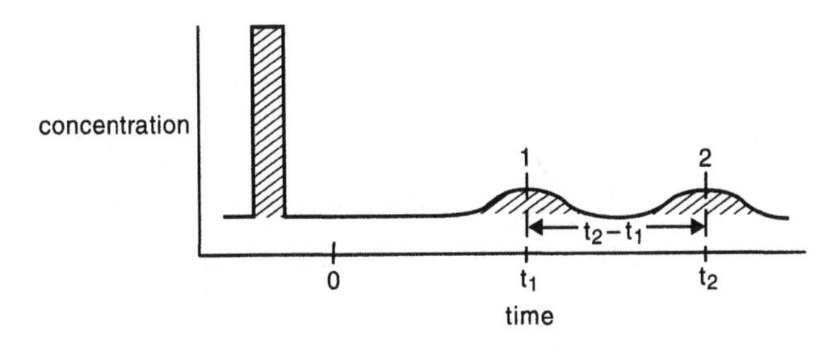

Figure 5.57 Peak broadening due to dispersion.

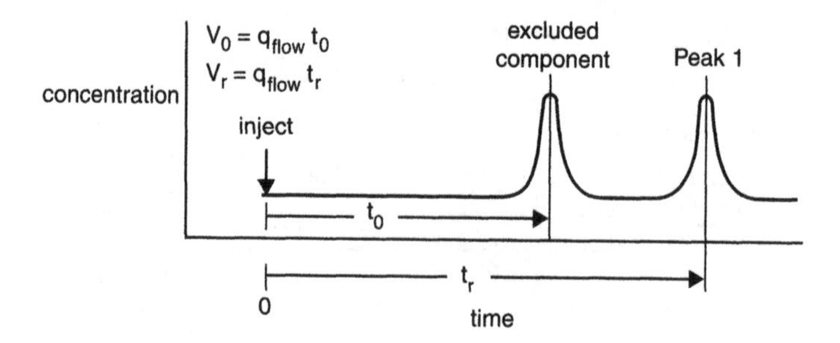

Figure 5.58 Schematic representation for retention times of two peaks.

$$K_D = \text{distribution coefficient} = \frac{q_{eq}}{C_{eq}} \text{ by Eq. (5.136).}$$

The capacity factor for liquid chromatography is defined as

$$k' = \frac{t_r - t_o}{t_o} = \frac{q_{flow}t_r - q_{flow}t_o}{q_{flow}t_o} = \frac{V_R - V_o}{V_o} \tag{5.139}$$

where t_r is the retention time of the peak and t_o is the retention time of an excluded component, q_{flow} represents flowrate, while V_R is the retention volume and V_o is the interparticulate void volume. These values are obtained by reading the chromatogram as in Figure 5.58.

The capacity factor represents the ratio of the moles of component 1 on or in the stationary phase to the moles of the same solute in the extraparticulate fluid volume. For an excluded component, $t_r = t_o$ and $k' = 0$. For a strongly retained component, for example, if $t_r = 10t_o$, the capacity factor will have a large value:

$$k' = \frac{(10t_o) - t_o}{t_o} = 9 \tag{5.140}$$

indicating that the stationary phase has a significant capacity for the component. Comparison of Eqs. (5.138) and (5.140) shows that the peak's retention time, and therefore the capacity factor k', should be independent of the concentration of the injected solute as long as the concen-

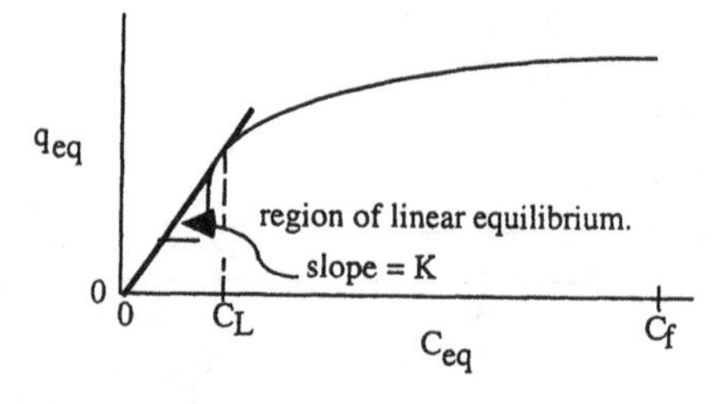

Figure 5.59 Schematic representation of an equilibrium isotherm.

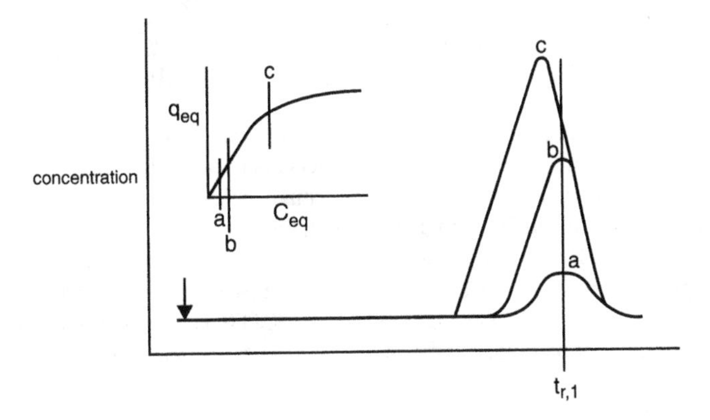

Figure 5.60 Schematic representation of peak shape and retention as a function of solute concentrations. Points a, b, and c in the equilibrium curve (upper left) correspond to peaks a, b, and c in the drawing.

tration is within the range of linear equilibrium. This region is shown schematically in Figure 5.59. This schematic applies to many types of liquid chromatographic separations, particularly where the solutes are at a low concentration.

In Figure 5.59, q_{eq} and C_{eq} represent concentrations of the solute on the solid and mobile phases at equilibrium. As long as the concentration is below C_L and ϕ is defined, it should be possible to obtain K, from the peak's chromatographic retention time, since the capacity factor is related to retention time through Eq. (5.139) and the equilibrium constant is related to capacity factor through Eq. (5.138). Linear equilibria are confirmed when samples containing different concentrations are injected into the liquid chromatography column, and all elute in the same retention time. If the concentration is above C_L (corresponds to peak *c* in Figure 5.60), the retention time (as measured by the peak center) will decrease to less than $t_{r,1}$, since a portion of the solute is less retained due to the equilibrium curve being in a nonlinear area (see the inset in Figure 5.60) where q_{eq} increases less proportionately than the increase in C_{eq}. Hence some of the solute will move ahead in the column and elute earlier than the rest of the peak. When this occurs, the peak becomes more triangular in shape. Components whose concentrations are in a range where the retention time is independent of concentration (i.e., linear equilibrium) can be used to determine the distribution coefficient using Eq. (5.138). However, this must be confirmed through measurement of k' over a range of flowrates. If the flowrate is too fast, equilibria may not be attained, and the values measured from retention time could be smaller than they actually would be if equilibrium occurs.

5.21 CHROMATOGRAPHIC SEPARATIONS ARE DEFINED BY DIVERGENCE OF PEAK CENTERS: CAPACITY FACTORS, PHASE RATIOS, AND RESOLUTION

The peak centers of the representations shown in Figures 5.56 and 5.57 are separated by $t_2 - t_1$. At a constant flowrate of q_{flow}, this difference corresponds to the difference in retention volumes of $V_{R,2} - V_{R,1}$, since $V_2 = q_{flow}\, t_2$ and $V_1 = q_{flow}\, t_1$. Separation of two peaks occurs because they have different capacity factors with respect to the stationary phase at the conditions of the separation. In the case of size exclusion chromatography, the maximum capacity factor will be given by

$$k' = \frac{V_{col} - (V_o + V_p)}{V_o} = \frac{V_{col} - (\varepsilon_b V_{col} + \varepsilon_p V_{col})}{\varepsilon_b V_{col}} = \frac{1 - (\varepsilon_b + \varepsilon_p)}{\varepsilon_b} = \phi_{bp} \tag{5.141}$$

since all of the solute must elute within one column volume, V_{col}. ϕ_{bp} denotes the phase ratio based on the total column void fraction. Retention occurs because the smaller molecules can explore more of the column's void volume than larger molecules which are not able to penetrate the small pores, as represented for a single particle in Figure 5.61.

The small molecules have a longer residence time in the column, on the average, since they penetrate the pores of the stationary phase particles. For a packing material with a particle size of 40 microns ($d_p = 0.04$ cm), the sieving occurs over 180,000 particles packed in a 10 mL column, since the number of particles per cm^3 of column can be estimated as:

$$n = \frac{6(1-\varepsilon_b)}{\pi d_p^3} = \frac{6(1-0.4)}{\pi (0.04)^3} = 17{,}904 \,\frac{\text{particles}}{\text{cm}^3}$$

Hence, the transport of the smaller solute is sufficiently hindered so that it may elute as a separate peak.

The fluid phase associated with the extraparticulate volume is given by $\varepsilon_b V_{col}$, as developed previously in sections 5.3 and 5.15. Hence, the maximum capacity factor for an excluded, non-adsorbed molecule in size exclusion chromatography also represents a definition of the phase ratio:

$$k' = \frac{V_{col} - \left(V_o + {}_o\mathcal{V}_p\right)}{V_o} = \frac{1 - \varepsilon_b}{\varepsilon_b} = \phi_b \tag{5.142}$$

The capacity factors for two components (named here as peak 1 and peak 2) must be sufficiently different so that the inlet pulse of sample diverges into its individual components. At the same time the capacity factors should not be too different so that the column length (and

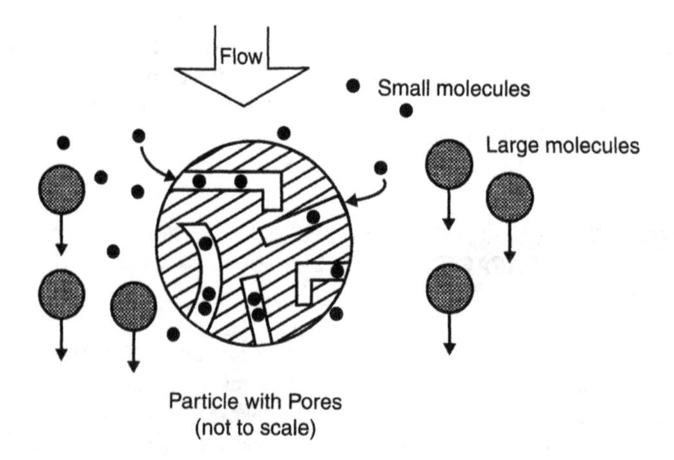

Figure 5.61 Schematic representation of size exclusion. Large molecules do not diffuse into small pores, and hence are said to be excluded. (Note: drawing is not to scale).

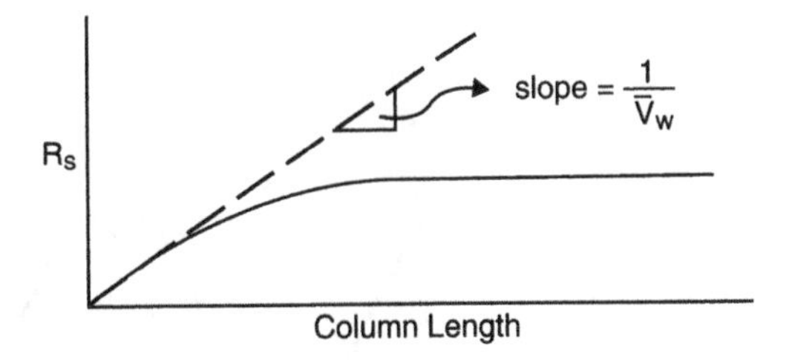

Figure 5.62 Resolution as a function of column length for ideal chromatography (straight line) compared to actual case (curve).

volume), while sufficient for the desired separation, is not so long as to result in extra cost and time. Conversely, if the capacity factors are similar because the distribution coefficients for the two components with respect to a given stationary phase are similar, only a partial separation may be achieved for given length of column. A longer column would be an intuitive choice for improving the difference of divergence of the peak centers. For ideal chromatography, the resolution would increase in proportion to the column's length as represented by the dashed line in Figure 5.62. However, due to peak dispersion, the peaks become wider and their maximum concentration less as the column length increases. Consequently the improvement in resolution, R_s, as defined by Eq. (5.143) is less than the ideal case represented in Figure 5.62. The difference between the peak centers is normalized to the average of the widths of the two peaks, since the two peaks will broaden as they pass through the column (represented in Figure 5.57) with the width of adjacent peaks affecting the separation of the two as represented in Figures 5.63 and 5.64. In the case of Figure 5.63, the peaks are considered to have a resolution of 1, even though there is overlap, while the resolution in Figure 5.64 is greater than 1.

The definition for resolution must therefore consider both divergence of peak centers and peak width:

$$R_s = \frac{\text{divergence of peak centers}}{\text{average of peak widths}} = \frac{V_{R,2} - V_{R,1}}{\frac{(V_{w,1} + V_{w,2})}{2}} \tag{5.143}$$

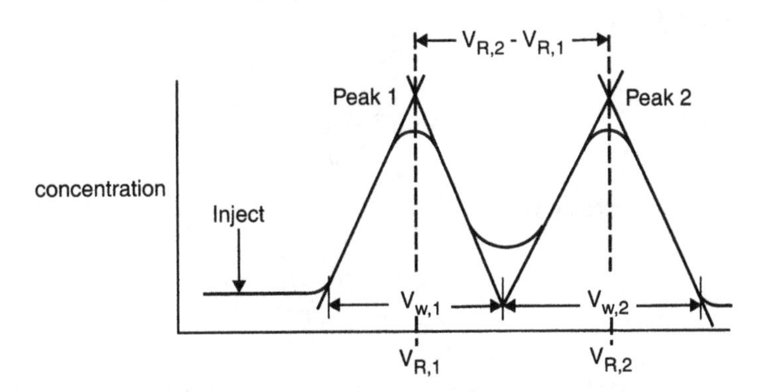

Figure 5.63 Schematic representation of divergence of peak centers (resolution of two peaks).

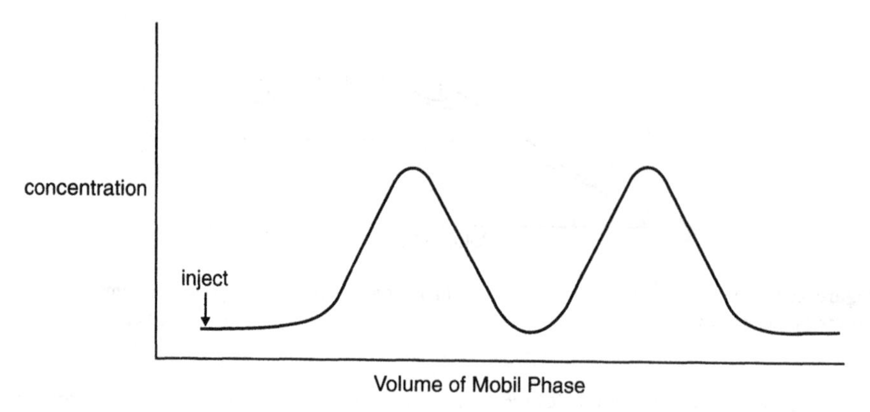

Figure 5.64 Schematic representation of completely resolved peaks, where resolution is greater than one.

where

$V_{R,2}$ = retention volume of peak 2, mL
$V_{R,1}$ = retention volume of peak 1, mL
$V_{w,1}$ = width of peak 1 as defined in Figure 5.38, mL
$V_{w,2}$ = width of peak 2, mL

In an ideal case the peak widths would not change as represented in Figure 5.56, and the resolution would increase with column length as shown by the straight line in Figure 5.62 with a slope of $1/V_w$. This would result in a constant average of peak widths and a linear relation, where resolution is directly proportional to column length:

$$R_s = \frac{1}{V_w}(V_{R,2} - V_{R,1}) \tag{5.144}$$

where V_w = the average of the two peak widths, a constant. Dispersion processes at a constant flowrate give a Gaussian peak (for a component that exhibits linear equilibrium) with a width that follows from Eqs. (5.29) and (5.33):

$$N = 16\left(\frac{V_R}{V_w}\right)^2 \tag{5.145}$$

which, upon rearrangement, gives

$$V_w = \frac{4}{\sqrt{N}} V_R = \frac{4V_R\sqrt{H}}{\sqrt{L}} \tag{5.146}$$

since $N = L/H$. The plate height H is a constant as discussed in section 5.12. As the column is made longer, the width of the peak changes as

$$\frac{V_{w,L_2}}{V_{w,L_1}} = \frac{\dfrac{4V_{R,L_2}\sqrt{H}}{\sqrt{L_2}}}{\dfrac{4V_{R,L_1}\sqrt{H}}{\sqrt{L_1}}} = \frac{\sqrt{L_1}\,V_{R,L_2}}{\sqrt{L_2}\,V_{R,L_1}} \tag{5.146}$$

where

L_1 = length of column 1
L_2 = length of column 2
V_{R,L_1} = retention volume of peak injected into column 1
V_{R,L_2} = retention volume of peak injected into column 2
V_{w,L_2} = width of peak eluting from column of length L_2
V_{w,L_1} = width of peak eluting from column of length L_1

If the column were doubled in length ($L_2 = 2L_1$) so that the retention volume is doubled ($V_{R,L_2} = 2V_{R,L_1}$), the width of the peak eluting from the column with length L_2, V_{w,L_2}, is given by

$$V_{w,L_2} = \sqrt{\frac{L_1}{2L_1}} \cdot \frac{2V_{R,L_1}}{V_{R,L_1}} \cdot V_{w,L_1} = \sqrt{2}\,V_{w,L_1} \tag{5.147}$$

In other words, doubling the length of the column increases the peak width by 1.44. For a two component mixture, the resolution for a column of length $2L_1$ is given by

$$R_{s,L_2} = \frac{2(V_{R,2} - V_{R,1})}{\sqrt{2}\left(\dfrac{V_{w,1} + V_{w,2}}{2}\right)} = \sqrt{2}\left(\frac{\Delta V_R}{\Delta V_w}\right) \tag{5.148}$$

where ΔV_R, ΔV_w, $V_{R,2}$, $V_{R,1}$, $V_{w,1}$, and $V_{w,2}$ refer to retention volumes and peak widths for components 1 and 2 eluting from the shorter column of length L_1 or longer column of length L_2. The ratio of resolutions for columns of lengths L_2 and L_1, respectively, is

$$\frac{R_{s,L_2}}{R_{s,L_1}} = \sqrt{2}\,\frac{\left(\dfrac{\Delta V_R}{\Delta V_w}\right)}{\left(\dfrac{\Delta V_R}{\Delta V_w}\right)} = \sqrt{2} \tag{5.149}$$

Hence doubling the length of the column increases the resolution by a factor of $\sqrt{2}$, or by a factor of 1.44. This effect is represented by the solid line in Figure 5.62, which illustrates the relation between column length and separation for a Gaussian peak under linear equilibrium conditions. This type of rationale forms the basis of some of the scale-up rules for chromatography carried out under conditions of linear equilibrium, as discussed in the next chapter.

REFERENCES

Anderson, L., M. Gulati, P. Westgate, E. Kvam, K. Bowman, and M. R. Ladisch, 1996, Synthesis and optimization of a starch based adsorbent for dehumidification of air in a pressure swing drier, *Ind. Eng. Chem. Res.*, 35, 1180–1187.

Athalye, A. M., S. J. Gibbs, and E. N. Lightfoot, 1992, Predictability of chromatographic protein separations—Study of size-exclusion media with narrow particle size distributions, *J. Chromatogr.*, 589, 71–85.

Barford, R. A., R. McGraw, and H. L. Rothbart, Large sample volumes in preparative chromatography, *J. Chromatogr.*, 166, 365–372.

Beery, K., M. Gulati, E. P. Kvam, and M. R. Ladisch, 1998, Effect of enzyme modification of corn grits on their properties as an adsorbent in Skarstrom pressure swing cycle drier, *Adsorption*, 4, 321–335.

Bienkowski, P. R., A. Barthé, M. Voloch, R. N. Neuman, and M. R. Ladisch, 1986, Breakthrough behavior of 17.5 mol % water in methanol, ethanol, isopropanol, and t-butanol vapors passed over corn grits, *Biotechnol. Bioeng.*, 28, 960–964.

Colin, H., P. Hilaireau, and J. de Tournemire, 1990, Dynamic axial compression columns for preparative high performance liquid chromatography, *Liquid Chromtography: Gas Chromatography*, 8(4), 302–312.

Collentro, W. V., 1999, The use of 0.1-μm membrane filters in USP purified water systems—Case histories, part I, *Pharmaceutical Technol.*, 23(3), 176–192.

Crank, J., 1975, *The Mathematics of Diffusion*, Oxford University Press, London, 89–91.

Cussler, E. L., 1997, *Diffusion: Mass Transfer in Fluid Systems*, 2nd ed., Cambridge University Press, Cambridge, 224.

Dean, J. A., ed., 1973, *Lange's Handbook of Chemistry*, McGraw-Hill, New York, 10-277–10-287.

Ding, H., M.-C. Yang, D. Schisla, and E. L. Cussler, 1989, Hollow-fiber liquid chromatography, *AIChE J.*, 35(5), 814–820.

Dullien, F. A. L., 1992, *Porous Media: Fluid Transport and Pore Structure*, 2nd ed., Academic Press, San Diego, CA, 487–557.

Foley, J. P., and J. G. Dorsey, 1983, Equations for calculating chromatographic figures of merit for ideal and skewed peaks, *Anal. Chem.*, 55, 730–737.

Gellert, W., S. Gottwald, M. Hellwich, H. Kastner, and H. Kuster, 1975, *Concise Encyclopedia of Mathematics*, van Nostrand Reinhold, New York, 586–591.

Gibbs, S. J., and E. W. Lightfoot, 1986, Scaling up gradient elution chromatography, *Ind. Eng. Chem. Fund.*, 25, 490–498.

Grushka, E., L. R. Snyder, and J. H. Knox, 1975, Advances in band spreading theories, *J. Chromatogr. Sci.*, 13, 25–37.

Hamaker, K. H., and M. R. Ladisch, 1996, Intraparticulate flow and plate height effects in liquid chromatography stationary phases, *Separation and Purification Methods*, 25(1), 47–83.

Hamaker, K., J. Liu, C. M. Ladisch, and M. R. Ladisch, 1998, Transport properties of rolled stationary phase columns, *Biotechnol. Progr.*, 14(1), 21–30.

Hjertén, S., J. Mohammad, and K. Nakazoto, 1991, Improvement in flow properties and pH stability of compressed, continuous polymer beds for high performance liquid chromatography, *J. Chromatogr.*, 646, 121–128.

Kearny, M., 1997, Engineered fractal cascades for fluid control applications, in *Proc. L'Ingenieur et Les Fractals*, Arcachon, France, Institut National De Recherche En Informatique et En Automatique (INRIA), June 25–27, 1997.

King, C. J., 1971, *Separation Processes*, McGraw-Hill, New York, 146–263.

Kiso, Y., K. Jinno, and T. Nagoshi, 1986, Liquid chromatography in a capillary packed with fibrous cellulose acetate, *J. High Resolution Chromatogr. Chromatogr. Commun.*, 9, 763–764.

Knight, P., 1990, Bioseparations: Media and modes, *Bio/Technol.*, 8, 200–201.

Ladisch, M. R., and G. T. Tsao, 1978, Theory and practice of rapid liquid chromatography at moderate pressures, using water as the eluent, *J. Chromatogr.*, 166, 85–100.

Ladisch, M. R., M. Voloch, J. Hong, P. Bienkowski, and G. T. Tsao, 1984, Cornmeal adsorber for dehydrating ethanol vapors, *Ind. Eng. Chem. Process Des. Dev.*, 23(3), 437–443.

Lee, J. Y., P. J. Westgate, and M. R. Ladisch, 1991, Water and ethanol sorption phenomena on starch, *AIChE J.*, 37(8), 1187–1195.

Levenspiel, O., 1972, *Chemical Reaction Engineering*, 2nd ed., Wiley, New York, 462–479.

Lin, J. K., B. J. Jacobson, A. N. Pereira, and M. R. Ladisch, 1988, Liquid chromatography of carbohydrate monomers and oligomers, in *Methods in Enzymology*, vol. 160, W. A. Wood and S. T. Kellogg, eds., Academic Press, New York, 145–159.

Martin, A. J. P., and R. L. M. Synge, 1941, A new form of chromatogram employing two liquid phases: 1. A theory of chromatography; 2. Application to the micro-determination of the higher monoamino-acids in proteins, *Biochem. J.*, 35, 1358–1368.

Martin, C. W., and Y. Shalon, 1985, An apparatus for slurry-packing preparative LC columns, *LC*, 3(6).

Monke, K., A. Velayudhan, and M. R. Ladisch, 1990, Characterization at the swelling of a size-exclusion gel, *Biotechnol. Progr.*, 6, 376–382.

News item, 1999, Chromatography market profile: LC detectors in pharmaceutical laboratories, *Liquid Chromatography: Gas Chromatography*, 17(1), 10.

Perry, R. H., C. H. Chilton, and S. D. Kirkpatrick, 1963, *Chemical Engineer's Handbook*, McGraw-Hill, New York, 3-228–3-231.

Pfeiffer, J. F., J. C. Chen, and J. T. Hsu, 1996, Permeability of gigaporous particles, *AIChE J.*, 42(4), 932–939.

Prochrom, *Industrial Continuous Liquid Chromatography, Preparative/Production Continuous Chromatography*, Brochure, Indianapolis, IN, March 1993.

Roe, S., 1989a, Separation based on structure, in *Protein Purification Methods: A Practical Approach*, E. L. V. Harris and S. Angal, eds., IRL Press, Oxford, 182–184.

Roe, S., 1989b, Separation based on structure, in *Protein Purification Methods: A Practical Approach*, E. L. V. Harris and S. Angal, eds., IRL Press, Oxford, 194–195.

Rudge, S. R., and M. R. Ladisch, 1986, Process considerations for scale-up of liquid chromatography and electrophoresis, in *Separation, Recovery, and Purification in Biotechnology*, ACS Symp. Ser., 314, J. A. Asenjo and J. Hong, eds., Am. Chem. Soc., Washingtion, DC, 122–152.

Schwartz, C. E., and J. M. Smith, 1953, Flow distribution in packed beds, *Ind. Eng. Chem.*, 45(6), 1209–1218.

Sepracor, 1992, Process Scale Glass Columns for Biochromatography, Lit. Code PB03.

Sherwood, T. K., R. L. Pigford, and C. R. Wilke, 1975, *Mass Transfer*, McGraw-Hill, New York, 32–37, 242–246, 565–589.

Sofer, G. K., and L. E. Nystrom, 1989, *Process Chromatography: A Practical Guide*, Academic Press, London, 73–80, 93–105.

Svec, F., and J. M. J. Fréchet, 1996, New designs of macroporous polymers and supports: From separation to biocatalysis, *Science*, 273(5272), 205–211.

Treybal, R. E., 1968, *Mass Transfer Operations*, McGraw-Hill, New York, 629–673.

Tyn, M. Y., and T. W. Gusek, 1990, Prediction of diffusion coefficients of proteins, *Biotechnol. Bioeng.*, 35, 327–338.

UpChurch Scientific, 1998, *Catalogue of Chromatography and Fluid Transfer Fittings*, 119.

van Deemter, J. J., F. J. Zuiderweg, and A. Klinkenberg, 1956, Longitudinal diffusion and resistance to mass transfer as causes of non-ideality in chromatography, *Chem. Eng. Sci.*, 5, 271–289.

van der Brekel, L. D. M., and E. J. de Jong, 1989, Hydrodynamics in packed textile beds, *Textile Res. J.*, 59(8), 433–440.

Velayudhan, A., 1989, Personal communication: The use of the Biorad 385 gradient former to generate linear gradients.

Wankat, P. C., 1990, *Rate-Controlled Separations*, Chapman and Hall, London.

Westgate, P. J., and M. R. Ladisch, 1993, Air drying using corn grits as the sorbent in a pressure swing adsorber, *AIChE J.*, 39(4), 720–723.

Westgate, P., J. Y. Lee, and M. R. Ladisch, 1992, Modeling of equilibrium sorption of water vapor on starch materials, *Trans. ASAE*, 35(1), 213–219.

Yang, Y., A. Velayudhan, C. M. Ladisch, and M. R. Ladisch, 1992, Protein chromatography using a continuous stationary phase, *J. Chromatogr.*, 598, 169–180.

SUGGESTED READING

Aiba, S., A. E. Humphrey, and N. F. Millis, 1973, *Biochemical Engineering*, 2nd ed., Academic Press, New York, 346–392.

Alltech Catalogue 450, *Injection Valves* (1999).

Bauman, W. C., R. M. Wheaton, and D. W. Simpson, 1956, Ion exclusion, in *Ion Exchange Technology*, F. C. Nachod and J. Schubert, eds., Academic Press, New York, 182–202.

Belter, P. A., E. L. Cussler, and W.-S. Hu, 1988, *Bioseparations: Downstream Processing for Biotechnology*, Wiley, New York, 1–5.

Bird, R. H., W. E. Stewart, and E. N. Lightfoot, 1960, *Transport Phenomena*, Wiley, New York, 196.

Carberry, J. J., 1958, Axial dispersion and void-cell mixing efficiency in fluid flow in fixed beds, *AIChE J.*, 4(1), 13M–21M.

Cavaco-Paulo, A., and L. Almeida, 1996, Cellulase activities and finishing effects. *Textile Chemist Colorist*, 28(6), 28–32.

Cohen, Y., and A. B. Metzner, 1981, Wall effects in laminar flow of fluids through packed beds, *AIChE J.*, 27(4), 705–880.

Colin, H., G. B. Cox, O. Dapremont, and P. Hilaireau, 1995, Optimization of Preparative Separations: Comparison of the Economics of Continuous and Batch Operation, Poster-Preptech '95, NJ (February 1995).

Colin, H., 1993, Large scale high performance preparative liquid chromatography, in *Preparative and Production Scale Chromatography*, G. Ganetsos and P. E. Barker, eds., Marcel Dekker, New York, ch. 2.

Collentro, W. V., 1995, USP purified water and water for injection storage system and accessories, part I, *Pharmaceutical Technol.*, 19(3), 78–94.

Cordes, A., J. Flossdorf, and M.-R. Kula, 1987, Affinity partitioning: Development of mathematical model describing behavior of biomolecules in aqueous two-phase systems, *Biotechnol. Bioeng.*, 30(4), 514–520.

Corn Refiners Association, 1996, *Annual Report*, Washington, DC.

Curling, J. M., and J. M. Cooney, 1982, Operation of large scale gel filtration and ion exchange systems, *J. Parenteral Sci. Technol.*, 36, 59–64.

Dolan, J. W., 1998, Extracolumn band broadening from injection and peak transfer, *Liquid Chromatography: Gas Chromatography*, 16(3), 248, 250.

Ganetsos, G., and P. E. Barker, 1993, *Preparative and Production Scale Chromatography*, Marcel Dekker, New York.

Goldstein, S. F. R. S., 1953, On the mathematics of exchange processes in fixed columns, I. Mathematical solutions, and asymptotic expansions, *Proc. Roy Soc. (London)*, A219, 171.

Gong, C. S., M. R. Ladisch, and G. T. Tsao, 1979, Biosynthesis, purification, and mode of action of cellulases of trichoderma reesei, in *Hydrolysis of Cellulose: Mechanisms of Enzymatic and Chemical Analysis*, Adv. Chem. Ser. 181, Am. Chem. Soc., Washington, DC, 261–287.

Goto, M., N. Hayashi, and S. Goto, 1983, Separation of electrolyte and non-electrolyte by an ion retardation resin, *Sep. Sci. Tech.*, 18, 475.

Greenkorn, R. A., and D. P. Kessler, 1972, *Transfer Operations*, McGraw-Hill, New York, 368–369, 488–489.

Hamaker, K. H., J. Liu, R. J. Seely, C. M. Ladisch, and M. R. Ladisch, 1996, Chromatography for rapid buffer exchange and refolding of secretory leukocyte protease inhibitor, *Biotechnol. Progr.*, 12, 184–189.

Ho, Sa. V., 1990, Strategies for large-scale protein purification, in *Protein Purification: From Molecular Mechanisms to Large-Scale Processes*, M. R. Ladisch, R. C. Willson, C.-D. C. Painton, and S. E. Builder, eds., Am. Chem. Soc., Washington, DC, 14–34.

Huber, J. F. K., 1973, Stofftransport and Stoffverteilung bei chromatographischen Prozessen, *Ber. Bunsen Ges.*, 77, 179.

Jageland, P., J. Magnusson, and M. Bryntesson, 1994, Optimization of industrial-scale high-performance liquid chromatography applications using a newly developed software, *J. Chromatogr. A.*, 658, 497–504.

Johansson, H.-O., G. Karlstrom, F. Tjerneld, and C. A. Haynes, 1998, Driving forces for phase separation and partitioning in aqueous two-phase systems, *J. Chromatogr. B,* 711, 3–17.

Johnson, R. D., 1986, The processing of biomolecules: A challenge for the eighties, *Fluid Phase Equilibria*, 29, 109–123.

Kipling, J. J., 1965, *Adsorption from Solutions of Non-electrolytes*, Academic Press, London, 1–31, 152–159.

Kirkland, J. J., W. W. Yau, H. J. Stoklasa, and C. H. Dilks Jr., 1977, Sampling and extra-column effects in high-performance liquid chromatography; influence of peak skew on plate calculations, *J. Chromatogr. Sci.*, 15, 303–316.

Kirschner, E. M., 1998, Soaps and detergents: New washing machines, enzymes, liquids churn the laundry detergent ingredients market, *Chem. Eng. News*, 76(4), 39–54.

Kovach, J. L., 1978, Gas-phase adsorption and air purification, in *Carbon Adsorption Handbook*, P. N. Cheremisinoff and F. Ellerbusch, eds., Ann Arbor Science Publishers, Ann Arbor, MI, 331–356.

Kucera, E., 1963, Contribution to the theory of chromatography: Linear non-equilibrium elution chromatography, *J. Chromatogr.,* 19, 237–248.

Ladisch, M. R., 1987, Separation by sorption, in *Advanced Biochemical Engineering*, H. R. Bungay and G. Belfort, eds., Wiley, New York, 219–237.

Ladisch, M. R., 1991, Fermentation derived butanol and scenarios for its uses in energy related applications, *Enz. Microb. Technol.*, 13(3), 280–283.

Ladisch, M. R., A. Emery, and V. W. Rodwell, 1977, Economic implications of purification of glucose isomerase prior to immobilization, *Ind. Eng. Chem. Process Des. Dev.*, 16(3), 309–313.

Ladisch, M. R., and A. Velayudhan, 1999, Scale-up techniques in bioseparation processes, in *Bioseparation Processes in Foods*, Rakesh K. Singh and Syed S. H. Rizi, eds., Marcel Dekker, New York, 113–138.

Ladisch, M. R., and P. C. Wankat, 1988, Scale-up of bioseparations for microbial and biochemical technology, in *Impact of Chemistry on Biotechnology*, M. Phillips, S. Shoemaker, R. Ottenbrite, and R. Middlekauf, eds., 72nd ACS Symp. Ser., 362, Am. Chem. Soc., Washington, DC.

Ladisch, M. R., M. Voloch, and B. Jacobson, 1984, Bioseparations: Column design factors in liquid chromatography, *Biotechnol. Bioeng. Symp.*, 14, 525.

Ladisch, M. R., R. L. Hendrickson, and K. L. Kohlmann, 1990, Anion exchange stationary phase for (-galactosidase, bovine serum albumin, and insulin: Separation and sorption characteristics, in *Protein*

Purification: From Molecular Mechanisms to Large-Scale Processes, M. R. Ladisch, R. C. Willson, C.-D. Painton, and S. E. Builder, eds., Am. Chem. Soc., ACS Symp. Ser., 427, 93–103.

Lapidus, L., and N. H. Amundson, 1952, Mathematics of adsorption in beds: VI. The effect of longitudinal diffusion in ion exchange and chromatographic columns, *J. Phys. Chem.*, 56, 984–988.

Lawlis, A. B., and H. Heinsohn, 1993, Industrial perspectives on the role of chromatography in the purification of biological molecules, *Liquid Chromatography: Gas Chromatography*, 11(10), 720–729.

Lochmüller, C. H., and M. Sumner, 1980, Estimation of extra-column dead volume effects using a mixing cell model, *J. Chromatogr. Sci.*, 18, 159–165.

Makulowich, J. S., 1993, Pharmacia unveils streamline and resource process systems, *GEN*, 13(9), 1, 27.

McKown, R., R. Tentonico, and S. Fox, 1996, Biotech manufacturing economics: Decreasing costs and increasing yields, *Genetic Eng. News*, 16(6), 6, 28.

Miller, N. T., and J. M. DiBusslo, 1990, Studies on the stability of *n*-alkyl-bonded silica gels under basic pH conditions, *J. Chromatogr.*, 499, 317–332.

Neuman, R. P., and M. R. Ladisch, 1986, Sulfuric acid recovery by ion exclusion, *Proc. 6th An. Solar, Biomass, and Wind Energy Workshop*, USDA, Atlanta, GA, February 25–27, 1986.

Neuman, R. P., S. R. Rudge, and M. R. Ladisch, 1987, Sulfuric acid–sugar separation by ion exclusion, *Reactive Polymers*, 5, 55.

Petsko, G. A., 1988, Protein engineering, in *Biotechnology and Materials Science*, M. L. Good, ed., Am. Chem. Soc., Washington, DC, 60.

Pieri, G., P. Piccardi, G. Muratori, and L. Caval, 1983, Scale-up for preparative liquid chromatography of fine chemicals, *La Chimica e. L'Industria*, 65, 331.

Reisman, H. B., 1988, *Economic Analysis of Fermentation Processes*, CRC Press, Boca Raton, FL, 69–77, 91, 92.

Rudge, S. R., and M. R. Ladisch, 1988, Electrochromatography, *Biotech. Progress*, 4, 123.

Ruthven, D. M., 1984, *Principles of Adsorption and Adsorption Processes*, Wiley, New York.

Siddiqi, S. F., N. J. Titchener-Hooker, and P. A. Shamlou, 1997, High pressure disruption of yeast cells: The use of scale down operations for the prediction of protein release and cell debris size distribution, *Biotechnol. Bioeng.*, 55(4), 642–649.

Snyder, L. R., 1980, Gradient elution, in *High Performance Liquid Chromatography—Advances and Perspectives*, C. Horvath, ed., Academic Press, New York.

Tice, P. A., I. Mazsaroff, and F. E. Regnier, 1987, Effect of large sample loads on column lifetime in preparative-scale liquid chromatography, *J. Chromatogr.*, 410, 43.

Tudge, A. P., 1961, Studies in chromatographic transport: I. A simplified theory, *Can. J. Phys.*, 39, 1600–1610.

Tudge, A. P., Studies in chromatographic transport: II. The effect of adsorption isotherm shape, *Can. J. Phys.*, 39, 1611–1618.

Velayudhan, A., and M. R. Ladisch, 1993, Plate models in chromatography: Analysis and implications for scale-up, *Adv. Biochem. Eng./Biotechnol.*, 49, 123–145.

Velayudhan, A., M. R. Ladisch, and J. E. Porter, 1993, Modeling of non-linear elution chromatography for preparative scale separations, 1, *AIChE Symp. Ser.*, 290.

Viswanth, D. S., and G. Natarajan, 1989, *Data Book on the Viscosity of Liquids*, Hemisphere Publishing, New York, 714–715.

Voloch, M., N. B. Jansen, M. R. Ladisch, G. T. Tsao, R. Narayan, and V. W. Rodwell, 1985, 2,3-Butanediol, in *Comprehensive Biotechnology*, M. Moo-Young, C. L. Cooney, and A. E. Humphrey, eds., Pergamen Press, Oxford, 933–947.

Wakao, N., and T. Furazkri, 1978, Effect of fluid dispersion coefficients on particle-to-fluid mass transfer coefficients in packed beds, *Chem. Eng. Sci.*, 33, 1375.

Wankat, P. C., and Y.-M. Koo, 1988, Scaling rules for isocratic elution chromatography, *AIChE J.*, 34, 1006.

Welinder, B. S., T. Kornfelt, and H. H. Sorenson, 1995, Stationary phases: The weak link in the LC chain?, *Today's Chemist at Work*, 4(8), 35–38.

Wilson, E. J., and C. J. Geankoplis, 1966, Liquid mass transfer at very low Reynolds number in packed beds, *Ind. Eng. Chem. Fund.*, 5, 9.

Yamamoto, S., 1995, Plate height determination for gradient elution chromatography of proteins, *Biotechnol. Bioeng.*, 48, 444–451.

Yamamoto, S., M. Nomura, and Y. Sano, 1986, Scaling up of medium-performance gel filtration chromatography of proteins, *J. Chem. Eng. Jpn.*, 19, 227.

Yamamoto, S., M. Nomura, and Y. Sano, 1987, Factors affecting the relationship between plate height and the linear mobile phase velocity in gel filtration chromatography of proteins, *J. Chromatogr.*, 394, 363.

PROBLEMS

5.1 The capacity factor, k', is defined as the ratio of moles of solute in the stationary phase to moles solute in the mobile phase. Use the expression for retention volume, V_r, as a function of the distribution coefficient, K, in order to derive the expression for the capacity factor, k'; given below.

t_r = retention time in minutes of a specified component, such as salt

t_o = retention time of an excluded component, such as blue dextran

V_r = retention volume

V_m = mobile phase volume

5.2 Molecular diffusion in the axial direction is given by Fick's second law (Cussler, 1997):

$$\frac{\partial c}{\partial t} = D\frac{\partial^2 c}{\partial x^2} \qquad \text{(a)}$$

where

x = axial distance

D = molecular diffusion coefficient

If D is replaced by an axial dispersion coefficient D, equation (a) can be interpreted to represent the degree of backmixing during flow. Show how equation (a) can be used to obtain a definition of the bed Peclet number

$$Pe = \frac{\mathrm{D}}{vL} \qquad \text{(b)}$$

5.3 Partial purification of insulin is carried out over a size exclusion chromatography column to remove peptide fragments. The empty column volume is 12 L and the column length is 60 cm. The particle size of the stationary phase is 100 microns (i.e., 0.01 cm). The chromatograms obtained with the column are shown in Figure 5.65.

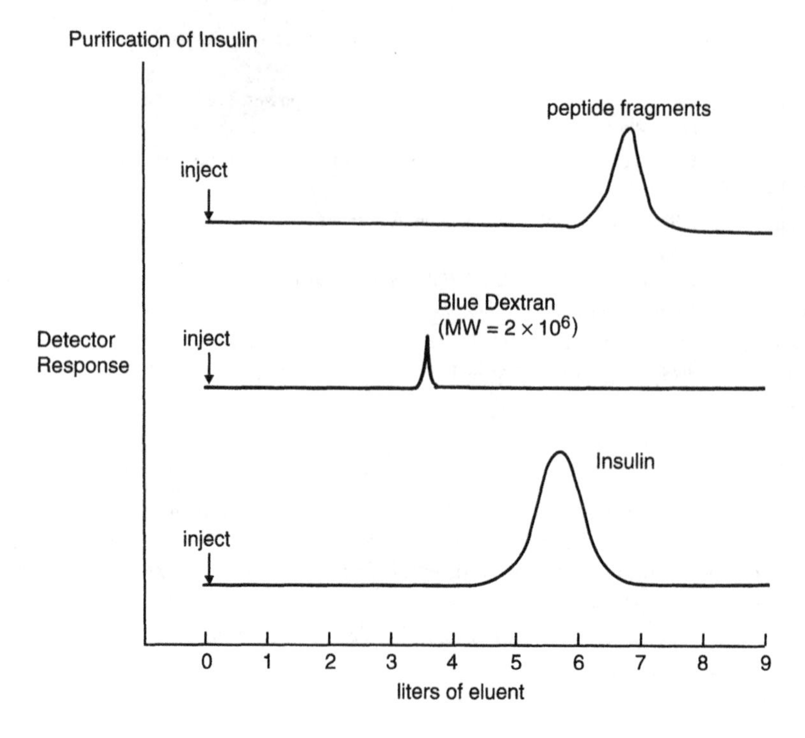

Figure 5.65 Hypothetical chromatogram for problem 5.3 showing a possible separation of insulin from peptides.

a. Calculate the ***distribution*** coefficient, K_D for insulin.

b. Calculate the number of stationary phase particles per theoretical plate. Assume that the particles touch end to end along the length of the column.

c. Estimate the length of column required to obtain a resolution of 0.95 between the insulin peak and the peptide fragments. Clearly state your assumptions. Show your work!

5.4 A polymeric stationary phase is identified for an antibiotic purification by liquid chromatography. Experiments with three columns of different diameters are carried out using the eluent flowrates as shown in Table 5.8. The retention time, t_o, of the *excluded* component, blue dextran, and the antibiotic (retention time of t_r) are also shown in Table 5.8. All columns are 25 cm in length.

Recall the basic definitions:

$$\varepsilon_{tot} = \frac{V_o}{V_T} = \frac{V_o}{V_p + V_o}$$

where

TABLE 5.8 Data for Problem 5.4

Inside Diameter mm	t_o min	t_r min
3.2	1.45	2.76
4.6	2.17	—
10.0	8.64	—

V_o = extraparticulate void volume
V_p = void volume due to pores
$N = L/H = 16(t_r/t_w)^2$ for peak width measured at the baseline

Based on these data, determine if there is an increase in the capacity factor with column diameter. Make this assessment as follows:

a. Derive an expression for k' as a function of K_D and ε_{tot} $(= \varepsilon_p + \varepsilon_b)$.

b. Calculate void fractions for the three columns having inside diameters of 3.2, 4.6, and 10.0 mm, respectively.

c. Calculate the capacity factors for each of the three columns. (Hint: The k' for each column is different.)

d. Give a brief explanation for the results calculated in parts b and c. Why does void fraction change for columns of different inside diameters?

e. What is anticipated retention time for the antibiotic in the column having the 10.0 mm inside diameter?

f. Does Table 5.8 give sufficient data to estimate K? If your answer is yes, explain how you would determine K. If your answer is no, explain what would be needed to determine K.

5.5 The pressure drop across a LC column packed with a dextran gel is given by

$$\Delta p = \frac{150\,(L\bar{V}_o \mu\,(1-\varepsilon_b)^2)}{g_c D_p \varepsilon_b^3}$$

where V_o is the average interstitial velocity. The dextran gel packing material consists of spheres having an average diameter of 120 microns. When packed in a bed having a 60 cm depth, the pressure drop is 100 psig. At identical conditions, a column of 100 cm length gives a Δp that exceeds 800 psig, and flow essentially stops. Inspection of the 100 cm column shows a compacted region near the column outlet, and a change in shape of the particles near the outlet from spherical to a shape which approximates a cylinder. The average void fraction, ε_b, of the 100 cm column is 0.33, while that of the 60 cm is 0.36.

Propose an explanation of why the pressure drop is so much higher in the 100 cm long column. Be specific. Does the change in ε_b account for the increase? Is it the change in shape, or is it something else?

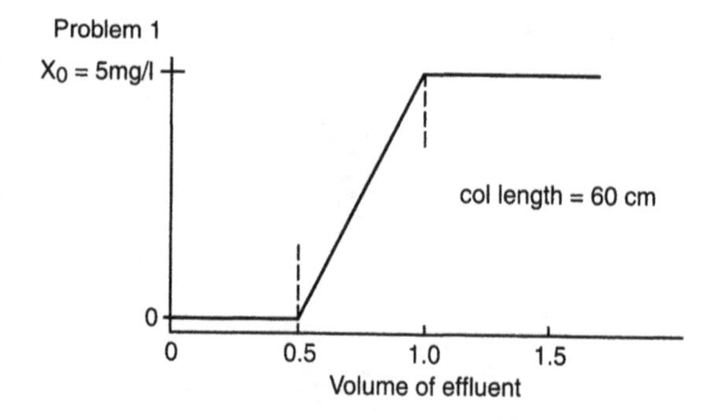

Figure 5.66 Hypothetical breakthrough curve for a 60 cm long affinity column (problem 5.6).

Note: $S_p/V_p = 6/d_p$ for spheres, cubes, and cylinders where S_p = surface area, V_p = volume, d_p = diameter or characteristic length.

5.6 The partial purification of an interferon is to be carried out using an affinity chromatography column. The interferon complexes with an affinity ligand immobilized on the sorbent and gives the breakthrough profile shown in Figure 5.66. The column length is 60 cm in this case, and the feed concentration, x_o, is 5 mg/L. The effluent volume is corrected for the column void volume.

A longer column is packed and gives the profile shown in Figure 5.67. The feed to the column is to be interrupted when $x = 0.12\ x_o$, and desorption started at this point. On this basis, and using the information given, calculate the length of column required so that 80% of the total capacity of the sorbent is utilized when $x = 0.12x_o$ (i.e., breakthrough occurs).

5.7 The separation of aspartame (i.e., Nutrasweet) from other amino acids is carried out using 10% ethanol as the eluent over a macroporous stationary phase. The separation occurs due to differences in adsorption of the amino acids. The other amino acids elute as one peak at the beginning of the chromatogram, with a retention time of 14 minutes. The aspartame peak elutes at 40 min with a peak width (at baseline) of 9 min.

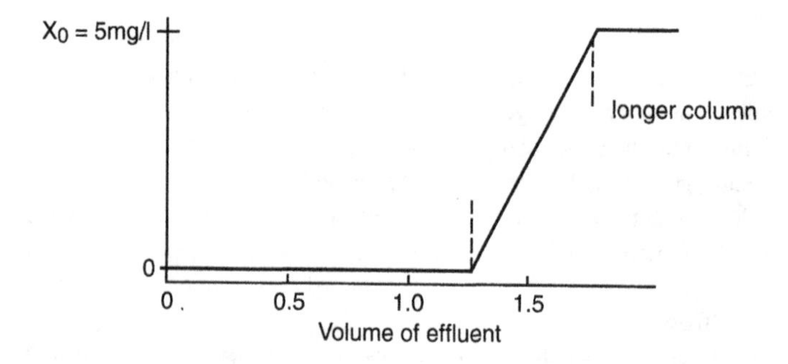

Figure 5.67 Hypothetical breakthrough curve for a longer column, packed with the same affinity stationary phase as in Figure 5.66 (problem 5.6).

When DNA ($MW = 5 \times 10^6$) is injected, DNA is not adsorbed by the stationary phase, and elutes at 7 min. When NaCl is injected, the chromatogram elutes at 12 min, with a peak width of 3 min.

a. Calculate the intraparticle and extraparticle void volumes.

b. Calculate the capacity factors for the amino acid peak and the aspartame peak.

c. What is the plate height based on NaCl? Based on aspartame?

d. Is the retention of aspartame due to a size exclusion mechanism, or by some other mechanism? Explain.

TABLE 5.9 Data for Problem 5.9

	BSA Separation	DNA Separation
Temperature, °	20 °C	60 °C
Column diameter	5 cm	5 cm
Column length	30 cm	30 cm
Retention volume, blue dextran	224 mL	224 mL
Retention volume, BSA	250 mL	224 mL
Retention volume, NaCl	442 mL	442 mL
Column area	19.6 cm^2	19.6 cm^2
Column volume	589 mL	589 mL
Interstitial area (= $A_o \varepsilon_b$)	7.47 cm^2	7.47 cm^2
Void fraction, ε_b	0.38 dimensionless	0.38 dimensionless
Flowrate	4.2 mL/min	8.4 mL/min
$m_{protein}$ or m_{DNA}	25 mg	25 mg
m_{NaCl}	100 mg	100 mg
Interstitial velocity	0.563 cm/min	1.125 cm/min
BSA or	no adsorption	
DNA		
K_D	0 by definition	0 by definition
ε_p^* BSA, Eq. 5.58	0.071 by Eq.	0.000
$x_{protein}$ by Eq. (5.57)	0.896 for BSA	1.000 for BSA
BSA radius of gyration	30.6 angstrom, Tyn and Gusel, 1990	30.6
NaCl K_D no adsorption	0 by definition	0 by definition
ε_p^* NaCl	0.597 by Eq. (5.58)	0.597 by Eq. (5.58)
x_{NaCl} by Eq. (5.57)	0.507 for NaCl	0.507 for NaCl
NaCl radius of gyration	20 angstrom, assumed	
	Stationary phase	
R, case 1 radius	0.002 cm = 20 microns	0.002
	Mobile phase	
Buffer (Water)		
Viscosity	1.006 cp = 10^{-2} g/(cm·s)	0.467
Density	1 g/mL	1
	Mass transfer parameters	
$m\, D_m/D_p^* \varepsilon_p$	35.10 if $D_p = D_m/2.5$	35.10[a]
γ Nu at large ReSc	2.87[a]	2.87
β Nu at lower ReSc	4	4
m^* [a]	0 for BSA	0

[a]Athalye, Gibbs, and Lightfoot, 1992.

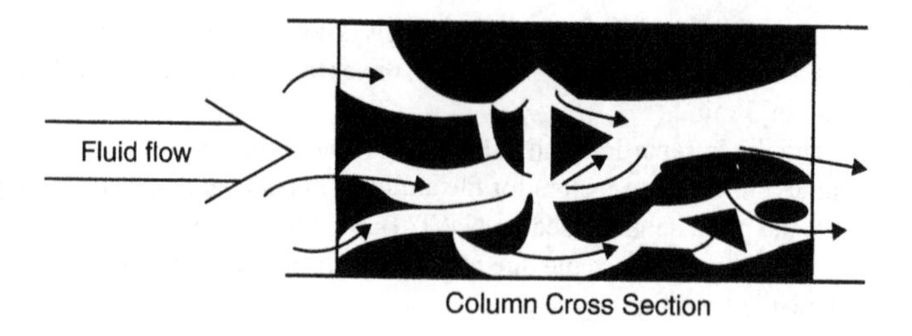

Figure 5.68 Schematic representation of fluid flow through pores.

5.8 Size exclusion chromatography separates molecules based on differences in size. For a completely excluded component (e.g., dextran – $MW = 10^6$), the retention volume, V_R, is

$$V_{R,e} = V_o \tag{a}$$

where V_o is the extraparticle void volume that corresponds to the extraparticulate void fraction, ε_b. For a completely included component, such as NaCl, the retention volume is

$$V_{R,i} = V_o + V_i \tag{b}$$

where V_i is the internal void volume. Using equations (a) and (b), obtain the expression for the distribution coefficient, K_D, for an intermediate sized molecule such as monomeric insulin ($MW = 6000$). Are there limits on the value of K_D? If so, give these limits. Explain your answers.

5.9 A column designed to carry out size exclusion chromatography of large proteins from salt has the characteristics shown in Table 5.9. This column is to be modified for the desalting of a DNA fragment that has the same radius of gyration as the blue dextran. The separation is to be carried out at a higher temperature of 60°C instead of the 20°C, since the DNA is stable at this temperature and since the column pressure drop is lower at the higher temperature. The flowrate is to be doubled from 4.2 mL/min used for the protein separation to 8.4 mL for the DNA separation. The operating specifications for the BSA and DNA separations are given in Table 5.9.

a. Calculate and graph the plate height curves for BSA and DNA for the two cases for flowrates ranging from 0.00084 to 42 mL/min. Graph the reduced plate height as a function of ReSc and then as a function of interstitial velocity.

b. How many 30 cm long column(s) will be needed to achieve 1000 plates based on the DNA peak? Based on the BSA peak? (Assume that the columns will be connected in series if more than one column is required.)

5.10 A column is designed that enables flow to occur through all of the pore spaces in the stationary phase as shown in Figure 5.68. At higher velocities, where ReSc > 20, the reduced plate height curve for an adsorbed solute found to change slowly but linearly with increasing flowrate (or ReSc). For an unadsorbed solute, the reduced plate height decreases with increasing flowrate, depending on the solute that is used. For this column $\varepsilon_b = 0.5$, show that these observations are consistent with equation (s) given in sections 5.17 and/or 5.18. Give a brief explanation of your rationale.

CHAPTER 6

LIQUID CHROMATOGRAPHY SCALE-UP

INTRODUCTION

Scale-up rules for chromatography are based on selecting the limiting condition (mass transfer or pore diffusion) and using dimensionless ratios of the resolution (R_s), plate count (N), and capacity factors to calculate column dimensions. The column's length and diameter for a given throughput, particle size, and degree of separation can be calculated using scaling rules for both linear chromatography and some types of nonlinear behavior. The procedures outlined in this chapter assume that the mechanism that dominates the separation at the bench scale does not change in proceeding from bench scale to process scale. This assumption must be verified on a case-by-case basis, particularly if particle size and/or interstitial velocity change. For example, pore diffusion is likely to be a dominating factor for porous stationary phases with an average particle size greater than 40 microns. Hence an elution profile for a stationary phase with a particle size of 10 microns, where mass transfer dominates, may not apply to a column packed with the same type of stationary phase consisting of 100 micron particles where pore diffusion dominates. Experimental data would be needed for the larger particle size stationary phase before scaling rules can be applied. The rules for linear chromatography, are from DeJong et al. (1981), Ladisch and Velayudhan (1992, 1999), Wankat and Koo (1988), Ladisch and Wankat (1988), Ladisch et al. (1984), and Snyder and Kirkland (1979).

Scale-up rules for nonlinear chromatography are based on application of local equilibrium theory to calculate elution time of the front of the peak, and mass transfer theory (Thomas solution) to estimate peak width. This approach would appear to invoke local equilibrium conditions, and then attribute band-spreading to mass transfer, even though local equilibrium theory assumes that mass transfer effects are small. Actually, the two effects—equilibrium and dispersion due to mass transfer—are calculated independently. The equilibrium calculation provides

The author thanks Juan Hong (University of California, Davis), Ajoy Velayudhan (Oregon State University), and Scott Rudge (FeRx) for their helpful inputs and contributions to subject matter in this chapter as students and colleagues.

a measure of the average elution time of the peak front analogous to retention time for linear equilibria.

6.1 SCALE-UP RULES ENABLE INITIAL SPECIFICATION OF CHROMATOGRAPHY COLUMNS

Development of a detailed mass transfer, pore diffusion, or local equilibrium-based model may not be practical in the early phases of purification development, even if approximations are used. Conversely, such a model, and the data for applying it, are needed when a separation has been developed and optimization of the process is the goal. However, the industrial process designer initially requires a first estimate of the size of chromatography equipment for cost estimating purposes, and will usually have only limited time and data[1] to generate such an estimate. Consequently scaling rules are needed. The utility of such rules lies in their ability to quickly obtain column dimensions and operating conditions for purposes of equipment sizing and process specification. Once a preliminary design is developed, and the proposed column appears to be feasible, further experimental work can be justified and a complete model developed. The application of scaling rules are only a first step to column scale-up.

Scale-Up Rules for Linear Chromatography

Linear chromatography plays a major role in the downstream processing of biological materials encountered in the manufacture of food, pharmaceutical, and biotechnology products (Ladisch and Velayudhan, 1999; Ganetos and Barker, 1993; Velayudhan et al., 1993; Ladisch and Wankat, 1988). Linear behavior is associated with isocratic gel permeation chromatography of proteins. It also occurs for analytical and process chromatography carried out at low solute concentrations, and in special cases where linear equilibria appear to exist at high solute concentrations.

Scale-up of a given chromatography separation has the objective of maintaining the separation between components. Product would constitute one peak, and all other components would elute on either side or both sides of the product peak. An example is desalting where components are separated from a product peak based on differences in their molecular weight. Sometimes the impurities will elute as several peaks that flank both sides of the product peak. In the case of scaling size exclusion chromatography of such a ternary system, all of the high molecular weight impurities would be lumped into the first peak and all of the low molecular weight impurities into the third peak. A product peak of intermediate molecular weight would elute between the two. Separation, for purposes of scale-up, is defined by resolution between the product peak and the component which elutes closest to the product peak, and is a function of plate counts and capacity factors of the two peaks.

A one-dimensional material balance, such as developed by Lapidus and Amundson (1952) and described later in this chapter or presented in Eq. (5.48) of Chapter 5, can facilitate calculation of elution profiles for linear chromatography as long as a lumped dispersion parameter can be readily determined. The representation of a chromatography column as a tank-in-series model provides rationale for relating axial dispersion to the column's plate count through an averaged Peclet number, and thereby provides an operational definition of a lumped dispersion

[1]The ease with which experimental data can be obtained is currently being greatly enhanced by the availability of instruments that rapidly and automatically carry out sequences of preprogrammed runs at different buffer, flowrate, and sample conditions. Hence the number and rate of replicate chromatograms that can quickly be attained for scale-up purposes is also expected to increase.

parameter. The solid, stationary phase particles (i.e., dark dots) in Figure 6.1 are shown to be vigorously mixed with the fluid or mobile phase in each tank to promote contact between the two phases (Uspensky, 1937). The dissolved solute equilibrates with the stationary phase particles so that local equilibrium (i.e., equilibrium in each tank) is achieved. The fluid phase with the remaining solute then overflows and moves to the next tank, where the equilibration process repeats itself. After n tanks, the mobile phase with the remainder of the solute leaves the system. This description is similar to the tanks-in-series model for determining residence distribution of a tracer discussed by Levenspiel (1972), or the model for ion exchange presented by Sherwood et al. (1975).

The plate count N is related to the number of tanks in series represented in Figure 6.1 by the equation

$$\frac{1}{N} = \frac{1}{n} + \frac{2v}{L}\frac{1}{k_{MT}}\frac{k'}{(1+k')^2} \tag{6.1}$$

This was proposed by Villermaux as discussed previously in a review (Velayudhan and Ladisch, 1993). The definitions of v (interstitial velocity), k' (capacity or retention factor), and L (column length) in Eq. (6.1) are the same as given previously, with k' from Eq. (5.141) in Chapter 5. The parameter k_{MT} gives the mass transfer coefficient.

Ruthven (1984) gives a relation between plate count and a rate expression that has a similar form to Eq. (6.1):

$$\frac{1}{N} = \frac{2D}{vL} + \frac{2v}{L}\frac{1}{k_{overall}}\frac{k'}{(1+k')^2} \tag{6.2}$$

The analogies between Eqs. (6.1) and (6.2) consist of

$$k_{MT} = k_{overall} \tag{6.3}$$

and

$$\frac{1}{n} = \frac{2D}{vL} \tag{6.4}$$

where $k_{overall}$ represents an overall mass transfer coefficient, and $2D/vL$ is an axial dispersion term. D represents an axial dispersion (mixing) coefficient that combines the various causes of

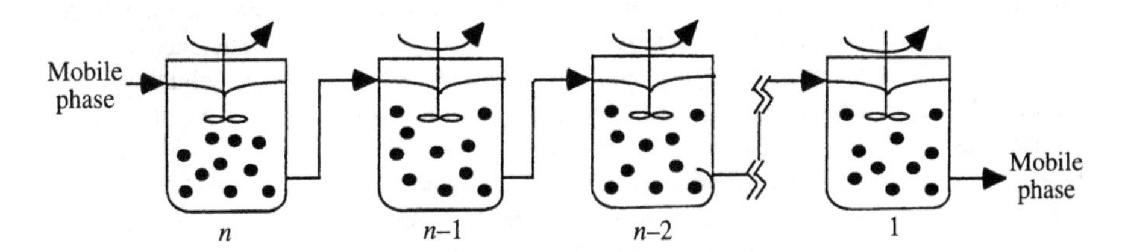

Figure 6.1 Tank-in-series representation.

dispersion in a packed bed. If axial dispersion (mixing) dominates mass transfer processes, then the axial dispersion will be larger than the mass transfer term:

$$\frac{2D}{vL} > \frac{2v}{L} \cdot \frac{1}{k_{\text{overall}}} \cdot \frac{k'}{(1+k')^2} \tag{6.5}$$

When axial dispersion dominates, Eq. (6.2) results in the approximation

$$\frac{1}{N} \cong \frac{2D}{vL} = \frac{2}{\text{Pe}} \tag{6.6}$$

or

$$\boxed{\overline{\text{Pe}} \cong 2N} \tag{6.7}$$

where $\overline{\text{Pe}}$ represents an average bed Peclet number. This also suggests an analogous equation for the mixing tank representation of dispersion in Figure 6.1 based on Eq. (6.4):

$$\overline{\text{Pe}} \cong 2n \tag{6.8}$$

where $\overline{\text{Pe}}$ represents an average Peclet number for a bed that is represented as a series of n mixing cells or tanks. The approximation in Eq. (6.7) is frequently used, since N can be determined from a chromatogram.

The attainment of equilibrium of the solute between the mobile phase and the stationary phase particles at each stage of Figure 6.1 assumes that either the residence time in each stage is long or that mass transfer is fast. If the residence time is long, the interstitial velocity, v, would have to be small and the inequality of Eq. (6.5) would be satisfied. If mass transfer is fast, the mass transfer parameter, k_{overall} would be large and the term on the right-hand side of Eq. (6.5) would again be small. Again, the inequality of Eq. (6.5) would be satisfied. The approximation of Eq. (6.7) provides a means for the bed Peclet number, $\overline{\text{Pe}}$, to bridge the concepts of plate count N and mass transfer by assuming that $\overline{\text{Pe}} = 2N$ (Velayudhan and Ladisch, 1993). However, generation and fitting of experimentally generated elution profiles must still be carried out so that the constants in equations such as Eqs. (5.87), (5.92), or (5.93) in Chapter 5 can be specified. If mass transfer and pore diffusion resistances are both significant, an equation such as that represented by Eqs. (5.33), (5.97), and (5.99) in Chapter 5 would be required to calculate the number of plates. Significant experimental effort or prior knowledge becomes necessary to determine the values of pore and mobile phase diffusivities, the distribution coefficient (assuming that linear equilibrium conditions apply), and other parameters that define the solute's properties when dissolved in the mobile phase or associated with the stationary phase. All of these approaches, as well as the scaling rules presented below for both linear and nonlinear chromatography, assume that the presence of one solute is not affected by another.

Resolution, R_s, is defined by Eq. (5.143) in Chapter 5 or approximated by an expression developed for analytical chromatography that utilizes averaged plate counts and capacity factors also given by Eqs. (5.27) and (5.139) in Chapter 5 (Snyder and Kirkland, 1979):

$$R_s = \frac{1}{4}(\alpha - 1)N_{1,2}^{1/2}\left[\frac{k'_{1,2}}{1 + k'_{1,2}}\right] \tag{6.9}$$

where

$\alpha = k'_2/k'_1$ = separation factor between peaks 1 and 2
$N_{1,2} = (N_1 + N_2)/2$ = average of plate counts for peaks 1 and 2
$k'_{1,2} = (k'_1 + k'_2)/2$ = average of capacity factors for peaks 1 and 2

Equation (6.9) represents an empirical definition of resolution, adopted from analytical chromatography, where the plate counts and capacity factors of adjacent peaks are assumed to be similar. If extra column band-broadening or dispersion for a solute that is in the linear range of its equilibrium curve occurs, chromatographic peaks may result that are significantly skewed. As a result the assumption of a simple Gaussian peak would not be valid. The plate count N can be calculated from the peak's assymmetry by measuring the parameters a and b, at a point 10% above the peak's base and then calculating the plate count N by Eq. (5.40), as described in Chapter 5, for exponentially modified Gaussian peaks.

Peak shape can be affected by solute concentration and the mobile phase/stationary phase equilibrium even if the sample volume is small and extra-column effects are absent. As described by Colin (1993): "the amount of broadening depends on the injected quantity and also some physicochemical parameters of the chromatographic system, such as the chemical nature of the mobile and stationary phases, the specific area of the stationary phase, and temperature." Of particular importance is the type of equilibrium isotherm that relates the concentration of the solute in the stationary phase to that in the mobile phase. Often the isotherm is Langmuirian, as illustrated in Figure 5.59 in Chapter 5 and in Figure 6.2.

A Langmuir isotherm has a plateau indicating the existence of a maximum solute concentration or saturation capacity that can be achieved on the stationary phase. The slope of the isotherm at a given concentration of the solute in the mobile phase is proportional to its capacity factor, k'. This is shown for the three points in Figure 6.2 for a favorable isotherm. The slopes decrease as the isotherm progresses from point 1 to 2 to 3. The capacity factors, and therefore elution times for increasing concentrations of a component injected in a constant volume of feed that would correspond to these points, would also decrease: $k'_{i,3} < k'_{i,2} < k'_{i,1}$. The subscripts 1, 2, and 3 correspond to points 1, 2, and 3 in Figure 6.2. The peak maximum moves to shorter retention times as mass of solute increases. This is illustrated in Figure 6.3. The peak becomes triangular as the concentration increases while the tailing edge of the peak appears at the same time as shown on the right-hand side of Figure 6.3.

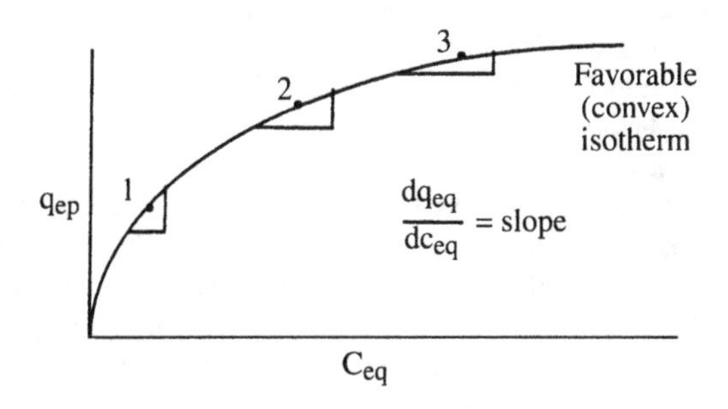

Figure 6.2 Non-linear adsorption isotherm.

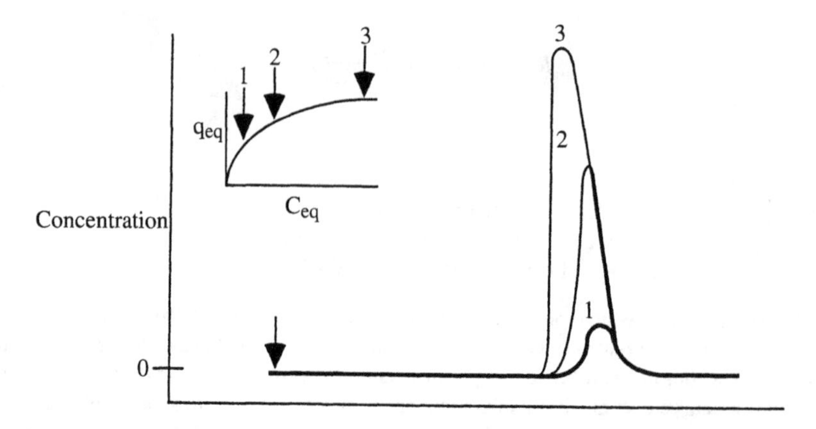

Figure 6.3 Retention times of peaks 1, 2 and 3 corresponding to points 1, 2 and 3 of adsorption isotherm.

Peaks 2 and 3 in Figure 6.3 occur at overloaded column conditions. If injection of constant feed volumes gives a peak shape that is affected by the concentration of the solute in the feed sample, linear equilibrium cannot be assumed. Therefore scale-up rules for linear chromatography presented here are not directly applicable for concentrations of solute (and peaks) that correspond to cases 2 and 3 in Figure 6.3. Rather, scale-up rules for nonlinear chromatography based on local equilibrium theory must be used, as discussed later in this chapter.

Linear equilibria and Gaussian peak shapes enable application of a systematic approach to estimating column length, diameter, and separation time to obtain a first estimate of column size for a given binary separation. Three cases are considered here for linear equilibria: pore diffusion alone, mass transfer alone, and both pore diffusion and mass transfer acting together as the primary determinants of solute dispersion. It is emphasized that the use of these rules is intended for solutes whose isotherms approximate a straight line and generally correspond to low solute concentrations.

6.2 SCALE-UP RULES FOR SIZE EXCLUSION CHROMATOGRAPHY (SEC) ASSUME PORE DIFFUSION CONTROLS

Since there is no sorptive mechanism in SEC, the equilibrium is inherently linear, and SEC lends itself to scale-up rules. Yamamoto et al. (1986, 1987) indicate that for 40 μm or larger particles, pore diffusion is a major contributor to band spreading. Then the assumptions used to simplify Equations (6.1) and (6.2) to give $\overline{\text{Pe}} = 2N$ would not be applicable, since pore diffusion rather than mass transfer controls peak width and therefore N. However, assuming that the controlling mechanism does not change upon scale-up, it is still possible to carry out scale-up calculations when the desired separation at bench scale has already been achieved and the bench-scale parameters are known (Ladisch and Wankat, 1988; Ladisch et al., 1984; Snyder, 1980; Wankat and Koo, 1988). If the ratio of interstitial velocities (v_x/v_b) can be specified from pressure drop considerations, scale-up reduces to specifying the column dimensions and the separation time on a large scale. The velocity, v, refers to interstial velocity while the subscript x refers to the large scale, and b refers to the bench-scale separation.

The ratio of particle diameters ($d_{p,x}/d_{p,b}$) is likely to be decided by commercial availability and cost of the stationary phase at a given particle size, while the ratio (v_x/v_b) may be constrained by pressure-drop considerations. The ratio of sample volumes ($V_{spl,x}/V_{spl,b}$) is specified by resolution based on the rationale presented in Eqs. (5.70) to (5.85) in Chapter 5. The plate count N is given by Eq. (6.10) when the plate count N is constrained to be the same at the bench and at large scale, and when pore diffusion is the controlling molecular process (Gibbs and Lightfoot, 1986; Ruthven, 1984):

$$N_{\text{pore diffusion}} \sim \frac{L}{v d_p^2} \tag{6.10}$$

For a constant plate count, the ratio of lengths of a large-scale column (length L_x) and bench-scale column (length L_b) is given by

$$\boxed{\left(\frac{L_x}{L_b}\right) = \left(\frac{v_x}{v_b}\right)\left(\frac{d_{p,x}}{d_{p,b}}\right)^2} \tag{6.11}$$

if the ratios on the right-hand side of Eq. (6.11) are known. The column diameter d_{col} is then specified by requiring that the ratio of the sample volume to the column volume be the same at the bench and the large scale. Thus

$$\frac{V_{col,\,x}}{V_{col,\,b}} = \frac{\pi\left(\frac{d_{col,\,x}}{2}\right)^2 L_x}{\pi\left(\frac{d_{col,\,b}}{2}\right)^2 L_b} = \frac{V_{spl,\,x}}{V_{spl,\,b}} \tag{6.12}$$

and hence

$$\boxed{\left(\frac{d_{col,\,x}}{d_{col,\,b}}\right) = \left[\left(\frac{L_b}{L_x}\right)\left(\frac{V_{spl,\,x}}{V_{spl,\,b}}\right)\right]^{1/2}} \tag{6.13}$$

where the length ratio is the reciprocal of that given in Eq. (6.11). The ratio of maximum separation times for size exclusion chromatography is obtained from the volume of mobile phase that must pass through the column to elute all of the peaks:

$$\boxed{\left(\frac{t_x}{t_b}\right)_{sec,\,max} = \left(\frac{L_x}{L_b}\right)\left(\frac{v_b}{v_x}\right)} \tag{6.14}$$

Equation (6.14) assumes that the interstitial velocities and interparticle void fractions are the same at both scales. Separation time is defined as the time that elapses between injection of the sample and elution of all of the peak, and corresponds to the sum of $V_{spl} + V_{R,f}$ in Figure 5.48 in Chapter 5. Separation time is a function of mobile phase flowrate and column length as indicated in Eq. (6.14), as well as a function of the value of K_D as incorporated into the $V_{tot,plate}$

term in Eq. (5.82) in Chapter 5. Equation (6.14) is a special case for size exclusion chromatography only.

A decrease in plate height H over a range of interstitial velocities, results when particle size of the stationary phase decreases. Equations (5.99) and (5.108) in Chapter 5 show that reduced plate heights fall into a range of values (i.e., a characteristic band) when plotted as a function of reduced velocity (Grushka et al., 1975). Hence the comparison of different types of stationary phases over a range of particle sizes and velocities for a given feed sample volume and concentration is justified. The benefits of a smaller column—sharper peaks, shorter separation time, and/or better resolution—may be offset by the higher cost of smaller particles and larger pressure drops, as discussed below (Grushka et al., 1975; Peskin and Rudge, 1992).

The cost of chromatography resins was tabulated and plotted by Peskin and Rudge (1992), as shown in Figure 6.4 (Peskin and Rudge, 1992). They found that the cost of the resins between 2 and 100 μm followed the relation in Eq. (6.15), which is indicated by the solid line in Figure 6.4:

$$C_{\text{cst,sp}} = 12{,}600\,(d_{\text{p}})^{-2.5} \tag{6.15}$$

where

$C_{\text{cst,sp}}$ = cost of stationary phase in \$/g stationary phase
d_{p} = average diameter of stationary phase particles, microns

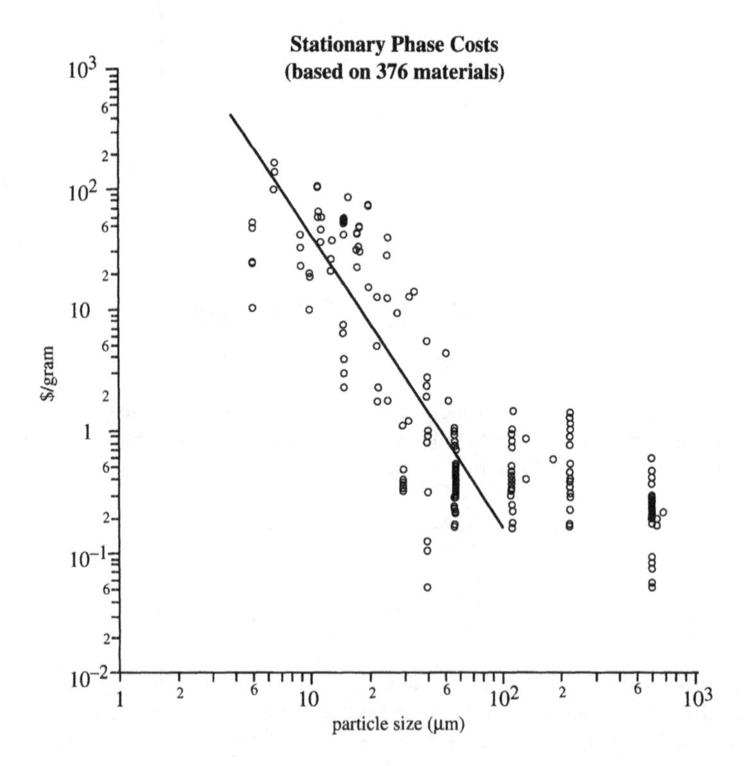

Figure 6.4 Cost of chromatography stationary phases as a function of particle size, based on data for 376 different stationary phases (from Peskin and Rudge, 1992, fig. 1, p. 51.)

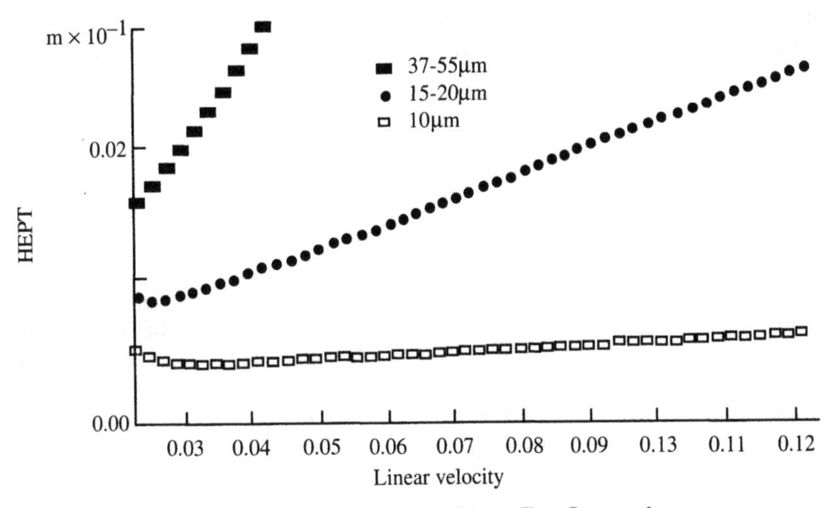

The Van Deemeter Plots For Several Bondapak C_{18} Particle Sizes

THE SMALLER PARTICLE SIZES ARE MORE EFFICIENT AT INCREASED FLOW RATE

Figure 6.5 Effect of particle size on plate height (HETP = H) as a function of linear velocity of mobile phase (cm/s) for a selected stationary phase (Bondapak C_{18}) with permission from MacLennan and Murphy, 1994, fig. 4, p. 34.

The stationary phase costs increase sharply while the plate height decreases and becomes less sensitive to flowrate when the stationary phase goes below 40 micron particle size (Figures 6.4 and 6.5). A low plate height, which is constant over a wide range of flowrates (see 10 μm line in Figure 6.5) is desirable, but it usually requires small particle size stationary phases that cause high-pressure drops, and which are costly. Many process scale resins or stationary phases have a particle size that exceed 100 μm and hence fall in a range where the cost begins to flatten out, though they exhibit larger plate heights as indicated in Figure 6.4. (Note that the data are plotted on a log plot.)

The pressure drop is given by (Grushka et al., 1975)

$$\Delta p = \varnothing \frac{\mu D_m \lambda \upsilon}{d_p^2} = \frac{\varnothing \mu D_m \frac{L}{d_p} \frac{v d_p}{D_m}}{d_p^2} = \frac{\varnothing \mu L v}{d_p^2} \tag{6.16}$$

where

$\varnothing$ = dimensionless, flow resistance parameter
$\lambda = L/d_p$ = reduced column length, dimensionless
D_m = diffusivity of solute in mobile phase, cm^2/s
μ = viscosity of the mobile phase, g/(cm · s)
$\upsilon = v d_p / D_m$ reduced velocity, dimensionless

Comparison of two columns of length L, operated at the same temperature, and mobile phase velocity shows the pressure drop to be inversely proportional to the square of the ratio of particle sizes:

$$\boxed{\frac{\Delta p_x}{\Delta p_b} = \frac{\varnothing_x \, d_{p,1}^2}{\varnothing_b \, d_{p,2}^2} \approx \left(\frac{d_{p,b}}{d_{p,x}}\right)^2} \tag{6.17}$$

In this case $\varnothing$ is assumed to be similar for the different particle size materials of the same stationary phase, and hence $\varnothing_x/\varnothing_b \approx 1$. Grushka et al. (1975) give an example that would justify this approximation. The value of $\varnothing$ ranged from 900 to 1000 when the particle size (of the same stationary phase) ranged from 6 to 44 microns.

The value of $\varnothing$ may be determined from the measurement of pressure drop of a column of given dimensions packed with a stationary phase of known particle size, by using a rearranged form of Eq. (6.16):

$$\varnothing = \frac{\Delta p d_p^2}{\mu L v} \tag{6.18}$$

This result leads to a correlation of elution time as a function of plate height and pressure drop:

$$t = \frac{\varnothing \mu N^2 H^2}{\Delta p d_p^2} \tag{6.19}$$

Equation (6.19) is obtained from Eq. (6.18) by substituting $v = L/t$ and $L = NH$, where H is the plate height and N the number of plates. The upper value of the interstitial velocity, v, for a given value of L and d_p is therefore set by the pressure drop and flow resistance parameter, $\varnothing$. This value then helps to define a plate height at the maximum flowrate at which the column may be run, thereby connecting a change in particle size to the hydrodynamics of the column, as well as to the column's dispersion characteristics. One approach to selecting the maximum flow velocity is to determine the upper limit of pressure drop that the equipment or stationary phase can withstand, and then use the measured pressure drop in Eq. (6.17) to select the smallest particle size that fits this constraint. An alternate criterion for selection of the most appropriate particle size is the cost of the stationary phase. In this case, particle size would be calculated based on an upper limit of cost using Eq. (6.15) for particle sizes that are less than 100 microns in diameter, or read from Figure 6.4 for sizes larger than 100 microns. Once the particle size is selected, the pressure drop would be checked to confirm that it does not exceed the equipment's limits. Equation (6.16) can then be used to calculate v, the interstial velocity for a given Δp.

Example 1. Determining Particle Size and Column Dimensions for Scale-up of Size Exclusion Chromatography

A size exclusion column that measures 10 mm i.d. × 60 cm long with a 42 micron stationary phase gives a baseline separation for a two component polypeptide mixture consisting of a high molecular weight target molecule and a low molecular weight polypeptide. The column has 700 plates, and it is run in the cold room at 4°C. The bench-scale column, with an average particle size of 42 microns gives a 30 psig (2 atm) pressure drop. The stationary phase cost is \$700/kg. The mixture is injected so that the volume of sample is equivalent to initially filling 1.2 cm of the column's length (377 μL). One run is completed in 35 minutes.

The separation is to be scaled-up to a process column packed with 100 L of the same type of stationary phase, except that its particle size is large in order to satisfy cost and availability constraints. Commercial quantities of the 42 micron material are not available. Both bench-scale and process columns have the same void fraction ($\varepsilon_b = 0.4$), but the larger column may be run at a lower flowrate—if this is needed to keep the pressure drop within range (i.e., 60 psig or less). The stationary phase particle size should be selected to meet the criterion that it cost no more than \$500/kg.

Calculate:

1. Cost of the stationary phase, per unit volume
2. The pressure drop, and the adjustment of the interstitial velocity that will be required, to satisify the pressure drop constraints
3. The diameter and length of the 100 L column
4. The time required to carry out a complete run for the 100 L column, and the maximum amount of mobile phase needed to carry out a single run
5. The impact of particle size to diameter ratio

Then comment on:

6. The need to carry out another round of calculations

Solution

We will first test whether the sample volume is considered to be within the range of a small sample volume. If it is small, a variation in the sample volume would not change the width of the eluting peaks, once a design is specified. Use Eq. (5.70) in Chapter 5:

$$V_{spl} < 0.5V_{o,plate} + K_D V_{s,plate} \tag{5.70}$$

if $K_D = 0$ (an excluded component), then $K_D V_{s,plate} = 0$. Calculation of $V_{o,plate}$ gives

$$V_{o,plate} = HA_o\,\varepsilon_b \qquad \text{by Eq. (5.71) in Chapter 5}$$

$$= \left(\frac{60 \text{ cm}}{700 \text{ plates}}\right)\left[\left(\frac{1}{2}\right)^2 \pi\right](0.40) = 0.027\,\frac{\text{cm}^3}{\text{plate}}$$

The column length, number of plates, and diameter are given. Consequently Eq. (5.70) simplifies to

$$V_{spl} = (0.5)\,(0.027) = 0.0135 \text{ mL} \qquad \text{for an excluded component}$$

If $K_D = 1$ (which would be the case for a completely included component), then

$$V_{spl} = (0.5)(0.027) + K_D[(1-\varepsilon_b)\,HA_o]$$

$$= 0.0135 + (1)(1-0.4)\left(\frac{60}{700}\right)\left(\frac{1}{2}\right)^2 \pi = 0.0539 \text{ mL}$$

The volume of stationary phase $HA_o\,(1-\varepsilon_b)$ would coincide with the volume not occupied by the moving fluid, i.e., mobile phase. Based on these definitions, a small sample size would correspond to 13.5 to 53.9 μL for the bench-scale column. The sample injected into the bench-scale column was 377 μL. Based on a void fraction of 0.4, this corresponds to 1.2 cm of the column's length. (Since the injected fluid would not initially penetrate the pores of the stationary phase, the length of column occupied by the sample is based on ε_b.) This would not be a small sample volume; neither does it fit the criteria of a large sample volume given by Eq. (5.73) in Chapter 5. A large sample size would be

$$V_{slp} > 3V_{tot,plate}\sqrt{N} = 3(A_o)(H)\sqrt{700} = (3)\left(\left(\frac{1}{2}\right)^2 \pi\right)\left(\frac{60}{700}\right)\sqrt{700} = 5.34$$

Consequently, if the sample size were increased relative to the column's volume, the peak width would be expected to change as a function of column size, and the peak width would be equivalent to V_w given by Eq. (5.74) in Chapter 5:

$$V_w = V_{spl} + V_{tot,plate}\sqrt{2\pi N} \tag{5.74}$$

The sample volume relative to column volume will be kept constant for this scale-up calculation based on the assumption that resolution is maintained if N is the same, and V_w follows Eq. 5.74.

1. The particle size based on the cost constraint would be calculated using Eq. (6.15):

$$d_p = \left(\frac{12{,}600}{C_{cost,sp}}\right)^{1/2.5} = \left(\frac{12{,}600}{\dfrac{\$500/\text{kg}}{1000\ \text{g/kg}}}\right)^{1/2.5} = 57.7 \text{ microns}$$

2. The calculated particle size is within the range of the cost and particle size data which are correlated by Eq. (6.15) and presented in Figure 6.4. Since particle size must be larger to satisfy the cost constraint of \$500/kg (= \$0.50/g), the column dimensions will also be different, subject to the constraint that $N = 700$. But first we will need to check the pressure drop using Eq. (6.16):

$$\Delta p = \frac{\varnothing \mu L v}{d_p^2}$$

Since the pressure drop may not exceed 60 psig, we will assume the buffer, temperature, and interstitial velocity are the same at both bench and process scales. Therefore

$$\frac{\Delta p_x}{\Delta p_b} = \frac{\dfrac{\varnothing \mu L_x v}{d_{p,x}^2}}{\dfrac{\varnothing \mu L_b v}{d_{p,b}^2}} \cong \frac{L_x}{d_{p,x}^2} \cdot \frac{d_{p,b}^2}{L_b} \tag{6.20}$$

The mobile phase, temperature, and packing characteristics are assumed to be the same as both scales so that $\varnothing\mu$ is eliminated from Eq. (6.20). We need to calculate the length before we can obtain the pressure drop. At constant plate count, and constant interstitial velocity by Eq. (6.11),

$$\left(\frac{L_x}{L_b}\right) = \underbrace{\left(\frac{v_x}{v_b}\right)}_{1}\left(\frac{d_{p,x}}{d_{p,b}}\right)^2 = \left(\frac{d_{p,x}}{d_{p,b}}\right)^2 = \left(\frac{57.7}{42}\right)^2 = 1.89 \tag{6.21}$$

Therefore the length of the 100 L column would be (1.89) (60) = 113 cm, which gives a reduced column length of $\lambda = 113/(57.7 \cdot 10^{-4}) = 1.96 \cdot 10^4$ compared to $\lambda = 60/(42 \cdot 10^{-4}) = 1.43 \cdot 10^4$ for the bench scale column.

Based on the result from Eqs. (6.20) and (6.21) the pressure drop of the large column is estimated to be the same as for the bench-scale column, if its length is scaled by Eq. (6.21):

$$\Delta p_x = \Delta p_b \frac{L_x}{L_b} \cdot \left(\frac{d_{p,b}}{d_{p,x}}\right)^2 = \Delta p_b \left(\frac{d_{p,x}}{d_{p,b}}\right)^2 \left(\frac{d_{p,b}}{d_{p,x}}\right)^2 = \Delta p_b = 30 \text{ psig} \tag{6.22}$$

The result from Eqs. 6.21 and 6.22 shows that the column with the larger particle size will be significantly longer than the bench-scale column but will have the same pressure drop. The reason is that the increase in column length will be directly proportional to the square of the ratio of particle sizes while the increase in pressure drop would be inversely proportional to this ratio. At constant plate count and fluid velocity, the pressure drops will be similar. This result is specific to the case for pore diffusion as the dominant cause of dispersion. The case for mass transfer as the dominant mechanism gives a different result, since

$$N \sim \frac{L}{v d_p^{1.5}}.$$

3. The diameter of the large column is:

$$r_{col} = \sqrt{\frac{V_{col,x}}{\pi L_x}} = \sqrt{\frac{100{,}000 \text{ cm}^3}{\pi(1.89)(60)}} = 16.7 \text{ cm} \quad \text{or} \quad d_{col} = 35.5 \text{ cm}$$

4. The maximum time required to carry out the run is given by Eq. (6.14), subject to the constraint that the separation is occurring by size exclusion:

$$\left(\frac{t_x}{t_b}\right)_{sec,max} = \left(\frac{L_x}{L_b}\right)\left(\frac{v_b}{v_x}\right) = 1.89$$

$$1$$

so that

$$t_x = 1.89 t_b = (1.89)(35) = 66 \text{ min}$$

The maximum volume of mobile phase in the two cases will be equivalent to 1 empty column volume, since this is size exclusion chromatography:

$$V_{col,b} = \left(\frac{1}{2}\right)^2 \pi (60) = 47 \text{ mL}$$

$$V_{col,x} = 100 \text{ L}$$

The volumes of mobile phase required in the two cases would be no more than 47 mL and 100 L, respectively.

The scale-up of the separation results in a column that has the proportions, as approximately indicated in Figure 6.6.

5. The separation on the larger scale may be enhanced in the larger-diameter column, since the number of particles per column cross section is

$$\frac{355 \text{ mm}}{0.0577 \text{ mm}} = 6152 \frac{\text{particles}}{\text{column diameter}}$$

compared to

$$\frac{10 \text{ mm}}{0.042 \text{ mm}} = 238 \text{ particles}$$

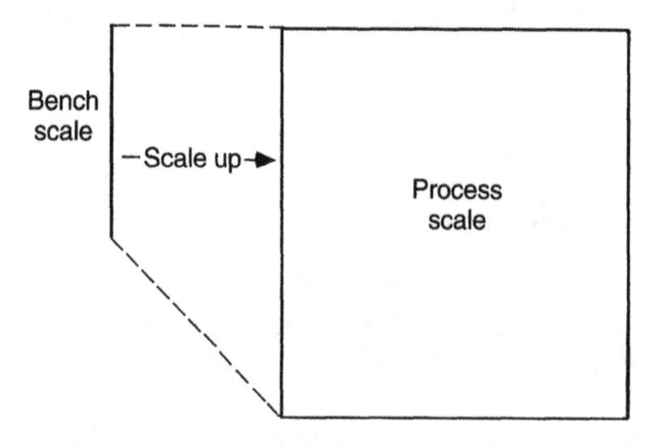

Figure 6.6 Approximate size of columns.

for the bench-scale column. However, in both cases, the number of particles per column cross section exceeds the generally accepted minimum of 80. Below 200 the irregularities in the void volumes induced at the column wall may be propagated into the bed and could result in channeling, which will decrease the plate count.

6. The dimensions of the column may not fit standard sizes that are available from the manufacturer. Hence further adjustments would be needed (e.g., the 113 cm column length may need to be 100 cm, and thereby require use of a stationary phase with a smaller particle size).

The take-home message from Example 1 is that the application of scaling rules enable a quick estimate of column dimensions to be generated using a hand calculator, since the column length in a size exclusion chromatography column scales in proportion to the square of the particle size ratio of the stationary phase, while the pressure drop is inversely proportional to this ratio. Once column dimensions are obtained, further experiments can be planned with the benefit of focusing the research plan onto a range of conditions that already meet predetermined criteria of cost, resolution, and availability of column equipment and stationary phase.

6.3 MASS TRANSFER CAN BE A LIMITING FACTOR AT SLOW FLOWRATES, OR FOR SOLUTES THAT HAVE SLOW DIFFUSION RATES

Diffusion rates generally increase with decreasing size of solute, although the transfer coefficient for an ion can still be small (10^{-5} cm/s) as was illustrated in Example 9 of Chapter 5. Hence small solutes, as well as large ones, may be subject to mass transfer limitation. The analysis of whether mass transfer or pore diffusion is a dominating factor requires that the importance of one relative to the other be discerned through experiments. For example, the plate counts for two different particle sizes of the same stationary phase could be measured for the same solute. If the plate count decreases in proportion to the inverse square of increase of particle size as given in Eq. (6.10), pore diffusion is likely to be a controlling resistance. If the decrease in N is in proportion to the inverse 1.5 power of the increase in particle size, mass transfer is indicated. Mass transfer resistance exhibits a different dependency than pore diffusion for the effect of particle size on plate count, and therefore has a different scaling relationship.

When mass transfer is dominant over other causes of dispersion, the plate count is proportional to the mass transfer coefficient k_{MT}, with the Sherwood number (Sh) giving the ratio of mass transfer velocity to diffusion velocity:

$$\mathrm{Sh} = \frac{k_{\mathrm{MT}} L}{D_{\mathrm{m}}} \tag{6.23}$$

where L represents a characteristic dimension. As pointed out by Cussler (1997), the characteristic length L in Eq. (6.23) is "the membrane thickness for membrane transport, but the sphere diameter for a dissolving sphere. The use of a dimensionless group requires that the physical system to which it is applied be clearly defined. The value of a Sherwood number of 2 for a membrane has a different meaning than the value of 2 for a dissolving sphere, just like 30% efficiency for a turbine has a different meaning than 30% efficiency for a running deer" (p. 224). Cussler also points out that the accuracy of the correlations "is typically on the order of thirty percent, but larger errors are not uncommon. Raw data can look more like the result of a shotgun blast than any sort of coherent experiment because the data include wide ranges of chemical and physical properties" (p. 225). In the case of the correlations presented here for liquid chromatography, L represents either column length or particle size (diameter) and k_{MT} an average mass transfer coefficient either across the bed or across a particle, respectively. This distinction will be made as needed.

The mass transfer coefficient can be estimated from the chromatographic elution profile of a solute that displays linear equilibrium. Sherwood et al. (1975) show that a pulse injection of a mass of solute Q (in g), is given by

$$C(z,t) = \frac{Q}{2\sqrt{\pi}}\left[\frac{(k_{MT}a)^2}{(v\varepsilon_b)^3 z\hat{t}\,(\rho_B K_D)^3}\right]^{1/4} \exp\left[-\left(\sqrt{\frac{k_{MT}az}{v\varepsilon_b}} - \sqrt{\frac{k_{MT}a\hat{t}}{K_D\rho_B}}\right)^2\right] \tag{6.24}$$

where

$C\,(x, t)$ = concentration of solute in the fluid, g/mL
k_{MT} = general mass transfer coefficient, cm/s
K_D = distribution coefficient = q_{eq}/c_{eq}
a = interfacial area of solid phase per volume of bed, cm^2/s^3
v = interstitial velocity, cm/s
ε_b = interparticle void fraction
z = axial distance along the bed, cm
$\hat{t} = t - z/v$ = elapsed time at point z after a solute molecule has arrived at that point, having entered the bed when $t = 0$ s
ρ_B = bulk density of solid phase = mass of solid phase/volume of bed, g/mL

If mass transfer between phases is very easy ($k_{MT}az > v\varepsilon_b$), the exponential term in Eq. (6.24) varies most rapidly as a function of the time and distance. The maximum peak concentration therefore occurs when

$$\frac{k_{MT}a\hat{t}}{K_D\rho_B} = \frac{k_{MT}az}{v\varepsilon_b} \tag{6.25}$$

at an elapsed time that corresponds to Eq. (6.26):

$$\hat{t}_{max} = t - \frac{z}{v} = \frac{K_D\rho_B z}{v\varepsilon_b} \tag{6.26}$$

The maximum concentration at $\hat{t}_{max}$ is given by

$$C_{max} \approx \left[\frac{Q}{2\sqrt{\pi}}\right]\left[\frac{(k_{MT}a)^2}{(v\varepsilon_b)^3 z\hat{t}_{max}\,(\rho_B K_D)^3}\right]^{1/4} \tag{6.27}$$

where Q is the mass of solute applied to the column. Different components have different times at which their peaks emerge from the column and are characterized as given by Sherwood et al., 1975 pp. 575–576. By ignoring the small changes of $(\hat{t})^{-1/4}$ in the pre-exponential term in Eq. (6.24), the times at which the peak passes through values of half of its maximum occur at

$$\sqrt{\frac{k_{MT}az}{v\varepsilon_b}} - \sqrt{\frac{k_{MT}a\hat{t}}{K_D\rho_B}} = \pm\sqrt{\ln 2} \tag{6.28}$$

The width of the peak, Δ, at half of its maximal concentration (or peak height) is

$$\Delta = 4\rho_B K_D \sqrt{\frac{z \ln 2}{v\varepsilon_b k_{MT} a}} = \sqrt{\frac{4\rho_B K_D z}{v\varepsilon_b}} \frac{\frac{1}{2}\ln 2}{\frac{1}{v\varepsilon_b} k_{MT} a} = \hat{t}_{max}\left(4\sqrt{\frac{\varepsilon_b v \ln 2}{k_{MT} az}}\right) \tag{6.29}$$

Since the number of transfer units per bed, n, is

$$n \equiv \frac{k_{MT}az}{v\varepsilon_b} \tag{6.30}$$

rearrangement of Eq. (6.29) and combination with Eq. (6.30) gives

$$n = (16 \ln 2)\left(\frac{\hat{t}_{max}}{\Delta}\right)^2 \tag{6.31}$$

Consequently injection of a small volume of sample at a concentration in which linear equilibrium is satisified, and where its elution occurs at a relatively slow flowrate, would enable determination of k_{MT} from the experimental parameters indicated in Figure 6.7 and given by Eq. (6.32). Equation (6.32) is obtained by solving Eq. (6.30) for $k_{MT}a$ and combining with Eq. (6.31):

$$k_{MT}a = n\frac{v\varepsilon_b}{z} = (16 \ln 2)\left(\frac{\hat{t}_{max}}{\Delta}\right)^2\left(\frac{v\varepsilon_b}{z}\right) \tag{6.32}$$

where $\hat{t}_{max}$ and Δ are read from the chromatogram in Figure 6.7, $v\varepsilon_b$ is the superficial velocity, and z is the length of the column. A large value of n would correspond to a narrow peak that elutes after a long retention time. A large value of n suggests, a large value of the mass transfer coefficient and rapid mass transfer.[2] This approach was presented by Sherwood et al. (1975) for gas chromatography, but it should also be applicable to liquid chromatography as long as mass transfer of the solute to the stationary phase ($k_{MT}\,az$) is rapid relative to the residence time of the solute in the column (represented by $v\varepsilon_b$). Equation (6.32) gives a value for the mass transfer coefficient lumped together with the interfacial area, a, since the interfacial area itself may not be known.

[2]The development of Eqs. (6.24) to (6.32) for chromatography extend to adsorption. The significance of n was concisely stated by Etzel (1995). The parameter n reflects the sharpness of the adsorption breakthrough curve and should be much greater than one for efficiently operated packed-column adsorbers. Etzel states, "The operating conditions that produce large values of n are long columns, low flowrates, high feed concentrations, and small bead diameters. An economic balance must be established between sharp breakthrough curves, which increase efficiency, and high column pressure drops which decrease throughput" (p. 406).

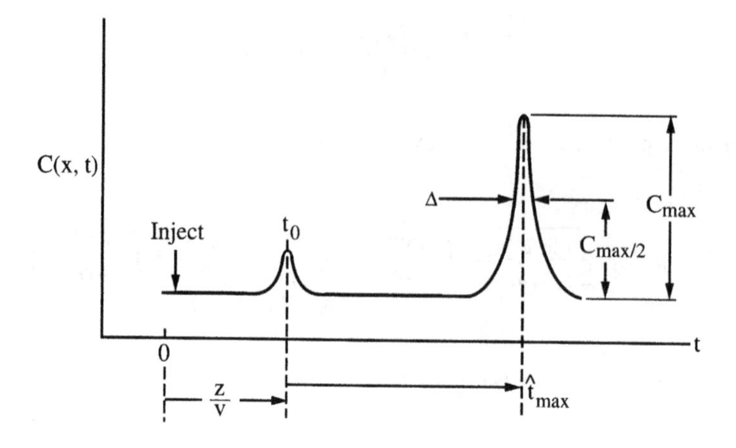

Figure 6.7 Parameters defined by Eqs. (6.26), (6.27), and (6.29).

A first step toward estimating the particle size and flowrates that achieves an economic balance when mass transfer is a dominating resistance is based on the definition of the plate count N as a function of the mass transfer coefficient:

$$N_{\text{mass transfer}} \cdot \frac{k_{\text{MT}}}{v} \frac{L}{d_{\text{p}}} = \frac{D}{vd_{\text{p}}} \text{Sh} = \frac{\varepsilon_{\text{b}}\text{Sh}}{\text{ReSc}} \tag{6.33}$$

where the ReSc number is defined based on the superficial velocity ($= \varepsilon_b v$) as given in Eq. (5.107) of Chapter 5. Here the plate count is denoted as $N_{\text{mass transfer}}$, since it represents a limiting condition (Ladisch and Velayudhan, 1999), D is a dispersion coefficient, v is the interstitial velocity, d_p the diameter of the stationary phase particles, and L column length.

Various correlations for the Sherwood number Sh, and therefore k_{MT} as a function of the ReSc ($= d_p \varepsilon_b v / D_m$) number, have been proposed. Goto et al. (1983) suggest that

$$\text{Sh} \sim \text{Re}^{1/2}\text{Sc}^{1/3} \tag{6.34}$$

where Re is the particle Reynolds number and Sc the Schmidt number. Another form is that of Wilson and Geankoplis (1966):

$$\text{Sh} \sim \text{Re}^{1/3}\text{Sc}^{1/3} \tag{6.35}$$

Wakao and Furazaki (1978) suggested a correlation of the form

$$\text{Sh} = 2.0 + 1.1\ \text{Re}^{0.6}\ \text{Sc}^{1/3} \tag{6.36}$$

However, none of these forms of the Sherwood number appear to be directly used for defining k_{MT} in Eq. (6.33) in order to obtain the proportional relationship between $N_{\text{mass transfer}}$ and column length, average diameter of the particle size, and interstitial velocity. Rather the form given in Eq. (6.37) is used:

$$N_{\text{mass transfer}} \sim \frac{L}{\nu^{1/2} d_p^{3/2}} \tag{6.37}$$

where the subscript "mass transfer" in $N_{\text{mass transfer}}$ denotes the plate count for the mass transfer limiting case. Equation (6.37) is obtained by combining definitions of the reduced velocity ($\upsilon = vd_p/D$) and reduced plate height ($h = H/d_p = (L/N)/d_p$) with an empirical equation from Grushka et al. (1975):

$$N = \frac{D^n}{m} \frac{L}{\nu^n d_p^{1+n}} \tag{6.38}$$

where D is an effective or averaged bed dispersion coefficient and m and n coefficients determined by fitting the data to the equation. Equation (6.38) follows from the Snyder equation

$$h = m\nu^n \tag{6.39}$$

which upon rearrangement gives Eq. (6.37), based on the definition of ν (Eq. 5.95 to 5.97) assumption that D and m are the same on both bench and process scales and that $n = 0.5$, since the dominating resistance (mass transfer in this case) is assumed to be the same on both the bench and process scales (Huber, 1973; Grushka et al., 1975; Pieri et al., 1983). Pieri et al. (1983) show that this approach is appropriate for scaling-up normal phase chromatography of a low molecular solute (insect phermone) using silica as the stationary phase. The porosity of the silica gel particles is small, and hence mass transfer is a rational choice as the dominating resistance.

Defining the plate count (and the resolution) to be the same at both bench and large scale gives

TABLE 6.1 Scaling Rules for Linear Equilibria for Mass Transfer or Pore Diffusion as the Dominant Resistance

Specify N and resolution to be sme on both process and large scale
1. Specify: $d_{P,B}/d_{p,x}$, $V_{spl,x}/V_{spl,B}$, v_x/v_B, L_B, t_B
2. Use: $k = 2$ if pore diffusion controls
$k = 1.5$ if mass transfer controls
3. Calculate large-scale parameters:
Length: $L_x = L_b\,(v_x/v_b)^{k-1}\,(d_{p,x}/d_{p,b})^k$
Column diameter: $d_{col,x} = d_{col,b}\,[(L_b/L_x)\,(V_{spl,x}/V_{spl,b})]^{1/2}$
Minimum run time: $t_x = t_b\,(L_x/L_b)\,(v_b/v_x)$

Source: Ladisch and Velayudhan (1999).

$$\frac{L_x}{[v_x d_{p,x}^3]^{1/2}} = \frac{L_b}{[v_b d_{p,b}^3]^{1/2}} \tag{6.40}$$

As before, we assume that the ratios (v_x/v_b), $(V_{spl,x}/V_{spl,b})$, and $(d_{p,x}/d_{p,b})$ are specified and that the pressure drop criteria of Eqs. (6.16) to (6.18) apply. The scaling rules are summarized in Table (6.1) with $k = 1.5$ $(= 1 + n = 1 + 0.5)$ for mass transfer control.

The assumption that one mechanism dominates band spreading allows the scale-up to be determined from simple equations involving operational parameters. Furthermore, if chromatographic profiles are available at a flowrate where mass transfer is known to be limiting, a value of $k_{MT}a$ can be estimated using the approach outlined by Eqs. (6.25) to (6.32). Once the value of $k_{MT}a$ is known, Eq. (6.24) may be used to generate an elution profile for a solute exhibiting linear equilibrium with a given distribution coefficient, K_D.

6.4 SCALE-UP RULES ARE SIMILAR FOR PORE DIFFUSION AND MASS TRANSFER LIMITING CASES

When either pore diffusion or mass transfer is the single controlling nonequilibrium mechanism, scaling rules can be used in a manner analogous to that for size exclusion chromatography. When pore diffusion is the controlling mechanism, the scaling equations are identical to those given for size exclusion chromatography in Eqs. (6.11) to (6.19). These rules are summarized in Table 6.1 with $k = 2$ for columns in which pore diffusion controls. The minimum time for a run is based on the assumption that at least one column volume will be passed through the system between sample injections. The expressions in Table 6.1 are valid only under the assumption that there is one dominant mechanism and that it does not change when moving from the bench- to the large-scale system. This assumption could be unrealistic, for instance, when the particle size at the pilot scale is much larger than at the bench scale. Mass transfer is usually assumed to be the controlling mechanism for a small particle size, while pore diffusion is much more likely to dominate for large particles. However, there are cases where film mass transfer and pore diffusion can both be dominant contributors to band spreading in linear chromatography. In addition another form of dispersion may arise from eddy diffusion (microscale mixing effects) that increases in importance as the particle size becomes larger. Under conditions of linear isocratic elution, slow binding kinetics are unlikely to be the controlling mechanism.

6.5 SCALE-UP WHEN MASS TRANSFER AND PORE DIFFUSION ARE OF COMPARABLE MAGNITUDE REQUIRES COMBINATION OF THESE RESISTANCES

If mass transfer and pore diffusion are of comparable magnitude so that neither is dominant, the effect of combined resistances must be considered. Since variances are additive under linear chromatography (Sherwood et al., 1975), the composite plate count should also be a combination of pore diffusion and mass transfer expressions (Ladisch and Velayudhan, 1999). One approach is suggested by

$$N_{\text{composite}} \sim A\frac{L}{v d_p^2} + B\frac{L}{v^{1/2} d_p^{3/2}} \tag{6.41}$$

Consider a case where pore diffusion becomes limiting when the particle size of the stationary phase is larger on a process scale than the bench scale. Also assume that the stationary phase on the bench scale exhibits a combination of mass transfer and pore diffusion resistance. If N is to be kept the same at both the large and bench scales, the equation for relating the bench-scale result, where mass transfer and pore diffusion are both important, to a case of pore diffusion controlling at the process scale, is

$$\underbrace{A\frac{L_x}{v_x (d_{p,x})^2}}_{\text{Process}} = \underbrace{A\frac{L_b}{v_b (d_{p,b})^2} + B\frac{L_b}{v_b^{1/2} (d_{p,b})^{3/2}}}_{\text{Bench scale}} \tag{6.42}$$

The value of A is assumed to be the same on both bench and process scales. The properties of the solute are embedded in the fitted parameters A and B. These are not explicit as is the case for the Eqs. (5.99) to (5.108) in Chapter 5 developed by Athalye et al. (1992). As before, the ratios (v_x/v_b), $(d_{p,x}/d_{p,b})$, and $(V_{spl,x}/V_{spl,b})$ are assumed to be specified, resulting in the expression:

$$\boxed{\left(\frac{L_x}{L_b}\right) = \frac{B' + A}{A}\left(\frac{v_x}{v_b}\right)\left(\frac{d_{p,x}}{d_{p,b}}\right)^2} \tag{6.43}$$

where $B' = B\,(v_b d_{p,b})^{1/2}$ and A and B are given by Ruthven (1984) as

$$A = 1.1\text{Sc}^{-1/6}\sqrt{D_M} \tag{6.44}$$

$$B = \frac{2}{15}\frac{\varepsilon_b}{(1+\varepsilon_b)\varepsilon_p}\frac{1}{D_p} \tag{6.45}$$

The parameters D_m, D_p, and ε_b in Eqs. (6.44) and (6.45) are the same as defined previously in Chaper 5. D_m is the diffusion coefficient for the solute in an unbounded mobile phase, and D_p is the diffusion coefficient for the solute within a pore. The equations for the diameter and the retention time of the large column are the same as given in Table 6.1.

Example 2. Derivation and Use of a Scaling Rule for Effect of Particle Size on Plate Count When Mass Transfer or Kinetic Effects Dominate

The plate height curve proposed by Snyder (Grushka et al., 1975), has the form

$$h = m\nu^n \tag{6.46}$$

if kinetic and mass transfer effects dominate and where

m, n = fitted parameters

h = reduced plate height = $\frac{L}{Nd_p}$

ν = reduced velocity = $\frac{vd_p}{D}$

and D is a dispersion coefficient. The other parameters are the same as defined previously.

1. Show the effect of particle size on plate count.
2. What is the expected change in plate count if the particle size of a stationary phase is increased from 0.005 cm (= 5 microns) to 0.010 cm (= 10 microns)?

Solution

1. Use Snyder's definition, and solve for N:

$$h = \frac{L}{Nd_p} = m\left(\frac{vd_p}{D}\right)^n \tag{6.47}$$

$$N = \frac{L}{d_p}\frac{D^n}{m}\frac{1}{v^n d_p^n} = \frac{D^n L}{mv^n d_p^{1+n}} \tag{6.48}$$

If the particle size doubles, and all other parameters have the same value, Eq. (6.48) shows that

$$\frac{N_{d_p}}{N_{2d_p}} = \frac{(2d_p)^{1+n}}{d_p^{1+n}} = 2^{1+n} \tag{6.49}$$

The size of the particle must still be known, so the dominating mechanism can be identified. The mechanisms on both the bench and large scale must also be confirmed to be the same. If they both prove to be limited by either a kinetic or a mass transfer effects, then Eq. (6.49) is valid.

2. The small particle size, in this case, would suggest that mass transfer is a limiting factor. This would justify the use of Eqs. (6.48) and (6.49) with $n = 0.5$:

$$\frac{N_5}{N_{10}} = \frac{(10)^{1+n}}{(5)^{1+n}} = \frac{10^{1.5}}{5^{1.5}} = 2.83$$

or

$$N_{10} = 0.354\, N_5$$

A significant decrease in plate count can be expected in this case. When a larger particle size is involved (e.g., for scale-up from $d_p = 5$ microns to $d_p = 100$ microns), this approach is not appropriate because pore diffusion rather than mass transfer could be dominant for the 100 micron particle size; the opposite would apply for the 5 micron particle.

Equation (6.48) has been used for the scale-up of the purification of phermones over silica gel (Pieri et al., 1983). The fitted constant m was the measure of column efficiency, and n had a value between 0.4 and 0.6.

The approach summarized in Table 6.1 gives a basis with which to anticipate increase in peak broadness (i.e., decrease in plate count) due to an increase in particle size and/or a change in the interstitial velocity (i.e., flowrate). If peak retention times of the two components are known as well, the average plate count and capacity factors can be estimated from Eq. (6.9). Equation (5.148) in Chapter 5 can be used if both retention volumes and peak widths are known. This way both an estimate of column length, and consequently column diameter for a given volume, can be obtained. The column diameter is specified by the volume of sample loaded. Once volumes and throughputs are known, costs can be estimated. Other forms of scaling rules, and guidelines for estimating costs for linear systems can be found in the literature (Wankat and Koo, 1988; Tice et al., 1987; Ladisch et al., 1984).

6.6 A MATERIAL BALANCE COMBINED WITH PLATE COUNT FACILITATES SIMULATION OF ELUTION PROFILES FOR LINEAR EQUILIBRIUM: SIZE EXCLUSION AND ION EXCLUSION EXAMPLES (WITH CONTRIBUTIONS BY SCOTT RUDGE)

The calculation of elution profiles based on a distribution function, such as using Eq. (5.30) in Chapter 5, is a first step toward modeling peak elution behavior, when the peaks have Gaussian characteristics. The development of models that are mechanistic in their origin, and that can be extrapolated to other conditions, must start with a material balance rather than a statistical function. The material balance (also referred to as the *continuum*) approach facilitates calculation of a complete elution profile for a solute, based on a mass balance on the solute, and either an equation of state describing the stationary phase capacity or a kinetic relationship describing the rate at which solute approaches equilibrium with the sorbent. This analysis is taken from a review by Ladisch and Velayudhan (1999). It is presented below with some modifications.

An early continuum model is that of Lapidus and Amundson (1952), which was presented in Chapter 5 as a mass balance on a single solute and given in Eq. (5.48) with the concentrations normalized for the phase ratio $(1 - \varepsilon_b)/\varepsilon_b$:

$$D\frac{\partial^2 C_m}{\partial z^2} = \frac{v\partial C_m}{\partial z} + \frac{\partial C_m}{\partial t} + \frac{1-\varepsilon_b}{\varepsilon_b}\frac{\partial q}{\partial t} \tag{5.48}$$

The loading on the stationary phase, given by q_{eq}, is defined using either the linear equilibrium relationship:

$$q_{eq} = K_D C_{eq}$$

or a kinetic relationship

$$\frac{\partial q}{\partial t} = k_1 C - k_2 q \tag{6.50}$$

where

v = interstitial velocity, cm/s
K_D = equilibrium constant = q_{eq}/c_{eq}
ε_b = extraparticulate void fraction
k_1, k_2 = rate constants; k_1 represents a mass transfer constant, k_2 a kinetic constant
C_m = concentration of a single solute as defined in Eq. (5.48)
D = convective axial, dispersion coefficient that lumps together second-order dispersion processes, without defining their origin, as a function of column length

Analytical solutions for the cases of adsorption and two types of pulse inputs were derived by Lapidus and Amundson (1952) based on the assumption of a semi-infinite column. The semi-infinite column is a column with a defined inlet, though it is infinitely long. This is a mathematical way of stating the assumption of no end effects, since the profile is always observed within the bed.

An initially clean column and a constant inlet concentration is represented by the initial and boundary conditions of

$$C = 0,\ t = 0,\ z > 0 \qquad \text{(pulse input)} \tag{6.51}$$

and

$$C = C_i,\ t > 0,\ z = 0 \qquad \text{(constant mobile phase concentration)} \tag{6.52}$$

gives the equation for adsorption breakthrough:

$$\boxed{\begin{aligned}\frac{C}{C_o} = 0.5\Bigg[1 + \operatorname{erf}\left[v\sqrt{\frac{t}{4\gamma D}} - z\sqrt{\frac{\gamma}{4tD}}\right]\\ + \exp\left(\frac{vz}{D}\right)\operatorname{erfc}\left[v\sqrt{\frac{t}{4\gamma D}} + z\sqrt{\frac{\gamma}{4tD}}\right]\Bigg]\end{aligned}} \tag{6.53}$$

where erf is the error function, erfc is the complementary error function, and γ is $1 + [(1 - \varepsilon_b)K_D/\varepsilon_b]$.

The lumped or averaged column dispersion coefficient, D, and plate number may be related through the bed's Peclet number, $\overline{\text{Pe}}$, as defined in Eqs. (6.2) to (Mood and Graybill, 1963)

$$\boxed{2N = \overline{\text{Pe}} = \frac{vL}{D}} \tag{6.54}$$

where v is the interstitial velocity and L is the length of the column. The definition of $\overline{\text{Pe}}$ in Eq. (6.54) differs from the definition for Pe given in Eq. (5.100) in Chapter 5. The definition of Pe in Chapter 5 is based on the ratio of convective mass transfer to diffusive mass transfer for a single particle of radius R. $\overline{\text{Pe}}$ in Eq. (6.54) is based on a bed of length L. Definition of a column Peclet number $\overline{\text{Pe}}$ represents a way of lumping together second-order dispersion effects into a parameter that can be adjusted to fit the data. $\overline{\text{Pe}}$ does not necessarily represent a ratio of convective to diffusive mass transfer.

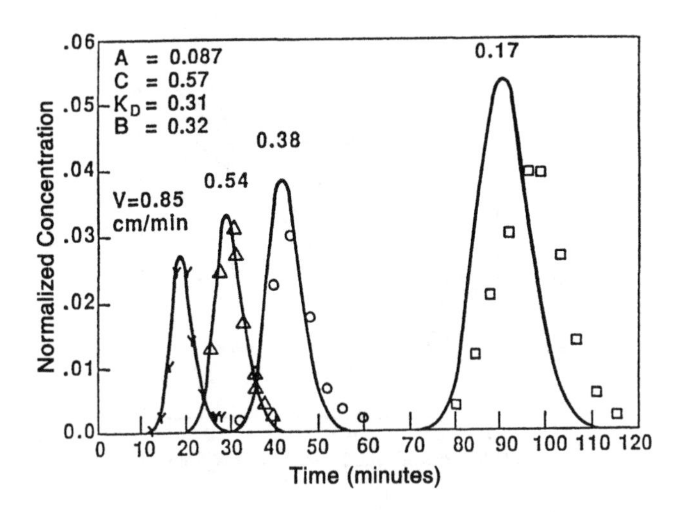

Figure 6.8 Elution profiles of the whey protein β-lactoglobulin B given by the Lapidus and Amundson equation, for different eluent velocities, with the dispersion coefficient estimated from the van Deemter equation. Solid lines are model predictions by Eq. 6.53 and 6.54: points are from column experiments on Sephadex Gl-75. (Reprinted with permission from American Chemical Society and American Institute of Chemical Engineers, Rudge and Ladisch, 1988, fig. 1, p. 125.)

The combination of the plate count Eq. (5.33) in Chapter 5 with the van Deemter Eq. (5.87) in Chapter 5 and the Peclet number as defined in Eq. (6.54), results in an empirical expression for a lumped bed dispersion coefficient:

$$D = \frac{vL}{2N} = \frac{1}{2} vH = \frac{1}{2} v \left(\frac{B}{v} + Cv + A \right) \tag{6.55}$$

The values of A, B, and C must be determined by fitting this equation to experimental measurements of plate height as a function of interstitial velocity for each column to be modeled.

Once the van Deemter coefficients are obtained, the dispersion coefficient can be interpolated for different flowrates. Equations (6.53) and (6.54) can be combined to generate an elution profile such as shown for a single protein, β-lactoglobulin B, for a Sephadex column (Figure 6.8) at eluent velocities ranging from 0.17 to 0.85 cm/min (Rudge and Ladisch, 1988). The individual peaks were obtained in separate runs but are superimposed in Figure 6.8 on a single graph. The calculated and measured elution profiles are similar at the higher flowrates, although significant deviation occurs when Eq. (6.53) is used at the lowest chromatographic velocity of 0.17 cm/min (represented by peak on right-hand side of Fig. 6.8). This approach only gives a first estimate of the expected elution profile shapes. These are nonetheless useful for initiating simulation of a chromatographic separation.

An alternate representation of this model was applied by Neuman et al. (1987) for sulfuric acid–sugar separation by ion exclusion. Ion exclusion takes its name from the exclusion of ionic species (sulfuric acid, in this case) from highly ionized cation exchange resins that are used as the stationary phase. Glucose concentrations in excess of 0.8 M (144 g/L) follow linear equilibria and are reversibly retained by the resin (Figures 6.9 and 6.10) (Neuman et al., 1987). The effluent concentration profile is given by

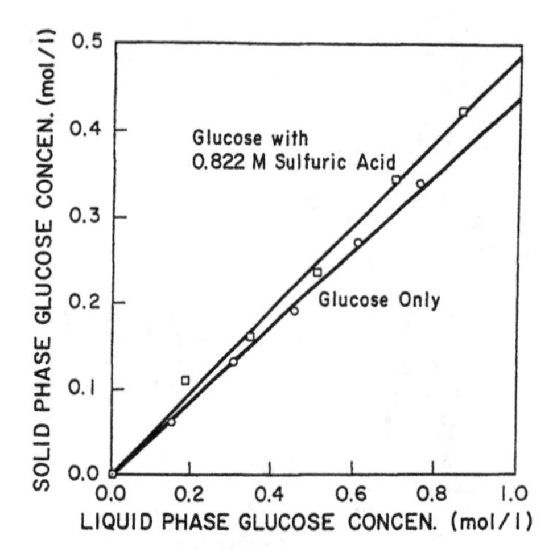

Figure 6.9 Equilibrium data for glucose in water and 0.822 M sulfuric acid on Amberlite IR-118 at 55°C. (Reprinted with permission from Neuman et al., 1987, fig. 4, p. 58, Reactive & Functional Polymers, Vol. 5, with permission from Elsevier Science).

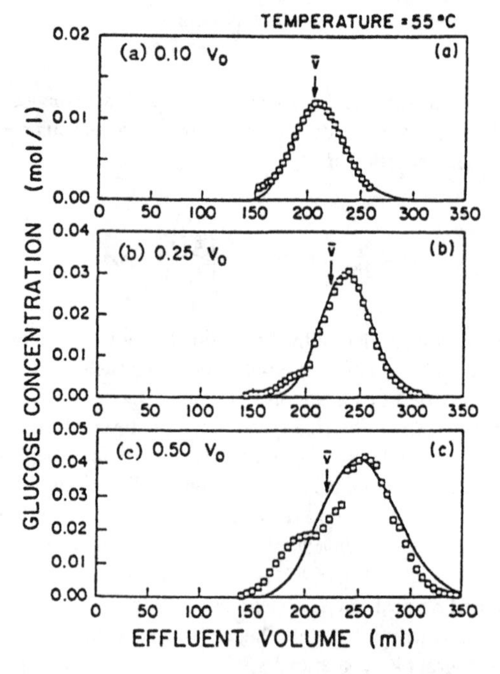

Figure 6.10 Experimental data with $0.10V_0$ sample volumes containing calculated values are summarized in Table 6.2. (Reprinted with permission from Neuman et al., 1987, fig. 5, p. 59; fig. 3, p. 58.); Experimental data with $0.25V_0$ sample volumes containing calculated values are summarized in Table 6.2. (Reprinted with permission from Neuman et al., 1987, fig. 5, p. 59; fig. 3, p. 58.); Experimental data with $0.50V_0$ sample volumes containing calculated values are summarized in Table 6.2. (Reprinted with permission from Neuman et al., 1987, fig. 5, p. 59; fig. 3, p. 58, Reactive & Functional Polymers, Vol. 5, with permission from Elsevier Science.)

TABLE 6.2 Comparison of Calculated Peclet Numbers and Dispersion Coefficients Corresponding to Figure 6.10a, b, and c

Figure	Temperature (°C)	V_o (% of V_o)	V (mL)	Pe	D (cm^2/min)
6.10a	55	10	204	138	0.271
6.10b	55	25	220	200	0.182
6.10c	55	50	219	109	0.343

$$\frac{C}{C_F} = \frac{1}{2}\left[1 + \text{erf}\left(\frac{\overline{\text{Pe}}^{1/2}(V - \overline{V})}{2(V\overline{V})^{1/2}}\right)\right] - \frac{1}{2}\left[1 + \text{erf}\left(\frac{\overline{\text{Pe}}^{1/2}(V - V_{\text{spl}} - \overline{V})}{2\left((V - V_{\text{spl}})\,\overline{V}\right)^{1/2}}\right)\right] \tag{6.56}$$

where V_{spl} is the sample volume, and $\overline{V}$ is the volume of feed required to saturate the column's capacity to take up the solute. The parameter $\overline{V}$ is related to the interparticle void volume V_o and the stationary phase volume for linear chromatography by

$$\overline{V} = V_R = V_o + K_D V_s = V_o + K_D \phi V_o = V_s + k' V_s \tag{6.57}$$

Equation (6.57) is a rearranged form of Eq. (5.139) in Chapter 5 based on the definition of capacity factor given in Eq. (5.137) in Chapter 5. The values of parameters which resulted in the fit of Eq. (6.56) to the data (Figure 6.10 are summarized in Table 6.2). The data and calculated result in Figure 6.10 show that a reasonable fit is possible with a sample volumes of up to 50% of the void fraction of the column. The column dimensions were 2.54 cm i.d. by 61 cm long, with a stationary phase bed volume of 309 mL. The stationary phase was Amberlite IR-118H, strong cation exchange resin from the Rohm and Haas Company (Philadelphia, Pa.). The mobile phase was distilled water that was pumped through the column at a flowrate of 3.0 to 3.2 mL/min.

Separation of the glucose from sulfuric acid is shown to give similar peak shapes for the glucose at temperatures ranging from 27 to 81°C (Figure 6.11(a) to (d)), with reduced skewing of the acid peak as the temperature increased. The effect of temperature on elution times for glucose is small, since glucose is retained due to gel permeation rather than adsorption. The average retention time for the acid peak is not affected by temperature either, since this peak is excluded from the stationary phase and does not interact with it. Significant dilution of the sample occurs as shown in Figure 6.11, which plot the fractional concentration at the outlet relative to the feed concentration (C/C_F), as a function of the number of void volumes of mobile phase eluted from the column.

Differences in retention occur, however, as the volume of sample that is loaded, increases (Figure 6.11e, f, and g). The retention of glucose is affected by the concentration of sulfuric acid. The feed sample placed onto the column contained 10 g/L glucose (equivalent to 0.056 M) or 10 g/L glucose in 7.7% (by weight) sulfuric acid. The increase in retention volume of the glucose is consistent with the equilibrium data, shown in Figure 6.9, where the presence of sulfuric acid increases the distribution coefficient. As the feed volume increases, the sulfuric acid peak undergoes less dilution (compare Figures 6.11e, f, and g). The increases in sulfuric acid concentration (maximum concentration of 0.38 at $0.10V_o$ verses 0.75 at $0.50V_o$) coincide with increased glucose retention. This provides an example where the presence of one solute affects the retention of the other.

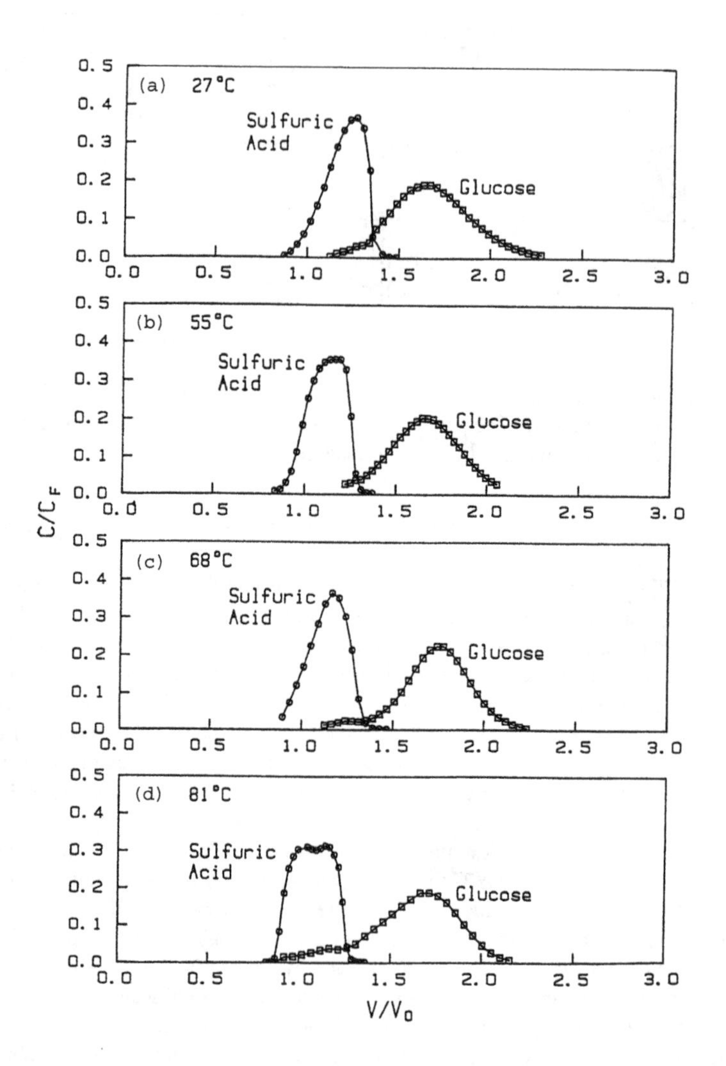

Figure 6.11 (a) Experimental results showing separation of sulfuric acid from glucose by IR-118H at temperatures of 27°C, with water as eluent. Sample volume of $0.1V_0$; effect of sample volume on peak retention and shape at 55°C. (b) Experimental results showing separation of sulfuric acid from glucose by IR-118H at temperatures of 55°C, with water as eluent. Sample volume of $0.1V_0$; effect of sample volume on peak retention and shape at 55°C. (c) Experimental results showing separation of sulfuric acid from glucose by IR-118H at temperatures of 68°C, with water as eluent. (d) Experimental results showing separation of sulfuric acid from glucose by IR-118H at temperatures of 81°C with water as eluent. Sample volume of $0.1V_0$; effect of sample volume on peak retention and shape at 55°C.

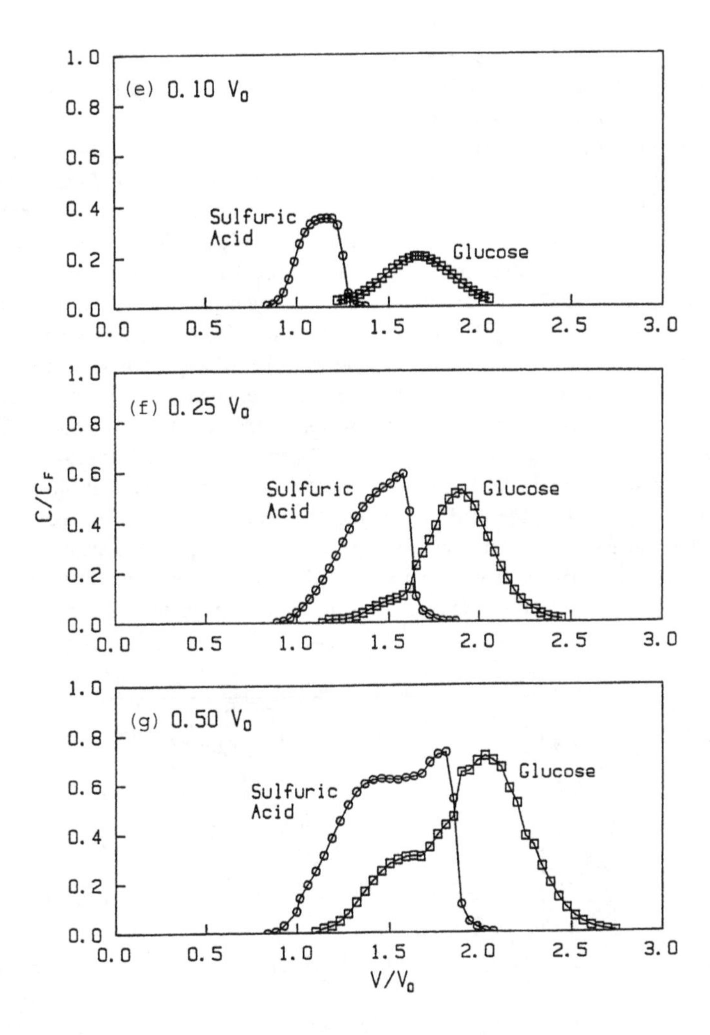

Figure 6.11 (Continued) Sample volume of $0.1V_o$; effect of sample volume on peak retention and shape at 55°C. (e) $0.10V_o$, where V_o denotes the interparticle void volume. Particle size of 297 to 1190 microns. Superficial eluent velocity of 0.59 to 0.63 cm/min ($\varepsilon_b v$). (f) $0.25V_o$, where V_o denotes the interparticle void volume. Particle size of 297 to 1190 microns. Superficial eluent velocity of 0.59 to 0.63 cm/min ($\varepsilon_b v$). (g) $0.50V_o$, where V_o denotes the interparticle void volume. Particle size of 297 to 1190 microns. Superficial eluent velocity of 0.59 to 0.63 cm/min ($\varepsilon_b v$). (Reprinted with permission from Neuman et al., 1987, fig. 2, p. 58, and fig. 3, p. 58.)

6.7 PHYSICAL PROPERTIES OF STATIONARY PHASE, MOBILE PHASE, AND FEED SAMPLE SHOULD NOT BE FORGOTTEN WHEN ANALYZING COLUMN PERFORMANCE

Both the sulfuric acid and glucose peaks become skewed as the sample volume increases. At $0.10V_0$, the glucose peak is close to Gaussian, while the sulfuric acid peak has the characteristic shape associated with ion exclusion (Figure 6.11e). However, as the feed volume increases to $0.25V_0$ (Figure 6.11f), the glucose peak becomes skewed, and at $0.50V_0$ this becomes even more noticeable, together with some deformation of the sulfuric acid peak (Figure 6.11f). The underlying mechanisms are complex, and they are not anticipated by the measurement of the equilibria alone. One contributing factor is the change in void fraction of the individual resin particles. Their void fraction (i.e., internal porosity) is 0.701 in 0.822 M H_2SO_4 and 0.766 in water (Neuman et al., 1987). The accompanying change in density from 1.12 ± 0.01 in water to 1.14 ± 0.01 g/mL in 0.822 M sulfuric acid is small, but also significant. Such a change in density could result in a change in bed packing characteristics, which in turn could cause a change in *D*, the lumped dispersion coefficient in Eq. 6.54. Another contributing factor is likely to be the decrease in the viscosity of the sample with increasing temperature, thus reducing fingering of the sample as it is layered on, or eluted by, water. When the sulfuric acid is passed downflow, its profile is skewed possibly due to channeling that may occur by a heavier layer of acid on top of water (Figure 6.12). The acid may finger into the lighter water phase below it. Conversely, when the acid is eluted upflow, a sharper and smoother profile is evident. A possible explanation again lies in the difference in densities. The lighter water layer is on top of the more dense acid layer, so the acid does not channel or finger. This example illustrates how physical

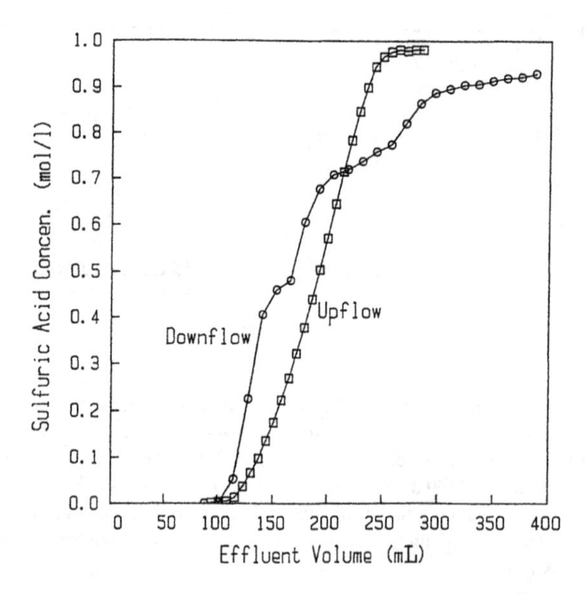

Figure 6.12 Effect of viscosity on shape of breakthrough curve for 0.822 M H_2SO_4 fed continuously into a column of Amberlite IR-118, in downflow and upflow directions. Sulfuric acid has a higher density than water, with skew caused by a sharp density gradient (from Neuman et al., 1987, fig. 6, p. 60, Reactive & Functional Polymers, Vol. 5, with permission from Elsevier Science).

characteristics of the bed at operational conditions have a strong impact on elution behavior that could easily be misinterpreted as nonideal mass transfer and/or equilibrium behavior.

Packing procedures and characteristics can also be an important and overlooked parameter in chromatographic separations, even though the effect of stationary phase packing characteristics on separations is well recognized. Comparison of elution profiles for glucose and sulfuric acid for the same system provide an example. Backflushing of the resin prior to a chromatography run may result in skewed peaks. The particular resin used in this work, IR 118H, had a particle size range of 297 to 1190 microns. Backflushing is a practice used in the water treatment industry where the mobile phase, water, is passed up the bed and the resin is gently fluidized. This causes fines to be removed from the bed but also causes the bed, once settled, to form a gradient of particle sizes, particularly when the resin has a broad particle size distribution (Ladisch et al., 1984). The largest particles settle to the bottom of the bed while the smallest ones are on top. This may reduce column performance as suggested by Figure 6.13. The column that was backflushed gave broader peaks with significant channeling as evidenced by elution of the first peak significantly earlier than the void volume of the bed. The expected elution time of an excluded component such as sulfuric acid would occur at about $V/V_0 = 1$ as indicated in Figure 6.14, which shows elution profiles for a column that was not backflushed. This example illustrates that care must be taken in identifying probable causes of peak skew: pore diffusion and mass transfer phenomena may not be the only explanation.

Example 3. Recycling a Waste Acid by Ion Exclusion Chromatography

Glucose and xylose are generated from sulfuric acid hydrolysis of biomass. The acid must be separated from the sugars before the sugars are sent to fermentation. If possible, a significant fraction of the acid should be recycled. Ion exclusion is considered, since it has the potential of returning the sulfuric acid to the hydrolysis process, while avoiding the generation of waste salts that would otherwise occur if the acid were simply neutralized. The process objective is to achieve as close to baseline separation of the sugars from the acid as possible, and to recycle 1000 kg of acid/day at a cost that is less than 2¢/kg sugar.

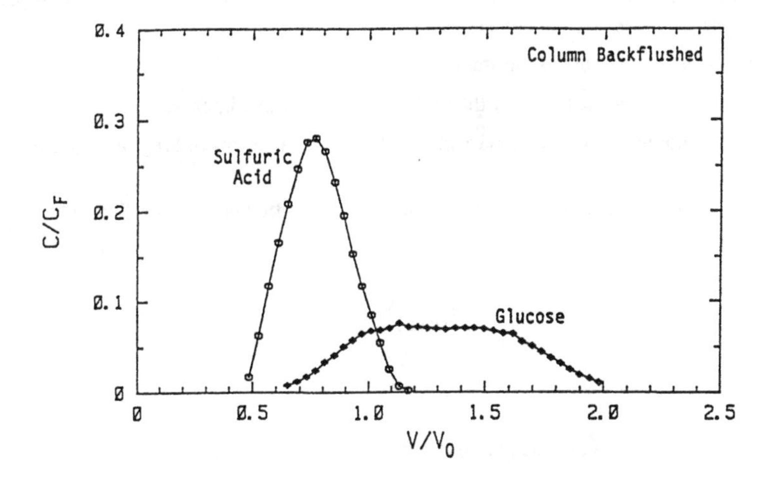

Figure 6.13 Example of channeling caused by backflushing of an ion exchange resin with a very broad particle size distribution (with permission from Neuman, 1986 thesis).

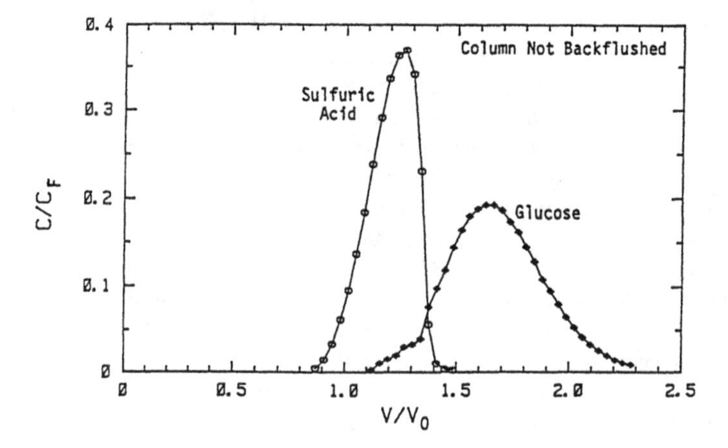

Figure 6.14 Peaks are broader and elute earlier than a column that was not backflushed (with permission from Neuman, 1986 thesis).

The first design calculation will be carried out based on $0.10V_0$ as the sample size, with the column to be run at 55°C. Assume that xylose and glucose co-elute, so that the chromatogram given in Figure 6.11e is applicable. The initial concentrations of glucose and xylose in the feed solution are 80 g/L and 60 g/L, respectively, while the sulfuric acid concentration is 77 g/L. The average particle size of the resin will be taken as 740 microns with a resin life of two years. The bench-scale run that gave the chromatogram in Figure 6.11e was carried out using a 2.54 cm i.d. × 61 cm long column at 55°C and an eluent flowrate of 3.1 mL/min of water. Sample volume is $0.10V_0$, where V_0 denotes the void volume between particles in the packed bed. The bed volume is 309 mL.

Calculate:

1. The length and diameter of the column(s) required to achieve 99% or greater separation of the monosaccharides (glucose and xylose) from H_2SO_4
2. The volume and cost of the stationary phase
3. The volume and total cost of the mobile phase consumed per day
4. The combined cost of stationary and mobile phase in terms of \$/kg of sugars recovered

Solution

1. Assume that the interstitial velocity of the mobile phase stays the same so that the plate height is the same:

$$\varepsilon_b v = \frac{3.1 \text{ cm}^3/\text{min}}{\pi\left(\frac{2.54}{2}\right)^2} = 0.611$$

Assume that $\varepsilon_b = 0.4$. Then the interstitial velocity is

$$v = \frac{0.611}{0.4} = 1.53 \frac{\text{cm}}{\text{min}}$$

The particle size is large, and the resin has a significant internal porosity, so assume that pore diffusion is the controlling resistance.

Use Eq. (6.9) to calculate the plate count, and therefore the column length. Since the temperature, flowrate, stationary phase, mobile phase, and sugars all stay the same, Eq. (6.9) reduces to

$$R_{s,b} = C N_{1,2,b}^{1/2}$$

where

$$C = \frac{1}{4}(\alpha - 1)\left(\frac{k'_{1,2}}{1 + k_{1,2}}\right)$$

Hence the ratio of resolutions at the process (subscript x) and bench (subscript b) scales should be

$$\frac{R_{s,x}}{R_{s,b}} = \left(\frac{N_{1,2,x}}{N_{1,2,b}}\right)^{1/2} \tag{6.58}$$

By convention, we set $R_{s,x} = 1.1$ in order to ensure baseline separation. Therefore Eq. (6.58) becomes

$$N_{1,2,x} = \left(\frac{1.1}{R_{s,b}}\right)^2 N_{1,2,b} \tag{6.59}$$

The resolution, $R_{s,b}$, and $N_{1,2,b}$ for the bench-scale column is estimated from Figure 6.12e, using the following procedure.

Assume that the sample is large so that Eq. (5.74) from Chapter 5 would apply:

$$V_w = V_{spl} + V_{tot,\ plate}\sqrt{2\pi N}$$

Solve for N:

$$N = \frac{1}{2\pi}\left[\frac{V_w - V_{spl}}{V_{tot,\ plate}}\right]^2 \tag{6.60a}$$

The sample size is given as $V_{spl} = 0.1 V_o$. The width of the acid peak, read from Figure 6.11e is

$$\frac{V_w}{V_o} = 1.4 - 0.8 = 0.6 \quad \text{or} \quad V_w = 0.6 V_o \tag{6.60b}$$

The effective plate volume is calculated using Eq. (5.71) in Chapter 5, where

$$V_{\text{tot, plate}} = HA_o(\varepsilon_b + K_D\varepsilon_p)$$

We assume that $\varepsilon_b = 0.4$, and from the problem statement $\varepsilon_p = (0.73)(1 - 0.4) = 0.44$. The particle porosity of 0.73 is an interpolated value. The distribution coefficient is zero, by definition, since the sulfuric acid is excluded from the resin. On this basis, Eq. (5.71) in Chapter 5 becomes

$$V_{\text{tot, plate}} = \frac{61}{N}(5.07)\,[0.4 + 0\,(0.44)] = \frac{124}{N} \tag{6.61}$$

since $H = L/N$ and the column length is 61 cm with a diameter of 2.54 cm ($A_o = 5.07\ \text{cm}^2$) (given in the text).

The combination of Eqs. (6.60a) and (6.61), together with expressions for V_{spl} and V_w, gives

$$N = \frac{1}{2\pi}\left[\frac{0.6V_o - 0.1V_o}{124/N}\right]^2 = \frac{N^2}{2\pi}\,[0.0040V_o]^2 \tag{6.62}$$

Since $V_o = (309)(0.4) = 123.6$, Eq. (6.62) becomes

$$\frac{1}{N} = \frac{1}{2\pi}\,[0.494]^2 = 0.0389$$

or $N = 26$ plates for surfuric acid.

Since we assumed the injected sample constitutes a large sample volume, it should satisfy the criterion

$$V_{\text{spl}} > 3V_{\text{tot, plate}}\sqrt{N}$$

where

$$V_{\text{tot, plate}} = \frac{124}{N} = \frac{124}{26} = 4.77$$

by Eq. (5.73) in Chapter 5. The sample volume is 12.4 mL, but

$$12.4 \not> 3\left(\frac{124}{26}\right)\sqrt{23} = 68.6\ \text{mL}$$

Since the sample injected does not satisfy the criterion that it is a large sample volume, the width of the peak should be given by Eq. (5.72) in Chapter 5, as defined for a small sample volume:

$$V_{w,o} = 4V_{\text{tot, plate}}\sqrt{N} = \text{constant}$$

So in this case

$$N = \frac{1}{16}\left(\frac{V_{w,o}}{V_{tot,\,plate}}\right)^2 = \frac{1}{16}\left(\frac{V_{w,o}}{124/N}\right)^2 = \frac{N^2}{16}\left(\frac{V_{w,o}}{124}\right)^2 \tag{6.63}$$

Substitution of Eqs. (6.61) and V_w into Eq. (6.63) results in

$$N = \frac{N^2}{16}\left(\frac{0.6V_o}{124}\right)^2$$

where $V_o = (309)\,(0.4) = 123.6$ mL. Then

$$\frac{1}{N} = \frac{1}{16}\left(\frac{0.6\,(123.6)}{124}\right)^2 = 0.0223 \tag{6.64}$$

Equation (6.64) gives the plate count of

$$\boxed{N = 44.8 \text{ plates}} \text{ for sulfuric acid}$$

A similar calculation for the glucose peak gives

$$\frac{V_w}{V_o} = 2.1 - 1.2 = 0.9$$

or

$$V_w = 0.9V_o \tag{6.65}$$

$$V_{tot,\,plate} = \frac{61}{N}\,(5.07)\,[0.4 + K_D\,0.44] \tag{6.66}$$

The distribution coefficient is obtained from Figure 6.9, and the plot for sugar in the presence of sulfuric acid is used. The distribution coefficient, K_D, is 0.49 on a weight per volume basis. Eq. (6.66) becomes

$$V_{tot,\,plate} = \frac{190}{N} \tag{6.67}$$

When results (6.65) and (6.67) are substituted into Eq. (6.63), the expression for the plate count gives

$$N = \frac{N^2}{16}\left(\frac{0.9\,V_o}{190}\right)^2 \tag{6.68}$$

which results in

$$\frac{1}{N}=\frac{1}{16}\left(\frac{111}{190}\right)^2=0.0214 \quad \text{or} \quad \boxed{N=46.7} \quad \text{for glucose}$$

A check on the applicability of the small sample equation, Eq. (5.70) in Chapter 5, shows that the sample volume satisfies the criterion of a small sample as would be expected from the result for sulfuric acid, calculated previously:

$$V_{\text{spl}} < 0.5V_{\text{tot, plate}}\sqrt{N}$$

$$12.4 < 0.5\left(\frac{190}{46.7}\right)\sqrt{46.7} = 13.9$$

We can now calculate the plate count that will be needed to achieve nearly complete separation of the two peaks, using Eq. (6.59) since

$$N_{1,2,\text{b}}=\frac{44.8+46.7}{2}=45.8$$

Then Eq. (6.59) gives

$$N_{1,2,\text{x}}=\left(\frac{1.1}{0.667}\right)^2(45.8)=124.6$$

By Eq. (5.143) in Chapter 5, and as seen from Figure 6.11e, the resolution at the bench scale is

$$R_{\text{s,b}}=\frac{(1.65-1.15)V_{\text{o}}}{\frac{(0.6+0.9)V_{\text{o}}}{2}}=0.667$$

Since the interstitial velocity will be held constant, the plate height is the same on both bench and process scales. The bench-scale column is given as being 61 cm long. Therefore

$$\frac{N_{\text{x}}}{N_{\text{b}}}=\frac{L_{\text{x}}/H_{\text{x}}}{L_{\text{b}}/H_{\text{b}}} \quad \text{and} \quad L_{\text{x}}=\frac{N_{\text{x}}}{N_{\text{b}}}L_{\text{b}}=\frac{124.6}{45.8}\cdot 61=166 \tag{5.75}$$

The length of the process scale column is estimated at

$$\boxed{L_{\text{X}} = 166 \text{ cm}}$$

2. The volume of stationary phase will depend on the throughput, which is 1000 kg acid per day. At the same loading as the bench scale this translates to the following values:

Bench scale: $0.1V_{\text{o}} = 0.1\ (309)\ (0.4) = 12.4$ mL for a throughput of (0.077) (12.4) = 0.95 g sulfuric acid per run.

Flowrate = 3.1 mL/min

Cycle time =

$$\frac{2.1 \text{ void volumes}}{\left(\dfrac{3.1 \text{ mL}}{\text{min}}\right)\left(\dfrac{1 \text{ void vol.}}{123.6 \text{ mL}}\right)} = 83.7 \text{ min}$$

Hence the bench-scale column could process

$$\frac{24 \cdot 60}{83.7} \cdot 0.95 = \boxed{16.3 \frac{\text{g } H_2SO_4}{\text{day}}} \text{ for bench scale}$$

This translates to

$$V_x = V_b \frac{(1000 \text{ kg/day})}{(0.0163 \text{ kg/day})} \cdot \frac{L_x}{L_b}$$

$$= (309)\,(61{,}350)\left(\frac{166}{61}\right)\left(\frac{1 \text{ L}}{1000 \text{ mL}}\right)$$

$$= \boxed{51{,}600 \text{ L}}$$

If all of the resin were packed in one (hypothetical) column, its diameter would be

$$\pi \left(\frac{d}{2}\right)^2 L = 51{,}600 \times 10^3 \text{ mL}$$

$$d = \left(\frac{4 \cdot 51{,}600\ 10^3 \text{ cm}^3}{(\pi)\,(167)}\right)^{1/2} = \boxed{627 \text{ cm}}$$

This would be too large for a single column, so the column would be divided into several smaller diameter columns.

The cost of the stationary phase on the bench and process scales would be estimated from Figure 6.4, since Eq. (6.15) is out of range. By visual inspection of Figure 6.4, the resin cost is estimated to be on the order of

$$C_{\$,\text{resin}} = \frac{0.10}{\text{g}} \quad \text{or} \quad \frac{\$100}{\text{kg}} \cdot (\text{dry})$$

If the material is at 50% moisture (x_{moisture}) the costs of the resins at the bench and process scales would be

$$V_{\text{bed}}\, \rho_{\text{resin}}\, x_{\text{moisture}}\, C_{\$,\text{resin}}$$

Bench: $0.309 \times 1.12 \times 0.50 \times 100 \cong \$17$

Process: $51{,}600 \times 1.12 \times 0.50 \times 100 \cong \$2.9 \times 10^6$

These costs could be much lower, since ion exchange resins (e.g., those used in home water softeners) may suffice for this application, resulting in lower costs.

A more exact estimate would adjust the cost figures for inflation. The data in Figure 6.4 are from about 1989. The Marshall-Swift index could be used for this purpose.

3. The volume of mobile phase in the interstitial void volume would be

$$V_o = (V_{col})\,(\varepsilon_b) = (51{,}600)\,(0.4) = 20{,}640 \text{ L}$$

At the same interstitial velocity, the number of void volumes per run would remain at 2.1. The run time for the bench scale system is $t_b = (L_b/v_b)\,(2.1) = (61/1.53)\,(2.1) = 83.8$ min so the number of runs per day would be $(24 \cdot 60)/83.8 \cong 17.2$.

Since the process scale column is longer,

$$t_x = \left(\frac{L_x}{v_x}\right)(2.1) = \left(\frac{166}{1.53}\right)(2.1) = 227\,\frac{\text{min}}{\text{run}}$$

where the interstitial velocity is kept the same (at 1.53 cm/min) at both bench and process scales.

Hence the number of runs per day on the process scale is

$$= \frac{(24)\,(60)\text{ min}}{227\text{ min/run}} = 6.34 \text{ runs}$$

The total mobile phase requirement per day would be

$$\left(6.34\,\frac{\text{runs}}{\text{day}}\right)\left(2.1\,\frac{\text{void vol.}}{\text{run}}\right)\left(20{,}640\,\frac{\text{L}}{\text{void vol.}}\right) \cong 275{,}900\,\frac{\text{L}}{\text{day}}$$

If the cost of water is \$1.00 per 1000 L, mobile phase costs alone would be about \$275 per day or \$91,700 per year if the system is operated at 8000 hrs a year.

There is significant opportunity for optimizing and improving the cost effectiveness of this process.

4. If the stationary phase is stable for 5 years, its annual cost would be on the order of \$500,000. This cost would be significantly lower (by a factor of 10 or more) if a commercial ion exchanger were to be used in place of chromatographic resins.

The total sugar processed would be 140 g/L for each 77 g/L acid. In this case the total sugar obtained over a year's time would be

$$\text{Throughput} = \frac{140 \text{ g sugars}}{77 \text{ g } H_2SO_4} \cdot \frac{1000 \text{ kg acid}}{\text{day}} \cdot \frac{8000 \text{ hr/yr}}{24 \text{ hr/day}}$$

$$= 606 \cdot 10^3\,\frac{\text{kg sugars}}{\text{yr}}$$

Cost:

$$\text{High:} \quad \frac{(500{,}000 + 88{,}700)\ \$/\text{yr}}{606 \cdot 10^3\,\frac{\text{kg sugars}}{\text{yr}}} = \frac{\$0.97}{\text{kg sugars}}$$

$$\text{Low:} \quad \frac{(50{,}000 + 88{,}700)}{606 \cdot 10^3} = \frac{\$0.23}{\text{kg sugar}}$$

5. Both the high and low costs for the current design are too high for this application. Optimization could include

i. Reducing particle size, if chromatographic grade resin is to be used, since there would appear to be only minimal price differential between 100 and 1000 micron particle size (see Figure 6.5). If plate height is reduced by use of a smaller particle size, length of column and volume of resin may also decrease, as will void volume.

ii. Relax specifications on recovery. Combined with step i, this would reduce solvent consumption.

iii. Examine increasing feed concentration and volume, or other methods to increase the amount of resin under use in the column. Figure 6.12e suggest that usage rate could be increased by 45%, since there is a volume gap of about 0.9 void volumes between injection and elution of the first peak.

iv. Use simulated moving bed chromatography (see Sections 6.35 and 6.39 of this chapter).

6.8 ION EXCLUSION HAS POSSIBLE APPLICATIONS FOR A GREENER CHEMICAL INDUSTRY

The distribution coefficient for HCl with respect to a cation exchange resin (Dowex 50-X8) ranges between 0.1 to 2.0 for acid concentrations between 1.0 and 5.0 M (Figure 6.15). A non-

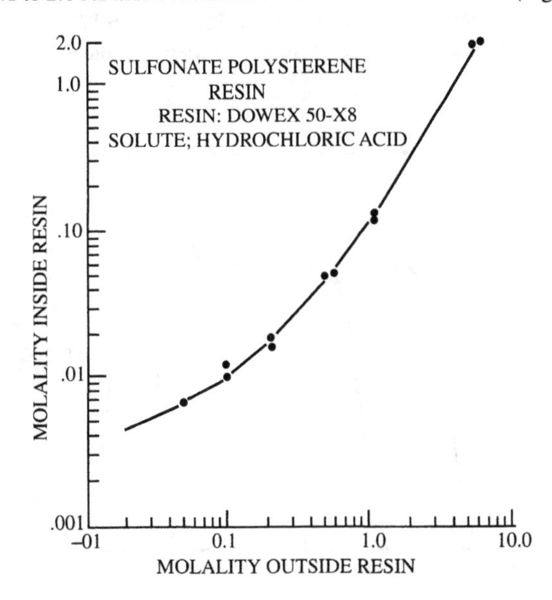

Figure 6.15 Equilibrium distribution curve for ionic solute (HCl) in water with respect to a gel type, sulfonated, polystyrene cation exchange resin. Note that the axis are log scales. (Reprinted from Bauman et al., 1956, fig. 7.4, p. 186.) The material is reproduced with permission of Academic Press.

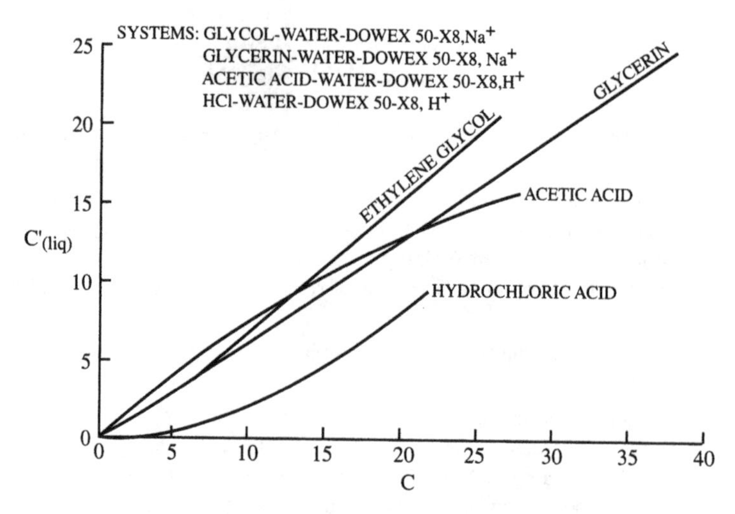

Figure 6.16 Equilibrium isotherms for ethylene glycol, glycerin, and acetic acid. Note that C_l refers to the concentration of solute in the liquid that is trapped within the resin. (Reprinted from Bauman et al., 1956, fig. 7.7, p. 188.) The material is reproduced with permission of Academic Press.

ionic species can be separated from an ionic species since the distribution coefficient for the organic or non-ionic species increases rapidly as the ionic species concentration (HCl, in this case) increases, while the isotherm for the HCl is not strongly affected by the concentration of the organic species. The differences in the equilibrium curves between ethylene glycol, glycerin, or acetic acid and HCl indicates separation is feasible between ionic and non-ionic species (Figure 6.16). The separation of ethylene glycol and HCl (Figure 6.17) shows that the non-ionic component can actually be concentrated, as well as purified, as indicated by the chromatogram

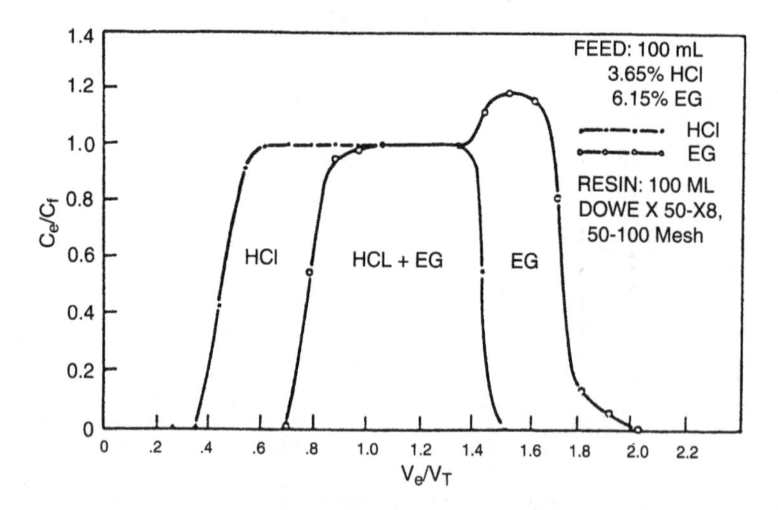

Figure 6.17 Ion-exclusion chromatogram showing concentration of ethylene glycol (from Bauman et al., 1956, fig. 7.9, p. 192.) The material is reproduced with permission of Academic Press.

in Figure 6.17. Probable explanations, offered by Bauman et al. (1956), are related to water activity and the displacement effect of the acid with respect to the glycol. This is reflected by the upward slope of the ethylene glycol equilibrium curve, shown in Figure 6.16. The shrink/swell characteristic of the resin was also proposed to be a contributing factor, with the resin swelling as the ion (HCl) concentration decreases and thereby incorporating additional water into the resin. Since the distribution coefficient for the glycol also decreases as the ion concentration decreases, the glycol is concentrated back into the mobile phase and gives the hump for peak EG (= ethylene glycol) in Figure 6.17.

Ion-exclusion chromatography was known and practiced as early as 1956. Its use may now expand as an environmentally friendly means of recovering non-ionic species from ion-containing fluids. This type of separation is likely to become important for the cost-effective processing of biomass[3] into value-added products.

6.9 LINEAR CHROMATOGRAPHY MAY DEPEND ON PARTICLE SIZE, TEMPERATURE, OR SOLUTE CONCENTRATION EFFECTS: CASE STUDY FOR LINEAR CHROMATOGRAPHY SCALE-UP

The elution profile for phenylalanine and aspartame shown in Figure 6.18 gives a good starting point for column design. However, more experimental measurements are still needed for the scale-up process because a number of non-ideal effects can alter the elution behavior of the same two components as conditions are changed (compare the data represented by circles and squares in Figure 6.18 to the lines). The separation of aspartic acid, phenylalanine, and the phenyl methyl ester of aspartic acid and phenylalanine (aspartame) has been studied because both the amino acids and the dipeptide were readily available and well characterized. The sequence of steps described here represents the approach generally used for chromatography development, but it includes insights into the types of unexpected results that could be obtained along the way. The case study is based on a previously published paper by Ladisch et al. (1991) and some of the paragraphs below are taken directly from this paper.

The stationary phase used in this example is a methacrylic, macroporous, polymeric sorbent with a dry surface area of 450 m^2/g, a pore size range of 20 to 600 Å, and average pore size of 200 to 300 Å, and a porosity of 58 to 63% (volume pores/volume spherical stationary phase particles). Water, 50% aqueous methanol, and 10, 25, and 50% (v/v) aqueous ethanol were tested as possible mobile phases. A 0.71 × 49.8 cm long column packed with 38.7 ± 6.9 μm particles was used to test for separation of 100 μl sample volumes in which Asp (1 g/L), Phe (6 g/L), and aspartame (4 g/L) were dissolved. Aspartame did not elute with water as the mobile phase. Resolution of all three components was achieved with both aqueous methanol and ethanol as the eluent, but 10% ethanol was chosen because it gave excellent resolution and higher solubility than the other concentrations of ethanol.

[3]Biomass refers to sources of cellulose-containing renewable materials including trees, sawdust, rice straw, paper, yard wastes, grasses, and sugar cane bagasse. At least 200 million tons of these materials could be available for conversion to glucose and other sugars, which in turn could be fermented to a number of products including fuel alcohol, precursors for the manufacture of biodegradable plastics, and organic acids. One of the processing approaches for converting biomass to fermentable sugars is based on use of H_2SO_4 as a hydrolysis catalyst to convert the cellulosic components to sugars (Scott et al., 1976). Although sulfuric acid, itself, is relatively inexpensive, its neutralization and disposal after its use is costly, thereby making the net cost of acid a significant factor in the economics of biomass conversion. Separation and recycle of the acid could reduce the cost of producing ethanol from cellulose. Cation exchange resin in the particle size range of 300 to 1200 microns could give 94% recovery of glucose while separating the glucose from the acid (Neuman et al., 1986; Neuman et al., 1987).

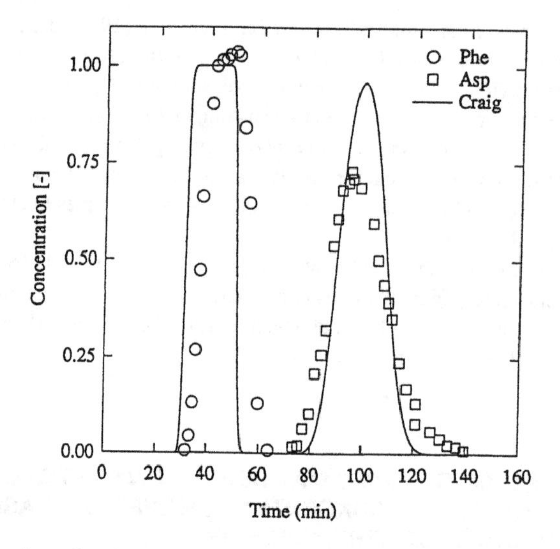

Figure 6.18 Comparison of a priori Craig segment simulation (using Eq. (6.116) and parameters from the literature) to elution profile data for aspartame and phenylalanine column dimensions 1.09 × 70 cm, operating temperature: 20°C; flowrate = 2 mL/min; $\varepsilon_b = 0.36$; $\varepsilon_T = 0.74$; $V_{spl} = 40$ mL of 5 mg/mL each of phenylalanine and aspartame; resin particle size of 60.3 ± 4.9 micrometers (from Velayudhan and Ladisch, 1993, fig. 5, p. 141.)

A retention map (i.e., a semilog plot of k' as a function of ethanol concentration in the mobile phase in Figure 6.19 shows 10% ethanol to be a good choice since the capacity factors for the three components differed by about a factor of 10. The capacity factors translate into the chromatogram shown in Figure 6.20, where an unknown peak between L-Asp and L-Phe was also detected. Retention times were then checked for columns packed with stationary phases having average particle sizes of 60.3 ± 4.9, 117 ± 15, and 164 ± 24 μm. The retention times for the different particle sizes are about the same (Figure 6.21). This would be consistent with Eq. (5.137) in Chapter 5 if ϕ [$= (1 - \varepsilon_b)/\varepsilon_b$] is constant, and linear equilibrium (K_D = constant) applies at the concentrations used. The width of the peaks increase with increasing particle size, however, and this diminishes the resolution as well as the average concentrations of the components that make up the peaks (Figure 6.21). This effect is consistent with Eq. (5.99) in Chapter 5 and the results of Example 8 in Chapter 5. The larger particle size might also contribute to additional extraparticulate dispersion and decrease overall mass transfer causing a decrease in $\overline{Pe}$ and N, as defined by Eq. (6.2). However, an increase in pore diffusion resistance with increasing particle size is likely to be a major contributing factor, since all three amino acids are small enough to penetrate the pores of the stationary phase.

Equilibrium measurements were carried out and a linear relation, up to the solubility limit of aspartame was found (Figure 6.22), with the slope ($= K_D$) decreasing as the temperature increased from 30° to 70°C. The equilibrium expression is given by

$$q = n = (n_o K)C_m = n_o \left[\kappa \exp\left(\frac{\Delta w}{k} \cdot \frac{1}{T} \right) \right] C_m \tag{6.69}$$

where

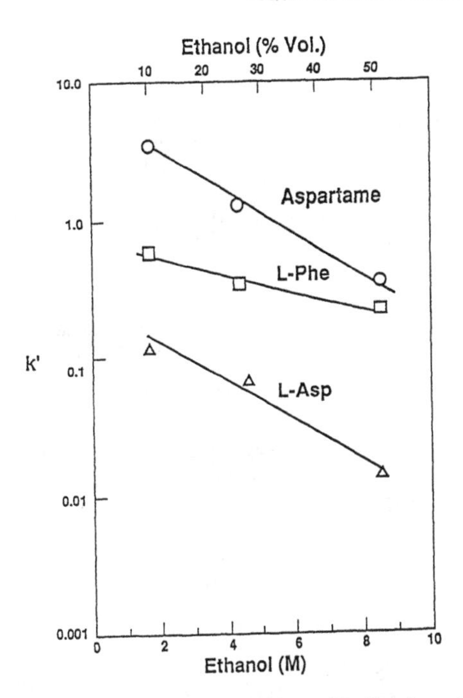

Figure 6.19 Plot of capacity factors for aspartame, L-Phe, and L-Asp determined at room temperature with respect to increasing ethanol concentration. ○ = aspartame; □ = L-Phe; △ = L-Asp; particle size = 38.7 ± 6.9 μm (from Ladisch et al., 1991, fig. 2, p. 89, J. Chromatogr., 540, 88–98, with permission from Elsevier Science).

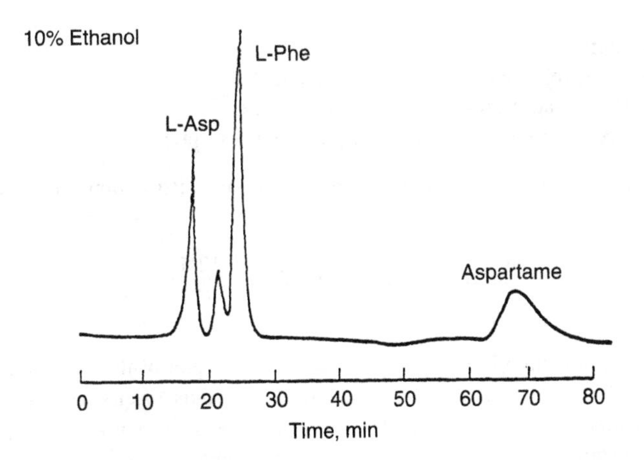

Figure 6.20 Chromatogram of the three components with 10% ethanol as the mobile phase; injection volume of 100 μL; flowrate at 0.95 mL/min; detection using differential refractometer; column was 0.71 × 49.8 cm, packed with 38.7 ± 6.9 μm material; run at ambient temperature (from Ladisch et al., 1991, fig. 1(b), p. 88).

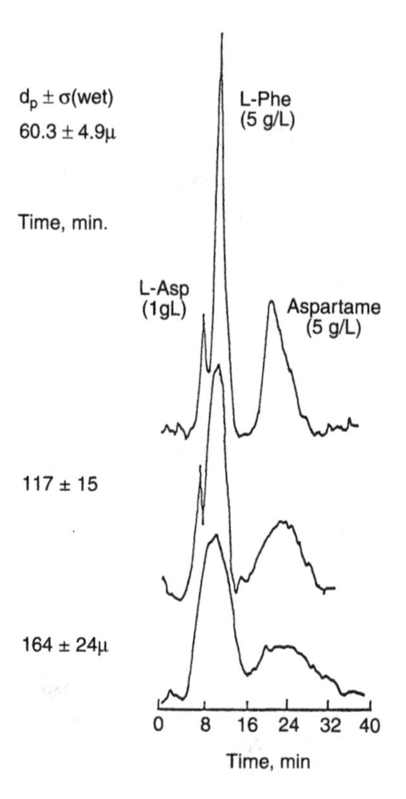

Figure 6.21 Separation as a function of particle size (from Ladisch et al., 1991, fig. 6, p. 96.)

T = temperature, K
k = a constant (proportional to gas law constant, R)
κ = ratio of solute adsorption rate to desorption rate
Δw = difference between energies of adsorption and desorption

This resulted in a temperature dependent correlation for the distribution coefficient,

$$K_{\mathrm{D,aspt}} = n_o K = 0.112 \exp\left[\frac{1240}{(T)}\right] \tag{6.70}$$

where T is the temperature in K. Figure 6.23 shows the replot of the temperature-dependent term in Eq. (6.69) to be linear, as is the analogous plot for the capacity factor for aspartame as a function of temperature (Figure 6.24). The capacity factor is given by $k' = \phi\, K_{\mathrm{aspt}}$. Equilibrium data for L-Phe showed that its adsorption was temperature independent, but still linear, with $K_{\mathrm{D,Phe}} = n_o\, K = 0.931$, where n_o represents the maximum surface coverage of the solute on the stationary phase.

The injection of a 40 mL sample of 5 mg/mL Phe and 5 mg/mL aspartame into a 1.09×70 cm column (volume = 65 mL) gave peaks that were close to Gaussian and had the charac-

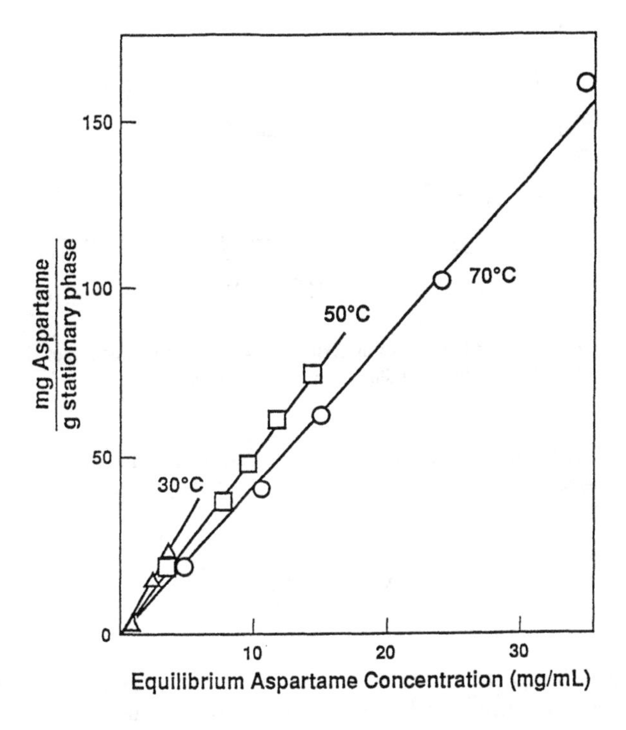

Figure 6.22 Equilibrium data for aspartame in 10% ethanol with respect to the polymeric stationary phase at 30, 50 and 70°C (from Ladisch et al., 1991, fig. 3, p. 91.)

teristics that would be expected from Eqs. (5.81) and (5.86) in Chapter 5. This includes a symmetrical shape, and a region of constant concentration at the peak maxima for Phe (Figure 6.25). However, when the sample concentration was increased 10-fold, to 50 mg/mL, the peaks lost their symmetry (Figure 6.26) and began to exhibit a shape that is associated with concentration overload, although the equilibrium data indicate linear equilibria (Figure 6.22). The separation was carried out at 82°C so that the amino acids would stay in solution at these high concentrations.

The plate count data were fitted with an expression that included the possible effect of temperature. The reduced velocity, ν (= vd_p/D_m), and reduced plate height, h (= H/d_p), were com-

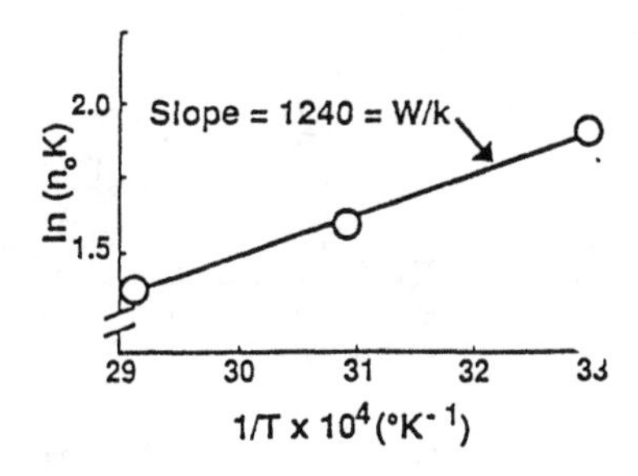

Figure 6.23 Distribution coefficient (from Ladisch et al., 1991, fig. 3, p. 91.)

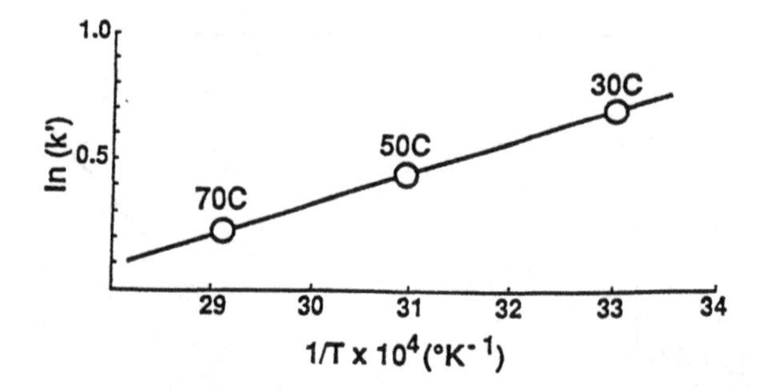

Figure 6.24 Capacity factor as a function of inversing temperature (from Ladisch et al., 1991, fig. 3, p. 91, J. Chromatogr., 540, 88–98, with permission from Elsevier Science.)

bined with the reduced form of the Snyder equation, as described earlier in the chapter in section 6.3, to obtain the expression

$$N = \frac{D_{\mathrm{m}}^{n}}{m} \frac{L}{v^{n} d_{\mathrm{p}}^{1+n}} \tag{6.38}$$

where H is the plate height, d_{p} the particle diameter, L the column length, N the plate count, v the interstitial (linear) velocity (cm/min), and D_{m} the diffusion constant (cm^2/min). The framework of Knox and Pyper, as expressed in Eq. (6.38), recommends that n = 1, and $m \cong 0.1$ at

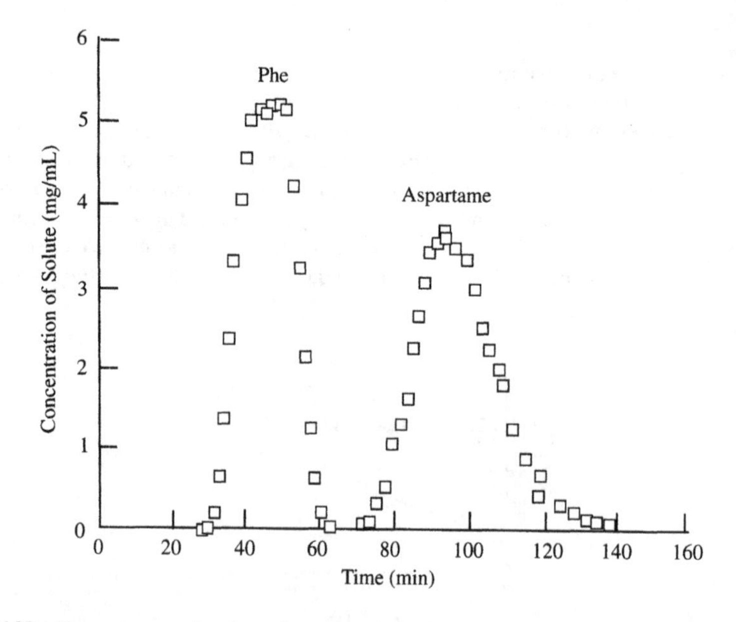

Figure 6.25 Chromatograms showing volume. Sample load was 40 mL; mobile phase of 10% ethanol was pumped at 2 mL/min. The run was done at ambient temperature (from Ladisch et al., 1991, fig. 5, p. 95, p. 96, J. Chromatogr., 540, 88–98, with permission from Elsevier Science.)

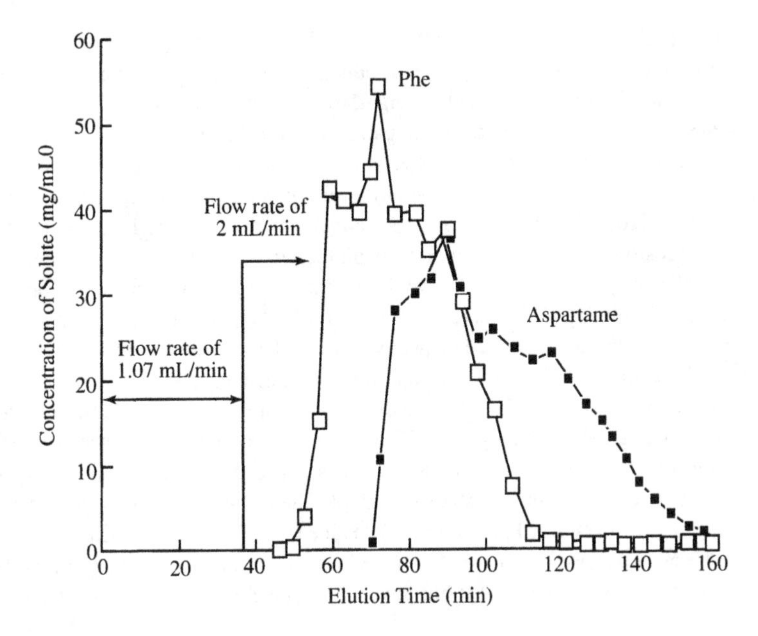

Figure 6.26 Chromatograms showing concentration overload conditions. Sample load was 40 mL; mobile phase of 10% ethanol was pumped at 2 mL/min. The run was carried out at 52°C (from Ladisch et al., 1991, fig. 7, p. 96, J. Chromatogr., 540, 88–98, with permission from Elsevier Science.)

large values of the reduced velocity, ν. These values of n and m are later shown to be applicable to the data presented here. The value of n may vary from 0.2 to 1.0, with pellicular packings having lower values of n and porous packings having values of n of 0.6 (Grushka et al., 1975) to 1.0 (Pieri et al., 1983). The first term in Eq. (6.71) gives the contribution to dispersion from axial diffusion. The Snyder Eq. (6.38) is a special case of the Knox equation (Eq. 6.71):

$$h = \frac{2\gamma}{\nu} + m\nu^n \approx m\nu^n \tag{6.71}$$

where $\nu > 30$ and the second term dominates. The second term of Eq. (6.71) gives the combined contributions of dispersion arising from flow of eluent over the packing, and the complex flow that occurs in the column (Grushka et al., 1975). Grushka et al. (1975), note that m and n cannot be assumed constant over a wide range of ν, so Eq. (6.71) may not always be reliable, although it is the basis of Eq. (6.38) and therefore the scale-up correlations in Table 6.1.

The temperature dependence of the reduced velocity is due to the temperature dependence of the diffusion coefficient, D_{Am}. The temperature dependence of the diffusion constant, D_{Am}, of dilute solute in multicomponent mixtures, is given by the modified Wilke-Chang equation (Pieri et al., 1983; Reid et al., 1977; Wilke and Chang, 1955):

$$D_{\text{Am}} = 7.4 \times 10^{-8} \frac{\left(\sum_{j=1}^{n} x_j \varnothing_j M_j \right)^{0.5} T}{\eta_{\text{m}} V_{\text{A}}^{0.6}} \tag{6.72}$$

where T is the absolute temperature, V_A is the molar volume of the solute, and η_m is the viscosity of the solvent mixture (in this case, aqueous ethanol, and where x_j, $\varnothing_j$, and M_j are the mole fraction, association factor, and molecular weight, respectively, of solvent component j where $j \neq A$. For the 10% (by volume) ethanol, the parameters for the mobile phase components of water ($j = 1$) and ethanol ($j = 2$) are, respectively, $x_1 = 0.9584$, $\varnothing_1 = 2.6$, and $m_1 = 18$; and $x_2 = 0.0416$, $\varnothing_2 = 1.5$ and $m_2 = 46$. The molar volumes of aspartame and phenylalanine were estimated to be $V_A = 320$ and 200 cm^3/gmole, respectively, by the method of LeBas (Reid et al., 1977). The viscosity of 10% (by volume) ethanol decreases with increasing temperature (*Lange's Handbook of Chemistry*, 1973) while the diffusion constant for aspartame in 10% ethanol, calculated from Eq. (6.72) increases from 2.5×10^{-4} to 6.0×10^{-4} cm^2/min as the temperature increases from 30 to 70°C. For phenylalanine, the estimated diffusion constant increases from 3.4×10^{-4} to 8.0×10^{-4} cm^2/min over the same temperature range.

Plate counts for aspartame and phenylalanine from columns packed with 60, 117, and 165 micron particle sizes are shown in Figures 6.27 and 6.28. The columns were run at 30° and 70°C with 10% ethanol as the eluent at superficial velocities, $v\varepsilon_b$, of 5.4 to 55 cm/min. Values of the reduced velocity were in the range of 50 to 700. Unlike the strongly adsorbed aspartame, a discernible temperature effect is lacking for weakly retained phenylalanine (compare Figures 6.27 and 6.28). The data were fitted to Eq. (6.38) for $n = 0.5$ and $n = 1.0$. While both the mass transfer ($n = 0.5$) and pore diffusion ($n = 1.0$) forms gave reasonable fits, the case for n = 1 was better.

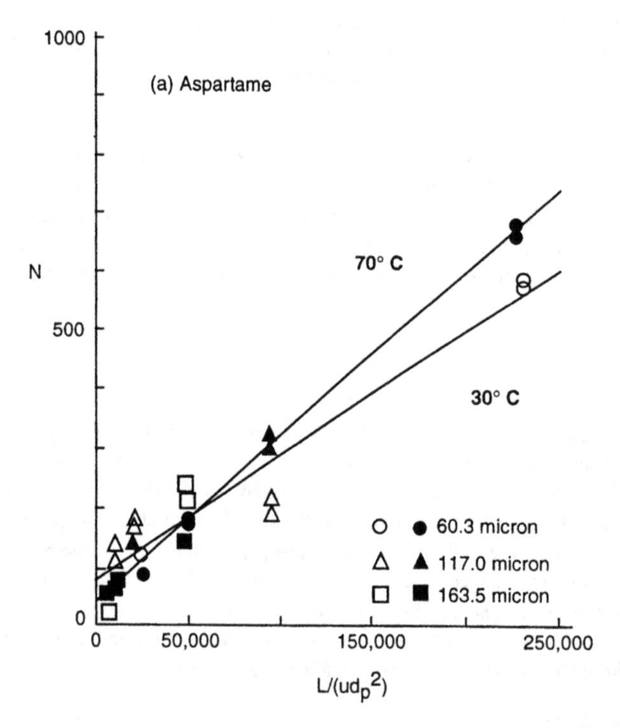

Figure 6.27 Relation of plate count to particle size, eluent superficial velocity, and column length by Eq. (6.122) for $n = 1$ for aspartame. Eluent is 10% ethanol. Values of L, u, and d_p have dimensions of cm, cm/min, and cm, respectively. Open symbols for temperature = 30. Dark symbols for temperature = 70 (from Ladisch et al., 1991, fig. 8, p. 98, J. Chromatogr., 540, 88–98, with permission from Elsevier Science).

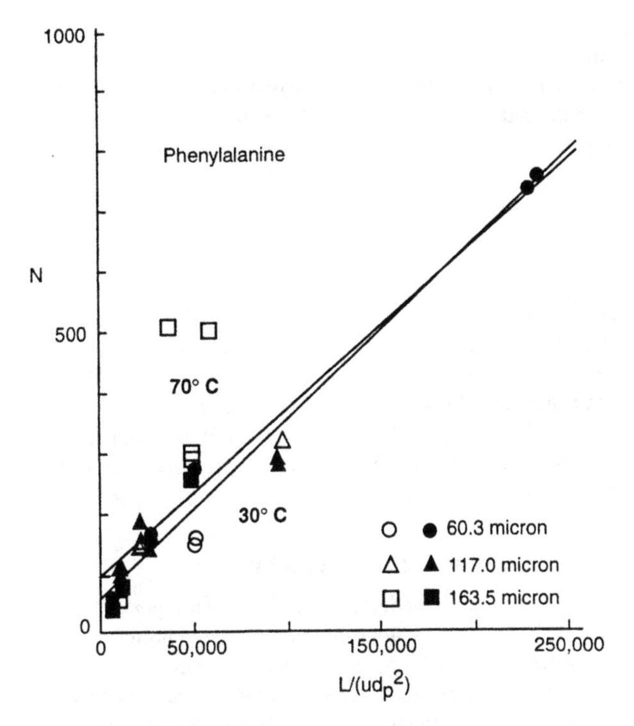

Figure 6.28 Relation of plate count to particle size, eluent superficial velocity, and column length by Eq. (6.122) for $n = 1$ for phenylalanine. Eluent is 10% ethanol. Values of L, u, and d_p have dimensions of cm, cm/min, and cm, respectively. Open symbols for temperature = 30. Dark symbols for temperature = 70 (from Ladisch et al., 1991, fig. 8, p. 98, J. Chromatogr., 540, 88–98, with permission from Elsevier Science).

On this basis, and in view of the large particle size, pore diffusion was chosen to be the more significant factor, although both mass transfer and pore diffusion are likley to contribute to the observed plate counts.

The values of m are calculated from the slopes ($= D_{Am}/m$) in Figures 6.27 and 6.28, with the values of D_{Am} estimated by Eq. (6.72). This gives $m = 0.11$ to 0.12 at 30°C and 0.28 to 0.29 at 70°C which is close to the expected $m \cong 0.1$ (Grushka et al., 1975; Pieri et al., 1983). The values of m calculated for these data are probably high, since values of D_{Am} represent free solution diffusion, rather than an effective hindered diffusion inside of the pores. The effective diffusion coefficient, D_{eff} is given by

$$D_{eff} = \frac{\Theta D_{Am}}{\tau} \tag{6.73}$$

where τ is a tortuosity factor for the pores and Θ is the fraction of free cross section for diffusion (Sherwood et al., 1975). Tortuosity varies inversely with porosity, and generally tortuosity values range from 2 to 6 (Ladisch, 1987). Since the stationary phase has a high porosity in the range of 58 to 63%, a value of $\Theta = 0.6$ and $\tau = 2$ was used. Equation (6.73) thus gives $D_{eff} = 0.3D_{Am}$. On this basis the free solution diffusivities and values of m would be multiplied by 0.3 to give

$m \cong 0.04$ and $m \cong 0.09$ at 30 and 70°C, respectively. These values are in the same range as those proposed by Knox and Pyper (1986).

The data show that $n = 1$ in Eq. (6.38) gives the best fit of N as a function of velocity, u, and particle size, d_p, for the particle sizes used here. Therefore, for a given temperature, changes in column length, due to changes in particle size, can be anticipated by the relation

$$L_1 = \left(\frac{N_1}{N_2}\right)\left(\frac{v_1}{v_2}\right)\left(\frac{d_{p,1}}{d_{p,2}}\right)^2 L_2 \tag{6.74}$$

where the subscripts 1 and 2 represent columns 1 and 2 which are packed with particle sizes 1 and 2, respectively, and pore diffusion is assumed to be the dominant cause of peak dispersion. Equation (6.74) is useful in providing a magnitude of order estimate of column length, since particle size, plate count, and eluent velocity usually change in scaling up a given separation (*Lange's Handbook of Chemistry*, 1973; Pieri et al., 1983; Wakao and Furazki, 1978). Column length is a key specification in estimating run time, solvent usage, and volume of stationary phase needed for a preparative separation.

Numerical Models for Linear Chromatography

The Craig and stirred-tank in series models are introduced at this point of the chapter, since they provide numerical methods for simulating multicomponent chromatography. Both types of models have the potential to be useful tools for simulating chromatographic separations if the distribution coefficient and void fractions are known for a given stationary phase, although the Craig model is not applicable to unretained solutes. While other methods can be used to simulate elution profiles, and are discussed in this book, the approach described in this section begins to address the need for carrying out a calculation that predicts elution behavior based on fundamental properties of the stationary phase. The other methods discussed, so far, simulate separation behavior after it has been achieved experimentally, and after an elution profile is available for fitting a model to it.

This discussion on the Craig and plate models also introduces the reader to some of the mathematics of finite difference methods as they apply to modeling of chromatographic separations. These can be used to develop more sophisticated models of chromatographic separations as the need arises. One disadvantage of both types of plate models is that they cannot satisfy the Danckwerts boundary conditions (Danckwerts, 1953). However, it is known that the choice of boundary conditions is only significant when the axial Peclet number is small and the column is relatively short (Ruthven, 1984). This combination does not occur very often in chromatographic practice.

6.10 THE CRAIG MODEL MAY BE USED TO PREDICT ELUTION PROFILES FOR STRONGLY RETAINED COMPONENTS ($k' >> 0$) (WITH CONTRIBUTIONS BY AJOY VELAYUDHAN)

The countercurrent contacting method, illustrated in Figure 6.29, is described by the Craig model. This is a hypothetical contacting scheme in which the solution and adsorbent are mixed, equilibrated, separated, and moved in opposite directions. Figure 6.29 illustrates how the liquid solution is enriched in component 2 while the adsorbent loading increases with component 1 being picked up to a greater extent than component 2. The representation of the separation sequence in Figure 6.29 may be modeled by a discontinuous contacting (tank-in-series) model as

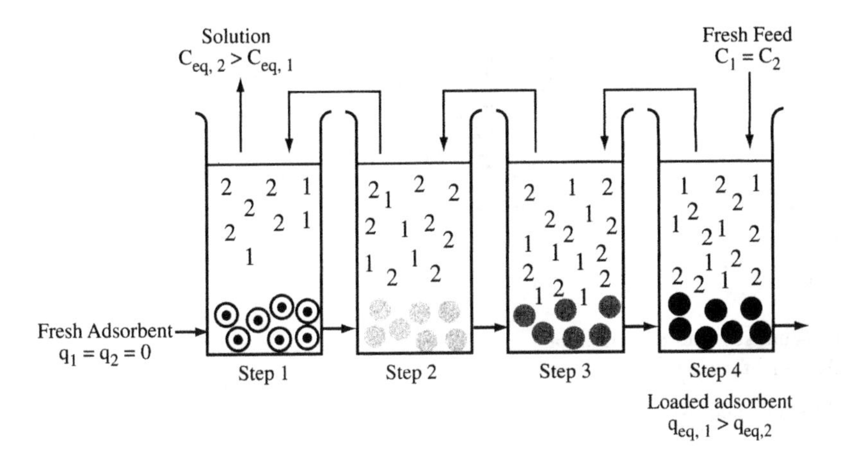

Figure 6.29 Representation of discontinuous, tank-in-series model.

well as the Craig model. An analysis of the Craig model and comparison to a stirred tank-in-series model by Velayudhan and Ladisch (1993) is presented here.

The Craig model is a discontinuous-flow plate model that represents the retention and peak-spreading behavior of retained components in linear chromatography. An unretained component cannot be simulated by the Craig model, since the model only predicts that the shape of the solute band is not altered by its passage through a chromatography column. The results for retained components are similar when the appropriate definition of plate count for the Craig model is used.

The Craig model is given by

$$C_m(i, j, k) + \phi q(i, j, k) = C_m(i, j-1, k-1) + \phi q_m(i, j, k-1) \tag{6.75}$$

where $c(i, j, k)$ is the mobile phase concentration of the ith species in the jth plate at the kth instant, $q(i, j, k)$ is the corresponding stationary phase concentration, and ϕ is the phase ratio. The development of the Craig model that follows, uses $q_{eq} = K_D C_{eq}$ and $k' = \phi K_D$ and is therefore restricted to linear equilibrium. The governing equation is

$$(1 + k_i')C_m(i, j, k) = c(i, j-1, k-1) + k'C_m(i, j, k-1) \tag{6.76}$$

Equation (6.76) is a linear partial difference equation, whose solution is outlined elsewhere (Jury, 1964; Uspensky, 1937). The analytical solutions to linear frontal chromatography and linear elution chromatography (when the pulse input fills several plates) are derived by the method of the two-dimensional z-transform (Velayudhan and Ladisch, 1993). The final forms of the solution are

$$\frac{C_m(i,j,k)}{C_o} = \begin{cases} (p_i')^j \sum_{m=o}^{k-j} \begin{bmatrix} j+m-1 \\ m \end{bmatrix} (q_i')^m, & k-j < M \\ (p_i')^j \sum_{m=k-j-M}^{k-j} \begin{bmatrix} j+m-1 \\ m \end{bmatrix} (q_i')^m, & k-j \geq M \end{cases} \tag{6.77}$$

where

C_o = initial solute concentration
M = input pulse width
$\begin{bmatrix} n \\ r \end{bmatrix}$ = binomial coefficient

and

$$p_i' = \frac{1}{1+k_i'} \tag{6.78}$$

$$q_i' = \frac{k_i'}{1+k_i'} \tag{6.79}$$

The experimentally obtained chromatogram for a column of N_i plates is obtained by substituting $j = N_i$ in Eq. (6.77).

The analogous solution for adsorption is

$$\frac{C_{mF}(i,j,k)}{C_o} = (p_i')^j \sum_{m=o}^{k-j} \begin{bmatrix} j+m-1 \\ m \end{bmatrix} (q_i')^m \tag{6.80}$$

The Craig distribution results in a binomial distribution for the peak profile in the column (Karger et al., 1973; Keulemans, 1957). Equations (6.77) and (6.80) represent the finite sums of the binomial distributions with respect to inputs of a pulse of finite width or input of a step, respectively. These results are expressed as a nonintegral J (plate count) or M (input pulse width) by making use of the identity

$$(p_i')^j \sum_{m=o}^{k-j} \binom{j+m-1}{m} (q_i')^m = I_{p_i'}(j, k-j) \tag{6.81}$$

where the right-hand side of Eq. (6.81) is an incomplete beta function (Abramowitz and Stegun, 1965).[4] The frontal (adsorption) solution can be represented as

$$\frac{C_F(i, N_i, k)}{C_o} = I_{p_i'}(N_i, k - N_i) \tag{6.82}$$

where p_i' is the probability of the ith species being in the mobile phase, N_i is the number of Craig segments or Craig plates, and k represents a time interval. The corresponding result for chromatography is

$$\frac{C(i, N_i, k)}{C_o} = I_{p_i'}(N_i, k - N_i) - I_{p_i'}(N_i, k - N_i - M) \tag{6.83}$$

where M represents the width of the input pulse.

Since the arguments of the incomplete beta function do not need to be integers, rational numbers may be used for N_i and M in Eqs. (6.82) and (6.83). Equation (6.81) is found in the theory of statistics (Pearson, 1934), and an analogous result can be seen in the early work of Stene (1945).

The calculation of moments for the Craig distribution (Fritz and Scott, 1983; Mood and Graybill, 1963) relate the plate count, N_{plate}, from an elution profile to the number of plates, N_i, in the Craig column:

$$N_i = N_{craig} = N_{plate}\left(\frac{k'}{1 + k'}\right) \tag{6.84}$$

Here the nomenclature of N_{plate} (instead of N) and N_i is used to differentiate the two different concepts. N_{plate} denotes the plate count as obtained from a chromatogram using Eq. (5.27) in Chapter 5, while $N_i = N_{craig}$ represents the Craig or discontinuous plate concept. Equation (6.84) shows how the two types of plate counts are related. If the capacity factor is zero (i.e., $k' = 0$ for an unretained component), the plate count for the Craig model would be zero and is therefore not usable for modeling unretained components. For this reason the Craig model will not accurately reflect the dispersion process for an unretained component. The values of the two definitions of plate count approach one another as k' becomes large.

[4]The incomplete beta function is given by

$$I_x(a, b) = \frac{\int_0^x t^{n-1}(1 - t)^{b-1}\,dt}{\int_0^1 t^{n-1}(1 - t)^{b-1}\,dt} \tag{c}$$

The plate count N_{plate}, is obtained from the ratio of the square of the first moment to the second central of a chromatographic peak, as described by Haynes (1975). The first moment (retention time) denoted by $\mu_{i,j}$ is given by

$$\mu_{i,j} = \frac{L}{v}(1 + \phi K_{D,i}) \tag{6.85}$$

where

v = interstitial (mobile phase) velocity cm/s
L = column length
$K_{D,i}$ = distribution coefficient for component i
ϕ = volumetric phase ratio

The interstitial phase velocity and phase ratio are defined *differently* here than was done previously in Chapter 5. These are given by

$$v = \frac{\nu}{\varepsilon_b + (1 - \varepsilon_b)\,\varepsilon_y + (1 - \varepsilon_b)\,(1 - \varepsilon_y)\,\varepsilon_x} \tag{6.86}$$

and

$$\phi = \frac{(1 - \varepsilon_b)\,(1 - \varepsilon_y)\,\varepsilon_x}{\varepsilon_b + (1 - \varepsilon_b)\,\varepsilon_y + (1 - \varepsilon_b)\,(1 - \varepsilon_y)\,\varepsilon_x} \tag{6.87}$$

where

v = superficial velocity as obtained by the ratio of flowrate to the column's cross-sectional area, cm/s; ν represents velocity of a single solute.
ϕ = phase ratio for a porous stationary phase
ε_b = interparticle bed porosity, cm^3 of voids between particles per m^3 bed
ε_x = micropore porosity
ε_y = macropore porosity

The plate count, N_{plate}, for the case in which rapid adsorption, rapid surface diffusion, and negligible bulk flow effects apply, is given by

$$\frac{1}{N_{plate}} = \frac{2}{Pe} + \frac{(1 - \varepsilon_b)\beta^2}{[\varepsilon_b + (1 - \varepsilon_b)\beta]^2}\left[\frac{2}{3}\,\frac{Pe_y\left(\frac{R_y}{L}\right)}{Sh} + \frac{Pe_y\left(\frac{R_y}{L}\right)}{15}\right] + \frac{(1 - \varepsilon_b)}{(1 - \varepsilon_y)}\,\frac{(\beta - \varepsilon_y)^2}{(\varepsilon_b + (1 - \varepsilon_b)\beta)^2}\left[\frac{Pe_x\frac{R_x}{L}}{15}\right] \tag{6.88}$$

where

$$\beta = \varepsilon_y + (1 - \varepsilon_y)\,\varepsilon_x\,(1 + K_i^d) \tag{6.89}$$

and where

$\overline{\text{Pe}}$ = bed Peclet number

$\text{Pe}_x = \dfrac{v\,(2R_x)}{D_x}$ = microparticle Peclet number defined in terms of the superficial velocity, v

Pe_y = particle or pellet Peclet number = $\dfrac{v\,(2R_y)}{D_y}$

L = column length, cm

$\text{Sh} = \dfrac{k_{\text{fMT}}\,(2R_y)}{D_y}$

k_{fMT} = external mass transfer coefficient, cm/s

R_x = microparticle radius, cm, in which microparticles, when agglomerated together, make up a larger particle or pellet with radius, R_y

R_y = particle or pellet radius, cm

D_x = effective micropore diffusivity, cm^2/s

D_y = effective macropore diffusivity, cm^2/s

Note that the distribution coefficient is denoted here using a different nomenclature, since

$$K^{\text{d}} = \frac{\rho s}{\varepsilon_x(1 - \varepsilon_y)}\,\frac{q_{\text{a}}}{C_x} \tag{6.90}$$

where

K^{d} = distribution coefficient based on surface concentration of solute on stationary phase

ρ = density of stationary phase, g/cm^3

s = surface area of stationary phase, cm^2/g

q_{a} = concentration of solute on the surface of the stationary phase, gmol/cm^2

C_x = concentration of solute in fluid that is present in the micropore, gmol/cm^3

The distribution coefficient, K_{D}, as defined elsewhere in this book is not directly interchangeable with the K^{d} that is used in Haynes and Sharma's (1973) definition of the plate count. The distribution coefficient K_{D} (= $q_{\text{eq}}/C_{\text{eq}}$) would need to be converted to K^{d} based on a more detailed knowledge of the particle's void fractions, density, and surface area.

The plate count N_{plate} defined by Eq. (6.88) is simplified when

$$(1 - \varepsilon_{\text{b}})\,\beta >> \varepsilon_{\text{b}} \tag{6.91}$$

and

$$\beta >> \varepsilon_{\text{y}} \tag{6.92}$$

The conditions in Eqs. (6.91) and (6.92) represent the case where the solute is strongly retained. In this case, K^{d} is large, and therefore $k' = \phi\,K^{\text{d}}$ is also large. The simplified form of Eq. (6.88) is

$$\frac{1}{N_{\text{plate}}} = \frac{2}{\overline{\text{Pe}}} + \frac{2}{3} \frac{\text{Pe}_y \left(\frac{R_y}{L} \right)}{(1 - \varepsilon_b)\,\text{Sh}} + \frac{1}{15} \frac{\text{Pe}_y \left(\frac{R_y}{L} \right)}{(1 - \varepsilon_b)} + \frac{1}{15} \frac{\text{Pe}_x \left(\frac{R_x}{L} \right)}{(1 - \varepsilon_b)\,(1 - \varepsilon_y)} \tag{6.93}$$

where the variables are the same as defined for Eq. (6.88). If there is no adsorption, and the component is unretained, $K^{\text{d}} = 0$, $\beta = \varepsilon_y$, and Eq. (6.88) becomes

$$\frac{1}{N_{\text{plate}}} = \frac{2}{\overline{\text{Pe}}} + \frac{(1 - \varepsilon_b)\,\varepsilon_y^2}{[\varepsilon_b + (1 - \varepsilon_b)\,\varepsilon_y]^2} \left[\frac{2}{3} \frac{\text{Pe}_y}{\text{Sh}} \frac{R_y}{L} + \frac{\text{Pe}_y}{15} \frac{R_y}{L} \right] \tag{6.94}$$

If the component is so large that it is not able to explore any of the pores or internal parts of the stationary phase, then $\varepsilon_y = 0$ for that component and Eq. (6.94) simplifies further, to give

$$\frac{1}{N_{\text{plate}}} = \frac{2}{\overline{\text{Pe}}} \tag{6.95}$$

or

$$\boxed{\overline{\text{Pe}} = 2N_{\text{plate}}} \tag{6.54}$$

which is the same result as obtained from Eqs. (6.2) to (6.6), although different assumptions were used in the case of Eqs. (6.2) to (6.6).

Equation (6.88) enables calculation of N_{plate}, the plate count for a specified component, and Eq. (6.84) enables calculation of N_i, the number of divisions in a Craig column for which the same bandspreading will be produced. Setting $j = N_i$ in Eqs. (6.77) or (6.80) gives the corresponding analytical expression for the chromatogram. As has been the case throughout most of this chapter, one of the restrictions of linear chromatography is that each component traverses the column independently of all others. (This is not always true, as was seen for separation of sulfuric acid from glucose at high loading, as discussed earlier in this chapter.) When one component is independent of the other, plate count $N_{\text{plate},i}$ or the corresponding Craig plate number N_i is determined independently for each component.

6.11 THE STIRRED-TANK-IN-SERIES MODEL OF CHROMATOGRAPHY IS BASED ON A MATERIAL BALANCE ($k' \geq 0$) (WITH CONTRIBUTIONS BY AJOY VELAYUDHAN)

The stirred-tank-in-series model handles the cases of retained as well as unretained components. This model is based on complete mixing and instantaneous equilibrium occurs. A material balance around the jth vessel is given by

$$V_{\text{T}}\varepsilon_{\text{T}} \frac{dC_{\text{m}}(i, j, t)}{dt} + V_{\text{T}}(1 - \varepsilon_{\text{T}}) \frac{dq(i, j, t)}{dt} = F[C_{\text{m}}(i, j - 1, t) - C_{\text{m}}(i, j, t)] \tag{6.96}$$

where

V_T = volume of one vessel (i.e., one mixing stage)
t = time since beginning of run
F = volumetric flowrate
$C_m(i, j, t)$ = concentration of solute i, in tank j, at time, t
$q(i, j, t)$ = concentration of solute i, in tank j, at time, t on the stationary phase
ε_T = total void fraction = $\varepsilon_b + \varepsilon_p$

The notation for concentration can be written as:

$$\frac{dC_m(i, j, t)}{dt} = \frac{N_i}{t_{r,i}}[C_m(i, j-1, t) - C_m(i, j, t)] \tag{6.97}$$

where

$$j = 1, 2, \ldots, N_i$$

$$t_{r,i} = t_o(1 + k') \tag{6.98}$$

$$t_o = \frac{L}{v}$$

with

L = column length
$v = \Delta x / \Delta t$
$\Delta x = \frac{L}{N}$ = length of the plate

The solution for Eq. (6.97) for adsorption (frontal chromatography) is given by

$$\frac{C_F(i, J, t)}{C_{i,o}} = 1 - \exp\left(\frac{-N_i t}{t_{r,i}}\right)\sum_{j=0}^{J-1}\frac{\left(\frac{N_i}{t_{r,i}}\right)^j t^j}{j!} = \exp\left(\frac{-N_i t}{t_{r,i}}\right)\sum_{j=J}^{\infty}\frac{\left(\frac{N_i}{t_{r,i}}\right)^j t^j}{j!} \tag{6.99}$$

where

J_i = $N_{plate,i}$ = plate count for component i
$C_{i,o}$ = concentration of component i at the inlet
$C_F(i, j, t)$ = outlet concentration of solute for a column with J plates
$t_{r,i}$ = retention time for component i as given in Eq. (6.98)

The elution profile for a chromatography column with an injection time, τ, for the pulse of solute is

$$\frac{C(i,J,t)}{C_{i,o}} = \exp\left[\frac{-N_i}{t_{r,i}}(t-\tau)\right]\left[\sum_{j=0}^{J-1}\frac{\left(\frac{N_i}{t_{r,i}}\right)^j (t-\tau)^j}{j!}\right] - \exp\left[\frac{-N_i t}{t_{r,i}}\right]\left[\sum_{j=0}^{J-1}\frac{\left(\frac{N_i}{t_{r,i}}\right)^j t^j}{j!}\right] \tag{6.100}$$

The result for adsorption given in Eq. (6.99) can be concisely expressed in terms of the incomplete gamma function. The result for the adsorption case is

$$\frac{C_F(i,J,t)}{C_{i,o}} = P\left(J_i, \frac{J_i t}{t_{r,i}}\right) \tag{6.101}$$

where

$P(a, x)$ = incomplete gamma function

$$= \int_0^x t^{a-1}\exp(-t)dt \Big/ \int_0^\infty t^{\alpha-1}\exp(-t)dt, \text{ where } a = J_i \text{ and } x = J_i t/t_{r,i}$$

The solution for elution of a pulse input given in Eq. (6.100), results in

$$\frac{C(i,J,t)}{C_{i,o}} = P\left(J_i, \frac{J_i t}{t_{r,i}}\right) - P\left[J_i, J_i\left(\frac{t-\tau}{t_{r,i}}\right)\right] \tag{6.102}$$

In both Eqs. (6.101) and (6.107), $t_{r,i}$, is the retention time of component i. The expression for retention time incorporates the distribution coefficient for component i between the mobile and stationary phases since $k_i' = \phi K_{D,i}$. In cases of linear equilibria both k' and $K_{D,i}$ are constants.

6.12 THE CRAIG AND STIRRED TANK IN SERIES MODELS GIVE SIMILAR RESULTS: GLUCOSE/FRUCTOSE SEPARATION EXAMPLE

The separation of glucose from fructose is a key step for the production of high fructose corn syrups. These syrups are widely used in food products ranging from donuts to soft drinks. The syrups are obtained from glucose derived from corn starch, with glucose being partially isomerized to fructose by the enzyme glucose isomerase. The resulting mixture must be enriched in fructose to achieve the proper sweetness. The hypothetical example below illustrates how the Craig or discrete models may be used to calculate the elution profile for separation of these sugars. Adsorption of one component through a single plate is used here to illustrate the Craig model and to contrast it to the continuous-flow plate model (stirred-tank-in-series). A liquid chromatography system will consist of numerous plates. In this respect the example is unrealistic. One plate is used here for the sake of illustration. This section is taken from Ladisch and Velayudhan (1993) with minor modifications.

First, the adsorption breakthrough characteristics of the two types of models are considered. From Eq. (6.80), the effluent history for the Craig model is given by

$$\frac{C_F^{disc}(k)}{C_o} = p' \sum_{m=o}^{k-1} (q')^m \tag{6.103}$$

where the subscript i has been dropped, since only one component is being considered, and j = 1 (exactly one plate). This expression involves the sum of a geometric series. When this is evaluated, the result is

$$\frac{C_F^{disc}(k)}{C_o} = 1 - (q')^k = 1 - (1 - p')^k \tag{6.104}$$

The superscript "disc," denoting "discrete flow" or "discontinuous flow," has been added to distinguish the Craig from the continuous-flow model, and k denotes a time step.

Under the same conditions, Eq. (6.99) gives

$$\frac{C_F^{cont}(t)}{C_o} = 1 - e^{-\alpha t} \tag{6.105}$$

where

$$\alpha = \frac{N}{t_R} = \frac{v}{(\Delta x)(1 + k')} = \frac{1}{(\Delta t)(1 + k')} \tag{6.106}$$

and Δx *is* L/N, or the length of a plate. In order to compare this to the result from the Craig model, we consider the output at the finite values $t_k = k\Delta t$, for which

$$\frac{C_F^{cont}(k)}{C_o} = 1 - e^{-p'k} \tag{6.107}$$

The value of p' is calculated from Eq. (6.78). Comparing the results for the two kinds of plate models, it is clear that $e^{-p'k} > 1 - p'$ and therefore $e^{-p'k} > (1 - p')^k$. Consequently $C_F^{disc}(k) > C_p^{cont}(k)$ for all $t = t_k$. The Craig effluent is always higher in solute concentration compared to that calculated from the continuous-flow plate. The Craig effluent also will reach the initial value faster, and the model is therefore said to be more efficient. These results are shown in Figure (6.30), for $k' = 1$. Since the argument will extend to a finite number of plates, it is clear that a column composed of a certain number of Craig plates will be more efficient than one composed of an equal number of continuous-flow plates. Thus Craig plates should be distinguished from continuous-flow plates. In fact it is better to avoid the term "plate" and speak of Craig segments and continuous-flow segments. This avoids confusion with the experimentally well-defined plate number:

$$N_{plate} = \left(\frac{t_R}{\sigma_t}\right)^2 \tag{6.108}$$

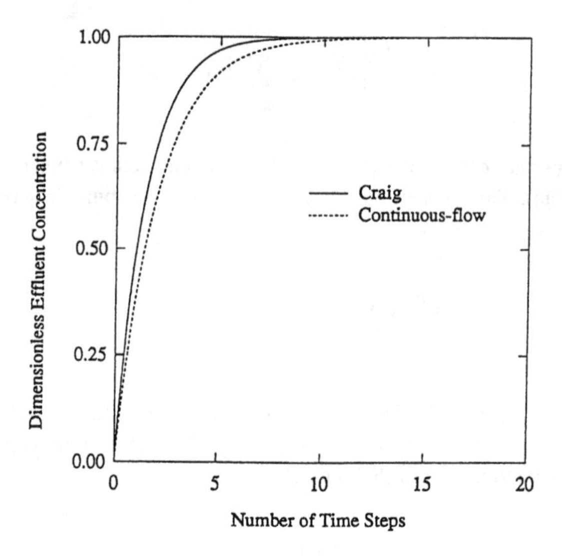

Figure 6.30 Graph of solution given by Eq. (6.82) for a single Craig segment and by Eq. (6.101) for a single continuous-flow segment for adsorption (from Velayudhan and Ladisch, 1993, fig. 1, p. 132.)

for the effluent peak of any component where t_R is the retention time (the time of emergence of its center of mass) and σ_t is its standard deviation (in consistent time units).

The Craig simulation requires data on column dimensions, and values of the various porosities in order to calculate the volumetric phase ratio from Eq. (6.87). If the distribution coefficient K^d is given in the form of moles of solute adsorbed per unit surface area, the corresponding retention factor k' can be calculated ($= \phi K^d$). The probabilities p' and q' can then be found from Eqs. (6.78) and (6.79). The chromatographic velocity can be calculated from Eq. (6.85); this, together with the column length, specifies the column hold-up time. The plate count, N_{plate}, can be calculated from Eq. (6.108). Sometimes data is reported in terms of dimensionless groups other than those used in Eq. (6.108), so the data must be transformed into the proper dimensions or formalism. The number of Craig segments is then calculated from Eq. (6.84):

$$N_{\text{craig}} = N_i = N^{\text{disc}} = N_{\text{plate}}\left(\frac{k'}{1+k'}\right) \tag{6.84}$$

The (discrete) input pulse width must also be specified. If the effluent history is in dimensionless time units, and is scaled to the column hold-up time, then the input pulse width must also be appropriately scaled, and the concentration reported in dimensionless terms, scaled to its input inlet value. With this information, the Craig simulation as expressed by Eq. (6.83) can be carried out. A FORTRAN program for this calculation is given by Velayudhan and Ladisch (1993) where the IMSL function for the incomplete beta function was used. The results of this simulation is comparable to that of Raghavan and Ruthven (1985) as shown in Figure 6.31.

Each component can be assigned its own plate number in linear chromatography. Thus plate models can also be used to describe multicomponent separations. Figure 6.32 shows the comparison of plate simulations with Carta's (1988) results on the separation of fructose from glucose. A third example is given by the previously described Figure 6.18 that shows a comparison

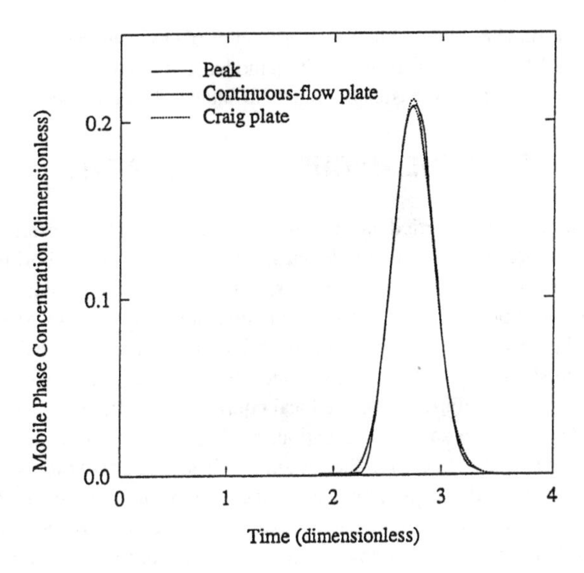

Figure 6.31 Comparison of Craig plate segment simulations and continuous flow segment simulations with numerical results of Raghavan and Ruthven (1985)—denoted as "peak" in the figure—for linear chromatography. The parameter values used with Eq. (6.83) for the Craig simulation and Eq. (6.102) for the continuous-flow plate are $\varepsilon_x = 0.42$, $\varepsilon_y = 0.32$, $\varepsilon_b = 0.41$, $K^d = 11.4$, $\overline{Pe} = \infty$, Sh = 2000, (r_y/L) $Pe_y = 3.66 \times 10^{-2}$, (r_x/L) $Pe_x = 5.06 \times 10^{-3}$, and $N_{plate} = 200$ (from Velayudhan and Ladisch, 1993, fig. 2, p. 135.)

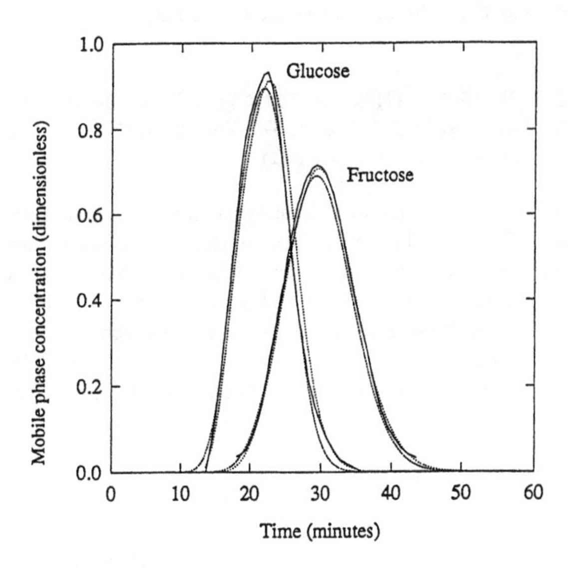

Figure 6.32 Simulation of glucose/fructose separation using Eqs. (6.83) or (6.102) and comparison to numerical results of Carta (1988). Parameters are $\varepsilon_b = 0.39$; *for glucose*: Sh = 4,545.5; (r_y/L) $Pe_y = 2.35$; $K^* = \varepsilon_b + (1 - \varepsilon_y)K^d = 0.26$; $N_{plate} = 46.5$; and *for fructose*: Sh = 1785.7; (r_y/L) $Pe_y = 0.92$; $K^* = 0.66$; $N_{plate} = 38.2$. The numerical results, the simulation of the Craig segment model, and the simulation of the continuous segment model are nearly the same (from Velayudhan and Ladisch, 1993, fig. 4, p. 137.)

of large-scale isocratic elution data for the separation of phenyl alanine from aspartame taken from Ladisch et al. (1991) to results from the Craig model. The Craig simulation used calculated values of parameters and gave a reasonable fit to the data (Velayudhan and Ladisch, 1993).

NONLINEAR CHROMATOGRAPHY

Nonlinear chromatography is carried out for solutes whose concentrations are high enough so that their ratios of stationary phase to mobile phase concentrations at equilibrium are no longer constant over a range of concentrations. However, if the combination of stationary phase properties, solute concentrations, and mobile phase velocity enables equilibrium to be approximated at each theoretical plate in the column, local equilibrium theory facilitates calculation of solute retention. Mass transfer and/or pore diffusion effects that result in peak dispersion are handled separately. This approach, referred to as the local equilibrium method or theory, requires that the equilibrium data between solutes and stationary phase be known or determined. In most cases the solutes are assumed to bind to the stationary phase and travel through the column independently of each other. If competition occurs, one component will hinder adsorption of the second and, when injected individually, will have different elution times (and shapes) then when combined into the same feed sample. A competition factor must therefore be determined and introduced into the adsorption isotherm equation for multiple solutes.

The simulation of elution profiles may be carried out using stage (discrete) models or the simultaneous solution of partial differential equations that give material balances across the column, around individual stationary phase particles, and within the pores for multiple interfering components (Gu et al., 1993). These approaches are not developed here. Rather, a method that enables estimation of particle size effects for cases where at least one peak exhibits nonlinear behavior is presented. This requires that the concepts of local equilibrium theory be familiar to the reader, and hence local equilibrium theory is discussed here.

6.13 LOCAL EQUILIBRIUM THEORY RELATES ELUTION PROFILES FOR AN ADSORBING OR DESORBING SOLUTE TO ITS EQUILIBRIUM ISOTHERM (WITH CONTRIBUTIONS BY JUAN HONG)

Adsorption and desorption isotherms are ultimately required for detailed modeling of chromatographic separations. There are three major types of isotherms as shown in Figure 6.33. A favorable isotherm is represented by a convex curve, while an unfavorable isotherm is given by a concave curve (Lightfoot et al., 1962). During adsorption the concentration profile for a solute with a favorable isotherm becomes increasingly sharp and approaches a step function, since the solute moves more rapidly through the column in regions where the solute's concentration on the stationary phase is higher. Solute in regions of higher concentration overtake the same sol-

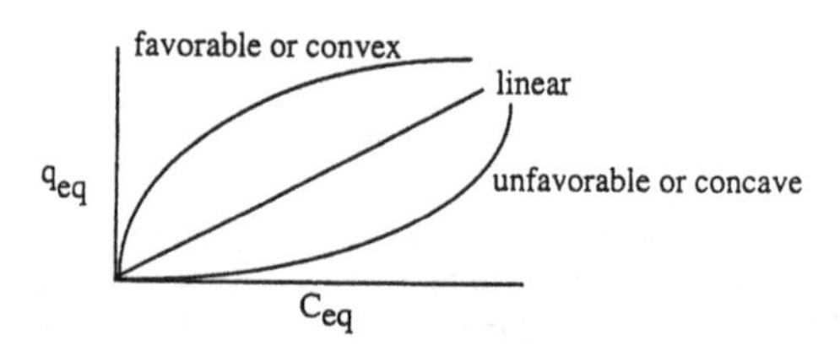

Figure 6.33 Types of isotherms.

ute in regions of lower concentration, and the front of the solute peak becomes sharper. For a linear isotherm, the solute travels at a rate that is independent of concentration, and there is neither a sharpening nor broadening effect. For an unfavorable isotherm, progressive broadening occurs as the solute moves more rapidly in regions of lower concentration. In the case of *desorption* or *elution*, a favorable isotherm results progressive broadening, while an unfavorable isotherm causes the solute front to become sharper. The idealized representations of the three cases are shown in Figure 6.34.

The representations of the adsorption and desorption fronts are not meant to include the effect of dispersive or diffusive processes. The profiles in Figure 6.34 represent a local equilibrium condition where the solute in the mobile phase is always in equilibrium with the stationary phase. This concept is referred to as local equilibrium, since it applies to short (i.e., local) segments of the column. Measurement of the elution time of the adsorbate from chromatography columns, together with the elution profile shape,[5] gives a qualitative measure of the type of isotherm and an indication of whether the adsorption and desorption isotherms are different. The benefit of local equilibrium theory is that the entire equilibrium isotherm is not needed to estimate expected breakthrough profile shape, as represented in Figure 6.35.

Conditions inherent to local equilibrium theory are that the adsorption process is isothermal, the vapor or liquid velocity is constant throughout the length of the column, radial velocity gradients are negligible, and dispersion due to channeling, diffusion or eddy mixing is negligible. These assumptions would, at first glance, seem to limit the utility of the theory in practice, to estimating retention times of peaks for an ideal adsorption column that does not exist. Actually local equilibrium theory gives the average retention time of an eluting front that has been subjected to dispersion processes much like retention time for a Gaussian peak that represents the center of mass of the peak.

The expression for the elution time is derived from a mass balance for a single solute, dissolved in the mobile phase that is passed through a fixed bed of adsorbent particles:

$$\varepsilon_b \frac{\partial C_m}{\partial t} + \rho_B \frac{\partial q}{\partial t} + \varepsilon_b v \frac{\partial C_m}{\partial z} = 0 \tag{6.109}$$

where

ε_b = extraparticulate void fraction, dimensionless (assumes solute does not penetrate adsorbent)
ρ_B = packing density of adsorbent, g adsorbent/(cm^3 column volume)
v = interstitial velocity, cm/s
C_m = solute concentration in vapor or mobile phase, moles/L
q = adsorbate concentration on adsorbent, moles/g
z = distance along column axis, cm
t = time, s

[5]Kipling credits Tiselius for showing that frontal analysis could be used to obtain an adsorption isotherm. A column is packed with the adsorbent and filled with pure solvent. The concentration of solute in the eluent is continuously measured as the solution is run through the column. For a two component solution consisting of solvent and a low concentration of *X* dissolved in it, a profile is obtained of the type shown in Figure 6.35.

In Figure 6.35 the volume is collected up to point A. This is the volume of solute which has been denuded of solute. Allowance has been made for the volume of liquid held mechanically in the column containing the adsorbent. Beyond the sharp rise the solution and absorbate in the column will be in equilibrium for a concentration of solute in the eluent that remains constant. (Kipling, 1965, p. 6). This rationale is generally applicable for linear or parabolic adsorption isotherms. Nonlinear isotherms with multiple inflection points give a number of different shapes as discussed by Goldstein (1953) and Tudge (1961a, b).

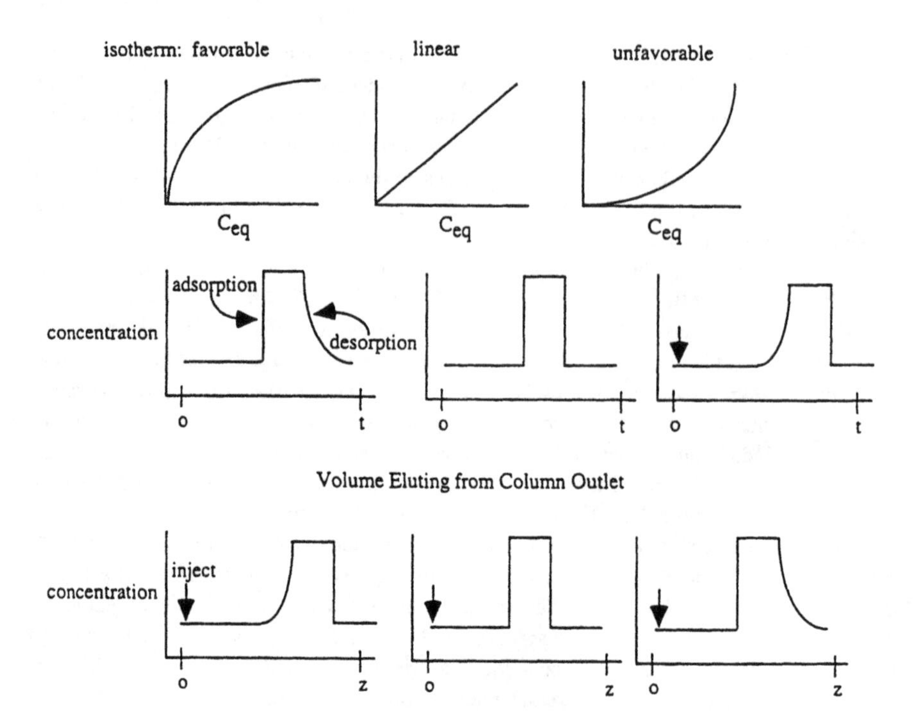

Figure 6.34 Elution profiles (ideal chromatography) correlated to types of isotherms.

If the solute concentration is given by an equilibrium isotherm having the general form:

$$q_{\text{eq}} = q(C_{\text{m,eq}}) \tag{6.110}$$

Equations (6.109) and (6.110), when combined, give the equation for local equilibrium:

$$\left(1 + \frac{\rho_{\text{B}}}{\varepsilon_{\text{b}}} q'(C_{\text{m}})\right) \frac{\partial C_{\text{m}}}{\partial t} + v \frac{\partial C_{\text{m}}}{\partial z} = 0 \tag{6.111}$$

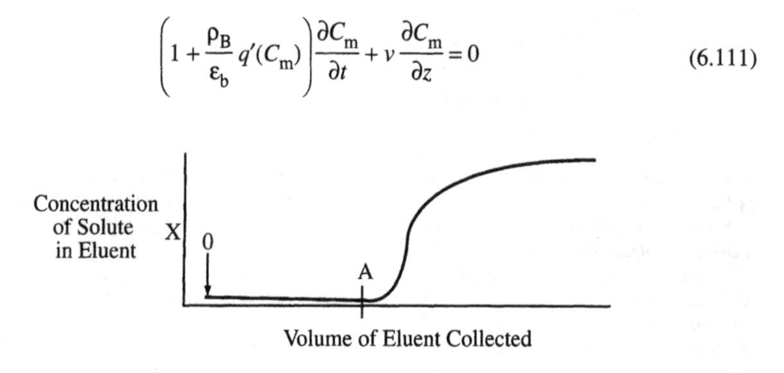

Figure 6.35 Breakthrough profile shape.

where

$$q'(C_m) = \frac{dq}{dC_m} \text{ and } \frac{\partial q}{\partial t} = \frac{\partial q}{\partial C_m}\frac{\partial C_m}{\partial t} \tag{6.112}$$

As described by Goldstein (1953), there are continuous and discontinuous solutions for Eq. (6.111). The continuous solution at $z > 0$, if it exists, is

$$C_m = f\left(z - \frac{vt}{1 + \frac{\rho_B}{\varepsilon_b} q'(C_m)}\right) \tag{6.113}$$

The function f in Eq. (6.113) is determined by initial conditions and the isotherm given by $q(C_m)$. If there is a discontinuous solution for C, the discontinuity originally at $z = 0$ must be propagated along the curve $z = z(t)$. Application of Stoke's theorem to Eq. (6.109) says that on the curve $z = z(t)$,

$$\frac{dz}{dt} = \frac{v}{1 + \frac{\rho_B [q]}{\varepsilon_b [C_m]}} \tag{6.114}$$

where $[q]$ and $[C_m]$ represent discontinuities in q and c across the curve. The discontinuous solution should be constructed whenever there are multiple solutions for the continuous solution.

Multiple solutions that would represent two different concentrations of the solute at the outlet of the column at the same time are not physically reasonable. Consider the example for a Type II isotherm[6] with hysteresis as illustrated in Figure 6.36. The continuous solution (given by the dotted line in Figure 6.36) represents the solution of Eq. (6.113) as represented by Eq. (6.115) below. When the argument is less than zero, the concentration of the solute in the mobile phase is the same as the inlet concentration, C_o: $C_m = C_o$. When the argument of the function of Eq. (6.113) is zero, C_m as a function of t is given by the dotted line in Figure 6.36. At some point greater than zero, $C_m = 0$. However, the portion of the dotted curve between $C_m = 0$ and $C_m = C_c$ is physically not possible. Thus the breakthrough curve is composed of a discontinuity which is generated by the chord between the origin and the tangential point, C_c, in Figure 6.37 and a diffusive front which is given by the equilibrium curve between the tangential point, C_c, and the inlet concentration, C_o denoted by the arrow in Figure 6.37. The time at which breakthrough occurs is

$$\boxed{t = \frac{z}{v}\left(1 + \frac{\rho_B}{\varepsilon_b} q_A'(C_C)\right) = \frac{z}{v}\left(1 + \frac{\rho_B}{\varepsilon_b}\frac{q_A(C_C)}{C_C}\right)} \tag{6.115}$$

[6]The five basic types of adsorption isotherms were classified by Branauer et al., as reviewed by Kovach (Kovach, 1978). These are Type I where adsorption does not proceed past a monolayer coverage of adsorbent surface, Type II for multilayer coverage, Type III where adsorption increases without limit, Type IV which is a variation of Type II where an adsorbing gas essentially condenses and completely fills the capillaries in the adsorbent particles, and Type V which is a variation of Type III with capillaries in adsorbent being filled with adsorbate. Type I isotherms are the most prevalent. The Langmuir and Freundlich isotherms used in modeling of liquid chromatography approximate a Type I isotherm. Since solute concentrations are often low, the isotherms also approximate linear adsorption equilibrium.

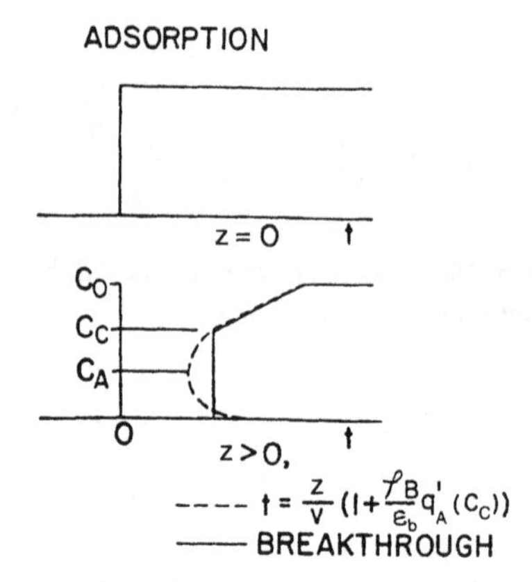

Figure 6.36 Schematic representation of local equilibrium theory applied to a Type II isotherm with hysteresis. A step increase in the solute concentration at the column's inlet ($z = 0$) and maintaining the solute concentration at a constant level gives the step shaped concentration profile as a function of time, as shown at the top, and a sharp increase followed by a more gradual increase near the outlet as shown by the solid line at the bottom. The dotted line gives the solution to Eq. (6.113) when the argument is zero.

The ratio z/v is column length divided by interstitial velocity, namely, the retention time of an excluded component. The subscript "A" denotes adsorption. The breakthrough time t is independent of concentration when the step input concentration (denoted by C_0) is greater than the concentration C_c, at the tangential point C in Figure 6.37. The quantity $q'_A\ (C_c)$ denotes the derivative for the adsorption curve at point C, as given by the expression in Eq. (6.112). The derivative $q_D\ (C_D)$ is for the desorption curve of point D (Figure 6.38). The time of appearance of the discontinuity for desorption is given by

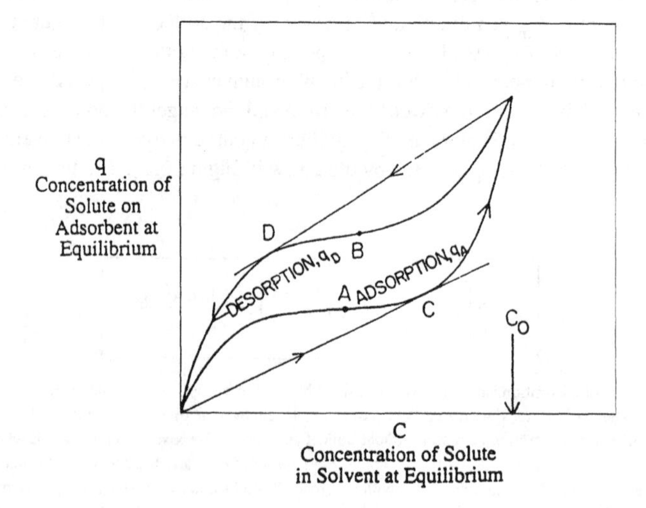

Figure 6.37 The general shape of the equilibrium isotherm for this example exhibits hysteresis.

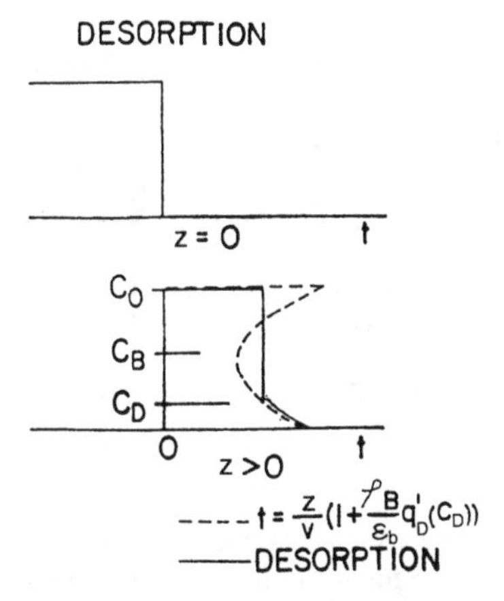

Figure 6.38 A step decrease of solute, back to a pure solvent (top) results in the desorption profile, some time later (bottom). The dotted line corresponds to the solution of Eq. (6.41).

$$t = \frac{z}{v}\left(1 + \frac{\rho_B}{\varepsilon_b} q'_D(C_D)\right) = \frac{z}{v}\left(1 + \frac{\rho_B}{\varepsilon_b}\left(\frac{q_D(C_o) - q_D(c_D)}{C_o - C_D}\right)\right) \tag{6.116}$$

where t is a function of the initial concentration of the solute, C_o. The subscript D for q_D denotes desorption, while the subscript D for C_D denotes point D. The discontinuity is followed by a diffusive trailing edge which corresponds to the desorption isotherm between point D and the origin of the isotherms at the intersection of the x- and y-axis in Figure 6.37. The same rationale applies to the adsorption and desorption profile for an initial solute concentration, C_o, that is less than the concentration corresponding to the tangential point C in Figure 6.37. In this case the breakthrough curve would consist solely of a step change, while the desorption curve would have a short discontinuity followed by a trailing edge as shown in Figure 6.38. C_D depends on the location of inflection point B.

6.14 DESORPTION ISOTHERMS MAY DIFFER FROM ADSORPTION ISOTHERMS: HYSTERESIS EFFECTS AND LOCAL EQUILIBRIUM THEORY

The concentration of the solute that is associated with the stationary phase may be higher for desorption than for adsorption at the same concentration of the solute in the mobile phase. This effect is usually associated with vapor phase adsorption processes, but it could be encountered in liquid phase adsorption processes. For example, the adsorption of some types of polymers can result in hysteresis (Kipling, 1965). The determination of desorption equilibria in a few cases may therefore be necessary. Strong binding and/or formation of weak chemical bonds between the solute and adsorbent can lead to a hysteresis effect in which the equilibrium between solute and adsorbent is different for adsorption than for desorption. This concept was illustrated for a Type II isotherm[6] (Figure 6.37) in the analysis of local equilibrium theory in the previous section, although hysteresis effects are not common in liquid chromatography. The graphical representations of the types of isotherms are given in Figure 6.39. While these isotherms were originally defined for gases, they have been also used for solutes adsorbing from liquids. For

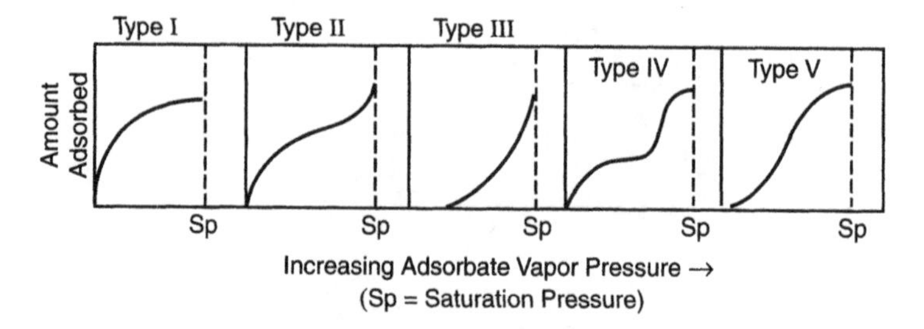

Figure 6.39 Representation of Type I, II, III, IV, and V isotherms (reproduced from Figure 9-1, pg. 338, of Kovach, 1978).

liquid equilibria of a solute with the stationary phase based on size exclusion, the equilibrium will be linear and will be the same for adsorption and desorption. Similarly solute adsorption and desorption equilibrium concentrations are the same for reversed phase or hydrophobic interaction chromatography. The emphasis in this chapter is on adsorption in the absence of hysteresis.

Local equilibrium analysis was developed for vapor phase adsorption, where diffusion of the solute will be relatively rapid and multi-layer adsorption is more likely to be observed than for adsorption of a solute from a liquid system. Nonetheless, local equilibrium theory applies equally well to liquid systems, according to Sherwood et al. (1975). This theory is an approximation since it represents solute behavior when k_{MT} is infinitely great and assumes that local equilibrium exists between the particles and the adjacent fluid at all points and all times (Sherwood et al., 1975, p. 555). The manner in which the theory is used for scale-up purposes is a matter of engineering judgement. If the elution profiles are fixed to have a similar shape, with retention characteristics that are similar on both bench and process scales when normalized for the column's length and mobile phase velocity, local equilibrium theory is useful, for it enables estimation of retention times for a large-scale column based on bench-scale results.

6.15 TRIANGULAR PEAKS ARE ASSOCIATED WITH NONLINEAR CHROMATOGRAPHY: OVERLOAD CONDITIONS

Nonlinear chromatography for a solute that follows the Langmuir isotherm occurs when

$$KC_{eq} \tilde{>} 1 \quad \text{for} \quad q_{eq} = \frac{q_m K C_{eq}}{1 + KC_{eq}} \tag{6.117}$$

If $KC_{eq} >> 1$, then $q_{eq} = q_m$. Elution profiles that are representative of nonlinear chromatography approximate a triangular shape as represented in an idealized drawing in Figure 6.40 (adopted from Eisenbess et al., 1985). The values of 1, 50, and 100 represent increasing solute concentrations dissolved in the same volume of sample injected into the column. The elution profile forms a triangular or trapezoidal shape if the concentration of solute in the sample is so large that a fraction of the solute does not adsorb.

The triangular shape of this type of peak is indicative of concentration or mass overload where the solute concentration corresponds to the nonlinear portion of the equilibrium curve. If the solute is at a sufficiently high concentration that it overloads the capacity of the stationary phase to adsorb it near the inlet of the column, the stationary phase becomes saturated or com-

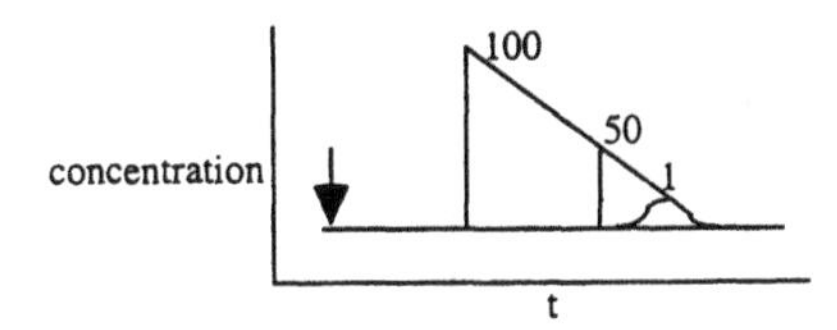

Figure 6.40 Triangular profiles at increasing sample concentration.

pletely covered with solute (i.e., $q_{eq} = q_m$). Most of the solute remains in the feed sample and is pushed ahead until it encounters fresh stationary phase where some more of the solute is adsorbed, and saturation is again established. The rest of the solute band continues to move ahead. The resulting concentration profile along the axis of the column, and the elution profile at the outlet, is characterized by the curves given in Figure 6.41. The profile will have a plateau or flat-top region where $C_m = C_o$, where C_o is the solute concentration in the feed (see left side of Figure 6.41). As the solute band travels further down the column, adsorption removes more of the solute. By the time the solute elutes from the column, the peak may have lost its plateau and elutes in a triangular shape.

Both equilibrium and dispersion effects result in broadening of the peak. Eventually $C_m < C_o$, so that the flat-top is lost. If the column is sufficiently long, the profile at the outlet of the column will resemble the shape shown at the right of Figure 6.41 or as represented by the idealized profile of where a sharp solute front elutes first, followed by a decreasing solute concentration as a function of time. Figure 6.42 gives a chromatogram where triangular profiles occur for the separation of two dipeptides using 15% aqueous acetonitrile as the mobile phase (Kim and Velayudhan, 1998). The analysis of Snyder et al. (1987) shows that small solutes, or solutes that exhibit the behavior shown in Figures 6.40 and 6.41, may be treated as right triangles for the purpose of scaling up a separation in which relative peak shape and spacing are held constant. The areas under the peak at the column inlet and outlet are the same, since mass is conserved.

An increasing solute concentration (also referred to as loading) decreases the apparent plate count as the load increases above a critical value. This is illustrated in Figure 6.43 (adapted from DeJong et al., 1981) that represents the curve for adsorption of 2,4-dimethylphenol on Lichrosorb S160 (DeJong et al., 1981). While loadings and plate counts are different for other types of solutes and stationary phases, the general trend is similar. At low sample concentration the peak shape at the column's outlet is Gaussian as illustrated at the left of Figure 6.43. As the concentration of the solute in the sample (feed) injection is increased, the peak broadens and

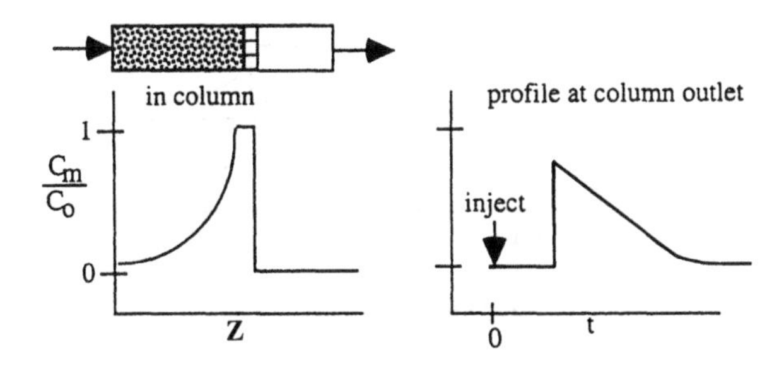

Figure 6.41 Elution profile for high concentration of solute.

The left-hand side of an elution profile has also been referred to as a shock wave (presumably since the front of the peak is sharp) or the front of the peak (since it elutes first). The velocity of the center of the mass transfer zone (abbreviated MTZ) is given by Eq. (6.119) and is obtained as explained by Eqs. (6.109) to (6.114):

$$u_{\mathrm{MTZ}} = \frac{v}{1 + \dfrac{1 - \varepsilon_{\mathrm{tot}}}{\varepsilon_{\mathrm{tot}}} \rho_{\mathrm{B}} \dfrac{\Delta q}{\Delta c}} = \frac{v}{1 + \phi \, \rho_{\mathrm{B}} \dfrac{\Delta q}{\Delta c}} \tag{6.119}$$

where $\varepsilon_{\mathrm{tot}}$ is defined by the molecule's size and the adsorbent's porosity, and

v = interstitial velocity of the mobile phase, cm/s
u_{MTZ} = velocity of solute as it passes through the column as measured from the midpoint of the peak shown in Figure 6.45, cm/s

Equation (6.119) assumes that the solute penetrates the pores of the stationary phase and adsorbs on the stationary phase. The velocity of MTZ will be slower than the interstitial velocity of the mobile phase since the solute is reversibly retained by the stationary phase. If no adsorption occurs, and the nonadsorbed solute is too large to enter pores of the stationary phase, the solute will travel at the same velocity as the mobile phase (i.e., $u_{\mathrm{MTZ}} = v$, since $\Delta q = 0$).

The width of the front (left side) of the peak spreads out because of mass transfer resistances or pore diffusional effects. The width of mass transfer zone is proportional to:

$$W_{\mathrm{MTZ}} \sim v^{k-1} d_{\mathrm{p}}^{k} \tag{6.120}$$

where $k = 2$ for pore diffusion controlling and $k = 1.5$ for mass transfer (film diffusion) control as discussed previously (Eqs. 6.10 and 6.37; Table 6.1). Particle size effects that cause changes in peak width due to either mass transfer or pore diffusion are incorporated into the scale-up rules through Eq. (6.120). The width of the mass transfer zone is given by $t_{\mathrm{w,MTZ}}$, which is the time required for MTZ (i.e., front of the peak) to elute relative to its average retention time as shown in Figure 6.45.

The width of the peak in dimensions of time is represented by

$$t_{\mathrm{w}} = t_{\mathrm{e}} - t_{\mathrm{MTZ}} + \frac{t_{\mathrm{w,MTZ}}}{2} \tag{6.121}$$

where

$t_{\mathrm{MTZ}} = L/u_{\mathrm{MTZ}}$ = time that has elapsed between injection of solute and elution of center of mass transfer zone for a flat-topped peak, s
$t_{\mathrm{w,MTZ}}$ = width of mass transfer zone on the left side of peak in Figure 6.45, s
t_{e} = time required for the entire peak to elute, s

The time required for the entire peak to elute is a function of the length of the column, L, and the velocity of the solute through the column:

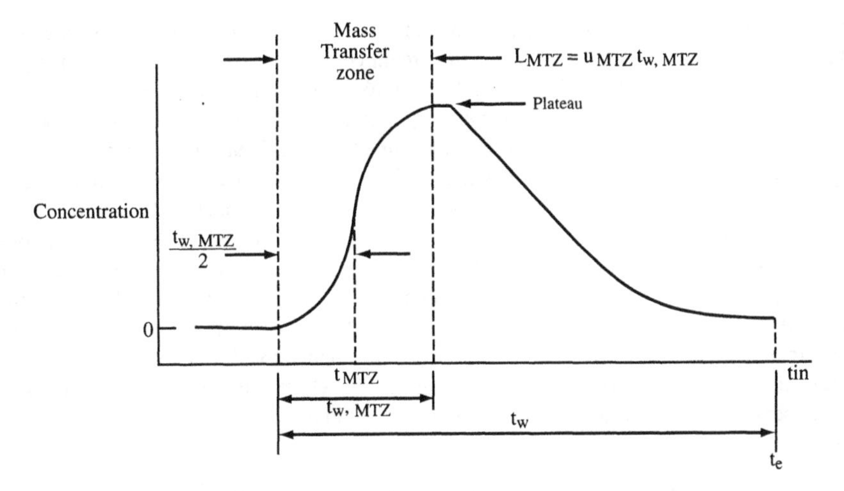

Figure 6.45 Schematic representation of peak and parameters corresponding to local equilibrium theory for peak that does not have a plateau.

$$t_{\mathrm{e}} = \frac{L}{u_{\mathrm{e}}} \tag{6.122}$$

The velocity, u_{e}, is obtained from the mobile phase flowrate, q_{flow}, elution volume, V_{e}, and column length:

$$u_{\mathrm{e}} = \frac{L}{(V_{\mathrm{e}}/q_{\mathrm{flow}})} \tag{6.123}$$

where V_{e} is the elution volume required for the entire peak to elute. The time at which the middle of the front of the peak (i.e., mass transfer zone) elutes is given by

$$t_{\mathrm{MTZ}} = \frac{L}{u_{\mathrm{MTZ}}} \tag{6.124}$$

where the velocity of the mass transfer zone, u_{MTZ}, is calculated from Eq. (6.119). The time required for the mass transfer zone (left side of peak) to pass the outlet of the column shown in Figure 6.45 is

$$t_{\mathrm{w,MTZ}} = \frac{L_{\mathrm{MTZ}}}{u_{\mathrm{MTZ}}} \tag{6.125}$$

where u_{MTZ} is the velocity of the mass transfer zone given in Eq. (6.119).

The definitions given in Eqs. (6.121) to (6.125) enable scale-up for a separation in which one or both of the components exhibits nonlinear behavior. An estimate for scaling up a column is carried out by specifying the particle size of the stationary phase, sample volume, and interstitial velocity of the eluent. The overlap of two peaks (as explained in Section 6.17) divided by their combined widths is assumed to be the same for both the large and bench scale columns. The elution time of a peak is based on the time required for the mass transfer zone of the peak to emerge from the column, and it is based on Eq. (6.119), that is, local equilibrium. Wankat and Koo (1988) use this approach in developing recommendations for scaling rules for non-linear systems. An example of skewed peaks obtained for chiral chromatography is shown in Figures 6.46 and 6.47. Figure 6.46 shows expected (dotted line) vs data, as discussed in Section 6.20. The two coincide when solute effects on each other are accounted for (Figure 6.47).

The width of the front of the peak (also referred to as the width of the mass transfer zone) will depend on the type of isotherm as well as solute concentrations and the type of stationary phase. For a favorable isotherm, such as a Langmuir isotherm, the shape and width of the front of the peak will be independent of column length and form a proportional or constant pattern (see Figure 5.54 in Chapter 5) once the column is long enough to allow a breakthrough pattern to fully develop (Sherwood et al., 1975). Wankat and Koo (1988) give an equation for the width of the mass transfer zone, for a solute exhibiting a Langmuir adsorption isotherm, as

$$t_{w,MTZ} = \frac{L_{MTZ}}{u_{MTZ}} = \rho_B (1 - \varepsilon_B) \frac{q_{eq,F}}{C_{eq,F}} \left[\frac{K_D}{\frac{q_{eq,F}}{C_{eq,F}} - K_D} \ln \left[\left(\frac{0.05}{0.95} \right) \left(\frac{g}{h} \right) \right] + \ln \left(\frac{g}{h} \right) \right] \tag{6.126}$$

where

$t_{w,MTZ}$ = width of mass transfer zone, s
L_{MTZ} = width of mass transfer zone, cm
u_{MTZ} = average velocity of mass transfer zone, cm/s
ρ_B = density of stationary phase, g/cm^3

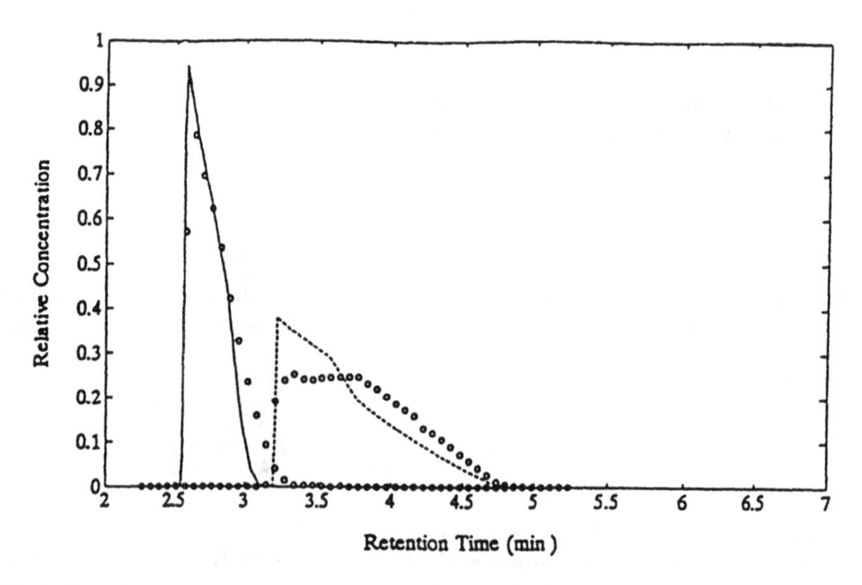

Figure 6.46 Chromatogram showing comparison of elution profile (circles) to simulated profile (solid line) using Langmuir Eq. (6.147) without the competition factor Γ (i.e., $\Gamma = 1$) (from Jageland et al., 1994, fig. 4, p. 499.)

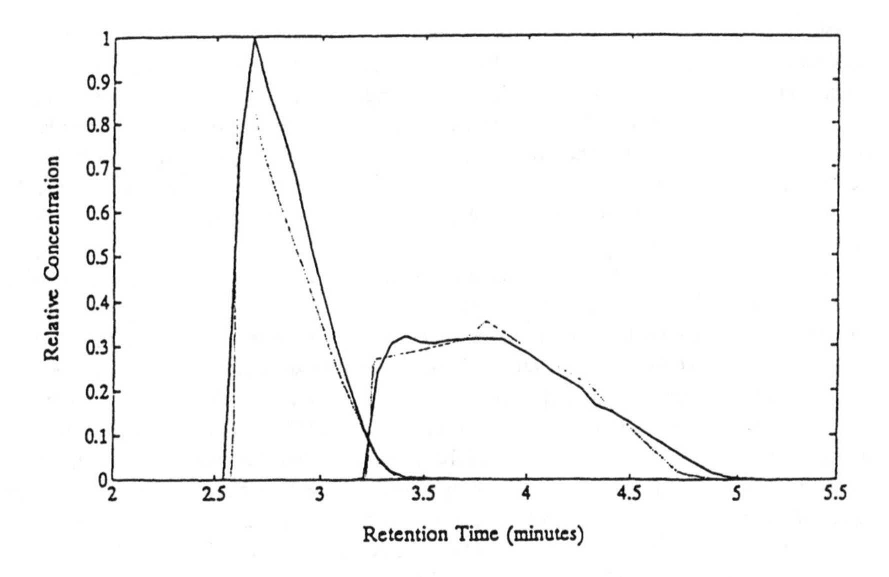

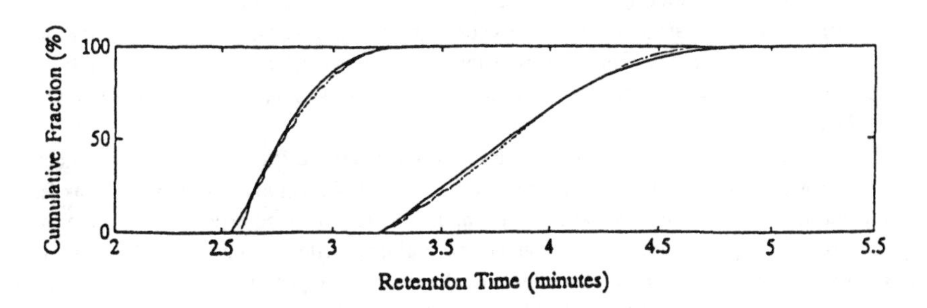

Figure 6.47 Comparison of data (dark line) to simulation (dotted line) for Langmuir isotherm (Eq. 6.148) that incorporates competition factor (the constants for the two peaks are $q_{m,1} = 340$ mg/g $\Gamma = 0.57$ and $q_{m,2} = 310$ mg/g $\Gamma_{2,1} = 1.32$). The upper trace gives a fractional elution profile, and the lower graph gives a cumulative elution profile. The eluent was 30% aqueous acetonitrile, and the stationary phase was a reversed phase C_{18} material (from Jageland et al., 1994, fig. 6, p. 501).

ε_B = interparticle void fraction

$K_D = q_m K$ = distribution coefficient in Langmuir equation, where q_m is the concentration of solute that corresponds to monolayer coverage and K is the adsorption (equilibrium) constant

$$g = q_m K - \frac{q_{eq,F}}{C_{eq,F}} - 0.95 K q_{eq,F}$$

$$h = q_m K - \frac{q_{eq,F}}{C_{eq,F}} - 0.05 K q_{eq,F}$$

$q_{eq,F}$ = solute concentration on stationary phase that is in equilibrium with the solute concentration in the feed, $C_{eq,F}$

Equation (6.126) represents the width of the mass transfer zone at the special conditions of a concentration which is large enough so that it falls in the nonlinear range of the adsorption isotherm and is able to give an elution profile whose range goes from 5% of the solute's concentration in the feed sample to 95% of its concentration. In effect, an elution profile with a plateau is needed to ensure that this condition is satisfied. In addition the constants (q_m and K) for the Langmuir adsorption isotherm equation must also be known. Note that K and K_D are not the same, and that the ratio of concentrations of $q_{eq,F}$ and $C_{eq,F}$ must be calculated from the Langmuir adsorption isotherm Eq. (6.117).

Local equilibrium theory enables calculation the average velocity (and from this, the elution time) of the front (left-hand side) of the peak. If there were no mass transfer or pore diffusion resistance, the front of the peak would be represented as a step-change, as illustrated by the bold line in Figure 6.48. However, diffusion and dispersion effects will cause the front to spread out, resulting in the formation of a gradual breakthrough curve which is also illustrated in Figure 6.48. Consequently an equation such as Eq. (6.126), is needed to calculate the width of this part of the profile, although Eq. (6.126) can only be used for a solute that exhibits a Langmuir adsorption isotherm and that has experimentally verified constant pattern elution profiles. The width of this zone, which is also referred to as an exchange zone, has historically been defined as the region of the breakthrough profile where the concentration of the solute changes from 5% to 95% of its influent value (Michaels, 1952).

An equation for calculating t_e requires that u_e be known, or calculated from theory. This is the velocity of the last part of the peak as its concentration approaches zero. The calculation of u_e requires measured values of the mass transfer coefficient as well as equilibrium data. If equilibrium data are available, a mathematical model could be used and a scaling rule would not really be needed.

The calculation of chromatographic velocity is carried out using Eq. (6.119) and assumes that the profile elutes as a step, as illustrated in Figure 6.48. In fact dispersion of the peak front will occur due to mass transfer, pore diffusion, and/or micromixing effects and yield a more gradual profile, which is superimposed on the equilibrium profile in Figure 6.48. Unlike the elution profile for the decreasing part of the peak, the front of the peak will form a pattern that is independent of column length, once column length exceeds a minimum value. This characteristic enables the width of the mass transfer zone to be used as a scale parameter if the equilibrium is known to follow a Langmuir isotherm.

The mass of solute that is retained on the stationary phase in the absence of mass transfer is reflected by the step change in Figure 6.48 and given by the area

$$\text{Mass} = \left(\frac{C_M}{C_{M,\text{feed}}}\right)(C_{m,\text{feed}})\, q_{\text{flow}}\,(t_{MTZ} - t_o) = (1)\, C_{M,\text{feed}}\,(V_{MTZ} - V_o) \tag{6.127}$$

where

V_{MTZ} = elution volume for mass transfer (or peak) front = $q_{flow}\, t_{MTZ}$

V_o = interparticle void volume = $q_{flow}\, t_o$

q_{flow} = volumetric flowrate, mL/min

and where $C_{M,feed}$ is the concentration of solute in the feed sample (mg/mL) and where t_o represents the elution time of fluid that is found in the spaces between the particles (i.e., the interparticle void volume). The area of the mass in Eq. 6.127 is given by the fill represented by diagonal lines in Figure 6.48. The ratio $C_M/C_{M,feed} = 1$, since the concentrations of solute at column inlet and outlet are equivalent at breakthrough for the ideal case of no dispersion. The

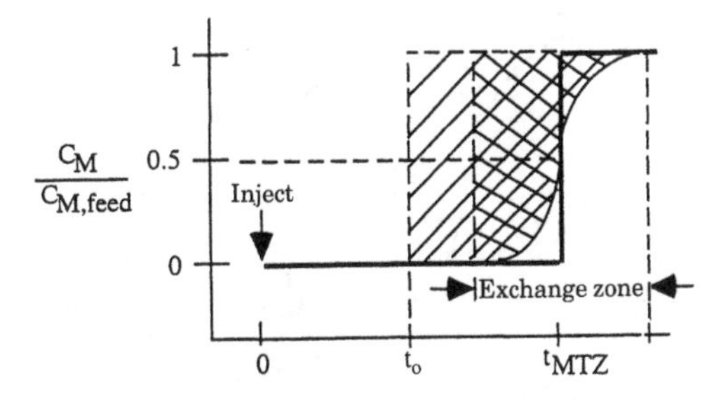

Figure 6.48 Relation of breakthrough curve to mass transfer zone for local equilibrium assumption.

quantity of solute adsorbed at breakthrough when resistance to mass transfer is significant is given by the cross-hatched area of Figure 6.48. Sherwood et al. (1975) show that the slope of the breakthrough curve, at its midpoint (where $C_M/C_{M,\mathrm{feed}} = 0.5$) as illustrated in Figure 6.48, is given by

$$\text{Slope} = (n)\left(\frac{1}{\dfrac{\rho_B z q_{\mathrm{eq,feed}}}{\varepsilon_b v C_{M,\mathrm{feed}}}}\right)\left(\frac{KC_{\mathrm{eq}}}{1 + KC_{\mathrm{eq}}}\right) \tag{6.128}$$

where

$$n = \frac{kaz}{\varepsilon_b v} = \text{number of transfer units} \tag{6.129}$$

and where

k = mass transfer coefficient, gmoles/s

a = interfacial area per unit volume of solid $= \dfrac{\pi d_p^2}{\pi d_p^3/6} = \dfrac{6}{d_p}, \dfrac{\mathrm{cm}^2}{\mathrm{cm}^3}$

z = axial distance from inlet of bed, cm

ε_b = void fraction of bed based on volume between particles

v = interstitial velocity, cm/s

$\varepsilon_b v$ = superficial velocity, cm/s

ρ_B = bulk density = mass of stationary phase divided by volume of bed, g/cm^3

$C_{M,\mathrm{feed}}$ = concentration of solute in the mobile phase that makes up the feed to the column

$q_{\mathrm{eq,feed}}$ = solid phase concentration of adsorbed material in equilibrium with the concentration of solute in the feed, $C_{M,\mathrm{feed}}$

K = adsorption constant in Langmuir equation

Equation (6.128) is limited to the case of nonporous particles, since it is based on the assumption that the major resistance to mass transfer is in the fluid phase outside of the particles. Such an equation might apply to silica gel particles used to carry out normal or reversed phase chromatography. Equations (6.128) and (6.129) provide a rationale for why the slope of the elution profile increases as the number of transfer units increases, causing the breakthrough profile to

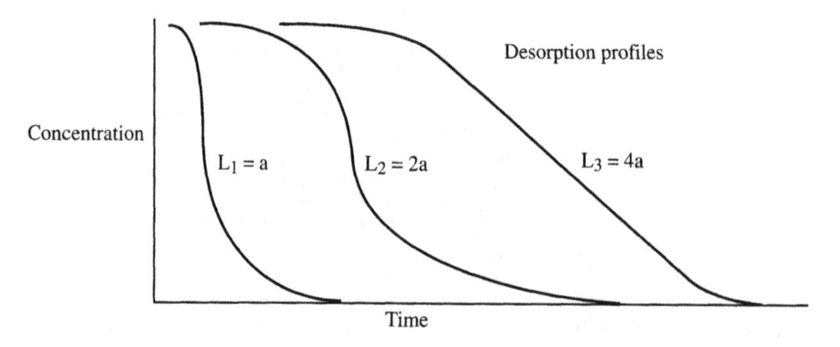

Figure 6.49 Desorption profiles vs. column length for local equilibrium.

become sharper. Equation (6.129) shows that the number of transfer units will increase with increasing mass transfer, decreasing particle size, increasing column length, decreasing void fraction, and decreasing flowrate when external mass transfer is the limiting resistance. The pattern itself will be independent of column length for a solute with a Langmuir isotherm, if all other parameters are held constant.

While the breakthrough profile forms a constant pattern, the desorption profile for a Langmuir isotherm does not. Its elution time is a function of column length, and the pattern for the desorption profiles is not constant, with the width of the profile increasing with column length ($L_1 < L_2 < L_3$) as illustrated in Figure 6.49.

One of the challenges in scaling-up the nonlinear chromatography is to estimate the time required for the desorption to be completed and the entire peak to elute, since column length will probably be changed upon scale-up. Analysis of the solution for the Thomas equation led Sherwood et al. (1975) to the conclusion that the proportional pattern desorption profile of a nonlinear system (as illustrated in Figure 6.49) will "scale in essentially the same way when mass transfer resistances are included as it does for the local equilibrium theory." Their statement is based on comparison of elution profile widths for the same solute but for stationary phases with particle sizes of 240 and 476 mm, respectively, and for two different cases where local equilibrium or pore diffusion were controlling, respectively. The bandwidths in both cases were almost the same.

6.17 SCALE-UP OF NONLINEAR CHROMATOGRAPHY IS BASED ON MAINTAINING THE RELATIVE PEAK POSITION AND OVERLAP TWO OR MORE PEAKS AT A FIXED RATIO

The challenge in scaling-up and modeling nonlinear chromatography is to identify simplified theories and equations that capture the behavior of peaks that occur at high concentration of the solute. These peaks are wider than peaks observed in analytical chromatograms, even though the sample volume may be small relative to the column volume (Velayudhan et al., 1992). The goal is to estimate column length and diameter, as well as fluid velocity, in order to reproduce a bench-scale separation on the process scale.

The rationale for determining scale-up ratios is based on the definition of peak width given in Eq. (6.121), and the hypothesis that each term is independent and therefore scales in the same proportion as every other term. Thus, for two solutes that exhibit nonlinear behavior such as illustrated in Figure 6.50, the width refers to the combined peaks. Figure 6.50 shows both the

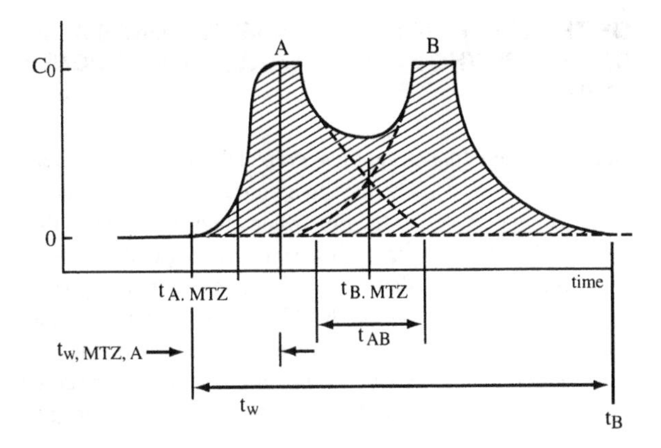

Figure 6.50 Overlapping peaks with plateaus showing parameters for local equilibrium theory.

profile as it would actually appear on a chromatogram (bold line) as well as the individual peaks (dotted lines) of components A and B, which give the chromatographic trace when combined.

The width of the two peaks is given by Eq. (6.130). The scale-up of bench-scale data to the process scale is based on the assumption that a proportional increase in t_{w} occurs due to the same portional change in each of the individual terms, on the right-hand side of Eq. (6.130), which are added to give t_{w}. Hence a doubling of each of the terms t_B, $t_{\mathrm{A,MTZ}}$, and $t_{\mathrm{w,MTZ},A}$ would result in a doubling of t_{w}. The other major assumption is that the interaction of one peak with the stationary phase does not affect interations of the other peak with the stationary phase. The overall width of the peak, t_{w}, is

$$t_{\mathrm{w}} = t_B - t_{\mathrm{A,MTZ}} + \frac{t_{\mathrm{w,MTZ},A}}{2} \tag{6.130}$$

where

$$t_B = \frac{L}{u_{B,\mathrm{o}}} + t_{\mathrm{F}} \tag{6.131}$$

$$t_{\mathrm{A,MTZ}} = \frac{L}{u_{\mathrm{A,MTZ}}} \tag{6.132}$$

and

$$t_{\mathrm{w,MTZ},A} = \frac{L_{\mathrm{A,MTZ}}}{u_{\mathrm{A,MTZ}}} \tag{6.133}$$

The value of the width of the feed pulse, t_{F}, in Eq. (6.131) is obtained by dividing the volume of the sample by the volumetric flowrate. The subscript A refers to peak A and subscript B refers to peak B in Figure 6.50. The velocity of the mass transfer zone $u_{\mathrm{A,MTZ}}$ in Eqs. (6.132) and (6.133), could be obtained from Eq. (6.119), but it requires equilibrium data to calculate $\Delta q/\Delta C$.

6.18 RATIOS OF THE WIDTH OF MASS TRANSFER ZONES AT PROCESS AND BENCH SCALES ARE THE BASIS OF SCALE-UP FOR NONLINEAR CHROMATOGRAPHY

The parameters that are changed upon scale-up are typically particle diameter, column length, and mobile phase velocity. These will change the number of transfer units, and therefore the width of the mass transfer zone as well as the time when the peak elutes. In the case of a single solute, the ratio of the width of the mass transfer zone to the length of columns for the process- and bench-scale cases follow relations that are similar to the equations given in Table 6.1 for linear chromatography. Equations (6.134) and (6.135), below, are for mass transfer ($k = 1.5$) or pore diffusion limiting ($k = 2$) conditions. These equations are based on the assumption that the plate height for nonlinear chromatography of a solute follows the empirical relation derived in Example 2 in Section 6.5. The ratios of the widths of the front of the peaks are given by

$$\boxed{\frac{L_{\mathrm{MTZ,x}}}{L_{\mathrm{MTZ,b}}} = \frac{u_{\mathrm{MTZ,x}}}{u_{\mathrm{MTZ,b}}} \cdot \left(\frac{v_{\mathrm{x}}}{v_{\mathrm{b}}}\right)^{k-1} \left(\frac{d_{\mathrm{p,x}}}{d_{\mathrm{p,b}}}\right)^{k}} \tag{6.134}$$

If L/L_{MTZ} (= ratio of length of column to the width of the mass transfer zone) is kept the same at both the process and bench scales, the ratio of the column lengths follows from the ratio of the widths of the mass transfer zone, Eq. 6.135, below, and Eq. 6.134 is equivalent to Eq. 6.11 or 6.40.

$$\boxed{\frac{L_{\mathrm{x}}}{L_{\mathrm{b}}} = \frac{L_{\mathrm{MTZ,x}}}{L_{\mathrm{MTZ,b}}}} \tag{6.135}$$

The subscript x denotes the process scale and b the bench scale. The scaling rules for two peaks with plateaus follow from Eqs. (6.132) and (6.133). When the elution time, $t_{A,\mathrm{MTZ}}$, relative to the peak width stays the same at both process and bench scales, then the length of the column will be proportional to

$$\left(\frac{t_{A,\mathrm{MTZ}}}{t_{\mathrm{w,MTZ},A}}\right)_{\mathrm{b}} = \left(\frac{t_{A,\mathrm{MTZ}}}{t_{\mathrm{w,MTZ},A}}\right)_{\mathrm{x}} \tag{6.136}$$

Combination of Eq. (6.136) with Eqs. (6.132) and (6.135), followed by rearrangement, gives

$$\frac{L_{\mathrm{b}}/u_{A,\mathrm{MTZ,b}}}{L_{A,\mathrm{MTZ,b}}/u_{A,\mathrm{MTZ,b}}} = \frac{L_{\mathrm{x}}/u_{A,\mathrm{MTZ,x}}}{L_{A,\mathrm{MTZ,x}}/u_{A,\mathrm{MTZ,x}}}$$

Simplification of the equation results in

$$\frac{L_{\mathrm{x}}}{L_{\mathrm{b}}} = \frac{\left(\dfrac{L_{A,\mathrm{MTZ,x}}}{u_{A,\mathrm{MTZ,x}}}\right)\left(\dfrac{u_{A,\mathrm{MTZ,x}}}{u_{A,\mathrm{MTZ,b}}}\right)}{\left(\dfrac{L_{A,\mathrm{MTZ,b}}}{u_{A,\mathrm{MTZ,b}}}\right)} = \frac{L_{A,\mathrm{MTZ,x}}}{L_{A,\mathrm{MTZ,b}}} \tag{6.137}$$

The scaling rules for nonlinear chromatography proposed by Wankat and Koo (1988) can be summarized as follows: The width of the elution profile consisting of one or more peaks as a fraction of column length is fixed to be the same at both bench and process scales. The contributions to the width as given in Eqs. (6.131) to (6.133) are assumed to scale in the same proportion as the change of the peak width. The width is the sum of the terms in Eqs. (6.131) to (6.133), as given in Eq. (6.130). Hence any or all of the individual terms have potential to give useful information for scale-up. The elution time, and therefore t_B given in Eq. (6.131), is a function of column length (Sherwood et al., 1975), and a mathematical model is needed to calculate it. As was the case for linear chromatography, it is not possible to estimate, a priori, the time required and mobile phase volume needed to be passed through the column for the peak to elute, unless sufficient information is given to model the profile. If this information is available, it may be more judicious to calculate the result using an appropriate model. However, this is often not the case during early stages of purification development. An approach for rapidly estimating column dimensions for nonlinear chromatography will depend on Eqs. (6.131), (6.132) and/or (6.133) and an elution profile from a bench-scale column.

The rules developed from Eqs. (6.132) and (6.133) are based on the width at the front of the peak, which is also referred to as the mass transfer zone. The ratios of the chromatographic and interstitial velocities and ratios of particle sizes at the process and bench scale are defined in Eq. (6.136). These ratios follow the same functionality as that developed for linear chromatography. The width of the mass transfer zone will scale as the square of the particle size when pore diffusion is the dominating resistance, and to the power of 1.5 for mass transfer. The lengths of the columns are assumed to be directly proportional to the ratio of the widths of the mass transfer zones for the process and bench scales as shown in Eq. (6.137). The presence of multiple solutes are assumed not to affect each other.

The scaling rules of Wankat and Koo are based on the plate height equation of Snyder rather than the more mechanistic approach of Sherwood et al. (1975) that gives an expression for the effect of mass transfer on the width of the mass transfer zone. These rules enable an estimate of the length and diameter of a process column, based on bench-scale data, to be carried out, as summarized in Table 6.3. The rules in Table 6.3 incorporate the assumption that the cross-sectional area of the column will increase in direct proportion to the sample volume, which is consistent with the recommendation of Sofer and Britten (1983). The length is assumed to increase in proportion to the width of the mass transfer zone.

A multicomponent system consisting of peaks without plateaus (as given in Figure 6.44) has different boundary conditions than peaks with plateaus such as represented in Figure 6.50. However, the rules summarized in Table 6.3 can still be used to estimate a minimum length of column, under the restrictions that pore diffusion is a controlling resistance, and the ratio of mass transfer zone to column length is the same at both bench and process scales. In the context of these restrictions, Wankat and Koo (1988) recommend that "systems with and without plateaus are essentially the same system except that a longer column is used in the latter case." As is the case for linear chromatography, and nonlinear chromatography of peaks with plateaus, the purpose of these rules should be viewed as a way of designing experiments to confirm separation characteristics that are to be achieved upon column scale-up. The necessary columns can then be set up and experiments carried out to obtain experimentally measured elution profiles. Once data are in hand, and equilibrium data are generated as described in the next section of this chapter, numerical modeling of the separation can—and should—be carried out (Gu et al., 1993). These rules should be applied carefully in the context of obtaining a first estimate of column size.

TABLE 6.3 Scaling Rules for Nonlinear Chromatography of Noninterfering Peaks

1. Specify $d_{p,b}/d_{p,x}$; $V_{spl,x}/V_{spl,b}$; v_x/v_b; L_b, t_b. Assume that $t_{w,MTZ}/t_b$ is constant.
2. Calculate chromatographic velocity of solute that elutes last

If local equilibrium,

$$u_{MTZ} = \frac{v}{1 + \phi\rho_B \frac{dq}{dC}}$$

If linear equilibrium,

$$u_{MTZ} = \frac{v}{1 + \phi\rho_B K_D}$$

Note that equilibrium data are needed to specify K_D or dq/dC over the concentration range of interest.

3. Calculate scale parameters:
   $k = 2$ for pore diffusion control
   $k = 1.5$ for mass transfer control

$$\text{Length} = L_x = L_b \frac{u_{MTZ,x}}{u_{MTZ,b}} \left(\frac{v_x}{v_b}\right)^{k-1} \left(\frac{d_{p,x}}{d_{p,b}}\right)^k$$

$$\text{Column diameter} = d_{col,x} = d_{col,b} \left(\frac{L_b}{L_x}\right)\left(\frac{V_{spl,x}}{V_{spl,b}}\right)^{1/2}$$

$$\text{Run time} = t_x = \frac{L_x}{u_x}$$

6.19 BATCH EQUILIBRIUM EXPERIMENTS ARE NEEDED FOR DETERMINING NONLINEAR EQUILIBRIA OR CONFIRMING EQUILIBRIUM CONSTANTS OBTAINED FROM COLUMN CHROMATOGRAPHY MEASUREMENTS

The determination of scale-up parameters to specify the column diameter and length, particle size, sample volume, and mobile phase flowrate as outlined in Tables 6.1 and 6.3 represent first steps. Equilibrium data are needed to further model the separation and optimize column design and operation. Equilibrium data must be independently determined so that nonlinear equilibrium expressions developed. The first step is to decide what the probable shape of the isotherm might be by using local equilibrium theory together with column chromatography data. This information can be used to plan and carry out equilibrium measurements. The equilibrium equations are then used together with numerical simulations such as the Craig or plate methods (discussed earlier in this chapter) in Sections 6.10 and 6.11 to model column chromatography processes. The characteristics of the stationary phases are reflected by both the form of the isotherm and the values of the constants. These characteristics are useful for selecting the most appropriate types of stationary phases, as discussed in Chapter 8, and for enabling numerical modeling of chromatographic separations.

Equilibrium measurements are usually first carried out for individual solutes dissolved in the buffer. This is followed by measurements of equilibria of one solute in the presence of others. Incubation of the stationary phase in a solution of the solute should result in the uptake of the solute on the stationary phase. The adsorbed solute is referred to as the adsorbate, and the amount of solute associated with the adsorbent defines the capacity of the adsorbent. The capacity is determined by measuring the solute's concentration in the solution containing the adsorbent over several time points. Once the solute in the liquid attains a constant concentration, (i.e., equilibrium has been reached), the amount of the solute associated with the solid phase is calculated by difference using a material balance based on the change in the adsorbate's concentration in the liquid phase:

$$q_{e,i} = \frac{[R_w C_{o,i} - C_{e,i}] W_{w,t}}{W_s} \tag{6.138}$$

where

$$R_w = \frac{W_w}{W_w + W_{w,s}} \tag{6.139}$$

and where

$q_{e,i}$ = amount of adsorbate (i.e., solute) asociated with the adsorbent (i.e., stationary phase) at equilibrium, g solute/g stationary phase
$C_{o,i}$ = initial concentration of the solute in the liquid added to the adsorbent g solute/(L water or buffer) for component *i*
$C_{e,i}$ = equilibrium concentration of the solute *i* when in contact with the stationary phase, g/L
W_s = dry weight of the adsorbent, g
W_w = weight of water or buffer added, g
$W_{w,t}$ = weight of water or buffer that is outside of stationary phase particles at time *t*, g
$W_{w,s}$ = weight of water or buffer associated with the adsorbent, g (water or buffer between particles, but does not include water or buffer within the pores of the adsorbent).

The subscript *i* is used in Eq. (6.138) because determination of equilibrium data for one solute ($i = 1$) may be carried out in the presence of a second solute ($i = 2$). The factor R_w accounts for dilution of the adsorbate when the water associated with the wet stationary phase dilutes the solution to which it is added as represented in Figure 6.51.

If initially dry stationary phase is used that has no external water associated with it, then initially $W_{w,s} = 0$. This is represented on the right-hand side of Figure 6.51. Later, when the water associated with the stationary phase or adsorbent is in equilibrium with the water that has been added, R_w is given by

$$R_w = \frac{W_w}{W_w - (W_{s,wet})(f)} \tag{6.140}$$

where

$W_{s,wet}$ = wet weight of adsorbent or stationary phase, g. The wet weight in this case includes both the weight of the water that is outside of the stationary phase and trapped between particles, as well as water within the stationary phase pores.
f = uptake of water by dry stationary phase expressed as fraction of its dry weight. For example, 50% water content, total weight basis, coincides with $f = 1$; a wet stationary phase with 20% water content coincides with $f = 0.25$; and 80% coincides with $f = 4$ (= 80 g water/20 g dry stationary phase).

In this case the factor R_w accounts for water uptake of dry stationary phases that rehydrate so rapidly that little, if any, solute is taken up initially. This causes the solute in the liquid phase

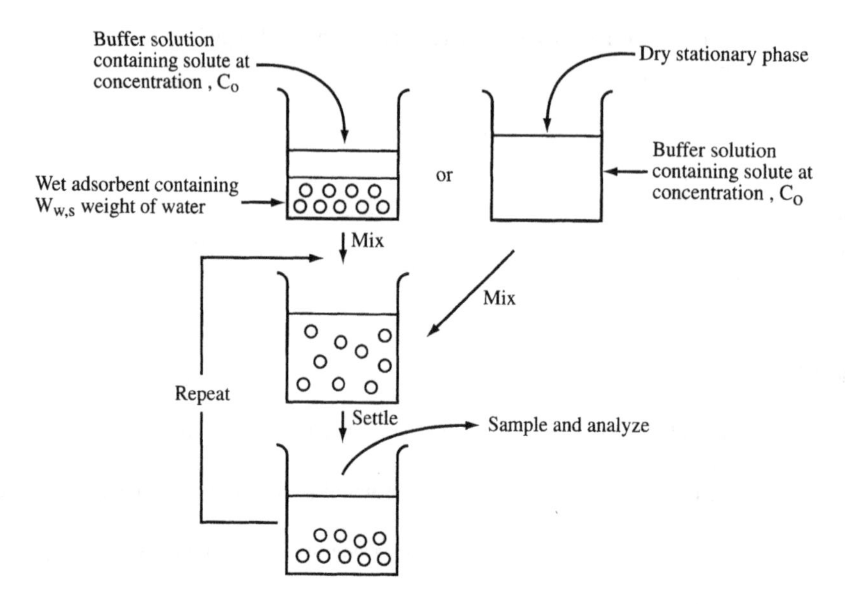

Figure 6.51 Representation of an experimental equilibrium measurement.

to increase in concentration over its initial level in the buffer added to the stationary phase. The solute concentration decreases again after this initial time, once the slower solute adsorption process takes place. Hence the concentration of solute will be measureably higher at time = 0, and this should be used to calculate the amount of solute that is taken up by the stationary phase once equilibrium is established.

The product $R_w C_{o,i}$ is the effective initial concentration of the solute or adsorbate, in the liquid phase, and $C_{e,i}$ is the measured equilibrium concentration that remains after some period of contact with the adsorbent. The value of $q_{e,i}$ is calculated by difference using Eq. (6.138), where R_w is given by either Eq. (6.139) or Eq. (6.140). The volume of solution added to the adsorbent must be large enough so that the change in volume due to withdrawal of samples is small and is typically less than 3% over the course of an experiment. If the change is more than this, the change in volume that occurs as samples are sequentially removed must be accounted for. Also the ratio of the volume of liquid solution added to the weight of adsorbent must be small enough so that the change in solute concentration due to its removal by adsorption is accurately measurable. A large volume of liquid added to a small amount of stationary phase would result in a small change in the solute concentration, because the total amount removed by the stationary phase would be small relative to the background solute concentration in the liquid.

Calculation of adsorbate loading, $q_{eq,i}$, for equilibrium concentration $C_{eq,i}$, plotted against $C_{eq,i}$, gives the equilibrium isotherm at temperature T_1 (Figure 6.52). If the adsorption is exothermic, a second set of measurements at a higher temperature, T_2, would give a reduced loading as shown by the curve for T_2 in Figure 6.52. The extent of adsorption may also be affected by pH and ionic strength of the buffer. Equilibrium loadings must therefore be reported together with their corresponding liquid phase equilibrium concentration, pH, ionic strength, and temperature to be meaningful. Measurements of solute uptake at different times will result in parabolic rate curves, in which the leveled-off part indicates that equilibrium has been obtained as illustrated on the left side of Figure 6.52.

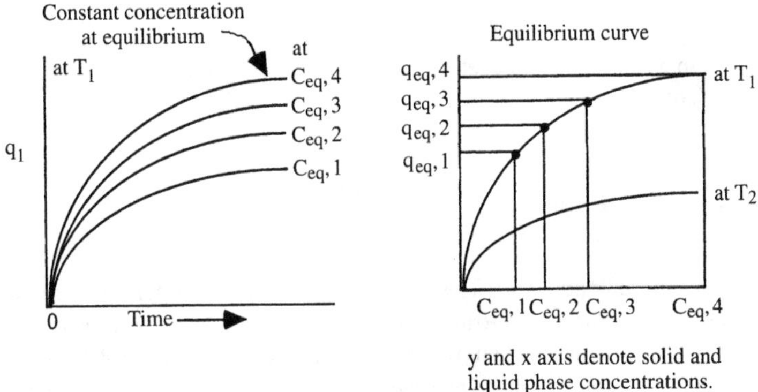

Figure 6.52 Adsorption time course (left side); non-linear equilibrium curve (right side).

The distribution of solute between the liquid and stationary phases, at equilibrium, is given by K as described by Sherwood et al. (1975):

$$K = \frac{(C_o - C_{eq})q_{eq}}{C_{eq}(q_m - q_{eq})} \tag{6.141}$$

where

- $K =$ an equilibrium parameter that reflects distribution of solute between liquid and stationary phases
- $C_o =$ initial concentration of solute in mobile phase (assumed to be a constant), g/mL liquid phase (where liquid phase is defined as the fluid outside the particles, and does not include fluid associated with or trapped inside particle).
- $C_{eq} =$ concentration of solute in mobile phase once equilibrium has been obtained, g/mL
- $q_m =$ concentration of solute on stationary phase corresponding to every site—on which the solute could adsorb-being filled, g/mL stationary phase.
- $q_{eq} =$ concentration of solute on stationary phase in equilibrium with mobile phase solute concentration, C_{eq}, expressed as g/mL stationary phase

The parameters q_{eq} and q_m are converted to a weight basis by dividing them by the density of the stationary phase. Sherwood et al. (1975) show that rearrangement of Eq. (6.141) results in Eq. (6.142) which has the same form as the Langmuir equation but a different physical interpretation since it represents a liquid, rather than a gas system. Hence, the equilibrium loading of a single solute at a specified temperature, pH and ionic strength, is given by an equation that has a form that is analogous to the Langmuir equation:

$$q_{eq} = \frac{q_m K C_{eq}}{1 + K C_{eq}} \tag{6.142}$$

At low solute concentrations, Eq. (6.142) simplifies to the linear equation that formed the basis of the discussion on principles of liquid chromatography in Chapter 5, as well as scale-up equations for linear chromatography in this chapter.

$$q_{eq} = q_m K C_{eq} = K_D C_{eq} \tag{6.143}$$

where K_D represents a distribution coefficient. Another type of nonlinear isotherm is given by the Freundlich isotherm:

$$q_{eq} = k C_{eq}^{1/n} \tag{6.144}$$

The forms of the equilibrium graphs represented by Eqs. (6.142) to (6.144) are summarized in Figure 6.53*a* to 6.53*c*. The linear isotherm is often a good approximation for many types of protein purification due to the low concentrations which are typical in the manufacture of biopharmaceutical proteins. For a strongly adsorbed species, adsorption of the solute from solution will leave essentially no solute remaining in the solution until the adsorption capacity of the adsorbent is completely used up. In this case, the isotherm equation is represented as a constant ($q_{eq} = K_D$) and the maximum capacity of the adsorbent is independent of the solute concentration (Figure 6.53*d*). This type of isotherm is representative of ion-exchange or reversed phase chromatography of proteins in which the solute is taken up by, and remains associated with, the stationary phase until the mobile phase composition is changed. For example, protein may be adsorbed onto the stationary phase from a buffer solution. Desorption of the protein would require that salt be added to the buffer so that the salt displaces the protein and causes the protein to desorb from the stationary phase. This type of separation is typical of gradient chromatography that is discussed in Chapter 7.

The adsorption of a solute from a multicomponent mixture that contains one or more additional solutes that adsorb on the stationary phase is represented as a Langmuir isotherm by

$$q_{eq,1} = \frac{q_m K_1 C_{eq,1}}{1 + \sum_{i=1}^{n} K_i C_{eq,i}} \tag{6.145}$$

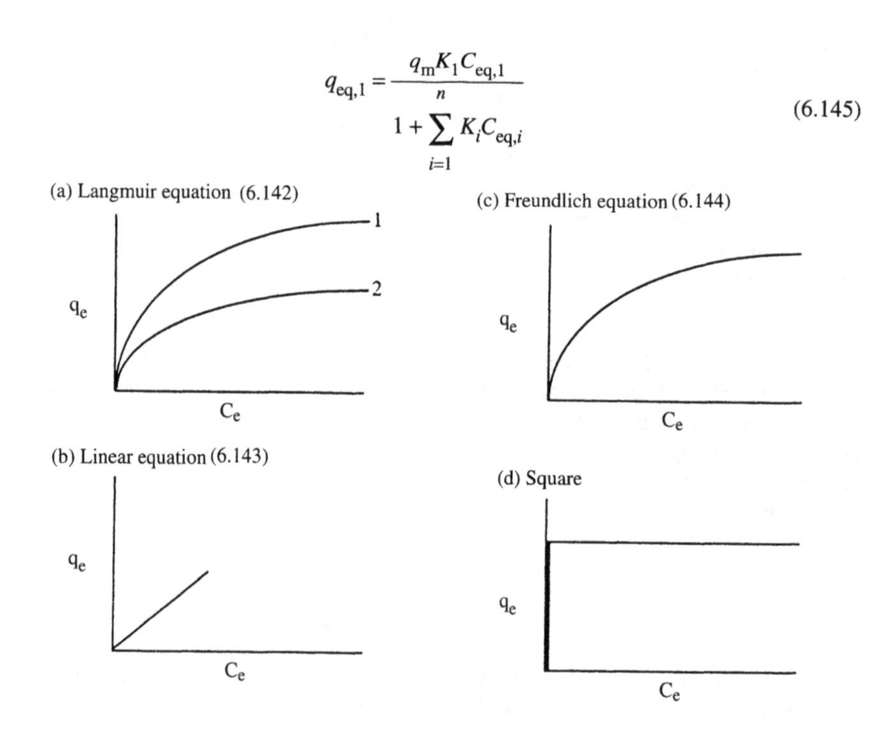

Figure 6.53 Shapes of isotherms.

for n components, where K_i represents an equilibrium adsorption constant for component i and the subscript 1 denotes component 1. For a two component mixture, the concentration of the second solute associated with the solid phase is given by

$$q_{eq,2} = \frac{q_m K_2 C_{eq,2}}{1 + K_1 C_{eq,1} + K_2 C_{eq,2}} \tag{6.146}$$

6.20 A COMPETITION FACTOR IN THE LANGMUIR EQUATION ACCOUNTS FOR CASES WHERE ADSORPTION OF ONE SOLUTE AFFECTS THE OTHER

The adsorption of two components from a binary solution may result in a competition effect where component 1 may hinder the adsorption of component 2. Hence the use of individual adsorption isotherms may not give the correct result. A competition factor is therefore introduced into the Langmuir isotherm, and results in the form given by Jageland et al. (1994):

$$q_{eq,i} = \frac{K_i C_{eq,i}}{1 + \Sigma_{j=1}^{n} \Gamma_{ij} K_j C_{eq,j} / q_{m,j}} \tag{6.147}$$

If there is a single component or if one component does not affect the other $\Gamma_{ij} = 1$. A test case for the separation of benzyl alcohol and 2-phenyl ethanol in a 50 : 50 mixture, with acetonitrile as the eluent was run by Jageland et al. (1994). The concentration of the solutes was high enough (about 400 g/L) so that the isotherm would be in a nonlinear range. Injection of this sample, and separation using a 30% aqueous acetonitrile mobile phase, gave the chromatogram in Figure 6.46. A better fit of the model was obtained when the Langmuir adsorption isotherm with the competition factor was used Figure 6.47. For component 1, the isotherm equation was

$$q_{eq,1} = \frac{K_1 C_{eq,1}}{1 + \dfrac{0.57 K_1 C_{eq,1}}{340} + \dfrac{1.32 K_2 C_{eq,2}}{310}} \tag{6.148}$$

where $q_m = 340$ mg/g and $\Gamma_{1,2} = 0.57$ (coefficient representing interference or blockage of access to the stationary phase for solute 1 by solute 2), and $q_{m,2} = 310$ mg/g and $\Gamma_{2,1} = 1.32$ (= coefficient representing blockage of access to the stationary phase for solute 2 by solute 1).

6.21 A DIFFERENCE IN EQUILIBRIUM CURVES OF TWO COMPONENTS INDICATES THAT SEPARATION IS POSSIBLE (LANGMUIR ISOTHERM)

A measurable difference between the equilibrium curves, such as shown in Figure 6.54, means that the relative uptake of the two components is different and that a partial separation would occur in a solution containing the adsorbent and both components. Removal of the liquid and transfer to another test tube containing fresh adsorbent will increase the separation of the two components, while transfer of the solid phase to fresh feed solution would increase the amount of solute associated with the adsorbent. Divergent equilibrium curves, such as illustrated in Figure 6.54, give large differences in the equilibrium concentrations of the two components and hence a large separation. A large difference in the equilibrium curves indicates that a substantial fractionation of each of the components will occur at each step, and that relatively few steps (or

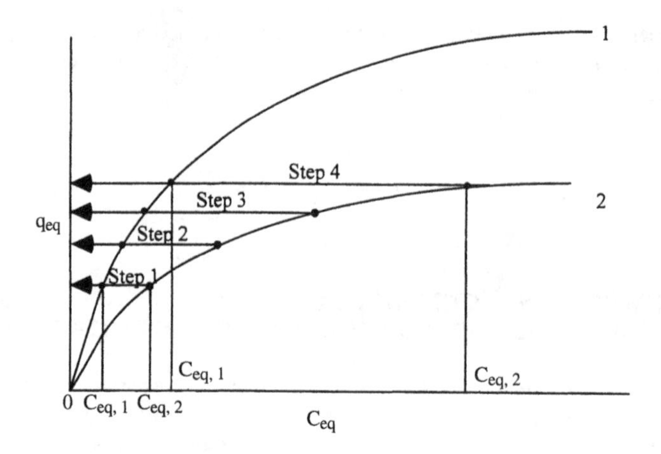

Figure 6.54 Illustration of difference in equilibria leading to separation.

stages) are needed to achieve the desired separation. When the difference in equilibrium between two or more components is small, the number of contact stages must be large; that is, a long column packed with adsorbent is needed to achieve separation.

6.22 DIFFERENCES IN RATES OF ADSORPTION MAY ENABLE A SEPARATION TO BE ACHIEVED WHEN THE EQUILIBRIUM ISOTHERMS FOR TWO COMPONENTS ARE SIMILAR

Equilibrium curves that approach each other, so that the differences in concentrations are small, do not always require a large number of equilibrium stages to achieve separation if the process is kinetically driven. An example of an industrial-scale kinetically driven separation is given in Figure 6.55 for the vapor phase removal of water from fermentation ethanol by ground corn.

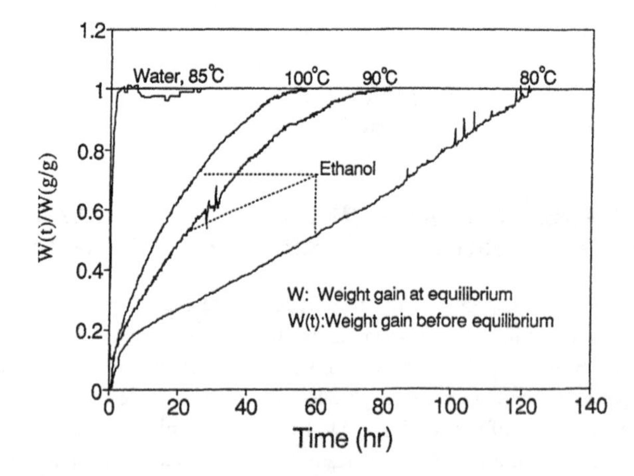

Figure 6.55 Adsorption rate curves for water and ethanol on corn that illustrate principle of kinetically driven separation (from Lee et al., 1991, fig. 6, p. 1193.) Reproduced with permission of the American Institute of Chemical Engineers, AIChE.

The rate of water adsorption by initially dry corn (i.e., fresh adsorbent) is 1000 × higher than that of the second component, ethanol (Ladisch et al., 1984b, Westgate and Ladisch, 1993). If allowed to come to equilibrium, the adsorbent will pick up similar amounts of water and alcohol. In order to achieve the desired separation, the adsorption step is stopped long before the maximum capacity of the adsorbent for ethanol is attained. This operational strategy takes advantage of the large differences in the rates of adsorption of water and ethanol to achieve the desired dryness in the alcohol product. Sometimes only a fraction of the adsorbent's capacity is used in this type of separation. In the case of corn, the adsorbent itself is sufficiently inexpensive (about 5 to 10¢/lb) to make such an approach economically practical. Hence small differences in equilibrium adsorption can be amplified through appropriate selection of operating conditions and design of the adsorption system itself. This type of separation driving force is the exception rather than the rule.

HYDRODYNAMICS

The basic configurations for analytical, preparative, and process-scale liquid chromatography (LC) systems are similar, except for the size of the equipment. The key part of a liquid chromatography system is the stationary phase and column: "The column is the heart of the LC system .... Chromatographers must be able to repeat the separation they developed and optimized yesterday for use today, tomorrow, and the day after tomorrow" (Welinder et al., 1995). However, the proper operation of the column requires that the surrounding equipment is properly configured, and that the hydraulic properties of the stationary phase are predictable and robust. The next section gives a case study of pressure drop characteristics of a gel-type stationary phase. System characteristics that may affect the column's operation are summarized here as well.

6.23 COMPRESSION OF GEL-TYPE STATIONARY PHASES IN PACKED BEDS MAY CAUSE INCREASED PRESSURE DROPS: CASE STUDY FOR STYRENE/DVB GEL-TYPE ION-EXCHANGER

Compressible media include some types of gel-based ion exchange resins used for sugar chromatography and dextran or agarose gels that are widely used for gel permeation (i.e., size exclusion) chromatography. While all are subject to deformation under pressure, gel-type ion exchange resins are more resistant to pressure than spherical dextran or agarose gel particles. A study with a 4% crosslinked styrene-divinyl benzene, sulfonated resin, used in sugar chromatography, illustrates consequences of stationary phase deformation due to loss of support from the walls as the diameter of the chromatography column becomes larger. This work was reported by Ladisch and Tsao (1978) and their development of the equation is repeated here.

Two 60 cm long columns were packed with the styrene/divinylbenzene type ion exchange resin. One column had a 6 mm inside diameter, while the other was 8 mm. The stationary phase packed in both columns had a particle size of 20 to 30 μm, was from the same batch of material, and was packed in the same manner using the slurry method described in Chapter 5. After being packed, capped with end-fittings, and run, the 6 mm column had an operational pressure drop of 130 psig at a mobile phase flowrate of 0.5 mL/min, while the 8 mm i.d. column had a pressure drop of 3600 psig at the same flowrate. Removal of the column endcaps did not show inlet void volume formation for either the 6 or 8 mm i.d. column. Initially it was assumed that once packed, the bed was incompressible. However, unpacking of the beds of both columns, and examination of the stationary phase by scanning electron microscopy, showed that some of the resin near the outlet of the 8 mm column had been compacted while the resin near the outlet of the 6 mm column appeared normal.

The bed of the 8 mm column consisted of two visible regions, a small compressed region at the column outlet and a large uncompressed one for the rest of the column. The existence of two regions suggested that the packed bed should be treated as two columns in series Figure 6.56: one column of void fraction $\varepsilon_{b,1}$ and length L_1 (= 59.5 cm) followed by another of void fraction $\varepsilon_{b,2}$ of length L_2 (= 0.5 cm). $\varepsilon_{b,1}$ and $\varepsilon_{b,2}$ represent the extraparticulate void fractions. A form of the Blake-Kozeny correlation for pressure drop across two packed beds in series is given in Eq. (6.149). This equation assumes that each of the two bed sections is uniform and incompressible once the column has been operated at steady-state conditions.

$$\Delta p = \frac{150\mu\upsilon_o}{d_p^2} \cdot \left(\frac{L_1(1-\varepsilon_{b,1})^2}{\varepsilon_{b,1}^3} + \frac{L_2(1-\varepsilon_{b,2})^2}{\varepsilon_{b,2}^3} \right) \tag{6.149}$$

where

L_1, L_2 = the lengths of the two column sections in Figure 6.56,
μ = viscosity of mobile phase at the operational temperature of the column, g/(cm · s)
d_p = average diameter of particles, cm (= $2.5 \cdot 10^{-3}$ for this example)
υ_o = superficial velocity, cm/s = $\varepsilon_b v$

The void fraction, ε_b, is defined by Eq. (5.1) in Chapter 5:

$$\varepsilon_b \equiv \frac{V_o}{V_{col}} \tag{5.1}$$

where

V_o = extraparticulate void volume, which is the volume between particles of the stationary phase, mL
V_{col} = volume of an empty column, mL

For two columns in series, an average apparent void volume, $\varepsilon_{b,app}$, for both columns can be defined as

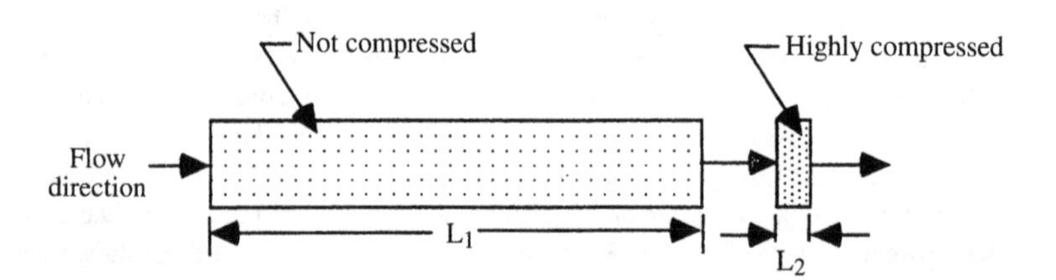

Figure 6.56 Schematic of packed column divided into compressed and non-compressed regions.

$$\varepsilon_{b,app} \equiv \frac{V_{o,1} + V_{o,2}}{V_{tot}} \tag{6.150}$$

where

$V_{o,1}, V_{o,2} =$ extraparticulate void volumes of columns 1 and 2
$V_{tot} =$ $V_{col,1} + V_{col,2}$ = the sum of volumes for empty columns 1 and 2

The volume of column 2 is a fraction, x_L, of column 1 when both columns have the same diameter:

$$V_{col,2} = x_L V_{col,1} \tag{6.151}$$

Equation (6.150) and (6.151) give

$$\varepsilon_{b,app} = \frac{V_{o,1} + V_{o,2}}{V_{col,1} + V_{col,2}} = \frac{V_{o,1} + V_{o,2}}{V_{col,1}(1 + x_L)} \tag{6.152}$$

where $x_L = L_2/L_1$ is the ratio of the lengths of the two beds of stationary phase, shown in Figure 6.56. Rearrangement of Eq. (6.152) results in

$$(1 + x_L)\varepsilon_{b,app} = \frac{V_{o,1} + V_{o,2}}{V_{col,1}} \tag{6.153}$$

The void fractions of columns 1 and 2, respectively, can be expressed individually as

$$\varepsilon_{b,1} = \frac{V_{o,1}}{V_{col,1}} \quad \text{and} \quad \varepsilon_{b,2} = \frac{V_{o,2}}{x_L V_{col,1}} \tag{6.154}$$

These may be summed to give

$$\varepsilon_{b,1} + x_L \varepsilon_{b,2} = \frac{V_{o,1} + V_{o,2}}{V_{col,1}} \tag{6.155}$$

Equating Eqs. (6.153) and (6.155) and rearranging results in

$$\varepsilon_{b,app} = \frac{\varepsilon_{b,1} + x_L \varepsilon_{b,2}}{(1 + x_L)} \tag{6.156}$$

Equation (6.156) relates the apparent void volume, $\varepsilon_{b,app}$, for a column with two distinct regions of different void fractions to the sum of individual void fractions for each region. In the case where the column is uniformly packed, $\varepsilon_{b,1} = \varepsilon_{b,2}$, Eq. (6.156) reduces to

$$\varepsilon_{b,app} = \frac{\varepsilon_{b,1}(1 + x_L)}{(1 + x_L)} = \varepsilon_{b,1} \tag{6.157}$$

For this example, $x_L = 0.5/59.5 = 0.0084$. Since $x_L << 1$, Eq. (6.156) simplifies to

$$\varepsilon_{b,app} = \varepsilon_{b,1} + x_L\, \varepsilon_{b,2} \tag{6.158}$$

A special case of Eq. (6.158) is the situation where $\varepsilon_{b,2} << \varepsilon_{b,1}$. Then for small values of x, Eq. (6.158) would simplify to

$$\varepsilon_{b,app} \cong \varepsilon_{b,1} \tag{6.159}$$

Hence the value of the apparent void fraction, ε_{app}, of two columns in series may be approximated by the void fraction of the first column, $\varepsilon_{b,1}$, if $L_1 >> L_2$ and if $\varepsilon_{b,2} << \varepsilon_{b,1}$, for $\varepsilon_{b,1}$, $\varepsilon_{b,2} < 0.5$. These criteria are consistent with the characteristics observed for the 8 mm × 60 cm column. This also provides an explanation of how the two different columns (8 mm × 60 cm vs. 6 mm × 60 cm) with apparently similar void fractions, could have significantly different pressure drops.

An example is given by the characteristics of an 8 mm i.d. column compared to a 6 mm i.d. column. Both columns were 60 cm long and were packed with a 4% crosslinked styrene-divinyl benzene ion exchange stationary phase in the Ca^{2+} form. The stationary phase was a gel-type ion exchanger that is compressed or deformed at pressures above several hundred psig. Calculation of the particle Reynolds number for the 8 mm i.d. column, Re_p, shows that the flow regime in the column is in creep flow. The pressure should be on the order of 64 psig. The pressure drop was actually 3600 psig. This was accounted for by a small compressed plug of stationary phase whose presence probably caused 98% of the observed pressure drop. This result illustrates the importance of column diameter in mechanically supporting the particles that make up the stationary phase and the flowrate limitations that must be experimentally determined in order to avoid pressure induced compression of the stationary phase (Ladisch and Tsao, 1978).

Example 4. Analysis of Pressure Drop in a Packed Bed of a Gel-Type Stationary Phase

A compressible, size exclusion packing material consists of spheres having an average diameter of 120 microns. When packed in a bed having a 15 mm i.d. and 60 cm depth, the pressure drop is 100 psig. At identical conditions, a column of 15 mm i.d. and 100 cm length gives a Δp of 260 psig. Inspection of the 100 cm column shows a compacted region near the column outlet, and a change in shape of the particles near the outlet from spherical to a shape that approximates a cylinder. Unpacking and examination of the stationary phase of the 60 cm column does not show any unusual characteristics.

The chromatograms of an excluded, nonadsorbed component for the two columns are shown in Figure 6.57. The peaks were obtained by collecting fractions and determining concentrations of a high molecular weight solute using a spectrophotometric method. The void volumes were estimated to be about 38 and 58 mL, respectively, for the 60 and 100 cm long columns, respectively.

Give an explanation of why the pressure drop is so much higher in the 100 cm long column. Be specific. Does a change in ε_b account for the increase? Is a change in the shape of the stationary phase a possible explanation or is another hypothesis more reasonable? Consider: (1) change in geometry; (2) change in ε_b, and (3) one other explanation. Of the three possibilities, which is the most likely hypothesis? (Note that $S_p/v_p = 6/d_p$ for spheres, cubes, and cylinders.)

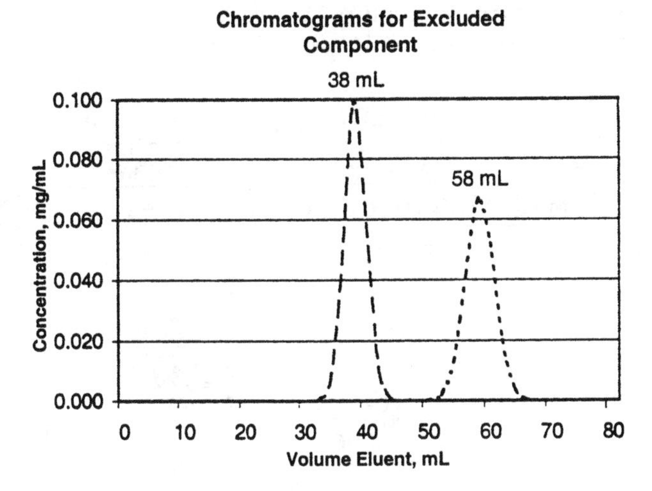

Figure 6.57 Superimposed chromatograms for example 4.

Solution

1. Characteristic dimension is unlikely to take shape of cylinder (Figure 6.58). That is, the d_p's will be the same magnitude of order

2. The second possibility is that a change in pressure drop is due to change in void fractions. The void fractions can be determined from the peaks of excluded components that are superimposed on the same chromatogram above.

The volumes of columns are

$$V_{col,60} = \left(\frac{1.5}{2}\right)^2 \pi\,(60) = 106 \text{ mL}$$

$$V_{col,100} = \left(\frac{1.5}{2}\right)^2 \pi\,(100) = 177 \text{ mL}$$

The corresponding, extraparticulate void fractions are given by substituting retention volumes given by the chromatogram shown in Figure 6.57 into Eq. (6.):

$$\varepsilon_{b,60} = \frac{38}{106} = 0.36$$

$$\varepsilon_{b,100} = \frac{58}{177} = 0.33$$

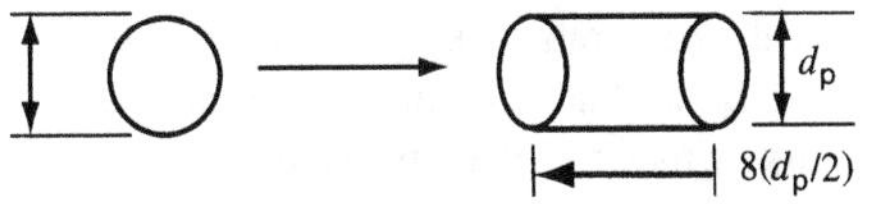

d_p

$8(d_p/2)$

Figure 6.58 Characteristic dimensions—example 4, part 1.

The void fractions differ by about 10%. The ratio of the pressure drops of the two columns can be estimated from Eq. (6.149), assuming that $L_2 = 0$. The ratio

$$\frac{\Delta p_{60}}{\Delta p_{100}} = \frac{\dfrac{(150)(60)(v_o\mu)(1-\varepsilon_{b,60})^2}{g_c d_p \varepsilon_{b,60}^3}}{\dfrac{(150)(100)(v_o\mu)(1-\varepsilon_{b,100})^2}{g_c d_p \varepsilon_{b,100}^3}} = \frac{60\dfrac{(1-\varepsilon_{b,60})^2}{\varepsilon_{b,60}^3}}{100\dfrac{(1-\varepsilon_{b,100})^2}{\varepsilon_{b,100}^3}}$$

$$\frac{\Delta p_{100}}{\Delta p_{60}} = \frac{100\dfrac{(1-0.33)^2}{(0.33)^3}}{60\dfrac{(1-0.36)^2}{(0.36)^3}} = \frac{1249}{527} = 2.4\times$$

3. Column plugs at bottom, at outlet distributor plate, or compacts in a way that gives reduced flow. Since the change in pressure drop (from 100 to 260 psig) is similar to the factor of 2.4 calculated from Eq. (5.71), solution 2 above is the most likely hypothesis.

The maximum flowrate is fixed by the compressibility characteristics of the stationary phase. However, there is also a minimum column diameter that must be specified relative to the average particle size of the stationary phase, in order to minimize channeling along the walls of the column.

6.24 COLUMN TO PARTICLE DIAMETER SHOULD EXCEED 80 IN ORDER TO MINIMIZE DISPERSION BY FINGERING

The axial dispersion and mixing due to void volumes and cells is enhanced when the number of particles per cross section of column is less than 40 (Carberry, 1958; Cohen and Metzner, 1981; Schwartz and Smith, 1953). Studies carried out for gases are discussed here since similar studies for liquid chromatography columns are not known to this author. The studies with gases provide insights of the types of channeling that may occur in any type of packed bed. The flow distribution of gases in 2 and 4 inch pipes packed with cylinders and spheres that ranged from 3/32 to 1/2 inch in diameter showed that the divergence of the flow from a uniform profile was less than 20% when the ratio of diameters of pipe to pellet exceeded 30 (Schwartz and Smith, 1953). However, flow near the wall could be double the velocity of the air at the center of the tube. The radial gradient in flow velocities for spherical particles is due to a gradient in the void fraction of the particles as a function of radial distance from the wall. The void fraction at the wall is close to 1, and then decreases to the average value of the bed at about 40 particle diameters away from the wall (Cohen and Metzner, 1981). Based on this observation, a suggested minimum column to particle diameter ratio is 80 (= 2×40 particles). In liquid chromatography the higher void fraction near the wall allows sample fingering, with sample movement in the axial direction being more likely to occur along the walls of the column than in the bulk phase of the column. This is another form of dispersion that decreases column efficiency since the sample no longer resembles a plug (Ladisch, 1987).

Example 5. Column to Particle Diameter Ratios

Assume that chromatography columns are to be packed with spherical stationary phase particles that range from 2 to 400 microns with average sizes of 2, 5, 10, 20, 40, 100, 200, or 300 microns. Standard column sizes used with the smallest three particle sizes have inside diameters of 2 mm or 4 mm, while the 20 micron particles are packed in 4, 6, or 8 mm columns. Particles above this size range will be packed in columns that have 10, 15, 25, or 50 mm inside diameters. Construct a plot that serves as a simple guide to determine whether the selected column diameters coincide with minimum recommended column diameter for a given particle size for a spherical stationary phase.

Solution

The plot of column diameter to column/particle ratio (Figure 6.59) is simple, but it serves as a reminder of the care that should be taken in generating data at the bench scale for purposes of scale-up. The larger particles require column diameters that begin to approach pilot scale in order to minimize possible fingering/channeling. A 4 mm i.d. column would be sufficient for testing the majority of liquid chromatography stationary phases used for analytical types of separations, since the particle diameters of the stationary phases will usually be 20 microns or less. However, materials in this size range are likely to be subject to mass transfer limitations as the dominant resistance. Use of data from such columns for scale-up should apply the appropriate scaling rules for mass transfer limiting cases. Figure 6.60 gives the plot for particles that are large enough that pore diffusion is likely to be the dominating resistance. An 8 mm i.d. column should be sufficient for testing candidate stationary phases for process liquid chromatography, if the diameters of these phases fall in the range of 50 to 100 microns. Figure 6.60 emphasizes that for process stationary phases, bench-scale profiles for specifying process chromatography columns should probably be obtained using columns with 8 mm i.d. or larger inside diameter based on practical considerations.

Once an appropriate diameter is selected, and the packing of the column is properly carried out to give reproducible plate counts, the column is ready for further evaluation. If the stationary phase is able to achieve the desired separation, it must then satisfy the criteria of being stable

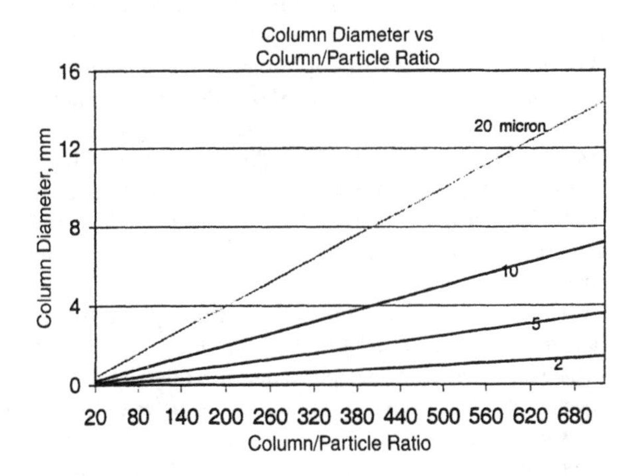

Figure 6.59 Column/particle ratio for 2 to 20 micron particles—example 5.

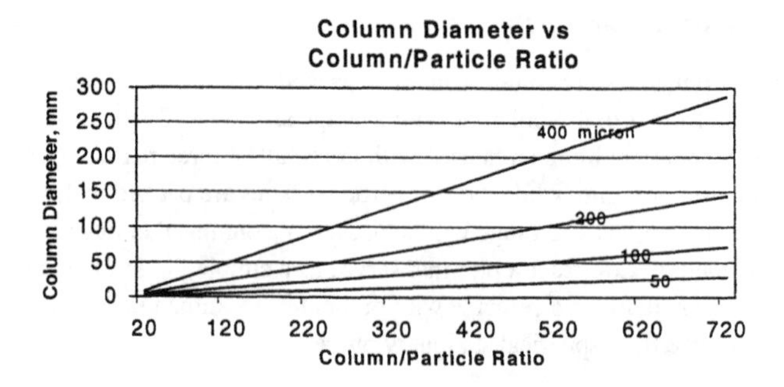

Figure 6.60 Column/particle ratio for 50 to 400 micron particles—example 5—Particle sizes of 50 to 400 micron.

and capable of withstanding reagents and conditions that may be required for process hygiene. The separation achievable with a column is also a function of being able to maintain plug flow as the sample passes from its injection point onto the column and travels from the column to the point where the separated peaks are collected. Sources of extra column band-broadening should be minimized. These process considerations are discussed at this point, since their impact on the separation, and modeling of column performance, is easy to overlook. If recognized, these factors can be controlled so that their effect on column performance is minimal.

6.25 MIXING AND DEAD VOLUMES MUST BE MINIMIZED IN LIQUID CHROMATOGRAPHY SYSTEMS: FITTINGS, INJECTORS, TUBING, AND FEED DISTRIBUTORS

Special care is exercised to minimize mixing and dilution of the sample as it moves from the injection valve to the column. This is accomplished by using low dead volume chromatography tubing, together with special fittings and unions that minimize dead zones or volumes where mixing might occur as discussed in Chapter 5. A low or zero dead-volume fitting, combined with low dead-volume tubing, minimizes band-spreading so that dilution of a pulse of sample volume is minimized. The diagram in Figure 6.61 shows the ideal situation where a pulse of sample remains a pulse, until it enters the column. After the sample passes into the column, unavoidable dispersion processes cause peak spreading, loss of the square wave character of the pulse, and a decreased maximum peak concentration. The peaks in Figure 6.61 illustrate these concepts. The variable x denotes distance between injector and column, z denotes distance from the column inlet to outlet, and t denotes the elapsed time since injection of the sample volume.

A dead volume between injector and column inlet effectively dilutes the sample before it enters the column and translates into reduced resolution as the components in the band begin to separate. This is analogous to injecting a large volume of a diluted form of the sample to the column. The consequence of using a high dead-volume tubing between injector and inlet of the column is mixing between separated peaks and decreased average concentration. A low dead-volume tubing/fitting between the injector and column inlet will maintain a sharp plug as represented in Figure 6.61 while a larger diameter tubing will cause unnecessary dispersion as

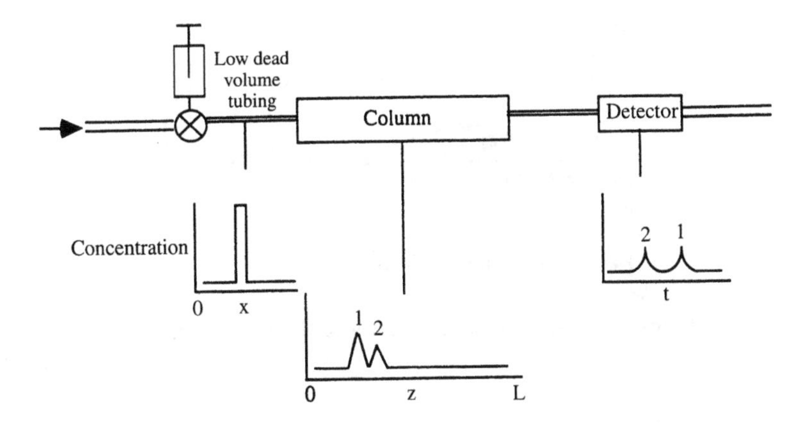

Figure 6.61 Dead volumes in a liquid chromatography system.

shown in Figure 6.62. Similarly a dead volume between column outlet and detector also causes dispersion and loss of resolution. Since dispersion within the chromatography column is unavoidable in most cases, the goal is to minimize the contribution of extra column band-spreading (Kirkland et al., 1977; Lochmüller and Sumner, 1980). The relative size of connecting tubings remain small compared to large diameter columns used in process chromatography. For example, a 15 cm i.d. chromatography column in our scale-up facility has an outlet tubing of 4 mm i.d. × 6.35 mm o.d.

Preservation of a tight band also requires that the feed be evenly distributed across the top of the column, and that zone spreading be minimized at the column outlet by making flow distribution as even as possible over the column outlet. An even distribution of flow

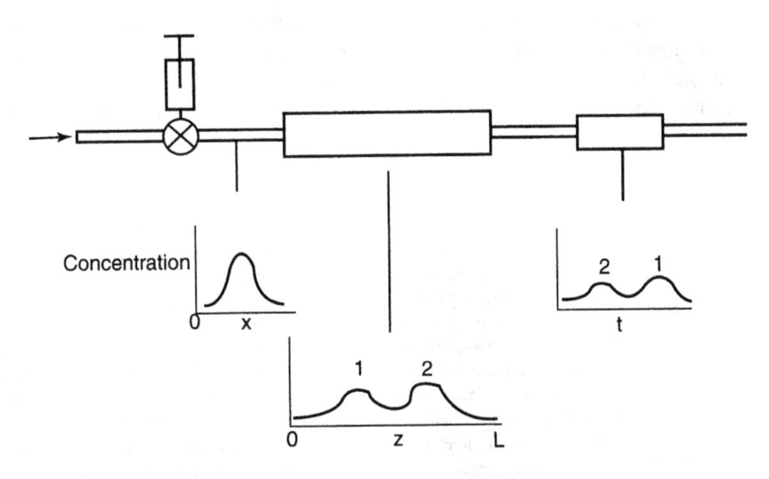

Figure 6.62 Effect of large dead volumes on peak shape.

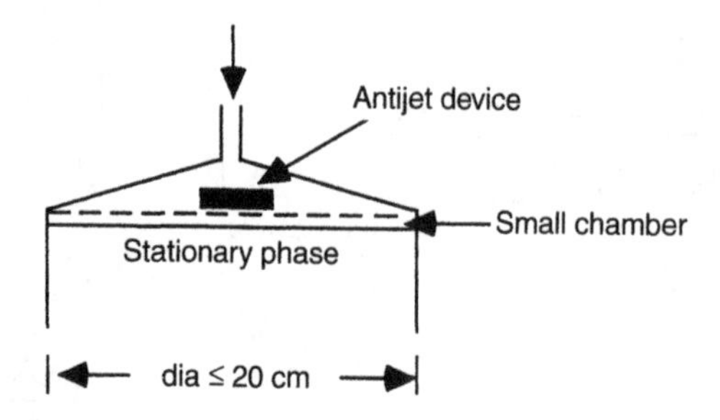

Figure 6.63 Representation of feed distributor with anti-jet device.

is achieved by designs in which the radial pressure drop in the distributor is negligible compared to the axial pressure drop. Sofer and Nystrom describe an appropriate distribution design for columns with a diameter of 20 cm or less. It consists of an anti-jet device to prevent the incoming fluid flow from disturbing the packed chromatographic media, and a small chamber to create a low radial pressure drop (Sofer and Nystrom, 1989). These concepts are illustrated in Figure 6.63.

Inlet distributors and outlet screens in columns that have diameters of more than 20 cm have a more complex design. For example, a multiple flow inlet distributor, shown in Figure 6.64, is used with process-scale columns for the chromatographic purification of insulin (Sofer and Nystrom, 1989). The distributor consists of a circular anti-jet plate and screens that act as spacers. This type of distributor is applicable to low pressure liquid chromatography carried out at pressures below 36 psig.[7] An alternate design for other applications uses flow distribution channels that extend radially from the center of the distribution plate (Figure 5.23) in Chapter 5 (Sepracor, 1992). Chromatography columns operated at higher pressures will utilize porous metal frits (i.e., filters) with pore size cutoffs of 10 to 20 microns, although these have the disadvantage that they may plug after long periods of use, unless the feed is subject to microfiltration prior to injection into the column.

Another type of distributor is based on a fractal design that is reported to avoid turbulence while forming a homogeneous surface (i.e., pulse of fluid that is evenly distributed across the cross-sectional area of the column) as shown in Figure 6.65. The characteristics of this distributor are amenable to both scale-up and scale-down since bifurcations (i.e., the patterns in Figure 6.65) are readily added or subtracted while maintaining symmetry. Fractal surface distributors

[7]Liquid chromatography in large glass columns must usually be carried out at pressures that range from 5 to 50 psig due to pressure limitations of the glass, seals, and materials of construction. Some smaller-diameter glass columns (15 to 25 mm i.d.) are capable of withstanding up to 700 psig pressure, although these columns are principally used for analytical or bench-scale research. The distinction between low-pressure chromatography (sometimes also referred to as open column chromatography when carried out at close to atmospheric pressure) and high-pressure liquid chromatography is not exact. High-pressure liquid chromatography usually uses small particle size, rigid, stationary phases, where the stationary phases will have average particle sizes of 5, 10, 20, 40, 75, or 100 microns, and cause pressure drops of 400 to 1000 psig/m, when packed in chromatography columns. Low-pressure liquid chromatography is usually associated with stationary phases that have a larger particle size (above 200 microns) and/or a gel or soft structure that is easily compressed. Pressures in these systems are typically 50 psig or less. Chromotography columns that are operated at pressures ranging from about 50 to 200 psig are sometimes referred to as moderate pressure systems.

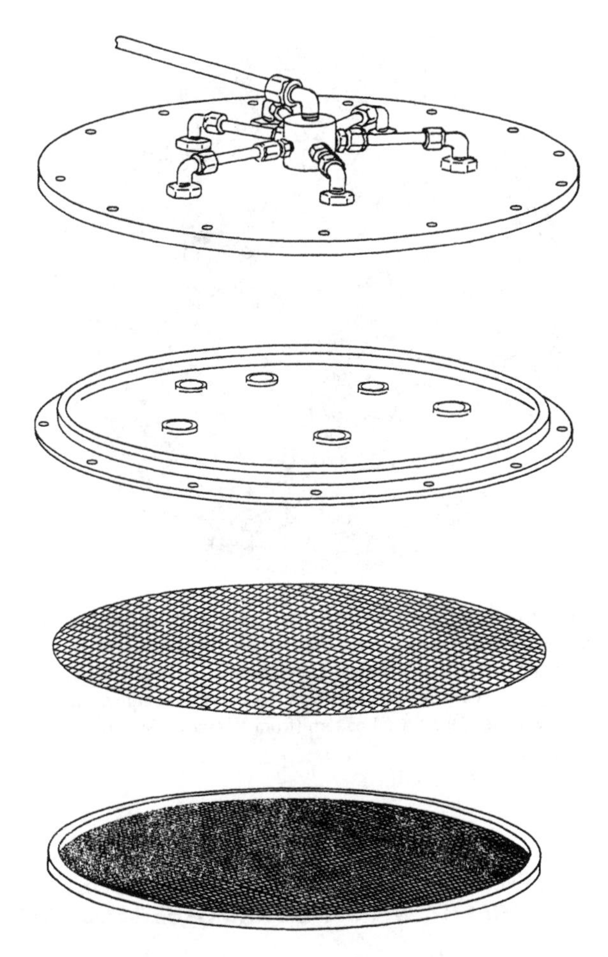

Figure 6.64 Schematic diagram of column distribution plate used in columns for low-pressure chromatographic purification of insulin (from Sofer and Nystrom, 1989, fig. 31, p. 79, with permission from Academic Press).

ranging from 15 cm to 6.7 m (used for industrial simulated moving bed chromatography) have been constructed (Kearny, 1997).

Laboratory and preparative scale columns, as well as some types of process columns are designed with adjustable plungers. These plungers are designed to enable the distributor plate to be moved to the top of the packed bed, thereby eliminating the void volume that forms when the bed settles during packing. An example of an adjustable plunger in a process scale column is given in Figure 5.23 in Chapter 5, where the plunger is positioned on top of the bed so that there are no voids. This type of column will typically only be packed to about half of its height, since the volume above the bed is needed during packing or unpacking to leave room for adding slurry, or expanding the bed, respectively. This head volume may sometimes serve as an in situ packer bulb in a manner that is analogous to the separate bulb that is shown in Figure 5.22 in Chapter 5 for packing analytical or preparative-scale chromatography columns.

Figure 6.65 Fractal surface distributor with 8064 pathways. Different sized conduits are on different planes and do not intersect. Fluid passing through the cascade exits at low turbulence and approximates a surface. The curved conduit is used to aid arranging the pattern on a circular surface while maintaining a close approximation of symmetry for all 8064 pathways. The number of equivalent paths is increased by a factor of four for each iteration of the fractal pattern (from Kearney, 1997, fig. 2, with permission from Amalgamated Research, Inc.).

6.26 THE MOBILE PHASE IS A MAJOR OPERATIONAL COST FOR PROCESS LIQUID CHROMATOGRAPHY: WFI WATER AND OTHER SOLVENTS

A major contributor to the high cost of industrial scale liquid chromatography is the cost of the mobile phase. The amounts of water and buffer components required for chromatography are large, and the water for purification of pharmaceutical molecules must often meet water-for-injection standards (abbreviated WFI). WFI water must be sterile, de-ionized, and pyrogen-free, with bacteria levels of less than 10 colony forming units (abbreviated cfu) per mL (Collentro, 1995). Organic reagents used to make up the mobile phases must also be of high purity, and this further adds to the expense. The contribution of solvent cost is up to 80% of the total cost of operating a chromatography system when acetonitrile is the mobile phase (Jageland et al., 1994)[8] or as little as 10% in the case of size exclusion chroma-

[8]This estimate is based on use of C18 silica gel as the stationary phase with an assumed cost of $487/kg, and a mobile phase cost of $1/L for 30% aqueous acetonitrile. The column packing material has an average particle size of 16 microns and a lifetime of 2000 hours. The chromatography system is utilized for 2000 h/yr, with 5 years of depreciation. The cycle time (i.e., the time required to complete one run) is defined as 1.2 times the retention time of the last component, to elute, or about 7 minutes per run for a flowrate of 860 L/hr. Costs are normalized to an hourly basis in this analysis. The basis is $1.70/(kg stationary phase • hr); system (equipment) cost of $65/hr; labor cost of $140/h; and a yield loss cost equivalent to $2/g of starting product (i.e., crude product) (Jageland et al., 1994). Use of simulated moving bed systems can reduce the cost by about one-third (Colin et al., 1995).

tography when the mobile phase is an aqueous buffer (Reisman, 1988).[9] In either case the cost of the mobile phase is significant.

6.27 SPECIAL OPERATING PROTOCOLS ARE REQUIRED FOR STORING THE MOBILE PHASE UNTIL IT IS USED

Recently published case studies show that filtration of the water through 0.1 micron filters combined with daily sanitization, eliminated biofilm development. Most water samples gave less than 1 cfu per 100 mL, compared to 10^4 to $5 \cdot 10^5$ cfu/100 mL in the starting water (Collentro, 1999). Stored water is continuously circulated through a 0.22 micron filter, and reintroduced to the storage tank in order to avoid stagnant water, according to one recommendation (Curling and Cooney, 1982). All buffers and cleaning solutions should be filtered through 0.45 micron, or preferably 0.22 micron filters. The feed solution that contains the components to be purified should also be filtered (Curling and Cooney, 1982; Sofer and Nystrom, 1989). The water tank itself should be periodically sanitized by circulating the water, heating it to 90°C and holding it for 1 hour at 90°C (Curling and Cooney, 1982). Stock solutions for making up aqueous buffers can be kept in forms that are resistant to bacterial contamination by maintaining the solutions at high concentration and extreme pH's. Examples of such stock solutions are glacial acetic acid, 20% sodium hydroxide and 40% sodium acetate.

6.28 PROCESS HYGIENE AFFECTS CHOICE OF MATERIALS OF CONSTRUCTION FOR COLUMN COMPONENTS (USE OF NaOH FOR CLEANING-IN-PLACE)

Components of both large- and small-scale chromatography systems that contact the feed stream can be cleaned-in-place using 250 mM NaOH (1% w/vol). NaOH at 0.1 M concentration is sufficient to kill vegative bacteria and yeast, while 0.5 to 1.0 M NaOH is needed for fungal contaminants. The times required to destroy 99% of the *B. subtilis* spores are 2 hours at 20°C in 1.0 M NaOH, 22 hours at 4°C, and 8 hours and 76 hours at 20° and 4°C, respectively, for 0.5 M NaOH (Figure 6.13) (Sofer and Nystrom, 1989). Sodium hydroxide is preferred over other possible disinfectants since it has a solubilizing effect on proteins that might otherwise serve as substrates for bacterial growth, as well as being capable of inactivating bacteria and destroying pyrogens. According to Sofer and Nystrom (1989), the cleaning of a packed column after a run for purifying recombinant protein may be achieved by a wash step using sterile filtered (through 0.22 or 0.45 micron membrane) 2 M NaCl to remove weakly bound proteins followed by contact with 0.5 M NaOH for time periods ranging from 30 minutes to 24 hours in order to solubilize irreversibly precipitated protein and adsorbed lipids. Filtered 0.5 M NaOH will also kill

[9] The purification of 1 unit of human albumin from plasma requires about 40 L of water. The water used to make up the mobile phase for size exclusion chromatography for pharmaceuticals must be pyrogen free, contain a low concentration of dissolved gases, and be available in large volumes at close to room temperature(15 to 20°C). The quality of water used for processing bioproducts that are intended for human use must satisfy standards for "water for injection" in the U.S. pharmacopoiae (USP). These and other applications require that the water be processed by ultrafiltration, reverse osmosis, and/or distillation before use as a chromatography mobile phase. The cost is about $0.002/L for water processed by reverse osmosis and $0.01 L for distilled water (1982 prices) (Curling and Cooney, 1982). Additional chemical costs associated with the operation of a gel permeation column bring the cost to about $0.40 out of a total of $4.42 per unit of albumin (where 1 unit = 100 mL of product at 25% weight/vol). Contributing components to the overall cost are chemicals ($0.35/unit of albumin), filters (1.55), membranes (0.15), electricity (0.02), process water (0.05), and stationary phase ($2.30) (Reisman, 1988).

TABLE 6.4 Example of Recommended Sanitization Protocol for a Dextran Gel Stationary Phase

Step	Column Volumes	Solution	Contact Time
Gram-negative bacteria, gram-positive bacteria (nonsporulating), and yeast			
1	2	Sterile-filtered 2 M NaCl	30 min
2	2	Filtered 0.5 M NaOH	30 min
3	2	Sterile-filtered H_2O^a	
4	2	Filtered 0.01 M $NaOH^b$	Storage
Molds and spores			
1	2	Sterile-filtered 2 M NaCl	30 min
2	2	Filtered 1 M NaOH	24 hr
3	2	Sterile-filtered H_2O	
4	2	Filtered 1 M NaOH	5 hr
5	2	Sterile-filtered H_2O^a	
6	2	Filtered 0.01 M $NaOH^b$	Storage

Source: From Sofer and Nystrom (1989), table 24, p. 103. The material is reproduced with permission of Academic Press.
[a]Column ready to use after equilibration.
[b]Solution for storage of column and stationary phase between runs.

gram-negative bacteria, nonsporulating gram-positive bacteria, and yeast, while 1 M NaOH is needed to kill molds and spores. NaOH acts more rapidly against gram-negative bacteria than gram-positive strains. The combination of aqueous NaOH treatment of chromatography system components and stationary phase, together with maintaining water for injection standards, should allow the system to be sufficiently free from microbial contamination so that it does not need to be autoclaved, according to Curling and Cooney (1982). Examples of recommended sanitization protocals are given in Table 6.4.

The materials of construction must be resistant to the reagents used for process hygiene, as well as to mobile phases used during the column's operation. If NaOH is to be used to clean the chromatography column and the stationary phase contained in it, both the materials of construction and the stationary phase must be resistant to corrosion by or be chemically stable against aqueous NaOH (Curling and Cooney, 1982; Sofer and Nystrom, 1989). Glass is a popular choice for this reason, but its use on a process scale is limited to pressures that are below about 6 bar for 9 cm diameter columns and 1.5 bar for 45 cm columns (1 bar = 14.5 psig). Metals or polymers that are resistant to the reagents are therefore used for high pressure applications. These may include stainless steels, titanium, platinum, teflon, and PEEK.[10] The separation media must also be resistant to periodic exposure to NaOH during clean-in-place procedures. Pyrogens bind to anion exchangers, and hence are removed from the injected sample as the sample moves through the column. Regeneration of the stationary phase must therefore be capable of removing pyrogens. Sodium hydroxide is preferred since it destroys pyrogens.[11]

Procedures for evaluating operational stability of stationary phase performance after it has been reused several hundred times has been outlined by Seely et al. (1994). They recommend that the physical and chemical states of the resin are assessed based on small-ion capacity (for

[10]The chemical reistances of materials used in process chromatography are summarized by Sofer and Nystrom (1989, Fig. 29, p. 75). Materials listed are PTFE, ETFE, CTFE, PFE, PVDF, PEEK, PI, PP, EPDM, titanium, platinum, and glass (borosilicate) with respect to over 35 solvents and aqueous solutions.

[11]Endotoxins (i.e., pyrogens) are detected in subnanogram amounts by the *Limulus* amoegocyte lysate (LAL) test, or in nanogram amounts when injected into a rabbit. Rabbits develop a fever if pyrogens are present.

ion-exchangers), pressure drops, changes in particle-size distribution, total organic carbon removed by high molarities of NaOH and H_2SO_4, the presence of microorganisms and endotoxins, and generation of leachable materials (Seely et al., 1994).

Some types of stationary phases will have backbones that are labile upon exposure to alkaline pH conditions or will consist of ligands (e.g., proteins or peptides used for affinity chromatography) that would be destroyed by contact with NaOH. If the stationary phase can not tolerate exposure to NaOH, bacteriostatic agents such as 1% formaldehyde or 70% ethanol must be used for maintaining column hygiene (Curling and Cooney, 1982). Other disinfectants include phenol, and benzyl alcohol. Treatment of a stationary phase with these disinfectants must be followed by extensive washing with sterile buffer. An excellent discussion of process considerations for chromatography column hygiene and methods suitable for addressing microbial challenge is given by Sofer and Nystrom (1989).

PRODUCTIVITY AND COSTS

The goal of a purification step is to increase purity of a given target molecule. Once the stationary phase, mobile phase, and operational conditions are initially selected to achieve this goal, the objective then becomes optimization of the chromtography system to maximize productivity per dollar spent (Jageland et al., 1994; Ladisch et al., 1984a; Nicaud and Colin, 1990). The amount of purified product produced per monetary unit, $P_{g/\$}$ (in g/$), is given by Jageland and Bryntesson (1994) as

$$P_{g/\$} = \frac{\text{throughput}}{C_{MP} + C_{SP} + C_{feed} + C_{RP} + C_{Cap} + C_{labor}} = \frac{\text{throughput}}{C_{tot}} \quad (6.160)$$

where

throughput = product recovered at a specified composition per hour, g/hr
C_{MP} = cost of mobile phase, $/hr
C_{SP} = cost of stationary phase, $/hr
C_{feed} = cost of feed or starting sample from which product is to be purified, $/hr
C_{RP} = cost of regeneration, $/hr
C_{Cap} = installed cost of equipment, $/hr
C_{labor} = cost of labor, $/hr
C_{tot} = total hourly cost, $/hr

The hourly cost of mobile phase depends on flowrate and throughput, where throughput is defined by the amount of purified product obtained per hour. The cost of the stationary phase is a function of its purchase cost and its longevity under operational conditions. The cost of the feed reflects inputs that are required to generate the target molecule through fermentation, cell culture, or chemical synthesis, prior to the purification of the molecule, as well as the cost of recovery or purification steps that precede the chromatography step in question. Equipment cost corresponds to its installed cost amortized over its lifetime, which will typically range from 3 to 10 years.

Jageland and Bryntesson (1994) present a hypothetical analysis for a low molecular weight product costing $1/g in an unpurified feed mixture; with a mobile phase cost of $0.50/g of product; a stationary phase cost of $1.80/g; a system cost of $150,000/yr; and labor cost of $80,000/yr per person. The molecule and purification system is not specified. The result is shown in Figures 6.66 and 6.67. Two key observations are presented in this figure. The first is that solvent cost is a major fraction of the overall purification cost. The fractional contribution

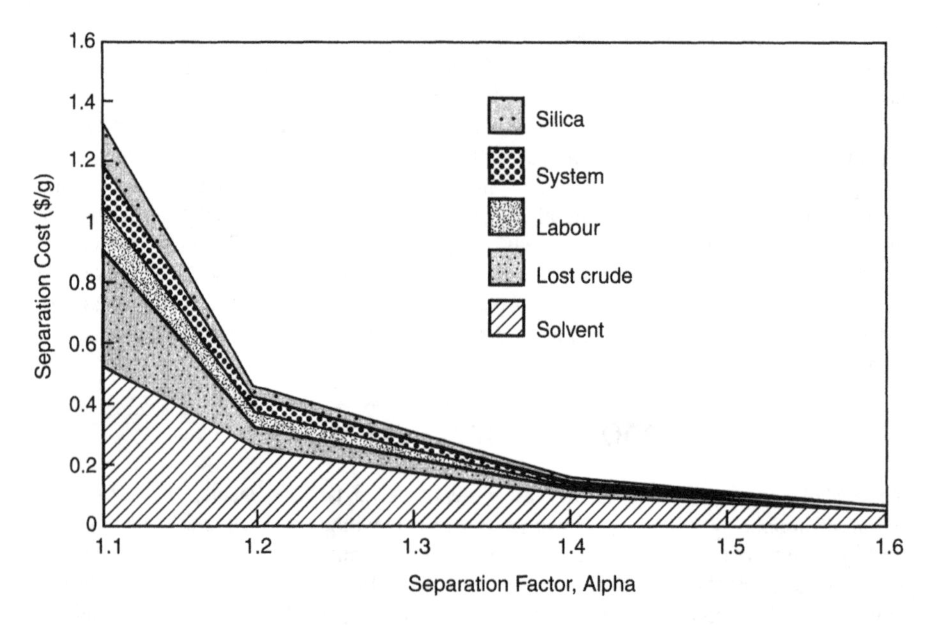

Figure 6.66 Effect of separation factor ($\alpha = k_2'/k_1'$) on cost. Conditions: 90 : 10 (mass ratio of product to impurity) feed purified to 99.5% in a 30 cm i.d. column on stationary phase with average particle size of 13 micron. Feed cost at $1/g, solvent cost of $0.50/L stationary phase (packing material) cost at $1.80/kg. (Reproduced from Jageland and Bryntesson, 1994, fig. 10, p. 11.)

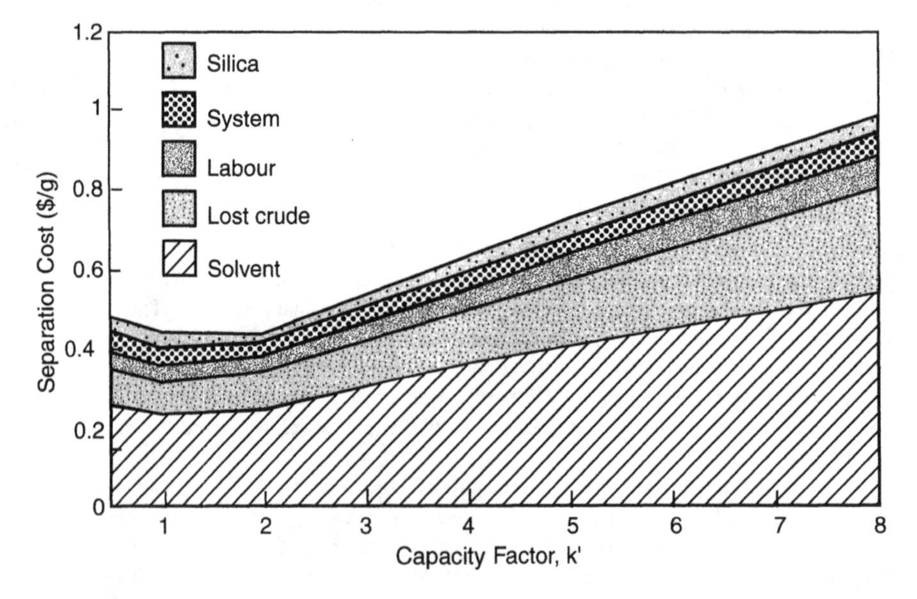

Figure 6.67 Effect of capacity factor on separation cost. Conditions: Same as in Figure 6. (Reproduced from Jageland and Bryntesson, 1994, fig. 11, p. 11.)

of solvent cost increases with increasing separation factor while the overall separation cost decreases with increasing separation factor. The second observation is that the optimum capacity factor is between 1 and 2 (Figure 6.67). Solvent cost increases significantly with increasing retention of the target molecule once the capacity factor exceeds 2, because the amount of mobile phase required to elute a product increases. A longer retention time for this peak relative to the others does not change its composition, but does require more mobile phase to wash it from the column. While the exact costs and optima will vary according to the type of stationary phase, target molecules and mobile phase, the general trends that are succinctly presented in Figure 6.67 give a first indication of the range of desired separation characteristics for process chromatography.

6.29 THE STATIONARY PHASE IS THE SINGLE MOST IMPORTANT FACTOR FOR PURIFICATION DEVELOPMENT

The stationary phase determines the range of separation and capacity factors that can be achieved (DePalma, 1996). While mobile phase compositions can be modified in order to adjust the separation characteristics during purification development, the stationary phase, once packed into the column, is fixed. The choice of stationary phase must match the type of molecule being separated so that the desired purity is achieved. The stationary phase must also be stable with respect to the chemicals and regeneration protocols associated with the purification procedures, as discussed in the previous section of this chapter. At the same time the retention characteristics of the stationary phase must be tuned so that mobile phase use is minimized.

The change in cost as a function of particle size in Figure 6.4 shows smaller particles are more expensive. Smaller particles exhibit higher efficiencies (correspond to smaller plate heights) than larger particles and less may be needed than a larger size material (Figure 6.5). The plate height is small and insensitive to linear velocity for the smallest material, and increases as particle size increases.

The value of the smaller particle size material is succinctly analyzed by MacLennan and Murphy (1994) from the perspective of scaling up an analytical separation to obtain larger amounts of purified material. Once a separation has been developed on an analytical scale, the goal is usually to increase the load so that more solute can be processed. "However, as load is increased, resolution is usually compromised . . . . Resolution can be increased by either decreasing the particle size, increasing the length of the column, or increasing the retention of the component of interest. In preparative chromatography increasing the retention of the component of interest significantly increases the volume of mobile phase consumed and results in the elution of a more dilute fraction, which makes isolation of the compound of interest more difficult" (MacLennan and Murphy, 1994, p. 29). From this perspective, a smaller particle size may be better, if it is available and if the amount of purified material required for the large scale can be readily obtained from a column having smaller dimensions. The cost of the stationary phase alone is insufficient to determine the cost of a preparative scale separation. The trade-off between cost of the mobile phase, volume of solute to be purified, availability of stationary phase, and regulatory requirements must all be considered. Examples of the impact of regulatory considerations are given in Chapter 8. The other factors are discussed here.

Quantitation of the characteristics of stationary phases under process conditions require that the separation be developed, and that tests be carried out under conditions that simulate process parameters. Accelerated tests can be devised such as continuously flowing the feed over the stationary phase to test for fouling effects, or incubating the stationary phase in a regenerating solution of 0.5 to 1 M NaOH at an elevated temperature for a month to test for possible alkaline degradation of the stationary phase. However, these tests require that the stationary phase and other conditions for a given separation already be selected. The need to estimate productivity and costs exists from the beginning of the development process. It is in this context that the concept of figures of merit become useful, if not essential, in guiding choices early on in the development process. The term "figures of merit" is credited to Dr. Jim Peterson of the National Renewable Energy Laboratory (NREL), and it refers to the calculation of best-case scenarios for the purpose of assessing their potential impact on transforming a new technology or process into an economically attractive venture. This approach could be used to organize calculations in a manner that enables ranges of costs to be assigned to best-case (or worst-case) scenarios for chromatographic separations. The equations and approach we give below are from previously published papers by Ladisch et al. (1984, 1987). These costs can then be added together and used to obtain a value of $P_{g/\$}$ from Eq. (6.160). Hence system productivity can be compared to cost ranges and prior operating experience in the industry. Practical choices can then be made based on the value of the purified final product, and the anticipated performance and costs of the scaled-up process that might be used to purify it.

6.30 YIELD REPRESENTS PRODUCT RECOVERY REGARDLESS OF ITS EXTENT OF PURIFICATION

The yield of component i after it has been through a chromatography column is represented by Eq. (6.161). While enrichment of the starting sample with respect to the product or target molecule is the goal of the separation step, the recovery of the product as represented by the yield, Y_i, must also meet minimum criteria. The yield is

$$Y_i = \frac{x_{P,i} \cdot V_{P,i} \cdot \rho_{M,P}}{(x_{s,i} \cdot V_{s,i} \cdot \rho_{M,s})} \tag{6.161}$$

where

$x_{p,i}$ = weight fraction of component i in product stream, g solute/g collected peak

$V_{p,i}$ = volume of product stream in which component i at weight fraction $x_{p,i}$ is contained, mL (= g, for water)

$x_{s,i}$ = weight fraction of component i in sample (feed), g solute i/g feed solution

$V_{s,i}$ = volume of sample in which component i at weight fraction $x_{s,i}$ is contained, mL (= g, for water)

$\rho_{M,p}$, $\rho_{M,s}$ = densities of mobile phases that make up product and sample volumes, respectively, g/mL. Usually the densities will be the same and have values close to 1.

The yield of product must ultimately be determined by operating experience in the plant, although bench scale data can provide guidance. Yields of 0.95 (i.e., 95%) would be considered to be excellent during early parts of a product recovery sequence, and necessary in the latter part of a purification sequence for an economical process. This figure of merit reflects the increase in value of the product as it becomes more purified.

6.31 THE PRODUCTIVITY, P_{prod}, OF A COLUMN FOR EACH CYCLE DEPENDS ON THE ACCEPTABLE EXTENT OF PURIFICATION OF THE PRODUCT

The productivity, $P_{prod,i}$, for one cycle (g product/(g stationary phase · cycle) is

$$P_{prod,i} = \frac{(x_{p,i} \cdot V_{p,i} \cdot \rho_{M,p})}{(W_s \cdot L_{cycle})} \tag{6.162}$$

where

L_{cycle} = number of cycles = 1

The weight of the stationary phase, W_s (in kg) is based on its wet weight although any other internally consistent basis could also be used. If two components constitute a desirable product, the productivity would, of course, be higher:

$$P_{prod,1} + P_{prod,2} = \frac{(x_{p,2} \cdot V_{p,2} + x_{p,1} \cdot V_{p,1})}{W_s L_{cycle}} \rho_{M,p} \tag{6.163}$$

The product fractions that are not of the desired purity must be reprocessed. If reprocessing is not an option, the product is lost and additional cost is incurred. Since the cost (or value) of the product is also a function of its purity, the value will increase as the product becomes more purified and the stream containing the product becomes more enriched in it. The cost of the feed that enters the purification step must therefore be carefully specified. The criterion for establishing its cost (or value) is process and company specific. Hence a simple predictive equation cannot be specified here.

The acceptable extent of product purification or enrichment is defined by the ratio of the weight of product to other solutes in the product stream divided by this ratio as calculated for the injected feed. Purification of component 1 from a mixture of n components can thus be expressed in terms of a purification coefficient given in

$$r_{pufn} = \frac{\left(\dfrac{x_{p,i} V_p}{\sum_{i=1}^{n} (x_{p,i} V_p)}\right)}{\left(\dfrac{x_{f,i} V_f}{\sum_{i=1}^{n} (x_{f,i} V_f)}\right)} = \frac{\dfrac{x_{p,i}}{\sum_{i=1}^{n} x_{p,i}}}{\dfrac{x_{f,i}}{\sum_{i=1}^{n} x_{f,i}}} \tag{6.164}$$

where

$x_{p,i}$ = fraction of component i in the product peak collected at the column's outlet
$x_{f,i}$ = fraction of component i in the feed solution injected into the column at the beginning of the run
V_f = volume of feed, L
V_p = volume of product peak, L
n = number of solutes in the feed mixture

The purification coefficient for a given step gives a measure of enrichment while the productivity for one run (cycle), calculated from Eq. (6.162), gives the amount of product obtained for a specified fractional concentration.

Example 6. Estimating Productivity

A 500 mL sample containing 4 components is injected into an ion exclusion column, having a bed volume of 10 L. Three of the components would ideally co-elute in peak 1 while the fourth component would elute later in a separate peak. However, neither the first nor the second peak, shown in Figure 6.68, is completely homogeneous. They both contain the four components indicated in Table 6.5.

Calculate:

1. The purification coefficient
2. The productivity for this separation

Assume that the density of the stationary phase is 1.05 g/mL packed volume and that the densities of the product and feed peaks are both 1.01. Assume further that one run requires two hours.

Solution

1. Calculate purification coefficient. Since 1 L = 1000 g, the corresponding weight fractions that follow from Table 6.5 are as follows:

	Feed	Peak 2
1	0.010	0.00033
2	0.004	0.00033
3	0.008	0.00066
4	0.020	0.00266
$\sum_{i=1}^{n}$	0.042	0.00398

Equation (6.164) gives

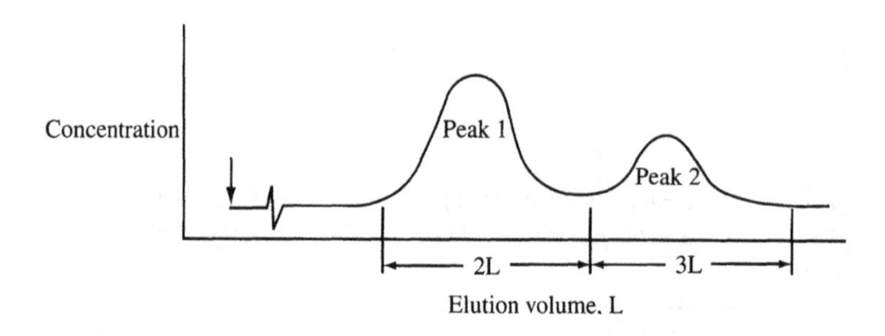

Figure 6.68 Schematic of a binary separation, example 6.

TABLE 6.5 Concentrations of Component n, g/L

Component	Feed	Peak 1	Peak 2
1	10	2.0	0.33
2	4	0.5	0.33
3	8	1.0	0.66
4	20	1.0	2.66

The volumes of feed, peak 1, and peak 2 are 500 mL, 2L, and 3L, respectively.

$$r_{\text{pufn}} = \frac{\left(\dfrac{x_{\text{p},4}}{\sum_{i=1}^{n} x_{\text{p},i}}\right)}{\left(\dfrac{x_{\text{f},4}}{\sum_{i=1}^{n} x_{\text{f},i}}\right)} = \frac{\dfrac{0.00266}{0.00398}}{\dfrac{0.020}{0.042}} = \boxed{1.4}$$

This example illustrates that two peaks do not necessarily reflect two components, and that purification is measured by the extent of enrichment of the product component with respect to the other components that are present.

2. The productivity per run is given by Eq. (6.162) with the volume of the product peak and weight fraction of the product (component 4) being obtained from Table 6.5.

$$P_{\text{prod},4} = \frac{(0.00266)\left(3\,\dfrac{\text{L}}{\text{run}}\right)\left(1.01\,\dfrac{\text{g}}{\text{mL}}\right)\left(1000\,\dfrac{\text{mL}}{\text{L}}\right)}{(10\text{ L})\,(1.05\text{ kg/L stationary phase})}$$

$$= 7.67 \cdot 10^{-1}\,\frac{\text{g}}{\text{kg stationary phase} \cdot \text{run}}$$

This productivity corresponds to 9.7 g/L per day since one run requires 2 hours. This level of productivity might be acceptable for a manufacturing process and represents a reasonable figure of merit for a high value product.

If the productivity does not change during the operational life of the support, the average productivity $P_{\text{prod},i}$, is the same as $P_{\text{prod},i}$. If productivity decreases as the support ages, the average productivity is based on the total weight of product, P_{tot}, obtained over L number of cycles. This is given by

$$P_{\text{prod},i} = \frac{P_{\text{tot},i}}{L_{\text{cycles}}} = \frac{P_{\text{tot}}}{\eta t} \tag{6.165}$$

where η represents the time required to complete one cycle and t is total time that the stationary phase is used. Hence

$$L_{\text{cycles}} = \eta t = \text{number of cycles of use of stationary phase} \tag{6.166}$$

One cycle, L_{cycle}, refers to the interval between when the sample is injected and the last eluent is collected, immediately prior to injecting the next sample. The operational life of the stationary phase is the time or number of cycles that have elapsed under operational conditions before the support is discarded. The turnaround time, η, refers to the fraction of the cycle (or number of cycles) completed in one hour. The average productivity for component i, $P_{\text{prod},i}$, is expressed in terms of dry weight of product obtained at a specified purity, per wet weight of stationary phase. The average g product, obtained per kg support, will be a function of the manner in which the support loses its operational capacity while giving a product of a specified composition and quantity. This is usually expressed as g product/(kg stationary phase-cycle).

One method of operating a column that is gradually losing resolution, is to gradually decrease the amount of product collected while keeping product composition at a constant level, and all other operating conditions the same. This is illustrated in Figure 6.69. This differs from analytical chromatography, where conditions of operation may be changed over a period of time, in order to maintain resolution. Changing of conditions at a process scale presents a challenge. It is difficult to gauge effects of changes in an upstream chromatography step on performance of chromatography columns that follow. Further only a narrow range of conditions would be allowed for a valid manufacturing process (as discussed in Chapter 8). These factors, combined with the higher loadings of process systems relative to analytical separations, result in less latitude to make changes.

If a sudden loss in separation efficiency occurs at a certain time, t_1, then the productivity of component i, $P_{\text{prod},i}$, is

$$P_{\text{prod},i} = P(l) = \frac{x_{\text{P},i} V_{\text{p}} \rho_{\text{M,P}}}{W_{\text{s}} L_{\text{cycle}}} \quad \text{constant at } t < t_1 \tag{6.167}$$

and

$$P_{\text{prod},i} = P(l) = 0 \quad \text{at } t > t_1 \tag{6.168}$$

where

$P_{\text{prod},i}$ = average productivity for component i, g product/kg stationary phase · cycle

The total weight of product obtained at weight fraction $x_{\text{p},i}$ is

$$P_{\text{tot},i} = P_{\text{prod},i} L_{\text{cycles}} \tag{6.169}$$

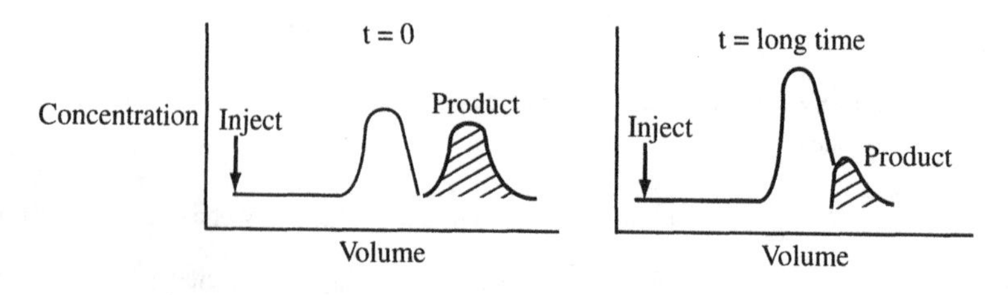

Figure 6.69 Loss of resolution results in less product at a specified purity.

where

$$P_{tot,i} = \text{total production of component } i, \frac{\text{g product}}{\text{kg stationary phase}}$$

where $P_{prod,i}$ is the same for each cycle until the stationary phase fails, Eqs. (6.167) and (6.168). If the separation efficiency decreases as a first order decay process, the productivity after each cycle can be approximated by

$$P_{prod,i}(l) = P_o \exp(-l/\tau) \tag{6.170}$$

where

$l =$ cycle number

$P_{o,i} =$ productivity when the column is first started up, $\frac{\text{g product}}{\text{kg stationary phase} \cdot \text{cycle}}$

$\tau =$ "time" constant (in cycles) for the loss in productivity

The average productivity obtained over a number of cycles, L_{cycle}, is

$$P_{prod,i} = P_o \int_0^{L_{cycle}} e^{-l/\tau} dl = P_o \tau\,(1 - e^{-L_{cycle}/\tau}) \tag{6.171}$$

The value of τ^{-1} is the slope of the line obtained by plotting $\ln[P(t)/P_o]$ against l, where l is treated as a continuous parameter as illustrated in Figure 6.70. This approach is approximate since the loss in productivity is expressed in terms of a discrete variable (i.e., l), rather than a continuous variable (i.e., t). This reflects the fact that the loss in productivity of a chromatographic process, if it occurs, would be measured after each cycle of operation rather than continuously.

If the stationary phase is periodically regenerated, the average productivity becomes

$$P_{prod,i} = \frac{P_{tot,R,i}}{L_R} \tag{6.172}$$

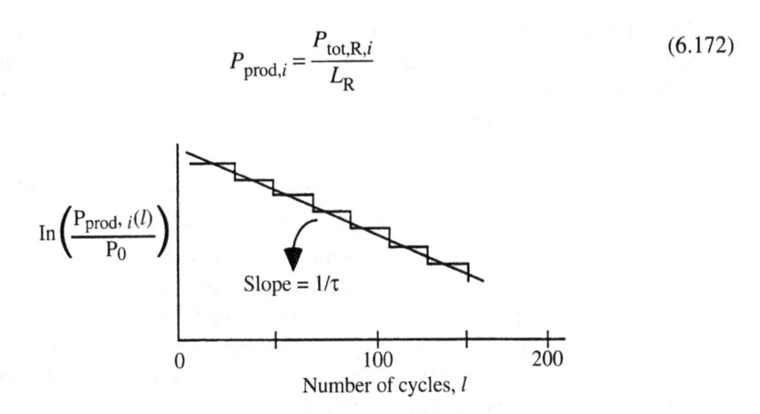

Figure 6.70 First order decay in column productivity.

where L_R is the number of cycles between regenerations, and $P_{tot,R,i}$ is the total product obtained (as g product/kg stationary phase) between regenerations. The product obtained between regenerations would be calculated from Eqs. (6.169) or (6.171). The productivity of the support is assumed to return to $P_{o,i}$ after each regeneration.

6.32 THE CALCULATION OF COSTS IS BASED ON THE PRODUCTIVITY OF THE SEPARATION

In the cases discussed below, the product is assumed to be a single component. Consequently the term "i" will be dropped for sake of simplicity, and $P = P_{prod,i}$. The cost of the stationary phase relative to the quantity of product obtained is given by

$$C_s = \frac{C_{sp}}{P \cdot \eta \cdot W_s} = \frac{S \cdot W_s/t}{P \cdot \eta \cdot W_s} = \frac{S}{P \cdot \eta \cdot t} \tag{6.173}$$

The hourly cost of stationary phase, as in Eq. (6.173), is

$$C_{SP} = \frac{S \cdot W_s}{L_{cycle}/\eta} = \frac{S \cdot W_s}{t} \tag{6.174}$$

where

C_S = product cost due to support, \$/g product
C_{SP} = hourly cost of stationary phase, \$/hr
S = direct cost of the stationary phase, \$/kg stationary phase
W_S = weight of stationary phase, kg
P = average productivity, g product/(kg stationary phase · cycle)
η = turnaround time, cycles/hr
t = operational life of the support (= L_{cycle}/η), not including time during storage or regeneration, hours
L_{cycle} = support life in cycles.

The cost, C_s, is asymptotic with respect to P, as shown in Figure 6.71. The relationship in Eq. (6.173) can be expressed in a linear form by plotting P^{-1} against C_s, as shown in Figure 6.72. The quantity $S/\eta t$ reflects the stationary phase cost per cycle where ηt is the support life expressed as the number of cycles of use as given in Eq. (6.166).

Example 7. Effect of Operational Life of Stationary Phase on Separations Lost

A stationary phase that costs \$1000/kg is used in a chromatography column where a sample is injected every 2.5 hour. The stationary phase was chosen because it exhibits high-equilibrium capacity (50 mg/g) for the product molecule based on laboratory runs. However, the amount of product in the feed sample loaded onto the column under process conditions is equivalent to only 2 mg per g stationary phase (= 2 g per kg stationary phase), because equilibrium is not attained. The product peak consists of 500 mL mobile phase. The peak contains both the product and some impurities from the feed, although the product is recovered in an enriched (purified) form. The weight fraction of the product molecule in the

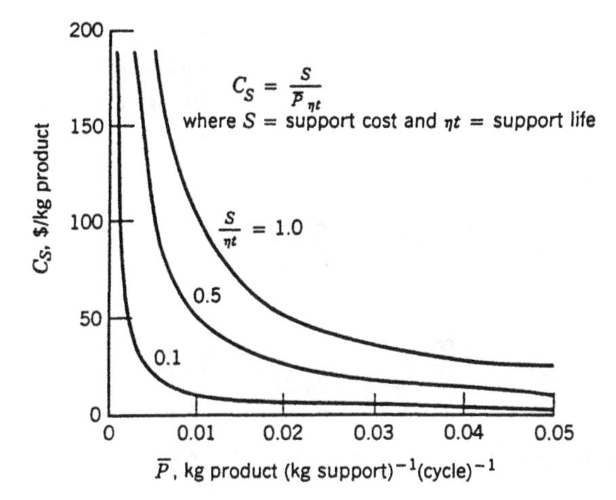

Figure 6.71 Cost functions showing effect of productivity P on stationary phase cost expressed in \$/kg product. Calculated from Eqs. (6.173) and (6.174). (Reproduced from Ladisch, 1987, fig. 9.4, p. 235, with permission from John Wiley and Sons, Inc.).

product peak is 0.008. The mobile phase density is 1.0 g/mL. Assume that the value of the product in the feed is \$200/g, before it is subjected to chromatography, and that the process column is packed with 4 kg of stationary phase.

Calculate the cost of stationary phase for operational lifetimes of

1. 250 hours
2. 2000 hours. Express the stationary phase cost in terms of \$/hr of operation and \$/g product.
3. The cost of the feed in \$/hr

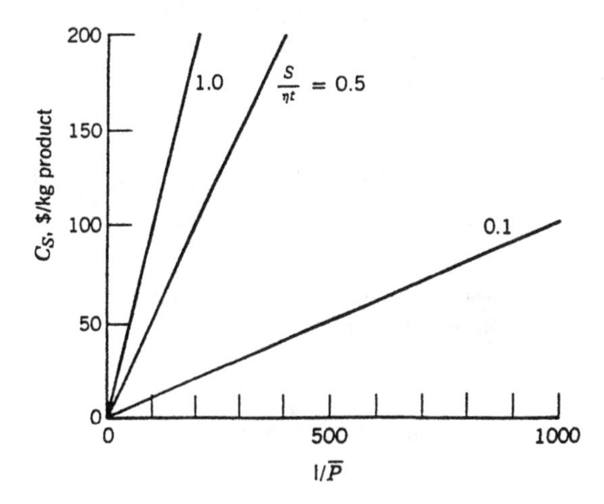

Figure 6.72 Cast as an inverse function of productivity (from Ladisch, 1987, with permission from John Wiley and Sons, Inc.).

Solution

1. In this case the average productivity is given by Eq. (6.167), since there is no decay in the performance of the stationary phase and the amount of product obtained per kg stationary phase per cycle is constant. The average productivity is

$$P_{\text{prod},i} = P = P(l) = \frac{x_{\text{P}} V_{\text{P}} \rho_{\text{M,P}}}{W_{\text{s}} \cdot L_{\text{cycle}}}$$

$$= \frac{(0.008)(500\ \text{mL})\ (1.0\ \text{g/mL})}{4\ \text{kg} \cdot 1\ \text{cycle}} = \boxed{1 \frac{\text{g product}}{\text{kg stationary phase} \cdot \text{cycle}}}$$

The number of cycles of use is calculated from Eq. (6.166):

$$L_{\text{cycles}} = \eta t = \left(\frac{1\ \text{cycle}}{2.5\ \text{hr}}\right)(250\ \text{hr}) = 100$$

The stationary phase cost on an hourly basis is given by Eq. (6.163):

$$C_{\text{sp}} = \frac{S \cdot W_{\text{s}}}{t} = \frac{(1000)\ (4)}{250} = \$16 \frac{1}{\text{hr}}$$

The cost of stationary phase per g product follows from Eqs. (6.162) and (6.173). Only 50% of the initial product is recovered in purified form (= (0.008)(500)(1)). From Eq. (6.173); and P above:

$$C_{\text{s}} = \frac{S}{P \eta t} = \frac{1000}{1\left(\frac{1}{2.5}\right)250} = \boxed{\$10 \frac{1}{\text{g product}}}$$

$$\frac{\left|\frac{\$}{\text{kg stationary phase}}\right|}{\left|\frac{\text{g product}}{\text{kg stationary phase} \cdot \text{cycle}}\right| \frac{\text{cycles}}{\text{hr}} \left| \text{hr} \right.}$$

2. At 2000 hours the calculation of costs use the same equations as in solution 1. The stationary phase cost on an hourly basis is

$$C_{\text{sp}} = \frac{\$1000 \cdot 4}{2000} = \$2 \frac{1}{\text{hr}}$$

and

$$C_s = \frac{1000}{1\,(1/2.5)\,(2000)} = \$1.25\,\frac{1}{g}$$

3. Cost of the feed is given by

$$C_{\text{feed}} = \frac{8 \text{ g product fed}}{\text{cycle}} \cdot \frac{\$200}{\text{g}} \cdot \frac{1 \text{ cycle}}{2.5 \text{ hr}} = \$640\,\frac{1}{\text{hr}}$$

This example illustrates how a relatively modest, hourly stationary phase cost (as given in solutions 1 and 2), can translate into a significant cost with respect to the amount of product processed (i.e., \$10 per gram product for a stationary phase life of 250 hours, and \$1.25/g for a stationary phase life of 2000 hours). The average loading of the stationary phase, compared to the equilibrium capacity of the stationary phase is a key determinant of this cost. A low average loading is due to the small fraction of the stationary phase volume that is in use at any one time, and to a lesser extent, the broadening of the product peak due to the product's strong retention on the stationary phase. This example shows that equilibrium capacity, alone, is not necessarily a good predictor of cost.

Another cost that contributes to purification cost is due to regeneration of a support (C_R in \$/(g product):

$$C_R = \frac{R}{P}\left(\frac{L_R}{L_{\text{cycle}}}\right) \tag{6.175}$$

and the regenerant cost in \$/hr is

$$C_{RP} = C_R \cdot P \cdot W_s \cdot \eta \tag{6.176}$$

where

C_{RP} = hourly regenerant cost, \$/hr
C_R = regenerant cost in \$/g product
R = the regenerant cost in \$/(kg support · cycle of regeneration)
L_{cycle} = number of cycles of operational use over life of stationary phase = ηt
L_R = number of regeneration cycles carried out over life of resin
W_s = weight of stationary phase, kg
η = cycle time, cycles/hr
P = average productivity, g product/(kg stationary phase · cycle)

The other parameters are as defined previously. The ratio L_{cycle}/L_R represents the frequency of regenerations carried out over the life of the support.

The cost of mobile phase is given by

$$C_M = \frac{M}{P} \tag{6.177}$$

where

C_M = mobile phase cost, \$/g
M = solvent cost, \$/kg support · cycle

and

$$C_{MP} = C_M \cdot P \cdot W_s \cdot \eta \tag{6.178}$$

where

C_{MP} = hourly mobile phase cost, \$/hr

and the other parameters are the same as defined above. The combined cost, C_{tot} (\$/hr) is,

$$C_{tot} = C_{MP} + C_{sp} + C_{feed} + C_{RP} + C_{cap} + C_{labor} \tag{6.179}$$

The capital, labor, and feed costs need to be specified separately, and are subject to accounting practices that vary on a case-by-case basis. The total cost is used in Eq. (6.179) to obtain the cost of a purified product, $P_{g/\$}$, in g/\$. This gives a first estimate of a separation cost once a product of satisfactory composition has been attained.

6.33 RECYCLE AND MOVING "STATIONARY" PHASE CHROMATOGRAPHY INCREASE PRODUCTIVITY

The efficient utilization of expensive stationary phases, and reduction of mobile phase costs can be achieved by increasing the fraction of stationary phase that is in use at any one time. The examples presented in this chapter so far are based on injection of a sample onto a column that is clean (q = o). The mobile phase itself is the only fluid used to move the sample through the column. Once a sample is injected, the run is carried out until all of the components in the sample have eluted from the column. This results in only a fraction of the stationary phase being used at any one time during the course of a run.

The simple remedy of injecting a second sample before the first one elutes can increase resin utilization while decreasing the quantity of mobile phase by a volume equal to the amount of sample injected. This approach of overlapping injections is illustrated in Figure 6.73. The first sample is injected at t = o, the second at $t = 2a$ (before any peaks have eluted), and the third at

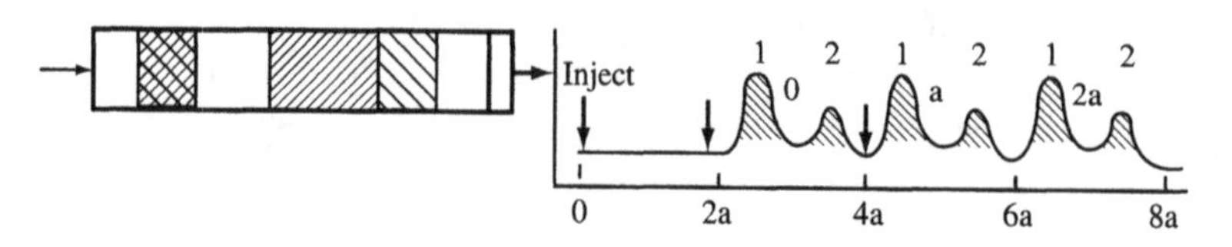

Figure 6.73 Overlapped injections.

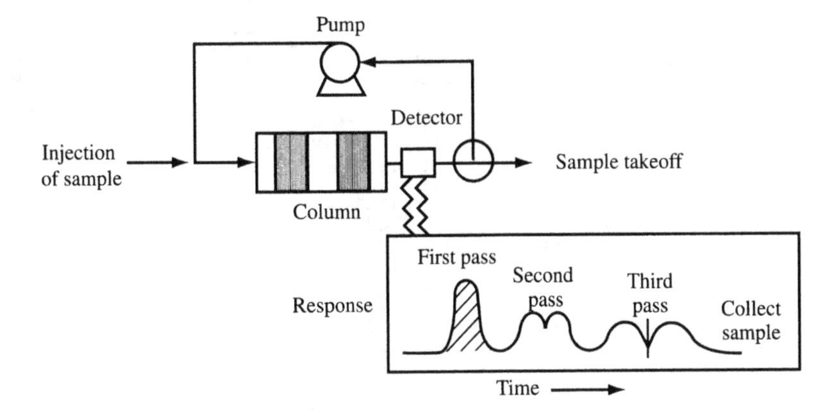

Figure 6.74 Recycle chromatography.

$t = 4a$, just as the sample injected at $t = o$ has eluted, and before the sample injected at $t = 2a$ begins to elute.

The resolved pair of peaks from the first injection begins to elute at $t = 2a$, the second pair at $t = 4a$, and the third at $t = 6a$, as illustrated in Figure 6.73. As a consequence more of the stationary phase is in contact with sample. Less mobile phase is required, since the fluid associated with the feed sample of subsequent injections pushes the mobile phase ahead of them through the column. In a single injection mode, mobile phase is used to push the sample through. Consequently properly timed injections improve the efficiency of use of the column. However, since the sample volumes are often less than 5% of the column volume, the benefits of reduced mobile phase volume are likely to be modest.

A variation of the approach illustrated in Figure 6.73 is the recycling of partially resolved peaks back onto the stationary phase, to improve separation (as illustrated in Figure 6.73). In this type of chromatography, the sample passes through the column more than once as illustrated in Figure 6.74. This makes better use of the stationary phase, since a shorter column can achieve the residence time required for the sample to separate. Savings in mobile phase also occur because the sample itself is used to push the fluid in front of it through the column. Both of the systems shown in Figures 6.73 and 6.74 require that the output be continuously monitored and that the separated components collected at the appropriate times. Recycle chromatography has been demonstrated for chromatographic resolution of racemates by chiral chromatography (Dingenen and Kinkel, 1994).

6.34 A MOVING BED SYSTEM MOVES THE "STATIONARY" PHASE TO ACHIEVE CONTINUOUS OPERATION

A continuous system known as the AVCO system was proposed in 1971, in which the "stationary" phase (i.e., ion exchange resin), itself is moved as a slurry. In this case the resin phase is moved downflow, by a slurry pump, while the feed water, wash, and regenerant are moved upflow as illustrated in Figure 6.75. Such a system was reported to be capable of deionizing or softening water at volumetric rates that were about 100 times higher than an equivalent batch system (Gold and Sonin, 1971). The feed and product takeoff points were kept at a constant location as illustrated in Figure 6.75. The movement of solid sorbent through such a system pre-

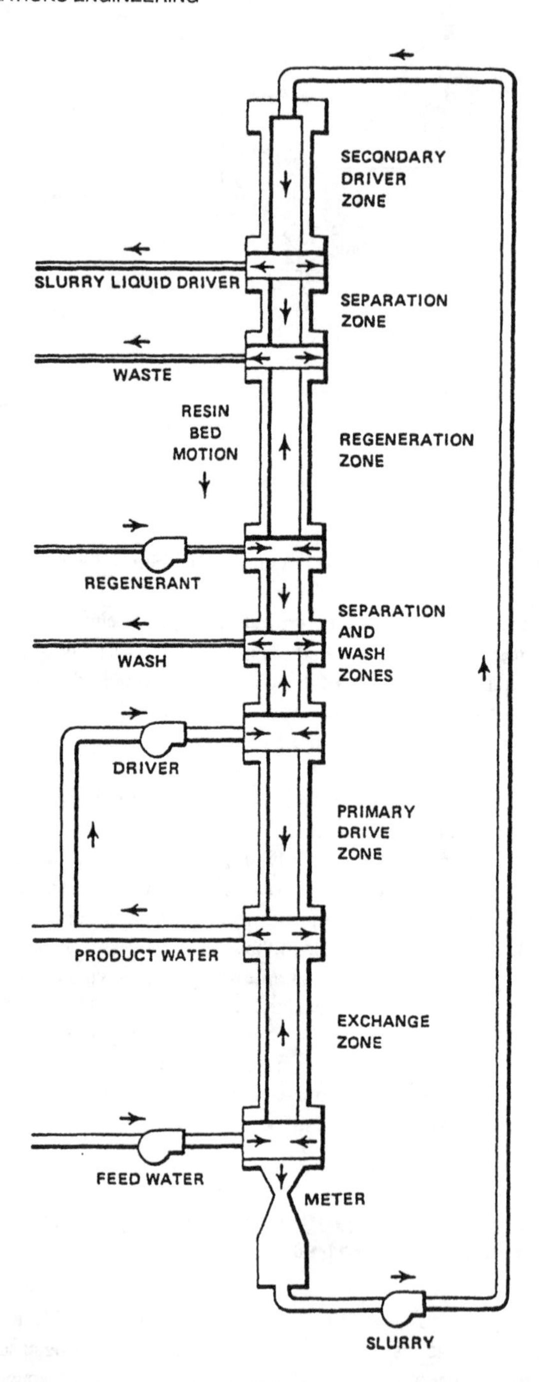

Figure 6.75 Schematic representation of an AVCO moving bed system, in which the feed ports are stationary, and the resin (adsorbent) actually moves. The system was developed for water de-ionization. The bed moves downward, as indicated by arrows in the column, while the fluids move upward. Part of the product (de-ionized) water is diverted for washing the bed (as indicated by arrows at the separation and wash zones, and primary driver zone). (Reproduced from Gold and Sonin, 1971, fig. 1, p. 2.)

sented hydraulic and materials handling problems despite its attractive benefit of reducing sorbent usage by a factor of 3 or 4. Alternate approaches simulate the movement of the adsorbent rather than physically moving the stationary phase. This is achieved through rotation of the column or column segments, or movement of fluid input and takeoff points, while keeping the stationary phase in a stationary bed.

6.35 ONE FORM OF CONTINUOUS CHROMATOGRAPHY MOVES THE ADSORBENT BY ROTATING THE COLUMN

The concept of continuous chromatography utilizing a rotating annulus was proposed about 50 years ago by Martin. The development of a pressurized system to achieve continuous chromatography on a large scale was pioneered by Scott et al., at Oak Ridge National Laboratory, and reported in 1976 for the separation of metal ions (Scott et al., 1976). A prototype of system is shown in Figure 6.76. The driving force for the mobile phase through the packed bed is provided by pressurizing the head space above the fluid at the column's inlet using a gas backpressure (Scott et al., 1976, Begovich, 1983). Gas pressures of up to 25 psig (274 kPa) could be used for columns fabricated from plexiglass, and 150 psig (1135 kPa) for columns made from metal. There would appear to be optimal annulus width for a given backpressure and particle size of stationary phase, since the interstitial velocity will be higher for a smaller annulus width than a wider one at a given back pressure. Rotation rates ranged from 15 to 275 deg/hr (Begovich et al., 1983). The stationary phase was a cation exchange resin (Dowex 50W-X8, 50–60 micron particle size).

Subsequent development of this technique demonstrated that the feed could be continuously introduced, and pH gradients could be used to achieve separations of four component feed solutions (Begovich et al., 1983). The concept was demonstrated for a system that had an annulus radius of 139.7 mm and column length of 114.3 cm so that 96.7% of the column's volume was packed with stationary phase. The largest radius tested was apparently 222.3 mm where 26.7% of the available cross-sectional area of the overall column diameter was used (Begovich et al., 1983).

A resolution of greater than one was achieved when the ratio of feed to mobile phase eluent was less than 0.02. The feed was continuously placed on the top of the rotating column through stationary feed nozzles. Mobile phase (eluent) is metered to the rest of the top of the bed through separate nozzles so that part of the bed not receiving feed is undergoing elution or regeneration. The separated components are continuously removed at the bottom of the device from stationary eluent exits.

The material balance for a rotating continuous annular chromatograph is analogous to equations developed in this chapter and Chapter 5 for column chromatography except that the angular displacement of the solute band, Θ, measured from the feed point (in radius) replaces time as the independent variable. This is given by (Scott et al., 1976; Begovich et al., 1983)

$$\Theta = \omega t = \frac{\omega \bar{\upsilon}}{\varepsilon_b v A} \tag{6.180}$$

where

ω = angular velocity or rotational speed, radians/sec, (analogous to velocity)

t = time for a reference annular segment to be displaced by an angle of Θ, from the feed point, s

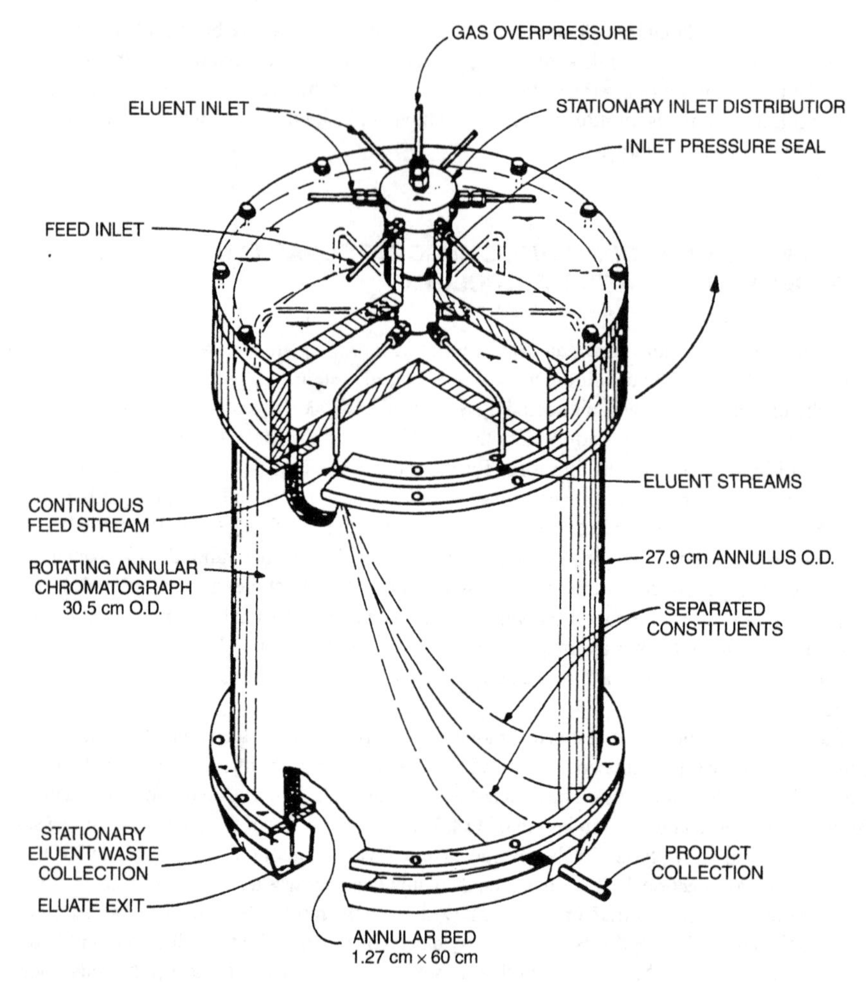

Figure 6.76 Drawing showing design of rotating annular chromatography system. This drawing is for a laboratory unit with an outside diameter of 27.9 cm, and a bed width of 1.27 cm. Other designs are for beds that have annular spaces up to 96% of the cross-sectional area of the bed diameter (drawing CEPS-X-329, shown in Scott et al., 1976, fig. 1, p. 383, J. Chromatogr., 126, 381–400, with permission from Elsevier Science).

$\overline{\upsilon}$ = volume of total eluent flow through an annular segment during time t and displacement Θ, cm^3

v = superficial velocity, cm/s

$\varepsilon_b v$ = interstitial velocity, cm/s

A = cross-sectional area of an annular segment, cm^2

Thus the material balance for the case of no dispersion becomes

$$\omega \frac{\partial C}{\partial \Theta} + \omega \phi \frac{\partial q}{\partial \Theta} + \frac{v}{\varepsilon_b} \frac{\partial C}{\partial z} = 0 \tag{6.181}$$

where

q = concentration of solute on the stationary phase
C = concentration of the solute in the mobile phase, g/cm^3
Θ = defined by Eq. (6.180)

and ω, ϕ, and ε_b are the angular velocity, phase ratio, and interparticle void fraction, respectively, and z is the column length as measured as the vertical distance from the feed point to exit in cm. Scott et al. (1976) develop the material balance for an annulus on the basis of:

Solute in – solute out = change in solute inventory

which is expressed by the equation

$$0 - C_1 d\overline{\upsilon} = d\left[\frac{V}{N}\varepsilon_b C_1 + \frac{V}{N}(1 - \varepsilon_b)K_D C_1\right] \tag{6.182}$$

where

C_1 = concentration of solute in the mobile phase that leaves the first theoretical plate
$\overline{\upsilon}$ = total eluent flow through an annular segment during time t and displacement Θ, mL
V = total volume of an annular segment corresponding to angle Θ, mL
N = total number of theoretical plates on an annular segment
K_D = distribution coefficient
ε_b = interparticle void fraction

The solute entering a column segment is zero in Eq. (6.182), since the sample is assumed to be injected as a pulse as given by the initial condition. Equation (6.182) represents the case for elution. Upon rearrangement, Eq. (6.182) gives the simple expression

$$\frac{dC_1}{C} = -B d\overline{\upsilon} \tag{6.183}$$

where

$$B = \frac{N}{V}\left[\frac{1}{\varepsilon_b + (1 - \varepsilon_b)K_D}\right] \tag{6.184}$$

with the boundary conditions of $\overline{\upsilon} = 0$ and $C_1 = C_f$ at $t = 0$, where C_f is the solute concentration in the feed pulse expressed as moles/mL. Integration of Eq. (6.183) subject to these boundary conditions gives the exponential function

$$C_1 = C_f \exp(-B\overline{\upsilon}) \tag{6.185}$$

or

$$\frac{C_1}{C_f} = \exp(-B\overline{\upsilon}) \tag{6.186}$$

Equation (6.186) provides the form for the equation that gives the mobile phase concentration, C_n, leaving any plate n:

$$\frac{C_n}{C_f} = [e^{-B\bar{\upsilon}}]\left[\frac{(B\upsilon)^{n-1}}{(n-1)!}\right] = [e^{-B'\Theta}]\left[\frac{(B'\Theta)^{n-1}}{(n-1)!}\right] \tag{6.187}$$

where

$$B' = \frac{N\varepsilon_b V}{\omega L}\left[\frac{1}{(\varepsilon_b + (1-\varepsilon_b)K_D}\right]$$

At the system's exit Eq. (6.187) becomes

$$\frac{C_N}{C_f} = \exp^{-B'\Theta}\left[\frac{(B'\Theta)^{N-1}}{(N-1)!}\right] \tag{6.188}$$

Equation (6.188) represents a Poisson distribution. This is approximated by the Gaussian distribution equation if N is large enough:

$$\frac{C_N}{C_f} = \frac{1}{\sqrt{2\pi B'\Theta}} \exp -\left[\frac{(B'\Theta - (N-1))^2}{2B'\Theta}\right] \tag{6.189}$$

The elution position (analogous to elution time) is obtained by taking the derivative of Eq. (6.189), setting it to zero, and solving for the elution position, $\bar{\Theta}$:

$$\bar{\Theta} = \frac{\omega L}{\varepsilon_b V}[\varepsilon_b + (1-\varepsilon_b)K_D] \tag{6.190}$$

Scott et al. (1976) also show that the bandwidth at the outlet of the column is obtained from

$$\sigma_l = \sqrt{\sigma_f^2 + \frac{\bar{\Theta}^2}{N}} \tag{6.191}$$

where

σ_l = standard deviation of the solute concentration across the solute band, rad (recall that 4σ = width of Gaussian peak as its base)
σ_f = standard deviation of solute peak in feed

The plate count is given by Eq. (6.191), which upon rearrangement results in the expression

$$N = \frac{\bar{\Theta}^2}{(\sigma_l^2 - \sigma_f^2)} = \frac{\bar{\Theta}^2}{\sigma_t^2} \tag{6.192}$$

where

$\overline{\Theta}$ = angular displacement of the peak maxima at the column outlet (corresponds to retention time)
σ_t^2 = variance of peak at the outlet of the column

The variance at the base of the peak is related to its width at half-height by

$$\sigma_t^2 = \frac{W^2}{8 \ln 2} \tag{6.193}$$

When corrected for the band-spreading contributed by the injection of the feed, Eq. (6.193) becomes

$$\sigma_t^2 = \frac{W_{1/2}^2 - W_{o,1/2}^2}{8 \ln 2} \tag{6.194}$$

where

$W_{1/2}$ = peak width at half height, measured at half height, rad
$W_{o,1/2}$ = initial feed band with measured at half height, rad
σ_t^2 = variance from measurement of peak width at half-height

The combination of Eqs. (6.192) and (6.194) give the definition of the plate count for the number of theoretical plates of the peak width measured at half-height:

$$\boxed{N_{1/2} = \frac{\overline{\Theta}^2 (8 \ln 2)}{W^2 - W_o^2}} \tag{6.195}$$

This initial work was followed by development of an annular column in which most of the volume of the cylinder was packed with stationary phase (Figure 6.77) (Begovich et al., 1983). In this case the separation was measured by monitoring solute concentration in the fluid collected from the stationary outlet ports at the column exit. The chromatogram in Figure 6.78 was obtained in this manner and showed baseline resolution was achieved similar to the prior designs. Since the peaks are Gaussian, Begovich et al. (1983) used Gaussian equations to quantitate the plate count and resolution. The variance of the peak (σ^2), which was obtained by measuring peak widths, is related to a vessel dispersion number, VDN, by the equation

$$\sigma^2 = 2 \cdot \text{VDN} = 2 \cdot \frac{\varepsilon_b D}{\varepsilon_b v L} \tag{6.196}$$

where

D = axial dispersion coefficient
$\varepsilon_b v$ = superficial velocity, cm/min
ε_b = void fraction between particles
L = column length

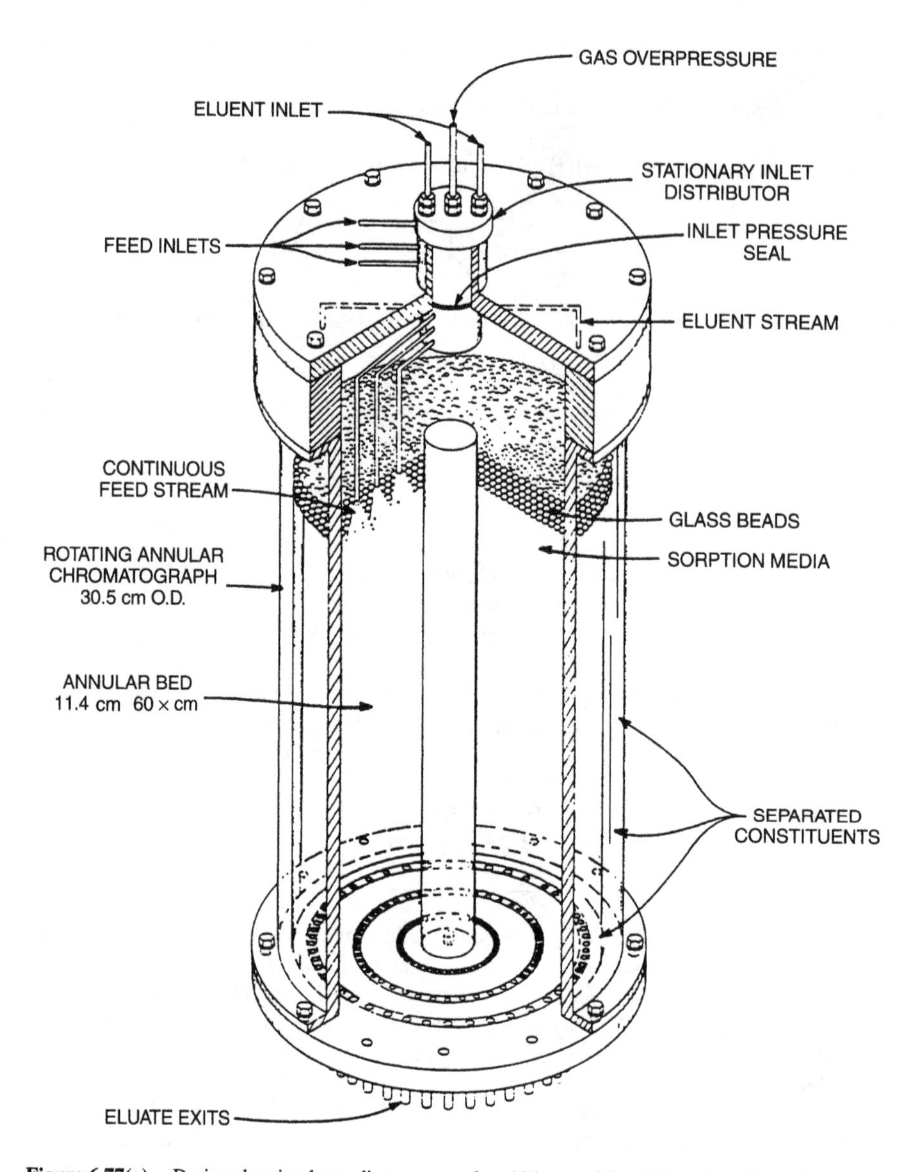

Figure 6.77(a) Design showing large diameter annulus, 114 mm wide × 60 cm long (from Begovich et al., 1983, fig. 2, p. 1174, and fig. 3, p. 1179, with permission from Marcel Dekker, Inc.). Continued to next page.

This relation was used to correct for band spreading and led to the definition of solute dispersion at the column outlet in terms of the particle Reynolds number:

$$\mathrm{VDN}_{\mathrm{solute}} = 1.06 \cdot 10^{-2}\,\mathrm{Re}_{\mathrm{p}}^{0.77} \tag{6.197}$$

where

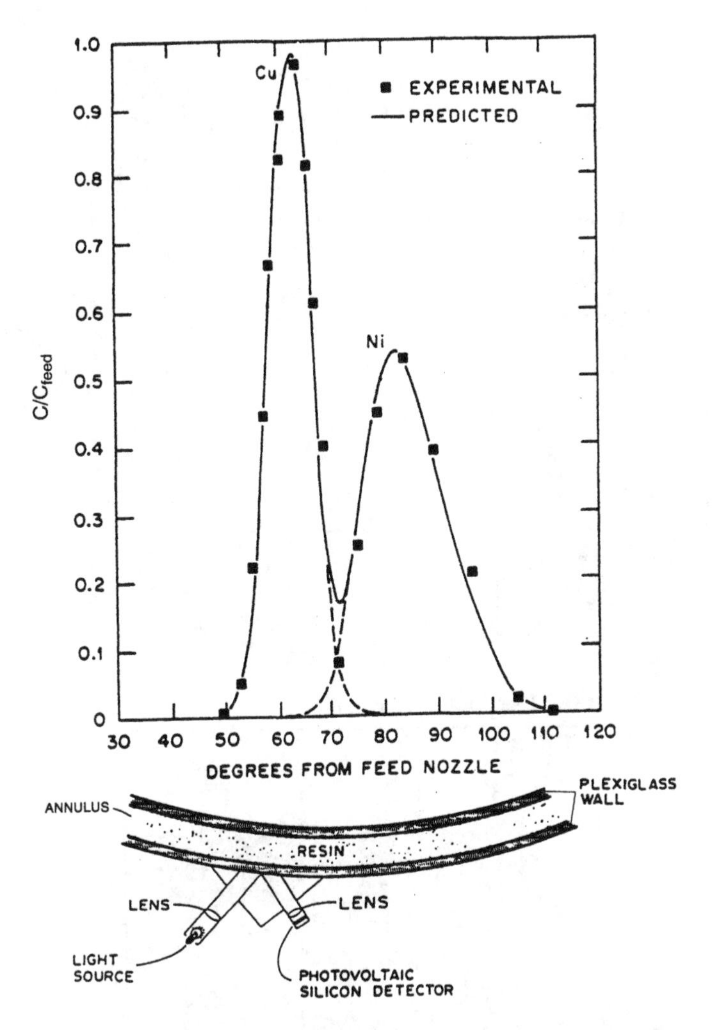

Figure 6.77(b) Chromatogram ot two ion peaks plotted as a function of the angular positon (Θ) instead of time (from Begovich et al., Figure 2, pg. 1174 and Figure 3, pg. 1179, ref. 190). Schematic diagram below chromatogram show detector at column wall (from Scott et al., Fig. 3, pg. 390, ref. 189).

VDN_{solute} = solute vessel dispersion number = $\overline{\Theta}D/\omega L^2$
Re_p = $\rho_M d_p v/\mu$ = particle Reynolds number

with

ρ_M = density of mobile phase, g/cm^3
d_p = diameter of stationary phase particle, cm
v = interstitial velocity of mobile phase through bed, cm/s
μ = viscosity of mobile phase (g · cm/s)

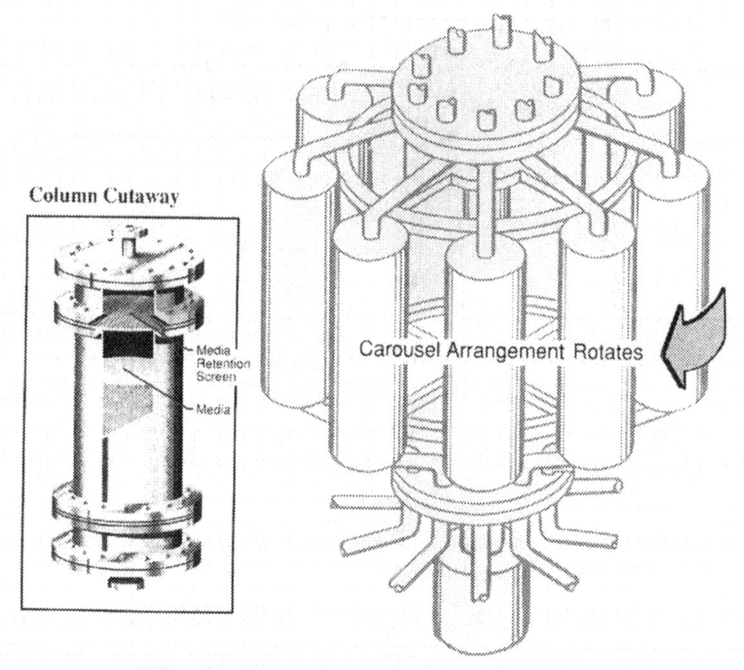

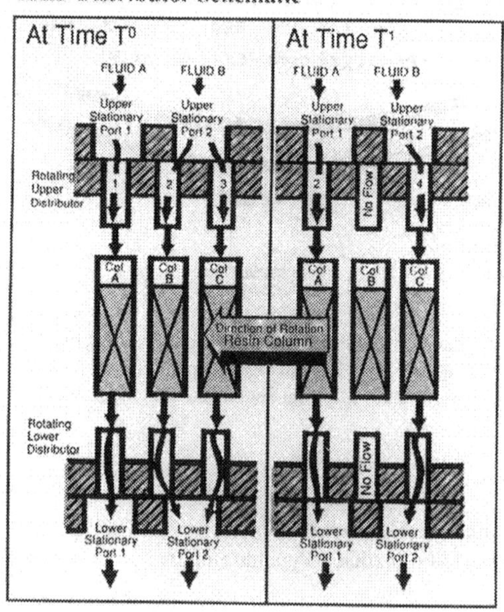

Figure 6.78 Schematic representation of ISEP system, showing arrangement of a carousel. The standard system consists of 20 stationary ports and 30 columns, although the drawing shows only 7 columns for purposes of illustration. The carousel rotates at a constant speed of 0.1 to 1.4 revolutions per hour, with fluid always flowing through the inlet or exit ports on the stationary distributor plates. There is no starting or stopping during the cycle (feed/eluent flows continuously). Ports are designed to withstand 50 psig pressure (from ISEP brochure, 1992).

and with

$\overline{\Theta}$ = solute peak position, deg
D = axial dispersion coefficient
ω = rotation rate, deg/s
L = bed length, cm

The resolution of peaks was originally monitored near the outlet of the column (but with the peaks still inside the column), using an optical sensor that scanned the side of the clear plexiglass column as indicated in Figure 6.78. The retention of the resulting peaks was thus represented in terms of angular displacement, Θ, instead of time (Scott et al., 1976). This convention was extended to cases where peaks were collected from outside of the column. The definition of resolution is also analogous to that for a stationary packed bed:

$$R_s = \frac{\Theta_2 - \Theta_1}{(W_1 + W_2)/2} \tag{6.198}$$

where peak retention parameters, Θ_1 and Θ_2, and peak widths, W_1 and W_2, are measured from a chromatogram of the type illustrated in Figure 6.82.

An extension of the method represented in Figure 6.74 is the combination of partial recycle of the column's effluent combined with a rotating annular ion exchange column. The feed distributor is fixed and a multichannel pump meters fresh feed, or recycled effluent to different injection ports while the annulus is rotating. Sample is removed from the bottom through stationary ports in a fixed sampling carousel. At steady-state operation, additional mobile phase is not required since the fluid required for the column consists of either the feed sample or recycled effluent. The sample is removed from the column, at about the same rate that feed is introduced to the column (Kitakawa, 1997). A mathematical model that anticipates separation of the binary system glutamic acid and valine is described by Kitakawa et al. (1997). While utilization of stationary phase is increased, another benefit of this type of system is the reduction in mobile phase volume needed to carry out the separation.

6.36 SIMULATED MOVING BEDS OPERATE THROUGH A SEQUENTIAL SWITCHING SCHEME TO MOVE THE FEED AND PRODUCT TAKEOFF POINTS

An industrial continuous chromatography system consists of stationary ports and a rotating carousel of stationary phase, is illustrated in Figure 6.78. The speed of rotation of the carousel is in the range of 0.1 to 1.4 revolutions an hour. The feed point moves while the adsorbent in each of the packed beds is held in a stationary configuration. This is similar to the continuous annular rotating chromatograph, except that movement of the separated components around the circumference of the column is blocked since the stationary phase is packed into discrete columns. This system is known as ISEP™ (trademark of Advanced Separations Technology Incorporated, Lakeland, FL), and consists of 30 fixed beds which are mounted on a support structure (i.e., the carousel). The upper and lower portions of the distributor rotate with the carousel, while the inlet and outlet manifolds are stationary, and contain 20 ports. Each port can direct feed into one or two columns at a time as illustrated in Figure 6.78. The zones that are associated with a chromatographic separation will be distributed among several columns as illustrated in Figure 6.79. The processes typically consist of either adsorption and elution zones (Figure 6.80)

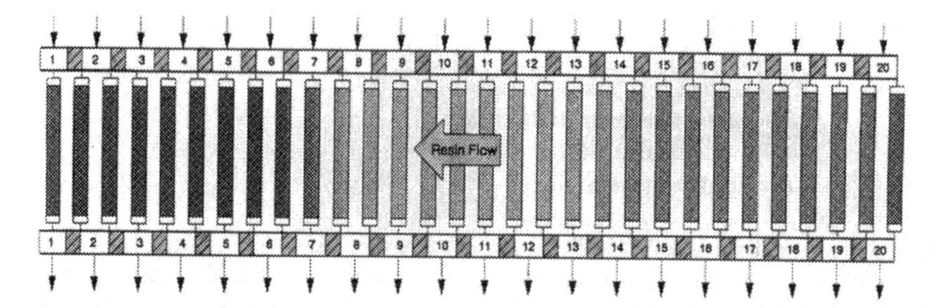

Figure 6.79 Sequencing of modules in an ISEP. Each module consists of a carousel assembly comprised of 30 columns that slowly rotate through 20 zones of operation. The zones are configured to define the treatment process. Typically the process will consist of an adsorption zone, a rinse zone, regeneration/backwash zone, and a wash zone. As the media-filled columns rotate with the carousel, they pass through each stationary port, thereby undergoing a complete process cycle in one revolution of the carousel. Fluid flow in each port is continuous.

or adsorption, rinse, regneration/backwash, and wash zones (Figure 6.80). The horizontal nomograph in Figure 6.80 illustrates how up to 50% of the stationary phase is utilized at one time. Biotechnology and food-processing applications of this system have included recovery and purification of fermentation products, sugar syrup purification, ion exclusion, antibiotic recovery, and optical isomer fractionation. Other industrial uses have been reported for the petrochemical, inorganic and organic chemicals, hydrometallurgy, wastewater, and pulp and paper industries.

Another major industrial system simulates continuous operation using stationary packed beds to which feed, rinse, eluting mobile phase, and/or regenerating solutions are continuously metered to the column through a rotating valve that directs the flows to different feed ports in a programmed, sequential manner. This concept is illustrated in Figure 6.81, and is referred to as the Sarex® or Sorbex® systems. Sarex® is a tradename associated with separation of sugars, while Sorbex® refers to separations of petrochemical streams. Another type of simulated moving bed system, developed by the Japan Organo, separates three components through the combination of simulated moving bed operation and intermittant addition of the feed (Japan Organo, 1999). A laboratory scale system for simulating this behavior (a Trezone unit) is available.

The Sorbex and Sarex systems can be represented by a series of column segments that are stacked one on top of the other, as represented in Figures 6.81 and 6.82. Figure 6.81 shows a hypothetical system in which the adsorbent moves as a dense bed, with the feed introduced as shown. The extract or product is withdrawn above the feed and is the strongly adsorbed component, which is desorbed in the mobile phase introduced at the top of the column. The less strongly adsorbed feed component is removed with the desorbent at the bottom of the column. The less strongly adsorbed component B moves to the bottom of the bed, since, by our definition, it has a higher affinity for the mobile phase than the solid. The less strongly adsorbed solute travels with the flow of the liquid. The opposite is true for A, so this component travels with the solid, up the column (Herber et al., 1991). This concept is the same as that of the AVCO bed shown in Figure 6.75, except that the bed itself is stationary and the liquid ports for the feed, desorbent, extract, and raffinate lines (shown in Figure 6.82) are moved in small steps around the bed. This simulates movement of the bed while maintaining a fixed distance between the liquid streams.

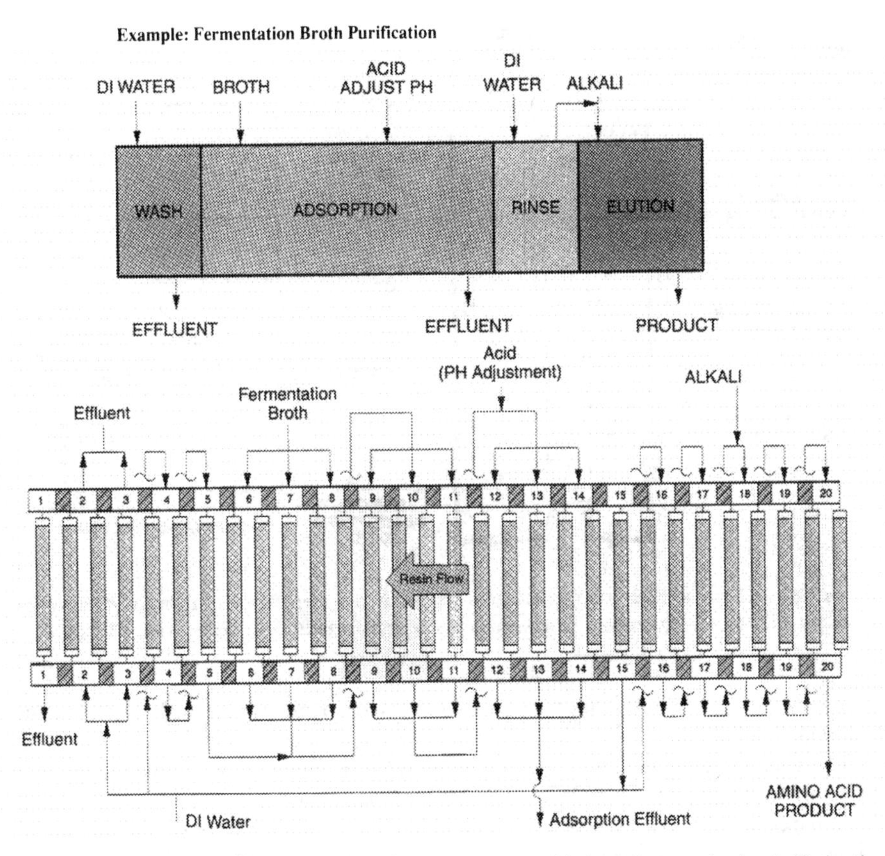

Figure 6.80 Sequence for recovery and separation of amino acids from fermentation broth. Notice the countercurrent flow in adsorption with pH adjustment in the middle zone. Also adsorption wash is upflow to fluidize the media each cycle and discharge suspended solids.

This idea is attributed to Donald Broughton of UOP who first tested it on a pilot plant scale in 1958 with an adsorbent volume of 40 L. The first patent being issued in 1961. A rotary valve that could withstand 200 psig pressure, and valve seal materials that were resistant to organic fluids at 90°C, had to be developed. This enabled the rotary valve concept illustrated in Figure 6.82 to be used (McCulloch et al., 1991). The adsorbent packed into the column is divided into beds of equal volume, with each bed section on chamber having multiple lines attached to it. The rotary valve (abbreviated RV in Figure 6.82) simultaneously directs the four process streams to different liquid transfer lines. Only four transfer lines are used at any one time. The drawing in Figure 6.82 shows lines 1, 5, 8, and 11 to be active, representing desorbent, extract, feed, and raffinate, respectively. The distance between each of the active transfer lines is fixed, and defines the distinct zones, represented in Figure 6.81. The mobile phase is circulated from the top to the bottom of the four zones, with the pump moving the liquid from the bottom of the column to the top, through the side loop. There are four different pumping rates that are used, with the pump being programmed to run at one flowrate as the raffinate zone passes through it, a second rate when the feed passes through it, and so forth. The pumping rate is adjusted to the

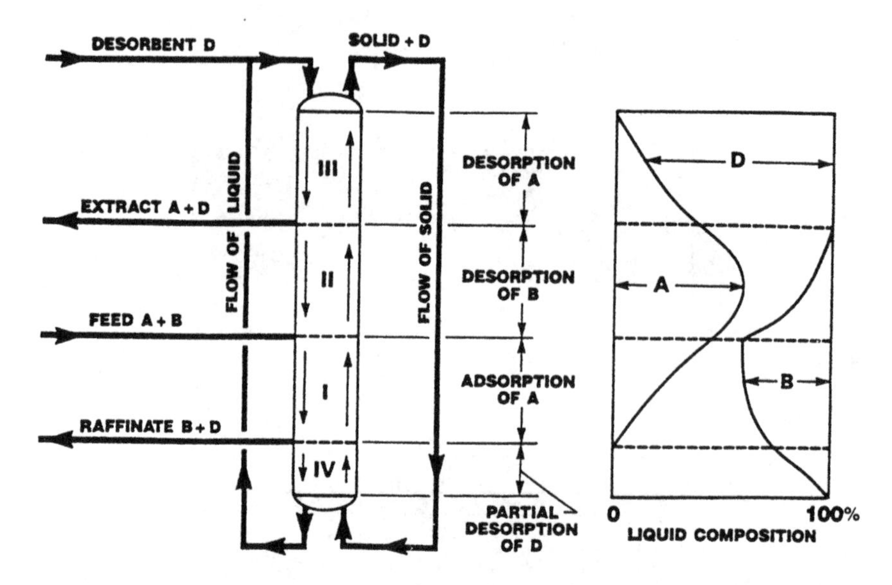

Figure 6.81 Schematic diagram of simulated moving bed system. Representation of hypothetical system in which the adsorbent is moved through adsorption and desorption zones (from Herber et al., 1991, fig. 2).

desired flowrate of the zone as the zone passes through the pump (Herber et al., 1991; McCulloch et al., 1991).

The other operating variable is the time required for complete rotation of the rotary valve. The purified components are found at either end of the column as indicated in Figure 6.82, where the concentration profiles of the two hypothetical components *A* and *B* are shown on opposing axes. These components will move as the valve cycles, although the proportionate shapes stay the same. Unlike analytical or process-scale diagrams viewed earlier in this chapter and in Chapter 5, the profile for this system resembles a single peak, where purified product is removed from the tail, as illustrated in Figure 6.83; with *A* and *B* representing purified components.

The initial concept was demonstrated between 1958 and 1961 for separating a C_5 from C_7 (petroleum) stream, although this particular separation was not economically viable. The first Molex unit was started up in 1964 for separation of normal praffins in the C_9 to C_{22} range. Flowrates were 10^2 L/min and the units were initially 3 to 6 feet in diameter. Later commercial units had adsorbent chambers that were up to 22 feet in diameter. The Molex process was then applied to separation of *p*-xylene (known as the Parex process) and the Sarex process to the separation of fructose from a mixture of glucose and fructose. In 1991 there were 5 Sarex units in operation (Herber et al., 1991), and a total of 89 Sorbex process units were licensed for a variety of industrial separations. These plants processed between 30,000 to 440,000 ton/yr of product. The combination of desorbent and adsorbent could differ for different units and would depend on the type of separation (McCulloch et al., 1991).

The modeling of simulated moving beds has been thoroughly described by Ching and Ruthven for glucose/fructose separation (Sarex process) as well as by Migliorini et al. (2000) for enantiomers. Their simulations show that 94% fructose recovery at 85% purity is possible (Ching and Ruthven, 1985a). Experimental studies, coupled with models for transient response and steady-state operation, describe the basic Sarex® process (Ching and Ruthven, 1985b, c;

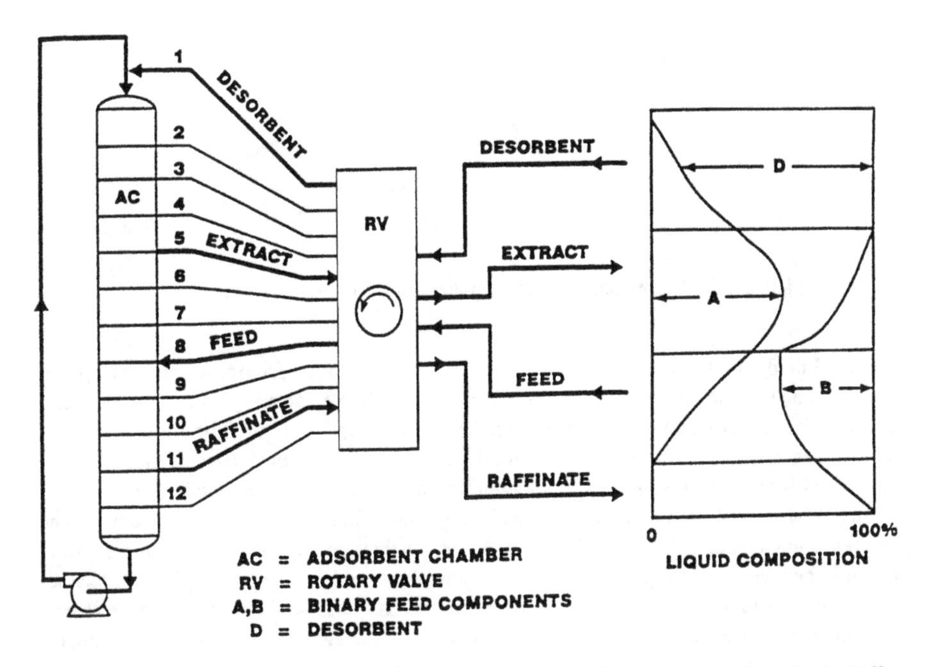

Figure 6.82 Schematic diagram of simulated moving bed system. Representation of moving bed effect obtained by moving the feed, extract, raffinate, and desorbent points using a rotary valve (RV) with a stationary adsorbent bed (from McCulloch et al., 1991, fig. 10.)

Ching et al., 1985). Equations for numerical simulation of this system are given by Hidajat et al. (1986). While glucose and fructose separations are characterized by linear equilibrium, other separations such as raffinose (in beet sugar) may be nonlinear. Systems with nonlinear equilibria have also been modeled by Ching et al. (1993) for the separation of raffinose from dextran. The modeling of simulated moving bed systems is complex, and would probably require another book for proper discussion of principles, practice and theory. The papers by Ruthven and Ching, Ching et al., and others are recommended in the absence of such a book.

Summary and Perspectives

The scale-up of separations for liquid chromatography systems discussed in this chapter addressed solutes that are characterized by either linear or nonlinear equilibria but for which the presence of one solute does not affect the adsorption behavior of the other. A heirarchy of scale-up rules was presented based on the specification that relative peak shapes and resolution would be the same on both bench and process-scales. These rules were summarized in Tables 6.1 and 6.3. While the approach outlined by Tables 6.1 and 6.3 constrains the generality of the scale-up procedures, it also reduces the amount of data that must first be obtained to generate an estimate of column dimensions, mobile phase requirements, and costs for purposes of scale-up.

Linear chromatography combined with measured peak dispersion characteristics enables analytical solutions to be specified that give elution profiles as a function of sample volume, column dimensions, and mobile phase flowrates. This was described in Eqs. (6.48) to (6.57), and (6.180) to (6.198). In cases where interference effects or nonlinear behavior are significant determinants of elution behavior, numerical models (rather than just scale-up rules) are needed.

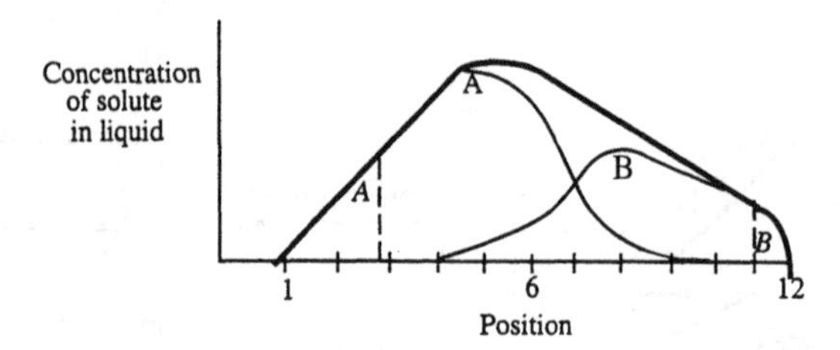

Figure 6.83 Representation of concentration profile in simulated moving bed.

Numerical modeling techniques and introduction of a competition factor into the appropriate isotherm equation and key concepts and mathematics of plate models for simulating were discussed for linear chromatographic separations (Section 6.20). Local equilibrium theory often provides an adequate approximation for chromatography at solute concentrations that are in the nonlinear portion of their isotherms. Both numerical and local equilibrium models are useful in calculating shapes of peaks and time dependent elution behavior at the column's outlet. The plate/rate approach, derived in Eqs. (6.75) to (6.108), enables prediction as well as interpolation of elution behavior if the equilibrium isotherm equations are known and the physical characteristics of the stationary phase quantitated. While this case was derived for a linear isotherm, the framework provided can be extended to nonlinear chromatography if the nonlinear isotherm equation is known.

Both the scaling rule and modeling approaches account for effects of particle size, flowrate, and sample volume on column design. The effect of viscosity of the mobile phase and its impact on pressure drop is also included. These approaches were discussed in sections 6.7 to 6.9.

Chemical as well as physical characteristics of the stationary phases affect operational ranges, with physical characteristics affecting pressure drop. This important consideration determines column hydrodynamics, and rates of chromatography. Pressure-drop equations are therefore developed in Eqs. (6.149) to (6.159) in section 6.23 in the context of a case study on the effect on column diamter on packing characteristics of a soft stationary phase. Column design requires consideration of process hygiene and the materials of construction needed to ensure appropriate chemical and corrosion resistance under cleaning-in-place conditions. These factors are discussed as well in sections 6.27 and 6.28.

The cost of chromatographic separations becomes important once product at a desired composition is obtained. For high-value products, cost is less of an issue than for higher-volume materials. However, in both cases measures of productivity (and therefore process volumes) are needed. Equations for this purpose were introduced in sections 6.29 and 6.30. Both mobile phase and stationary phase costs are important, and both can be reduced by making more efficient use of the stationary phase. This can be accomplished by simulating moving bed chromatography. The concepts of this approach were also introduced.

REFERENCES

Abramowitz, M., I. Stegun, 1965, *Handbook of Mathematical Functions with Formulas, Graphs and Mathematical Tables*, Dover, New York.

Advanced Separation Technologies, 1992, *The ISEP Principle*, Technical Brochure, Florida, 6.

Athalye, A. M., S. J. Gibbs, and E. N. Lightfoot, 1992, Predictability of chromatographic protein separations—Study of size-exclusion media with narrow particle size distributions, *J. Chromatogr.*, 589, 71–85.

Bauman, W. C., R. M. Wheaton, and D. W. Simpson, 1956, Ion exclusion, in *Ion Exchange Technology*, F. C. Nachod and J. Schubert, eds., Academic Press, New York, 182–202.

Begovich, J. M., C. H. Byers, and W. G. Sisson, 1983, A high-capacity pressurized continuous chromatograph, *Sep. Sci. Technol.*, 181(12–13), 1167–1191.

Carberry, J. J., 1958, Axial dispersion and void-cell mixing efficiency in fluid flow in fixed beds, *AIChE J.*, 4(1), 13M–21M.

Carta, G., 1988, Exact analytic solution of a mathematical model for chromatographic operations, *Chem. Eng. Sci.*, 43, 2877–2883.

Ching, C. B., and D. M. Ruthven, 1985a, An experimental study of a simulated counter-current adsorption system: I. Isothermal steady state operation, *Chem. Eng. Sci.*, 40(6), 877–885.

Ching, C. B., and D. M. Ruthven, 1985b, An experimental study of a simulated counter-current adsorption system: II. Transient response, *Chem. Eng. Sci.*, 40(6), 887–891.

Ching, C. B., and D. M. Ruthven, 1985c, Separation of glucose and fructose by simulated counter-current adsorption, in *Adsorption and Ion Exchange: Recent Developments*, AIChE Symp. Ser., 242, 81, J. P. Ausikaitis, A. L. Myers, and N. H. Sweed, eds., 1–8.

Ching, C. B., D. M. Ruthven, and K. Hidajat, 1985, Experimental study of a simulated counter-current adsorption system: III. Sorbex operation, *Chem. Eng. Sci.*, 40(8), 1411–1417.

Ching, C. B., K. H. Chu, K. Hidajat, and D. M. Ruthven, 1993, Experimental study of a simulated counter-current adsorption system: VII. Effects of non-linear and interacting isotherms, *Chem. Eng. Sci.*, 48(7), 1343–1351.

Cohen and Metzner, 1981, Wall effects in laminar flow of fluids through packed beds, *AIChE J.*, 27(4), 705–715.

Colin, H., 1993, Large scale high performance preparative liquid chromatography, in *Preparative and Production Scale Chromatography*, G. Ganetsos and P. E. Barker, eds., Chapter 2, Marcel Dekker, New York, ch. 2.

Colin, H., G. B. Cox, O. Dapremont, and P. Hilaireau, 1995, Optimization of preparative separations: Comparison of the economics of continuous and batch operation, Poster-Preptech '95, NJ (February).

Colin, H., P. Hilaireau, and J. de Tournemire, 1990, Dynamic axial compression columns for preparative high performance liquid chromatography, *LC-GC*, 8(4), 302–312.

Collentro, W. V., 1999, The use of 0.1-μm membrane filters in USP purified water systems—Case histories, part I, *Pharmaceutical Technol.*, 23(3), 176–192.

Collentro, W. V., 1995, USP purified water and water for injection storage system and accessories, part I, *Pharmaceutical Technol.*, 19(3), 78–94.

Curling, J. M., and J. M. Cooney, 1982, Operation of large scale gel filtration and ion exchange systems, *J. Parenteral Sci. Technol.*, 36, 59–64.

Cussler, E. L., 1997, *Diffusion: Mass Transfer in Fluid Systems*, 2nd ed., Cambridge University Press, Cambridge, 15–35, 222–226.

Danckwerts, P. V., 1953, *Chem. Eng. Sci.*, 2, 1–.

Dean, J. A., ed., 1973, *Lange's Handbook of Chemistry*, McGraw-Hill, New York, 10-277–10-287.

DeJong, A. W. J., H. Poppe, and J. C. Kraak, 1981, Column loadability and particle size in preparative liquid chromatography, *J. Chromatogr.*, 209, 432–436.

DePalma, A., 1996, Scale-up remains the critical for HPLC bioprocessing operations, *GEN*, 16(11), 6–7.

Dingenen, J., and J. N. Kinkel, 1994, Preparative chromatographic resolution of racemates on chiral stationary phases on laboratory and production scales by closed-loop recycling chromatography, *J. Chromatogr. A*, 666, 627–650.

Eisenbeiss, F., S. Ehlerding, A. Wehrli, and J. F. K. Huber, 1985, *Chromatographia*, 20, 657–.

Etzel, M. R., 1995, Whey protein isolation and fractionation using ion exchangers, in *Bioseparation Process in Foods*, R. K. Singh and S. S. H. Rizvi, eds., Marcel Dekker, New York, ch. 12.

Fritz, J. S., and P. M. Scott, 1983, *J. Chromatogr.*, 271, 193–.

Ganetsos, G., and P. E. Barker, 1993, *Preparative and Production Scale Chromatography*, Marcel Dekker, New York.

Gibbs, S. J., and E. W. Lightfoot, 1986, Scaling up gradient elution chromatography, *Ind. Eng. Chem. Fund.*, 25, 490–498.

Gold, H., and A. A. Sonin, 1971, The AVCO continuous moving bed ion exchange process, Report No. AVSD-WP-71-2, 32nd Annual Meeting, The International Water Conference, The Engineer's Society of Pennsylvania, Pittsburgh, PA, November 2–4.

Goldstein, S. F. R. S., 1953, *Proc. Roy Soc. (London)*, A219, 171.

Goto, M., N. Hayashi, and S. Goto, 1983, Separation of electrolyte and non-electrolyte by an ion retardation resin, *Sep. Sci. Tech.*, 18, 475.

Grushka, E., L. R. Snyder, and J. H. Knox, 1975, Advances in band spreading theories, *J. Chromatogr. Sci.*, 13, 25–37.

Gu, T., G.-J. Tsai, and G. T. Tsao, 1993, Modeling of nonlinear multicomponent chromatography, in *Adv. Biochem. Eng. Biotechnol.*, 49, 45–71.

Haynes, H. W., and Sharma, P. N., 1973, *AIChE J.*, 19, 1043.

Haynes, H. W., Jr., 1975, *Chem. Eng. Sci.*, 30, 955.

Herber, R. R., J. A. Johnson, and J. W. Priegnitz, 1991, Development of the UOP Sorbex process, AIChE Spring 1991 National Meeting, April 7–11.

Hidajat, K., C. B. Ching, and D. M. Ruthven, 1986, Numerical simulation of a semi-continuous counter-current adsorption unit for fructose-glucose separation, *Chem. Eng. J.*, 33, B55–B61.

Huber, J. F. K. 1973, Stofftransport and Stoffverteilung bei chromatographischen Prozessen, *Ber. Bunsen Ges.*, American Chemical Society, Washington, D.C., 77, 179.

ISEP Brochure, 1992, Advanced Separation Technologies Brochure, The ISEP Principle of Continuous Adsorption.

Jageland, P. T., and L. M. Bryntesson, 1994, Economical aspects and optimization in industrial HPLC, Poster Presented at PrepTech 94, Secaucus, NJ, March 22–24.

Jageland, P. T., L. M. Bryntesson, and P. L. Möeller, 1995, Preparative Chiral HPLC separation: modeling, optimization, and scale-up, *Tech. Bull.*, 14.

Jageland, P., J. Magnusson, and M. Bryntesson, 1994, Optimization of industrial-scale high-performance liquid chromatography applications using a newly developed software, *J. Chromatogr. A.*, 658, 497–504.

Japan Organo, 1999, *New Simulated Moving-Bed System*, Catalogue No. EA-53 (available at: *www.organo.co.jp*).

Jury, E. I., 1964, *Theory and Application of the z-Transform Method*, Wiley, New York.

Karger, B. L., L. R. Snyder, and C. S. Horvath, 1973, *An Introduction to Separation Science*, Wiley, New York.

Kearny, M., 1997, Engineered fractal cascades for fluid control applications, in *Proc. L'Ingenieur et Les Fractals*, Arcachon, France, Institut National De Recherche En Informatique et En Automatique (INRIA), June 25–27.

Keulemans, A. I. M., 1957, *Gas Chromatography*, Reinhold, New York.

Kim, B., and A. Velayudhan, 1998, Preparative gradient elution chromatography of chemotactic peptides, *J. Chromatogr. A.*, 796, 195–209.

Kipling, J. J., 1965, *Adsorption from Solutions of Non-electrolytes*, Academic Press, London, 1–31, 152–159.

Kirkland, J. J., W. W. Yau, H. J. Stoklasa, and C. H. Dilks Jr., 1977, Sampling and extra-column effects in high-performance liquid chromatography; influence of peak skew on plate calculations, *J. Chromatogr. Sci.*, 15, 303–316.

Kitakawa, A., 1997, Complete separation of amino acids using continuous rotating annular ion exchange chromatography with partial recycle of effluent, *Ind. Eng. Chem. Res.*, 36, 3809–3814.

Knox, J. H. and H. M. Pyper, 1986, Framework for Maximizing Throughput in Preparative Liquid Chromatography, *J. Chromatogr.,* 363, 1–30.

Kovach, J. L., 1978, Gas-phase adsorption and air purification, in *Carbon Adsorption Handbook*, P. N. Cheremisinoff and F. Ellerbusch, eds., Ann Arbor Science Publishers, Ann Arbor, MI, 331–356.

Lachman, Shih, and Brown, 1985, Interleukin 1 from Human Leukemic Monocytes, *Methods in Enzymology*, 116, 467–479.

Ladisch, M. R., 1987, Separation by sorption, in *Advanced Biochemical Engineering*, H. R. Bungay, and G. Belfort, eds., Wiley, New York, 219–237.

Ladisch, M. R., 1991, Fermentation derived butanol and scenarios for its uses in energy related applications, *Enz. Microb. Technol.*, 13(3), 280–283.

Ladisch, M. R., and A. Velayudhan, 1999, Scale-up techniques in bioseparation processes, in *Bioseparation Processes in Foods*, Rakesh K. Singh and Syed S. H. Rizi, eds., Mercel Dekker, New York, 113–138.

Ladisch, M. R., and G. T. Tsao, 1978, Theory and practice of rapid liquid chromatography at moderate pressures, using water as the eluent, *J. Chromatogr.*, 166, 85–100.

Ladisch, M. R., and P. C. Wankat, 1988, Scale-up of bioseparations for microbial and biochemical technology, in *Impact of Chemistry on Biotechnology*, M. Phillips, S. Shoemaker, R. Ottenbrite, and R. Middlekauf, eds., 72. ACS Symp. Ser., 362. Am. Chem. Soc., Washington, DC.

Ladisch, M. R., R. L. Hendrickson, and E. Firouztale, 1991, Analytical- and preparative-scale chromatographic separation of phenylalanine from aspartame using a new polymeric sorbent, *J. Chromatogr.*, 540, 85–101.

Ladisch, M. R., M. Voloch, and B. Jacobson, 1984, Bioseparations: Column design factors in liquid chromatography, *Biotechnol. Bioeng. Symp. Ser.*, 14, 525–540.

Ladisch, M. R., M. Voloch, J. Hong, P. Bienkowski, and G. T. Tsao, 1984, Cornmeal adsorber for dehydrating ethanol vapors, *Ind. Eng. Chem. Process Des. Dev.*, 23(3), 437–443.

Lapidus, L., and N. R. Amundson, 1952, Mathematics of adsorption in beds: VI. The effect of longitudinal diffusion in ion exchange and chromatographic columns, *J. Phys. Chem.*, 56, 984–988.

Lee, J. Y., P. J. Westgate, and M. R. Ladisch, 1991, Water and ethanol sorption phenomena on starch, *AIChE J.*, 37(8), 1187–1195.

Levenspiel, O., 1972, Chemical Reaction Engineering, 2nd ed., Wiley, New York, 290–293, 462–479.

Lightfoot, E. N., R. J. Sanchez-Palma, and D. O. Edwards, 1962, Chromtography and allied fixed-bed separations processes, in *New Chemical Engineering Separation Techniques*, H. M. Schoen, ed., Wiley Interscience, New York, 100–181.

Lochmüller, C. H., and M. Sumner, 1980, Estimation of extra-column dead volume effects using a mixing cell model, *J. Chromatogr. Sci.*, 18, 159–165.

MacLennan, J., and B. Murphy, 1994, Balance load vs. resolution when selecting column packing, *Today's Chemist at Work*, 3(2), 29–34.

McCulloch, B., J. A. Johnson, and A. R. Oskar, 1991, Zeolites for petrochemical separation, AIChE 1991 National Meeting, Houston, TX, April 7–11.

Michaels, A. S., 1952, Simplified Method of Interpreting Kinetic Data in Fixed-Bed Ion Exchange, *Ind. Eng. Chem.*, 44(8), 1922–1930.

Migliorini, C., and M. Mazzotti, 2000, Modeling chromatographic chiral separation under nonlinear competitive conditions, *AIChE J.*, 46(8).

Mood, A. M., and F. A. Graybill, 1963, *Introduction to the Theory of Statistics*, McGraw-Hill, New York.

Neuman, R. P., and M. R. Ladisch, 1986, Sulfuric acid recovery by ion exclusion, *Proc. 6th Annual Solar, Biomass, and Wind Energy Workshop*, USDA, Atlanta, GA, February 25–27.

Neuman, R. P., S. R. Rudge, and M. R. Ladisch, 1987, Sulfuric acid–sugar separation by ion exclusion, *Reactive Polymers*, 5, 55.

Neuman, R. P., 1986, *Ion Exclusion Separation of Glucose and Sulfuric Acid*, M.S. Thesis, p 54, 73.

Nicaud, R. H., and H. Colin, 1990, Some simple considerations for the optimization of economic factors in preparative liquid chromatography, *LC-GC*, 8(1), 24–32.

Pearson, K., 1934, *Tables of Incomplete Beta Function*, Biometrika Office, London.

Peskin, A. P., and S. R. Rudge, 1992, Optimization of large-scale chromatography for biotechnological applications, *Appl. Biochem. Biotechnol.*, 34–35, 49–59.

Pieri, G., P. Piccardi, G. Muratori, and L. Caval, 1983, Scale-up for preparative liquid chromatography of fine chemicals, *La Chimica e L'Industria*, 65, 331.

Raghavan, N. S., and D. M. Ruthven, 1985, *Chem. Eng. Sci.*, 40, 699–.

Reid, R. C., J. M. Prausnitz, and T. K. Sherwood, 1977, *The Properties of Liquids and Gases*, 3rd ed., McGraw-Hill, New York, 56–66, 567–590.

Reisman, H. B., 1988, *Economic Analysis of Fermentation Processes*, CRC Press, Boca Raton, FL., 69–77, 91, 92.

Rudge, S. R., and M. R. Ladisch, 1988, Electrochromatography, *Biotech. Progress*, 4, 123.

Ruthven, D. M., 1984, *Principles of Adsorption and Adsorption Processes*, Wiley, New York.

Schwartz, C. E., and J. M. Smith, 1953, Flow distribution in packed beds, *Ind. Eng. Chem.*, 45(6), 1209–1218.

Scott, C. D., R. D. Spence, and W. G. Sisson, 1976, Pressurized annular chromatograph for continuous separations, *J. Chromatogr.*, 126, 381–400.

Seely, R. J., H. D. Wight, H. M. Fry, S. R. Rudge, G. F. Slaff, G. Sofer, J. McEntire, and J. Akers, 1994, Biotechnology product validation, part 7: Validation of chromatography resin useful life, *BioPharm*, 7(9), 41–48.

Sepracor, 1992, Process Scale Glass Columns for Biochromatography, Lit. Code PB03.

Sherwood, T. K., R. L. Pigford, and C. R. Wilke, 1975, *Mass Transfer*, McGraw-Hill, New York, 39–43, 553.

Snyder, L. R., and J. J. Kirkland, 1979, *Introduction to Modern Liquid Chromatography*, Wiley, New York, 15–82.

Snyder, L. R., G. B. Cox, and P. E. Antle, 1987, *Chromatographia*, 24, 82–.

Snyder, L. R., 1980, Gradient elution, in *High Performance Liquid Chromatography—Advances and Perspectives*, C. Horvath, ed., Academic Press, New York.

Sofer, G., and V. J. Britton, 1983, Designing an optimal chromatographic purification scheme for proteins, *BioTechniques*, (November–December), 198–203.

Sofer, G. K., and L. E. Nystrom, 1989, *Process Chromatography: A Practical Guide*, Academic Press, London, 73–80, 93–105.

Stene, S., 1945, *Arkiv Kemi*, 18A(18), 1.

Terfloth, G., 1999, From nanograms to tons: Chiral stationary phases in the pharmaceutical industry, *LC-GC*, 17(5), 400–406.

Tice, P. A., I. Mazsaroff, and F. E. Regnier, 1987, Effect of large sample loads on column lifetime in preparative-scale liquid chromatography, *J. Chromatogr.*, 410, 43.

Tudge, A. P., 1961a, Studies in chromatographic transport: I. A simplified theory, *Can. J. Phys.*, 39, 1600–1610.

Tudge, A. P., 1961b, Studies in chromatographic transport: II. The effect of adsorption isotherm shape, *Can. J. Phys.*, 39, 1611–1618.

Uspensky, J. V., 1937, *Introduction to Mathematical Probability*, McGraw-Hill, New York.

Velayudhan, A., and M. R. Ladisch, 1993, Plate models in chromatography: Analysis and implications for scale-up, *Adv. Biochem. Eng./Biotechnol.*, 49, 123–145.

Velayudhan, A., M. R. Ladisch, and J. E. Porter, 1993, *Modeling of Non-linear Elution Chromatography for Preparative Scale Separations*, AIChE Symp. Ser., 290.

Velayudhan, A. and M. R. Ladisch, 1992, Effect of Modulator Sorption in Gradient Elution Chromatography: Gradient Deformation, *Chem. Eng. Sci.*, 47(1), 233–239.

Velayudhan, A., M. R. Ladisch, and J. E. Porter, 1992, Modeling of non-linear elution chromatography for preparative-scale separations, in *New Developments in Bioseparation*, 290 (88), 1–11.

Wakao, N and T. Furazki, 1978, Effect of fluid dispersion coefficients on particle-to-fluid mass transfer coefficients in packed beds, *Chem. Eng. Sci.*, 33, 1375.

Wankat, P. C., and Y.-M. Koo, 1988, Scaling Rules for Isocratic Elution Chromatography, *AIChE J.*, 34(6), 1006–1019.

Welinder, B. S., T. Kornfelt, and H. H. Sorenson, 1995, Stationary phases: The weak link in the LC chain?, *Today's Chemist At Work*, 4(8), 35–38.

Westgate, P. J., and M. R. Ladisch, 1993, Air drying using corn grits as the sorbent in a pressure swing adsorber, *AIChE J.*, 39(4), 720–723.

Wilke, C. R., and P. Chang, 1955, *AIChE J.*, 1(2), 264.

Wilson, E. J., and C. J. Geankoplis, 1966, Liquid mass transfer at very low Reynolds number in packed beds, *Ind. Eng. Chem. Fund.*, 5, 9.

Yamamoto, S., M. Nomura, and Y. Sano, 1987, Factors affecting the relationship between plate height and the linear mobile phase velocity in gel filtration chromatography of proteins, *J. Chromatogr.*, 394, 363.

Yamamoto, S., M. Nomura, and Y. Sano, 1986, Scaling up of medium-performance gel filtration chromatography of proteins, *J. Chem. Eng. Jpn.*, 19, 227.

SUGGESTED READING

Aiba, S., A. E. Humphrey, and N. F. Millis, 1973, *Biochemical Engineering*, 2nd ed., Academic Press, New York.

Alltech, 1999, *Catalogue 450*, Cleveland, Ohio *Injection Valves*.

Anderson, L., M. Gulati, P. Westgate, E. Kvam, K. Bowman, and M. R. Ladisch, 1996, Synthesis and optimization of a starch based adsorbent for dehumidification of air in a pressure swing drier, *Ind. Eng. Chem. Res.*, 35, 1180–1187.

Barder, T. J., P. J. Wohlman, C. Thrall, and P. D. Dubois, 1997, Fast chromatography and nonporous silica, *LC-GC*, 15(10), 918–926.

Beery, K., M. Gulati, E. P. Kvam, and M. R. Ladisch, 1998, Effect of enzyme modification of corn grits on their properties as an adsorbent in Skarstrom pressure swing cycle drier, *Adsorption*, 4, 321–335.

Belter, P. A., E. L. Cussler, and W.-S. Hu, *Bioseparations: Downstream Processing for Biotechnology*, Wiley, New York.

Bienkowski, P. R., A. Barthé, M. Voloch, R. N. Neuman, and M. R. Ladisch, 1986, Breakthrough behavior of 17.5 mol % water in methanol, ethanol, isopropanol, and t-butanol vapors passed over corn grits, *Biotechnol. Bioeng.*, 28, 960–964.

Bird, R. H., W. E. Stewart, and E. N. Lightfoot, 1960, *Transport Phenomena*, Wiley, New York.

Cavaco-Paulo, A., and L. Almeida, 1996, Cellulase activities and finishing effects, *Textile Chemist Colorist*, 28(6), 28–32.

Colby, C. B., B. K. O'Neill, F. Vaughn, and A. P. J. Middleberg, 1996, Simulation of compression effects during scale-up of a commercial ion exchange process, *Biotechnol. Progr.*, 12, 662–681.

Cordes, A., J. Flossdorf, and M.-R. Kula, 1987, Affinity partitioning: Development of mathematical model describing behavior of biomolecules in aqueous two-phase systems, *Biotechnol. Bioeng.*, 30(4), 514–520.

Corn Refiners Association, 1996, *Annual Report*, Washington, DC.

Crank, J., 1975, *The Mathematics of Diffusion*, Oxford University Press, Oxford, 89–91.

Ding, H., M. C. Yang, D. Schisla, and E. L. Cussler, 1989, Hollow-fiber liquid chromatography, *AIChE J.*, 35(5), 814–820.

Dolan, J. W., 1998, Extracolumn band broadening from injection and peak transfer, *LC-GC*, 16(3), 248, 250.

Dullien, F. A. L., 1992, *Porous Media: Fluid Transport and Pore Structure*, 2nd ed., Academic Press, San Diego, CA, 487–557.

Foley, J. P., and J. G. Dorsey, 1983, Equations for calculating chromatographic figures of merit for ideal and skewed peaks, *Anal. Chem.*, 55, 730–737.

Gellert, W., S. Gottwald, M. Hellwich, H. Kastner, and H. Kuster, 1975, *Concise Encyclopedia of Mathematics*, van Nostrand Reinhold, New York, 586–591.

Glueckauf, E., 1955, *Trans. Faraday Soc.*, 51, 34–.

Gong, C. S., M. R. Ladisch, and G. T. Tsao, 1979, Biosynthesis, purification, and mode of action of cellulases of trichoderma reesei, *Hydrolysis of Cellulose: Mechanisms of Enzymatic and Chemical Analysis*, Adv. Chem. Ser., 181, Am. Chem. Soc., Washington, DC, 261–287.

Greenkorn, R. A., and D. P. Kessler, 1972, *Transfer Operations*, McGraw-Hill, New York, 368–369, 488–489.

Hamaker, K. H., and M. R. Ladisch, 1996, Intraparticulate flow and plate height effects in liquid chromatography stationary phases, *Separation and Purification Methods*, 25(1), 47–83.

Hamaker, K. H., J. Liu, R. J. Seely, C. M. Ladisch, and M. R. Ladisch, 1996, Chromatography for rapid buffer exchange and refolding of secretory leukocyte protease inhibitor, *Biotechnol. Progr.*, 12, 184–189.

Hamaker, K., J. Liu, C. M. Ladisch, and M. R. Ladisch, 1998, Transport properties of rolled stationary phase columns, *Biotechnol. Progr.*, 14(1), 21–30.

Hjertén, S., J. Mohammad, and K. Nakazoto, 1986, Improvement in flow properties and pH stability of compressed, continuous polymer beds for high performance liquid chromatography, *J. Chromatogr.*, 646, 121–128.

Ho, Sa. V., 1990, Strategies for large-scale protein purification, in *Protein Purification: From Molecular Mechanisms to Large-Scale Processes*, M. R. Ladisch, R. C. Willson, C.-D. C. Painton, and S. E. Builder, eds., 14–34.

Johansson, H.-O., G. Karlstrom, F. Tjerneld, and C. A. Haynes, 1998, Driving forces for phase separation and partitioning in aqueous two-phase systems, *J. Chromatogr. B*, 711, 3–17.

Johnson, R. D., 1986, The processing of biomolecules: A challenge for the eighties, in *Fluid Phase Equilibria*, 29, 109–123.

King, C. J., 1971, *Separation Processes*, McGraw-Hill, New York, 146–263.

Kirschner, E. M., 1998, Soaps and detergents: New washing machines, enzymes, liquids churn the laundry detergent ingredients market, *Chem. Eng. News*, 76(4), 39–54.

Kiso, Y., K. Jinno, and T. Nagoshi, 1986, Liquid chromatography in a capillary packed with fibrous cellulose acetate, *J. High Resolution Chromatog. Chromatog. Commun.*, 9, 763–764.

Knight, P., 1990, Bioseparations: Media and modes, *Bio/Technol.*, 8, 200–201.

Kucera, E., 1963, Contribution to the theory of chromatography: Linear non-equilibrium elution chromatography, *J. Chromatogr.*, 19, 237–248.

Ladisch, M. R., 1989, Hydrolysis, in *Biomass Handbook*, C. W. Hall and O. Kitani, eds., Gordon and Breach, London, 434–451.

Ladisch, M. R., A. Emery, and V. W. Rodwell, 1977, Economic implications of purification of glucose isomerase prior to immobilization, *Ind. Eng. Chem. Process Des. Dev.*, 16(3), 309–313.

Ladisch, M. R., R. L. Hendrickson, and K. L. Kohlmann, 1990, Anion exchange stationary phase for (-galactosidase, bovine serum albumin, and insulin: Separation and sorption characteristics, in *Protein Purification: From Molecular Mechanisms to Large-Scale Processes*, M. R. Ladisch, R. C. Willson, C-d. Painton, and S. E. Builder, eds., Am. Chem. Soc., ACS Symp. Ser., 427, 93–103.

Lawlis, A. B., and H. Heinsohn, 1993, Industrial perspectives on the role of chromatography in the purification of biological molecules, *LC-GC*, 11(10), 720–729.

Lin, J. K., B. J. Jacobson, A. N. Pereira, and M. R. Ladisch, 1988, Liquid chromatography of carbohydrate monomers and oligomers, in *Methods in Enzymology*, vol. 160, W. A. Wood and S. T. Kellogg, eds., Academic Press, 145–159.

Majors, R. E., 1996, New chromatography columns and accessories at the 1996 Pittsburgh conference, part I, *LC-GC*, 14(3), 190–201.

Makulowich, J. S., 1993, Pharmacia unveils streamline and resource process systems, *GEN*, 13(9), 1, 27.

Martin, A. J. P., and R. L. M. Synge, 1941, A new form of chromatogram employing two liquid phases: 1. A theory of chromatography; 2. Application to the micro-determination of the higher monoamino-acids in proteins, *Biochem. J.*, 35, 1358–1368.

Martin, C. W., and Y. Shalon, 1985, An apparatus for slurry-packing preparative LC columns, *LC*, 3(6).

Mayer, S. W., and E. R. Tompkins, 1947, *Am. Chem. Soc.*, 69, 2866–.

McKown, R., R. Tentonico, and S. Fox, 1996, Biotech manufacturing economics: Decreasing costs and increasing yields, *Genetic Eng. News*, 16(6), 6, 28.

Miller, L., D. Handa, R. Fronek, and K. Howe, 1994, Examples of preparative chromtography on an amylose-based chiral stationary phase in support of pharmaceutical research, *J. Chromatogr. A*, 658, 429–435.

Miller, N. T., and J. M. DiBusslo, 1990, Studies on the stability of *n*-alkyl-bonded silica gels under basic pH conditions, *J. Chromatogr.*, 499, 317–332.

Monke, K., A. Velayudhan, and M. R. Ladisch, 1990, Characterization at the swelling of a size-exclusion gel, *Biotechnol. Progr.*, 6, 376–382.

News item, 1999, Chromatography market profile: LC dectectors in pharmaceutical laboratories, *LC-GC*, 17(1), 10.

News item, 1999, Chromatography market profile: Process-scale chromatography, *LC-GC*, 15(7), 600.

Perry, R. H., C. H. Chilton, and S. D. Kirkpatrick, 1963, *Chemical Engineer's Handbook*, McGraw-Hill, New York, 3-228–3-231.

Petsko, G. A., 1988, Protein engineering, in *Biotechnology and Materials Science*, M. L. Good, ed., Am. Chem. Soc., Washington, DC, 60.

Pfeiffer, J. F., J. C. Chen, and J. T. Hsu, 1996, Permeability of gigaporous particles, *AIChE J.*, 42(4), 932–939.

Prochrom, 1993, *Industrial Continuous Liquid Chromatography, Preparative/Production Continuous Chromatography*, Brochure, March.

Roe, S., 1989, Separation based on structure, in *Protein Purification Methods: A Practical Approach*, E. L. V. Harris and S. Angal, eds., IRL Press, Oxford, 182–184.

Rudge, S. R., and M. R. Ladisch, 1986. Process considerations for scale-up of liquid chromatography and electrophoresis, in *Separation, Recovery, and Purification in Biotechnology*, ACS Symp. Ser., 314, J. A. Asenjo and J. Hong, eds., 122–152.

Siddiqi, S. F., N. J. Titchener-Hooker, and P. A. Shamlou, 1997, High pressure disruption of yeast cells, the use of scale down operations for the prediction of protein release and cell debris size distribution, *Biotechnol. Bioeng.*, 55(4), 642–649.

Snyder, L. R., 1972, A rapid approach to selecting the best experimental conditions for high-speed liquid column chromatography: Part II. Estimating column length, operating pressure and separation time for some required sample resolution, *J. Chromatogr. Sci.*, 10, 369–379.

Svec, F., and J. M. J. Fréchet, 1996, New designs of macroporous polymers and supports: From separation to biocatalysis, *Science*, 273(5272), 205–211.

Treybal, R. E., 1968, *Mass Transfer Operations*, McGraw-Hill, New York, 629–673.

Tyn, M. Y., and T. W. Gusek, 1990, Prediction of diffusion coefficients of proteins, *Biotechnol. Bioeng.*, 35, 327–338.

UpChurch Scientific, 1998, *Catalogue of Chromatography and Fluid Transfer Fittings*, 119, Oak Harbor, WA.

van Deemter, J. J., F. J. Zuiderweg, and A. Klinkenberg, 1956, Longitudinal diffusion and resistance to mass transfer as causes of non-ideality in chromatography, *Chem. Eng. Sci.*, 5, 271–289.

van der Brekel, L. D. M, and E. J. DeJong, 1989, Hydrodynamics in packed textile beds, *Textile Res. J.*, 59(8), 433–440.

Velayudhan, A., 1989, Personal communication: The use of the Biorad 385 gradient former to generate linear gradients.

Viswanth, D. S., and G. Natarajan, 1989, *Data Book on the Viscosity of Liquids*, Hemisphere Publishing, New York, 714–715.

Voloch, M., N. B. Jansen, M. R. Ladisch, G. T. Tsao, R. Narayan, and V. W. Rodwell, 1985, 2,3-Butanediol, in *Comprehensive Biotechnology*, M. Moo-Young, C. L. Cooney, and A. E. Humphrey, eds., Pergamen Press, Oxford, 933–947.

Wang, W. K., S.-P. Lei, H. G. Monbouquette, and W. C. McGregor, 1995, Membrane adsorber process development for the isolation of a recombinant immunofusion protein, *BioPharm.*, 8(5), 52–59.

Wankat, P. C., *Rate-Controlled Separations*, Chapman and Hall, London.

Westgate, P., J. Y. Lee, and M. R. Ladisch, 1992, Modeling of equilibrium sorption of water vapor on starch materials, *Trans. ASAE*, 35(1), 213–219.

Yamamoto, S., 1995, Plate height determination for gradient elution chromatography of proteins, *Biotechnol. Bioeng.*, 48, 444–451.

Yang, Y., A. Velayudhan, C. M. Ladisch, and M. R. Ladisch, 1992, Protein chromatography using a continuous stationary phase, *J. Chromatogr.*, 598, 169–180.

PROBLEMS

6.1 Give the expression for the time required to elute all of the peaks for a process-scale size exclusion chromatography column, if the time for a smaller, bench-scale column is known, and the expression has the form:

$$t_x = f(t_b)$$

Assume that the fractional sample volume, type of stationary phase, the plate count, and interstitial velocity are the same on both the bench and process scales. The size of the particles used on the bench and process scales are different, while the void fractions are the same.

6.2 The Sherwood number is proportional to the ReSc (reduced velocity). Use the Sh expression of the Wakao and Furazaki Eq. (6.38) together with the definition of the mass transfer coefficient based plate count

$$N_{\text{mass transfer}} \sim \frac{k_{\text{MT}}L}{vd_{\text{p}}}$$

to develop a scale-up correlation that would enable column length to be calculated for a different particle size than used to obtain the laboratory data.

6.3 Refer to Example 3. Calculate the size of the column if the average particle size is decreased from 740 to 100 microns. Give the length and diameter of the column(s) and the volume of stationary phase needed. Also calculate:

a. Cost of stationary phase

b. Volume and cost of mobile phase

c. Combined cost of stationary phase and mobile phase in terms of $1/kg sugar

6.4 A column packed with 40 micron particles is to be characterized with respect to plate count and retention characteristics. The extraparticulate void fraction is determined using blue dextran while acetone, which absorbs UV light and can therefore be detected using a spectrophotometer, is used to obtain a measure of the internal porosity of the stationary phase. Phenylalanine, which adsorbs onto the stationary phase, has a known partitioning property with $K^* = 0.2$.

The properties of the column are

$V_{\text{col}} = 1$ L

$R_{\text{col}} = 2.5$ cm

The mobile phase has a flowrate of 10 mL/min. The known properties of the column, which have previously been determined through experimental measurements, are

$\varepsilon_{\text{b}} = 0.36$

$\varepsilon_{\text{p}} = 0.34$

$K_{\text{D}} = 0$ for blue dextran

$H = 2$ mm (assumed to be same for all 3 components)

Sample volumes of 100 mL blue dextran (at 5 mg/mL), acetone (10 mg/mL), and phenylalanine (2 mg/mL) are injected. In order to begin to characterize this column calculate the following from the data given above:

a. Peak width (V_{w}) and elution volumes at the center of the peak maximum (V_{max}) for all three components.

b. Elution volume that passes through the column between the times that the sample injection is initiated, and the last peak elutes.

c. Apparent plate count for the phenylalanine peak, that would be measured for the 100 mL injection volume.

6.5 A chromatography column has dimensions of 10 mm i.d. and 30 cm length. It is packed with a stationary phase to be used for size exclusion chromatography. An excluded component (a large protein) is injected and found to give a plate height of 2 mm. The void fractions of the column are $\varepsilon_{\text{b}} = 0.40$ and $\varepsilon_{\text{p}} = 0.30$. Given these limited data, would the use of a Gaussian distribution function be appropriate to model the peak, or should a Poisson distribution be used instead? Show your work that supports your answer.

6.6 The fractionation of two solutes having molecular weights of 4000 and 6000, respectively, is to be carried out by size exclusion chromatography using a macroreticular stationary phase. (Note: A macroreticular stationary phase is the same as a gigaporous particle, as shown in Figure 5.29 in Chapter 5). The 100 micron diameter stationary phase with the properties, as listed below, gives a partial separation of the two solutes when it is packed in a 180 cm long column. Due to pressure-drop considerations, the particle size (radius = R_y) must be kept constant, while other unspecified constraints require that the column be kept at a length of 180 cm. However, the vendor of the 100 micron stationary phase has a second stationary phase with an identical chemical structure, but with a microparticle size that is 10 times smaller ($R_x = 0.1$) while keeping all other parameters the same.

The column is run at a superficial velocity of:

$\upsilon = 2$ cm/min (superficial velocity)

Properties of stationary phase 1:

$\varepsilon_b = 0.4, \varepsilon_x = 0.2, \varepsilon_y = 0.3$

$D_y = D_m/z, D_x = D_m/1.5$
$q_a/C_x = 0.5$ mole/cm
$k_{fMT} = 10^{-4}$ cm/s
$\rho = 1.1$
$R_y = 100$ microns, $R_x = 1$ micron
$S = 300$ m^2/g

Properties of stationary phase 2: Same as for stationary phase 1 except that $R_x = 0.1$ micron. The change in surface area is minimal.

The goal is to increase the resolution from 0.8 to 1.1. Can this be achieved if stationary phase 1 is replaced by stationary phase 2 and all other conditions are kept the same? Show your work and clearly state any assumptions.

6.7 An ion exchange resin having 6% crosslinking, with a styrene-divinyl benzene backbone, is *found* to give satisfactory separation of a variety of carbohydrates. This resin is a cation exchange resin, in the H+ form. The eluent is pH 3.5 H_2SO_4 (very dilute acid).

You are asked to obtain a first, *very* rough estimate of a column capable of separating 100 kg per cycle (dry basis) of sugars *A* and *B* in a 10% aqueous solution. You run a bench-scale run with a 20 μ*l* injection of these sugars; the column is 60 cm long × 6 mm i.d. packed with the resin of 22 micron particle size. The large-scale run is to be carried out at the same temperature and linear velocity. However, the average particle size of the resin is 150 micron, instead of the 22 micron size used on the bench scale. Using the data you have obtained, obtain a first estimate of:

a. The maximum sample volume which can be injected. Express your solution in terms of V. Assume that $R_s \cong 1$.
b. The plate count *N*.
c. Column length for large-scale column.
d. Column diameter for large-scale column.
e. Product cost due to support (C_s). Assume resin costs \$200/ft^3 (with $\rho = 50$ lb/ft^3) and has a one year life. Also assume that P = constant.
f. Time required to completely elute one sample of 100 kg (dry) in water.

State your assumptions. Show your work!

6.8 **a.** Show that Eq. (6.190) can be obtained from Eq. (6.188).

b. Assume that the plate count for the column is N. What is the maximum fractional concentration of the eluting peak for a small sample? Express your answer as the ratio of concentration at the column's outlet to that in the feed.

6.9 The separation of a two component mixture gives the chromatogram shown below. Given the equilibrium curves for the two components, calculate:

a. The volume of mobile phase required to elute both peaks. (30 points)

b. The maximum injection (sample) volume that can be used before the two peaks overlap. (20 points)

To help solve (a) and (b), use the information given below.

Empty column volume = 1 L

Elution volume of DNA (MW = 2,000,000) = 0.40 L

Elution volume of NaCl = 0.75 L

Width of peak 1 = 0.1 L

Width of peak 2 = 0.3 L

6.10 Size exclusion chromatography.

a. The distribution coefficient in size exclusion chromatography, K_{sec}, is defined as:

$$K_{sec} = \frac{V_r - V_o}{V_t - V_o}$$

where

V_o = extraparticulate void volume

V_t = total void volume

V_r = retention volume of a given solute

What are the minimum and maximum values for K_{sec} for size exclusion. (5 points)

What is the physical significance of these extreme values? (5 points)

b. The following results were obtained on a size exclusion stationary phase:

Dextran (molecular weight = 2 million) gave a retention volume of 20 mL.

Deuterated water (D_2O) gave a retention volume of 45 mL.

A dissolved solute gave a retention volume of 35 mL.

What is the size exclusion distribution coefficient for the solute?

c. The "solute" actually represents a number of different low molecular components left after ion-exchange chromatography and reverse phase chromatography have been carried out on refolded insulin. The last step in this sequence, size exclusion, has the objective of removing the salts from the proteins (i.e., desalting). The protein peak is found to elute at a retention volume that is almost the same as the interstitial void volume of the column. The chromatogram for a 5 mL sample appears as shown in Figure 6.84, and it represents a satisfactory separation on a column having dimensions of 25 mm i.d. × 50 cm long.

It is now desired to obtain the same separation for a sample volume of 50 mL at the same linear eluent velocity and column conditions. Calculate the column dimensions required to achieve the scaled-up separation.

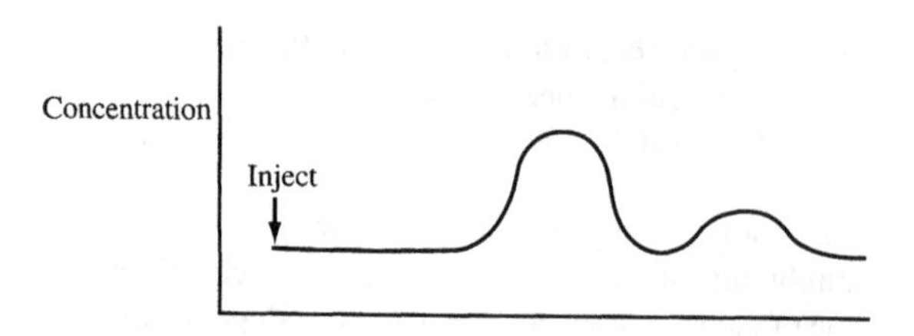

Figure 6.84 Chromatographic trace for Problem 6.10(c).

6.11 Chromatography optimization. Figure 6.85 shows a chromatographic separation of aspartic acid and phenylalanine and aspartame (the phenyl methyl ester of aspartic acid and phenylalanine known as Nutrasweet®). An excluded component elutes at about 30 minutes, while the elution times of the other species are as indicated in the figure. The peak widths for the Phe and aspartame peaks are 15 and 20 min respectively. Based on this chromatogram and the data given, estimate the minimum column volume which will give $R_s = 1$ and the amount of mobile phase which will be saved per run by shortening the column. Assume that the column diameter is kept constant and that the volume is changed by changing the bed height of the stationary phase. (50 points)

6.12 Interleukin 1 (IL-1) is a hormone like peptide that affects many organs. Lachman, Shih, and Brown (1985) report a separation of IL-1 from other constituents over a 5 micrometer hydrophobic, polypropyl A column of dimensions 4.6×200 mm. The sample size was 2 ml. The profile obtained is as given in Figure 6.86.
It is desired to recover the shaded peak in a larger-scale separation using a 120 micrometer particle size packing, with other conditions, including eluent linear velocity, being the same.

a. Estimate the length of column required to attain a similar separation for the 120micron particle size packing.

b. What is the capacity factor of component x? Of the IL-1 peak?

c. What diameter column would be required for a 200 ml sample size being separated over the 120 mciron particle size support?

Clearly state your assumptions.

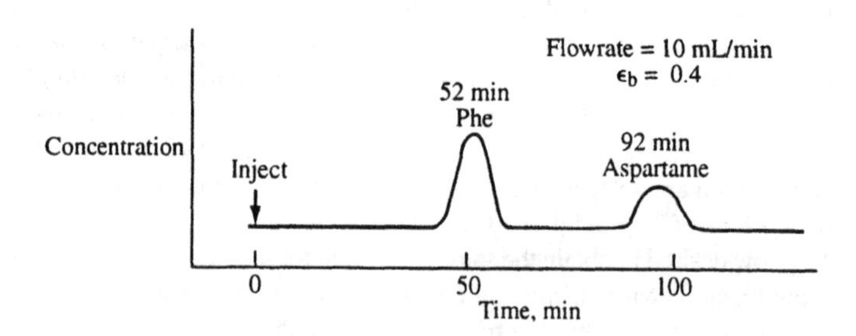

Figure 6.85 Chromatographic trace for Phe from Aspartame separation for Problem 6.11.

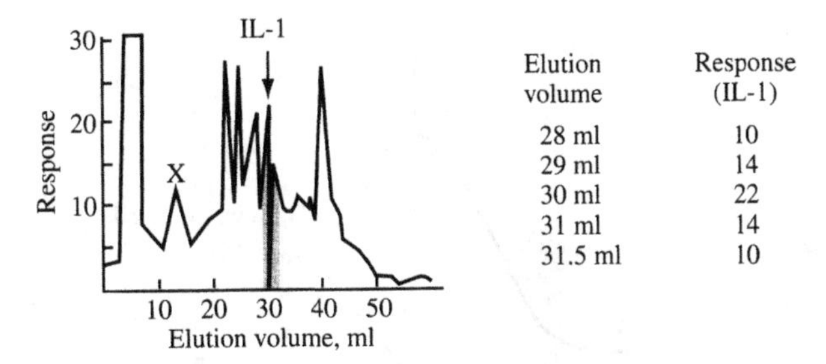

Elution volume	Response (IL-1)
28 ml	10
29 ml	14
30 ml	22
31 ml	14
31.5 ml	10

Figure 6.86 Separation of IL-I (from Lachman et al., 1985).

6.13 The determination of an equilibrium isotherm by frontal analysis is illustrated by Terfloth (1999) for chiral chromatography, where liquid CO_2 (under pressure) is used as the mobile phase. Unlike batch equilibrium measurements carried out at atmospheric pressure, the column can readily be run under pressure. This is particularly useful in cases where the mobile phase must be pressurized to keep it in a fluid phase for chromatographic separations. Terfloth suggests that the retention time of the front of a peak be used to estimate the equilibrium loading as illustrated in Figure 6.43. The first peak at time, t_o, corresponds to the elution time for solute that elutes within one void volume that corresponds to the interparticle void fraction (ε_b). The retention times t_1, t_2, t_3, and t_4 in Figure 6.87 elution of the front of the peaked high concentrations of solute that were injected in small volumes. The solutes are not completely adsorbed, but equilibration is indicated because the peaks are not flat-topped. If a flat top occurs, it corresponds to the part of the solute that

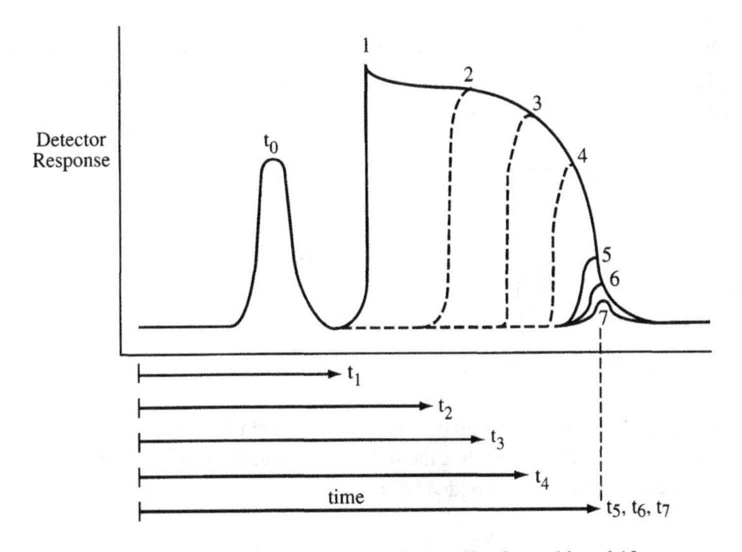

Figure 6.87 Superimposed elution profiles for problem 6.13.

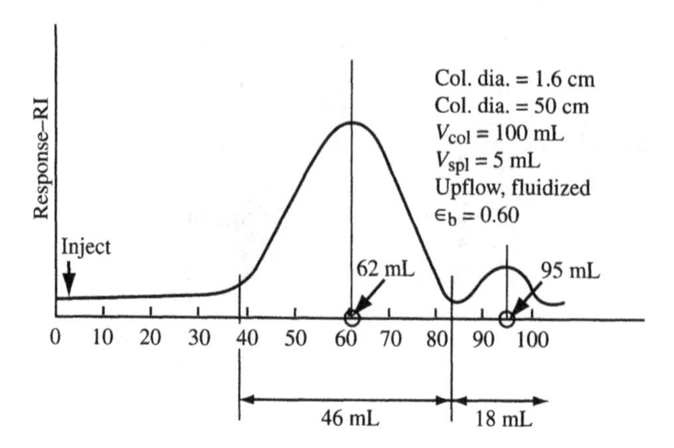

Figure 6.88 Inclusion body chromatogram for problem 6.14.

makes up a nonadsorbed fraction. Show how you would use the chromatograms in Figure 6.87 to calculate an equilibrium isotherm.

6.14 A novel stationary phase is proposed to separate the inclusion bodies from the soluble broth components. The bench-scale system consists of an upflow, fluidized bed system, where the void fraction of the column is 0.60 during operation. This allows the submicron size inclusion bodies to flow through without plugging the column, while separating out the dissolved broth components. The chromatogram obtained is shown in Figure 6.88, for a sample (injected volume) of 5 mL sample for a 100 mL column volume, with a column length of 50 cm. The column separation is to be scaled up for a trial in the pilot plant where the column diameter is fixed at 10 cm, and the plate count is to be double that of the laboratory column.

What is the plate height of the peak that corresponds to inclusion bodies? What is the height of the pilot scale column if it has double the plates of the bench-scale column? How much sample could it process per run?

6.15 An ion-exchange column has the following characteristics:

Column:	Diameter	2.54 cm
	Length	19.7 cm
Packing:	dp (characteristic particle dia.)	200 microns
Flowrate:	10 mL/min	(= 0.02 cm)
Chromatograms	(see Figure 6.89)	

a. 2 mg/mL BSA in 2.0 M NaCl
b. 2mg/mL D_2O in 2.0 M NaCl
c. Retention of proteins and related chromatograms in Figure 6.89.
Given this information, calculate the following parameters. Show your work and give your answers in the table provided below.

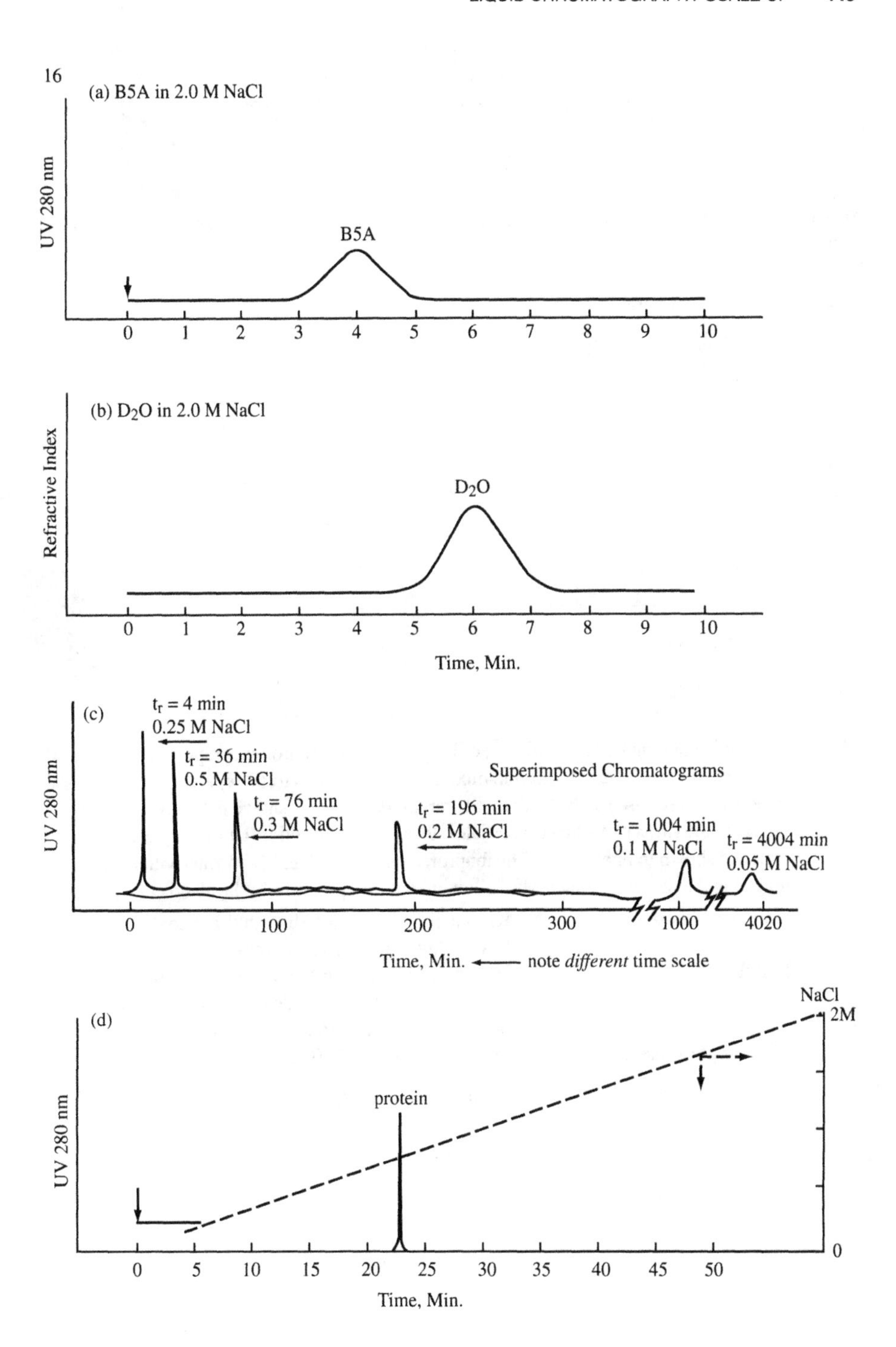

Figure 6.89 Chromatogram for problem 6.15.

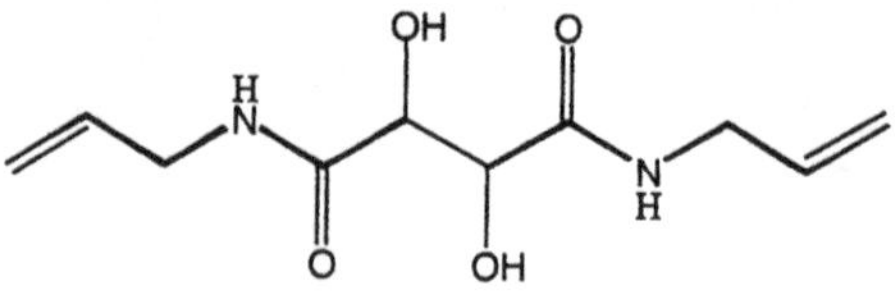

Figure 6.90 Structure of DATD—used to prepare various types of chiral stationary phases—problem 6.16.

Column Characteristics

Column Volume	______mL
ε_T	______
ε_b	______
ε_p	______
ϕ	______
N_{BSA}	______
N_{D_2O}	______
$d_{col}/d_p \left(\frac{\text{column dia.}}{\text{particle dia.}}\right)$	______
Chromatographic velocities	______
For BSA	______ $\frac{\text{cm}}{\text{min}}$
For D2O	______ $\frac{\text{cm}}{\text{min}}$

6.16 Cost for Separating Racemic Mixture of Ibuprofen. (chiral chromatography example based on Jageland et al. 1995). A racemic mixture of ibuprofen is to be separated using a chiral stationary phase based on N, N′-diallyl-L-tartamide (DATD), shown in Figure 6.91, polymerized on the surface of silica gel particles. Ibuprofen is an anti-inflammatory drug with the structure shown in Figure 6.95. The ibuprofen exists in (–) and (+) forms, with (+) ibuprofen being the desired product. The following data are provided by Jageland et al. (1995).

Stationary phase	Kromasil CSP, di -(3,5-dimethyl-benzoyl)-L-DATD (available in kg quantities)
Mobile phase	Hexane: MTBE: TFA volumetric ratio of 90:10:0.5 (note MTBE denotes methyl tert butyl ether; TFA denotes triflouroacetic acid)
Plate height relation	(Knox parameters A, B, C)

$$(+)\ \text{ibuprofen: } h = 0.79 + \frac{2.0}{V} + 0.48V$$

$$(-)\ \text{ibuprofen: } h = 0.74 + \frac{2.0}{V} + 0.58V$$

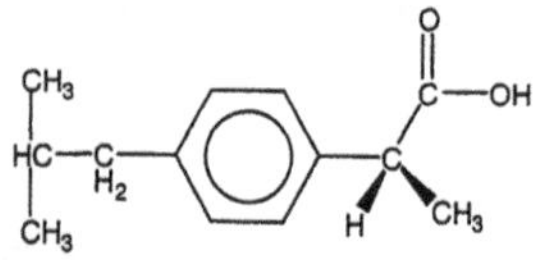

Figure 6.91 Structure of ibuprofen—problem 6.16.

Capacity factors (+) ibuprofen: $k_1' = 1.7$ ($i = 1$)

(−) ibuprofen: $k_2' = 2.3$ ($i = 2$)

Equilibrium relation
(with competition factor)

where

$$C_{i,1} = C_{i,1} = 0.60 = \text{competition factor}$$

$$q_{m,1} = 13.9 \sim$$

$$q_{m,2} = 13.3$$

Operational and cost parameters:

	Laboratory	Process
Feed cost (hypothetical), $/g	1.00	1.00
Product specification, % (+) ibuprofen	99.5	99.5
Column diameter, mm	4.6	300.0
length, mm	250.	—
Mobile phase flowrate, mL/min	0.7	—
Stationary phase particle size, microns	10.	10.
Stationary phase cost, $/(kg · hr)	4.	4.
Mobile phase cost, $/L	0.30	0.30
System cost $ (installed)	NA	300,000.
Maximum pressure drop, bar	60	60
System operational time hr/yr	8000.	8000.
Cycle time (hr/cycle)	0.17	0.6
Labor cost (at $80,000/(person · yr))	80,000.	240,000.

The elution profile for a 0.020 mg sample of racemic ibuprofen from the laboratory column gives a chromatogram that resembles the diagram shown in Figure 6.92.
A second set of runs using the same column, mobile phase and flowrate but at a loading of 1.6 mg ibuprofen/g stationary phase gives a profile that resembles the diagram

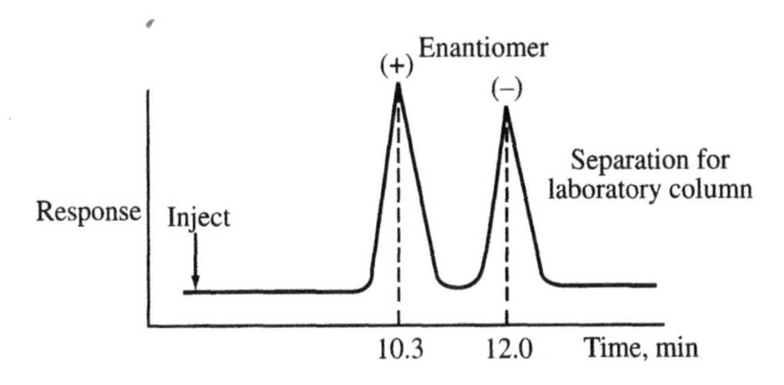

Figure 6.92 Sample chromatogram—problem 6.16.

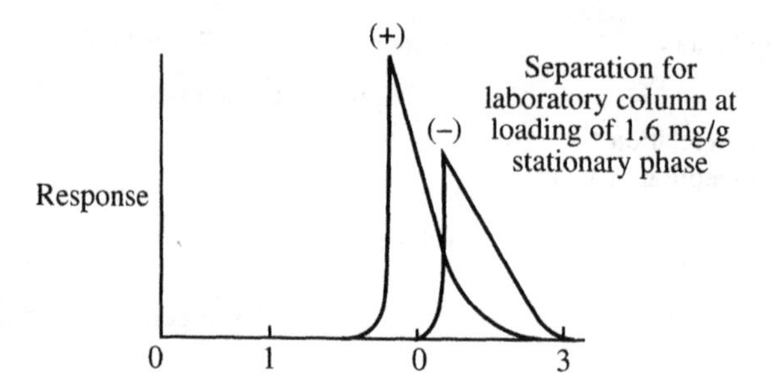

Figure 6.93 Profile indicating existence of column overload conditions.

shown in Figure 6.93.

Assume that a process-scale column of 300 mm inside diameter, with an adjustable head (or plunger) that can be used to adjust the bed height to between 30 and 100 cm, is available in the plant. Preliminary laboratory trials show that a loading of 2.1 mg ibuprofen/g stationary phase with 72% recovery of 99.5% (+) ibuprofen is possible. Due to time constraints, the engineer on the project is requested to immediately scale up the laboratory separation using the existing process column. The goal is to produce 2740 kg of (+) ibuprofen a year. The design is to specify the parameters in (a) through (f) below and to determine whether this goal is attainable with the 300 mm, i.d. process column. Calculate:

a. Column length and expected separation factor

b. Particle size of stationary phase

c. Mobile phase flowrate, and volume consumed on an annual basis

d. Mass of racemic mixture required in the feed solution

e. Concentration of racemic ibuprofen in the feed solution

f. Cost in \$/(g (+) ibuprofen product)

Solvent	______
Lost product	______
Labor	______
System	______
Stationary phase	______
Total	______

CHAPTER 7

PRINCIPLES OF GRADIENT ELUTION CHROMATOGRAPHY

INTRODUCTION

Gradient elution chromatography separates multiple solutes when a component added to the mobile phase alters retention of the solutes as they pass through a bed of adsorbent particles. The additive to the mobile phase is referred to as the modulator or modifier, since it modulates or modifies the retention of the feed components. The modulator's concentration at the column's inlet typically increases or decreases as a linear function of time. This change propagates through the column as the mobile phase travels through the column, thus forming a concentration gradient of the modulator along the length of the column. Commonly used modulators are NaCl in ion exchange chromatography and ammonium sulfate in hydrophobic interaction chromatography. Acetonitrile, methanol, ethanol, or other organic compounds are used in reversed phase or normal phase chromatography. Sugars, salts, peptides, proteins, or metal ions are modulators for affinity chromatography. The mobile phase in which the modulator is dissolved is usually an aqueous buffer. Proteins separated by gradient chromatography, and other types of chromatography will have molecular weights that range from 5 to 150 kD.

Gradient, ion exchange protein chromatography is widely used. The protein or other solutes to be separated by ion exchange chromatography are typically loaded onto the column in a buffer containing a low level of the modulator. This results in an initially high loading of the proteins onto the stationary phase near the inlet. An increase in the modulator concentration due to the application of a gradient causes the protein to be displaced. More strongly retained components move down the column more slowly than weakly retained ones. Differences in retention of different components at the same salt concentration result in the separation of the solutes.

Increasing concentration gradients of modulator are used not only in ion exchange, but also in reversed phase or affinity chromatography (Ladisch and Velayudhan, 1997; Velayudhan et al., 1992). Gradient will cause these components to elute into separate bands as illustrated in

The author thanks Ajoy Velayudhan (Oregon State University), Shuichi Yamamoto (Yamaguchi University), and Mark Etzel (University of Wisconsin) for helpful inputs and contributions to the subject matter of this chapter.

Figure 7.8 and in Example 1 of Chapter 5. However, in hydrophobic interaction chromatography, the sample is loaded at a *high* concentration of the salt (e.g., ammonium sulfate). The high salt concentration promotes a high loading of the protein onto the stationary phase. Elution is achieved in a decreasing gradient. The rationale for this type of behavior is presented in Chapter 3 in terms of surface tension and the action of chaotropic and kosmotropic salts.

If hydrophobic interaction chromatography is used, it is likely to precede the ion-exchange step. An example of hydrophobic interaction chromatography followed by ion exchange chromatography is given for human interleukin-3[1] produced using industrial microorganisms. Unglycosylated human IL-3 is derived from recombinant *Bacillus licheniformis* that secretes it. The broth is filtered and the cell-free filtrate containing the 15 kD IL-3 is passed over a hydrophobic interaction stationary phase that retains the IL-3. Most of the other contaminating proteins flow through. The IL-3 is then eluted from the hydrophobic interaction column by a decreasing gradient of ammonium sulfate. Fractions containing the protein are collected and pooled, concentrated by ammonium sulfate precipitation, and subsequently processed by anion exchange and size exclusion chromatography. The principles of gradient elution chromatography are presented in this chapter based on ion exchange gradient elution chromatography, and reversed phase chromatography. The concepts and equations for both ion exchange and reversed phase chromatography are similar. These principles of gradient elution chromatography discussed in this chapter are followed by the chemical basis of the different types of chromatography, criteria for selecting one type of chromatography over another, and case studies of multi-step chromatography processes in Chapter 8.

7.1 THE SYSTEM FOR CARRYING OUT GRADIENT CHROMATOGRAPHY IS SIMILAR TO THAT FOR ISOCRATIC CHROMATOGRAPHY

There are two or more reservoirs for the different mobile phases used to make up the gradient as shown in Figure 7.1. The system shown is for ion exchange chromatography, with the gradient ranging from 0 to 1 M NaCl. Buffer *A* is the equilibration buffer used to flush the column, prior to the start of a run. Buffer *B* is the buffer used to initiate the gradient, while buffer *C* is the final buffer composition that is passed through the column.

The peak often emerges from the column at a modulator concentration at which the protein's retention time and peak shape is close to that which would be expected if the protein were in the linear range of its adsorption isotherm. This behavior justifies the assumption that the solutes in the feed sample are in the linear region of their isotherms once they begin to move down the column. This enables the calculation of both retention time and peak width to be carried out in a manner that is analogous to chromatography carried out under isocratic elution (Velayudhan et al., 1992).

The elution volume that corresponds to the beginning of the appearance of the gradient at the column outlet is denoted V_s. The corresponding time delay, t_o, in Figure 7.1 corresponds to the extraparticulate column void volume, which is the volume between the stationary phase particles. While the volume that is accessible to the modulator (which is typically a low molecular weight species) is greater than the extraparticle void volume, the concentration of the modulator

[1]New blood cells come from stem cells located in the bone marrow, where glycoproteins called colony stimulating factors play a role in the differentiation of a stem cell to different types of blood cells. IL-3 is capable of increasing production of platelets and both red and white blood cells from stem cells and hence is of interest as a therapeutic protein (van Leen et al., 1991). While T-cells produce Interleukin-3, IL-3 derived from an industrial microorganism is of interest because the amount of IL-3 that could be derived from blood is exceedingly small. T-cells are part of a pool of mature circulating blood cells that are continuously re-supplied by the body since these cells have a short life-span.

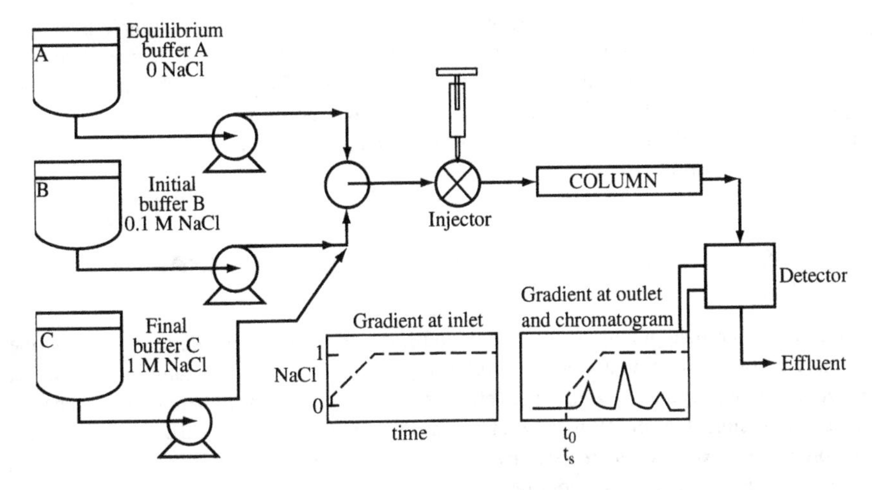

Figure 7.1 Representation of a gradient chromatography system.

is large enough so that most of it will begin to elute after one column void volume, V_o (= $\varepsilon_b V_{col}$), that corresponds to t_s (in Figure 7.1). The modulator may elute later, in which case $V_s > V_o$. The discussion of the principles of gradient chromatography in this section will assume that $V_s = V_o$ or that $t_s = t_o$.

The conceptual representation of linear gradient chromatography given by Figure 7.1 assumes that the gradient shapes at the inlet and outlet of the column are the same. However, the modulator itself can be present at high enough concentrations so that it is in the nonlinear portion of its own isotherm with respect to the stationary phase. When this occurs, the modulator may undergo self-interference resulting in a deformation of the gradient within the column, so the gradient at the outlet of the column may be different than the gradient at the column's inlet, as illustrated in Figure 7.2. When this occurs, the gradient is said to be deformed. This behavior is consistent with local equilibrium theory, discussed in Section 6.13 of Chapter 6. The modulator is fed to the column in a continuous manner, just as if the column were being run as an

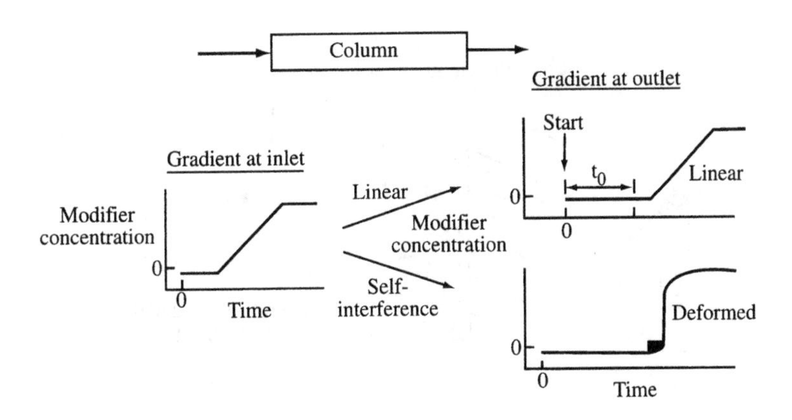

Figure 7.2 Linear vs deformed gradients.

adsorber with respect to the modulator. In the case of a step gradient, the column's operation would be the same as an adsorption system with the breakthrough profile of the modulator following the idealized representation illustrated in Figure 6.38 in Chapter 6. In the case of a time-varying gradient, illustrated in Figure 7.2, deformation will occur by the same principles, but part of the shape of the modulator's breakthrough curve will be spread out by the continual change of its concentration at the column's inlet.

ION EXCHANGE GRADIENT CHROMATOGRAPHY

Ion exchange chromatography is used for the purification of enzymes, proteins from murine cell lines, recombinant proteins from *E. coli*, and gene therapy vectors. Examples of how ion exchange chromatography is used for several applications are presented here to place the subsequent discussion of the modeling of gradient chromatography in a practical context. The presentation of these examples is preceded by a short summary of the principles of operation of an anion exchange gradient elution column.

7.2 LINEAR GRADIENT ELUTION IN ION EXCHANGE CHROMATOGRAPHY IS BASED ON EXCHANGE OF A MULTIVALENT PROTEIN FOR A MONO- OR DI-VALENT SALT

A linearly increasing salt gradient is initiated at the same time or shortly after the sample is injected. The various proteins in the sample first bind to the stationary phase near the column inlet, as illustrated in Figure 7.3. The binding may help to concentrate the protein onto the stationary phase. Hence this type of chromatography may purify and concentrate a peak at the same time.

The ions on the charged binding sites on the stationary phase near the inlet of the column exchange with the protein. The protein is held there until an appropriately high modulator concentration (NaCl in this case) is reached as illustrated in Figure 7.4. When the modulator (an ion) exchanges for the protein, the protein band moves down the column at the velocity of the modulator concentration that causes it to desorb.

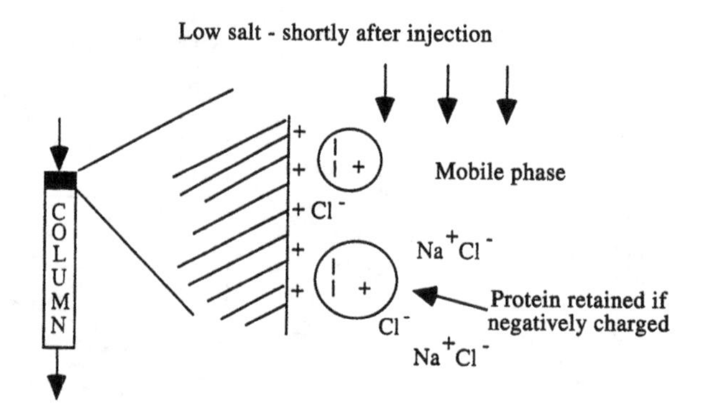

Figure 7.3 Protein ion exchange before start of gradient.

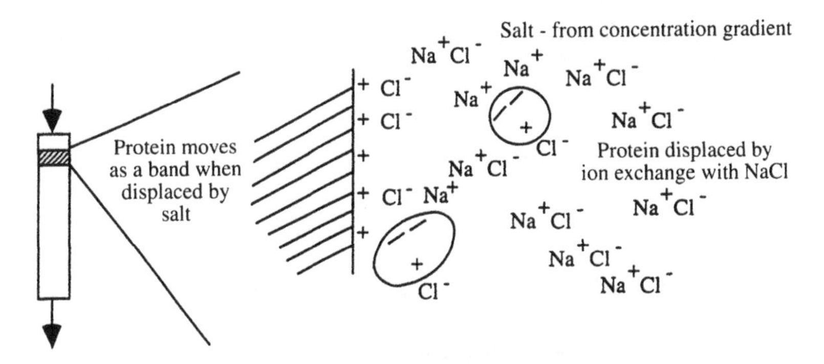

Figure 7.4 After start of gradient, protein begins to move down column.

7.3 SEPARATION IN ION EXCHANGE GRADIENT CHROMATOGRAPHY IS DRIVEN BY THE TIME-DEPENDENT INCREASE IN SALT CONCENTRATION IN THE MOBILE PHASE

A change in the initial salt concentration, or the time at which the salt concentration is increased, will alter the retention time of the protein. Figure 7.5 superimposes the same protein peak at different elution times on the same graph. This figure illustrates how the same protein elutes at approximately the same salt concentration $C_{salt,1}$ but at different retention volumes (denoted $V_{1,a}$ and $V_{1,b}$) or retention times ($t_{1,a}$ or $t_{1,b}$) when the starting concentrations of the salt introduced with start of the gradient (denoted by $C_{salt,a}$ and $C_{salt,b}$), are different. The shape of the gradient is assumed to be the same at the column inlet and outlet. As shown in Figure 7.5, the salt gradient initiated at concentration $C_{salt,a}$ causes the protein peak to elute at volume $V_{1,a}$ since the concentration $C_{salt,1}$ is attained sooner than when the starting salt concentration is $C_{salt,b}$. The volume V_o denotes the interparticle void volume in the column that corresponds to ε_b. The salt gradient does not begin to emerge until V_o is attained, since the salt solution pumped into the column must first displace the buffer (that contains salt concentration $C_{salt,o}$) from the column. The times that correspond to the elution volumes in Figure 7.5 are indicated by t_o, $t_{1,a}$, and $t_{1,b}$, while the elution volume and time for the modulator peak are given by V_S and t_s, respectively.

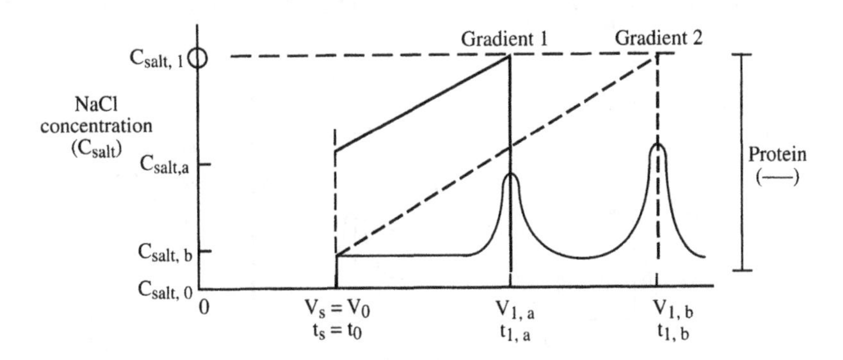

Figure 7.5 Effect of initial gradient concentration ($C_{salt,a}$ vs $C_{salt,b}$) on peak elution ($V_{1,a}$ vs $V_{2,b}$).

The instantaneous concentration of salt at the outlet of the column, C_{salt} in Figure 7.5 is a function of the elution volume, V, as given by Eq. (7.1) when the gradient is initiated at the same time as when the sample is injected:

$$C_{salt}(V) = M(V - V_o) + (C_{salt}(V_o)) \quad \text{when } V \geq V_o \tag{7.1}$$

where

M = slope of salt gradient (g/mL^2 or gmole/L^2) at inlet (and outlet) of column = $\Delta I/\Delta V$

ΔI = change in salt concentration, g/mL or gmole/L

V = volume of mobile phase passed through the column between injection of the sample (V = 0) at start of gradient and the volume at which the salt concentration is measured at the column's outlet, mL or L

V_o = void volume of column—corresponds to volume of mobile phase between particles, mL or L, as well as the volume when the salt gradient first begins to emerge from the column

ΔV = change in (difference of) elution volume, mL or L

$C_{salt}(V)$ = salt concentration at outlet of column as a function of volume of mobile phase passed through the column since initiation of gradient, g/mL

$C_{salt}(V_o)$ = concentration of salt in buffer at column inlet, immediately after gradient is initiated, g/mL (corresponds to buffer SB or C in Figure 7.1)

Otherwise, $C_{salt} = C_{salt,o}$ when $V < V_o$ and where $C_{salt,o}$ is the initial concentration of salt in the buffer at the column inlet immediately before gradient is initiated (usually $C_{salt,o}$ = 0, g/mL). $C_{salt,o}$ would correspond to the salt concentration in buffer A in Figure 7.1.

Separation between two different proteins is achieved when the two components elute at different concentrations of the modulator, as represented in Figure 7.6 for peaks 1 and 2 that elute at salt concentrations $C_{salt,1}$ and $C_{salt,2}$. The salt gradient is linear at both inlet and outlet and is generated at a constant flowrate in this example. The interaction at the surface of an anion exchange stationary phase as illustrated in Figure 7.3 results in a separation if the proteins have different charges at the same pH. In Figure 7.6 protein 1 is less charged than protein peak 2. Therefore protein 1 elutes at a lower salt concentration ($C_{salt,1}$) than protein 2 ($C_{salt,2}$). As is the case for a size exclusion support, large molecules may not be able to penetrate the stationary phase. Therefore they will not interact with ion exchange groups

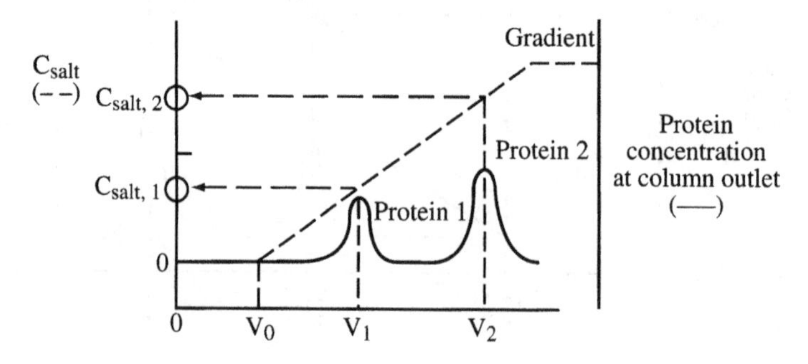

Figure 7.6 Separation of proteins in a gradient.

that are located on the internal surfaces of the stationary phase particles. Binding requires that the charged sites be accessible. This is discussed in Section 8.26 of Chapter 8.

Linear gradient elution chromatography enhances the resolution of multiple peaks by reducing the slope of the gradient (Yamamoto, 1987). However, the concentration of the salt (i.e., ionic strength) at which the peak elutes decreases and the peak becomes wider when the slope of the gradient is decreased (Yamamoto et al., 1983a,b).

7.4 GRADIENT CHROMATOGRAPHY IS OFTEN CARRIED OUT IN THE MIDDLE OF A PURIFICATION SEQUENCE

The positioning of an ion exchange chromatography step in a protein purification sequence is illustrated in Figure 7.7, for a hypothetical protein (an enzyme) (Burgess, 1987). The enzyme

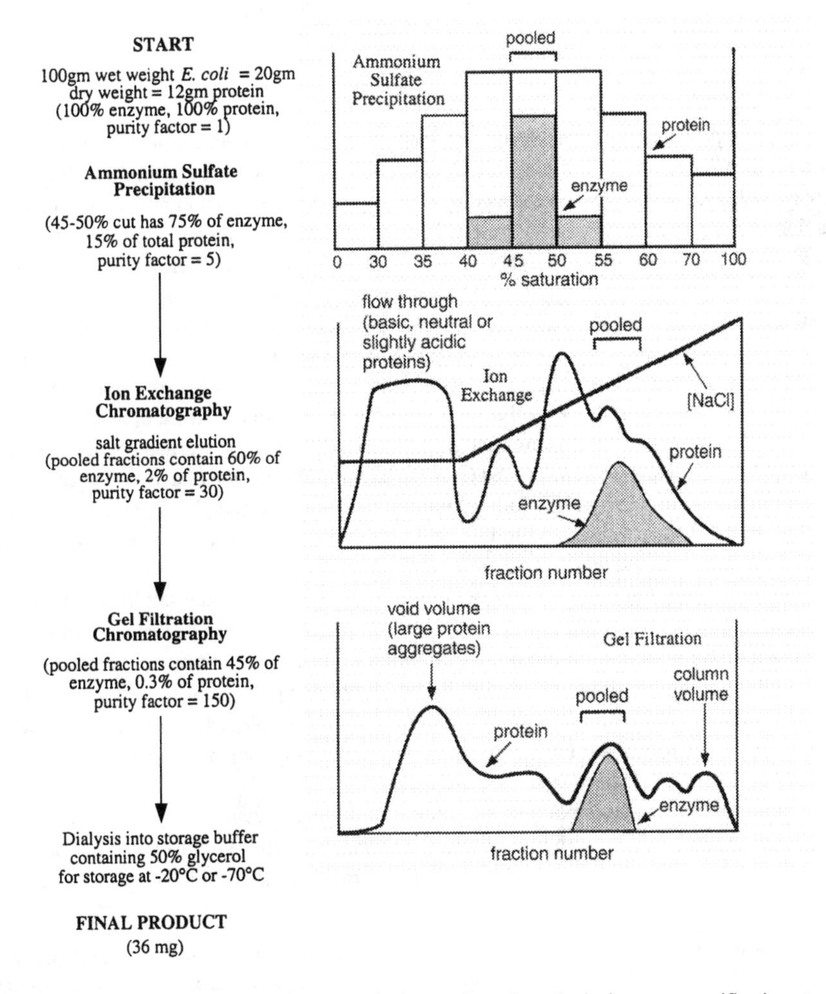

Figure 7.7 Staging of chromatography separation steps for a hypothetical enzyme purification step. (Reproduced from Burgess, 1987, fig. 2, p. 76.) Reprinted with permission of John Wiley & Sons, Inc.

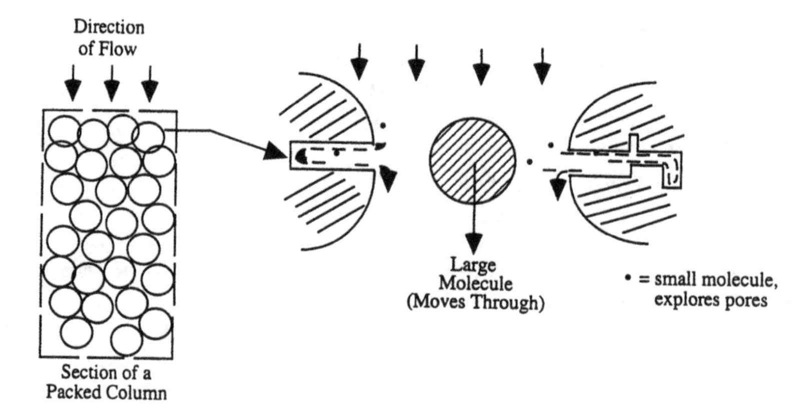

Figure 7.8 Size affects molecule retention.

is initially separated by ammonium sulfate precipitation, with the desired protein being precipitated in a solution that is 40 to 50% saturated with ammonium sulfate (see Section 4.3). The fractions that are pooled (grey area) are then separated by centrifugation, redissolved in buffer, and run through an ion exchange chromatography column. The enzyme adsorbs onto the stationary phase, while other proteins, nonprotein contaminants, and ammonium sulfate flow through the column, as indicated in the middle panel of Figure 7.7. The enzyme that is eluted by the salt gradient (grey-shaded peak) is then collected, pooled, and passed over a size exclusion column. In this example, 36 mg of final product is obtained from a starting weight of 12 g protein from 20 g dry weight of cells.

Proteins that have different molecular weights are separated by gel filtration due to differences in the diffusional pathways that are accessible to the different sized molecules, as shown in Figure 7.8. A size exclusion chromatography step will often follow gradient chromatography so that NaCl ions can also be separated from the larger molecule that eluted in the gradient. Size exclusion chromatography often follows ion exchange or reversed phase gradient chromatography steps in order to separate modulator from the product molecule as well as the product molecule from other proteins (Ogez et al., 1989).

The example in Figure 7.7 shows that only partial separation of the salt from the enzyme is achieved in the size exclusion step, and the enzyme is one of the collected fractions. For this example, dialysis (the last step indicated in Figure 7.7) removes most of the remaining salt and low molecular species (Burgess, 1987). The principles and practice of size exclusion chromatography as well as criteria for selecting different stationary phases are explained in Sections 8.31 to 8.39 in Chapter 8. Methods for removing salts using membranes and dialysis are discussed in Sections 3.3 to 3.8, and 3.11 in Chapter 3.

Combinations of anion followed by cation exchange chromatography, and vice versa, are also used. An example is given by recombinant human prolactin[2] produced in a murine cell expression system. In this case a cation exchange chromatography step precedes the anion ex-

[2]Prolactin is a hormone associated with pregnancy and lactation. It is a 199 amino-acid polypeptide (molecular weight of 23 to 25 kD) that undergoes post-translational modifications of glycosylation, phosphorylation, dimerization, and enzymatic cleavage. More than 85 diverse effects are attributed to this hormone (Cole et al., 1991; Price et al., 1995).

change step (7). The study of the microheterogeneity of this polypeptide, and the role of modified forms of prolactin on its in vivo properties, is facilitated by the availability of gram quantities of the purified protein (Cole et al., 1991).

7.5 PURIFICATION OF RECOMBINANT PROTEINS FROM *E. COLI* REQUIRES STEPS THAT FIRST DISSOLVE AND THEN REFOLD THE PROTEINS

Proteins that are obtained from recombinant *E. coli* would be purified using chromatography steps that are similar to the example given in Figure 7.7, although a number of recovery steps precede chromatography. Intracellular particles of precipitated protein known as inclusion bodies form in 80% or more of the cases where the protein has been overproduced in genetically engineered *E. coli*. (Burgess, 1987) gives an example of recovery of the sigma subunit of *E. coli* RNA polymerase that was overproduced 200-fold in a recombinant *E. coli*. The protein has a molecular weight of 70,000 (= 70 kD) and is considered to be acidic, since it carries a net negative charge. The sequence for recovering and purifying this protein consists of the following steps (Burgess, 1987).

1. Cells are lysed on ice using Tris, lysozyme, and EDTA followed by 0.05% sodium deoxycholate (10% of the protein in this extract is the target molecule).
2. Cell debris and insoluble target molecule in the inclusion bodies are centrifuged.
3. The pellet (centrifuged solids) from step 2 is washed to remove impurities (70% of the pellet is the target molecule).
4. The pellet is dissolved in 6 M GuHCl and 0.1 mM dithiothreitol. This also denatures (unfolds) the protein. The protein is renatured when buffer is added to the GuHCl in order to dilute it by a factor of 60.
5. The precipitate that forms during the dilution step is removed by centrifugation leaving a solution (supernate) in which the protein is 90% target molecule.
6. The supernate from step (5) that contains 90% of the target protein is purified by anion exchange chromatography followed by size exclusion (gel filtration) chromatography.

This sequence illustrates how the ion exchange chromatography step typically occurs later in a purification sequence, and is followed by size exclusion.

7.6 ANION EXCHANGE CHROMATOGRAPHY IS PROMINENT IN THE PURIFICATION OF BLOOD PRODUCTS: PRO- AND ANTICOAGULANT FACTORS

Many blood products may be derived through human plasma fractionation. These include procoagulant factors VIII and IX, anticoagulants antithrombin III, and protein C (Clark et al., 1989). Another example of protein purification using gradient chromatography is given by protein C from the milk of transgenic pigs. Anion exchange chromatography of this protein employs a NaCl and $CaCl_2$ gradient (Drohan et al., 1994). Protein C is a member of the plasma vitamin K–dependent protein family that undergoes a complex sequence of co- and post-translation modification steps before it is transformed to its anticoagulant, active form. This has made it difficult to produce this protein by cell culture systems. An alternate approach has been proposed in which the protein is expressed in porcine milk from transgenic pigs at levels ranging from 0.1 to 1 mg/mL. The protein would then be recovered and purified from relatively large quantities of milk. The projected demand for this protein is 100 kg/year. A key to success is the ability to purify the protein.

7.7 GENE THERAPY VECTORS CAN BE PURIFIED BY ANION EXCHANGE CHROMATOGRAPHY USING PHOSPHATE BUFFER AND KCl GRADIENTS

Recent advances in gene therapy research and about 100 approved gene transfer clinical protocols have promoted the need for manufacture and purification of human gene therapy products.[3] One form of gene therapy could be based on use of DNA that is delivered in the form of a plasmid. This requires that manufacturing processes be developed for pharmaceutical-grade plasmid DNA. A standard laboratory method for purification of plasmids has been cesium chloride/ethidium bromide ultracentrifugation. This approach can not be used for therapeutic plasmid DNA, since the reagents are mutagenic (Marquet et al., 1995). Analytical scale anion exchange chromatography using a polyacrylate (polymeric) DEAE stationary phase and a NaCl in phosphate buffer gradient has been successfully employed (Lahijani et al., 1996).

The ability for ion exchangers to purify plasmid DNA's was noted as early as 1978. Colman et al. (1978) showed that the mixed mode exchanger, hydroxyapetite (discussed in Chapter 8), removed all detectable protein and RNA in a solution containing plasmid DNA from *E. coli* without affecting the conformation of the plasmid (Colman et al., 1978). The plasmid DNA was eluted in 0.3 M phosphate buffer while larger linear DNA molecules (which must be removed) are retained by the hydroxyapetite (Colman et al., 1978). Small amounts of chromosomal DNA co-purified with the plasmid DNA. The method was found to work for plasmids over the molecular weight range of 3,000 to 14,000 kD.

Selective and reversible retention of small double-stranded DNA molecules (i.e., plasmid DNA) occurs at conditions known to prevent binding of RNA and protein (Colman et al., 1978). DEAE chromatography of plasmids derived from *E. coli* K12 cells showed the order of elution to be tRNA followed by rRNA, and finally the plasmid peak. The separation was obtained by a linear gradient of 0.225 to 1.5 M KCl, in 5 M urea, 20 mM phosphate buffer at pH 6.9, with the chromosomal DNA co-eluting with the plasmid DNA. Chromosomal DNA could be removed prior to the DEAE chromatography by precipitating the DNA by boiling the solution for 40 seconds (Copepella et al., 1987a,b).

[3]Clinical focus for gene therapy products has been on viral vector-based delivery systems. The ex vivo approach is based on removal of cells from a patient, culturing the cells, inserting the therapeutic gene, and then returning the cells containing the therapeutic gene to the patient. This approach is patient specific and labor-intensive (Horn et al., 1995). In vivo injection of a viral vector is an alternate approach that is appealing since it is patient generic, but safety concerns must be addressed, since the gene introduced by the virus may be integrated into the cell's genome and cause unexpected side effects such as activation of cancer-causing oncogenes (Marquet et al., 1995). Vectors derived from adeno-associated virus (AAV) appear to hold promise (Marquet et al., 1995). Alternate nonviral vectors for gene therapy have also developed out of safety concerns associated with viral-vectors. Nonviral vectors are receptor mediated, directly injected as naked DNA, or cytofectin-mediated. Direct injection of naked DNA into muscle tissue represents form of vaccine delivery. Direct DNA injection may be useful in cancer therapy in which transient expression in tumor cells may be sufficient to induce destruction of the tumor cells.

7.8 PROCESS-SCALE PURIFICATION OF PLASMID DNA EMPLOYS A SEQUENCE OF ANION EXCHANGE AND SIZE EXCLUSION CHROMATOGRAPHY

Marquet et al. (1995, pg. 32) points out that "it is prudent to include two chromatography steps of contrasting chemistry. Anion exchange may be used to remove RNA from the plasmid DNA product and to reduce the overall level of trace host cell contaminants such as chromosomal DNA." Consequently RNA is removed by the anion exchange column, and then the plasmid DNA can be subjected to a polishing step when processed through the size exclusion chromatography column. Since plasmid DNA, host cell DNA, and endotoxins have similar affinities for an anion exchange stationary phase, these components must be resolved by other means, namely by size exclusion chromatography which is able to differentiate these molecules on the basis of differences in their molecular weight. Indeed, size exclusion chromatography is the only chromatography step in a purification sequence that removes higher molecular weight RNA and other nucleic acid impurities from the plasmid DNA (purification sequence in Figure (7.9). The higher molecular weight plasmid DNA (MW = 3000 to 6000 kD) elutes ahead of the lower molecular weight impurities (as shown in Figure (7.10) that include endotoxins (MW = 10 to 300 kD), as well as some RNA fragments. The purification obtained is shown by the electrophoresis patterns (Figure 7.11) and by the comparison of analytical chromatograms of the load and eluted samples (Figure 7.12) (Horn et al., 1995; Marquet et al., 1995).

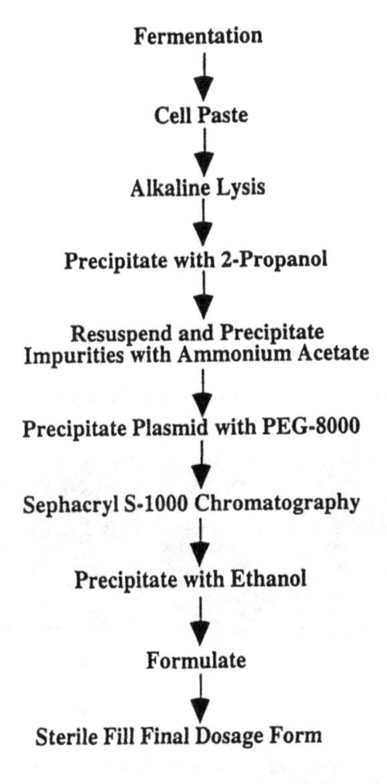

Figure 7.9 Process flow diagram showing position of size exclusion chromatography using Sephacryl S-1000 relative to other processing steps (from Horn et al., 1995, fig. 1, p. 569).

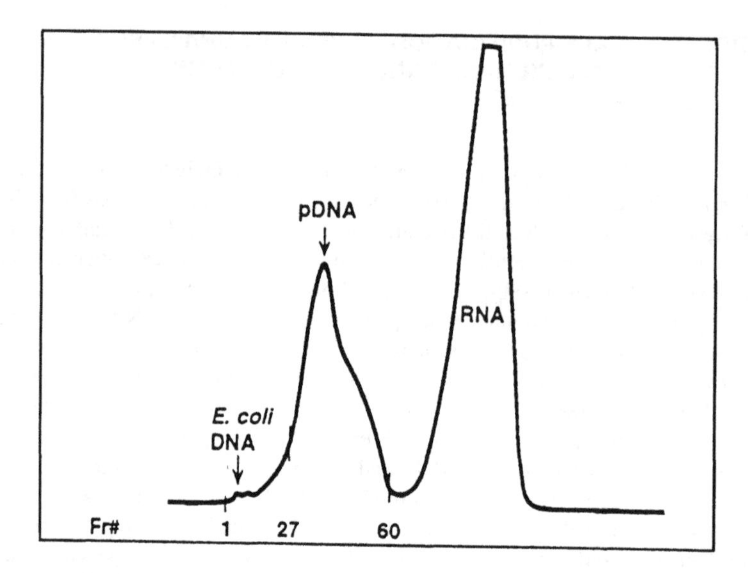

Figure 7.10 Elution profile from Sepharyl S-1000 chromatography column conditions: Column volume of approximately 900 mL, with a bed height of 85 cm. Sample volume of 10 mL. Sample contained plasmid from *E. coli* strain DH10B dissolved in 10 mM Tris, 1 mM EDTA, 150 mM NaCl at pH 8.0. pDNA denotes plasmid DNA (from Horn et al., 1995, fig. 3, p. 569).

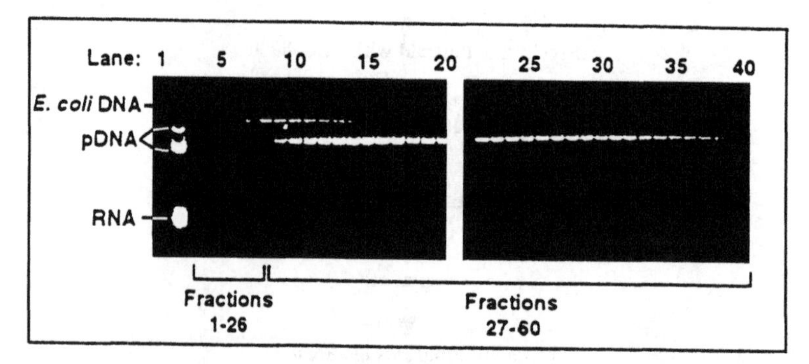

Figure 7.11 Gel analysis (0.8% agarose) of resulting 5 mL fractions collected at the column's outlet are shown underneath the chromatogram. 1 and 2 is column load; lanes 3 to 8, every 3rd to 4th fraction from fractions 1 to 26; lanes 9 to 40; every fraction from 27 to 60 (from Horn et al., 1995, fig. 2, p. 569).

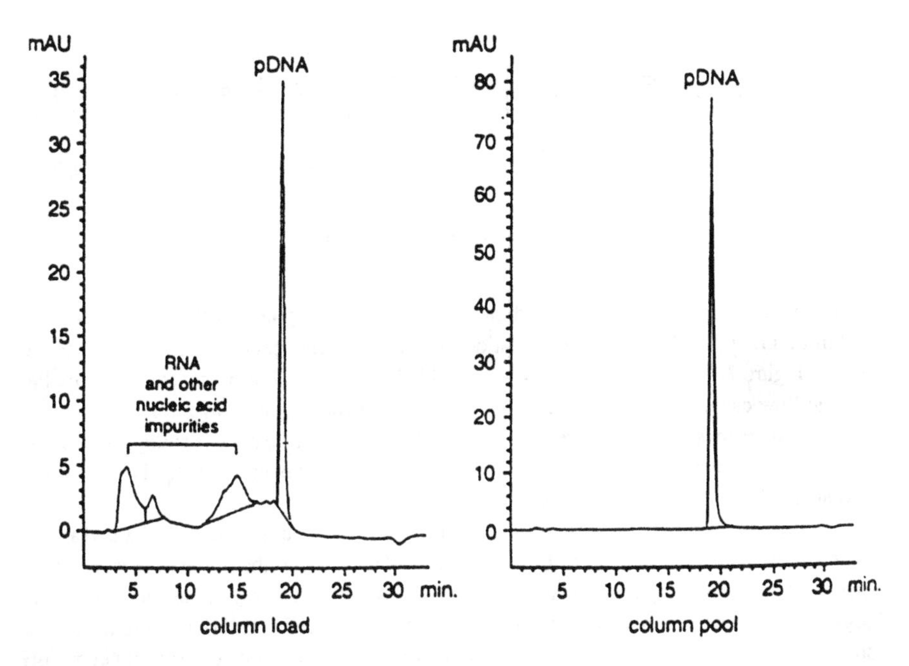

Figure 7.12 Comparison of load and eluted sample (pDNA peak) by analytical chromatography (from Horn et al., 1995).

7.9 AN ION-EXCHANGER IS A SOLID MATERIAL THAT CARRIES EXCHANGEABLE CATIONS OR ANIONS

A cation exchanger has a negatively charged surface and replaces one positively charged cation with another. An anion exchanger is positively charged, and exchanges negatively charged anions. Ion-exchangers are represented in Figures 7.13 through 7.16 as a solid matrix to which ionizable groups are attached. An acidic or cation exchange material is a weak ion-exchanger if it contains hydroxyl or carboxylic acid groups on the surface as represented in Figure 7.13 and a strong ion-exchanger if it has sulfonic acid groups as represented in Figure 7.14. A strong cation exchanger is represented by a sulphono (5) sulphonomethyl (SM) and sulphonopropyl (SP) groups.

Similarly basic or anion-exchangers can be either weak or strong. Weakly basic ion-exchangers are associated with the aminoethyl (AE) and diethylaminoethyl (DEAE) functionalities shown in Figure 7.15. Strong anion-exchangers are represented by triethylaminomethyl

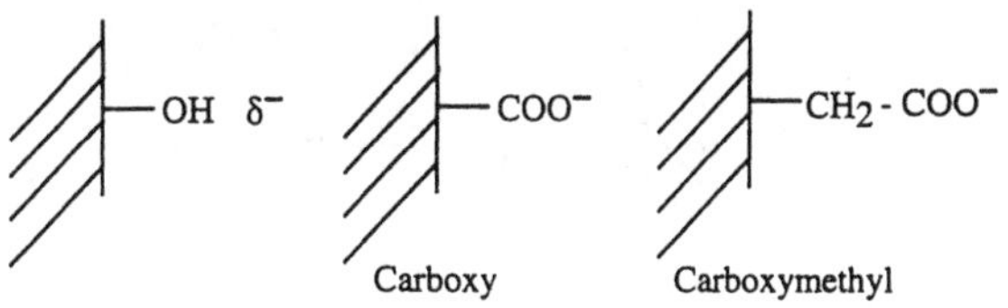

—OH δ^- —COO^- —CH_2 - COO^-

Carboxy Carboxymethyl

Figure 7.13 Weak cation exchangers.

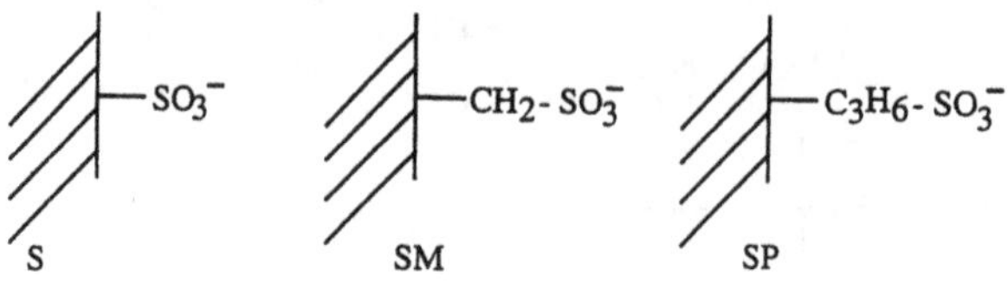

Figure 7.14 Strong cation exchangers.

(TAM), triethylaminoethyl (TEAE), and diethyl-2-hydroxypropylaminoethyl (QAE) as illustrated in Figure 7.16. The capacities of cation- and anion-exchangers as a function of pH are shown in Figure 7.17. The weakly acidic and weakly basic materials are ionized over a smaller pH range (low capacity is equivalent to low extent of ionization), as shown in Figure 7.17. A weakly acidic resin is charged above a pH of 6, while a weakly basic resin is charged below the pH of 6.0. The strong cation and strong anion resins are charged over a broad pH range, while the weakly acidic or anionic resins are not (Figure 7.17).

The charge on a protein is determined by its sum of positively and negatively charged amino acids residues. This is a function of the pH of the buffer in which the protein is dissolved. A protein that contains many aspartic and glutamic acid residues is likely to have a net negative charge at close to neutral pH and is referred to as an acidic protein. If lysine, arginine, and/or histidine residues are dominant, the charge on the protein is likely to be positive at neutral pH and is referred to as a basic protein. The pH at which the net charge is zero is referred to as the isoelectric point. The isoelectric points of various proteins are generally found in the range of pH 4.5 to 8.5. The effect of salt concentration on the protein's charge is generally small as compared to the effect of pH. This is illustrated by the titration curve for BSA shown in Figure 7.18 (Yamamoto et al., 1988).

Proteins are heterogeneous polymers of amino acids. The amino acids give the proteins zwitter ionic character so that the same protein can bind to either a cation or an anion exchanger. The charge on a protein changes with pH, with adsorption of the protein to a cation-exchanger occurring at a pH below the protein's isoelectric point and to an anion-exchanger above the protein's isoelectric point, as represented in Figure 7.19 (Yamamoto et al., 1988).

Discrimination between proteins by ion exchange chromatography is a function of both the charge and the manner in which the charges are clustered on the protein's surface. Charged

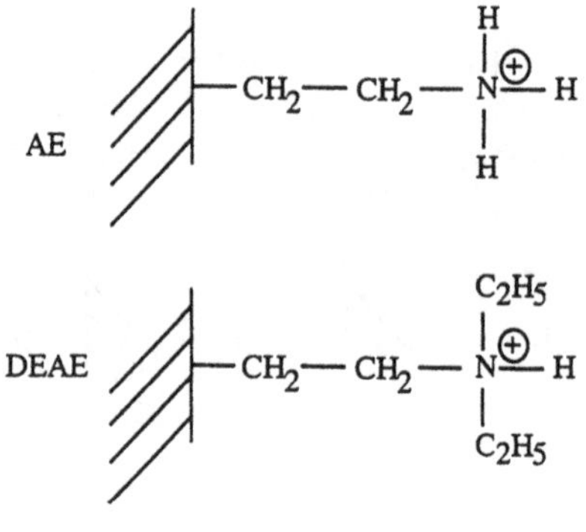

Figure 7.15 AE and DEAE (weak) anion exchangers.

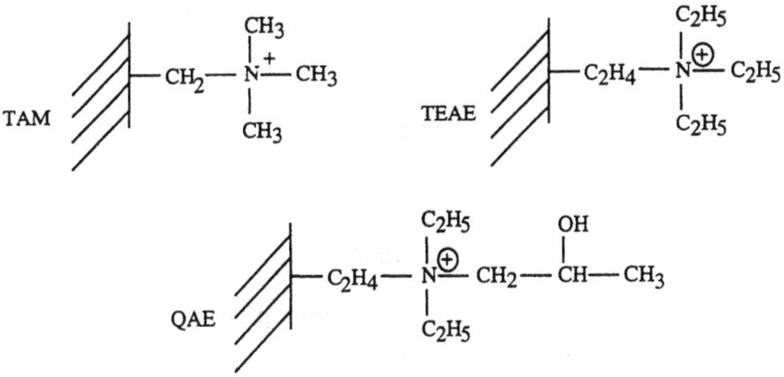

TAM: $-CH_2-N^{+}(CH_3)_3$

TEAE: $-C_2H_4-N^{\oplus}(C_2H_5)_3$

QAE: $-C_2H_4-N^{\oplus}(C_2H_5)_2-CH_2-CH(OH)-CH_3$

Figure 7.16 TAM and QAE strong cation exchangers.

amino acid residues may be distributed uniformly on the surface, or in regions of highly positive or negative charge. Hence the net charge on a protein can be different from the binding charge. Consider a globular protein, such as represented in Figure 7.20, that has a binding charge of –3 even though the net charge is –4. The protein will orient itself so that the free energy decrease upon binding to the stationary phase is minimized. The protein itself is likely to be too large for all of its surface charges to interact (Kopaciewicz et al., 1983). Not all adsorbed ions at the stationary phase's surface (denoted by S in Figure 7.20) are available for exchange, since the previously adsorbed protein will shield a number of them (Bosma and Wesselingh, 1998). Studies with BSA have indicated that at its isoelectric point, this protein has about 90 positive charges and 90 negative charges, but only 15 charged groups are located in the contact area of the protein. Most researchers have found that the binding charge for a variety of proteins ranges from about 2 to 8 (Bosma and Wesselingh, 1998).

While the net charge of the protein at its isoelectric point is zero, the protein at its isoelectric point can still be retained by a strong anion exchange or strong cation exchange column. This was shown by Kopaciewicz et al. (1983) for β-lactoglobulin (MW = 35 kD, pI = 5.13), BSA (MW = 69 kD with pI = 4.98, 5.07, or 5.18 depending on its source), ovalbumin (MW = 43.5 kD, pI = 4.7), conalbumin (chicken egg white, MW = 77 kD, pI = 6.0, 6.3, 6.6), chymotrypsin (MW = 25 kD, pI = 8.8, 9.2, 9.6), and ribonuclease (MW = 13.5 kD, pI = 8.7, 8.8). Zero retention, as measured by elution of the protein near the column void volume, was observed only for 3 of the 14 proteins studied: carbonic anhydrase, myoglobin, and immunoglobulin G (Kopaciewicz et al., 1983).

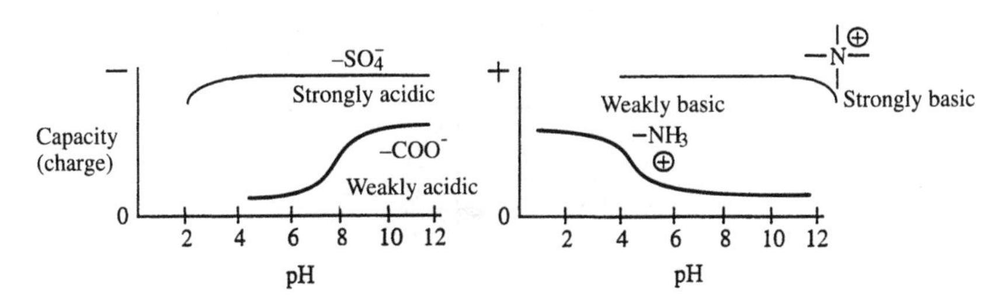

Figure 7.17 Charge vs pH for ion exchangers.

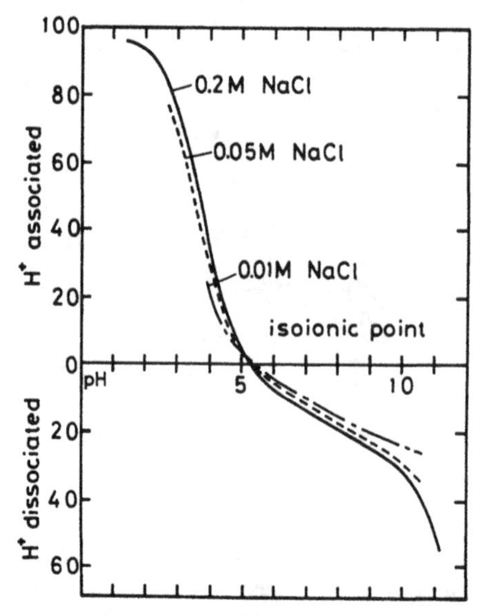

Figure 7.18 Titration curves for bovine serum albumin at NaCl concentrations of 0.01 to 0.2 M at 25°C. The dissociated or associated H^+ was calculated on the basis of a bovine serum albumin molecular weight of 65,000. (Adapted with permission from Marcel Dekker, Inc., from Yamamoto et al., 1988, fig. 3.10, p. 135.)

When adsorption was studied for BSA with respect to a strong anion exchanger that is charged over the range of pH's used to measure protein adsorption, it was found that the maximum adsorption capacity of the ion exchanger for BSA did not depend on pH (Bosma and Wesselingh, 1998). As pointed out by Kopaciewicz et al. (1983), strong ion exchangers (quaternary amines and sulfonic acids) are essentially always charged (as represented in Figure 7.17), and they retain the same charge density regardless of operating pH. Weak ion exchangers are seldom completely ionized and have a ligand density that is a function of pH of the mobile phase. Hence a fivefold variation of ligand density may occur over the normal operating range of the weak ion exchanger and hardly any for the strong ion exchanger. Retention properties are difficult to predict because both the net charge on the ion exchanger and the protein vary at the

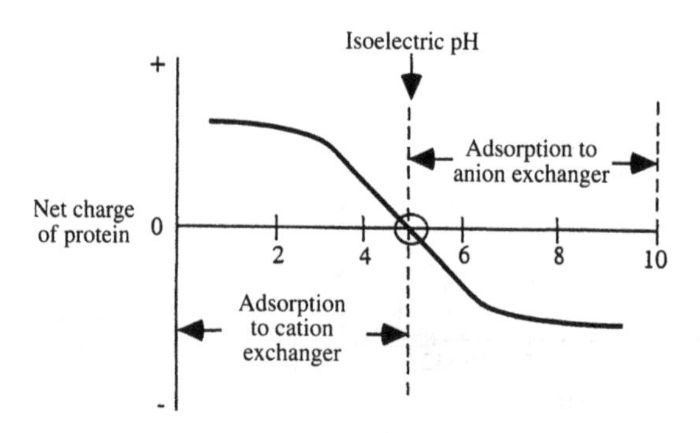

Figure 7.19 Charge vs pH for proteins (adopted from Yamamoto et. al., 1988).

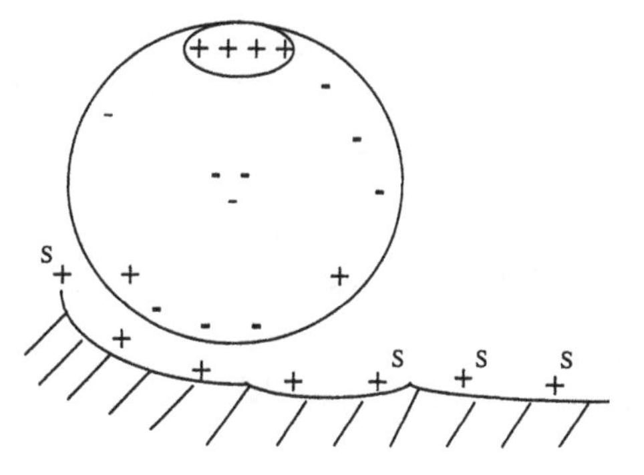

Figure 7.20 Globular protein (circle) at surface of stationary phase.

same time (compare Figures 7.17 and 7.19). Consequently, in modeling ion exchange chromatography, one must consider the type of ion exchanger, the number of charges on the protein's surface that interact with the ion exchanger, steric shielding effects, and charge distribution on the protein's surface.

7.10 RETENTION TIMES AND CAPACITY FACTORS OF CHARGED SPECIES IN ION EXCHANGE CHROMATOGRAPHY ARE PROPORTIONAL TO THEIR CHARGE (WITH CONTRIBUTIONS BY A. VELAYUDHAN)

The ion exchange phenomenon for the single ionic species A^+ and B^+ is represented by

$$\overline{A}^+ + B^+ \Leftrightarrow \overline{B}^+ + A^+ \tag{7.2}$$

where

$\overline{A}^+$ = positively charged counterion adsorbed onto the surface of a cation exchange resin; the bar above A^+ denotes that this species is associated with the resin phase

A^+ = positively charged ion A in the mobile or fluid phase

$\overline{B}^+$ = positively charged ion B^+ associated with the resin phase

B^+ = positively charged ion in the mobile or fluid phase

The distribution coefficients for the two species are defined in the same way as previously developed in Chapter 6:

$$K_{D,A} = \frac{\overline{A}^+}{A^+}, \quad K_{D,B} = \frac{\overline{B}^+}{B} \tag{7.3}$$

where $K_{D,A}$ and $K_{D,B}$ represent the distribution of either species A or B on the solid phase with respect to their respective equilibrium concentrations in the fluid or mobile phase. The ratio of the distribution coefficients given in equation (7.3) results in the definition of the separation factor for the two species, when both are found in the same solution as represented by the sequence (7.2). The separation factor of A with respect to B, α_{AB}, is given by

$$\alpha_{BA} = \frac{K_{D,B}}{K_{D,A}} = \frac{\dfrac{\overline{B}^+}{B^+}}{\dfrac{\overline{A}^+}{A^+}} = \frac{\overline{B}^+}{B^+} \cdot \frac{A^+}{\overline{A}^+} \tag{7.4}$$

This is equivalent to the distribution coefficient for A with respect to B:

$$K_{D,AB} = K_{D,B}\left(\frac{A^+}{\overline{A}^+}\right) \tag{7.5}$$

This representation can be given in a more general form by defining S as the counterion initially associated with the resin, before the resin is brought in contact with the solution containing A. The coefficient a represents the net charge on species A. The bar above the letter $\overline{A}$ or $\overline{S}$ denotes the association of the counterion S or ion A with the resin phase:

$$a\overline{S} + A \Leftrightarrow \overline{A} + aS \tag{7.6}$$

A distribution coefficient for the ion A relative to the counterion S can be defined by the same rationale as presented in Eqs. (7.2) to (7.4):

$$K_{D,AS} = K_{D,A}\frac{[S]^a}{[\overline{S}]^a} \tag{7.7}$$

where

$a =$ net charge on species a
$K_{D,A} =$ distribution coefficient for A between resin and fluid phases as defined in Eq. (7.3)
$K_{D,AS} =$ equilibrium constant for species A with respect to S in the presence of the ion exchanger

The concentration of the charged sites (i.e., binding sites) on the resin will be the sum of the number or concentration of the counterions on the surface and the number or concentration of the exchanged species $\overline{A}$ with charge a as given in

$$\Lambda = a[\overline{A}] + [\overline{S}] \tag{7.8}$$

where

$\Lambda =$ number or concentration of binding (charged) sites on the resin
$[\overline{S}] =$ number or concentration of counterions associated with the resin

A special case for Eq. (7.8) is given by $[\overline{S}] >> A^+$ and $[\overline{S}] >> a[\overline{A}]$. This occurs when a small sample is loaded onto the column as would be the case for analytical liquid chromatography, or if the loading of the stationary phase is low. The total charge is therefore approximated by the total amount of counterions associated with the stationary phase:

$$\Lambda = a[\bar{A}] + [\bar{S}] \cong [\bar{S}] \tag{7.9}$$

Equation (7.7) becomes

$$K_{D,AS} = K_{D,A} \frac{[S]^a}{[\Lambda]^a} \tag{7.10}$$

which upon rearrangement gives a distribution coefficient for species A in analytical, ion exchange chromatography:

$$K_{D,A} \cong K_{D,AS} \frac{[\Lambda]^a}{[S]^a} \tag{7.11}$$

This enables a capacity factor to be defined for ion exchange chromatography, since

$$\boxed{k' = \phi K_{D,A} = \phi K_{D,AS} \left[\frac{\Lambda}{S}\right]^a} \tag{7.12}$$

where

ϕ = phase ratio, as defined in Section 5.21, $= (1 - \varepsilon_B)/\varepsilon_B$
k′ = capacity factor as defined in Section 5.21

For the special case where $[\bar{S}] >> A$, the retention plot can be defined by rearranging Eq. (7.12):

$$\boxed{\log k' = -a \log[S] + \log(\phi K_{D,AS}[\Lambda]^a)} \tag{7.13}$$

These simplications provide a useful approach for capturing the change in the distribution coefficient of the protein $K_{D,A}$ (where component A is a protein) as a function of the salt concentration. This idealized representation is illustrated in Figure 7.21 based on the data of Yamamoto et al. (1988, pp. 129–135).

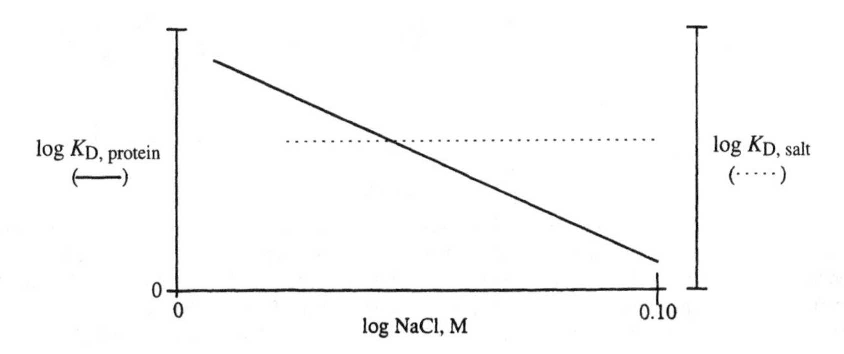

Figure 7.21 Representation of equilibrium vs salt concentration.

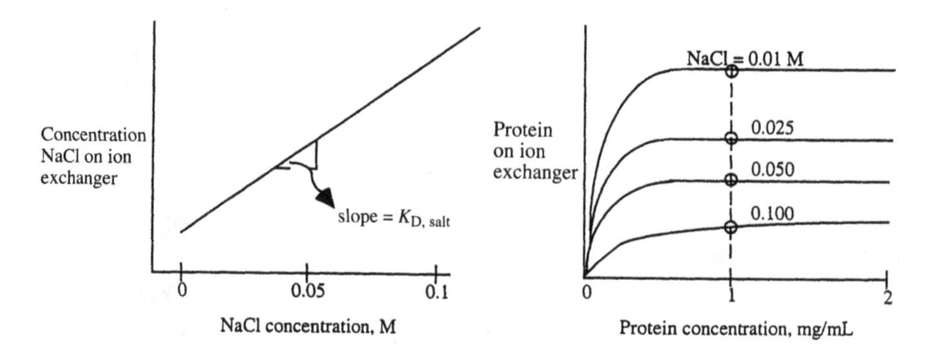

Figure 7.22 Representation of loading of salt or protein on ion exchanger vs. liquid concentration.

The distribution coefficient of the salt is represented as a constant, since the equilibrium between salt and the stationary phase is assumed to be linear as represented on the left side of Figure 7.22. The distribution coefficient for the protein decreases as a function of increasing salt concentration and follows the equilibrium curves represented on the right side of Figure 7.22. The distribution coefficient is also a function of the stationary phase, as illustrated in Figure 7.23. This figure represents the effective distribution coefficient as the difference between the measured coefficient at a given salt concentration, and the coefficient at a very large salt concentration (denoted as K_{crt} in Figure 7.23). K_{crt} represents the distribution coefficient of the pro-

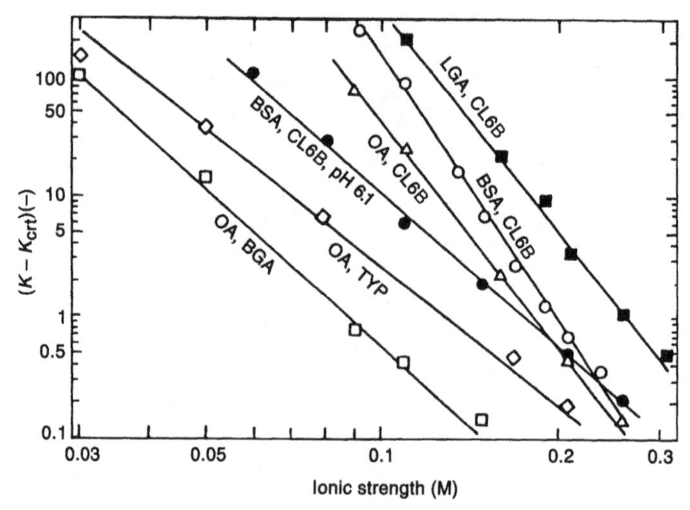

Figure 7.23 Distribution coefficients of bovine serum albumin (BSA), ovalbumin (OA), and lactoglobulin (LGA) vs. ionic strength for different anion exchangers at pH 7.9 unless otherwise indicated. The anion exchangers are: CL6B (DEAE Sepharose CL6B); BGA (DEAE Bio-gel A); and TYP (DEAE Toyopearl 650). K represents $K_{D,protein}$, and K_{crt}, is the distribution coefficient of the protein at a very large salt concentration where separation occurs essentially as size exclusion (ion exchange is negligible) (from Yamamoto et al., 1983, fig. 2, p. 1378, with permission of Wiley-Liss, Inc., a subsidiary of John Wiley and Sons, Inc.).

tein if ion exchange is almost completely suppressed by a high salt concentration. At this condition, protein retention would be due to size exclusion effects, only. If the protein is excluded, $K_{crt} \cong 0$, and $K - K_{crt}$ is equivalent to $K_{D,protein}$ (Yamamoto et al., 1983).

Equation (7.13) represents a useful result. It provides a relation whereby retention times of small injections of a charged species enable estimation of the charge on the molecule being adsorbed (denoted by A in our derivation). An experiment would consist of injection of small samples into the column and eluting each sample with a different concentration of counterion (e.g., NaCl where the Cl^- would compete with the anionic species A for binding sites on the resin). The higher the salt concentration, the less would be the expected retention of the injected species as represented in Figure 7.24 where the salt concentrations are $S_3 > S_2 > S_1$.

The capacity factors for the three salt concentrations would be calculated using Eqs. (5.137) or (5.141) in Chapter 5. Thus, for component A, the capacity factor at salt concentration S_1 would be given by

$$k'_{S1} = \frac{t_r - t_o}{t_o} = \frac{t_1 - t_o}{t_o} \tag{7.14}$$

where t represents the retention time of this component at salt concentration S_1 and t_o is the retention time of an excluded component. Capacity factors at salt concentrations S_2 and S_3 are calculated in the same manner. Their values will follow the order

$$k'_{S3} < k'_{S2} < k'_{S1}$$

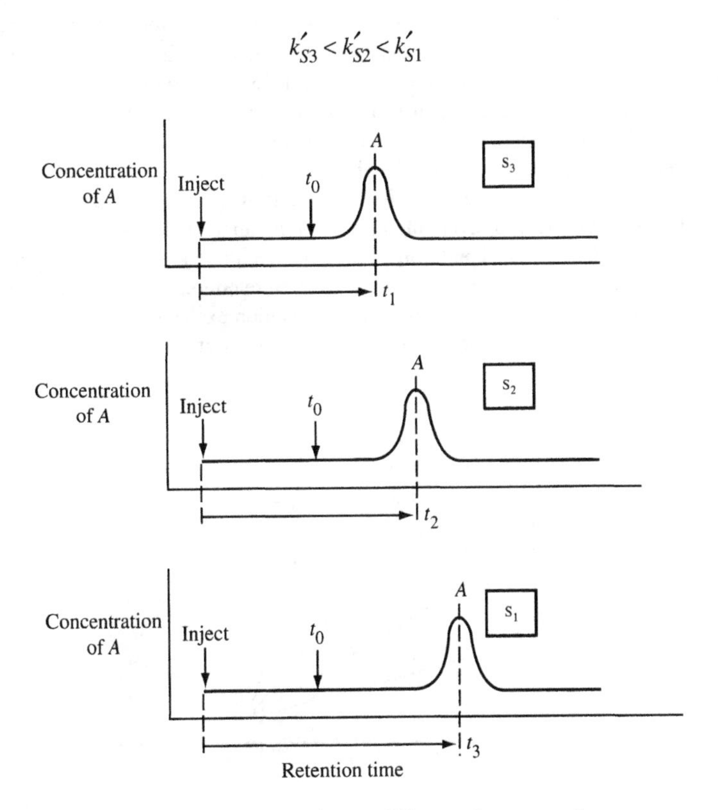

Figure 7.24 Retention times at different salt concentrations.

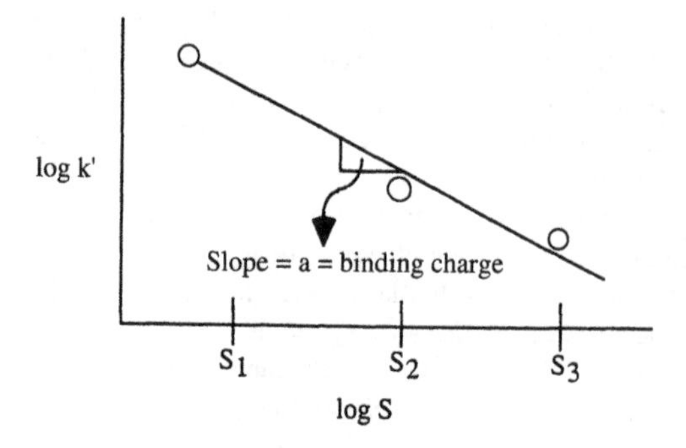

Figure 7.25 Schematic representation of capacity factor as functions of salt concentration.

A plot of log k' against log $[S]$ should follow Eq. (7.13). If a straight line is obtained, its slope will give the binding charge on the species (component A) as indicated in Figure 7.25.

The slope of the curve will be larger if the characteristic binding charge on the species A is larger. This is represented in Figure 7.26 where the slope increases as a increases. As the charge on A increases (e.g., as might occur if the pH is increased above the isoelectric point of the protein as shown in Figure 7.19), the amount of salt required to elute the protein at a given retention time would also increase.

A study on the pH dependence of bovine serum albumin (BSA) showed that the number of binding charges for this protein with respect to a strong anion exchanger was independent of pH between pH 5.2 and 7.9 and ionic strength between 18 and 440 mM acetate or tris buffer. The concept of binding charge provides the basis for defining capacity factors for ion exchange chromatography. The net charge of the protein is not the best indicator since it is much larger than the binding charge (Bosma and Wesselingh, 1998; Kopaciewicz et al., 1983). The proteins will separate if their binding charges are different even if the pI's of the proteins are the same. While small molecules undergo reversible binding in an analytical chromatography column, and can be eluted under isocratic conditions, a macromolecule (e.g., a protein) requires salt for elution to occur, with the type of salt having an effect on retention time in some cases. For example, the relative retention of ovalbumin on a strong anion exchange column decreases with the sequence of the chloride salts of the cations (Kopaciewicz et al., 1983):

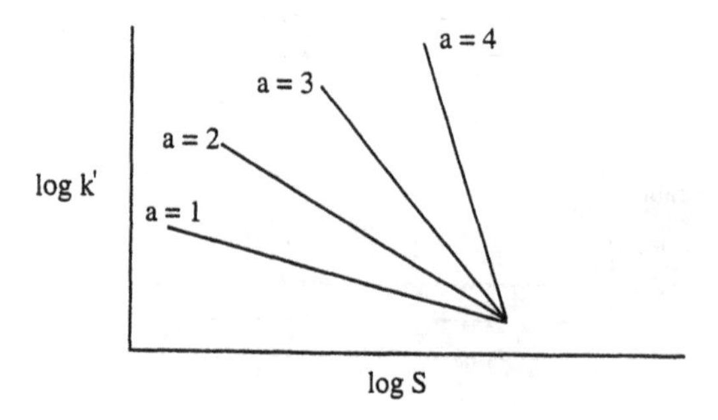

Figure 7.26 Representation of slope as a function of binding charge.

$$Li > Na^+ > K^+ > NH_4^+ > Mg^2 + > Ca^{2+}$$

The acetate salts of sodium and magnesium further decrease retention. (Also see Sections 4.3.)

Example 1. Effect of NaCl Concentration on Retention of a Protein (example contributed by A. Velayudhan)

The retention map of a protein is given by the graph shown in Figure 7.27. The mobile phase is either 0.01 M salt or 1 M salt. Calculate the retention time for the two cases, if the interparticle void volume is 2 mL.

Solution

For 10 mM (0.01 M) salt, and using the definition of the capacity factor (Section 5.21),

$$\log k' = 3 = \log\left(\frac{V_R}{V_o} - 1\right) \tag{7.15}$$

where

V_R = retention volume
V_o = interparticle void volume
k' = capacity factor (read from Figure 7.27 for purposes of this example, where $\log k' = 3$ or $k' = 1000$)
$V_R/V_o = 10^3 + 1 = 1001$
$V_R = V_o\,(1001) = (2)\,(1001) = 2002$ mL

At a flowrate of 1 mL/min, the elution time is 2002 min or about 33.4 hours.

If the salt concentration is 1 M, then $\log k' = 0$ as indicated by Figure 7.27, and Eq. 7.15 becomes

$$\log k' = 0 = \log\left(\frac{V_R}{V_o} - 1\right) \tag{7.16}$$

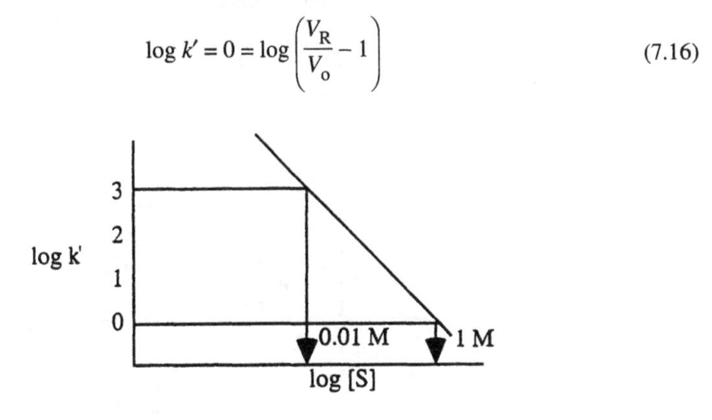

Figure 7.27 Hypothetical plot of capacity factor vs. salt concentration, example 1.

$$10^{o} = 1 = \frac{V_R}{V_o} - 1$$

$$V_R = 2V_o = (2)(2) = 4 \text{ mL}$$

At a flowrate of 1 mL/min, the protein elutes at 4 minutes. The plot in Figure 7.27 also shows that a salt concentration does not exist where there is no retention.

7.11 THE DEFINITION OF THE SEPARATION FACTOR IS BASED ON DIFFERENCES OF BINDING CHARGES OF THE TWO COMPONENTS *A* AND *B*

Different binding charges for components *A* and *B* will result in different slopes for the capacity factor plots as shown in Figure 7.28. By Eq. (7.12) the capacity factors for components *A* and *B* are

$$k'_A = \phi K_{D,AS} \left[\frac{\Lambda}{S}\right]^a \quad \text{and} \quad k'_B = \phi K_{D,BS} \left[\frac{\Lambda}{S}\right]^b$$

where ϕ is the phase ratio. The chromatogram for components *A* and *B* is shown in Figure 7.29, where the feed sample containing both components is loaded at salt concentration S_1. The mobile phase is then changed to consist of the same buffer with salt concentration S_2, followed by a step change to salt concentration S_3. Component *A* would elute first followed by *B*. The capacity factor never attains a value of zero (i.e., $k' \neq 0$), by definition. Hence peak *A* contains minute amounts of *B* and peak *B* trace amounts of *A*, even though the divergence of peak centers is large. Assurance of complete removal of components *A* and *B* from the column requires a regeneration step, such as use of NaOH, in a cleaning step between runs. Rearrangement of Eq. (7.12), gives an expression in terms of a distribution coefficient:

$$\log K_{D,A} = \log\,[K_{D,AS}[\Lambda]^a] - a \log\,[S] \tag{7.17}$$

where $K_{D,AS}$ represents the equilibrium constant between adsorbed species *A* and counterion *S*. Similarly an equation for the second component B can be written as

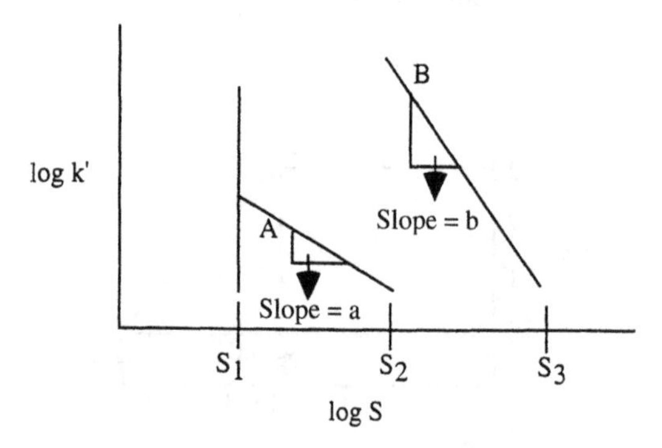

Figure 7.28 Capacity factor plot for proteins with different binding charges.

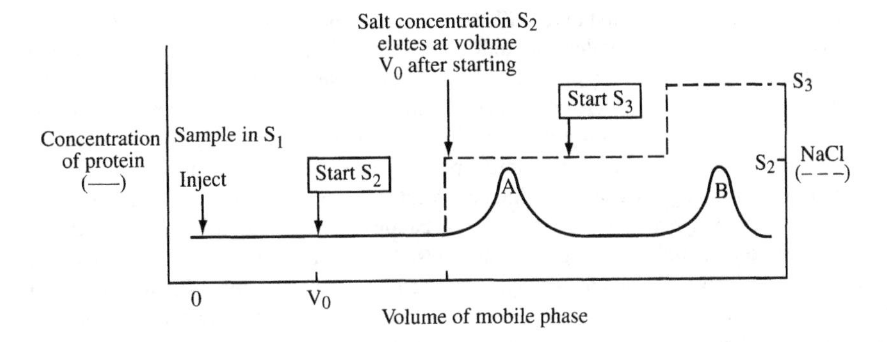

Figure 7.29 Hypothetical chromatogram for components A and B.

$$\log K_{\mathrm{D},B} = \log (K_{\mathrm{D},BS}[\Lambda]^{b}) - b \log [S] \tag{7.18}$$

where b is the binding charge on species B. The separation factor, α_{AB}, between A and B is obtained from the ratio of distribution coefficients (ratio is less than one) or by difference of the log of the distribution coefficients given in Eqs. (7.17) and (7.18):

$$\alpha_{AB} = \frac{K_{\mathrm{D},A}}{K_{\mathrm{D},B}} \quad \text{or} \quad \log \alpha_{AB} = \log \frac{K_{\mathrm{D},A}}{K_{\mathrm{D},B}} = \log K_{\mathrm{D},A} - \log K_{\mathrm{D},B} \tag{7.19}$$

where A and B denote the two components to be separated with A eluting before B. Combination of Eq. 7.17 and 7.18 with equation (7.19) results in an expression for the separation factor:

$$\log \alpha_{AB} = [\log [K_{\mathrm{D},AS}\Lambda^{a}] - a \log S] - [\log [K_{\mathrm{D},BS}\Lambda^{b}] - b \log S] \tag{7.20}$$

which upon rearrangement gives

$$\log \alpha_{AB} = \log \left[\frac{K_{\mathrm{D},AS}[\Lambda]^{a}}{K_{\mathrm{D},BS}[\Lambda]^{b}} \right] + (-a + b) \log [S] \tag{7.21}$$

or

$$\boxed{\log \alpha_{AB} = \log \left[\frac{K_{\mathrm{D},AS}}{K_{\mathrm{D},BS}} [\Lambda]^{a-b} \right] - (a - b) \log [S]} \tag{7.22}$$

Equation (7.22) applies to the case where the number of binding sites on the ion exchanger are much greater than the sites taken up by A^+ or B^+, namely, $[\bar{S}] >> a[A^+], b[B^+]$, and is related to the binding charge based on the concept represented in Figure 7.25, where the slope of the log of the capacity factor as a function of the log of the salt concentration gives the value of the binding charge. A line that represents a difference of binding charges is obtained by the plot of Eq. (7.22), where the separation factor is a function of the log of the salt concentration.

The framework presented in the Eqs. (7.13) and (7.22) and Figures 7.26 and 7.28 represent ideal cases, not only because of the assumption that $[\overline{S}] >> \overline{A}^{+}A^{+}$ but also because the effect of indirect interactions between the protein and stationary phase are ignored. Chicz and Regnier (1989) studied the effect of single amino acid substitutions in genetically engineered variants of subtilisin from *Bacillus amyloliquifacians* with respect to their chromatographic retention on a strong cation exchange resin. (Subtilisin is a proteolytic enzyme used in laundry detergents.) They found that a change in a single amino acid residue in a protein consisting of 275 residues was sufficient to change the protein's retention in cation exchange chromatography. They also reported that both neutral and charged amino acids affected separation. In addition to ion exchange, indirect effects include both electrostatic changes that affect the pK_a values of neighboring amino acids in the protein, as well as steric effects of hydrogen-bonded water molecules associated with the protein and stationary phase (Chicz and Regnier, 1989). Nonetheless, Eqs. (7.13) and (7.22) give a good starting point for characterizing retention behavior for experimentally measured protein peaks eluted at different salt concentrations.

7.12 DEFINITIONS OF PLATE HEIGHT AND RESOLUTION FOR LINEAR GRADIENT CHROMATOGRAPHY ARE ANALOGOUS TO THOSE FOR ISOCRATIC CHROMATOGRAPHY

A fundamental measure of chromatography column performance is the height of a theoretical plate, H, as described in Sections 5.12 to 5.17, in Chapter 5. These definitions were given for a mobile phase having a constant composition. Since the composition of the mobile phase changes in linear gradient elution chromatography, a modified definition of plate height was proposed by Yamamoto (1995):

$$H_{LG} = \frac{L}{N_{LG}} = \frac{L}{\left(\frac{W_{LG}}{W}\right)^2}\left[\frac{\left(\frac{W_{LG}}{4}\right)}{V_{R,LG}}\right]^2 \tag{7.23}$$

where

H_{LG} = plate height for a protein peak in linear gradient elution chromatography, cm
N_{LG} = total number of plates for linear gradient elution, dimensionless
L = column length, cm
W = peak width, at isocratic (constant salt concentration) conditions, mL ($\sigma_V = W/4$)
$V_{R,LG}$ = retention volume of peak in linear gradient elution at a specified pH and salt concentration of the mobile phase, mL
W_{LG} = width of the peak in a linear salt gradient, mL = $4\sigma_{V,LG}$
W_{LG}/W = zone-sharpening ratio, dimensionless
N_{LG} =

$$16\left[\frac{V_{R,LG}}{W_{LG}}\right]^2 \text{ = plate count for a protein peak in linear gradient chromatography} \tag{7.24}$$

The plate count can be obtained by direct experimental measurement of peaks obtained under linear gradient elution and isocratic salt conditions as illustrated in Figure 7.30. The dotted lines indicate salt concentration. The peak widths, W_{LG} and W, are often similar, and consequently, their ratio is close to one. The plate count can be estimated directly from the chromatogram for linear gradient elution. Zone sharpening, if it occurs, is defined by the ratio of the peak width in a gradient, W_{LG}, to the width, W at a constant salt concentration.

Peak width is independent of the volume of the feed, for protein concentrations that exhibit linear equilibria at the modulator concentration at which the peak elutes otherwise peak width will change. If zone-sharpening effects are small, the peak width in a gradient is almost the same as the peak width at a constant concentration of the modulator. This can be confirmed by measuring the ratio of $(W_{LG}/W)^2$ from experimental elution profiles, or estimating it from the distribution coefficients for the salt and the protein and the slope of the gradient. The distribution coefficient ratio, R_K, is given by (Yamamoto, 1995):

$$R_K = \frac{1 + \phi K_{D,\text{protein}}}{2\,G_{SP}\left|\dfrac{dK_{D,\text{protein}}}{dI}\right|(1 + \phi K_{crt})} \tag{7.25}$$

where

R_K = distribution coefficient ratio for estimating W_{LG}/W from Eqs. 7.27, 7.28, or 7.29

G_{SP} = slope of gradient expressed as a function of stationary phase volume (Eq. (7.26), moles/L

ϕ = phase ratio, dimensionless

$K_{D,\text{protein}}$ = distribution coefficient for protein at salt concentration or ionic strength that coincides with the elution of the peak's maxima (i.e., center of peak), dimensionless; K_{crt} is for protein eluting at very high salt concentration. (see Furasaki et. al, 1987)

$\left|\dfrac{dK_{D,\text{protein}}}{dI}\right|$ = absolute value of slope of plot of $K_{D,\text{protein}}$ against salt concentration (ionic strength)

The parameter G_{SP} is defined in Eq. (7.26) and represents the gradient normalized with respect to the volume of the stationary phase.

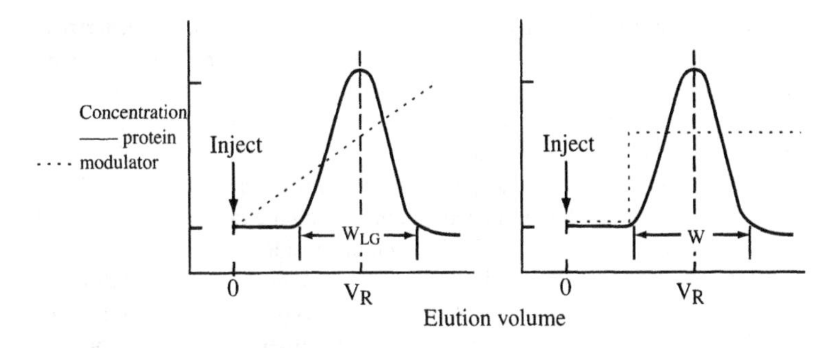

Figure 7.30 Linear vs step gradient across a protein peak.

The gradient concentration with respect to the column's stationary phase volume is given by

$$G_{SP} = M\phi V_o = M\left(\frac{1-\varepsilon_b}{\varepsilon_b}\right)V_o = M\left(\frac{V_o}{\varepsilon_b} - V_o\right) = M(V_{col} - V_o) \tag{7.26}$$

where

G_{SP} = slope of gradient concentration expressed as a function of stationary phase volume, moles/L
V_{col} = total volume of column, mL
$V_{col} - V_o$ = volume of column that is displaced by the stationary phase
M = slope of gradient as defined in Eq. (7.1)
ϕ = $(1 - \varepsilon_b)/\varepsilon_b$ = phase ratio

The distribution coefficient ratio, R_K, gives an indication of the ratio of the width of the peak in linear gradient elution to the same peak at isocratic conditions. The values of W_{LG}/W will be less than or equal to 1, with the ranges specified by Yamamoto (1995) increasing with increasing R_K. For $R_K < 0.25$,

$$\frac{W_{LG}}{W} = R_K^{0.5} \tag{7.27}$$

for $0.25 < R_K < 12$,

$$\frac{W_{LG}}{W} = \frac{3.22R_K}{1 + 3.13\,R_K} \tag{7.28}$$

and for $R_K > 12$,

$$\frac{W_{LG}}{W} = 1 \tag{7.29}$$

Example 2. Calculation of Plate Height for Linear Gradient Elution Chromatography (data from Yamamoto et al. 1988).

The protein ovalbumin (abbreviated Ova or OA) is used as a probe to characterize the plate height of an anion exchange chromatography column packed with CL6B stationary phase. An elution curve is generated in a salt gradient at pH 7.9 and 20°C. The gradient had an initial salt concentration of 0.11 and a final concentration of 0.24 M after 5.8 void volumes mobile phase had passed the outlet of the column (equivalent to 4.8 void volumes salt at the inlet). The protein concentration was 2 mg/mL in an injection volume of 6 mL. The peak maximum eluted in about 4.1 column void volumes, with the salt concentration at the peak maxima corresponding to 0.19 M NaCl. Column dimensions were 1.5 cm inside diameter by 10 cm long. The flowrate of mobile phase was 18 mL/hr. The column's external (extraparticulate) void fraction was $\varepsilon_b = 0.38$. At very high salt concentration, assume that OA has $K_{crt} = 0.95$ for this particular stationary phase and follows the relations shown in Figure 7.23.

1. For the conditions given, is the width of the peak under linear gradient elution going to be less than the width of the peak for the same protein at isocratic conditions? Will the peak elute at a higher average concentration of salt under linear gradient elution compared to isocratic conditions?
2. The slope of the gradient is decreased by a factor of 2 times to give a shallow gradient. If all other conditions are the same, calculate the expected ratio of peak widths W_{LG}/W. Assume that $dK_{D,protein}/dI = 30$.
3. The width of the peak under linear gradient conditions is equivalent to one column void volume for the 1.5 × 10 cm column. Calculate the expected plate height for linear gradient elution conditions for the gradient in part 2.

Solutions

1. First calculate distribution coefficient ratio using Eq. (7.25):

$$R_K = \frac{1 + \phi K_{D,protein}}{2G_{sp}\left|\frac{dK_{D,protein}}{dI}\right|(1 + \phi K_{crt})} \tag{7.30}$$

where

$$\phi = \frac{1 - \varepsilon_b}{\varepsilon_b} = \frac{1 - 0.38}{0.38} = 1.63$$

$K_{D,protein}$ = distribution coefficient that corresponds to salt concentration that coincides with elution of peak maximum at about 1.5 (read from Figure 7.23 for ionic strength of 0.19 for CL6B, $K - K_{crt} \cong 0.55$), since $K_{crt} \cong 0.95$ at very large salt concentration, for purposes of this example.

An approximation for the slope of the distribution coefficient as a function of ionic strength is obtained from the log-log plot given in Figure (7.23):

$\left|\frac{dK_{D,protein}}{dI}\right|$ = absolute value of change in ovalbumin's distribution coefficient measured at ionic strengths between 0.2 and 0.18 for Figure 7.23, CL6B:

$$\cong \frac{\Delta K_{D,protein}}{\Delta I} = \left|\frac{0.7 - 2.0}{0.2 - 0.18}\right| = \left|\frac{-1.3}{0.02}\right| = 65$$

Selection of a second set of coordinates at ionic strengths of 0.19 and 0.18 on Figure 7.23 for CL6B shows that

$$\left|\frac{1.0 - 2.0}{0.19 - 0.18}\right| = \left|\frac{-1.0}{0.01}\right| = 100$$

and selection of a third set of coordinates shows for ionic strengths between 0.2 and 0.19, the slope is

$$\left|\frac{0.7-1.0}{0.2-0.19}\right| = \left|\frac{-0.3}{0.01}\right| = 30$$

In view of the variation of $\Delta K_{D,protein}/\Delta I$ with ionic strength more detailed data are needed around each range. However, to obtain a first estimate, calculate the slope of the gradient as a function of concentration of salt using Eq. (7.26) and then use this to obtain the magnitude of order of R_K. The gradient slope is

$$G_{sp} = M\phi V_o = (0.00403)\,(1.63)\,(6.715) = 0.0441$$

where $V - V_o = 5.8V_o - V_o = 4.8V_o$ and $\phi = 1.63$ with

$$V_o = \varepsilon_B V_{col} = (0.38)\left(\frac{1.5}{2}\right)^2 \pi\,(10.0) = 6.715 \text{ mL}$$

$$M = \frac{\Delta I}{\Delta V} = \frac{0.24-0.11}{(4.8)(6.715)} = 0.00403\,\frac{\text{moles}}{\text{L}\cdot\text{mL}}$$

Substitution of values of ϕ, $K_{D,protein}$, G_{sp}, and $\left|\frac{dK_{D,protein}}{dI}\right|$ into equation (7.30) yields

$$R_K = \frac{1+(1.63)(1.5)}{2(0.0441)(65)(1+(1.63)(0.95))} = \frac{3.445}{14.6} = 0.236 \tag{7.31}$$

Based on the criteria in Eq. (7.27), the ratio W_{LG}/W would have an expected value of

$$\frac{W_{LG}}{W} = R_K^{0.5} = (0.236)^{0.5} = 0.486$$

In other words, the width of the peak at linear gradient conditions would be significantly less than the peak for the same protein at isocratic conditions. The conclusion is the same but relative peak width is different if a value of 30 is used for $(dK_{D,protein}/dI)$ in Eq. (7.31), R_K is

$$R_K = \frac{1+(1.63)(1.5)}{((2)(0.0441)(30)(1+(1.63)(0.95))} = 0.510$$

In this case, Eq. (7.28) is applicable, and

$$\frac{W_{LG}}{W} = \frac{3.22R_K}{1+3.13R_K} = \frac{(3.22)(0.510)}{1+(3.13)(0.510)} = \frac{1.64}{2.59} = 0.633$$

Since the total mass injected in both cases is the same, the area under the peak (i.e., the total mass) would be conserved. Hence a more narrow peak would result in a higher concentration, with linear gradient elution giving a narrower peak than isocratic elution.

2. Calculate G_{sp} using Eq. (7.26):

$$G_{sp} = M\phi V_o = \left(\frac{0.00403}{2}\right)(1.63)(6.715) = 0.022$$

Assume that $dK_{D,\mathrm{protein}}/dI = 30$, then

$$R_K = \frac{1 + (1.63)(0.95)}{2(0.022)(30)(1 + (1.63)(0.95))} = 0.758$$

Since $R_K < 12$ but > 0.25, correlation (7.28) still applies:

$$\frac{W_{LG}}{W} = \frac{3.22(0.758)}{1 + (3.13)(0.758)} = 0.724$$

3. The expected plate height at these conditions is calculated from Eq. (7.23), and the data given:

$$V_{col} = \left(\frac{1.5}{2}\right)^2 \pi(10) = 17.67$$

$$\frac{V_{R,LG}}{V_o} = 4.1 \text{ (peak maximum elutes after 4.1 column void volumes)}$$

$$V_{R,LG} = (4.1)(0.38)(17.67) = 27.5 \text{ mL}$$

The width is one void volume. The width ratio of the peak from case 2 is $W_{LG}/W = 0.724$. Hence, by Eq. (7.23),

$$H_{LG} = \frac{10}{(0.724)^2}\left[\frac{\frac{(0.38)(17.67)}{4}}{27.5}\right]^2 = 0.071 \text{cm}$$

7.13 THE PLATE HEIGHT INCREASES WITH INCREASING INTERSTITIAL VELOCITY IN LINEAR GRADIENT ELUTION CHROMATOGRAPHY

An example is given in Figure 7.31 for the protein lysozyme eluted at gradients equivalent to $G_{sp} = 0.0955$ M, with the initial salt concentrations ranging from 0.03 to 0.2 M. This plot suggests that the plate heights may increase (peaks get wider) with increasing salt concentration (see points for 0.05, 0.10, and 0.20 M salt at a linear velocity of about 18 cm/min), although this variation may lie within the variance of the experimental measurements. More important, the plate heights increase with increasing velocity over a range of salt concentrations. The plate height in linear gradient elution chromatography follows the form:

$$H_{LG} = A + Bv \tag{7.32}$$

as indicated by the example given in Figure 7.31. If plate height as a function of velocity is known, the resolution of a two component mixture in two different columns operated a different linear velocities can be calculated. This is based on the ratio given in Eq. (7.33) being the same

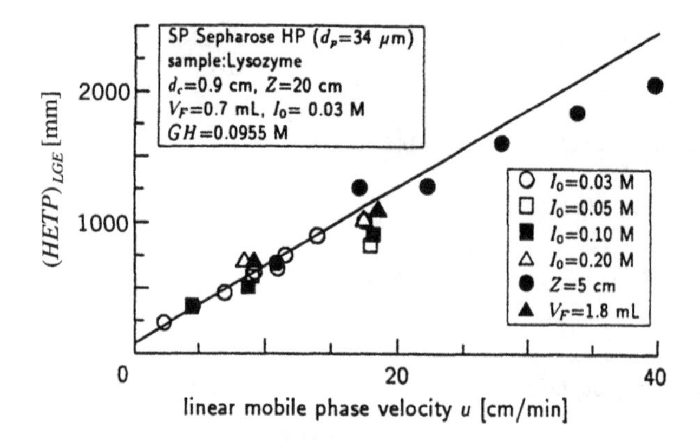

Figure 7.31 Plate height as a function of interstitial velocity, v, at initial ionic strengths of 0.03 M to 0.20 M for G_{sp} = 0.0995 M. Base conditions were a feed volume of 0.7 mL and I_o = 0.03 mL. Other runs were carried out at 1.8 mL feed volume and initial ionic strengths as indicated. Column dimensions of either 0.9 × 20 cm (base case) or 0.9 × 5 cm (for runs at other ionic strengths). Lysozyme concentration in feed was 1 mg/mL (from Yamamoto, 1995, fig. 2, p. 447, by permission of Wiley-Liss, Inc., a subsidiary of John Wiley & Sons, Inc.).

for the two columns (Yamamoto, 1995). Equation (7.33) also assumes that the particle size of the stationary phase is the same in both cases:

$$\left(\frac{L}{G_{V_o} H_{LG}}\right)_1 = \left(\frac{L}{G_{V_o} H_{LG}}\right)_2 \tag{7.33}$$

where

L = column length, cm
H_{LG} = plate height, cm
G_{V_o} = MV_o = gradient slope with respect to extra-particulate column void volume, moles/L. The slope M is given by Eq. (7.1).

The resolution, R_s, is proportional to (Yamamoto, 1995):

$$R_s \alpha \frac{D_m L}{G_{sp}\, v d_p^2} \tag{7.34}$$

where

d_p = average diameter of stationary phase particles, microns
G_{sp} = $M\phi V_o$ gradient slope with respect to stationary phase volume as defined in Eq. (7.26), moles/L
v = interstitial velocity, cm/s
D_m = diffusion coefficient of protein or solute in an unbounded solution, cm^2/s

Hence the particle size effect on resolution on two columns can be factored into a scale-up procedure by using the ratio of $R_{s,1}$ to $R_{s,2}$. This approach is credited to Snyder and Stadalius by

Yamamoto (1995). The proportionality constant for the two columns as well as the value of D_m is assumed to be the same. Therefore, resolution will vary as given in Eq. (7.35):

$$\frac{R_{s,1}}{R_{s,2}} = \frac{\left(\dfrac{L_1}{G_{sp,1}v_1 d_{p,1}^2}\right)}{\left(\dfrac{L_2}{G_{sp,2}v_2 d_{p,2}^2}\right)} = \frac{v_2 L_1 G_{sp,2}}{v_1 L_2 G_{sp,1}}\left(\frac{d_{p,2}}{d_{p,1}}\right)^2 \tag{7.35}$$

The ratio of particle sizes is squared, thus indicating that pore diffusion resistance (rather than mass transfer) is assumed to be the dominant contributor to peak dispersion. The use of Eq. (7.35) is analogous to the scale-up procedures described for isocratic chromatography in Section 6.2 of Chapter 6.

7.14 SCALE-UP OF GRADIENT CHROMATOGRAPHY IS MORE CHALLENGING THAN ISOCRATIC CHROMATOGRAPHY SINCE PEAK RETENTION AS A FUNCTION OF GRADIENT CHARACTERISTICS MUST EITHER BE KNOWN OR CALCULATED

Yamamoto et al. (1983a,b), Gibbs and Lightfoot (1986), Gallant et al. (1996), and others, have carried out thorough experimental and theoretical studies on the scale-up of gradient chromatography. Gibbs and Lightfoot derived explicit equations for extrapolating isocratic results to gradient elution, while Gallant et al. used a steric mass action model of ion exchange together with mass balance equations in order to model protein band movement under gradient elution conditions in a strong cation exchange system. Their model is presented in a form that can be programmed into a spreadsheet (Yamamoto et al., 1983a). Natarajan and Cramer (1999) present a model for simulating effects of dispersion of shock layers on resolution, while Bates and Frey (1998) have studied the complex behavior induced by decreasing pH gradients. Models of gradient elution chromatography by Yamamoto and his co-workers that are validated by examples from both Yamamoto (Velayudhan et al., 1995; Yamamoto et al., 1983a,b) and Furasaki et al. (1987) begin to address the complex interactions among stationary phase, modulator, adsorbate, particle size, and dispersion processes. Guichon and Zhu (1995) present an analysis of nonlinear systems in chromatography. They have developed models that capture non-ideal effects in gradient chromatography, while Gu, Tsai, and Tsao (1991a,b) provide models for multicomponent elution in both axial and radial flow chromatography columns. Other papers discuss electrical gradients (Keim and Ladisch, 2000; Basak and Ladisch, 1995; Rudge et al., 1993).

These and other publications provide depth on the complex factors that affect modeling of chromatographic processes and adsorptive phenomena in gradient chromatography. An underlying common factor in all of these analyses is the interaction between modulator concentration and the protein (or other solute) retention in the dynamic system where modulator concentration changes within the chromatography column as a function of time, distance, and the modulator concentration itself. Since the purpose of this chapter is to introduce the reader to the principles of gradient elution, the interactions among modulator gradients, protein distribution coefficients, and chromatographic elution profiles are presented in the context of linear systems. The scale-up of gradient chromatography may use rules in a scale-up approach that are similar to those for isocratic systems discussed in Chapter 6. Hence performance of the bench scale can be simulated at the process scale with minimal knowledge of the microscopic processes that are involved. While useful, this approach limits the chromatographic configurations that can be employed. If a greater amount of information is known, such as equilibrium relationships for both the modulator and the solute (i.e., protein) with respect to the stationary phase and the interaction between the two, scale-up can also encompass estimation of the effects in changes of the gradient characteristics. Both approaches are discussed in the remainder of this chapter. The discussion presents principles that will prepare the reader to study in a more comprehensive manner the excellent literature, cited above, that addresses gradient chromatography.

7.15 ONE APPROACH TO SCALE-UP OF GRADIENT CHROMATOGRAPHY IS BASED ON MAINTAINING A CONSTANT GRADIENT DURATION

The scale-up of gradient chromatography has the goal of achieving the same separation for a large-scale column as was obtained for the laboratory system. If the same ion exchanger, and particle size are used, and gradient conditions are identified that give the desired separation, scale-up may be based on gradient duration. A simple and clearly stated rule is given by MacLennan and Murphy (1994):

$$q_X G_X = \frac{V_{o,X}}{V_{o,B}} q_B G_B \tag{7.36}$$

where

G_X = gradient duration on the process scale, min
G_B = gradient duration on the bench scale, min
$V_{o,X}$ = void volume on process scale, mL
$V_{o,B}$ = void volume on bench scale, mL
q_X = flowrate at the process scale, mL/min
q_B = flowrate at the bench scale, mL/min

Equation (7.36) simply states that the volume of the gradient (given by $q_X G_X$ or $q_B G_B$) should be increased in proportion to the void volume of the column. This scale-up rule assumes that the slope of the gradient and its initial and final concentrations, as well as sample composition, are the same on both bench and process scales. These constraints mean that scale-up is achieved by increasing the cross-sectional area of the column in proportion to the increase in sample volume. Hence the rule given by Eq. (7.36) is constrained to the assumption of a constant linear velocity and proportional scaling of sample volume. The flowrate would then be scaled according to

$$q_X = q_B \left[\frac{D_X}{D_B}\right]^2 \tag{7.37}$$

where

D_X = diameter of process column, cm
D_B = diameter of bench-scale column, cm
q_X, q_B = flowrates for process- and bench-scale columns, respectively, in mL/min

The linear velocity is kept constant when the flowrate is scaled in this manner. Sample load is measured in terms of sample volume V_{spl} when the composition of the sample is constant. Sample volume is assumed to be a constant proportion of the column volume, as given by Eq. (7.38):

$$V_{spl,x} = V_{spl,B}\left[\frac{D_x}{D_B}\right]^2\left[\frac{L_x}{L_B}\right] \quad (7.38)$$

where

$V_{spl,x}$, $V_{spl,B}$ = sample volumes for the process- and bench-scale columns, respectively, mL

L_x, L_B = bed lengths of process and bench scale columns, respectively, cm

and the other variables are the same as defined previously. The rules given in Eqs. (7.36) to (7.38), presented by Maclennan and Murphy, comprise a rapid and conceptually understandable method by which gradient volume, and therefore peak retention volume, can be estimated from a small amount of experimental data. The peak retention volume for ion exchange chromatography is related to the capacity factor of the peak, and is defined for a specified salt concentration as given in Eqs. (7.2) to (7.13). In terms of retention volume, the capacity factor, as given in Section 5.21 in Chapter 5 is

$$k' = \frac{V_R - V_o}{V_o} \quad (7.39)$$

or

$$V_R = V_o\,(1 + k') \quad (7.40)$$

Hence, if the gradient duration and slope is the same, and the initial and final modulator concentrations are also the same, the retention volumes of the peaks on the process and bench scales will be proportional to the column void volumes. This follows from Eq. (7.40), which for a ratio of process to bench scales gives the equation

$$\frac{V_{R,x}}{V_{R,B}} = \frac{V_{o,x}\,(1 + k'_x)}{V_{o,B}(1 + k'_B)} \quad (7.41)$$

When the capacity factors on the process and bench scales are the same, the retention volume on the process scale would correspond to

$$\boxed{V_{R,x} = V_{R,B}\left(\frac{V_{o,x}}{V_{o,B}}\right)} \quad (7.42)$$

where the variables are defined in the same manner as for Eqs. (7.36) to (7.38).

The rules given by Eqs. (7.36) to (7.42) are approximations, and they should be carefully applied within the context of the assumptions used to develop them. Changes in gradient slope, gradient shape, and/or mobile phase flowrate can change peak retention and peak shape. Yamamoto et al. (1988) address these effects for linear gradient elution for ion exchange chromatography.

7.16 MATERIAL BALANCES ON BOTH MODULATOR AND PROTEIN ARE NEEDED TO SCALE UP LINEAR GRADIENT ELUTION WHEN COLUMN LENGTH AND/OR GRADIENT SLOPE ARE CHANGED

The modeling of linear gradient, ion exchange elution chromatography for a stationary phase consisting of spherical particles is based on the assumptions that the liquid-film mass transfer is negligible and that the equilibria of both protein and salt between stationary and mobile phase is linear. Furasaki et al. (1987) give a clear summary of the equations that model the elution behavior of the protein peaks at different salt concentrations. The material balances for the protein as it moves through the column is given by

$$\frac{\partial C_{\text{m}}}{\partial t} + v\frac{\partial C_{\text{m}}}{\partial Z} + \frac{6}{d_{\text{p}}}\phi D_{\text{p}}\left(\frac{\partial C_{\text{SP}}}{\partial r}\right)_{r=R} = D\frac{\partial^2 C_{\text{m}}}{\partial Z^2} \tag{7.43}$$

The material balance for a stationary phase particle is

$$\frac{\partial C_{\text{SP}}}{\partial t} = D_{\text{p}}\left(\frac{\partial^2 C_{\text{SP}}}{\partial r^2} + \frac{2}{r}\frac{\partial C_{\text{SP}}}{\partial r}\right) \tag{7.44}$$

with the boundary and initial conditions that represent an initially clean column

$$t = 0 \quad \text{and} \quad Z > 0$$

$$C_{\text{SP}} = 0 \quad \text{and} \quad C_{\text{m}} = 0$$

to which a pulse input of sample is injected at

$$Z = 0, \quad C = \delta(t) \tag{7.45}$$

where linear equilibrium applies:

$$t > 0 \quad \text{and} \quad r = R;\ C_{\text{SP}} = K_{\text{D}}C_{\text{m}} \tag{7.46}$$

The adsorption occurs in a symmetrical manner as given by $\partial C_{\text{SP}}/\partial r = 0$ for $r = 0$, where

C_{m} = concentration of protein in mobile phase g/mL
C_{SP} = concentration of protein associated with the stationary phase, g/mL
D = axial dispersion coefficient, for protein, cm^2/s
Z = distance from inlet of bed, cm
ϕ = phase ratio = $(1 - \varepsilon_b)/\varepsilon_b$, dimensionless
v = interstitial velocity cm/s
d_{p} = diameter of stationary phase particle
r = radial position in particle, cm
$R = d_{\text{p}}/2$ = particle radius, cm
D_{p} = intraparticle diffusivity of protein, cm^2/s

$\delta(t)$ = dirac delta input at time = t
K_D = distribution coefficient of protein = $K_{D,\text{protein}}$

Since linear conditions are assumed in Eq. (7.46), the retention time of the protein peak is given by

$$t_R = \frac{L}{v}(1 + \phi K_D) \tag{7.47}$$

L = column length

and the other parameters are as defined above. The peak width is related to the second moment by

$$W = 4\sigma = 4V_{col}\left[\mu_2'\left(\frac{\varepsilon_b v}{L}\right)^2\right]^{1/2} \tag{7.48}$$

where

V_{col} = volume of the column, mL

The second central moment, μ_2', is given by

$$\mu_2' = \frac{2L}{v}\left[\frac{D}{v^2}(1 + \phi K_D)^2 + \phi \frac{K_D d_p^2}{60D_p}\right] \tag{7.49}$$

where the parameters are as defined above. Equations (7.48) and (7.49) are for elution of the protein under linear, isocratic conditions. Equation (7.47), when differentiated with respect to time where Z_p denotes axial position of the protein, gives

$$\frac{d}{dt_R}\left(t_R = \frac{Z_p}{v}(1 + \phi\, K_D\right) = 1 = \frac{dZ_p}{dt}\left(\frac{1 + \phi K_D}{v}\right) \tag{7.50}$$

Equation (7.50) can be rearranged to give an expression for the position of the protein peak, under local equilibrium conditions:

$$\frac{dZ_p}{dt} = \frac{v}{1 + \phi K_D} \tag{7.51}$$

However, in gradient elution chromatography, the distribution coefficient is a function of ionic strength. A relation is needed that gives the peak position as a function of ionic strength. This correlation is presented by Yamamoto et al. (1983a,b, 1987a,b; see also Yamamoto, 1995) based on local equilibrium theory and linear equilibrium for the salt. The concentration of the salt is expressed in terms of ion strength, I, by a material balance that is analogous to Eq. (7.43):

$$\frac{\partial I_m}{\partial t} + \phi \frac{\partial I_{SP}}{\partial t} + v \frac{\partial I_m}{\partial Z} = D_I \frac{\partial^2 I_m}{\partial Z^2} \tag{7.52}$$

where

I_m = ionic strength (concentration) of salt in the mobile phase, mmoles/cm^3
I_{SP} = ionic strength (concentration) of salt in the stationary phase, mmoles/cm^3
D_I = axial dispersion coefficient of salt, cm^2/s

The equilibrium of the salt between mobile and stationary phases, is also assumed to be linear in this derivation of Furusaki et al. (1987):

$$I_{SP} = K_{D,salt} I_m \tag{7.53}$$

In addition, the gradient is assumed to be linear (i.e., constant slope) along the length of the column. Hence the axial gradient is a constant while salt concentration changes with time:

$$\frac{\partial I_m}{\partial Z} = \text{constant and } \frac{\partial I_{SP}}{\partial t} = K_{D,salt} \frac{\partial I_m}{\partial t} \tag{7.54}$$

Consequently $(\partial^2 I_m/\partial Z^2) = 0$. For linear equilibrium, Eq. (7.52) can be combined with Eq. (7.53) and (7.54) and then rearranged in the form

$$\frac{\partial I_m}{\partial t} + \phi \frac{\partial I_{SP}}{\partial t} = -v \frac{\partial I_m}{\partial Z}$$

$$(1 + \phi K_{D,salt}) \frac{\partial I_m}{\partial t} = -v \frac{\partial I_m}{\partial Z}$$

$$\frac{\partial I_m}{\partial t} = -\left(\frac{v}{1 + \phi K_{D,salt}}\right) \frac{\partial I_m}{\partial Z} \tag{7.55}$$

Based on Figure 7.5, the ionic strength at the inlet of the column is

$$I = I_o + \frac{\Delta I}{\Delta V} Qt \tag{7.56}$$

where

I = concentration of salt as a function of mobile phase volume, moles/L = I_m
I_o = initial concentration of salt at the beginning of the gradient, moles/L
$\Delta I/\Delta V$ = slope of gradient, moles/L/mL
ΔV = increment in volume of mobile phase, mL
Q = flowrate, mL/s
t = time elapsed since the initiation of the gradient, s

The ionic strength of the gradient at a distance Z from the inlet of the column is given by

$$I = I_o + \frac{\Delta I}{\Delta V} Q \left(t - \frac{(1 + \phi K_{D,salt}) Z_p}{v} \right) \quad (7.57)$$

Equation (7.57) enables the concentration of the modulator to be calculated anywhere along the column's length subject to Eq. 7.54. Yamamoto et al. (1983a, 1987b) showed that integration of the derivative of Eq. (7.57) gives a correlation by which the salt concentration corresponding to the elution time or volume of the peak can be determined. The derivative of Eq. (7.57) with respect to the axial position of the protein peak (denoted by Z_p) is

$$\frac{dI}{dZ_p} = \frac{\Delta I}{\Delta V} Q \left(\frac{dt}{dZ_p} - \frac{(1 + \phi K_{D,salt})}{v} \frac{dZ_p}{dZ_p} \right) \quad (7.58)$$

Rearrangement of Eq. (7.58), and combination with Eq. (7.51), gives

$$\frac{dI}{dZ_p} = \frac{\Delta I}{\Delta V} Q \left[\left(\frac{1 + \phi K_D}{v} \right) - \frac{(1 + \phi K_{D,salt})}{v} \right] \quad (7.59)$$

which simplifies to

$$\frac{dI}{dZ_p} = \frac{\Delta I}{\Delta V} \frac{Q\phi}{v} (K_D - K_{D,salt}) \quad (7.60)$$

where v is the interstitial velocity. The integral of Eq. (7.60) is given by

$$\int_{I_o}^{I_p} \frac{dI}{K_D - K_{D,salt}} = \frac{\Delta I \phi Q}{\Delta V v} \int_o^L dZ_p \quad (7.61)$$

or

$$\int_{I_o}^{I_p} \frac{dI}{K_D - K_{D,salt}} = \frac{\Delta I}{\Delta V} \frac{1 - \varepsilon_B}{\varepsilon_B} \frac{Q}{v} L = \frac{\Delta I}{\Delta V} (1 - \varepsilon_B) V_{col} \quad (7.62)$$

Since $Q/(\varepsilon_B v)$ is the cross-sectional area of the column, $V_{col} = [Q/(\varepsilon_B v)]\, L$, where L is the length of the bed. A solution to Eq. (7.62) is possible if the value of the distribution coefficient for the protein at an infinitely high concentration is known.

The measurement of adsorption equilibria data by Furasaki et al. (1987) over a range of salt concentrations for bovine serum albumin and hemoglobin gave the results shown in Figures 7.32 and 7.33. The data took the form

$$K_D = AI^{-B} + K_{D,\infty} \quad (7.63)$$

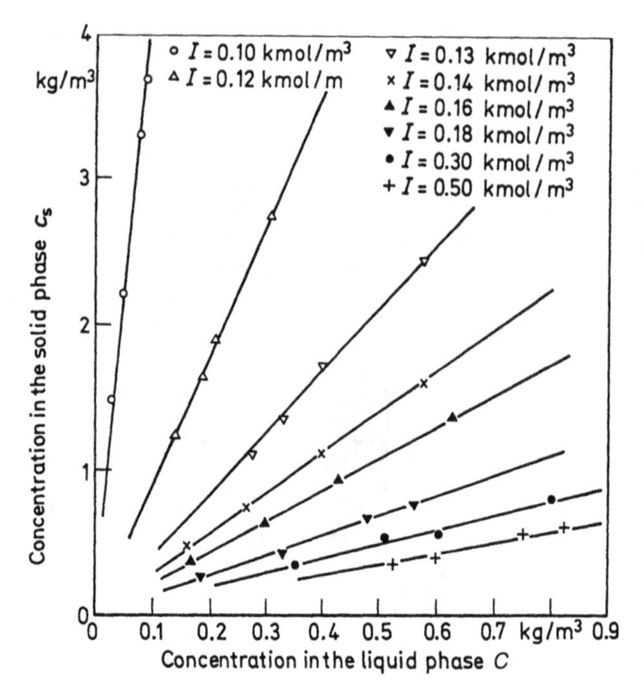

Figure 7.32 Equilibrium curves for adsorption of bovine serum albumin (BSA) on DEAE (anion) exchanger (from Furasaki et al., 1987, fig. 1, p. 51, with permission by Springer-Verlag).

or

$$K_D - K_{D,\infty} = AI^{-B}$$

where

K_D = distribution coefficient of protein at a given ionic strength
$K_{D,\infty}$ = distribution coefficient of protein at a very high ionic strength = K_{crt}
A, B = fitted constants determined from equilibrium data

The data in Figure 7.33 show that the distribution coefficients for BSA in the absence and presence of hemoglobin are the same and that the bovine serum albumin does not influence adsorption of hemoglobin. As the ionic strength of the mobile phase increases, the extent of adsorption of protein, and therefore its distribution coefficient, decreases. The parameters that correspond to the fitted lines for BSA and hemoglobin (abbreviated Hb) are

$$K_{D,BSA}:\quad A = 1.41 \cdot 10^{-8}, B = 9.5 \qquad \text{for } I < 0.14$$

$$A = 2.00 \cdot 10^{-3}, B = 3.5 \qquad \text{for } I \geq 0.14$$

$$K_{D,Hb}:\quad A = 1.95 \cdot 10^{-2}, B = 1.5$$

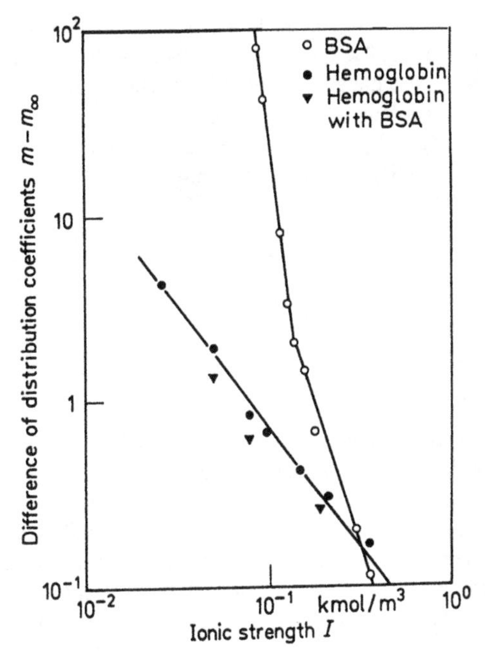

Figure 7.33 Distribution coefficient plot for hemoglobin and bovine serum albumin as a function of ionic strength. Note that both axis are logarithmic (from Furasaki et al., 1987, fig. 3, p. 51, with permission of Springer-Verlag).

The corresponding values for $K_{D,\infty}$ were 0.73 for BSA and 0.90 for Hb. Equation (7.63) for proteins, when rearranged, fits the form of Eq. (7.62) that was derived for salt (not protein) to give:

$$\int_{I_o}^{I_p} \frac{dI}{AI^{-B}} = \frac{\Delta I}{\Delta V}(1 - \varepsilon_B)\, V_{col} \tag{7.64}$$

where

$\Delta I/\Delta V$ = slope of the gradient
ε_B = void fraction based on mobile phase volume between particles
V_{col} = column volume, mL
I_o = ionic strength (salt concentration) upon initiating the salt gradient, moles/L
I_p = ionic strength (salt concentration) at which the peak elutes, moles/L

The integrated form of Eq. (7.64) is

$$I_p = \left[I_o^{(B+1)} + A\,(B+1)\,\frac{\Delta I}{\Delta V}(1 - \varepsilon_b)V_{col} \right]^{-(B+1)} \tag{7.65}$$

Once the value of I_p is known from numerical integration of Eq. (7.64), the distribution coefficient can be determined from Eq. (7.63) (with $I = I_p$). Once the distribution coefficient is known, t_R, and therefore V_R, can be calculated from Eq. (7.47). This approach enables the calculation of resolution, R_S of two peaks. The retention volumes for each of the two peaks are calculated from Eq. (7.47) and the relation $V_R = Qt_R$, and then the widths are obtained by Eqs. (7.48) and (7.49). Resolution is given by the operational definition

$$R_S = \frac{V_{R,2} - V_{R,1}}{(W_1 + W_2)/2} \tag{7.66}$$

where

$V_{R,1}$, $V_{R,2}$ = retention volumes of peaks 1 and 2, mL
W_1, W_2 = peak widths of components 1 and 2

Yamamoto et al. (1983,a,b, 1987b) showed that this approach is able to anticipate elution behavior for both ion exchange and hydrophobic interaction chromatography. Furasaki et al. (1987) gave an example for BSA and hemoglobin, where excellent agreement between theory and data was obtained, Figure 7.34. The degree of separation for BSA and hemoglobin is also shown to decrease as the slope of the gradient increased, Figure 7.35. The calculation of the resolution based on peak retention is based on the relative sample size being kept the same. However, increases in sample volume or protein concentration can change both peak retention

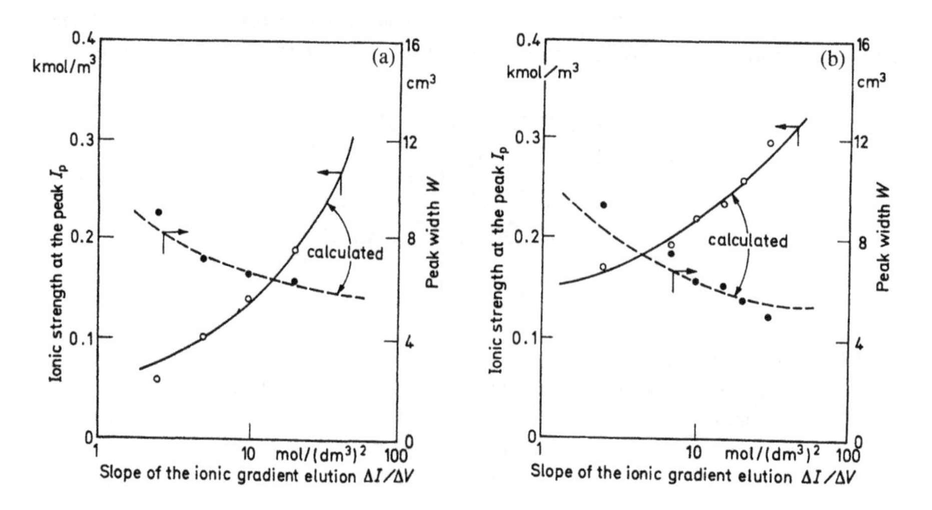

Figure 7.34 Ionic strength of gradient at peak maxima for hemoglobin and BSA as a function of gradient slope. (*a*) Peak position and width for hemoglobin (Hb) as functions of the slope of gradient elution; $Z = 8.6$ cm, $Q = 0.0167$ cm^3 s^{-1}. (*b*) Peak position and width for bovine serum albumin (BSA) as functions of the slope of gradient elution; $Z = 8.6$ cm, $Q = 0.0167$ cm^3 s^{-1} (from Furasaki et al., 1987, figs. 7 and 5, p. 52 with permission by Springer-Verlag GmbH & Co. KG, from Separations of Proteins by Ion-Exchange Chromatography Using Gradient Elution, Copyright notice of Springer-Verlag).

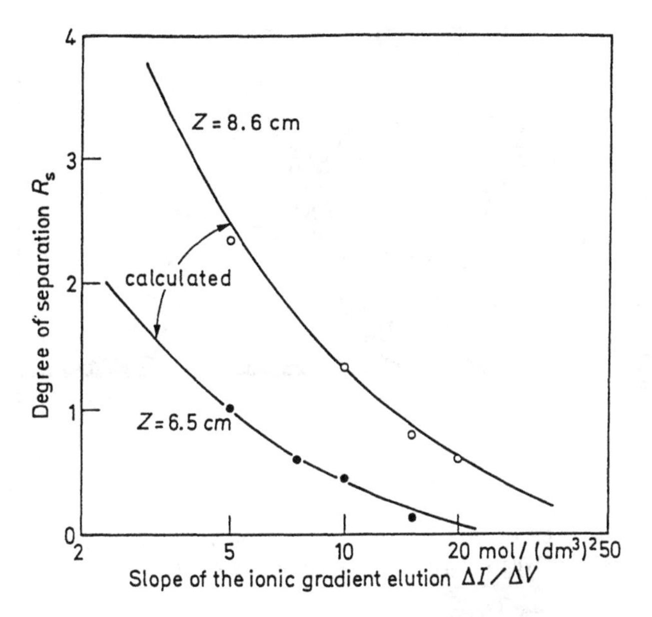

Figure 7.35 Resolution of bovine serum albumin from hemoglobin as the function of the slope of the gradient, $\Delta I/\Delta V$ (from Furasaki et al., 1987, fig. 8, p. 53 with permission of Springer-Verlag).

and peak shape (Yamamoto et al., 1983a) if the protein is in a nonlinear portion of its isotherm. When the volume is increased, peak width and skew increase as is shown for BSA (Figure 7.36). When protein concentration is increased, peak retention decreases and tailing increases (Figure 7.37).

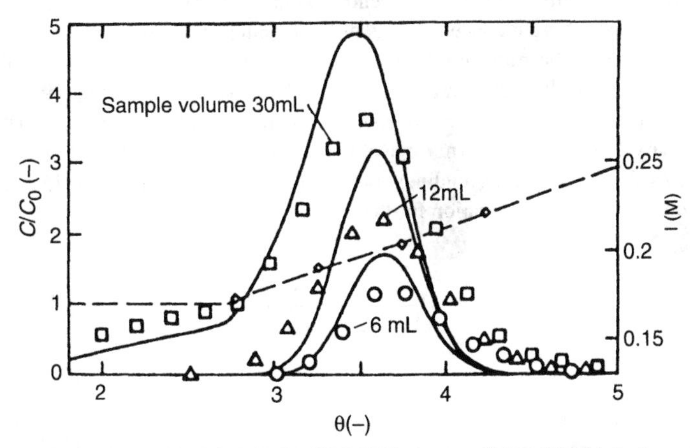

Figure 7.36 Effect of sample volume for 0.3 mg/mL BSA, for a gradient with initial salt concentration of 0.17 moles/L (abbreviated M) with a gradient slope $\Delta I/\Delta V = 10^{-3}$ M/mL at pH 7.9 and 20°C (from Yamamoto et al., 1983, figs. 5 and 8, pp. 1381 and 1384 by permission of Wiley-Liss, Inc., a subsidiary of John Wiley & Sons, Inc.).

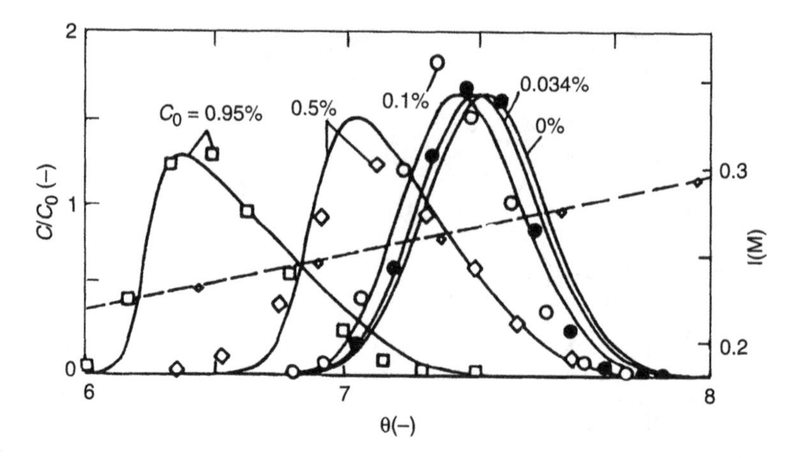

Figure 7.37 Effect of concentration of β-lactoglobulin A for an initial salt concentration of 0.11 M, a gradient of 5×10^{-3} M/mL and pH 7.9 and 20°C; both runs carried out in a 1.5 × 10 cm column packed with DEAE Sepharose CL6B at a flowrate of 18 mL/hr (from Yamamoto et al., 1983, figs. 5 and 8, pp. 1381 and 1384 by permission of Wiley-Liss, Inc., a subsidiary of John Wiley & Sons, Inc.).

7.17 ADSORPTION OF THE MODULATOR ON THE STATIONARY PHASE MAY CAUSE DEFORMATION OF THE GRADIENT

The modulator alters solute adsorption by competing for binding sites on the stationary phase. Since protein and peptide solutes are more strongly retained than the modulator, the modulator must usually be at high concentrations to compete with the binding sites for the macromolecules, thereby causing them to desorb. The modulator itself is likely to be in the nonlinear region of its adsorption isotherm.

At high concentration some of the modulator will adsorb while the remainder moves down the column. This causes the modulator concentration in the mobile phase to change once it enters the column. The different concentrations of the modulator will move down the column at different speeds. Hence the gradient introduced at the column's inlet will be distorted as it travels down the column, as illustrated in Figure 7.38. Velayudhan and Ladisch (1991, 1992) illustrated this effect for acetonitrile in an aqueous mobile phase with respect to a C-18 (reversed phase) stationary phase. The isotherm for acetonitrile in water (Figure 7.39) fitted to the Langmuir isotherms shows significant nonlinear behavior at concentrations above 15% acetonitrile concentration. The Langmuir equation for the acetonitrile is

$$q_{\mathrm{m}} = \frac{a_{\mathrm{m}} C_{\mathrm{m}}}{1 + K C_{\mathrm{m}}} = \frac{a_{\mathrm{m}} C_{\mathrm{m}}}{1 + b_{\mathrm{m}} c_{\mathrm{m}}} \tag{7.67}$$

where

q_{m} = stationary phase concentration of adsorbate, mmol/g stationary phase
a_{m} = 1.48 mmol/(g stationary phase · M)
K = 0.26 M^{-1}
C_{m} = concentration of solute in mobile phase, mole/L (equivalent to M)

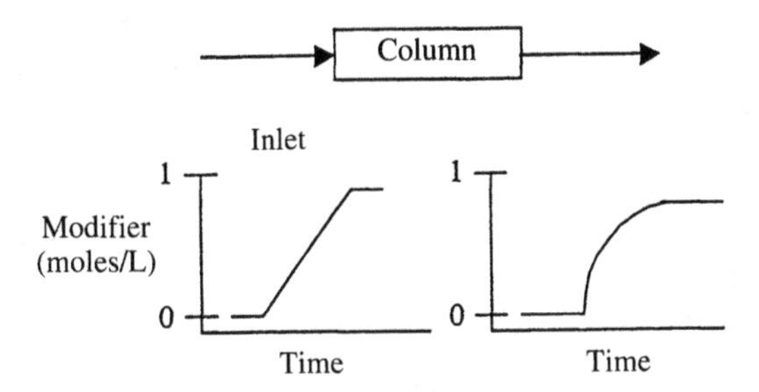

Figure 7.38 Linear vs. deformed gradient.

The modeling of this behavior in gradient chromatography, as well as its effect on generating what might appear to be anomalous behavior, was discussed by Velayudhan et al. (1995; see also Velayudhan and Ladisch, 1991, 1992). The theory described below was taken from one of these papers (Velayudhan and Ladisch, 1992) and is repeated below with some modification. The mass balance on the modulator (not the protein adsorbate) is given by Eq. (7.68). Dispersive effects are assumed to be small for this derivation. The equation is

$$\left(1+\bar{\phi}\frac{dq_{\mathrm{m}}}{dC_{\mathrm{m}}}\right)\frac{\partial C_{\mathrm{m}}}{\partial t}+v\frac{\partial C_{\mathrm{m}}}{\partial Z}=0 \tag{7.68}$$

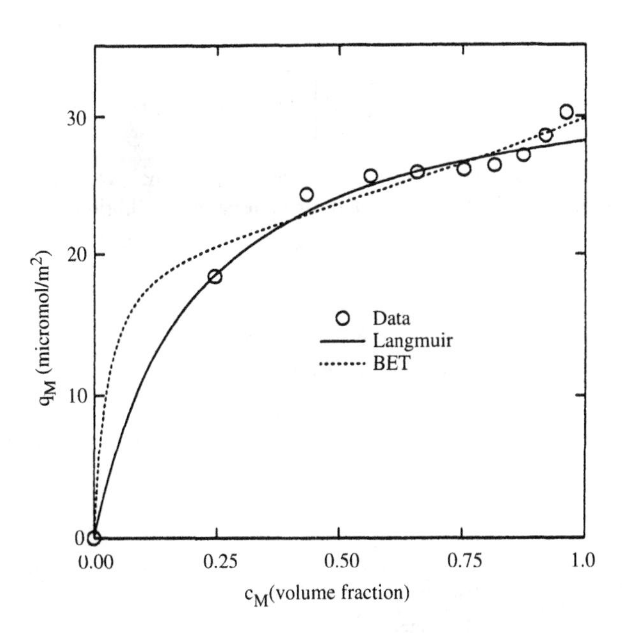

Figure 7.39 Isotherm for acetonitrile equilibrium with C-18 type stationary phase with surface area of 167 m^2/g (from Velayudhan and Ladisch, 1992, fig. 2, p. 235, with permission from Chem. Eng. Sci., 47(1), Elsevier Science).

where

q_m = modulator loading on stationary phase at equilibrium
t = time, s
Z = distance from inlet of bed
C_m = concentration of modulator in mobile phase after equilibrium has been attained, M (i.e., moles/L)
$\bar{\phi}$ = volumetric phase ratio = $[(1 - \varepsilon_b)(1 - \varepsilon_p)]/[\varepsilon_b + (1 - \varepsilon_b)\varepsilon_p]$

Note that the volumetric phase ratio, as defined here, is based on the volume displaced by the nonporous fraction of the stationary phase, and hence considers both interparticle (ε_b) and intraparticle (ε_p) void fractions. Since this differs from the definition used previously in this book which is based on ε_b, only, the phase ratio is denoted with a bar above the symbol $\bar{\phi}$.

The solute velocity, ν, is related to the superficial velocity, v, by the equation

$$\nu = \frac{\text{v}}{\varepsilon_b + (1 - \varepsilon_b)\varepsilon_p} \tag{7.69}$$

where

ε_b = extraparticulate void fraction
ε_p = interparticle void fraction

When Eq. (7.69), together with the Langmuir equation, is substituted into Eq. (7.68), the resulting equation is

$$\left[1 + \frac{\bar{\phi} a_m}{(1 + KC_M)^2}\right] \frac{\partial C_M}{\partial t} + \frac{\partial C_M}{\partial x} = 0 \tag{7.68}$$

Equation (7.68) is transformed into Eq. (7.69) by defining distance, velocity, and the equilibrium in terms of dimensionless variables, and substituting the equilibrium expression given in Eq. (7.68):

$$\left[1 + \frac{\bar{\phi} a_m}{(1 + \chi)^2}\right] \frac{\partial \chi}{\partial \tau} + \frac{\partial \chi}{\partial \xi} = 0 \tag{7.69}$$

where

$\chi = KC_M = b_m C_M$ (we assume $b_m = K$)
$\xi = Z/L$
$\tau = vt/L$
$\bar{\phi}$ = phase ratio = $\dfrac{(1 - \varepsilon_b)(1 - \varepsilon_p)}{\varepsilon_b + (1 - \varepsilon_b)\varepsilon_p}$

when Z is the distance from the inlet of the column. For a column at the beginning of a gradient run, the initial condition is given by

$$C_M(x,t) = C_{M,o} \quad \text{at } t = 0 \tag{7.70}$$

or

$$\chi = b_m C_{M,o} = \chi_o \text{ at } \tau = 0 \tag{7.71}$$

The boundary conditions for a linearly increasing gradient at $Z = 0$ (i.e., the column inlet) are

$$C_M = \begin{cases} C_{M,o}, & 0 \le t \le t_{inj} \\ C_{M,o} + M(t - t_{inj}) & t \ge t_{inj} \end{cases} \tag{7.72}$$

where M is the slope of the gradient as defined in Eq. (7.1). The gradient begins after the sample has been completely injected into the column over the time, t_{inj}. The dimensionless form of boundary conditions of Eq. (7.72) are given by

$$\xi = 0, \quad \chi = \begin{cases} \chi_o, & 0 \le \tau \le \tau_{inj} \\ \chi_o + \beta(\tau - \tau_{inj}) & \tau_{inj} \le \tau \end{cases} \tag{7.73}$$

where

$$\beta = \frac{MLK}{v} = Mt_oK \tag{7.74}$$

$$\tau_{inj} = \frac{v}{L} t_{inj} = \frac{t_{inj}}{t_o} \tag{7.75}$$

where $t_o = L/v$ = retention time that corresponds to ε_b and where

L = column length, cm
K = constant in denominator of Langmuir equation, 1/M
v = interstitial velocity, cm/s
$t_o = L/v$ = residence time, of an unretained component, s

When the method characteristics is applied to Eq. (7.69) the resulting expression is

$$\frac{d\tau}{ds} = 1 + \frac{\bar{\phi} a_M}{(1+\chi)^2} \tag{7.76}$$

where

$$\frac{d\xi}{ds} = 1 \tag{7.77}$$

and

$$\frac{d\chi}{ds} = 0 \tag{7.78}$$

The initial conditions are expressed using this nomenclature are

$$s = 0, \quad \xi = 0, \quad \text{and} \quad \tau = \eta \tag{7.79}$$

with χ being given by a modified form of Eq. (7.73) where $\tau = \eta$:

$$\chi = \begin{cases} \chi_o, & 0 \le \eta \le \tau_{inj} \\ \chi_o + \beta(\eta - \tau_{inj}), & \tau_{inj} \le \tau \end{cases} \tag{7.80}$$

The resulting expression for the dimensionless residence time $\tau(= vt/L)$ is given by

$$\tau = \eta + \xi\left[1 + \frac{\bar{\phi}a_m}{(1+\chi)^2}\right] \tag{7.81}$$

and

$$\chi = \chi_o + \beta\left\{\tau - \tau_{inj} - \xi\left[1 + \frac{\bar{\phi}a_m}{(1+\chi)^2}\right]\right\} \tag{7.82}$$

where

η = initial value of the dimensionless residence time, τ
$\beta = MKt_o$, where $\gamma = M$ = slope of gradient at *inlet* of column

The solutions of Eqs. (7.81) and (7.82) enable calculation of characteristic curves for χ for specified values of dimensionless distance ξ and calculated values of τ (from Eq. (7.81)). However, if the characteristic curves intersect, a physically impossible condition would exist, since there cannot be two different concentrations of the modulator at the same distance and time (see Section 6.13). Consequently this indicates a condition where a step change in the concentration of the modulator occurs, as discussed in Sections 6.13 and 6.14. This step change is referred to as a shock wave. The equations for the point and time at which the shock waves reach constant strength are given by $\xi_f\ (= Z_f/L)$ and τ_f:

$$\xi_f = \frac{(1+\chi_o)(1+\chi)^2}{2\bar{\phi}a_m\beta} \tag{7.83}$$

and

$$\tau_f = \tau_{inj} + \frac{(1+\chi)+(\chi-\chi_o)}{2\beta} + \frac{(1-\chi_o)(1+\chi)^2}{2\bar{\phi}a_m\beta} \tag{7.84}$$

for $\chi = \chi_f$ and a constant value of the inlet gradient at $t \ge t_f$ where

χ_o = concentration at which the shock wave begins to occur = $KC_{M,o}$
$\chi = KC_m$
$\bar{\phi}$ = phase ratio = $[(1 - \varepsilon_b(1 - \varepsilon_p)]/[\varepsilon_b + (1 - \varepsilon_b)\varepsilon_p]$
a_m = Langmuir constant, Eq. (7.67)
$\beta = \gamma K t_o = MKt_o$
$\gamma = M$ = slope of gradient at the inlet

Equation (7.83) can be used to calculate the axial distance at which a discontinuity (i.e., shock) in the gradient may occur for a specified modulator concentration, $C_{M,o}$, by setting $\chi = \chi_o$ in Eq. (7.83):

$$\xi^* = \frac{(1+\chi_o)^3}{2\bar{\phi} a_m \beta} < 1 \tag{7.85}$$

When ξ^* is less than 1, the condition for a shock wave to occur is present in the column. Equation (7.86) gives the dimensionless time at which this condition occurs:

$$\tau^* = \tau_{inj} + (1+\chi_o)\left[\frac{\bar{\phi} a_m + (1+\chi_o)^2}{2\bar{\phi} a_m \beta}\right] \tag{7.86}$$

Many columns will be too short to form a shock wave, and hence a linear gradient is likely to remain linear in many analytical scale columns whose lengths are typically in the 10 to 20 cm range. The scale-up rules given in Eqs. (7.36) to (7.39) assume that the length and gradient duration stays the same. In this special case, if gradient deformation is not observed at an analytical (or bench) scale, it is unlikely to be observed on a larger scale. However, scale-up will likely result in use of a longer column, and Eq. (7.87) shows that if a column is long enough, any gradient that is linear will become a discontinuous step by the time it reaches the outlet:

$$\xi^* = \frac{(1+\chi_o)^3}{2\bar{\phi} a_m \beta} = \frac{(1+\chi_o)^3}{2\bar{\phi} a_m \left(\frac{MLK}{v}\right)} \tag{7.87}$$

The column length, L, in the denominator of Eq. (7.87) will assure this result. Equation (7.87) represents the effects of modulator adsorption (through a_m and β), the porosity and packing of the stationary phase (through the phase ratio, $\bar{\phi}$), and initial modulator concentration and gradient slope (through the β and χ_o terms). Plate simulations illustrate the gradient shapes at the outlet that might result from a linear gradient at the column inlet (Figure 7.40).

Example 3. Calculation of Gradient Steepness That Results in Deformation of the Gradient and Shock Wave Formation

A 20 cm chromatography column with a phase ratio, $\bar{\phi}$, of 0.67 with the isotherm parameter of $a_m = 1.85$ and $K = 0.26$, and an initial modulator concentration of zero, are run at a scaled gradient steepness of $\gamma = M = 0.5$ M/min and a mobile phase linear velocity of 2 cm/min. The void volume corresponds to a retention time of 10 minutes, and $\chi_o = 1.13$.

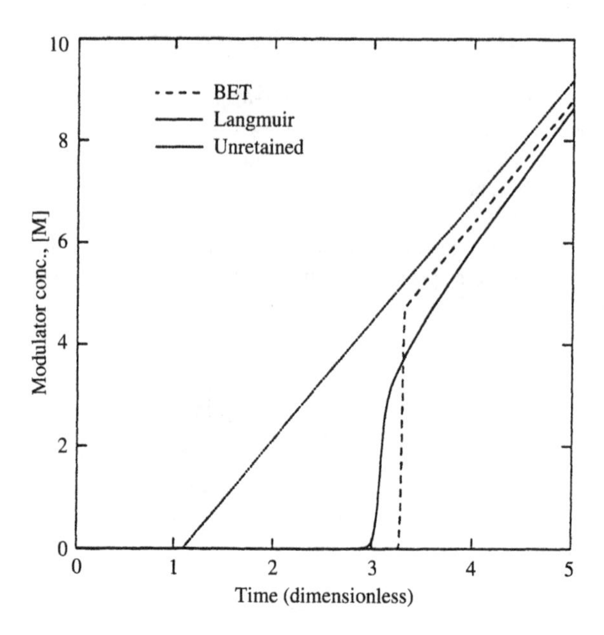

Figure 7.40 Possible types of gradient shapes at the column outlet for a linear inlet gradient with a slope of 2.35 M/min, with gradient initiated at a dimensionless time ($\tau = vt/L$) of 0.1. (Reprinted from Chem. Eng. Sci., 47(1), from Velayudhan and Ladisch, 1992, fig. 3, p. 238 with permission from Elsevier Science.)

1. Is it likely that deformation of the gradient will occur for the 20 cm long column?
2. If the column is 60 cm long, what is the distance from the inlet of the bed at which the shock wave attains a constant strength of KC_M if C_o = 1.1 M the modulator concentration at the inlet is brought to 5 M acetonitrile and then held at the concentration? Assume $\tau_{inj} = 0.1$.

Solutions

1. Use Eq. (7.85) with $\chi_o = 1.13$ in order to estimate if shock wave formation is likely to occur by the time that the gradient reaches the column outlet when $\chi_o = KC_m = 1.13$.

$$\xi^* = \frac{(1+\chi_o)^3}{2\bar{\phi}a_m\left[\dfrac{MLK}{v}\right]} = \frac{(1+1.13)^3}{2(0.67)(1.85)\left\{\dfrac{(0.5)(20)(0.26)}{2}\right\}} = \frac{(2.13)^3}{3.22} = 3.00$$

Since $\xi^* > 1$, shock wave formation is not likely.

2. First, use Eq. (7.85) to confirm that a shock wave could form:

$$\xi^* = \frac{(1+1.13)^3}{2(.67)(1.85)\{[(0.5)(60)(0.26)]/2\}} = 0.999$$

Tripling the length of the column may lead to gradient deformation at the outlet. Calculation of the distance at which the shock wave will attain a constant strength is based on knowledge of the modulator concentration at which shock wave begins to occur, combined with the application of Eqs. (7.83) and (7.84).

The concentration at which the shock wave will begin form is given by Eq. (7.85) when $\xi^* = 1$:

$$(1 + \chi_o)^3 = 2\bar{\phi} a_m \beta \tag{7.88}$$

$$\chi_o = (2\bar{\phi} a_m \beta)^{1/3} - 1$$

$$\chi_o = \left[2(0.67)(1.85)\left[\frac{(0.5)(60)(0.26)}{2}\right]\right]^{1/3} - 1 = 1.13$$

By definition, $\chi_o = 1.13 = KC_M$, and calculation shows a shock wave is expected at:

$$C_{M,o} = \frac{1.13}{K} = \frac{1.13}{0.26} = 4.35 \text{ moles/L} \tag{7.89}$$

The fractional distance from the inlet of the bed at which the shock wave forms is given by Eq. (7.83):

$$\xi_f = \frac{(1 + \chi_o)(1 + \chi_f)^2}{2\bar{\phi} a_m \beta} \tag{7.90}$$

where

$\chi_f = KC_{m,f} = (0.26)\ (5.00) = 1.3$ (since $C_{m,f}$ is 5.00 moles/L)
$\chi_o = KC_o = (0.26)(1.1) = 0.286$ (since $C_{M,o}$ is given at 1.1)

C_o is the initial modulator concentration at the column inlet and C_f is the concentration at the column inlet at which a shock wave begins to form. The other variables in Eq. (7.89) are

$\bar{\phi} = 0.67$ (given)
$a_m = 1.85$ (given)

$$\beta = MLK/v = \frac{[(0.5)(60)(0.26)]}{2} = 3.9 \text{ for the 60 cm long column.}$$

Therefore

$$\xi_f = \frac{(1 + \chi_o)(1 + \chi_f)^2}{2\phi a_m \beta} = \frac{(1 + 0.286)(1 + 1.3)^2}{2(0.67)(1.85)(3.9)} = 0.704 \tag{7.91}$$

Since

$$\xi_f = \frac{Z_f}{L} = \frac{Z_f}{60} \quad \text{and} \quad Z_f = (60)(0.704) = 42.2 \text{ cm}$$

This calculation indicates that the characteristic curves will travel at constant speed that correspond to the final concentration $\chi_f (= KC_{M,f})$.

The time at which this occurs is calculated from Eq. (7.84) with ξ_f from Eq. (7.91):

$$\tau_f = \tau_{inj} + \frac{(1+\chi_f)+(\chi_f-\chi_o)}{2\beta} + \xi_f = 0.1 + \frac{(1+1.3)+(1.3-0.286)}{2(3.9)} + 0.704 = 1.23$$

Hence the time at which it occurs is given by the definition of dimensionless time:

$$\tau_f = \frac{vt_f}{L} \tag{7.92}$$

Rearrangement of Eq. (7.92) gives the time at which the shock wave forms:

$$t_f = \tau_f \frac{L}{v} = 1.23\left(\frac{60}{2}\right) = 36.9 \text{ min}$$

Hence the shock wave forms about 37 minutes after the gradient is initiated at the column inlet. If the column is short enough so that the gradient emerges in 37 minutes or less, and all other conditions are the same, no shock wave would form.

Example 3 illustrates rules for anticipating deformation of the modulator. These rules were summarized by Velayudhan and Ladisch (1992) and are repeated here

1. A gradient that is linear at the column's inlet will become a discontinuous step by the time it reaches the column's outlet if the column is long enough. Hence, for the conditions in the second part of the example, the gradient would appear as a linear gradient at the column outlet if the column is 42.2 cm long (as calculated from Eq. 7.91), but a discontinuous one if the column is longer than this length.
2. Discontinuities in the gradient are promoted by a larger initial slope of the gradient, a modulator isotherm with a larger curvature (i.e., a large value of K), and a larger phase ratio $\bar{\phi}$. This is reflected by the form of Eq. (7.85).
3. The initial modulator concentration can be controlled, and it has a dominant effect on shock formation, since its effect varies as the square (in Eqs. (7.83) to (7.86). Increasing the initial modulator concentration while keeping the endpoint and gradient slope the same will lessen the likelihood of shock formation. However, for a sufficiently long column, a discontinuity can still form even for a high initial modulator concentration.

Effects that could lead to shock formation are important for scale-up, and they should be recognized as part of the purification development process. "If the separation is optimized at laboratory scale under conditions where no modulator shock forms, and then scaled-up without regard for [Eq. (7.85)], a shock might occur in the large column and affect separation significantly" (Velayudhan and Ladisch, 1992, p. 238). One possible effect is the formation of leading or tailing shoulders on the solute (protein) peak, depending on its position relative to the discontinuity in the gradient. This is illustrated in Figure 7.41 where a numerical simulation shows the effect of the gradient on peak shape. If the peak elutes under the gradient (peak with dotted line), it will have a Gaussian shape. However, if the same peak were to straddle the discontinuity, the peak would be both sharper and skewed (peak with solid line). In some cases a single

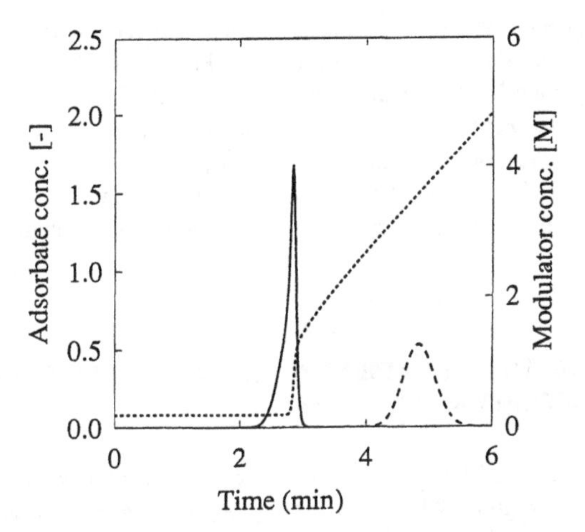

Figure 7.41 Numerical simulation showing possible effect in discontinuity of the gradient on peak shape. (Reprinted with permission from Velayudhan and Ladisch, 1991, fig. 1, p. 2030, American Chemical Society.)

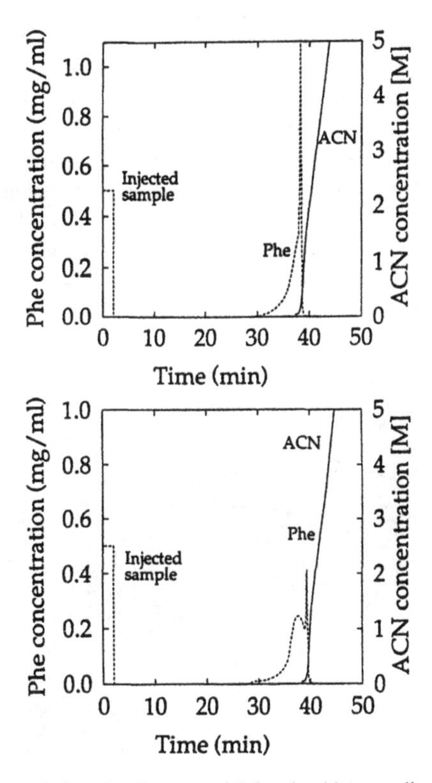

Figure 7.42 Numerical simulation showing potential for shoulder or split peak formation if emerging peak straddles a discontinuity (or shock wave). (*a*) Shoulder is formed (from Figure 7.18) and (*b*) two peaks form from a homogeneous component (from Velayudhan et al., 1995, fig. 4, p. 1190). Reproduced with permission of the American Institute of Chemical Engineers. AIChE. All rights reserved.

homogeneous component could form a shoulder or even two peaks at the point at which it straddles a discontinuity (i.e., shock wave) of the gradient Figure 7.42. Timing of the onset of the gradient relative to sample injection could have a beneficial effect by enabling simultaneous purification and concentration of the sample as shown by Velayudhan et al. (1995) for amino acids and peptides in reversed phase chromatography, and proteins in ion exchange chromatography (Velayudhan and Ladisch, 1995). In the limit a step gradient could also be used to intentionally introduce a discontinuity that would both purify and concentrate proteins (Yamamoto et al., 1992).

7.18 THE CONCEPTS OF GRADIENT CHROMATOGRAPHY CAN BE EXTENDED TO AFFINITY MEMBRANES

Affinity membranes consist of a stack of porous sheets that are fixed into a column configuration as represented in Figure 7.43. Ion exchange groups, as represented earlier in this chapter in Figures 7.13 to 7.14, or affinity ligands, as described in Chapter 9, are attached to the surfaces of the pores that permeate the membrane. Since these groups can strongly and selectively bind proteins that pass through the membrane, the membrane is said to have an affinity for the proteins. Hence the membranes are referred to as affinity membranes.

These membrane systems were developed to circumvent limitations of packed beds including pressure drops and intraparticle diffusion limitations, particularly at high flowrates. The potential benefit of an affinity membrane system is the ability of such a system to tolerate an increased linear velocity of the mobile phase to the point that association kinetics of solute with ligand become the limiting factor rather than pressure drop. While such a membrane can have high volumetric loading (measured as g/mL), the total loading (measured as g) may still be low. Furthermore axial diffusion effects are much more of a factor in a membrane device than a long packed bed.

Suen and Etzel (1992) and Kula et al. (Gebauer et al., 1997) presented a definitive analyses of such membrane systems for bioseparations. They showed their utility to rapid separation of biological molecules and provided a quantitative framework by which such membranes could

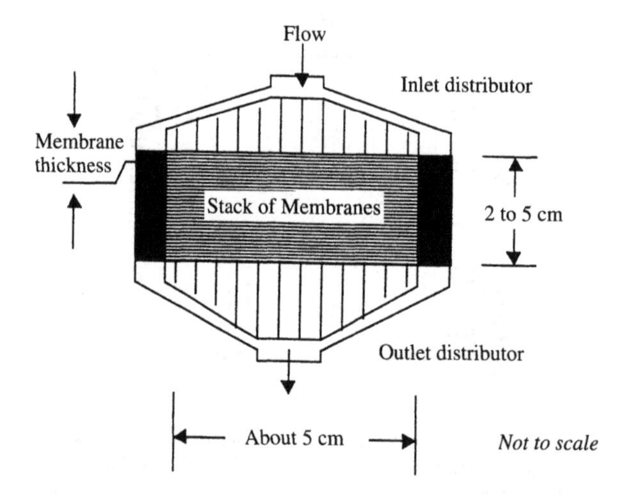

Figure 7.43 Stacked membrane column.

be properly designed and operated. Suen and Etzel calculated effects of axial diffusion, flow velocity, association kinetics, and nonuniformities in membrane thickness and porosity, on membrane performance. If the membrane is run in the adsorption mode, their calculations showed that the variation in membrane thickness must be less than 3% if the effect of the thickness on the breakthrough curve is to be insignificant (Figure 7.44). The breakthrough curves are also broad if the plate count (proportional to the Peclet number) is small, the association rate of solute with ligand is low, and the axial diffusion coefficient is large. Even if the plate height is small, the total plate count is still likely to be small since the membranes are thin and the total column length ($L = NH$) is relatively short compared to a packed bed. The effect of the Pe ($\approx 2N$) is shown in Figure 7.45. The other major contributor was shown to be porosity. As little as 3% variation resulted in major dispersion of the breakthrough curve (Figure 7.46). A thorough analysis of the flow properties and kinetics of affinity membranes showed that kinetics rather than pressure drop determine the best flow velocity.

Resolution and plate count for the profiles for affinity membranes shown in Figures (7.44) and (7.45) were given by Suen and Etzel as shown in Eqs. (7.93) and (7.94). Resolution, R_s, is

$$R_s = 1 + \frac{C_{M,o}}{K_{Dis}} \tag{7.93}$$

where

$C_{M,o}$ = initial concentration of protein in entering mobile phase, M (moles/L)
K_{Dis} = dissociation equilibrium constant = k_d/k_a, M (moles/L)
k_d = dissociation rate constant, 1/s
k_a = association rate constant, 1/M·s

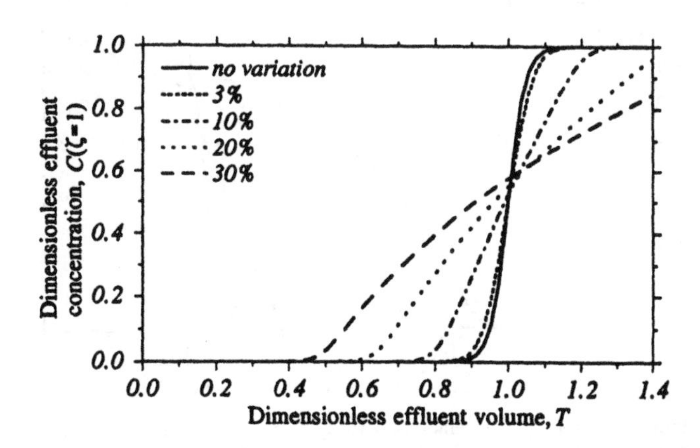

Figure 7.44 Effect of variation on membrane thickness on the shape of the breakthrough curve for R_s = 11, and $n \approx 48$. The plate count n varies with membrane thickness. Curves calculated using the Thomas model (Pe = ∞) as described by Suen and Etzel (1992). (Reprinted from Chem. Eng. Sci., 47(6), Suen and Etzel, 1992, fig.7, p. 1361 with permission from Elsevier Science.)

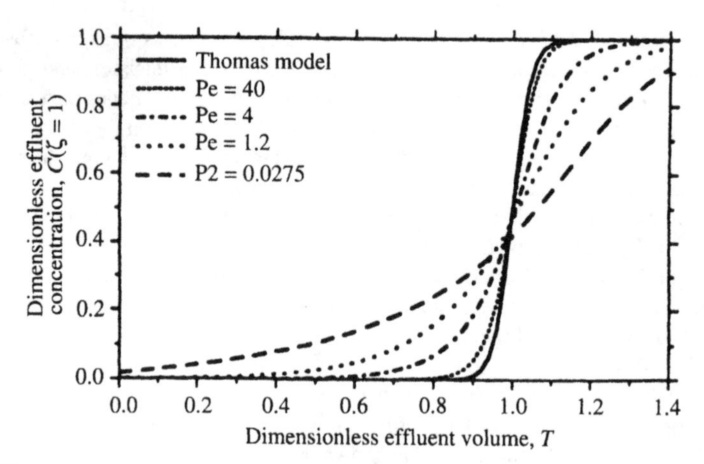

Figure 7.45 Influence of axial diffusion on the shape of the breakthrough curve when $R_s = 11$ and $n = 48.3$. The breakthrough curves are calculated as described by Suen and Etzel (1992). Axial diffusion dominates the loading process for small values of the axial Peclet number, Pe, and results in broad breakthrough curves. Pe is small when the membrane is thin, the association rate is low, and the axial diffusion coefficient is large. As Pe increases, the breakthrough curve becomes sharper. For Pe greater than 40, the effect of axial diffusion is insignificant. At this point the breakthrough curve is very close to the Thomas model (no axial diffusion, $Pe = \infty$). (Reprinted from Chem. Eng. Sci., 47(6), Suen and Etzel, 1992, fig. 3, p. 1361 with permission from Elsevier Science.)

The concept embodied in Eq. (7.93) is that the protein will reversibly, but strongly, bind with the affinity ligands fixed on the surfaces of the pores of the membranes. The goal is to selectively bind as much of the target molecule as possible, while all the other species pass through the membrane, as shown at the left in Figure 7.47. However, leakage of the target molecule, due

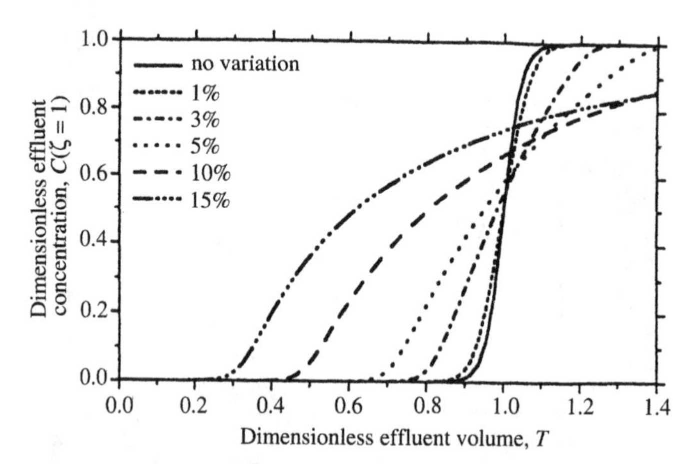

Figure 7.46 Influence of variation in membrane porosity on the shape of the breakthrough curve when $R_s = 11$. The parameter n varies with the membrane porosity, and has a value centered around 48.3. The predictions are from the Thomas model (no axial diffusion, $Pe = \infty$). The pressure drop across the membrane is held constant. The membrane porosity variation must be less than 1% for broadening of the breakthrough curve to be insignificant. (Reprinted from Chem. Eng. Sci., 47(6), Suen and Etzel, 1992, fig. 9, p. 1362 with permission from Elsevier Science.)

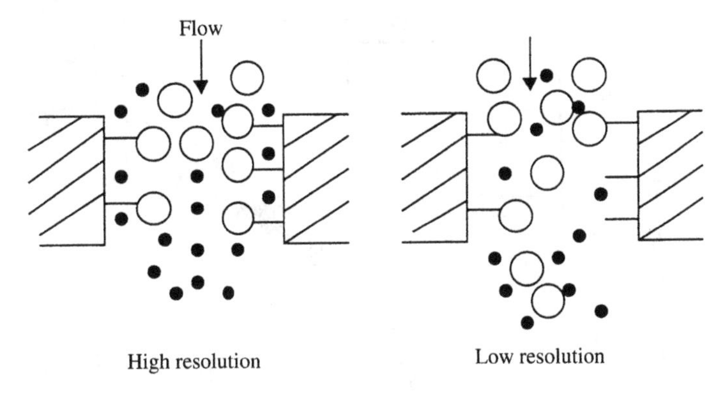

Figure 7.47 Capture of large molecules (large circle) at high and low resolution.

to dispersion or other mechanisms reduces resolution, since some of the target protein leaves the column together with the unbound impurities (as shown at the right-side of Figure 7.47).

The dimensionless number of transfer units, n, gives the plate count for such a system:

$$n = \frac{(1 - \varepsilon_b)C_{s,\text{ligand}}k_a L}{\varepsilon_b v} \tag{7.94}$$

where

ε_b = interpore (membrane) void fraction
$C_{s,\text{ligand}}$ = average concentration of ligand capacity based on the solid (membrane) volume, M (moles/L)
k_a = association constant of protein with ligand, 1/M·s
L = membrane thickness, cm
v = interstitial velocity, cm/s

A thorough case study of a high-capacity ion exchange membrane is presented by Gebauer et al. (1997) based on behavior of lysozyme and bovine serum albumin. Binding kinetics, film and pore diffusion, and solid diffusion were analyzed. A rate limitation was identified to result from the slow transport of proteins into the three-dimensional adsorbing layer at the surfaces of the membrane.

7.19 ELECTRICAL GRADIENTS MAY ALSO BE USED FOR CHROMATOGRAPHIC SEPARATIONS

Electric fields that are applied as an external force field may drive or enhance separation processes for changed molecules (Rudge and Todd, 1990). Electrophoresis is an example of the use of an electrically driven separation in which biological molecules are fractionated on the basis of differences of molecular weights, mobilities, and/or isoelectric points (Ivory et al., 1990). Thorough and definitive analyses of electrically driven separations are given by Ivory et al. (1990) and Rudge and Todd (1990). In the case of proteins, electrophoretic mobility (i.e., movement in an electric field) is related to the charge/mass ratio as summarized in Figure 7.48. The

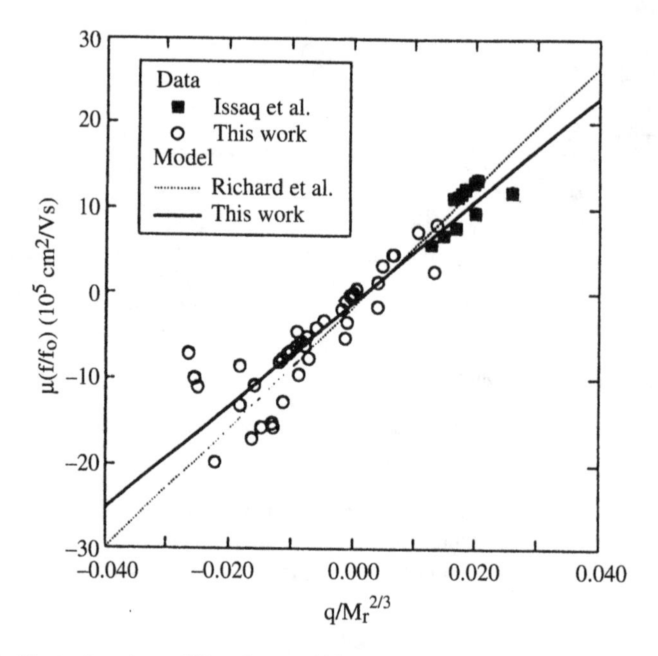

Figure 7.48 Electrophoretic mobility for peptides and proteins versus charge-to-size parameter, $q/M_r^{2/3}$. The equation and correlation coefficient by linear regression is $y = 6.048 \times 10^{-3}x - 1.13 \times 10^{-5}$, $r = 0.94$ (from Basak and Ladisch, 1995, fig. 3, p. 56 with permission from Academic Press).

electrophoretic mobility of a macro-molecule or colloidal particle is given by (Basak and Ladisch, 1995b)

$$\mu = \frac{q}{6\pi\eta r} \cdot \frac{f(kr)(1 + \kappa r_i)}{(1 + \kappa r + \kappa r_i(f/f_o)} \tag{7.95}$$

where

μ = electrophoretic mobility
q = charge
η = viscosity (assumed to be 0.89 Pa at 25°C)
$f(\kappa r)$ = function that corrects for electrical field distortion by large, nonconducting particles
r = effective radius of colloidal particle, Å
r_i = average radius of buffer ions (Å)

The Debye constant κ is

$$\kappa = \left(\frac{2000\text{Ne}^2 I}{\varepsilon_o \varepsilon_r kT}\right)^{0.5} \tag{7.96}$$

where

N = Avogaldro's number (6.023×10^{23})
e = electronic charge ($1.602 \cdot 10^{-19}$ coulombs)
I = ionic strength, moles/L
ε_o = permittivity of free space ($8.85 \cdot 10^{-12}$ C^2/J · meter)
ε_r = dielectric constant or relative permittivity of the medium (78.54 for water at 25°C)
k = Boltzmann constant ($1.38 \cdot 10^{-23}$ J/K)
T = absolute temperature (298 k)

The ratio f/f_o is a function that corrects for protein asymmetry. A tabulation of f/f_o for different molecular weight proteins together with models that relate μ to the charge/mass ratio of proteins is summarized in Basak and Ladisch (1995b). The data from the literature as well as other experimental measurements showed that μ (f/f_o) as a function of $(q/M_r)^{2/3}$ gives the plot shown in Figure 7.48, and follows the correlation

$$\mu\left(\frac{f}{f_o}\right) = 6.048 \cdot 10^{-3} \frac{q}{M_r^{2/3}} - 1.13 \cdot 10^{-5} \tag{7.97}$$

where

μ = electrophoretic mobility, cm^2/volt·s
M_r = molecular weight, Daltons

The frictional forces opposing electrophoretic migration are proportional to the surface area of the colloidal particles. This plot shows that if dispersion forces can be controlled, separation in an electric field may occur (Mosher et al., 1989).

When an electric field is superimposed on a chromatography column, the relative contributions of hydrodynamic compared to electrophoretic dispersion become important. As pointed out by Rudge and Todd (1990), electrophoretic mobilities of proteins are on the order of 10^{-4} cm^2/volt·s. Based on these characteristic values the authors show that the transport rate in electrophoresis may be 1000 to 10,000 higher than transport rates in chromatography, when the same extent of equilibrium is attained in both cases. Hence rapid separations with minimal dispersion are possible by electrophoresis, as long as electrically induced heating (i.e., Joule heating) is controlled.

When an axial electrical potential is applied across a chromatograph column as represented in Figure 7.49, simultaneous fractionation and concentration of proteins is a possibility (Ivory et al., 1990; Rudge et al., 1993; Basak and Ladisch, 1995a, 1995c) as indicated by both theory and experiment. However, the specific interactions between stationary phase, protein, electric gradient, buffer composition, heating effects, and electrode design are still a subject of current research (as of 2000).

The application of an electric potential as shown in Figure 7.49 can be used to increase retention of a negatively charged protein, while movement of a positively charged protein towards the column outlet would be enhanced. Application of the reverse gradient (negative electrode at the column inlet and positive charge at outlet) can be achieved almost instantaneously using an electric gradient. The positive protein would be more retained, while the negative one would move more rapidly. As long as there are sufficient differences in the charge to mass ratios (see Figure 7.48), separation is possible.

Improved electrochromatography column and electrode designs have made it possible to more readily carry out extended electrochromatography runs (Rudge et al., 1995; Basak and

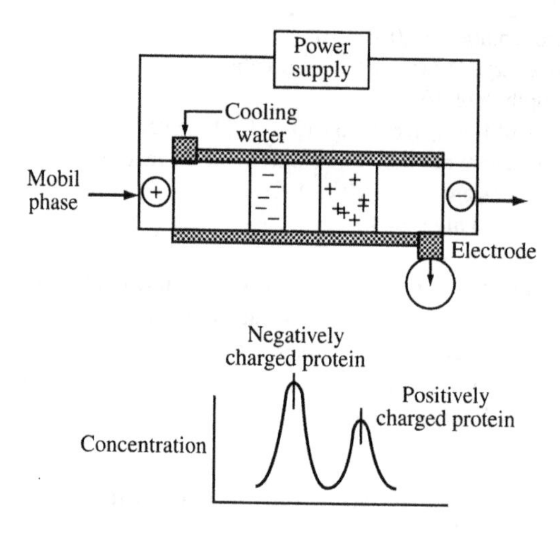

Figure 7.49 Electrochromatography system.

Ladisch, 1995b; Cole and Cabezas, 1997). As pointed out by Ivory et al. (1990 p. 227), "discrepancies between theory and observation stem from differences in column construction, operation, choice of granular gels (i.e., stationary phases) and mechanisms accounted for in models." Improvements in the technology, models, and robust process electrodes are likely to occur in the coming years. This will lead to process-scale separation systems where electrical gradients are used. These gradients avoid the need for additional chemicals to achieve a gradient and can be imposed or removed with the "flick of a switch." However, this vision has yet to be achieved for process-scale applications.

Summary and Perspectives

The application of chemical and other gradients across a chromatography column enhances the separation capabilities of the method but also the complexity of modeling and operating it. Current systems use salts or organic molecules as modifiers, and they are used in the industry for purification of peptides and proteins. The scale-up of gradient chromatography using bench-scale column results is achievable using scaling rules as long as the operating conditions and gradient duration are the same on the bench and process scales. If this is not the case, a more sophisticated method is needed, and equilibrium data that relate modulator to protein concentration within the column must be available, over at least part of the isotherm. A framework for carrying out scale-up, based on knowledge of the ionic gradient, was summarized here. More complete modeling approaches increase the complexity and intensity of computational requirements. References to these approaches are given at the end of this chapter. However, it is hoped that the principles presented here are sufficient to provide a starting point for the simulation, scale-up, and engineering of gradient chromatography.

REFERENCES

Basak, S., and M. R. Ladisch, 1995a, Mechanistic description and experimental studies of electrochromatography of proteins, *AIChE J.*, 41(11), 2499–2507.

Basak, S. K., and M. R. Ladisch, 1995b, Correlation of electrophoretic mobilities of proteins and peptides with their physiochemical properties, *Anal. Biochem.*, 226, 51–58.

Basak, S. K., A. Velayudhan, K. Kohlmann, and M. R. Ladisch, 1995c, Electrochromatographic separation of proteins, *J. Chromatogr. A*, 707, 69–76.

Bates, R. C., and D. D. Frey, 1998, Quasi-linear pH gradients for chromatofocusing using simple buffer mixtures: Local equilibrium theory and experimental verification, *J. Chromatogr. A*, 814, 43–54.

Bosma, J. C., and J. A. Wesselingh, 1998, pH dependence of ion exchange equilibrium of proteins, *AIChE J.*, 44(11), 2399–2409.

Burgess, R. R., 1987, Protein purification, in *Protein Engineering*, Alan R. Liss, Inc., New York, 71–82.

Chicz, R. M., and F. E. Regnier, 1989, Single amino acid contributions to protein retention in cation exchange chromatography: Resolution variants, *Anal. Chem.*, 61, 2059–2066.

Clark, D. B., W. N. Drohan, S. I. Miekka, and A. J. Katz, 1989, Strategy for purification of coagulation factor concentrates, *An. Clin. Lab. Sci.*, 19(3), 196–207.

Cole, K. D., and H. Cabezas, 1997, Improved preparative electrochromatography column design, *J. Chromatogr. A*, 760, 259–263.

Cole, E. S., E. H. Nichols, K. Lauziere, T. Edmunds, and J. M. McPherson, 1991, Characterization of the microheterogeneity of recombinant primate prolactin: Implications for post-translational modifications of the hormone, *in vivo*, *Endocrinol.*, 129(5), 2639–2646.

Colman, A., M. J. Byers, S. B. Primrose, and A. Lyons, 1978, Rapid purification of plasmid DNAs by hydroxyapatite chromatography, *Eur. J. Biochem.*, 91, 303–310.

Coppella, S. J., C. M. Acheson, and P. Dhurjati, 1987a, Isolation of high-molecular weight nucleic acids for copy number analysis using high-performance liquid chromatography, *J. Chromatogr.*, 402, 189–199.

Coppella, S. J., C. M. Acheson, and P. Dhurjati, 1987b, Measurements of copy number using HPLC, *Biotechnol. Bioeng.*, 29, 646–647.

Drohan, W. N., T. D. Wilkins, E. Latimer, D. Zhou, W. Velander, T. K. Lee, and H. Lubon, 1994, A scalable method for the purification of recombinant human protein C from the milk of transgenic swine, in *Advances in Bioprocess Engineering*, Kluwer Academic, Dordrecht, The Netherlands, 501–507.

Furasaki, S., F. Haruguchi, and T. Nozawa, 1987, Separation of proteins by ion-exchange chromatography using gradient elution, *Bioprocess Eng.*, 2, 49–53.

Gallant, S. R., S. Vunnum, and S. M. Cramer, 1996, Modeling gradient elution of proteins in ion-exchange chromatography, *AIChE J.*, 42(9), 2511–2520.

Gebauer, K. H., J. Thömmes, and M. R. Kula, 1997, Breakthrough performance of high-capacity membrane adsorbers, in protein chromatography, *Chem. Eng. Sci.*, 52(3), 405–419.

Gibbs, S. J., and E. N. Lightfoot, 1986, Scaling-up elution chromatography, *Ind. Eng. Chem. Fundam.*, 25, 490–498.

Gu, T., G-J. Tsai, and G. T. Tsao, 1991a, Simulation of multicomponent elution with the mobile phase containing competing modifiers, *Separations Technol.*, 1, 184–194.

Gu, T., G-J. Tsai, and G. T. Tsao, 1991b, A theoretical study of multicomponent radial flow chromatography, *Chem. Eng. Sci.*, 46(5), 1279–1288.

Guichon, G., and J. Zhu, 1995, Production rate of an isotactic train in displacement chromatography, *AIChE J.*, 41(1), 45–57.

Horn, N. A., J. A. Meek, G. Budahazi, and M. Marquet, 1995, Cancer gene therapy using plasmid DNA: Purification of DNA for human clinical trials, *Hum. Gene Ther.*, 6, 565–573.

Ivory, C. F., W. A. Gobie, and T. P. Adhi, 1990, Analytical, preparative and large-scale zone electrophoresis, in *Protein Purification: From Molecular Mechanisms to Large-Scale Processes*, M. R. Ladisch, R. C. Willson, C. C. Painton, and S. E. Builder, eds., ACS Symp. Ser., 427, AIChE, New York, 210–243.

Keim, C. and M. R. Ladisch, 2000, New system for preparative electrochromatography of proteins, *Biotechnol. Bioeng.*, 70(1), 72–81.

Kopaciewicz, W., M. A. Rounds, J. Fausnaugh, and F. E. Regnier, 1983, Retention model for high-performance ion-exchange chromatography, *J. Chromatogr.*, 266, 3–21.

Ladisch, M. R., and A. Velayudhan, 1997, Scale-up techniques in bioseparation processes, in *Bioseparation Processes in Foods*, R. K. Singh and S. S. H. Rizvi, eds., Marcel Dekker, New York, 113–138.

Lahijani, R., G. Hulley, G. Soriano, N. A. Horn, and M. Marquet, 1996, High yield production of pBR322-derived plasmids intended for human gene therapy by employing a temperature-controllable gene mutation, *Hum. Gene Ther.*, 7, 1971–1980.

Maclennan, J., and B. Murphy, 1994, Balance load vs. resolution when selecting column packing, *Today's Chemist at Work*, 3(2), 29–34.

Marquet, M., N. A. Horn, and J. A. Meek, 1995, Process development for the manufacture of plasmid DNA vectors for use in gene therapy, *BioPharm.*, 8(7), 26–37.

Mosher, R. A., D. Dewey, W. Thormann, D. A. Saville, and M. Bier, 1989, Computer simulation and experimental evaluation of the electrophoretic behavior of proteins, *Anal. Chem.*, 61, 362–366.

Natarajan, V., and S. M. Cramer, 1999, Modeling shock layers in ion exchange displacement chromatography, *AIChE J.*, 45(1), 27–37.

Ogez, J. R., J. C. Hodgdon, M. P. Beal, and S. E. Builder, 1989, Downstream processing of proteins: Recent advances, *Biotech. Adv.*, 7, 467–488.

Rudge, S. R., and P. Todd, 1990, Applied electric fields for downstream processing, in *Protein Purification: From Molecular Mechanisms to Large-Scale Processes*, M. R. Ladisch, R. C. Willson, C. C. Painton, and S. E. Builder, eds., ACS Symp. Ser., *427*, AIChE, New York, 210–243.

Rudge, S. R., S. Basak, and M. R. Ladisch, 1995, Solute retention in electrochromatography by electrically induced sorption, *AIChE J.*, 39(5), 797–808.

Rudge, S. R., S. Basak, and M. R. Ladisch, 1993, Solute retention in electrochromatography by electrically induced sorption, *AIChE J.*, 39(5), 797–808.

Suen, S.-Y., and M. Etzel, 1992, A mathematical analysis of affinity membrane separations, *Chem. Eng. Sci.*, 47(6), 1355–1364.

van Leen, R. W., J. G. Bakhuis, R. F. W. C. van Beckhoven, H. Burger, L. C. J. Dorssers, R. W. J. Hommes, P. J. Lemson, B. Noordam, L. M. Persoon, and G. Wagemaker, 1991, Case study in process development, production of human interleukin-3 using industrial microorganisms, *Bio/Technol.*, 9, 47–52.

Velayudhan, A., M. R. Ladisch, and J. E. Porter, 1992, Modeling of non-linear elution chromatography for preparative-scale separations, in *New Developments in Bioseparation*, AIChE Symp. Ser. 88(290), M. M. Ataai and S. K. Sikdar, eds., AIChE, New York, 1–11.

Velayudhan, A., R. Hendrickson, and M. R. Ladisch, 1995, Simultaneous concentration and purification through gradient deformation chromatography, *AIChE J.*, 41(5), 1184–1193.

Velayudhan, A., and M. R. Ladisch, 1995, Effect of modulator sorption on gradient shape in ion exchange chromatography, *Ind. Eng. Chem. Res.*, 34(8), 2805–2810.

Velayudhan, A., and M. R. Ladisch, 1992, Effect of modulator sorption in gradient elution chromatography: Gradient deformation, *Chem. Eng. Sci.*, 47(1), 233–239.

Velayudhan, A., and M. R. Ladisch, 1991, Role of the modulator in gradient elution chromatography, *Anal. Chem.*, 63, 2028–2032.

Yamamoto, S., 1995, Plate height determination for gradient elution chromatography of proteins, *Biotechnol. Bioeng.*, 48, 444–451.

Yamamoto, S., K. Nakanishi, R. Matsuno, and T. Kamibuko, 1983a, Ion exchange chromatography of proteins: Prediction of elution curves and operating conditions, I. Theoretical considerations, *Biotechnol. Bioeng.*, 25, 1465–1483.

Yamamoto, S., K. Nakanishi, R. Matsuno, and T. Kamibuko, 1983b, Ion exchange of proteins: Prediction of elution curves and operating conditions, *Biotechnol. Bioeng.*, 25, 1373–1391.

Yamamoto, S., K. Nakanishi, and R. Matsuno, 1988, *Ion Exchange Chromatography of Proteins, Chromatographic Sci. Ser.*, Marcel Dekker, New York.

Yamamoto, S., M. Nokomura, and Y. Sano, 1987a, Resolution of proteins in linear gradient ion-exchange and hydrophobic interaction chromatography, *J. Chromatogr.*, 409, 101–110.

Yamamoto, S., M. Nomura, and Y. Sano, 1987b, Adsorption chromatography of proteins, *AIChE J.*, 33(9), 1426–1434.

Yamamoto, S., M. Nomura, and Y. Sano, 1992, Stepwise elution chromatography as a method for both purification and concentration of proteins, *Chem. Eng. Sci.*, 47(1), 185–188.

SUGGESTED READING

Chicz, R. M., and F. E. Regnier, 1988, Surface-mediated Retention Effects of Subtilisin Site-Specific Variants in Cation-Exchange Chromatography, *J. Chromatogr.*, 443, 193–203.

Chu, A. H. T., and S. H. Langer, 1985, Characterization of a chemically bonded stationary phase with kinetics in a liquid chromatographic reactor, *Anal. Chem.*, 57(12), 2197–2204.

Fausnaugh-Pollitt, J., G. Thevedon, L. Janis, and F. E. Regnier, 1988, Chromatographic resolution of lysozyme variants, *J. Chromatogr.*, 443, 221–228.

Kim, B., and A. Velayudhan, 1998, Preparative gradient elution chromatography of chemotaltic peptides, *J. Chromatogr. A.*, 796, 195–209.

Price, A. E., K. B. Logvinenko, E. A. Higgins, E. S. Cole, and S. M. Richards, 1995, Studies on the microheterogeneity and *in vitro* activity of glycosylated and non-glycosylated recombinant human prolactin separated using a novel purification process, *Endocrinol.*, 136(11), 4827–4833.

PROBLEMS

7.1 An unknown (i.e., target) protein has the retention behavior shown in Figure 7.50. Calculate the characteristic charge of this protein. The protein capacity factors and retention data are given below where

$$\text{Protein capacity factor} = k' = \frac{t_{R,BSA} - t_{r,o}}{t_{r,o}} = \frac{t_{r,BSA}}{4} - 1$$

M NaCl	$t_{r,BSA}$	k'	$\log k'$
0.5	4	0	
0.4	36	10	1
0.3	76	1.8	1.25
0.2	196	4.8	1.68
0.1	1004	250	2.4
0.05	4004	1000	

7.2 Two adjacent characteristic curves (lines) emanating from the axis for dimensionless time, τ (= vt/L), have the form

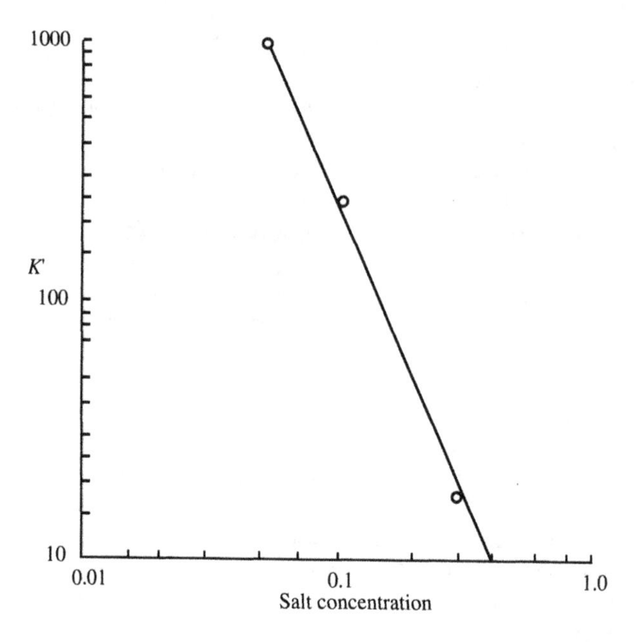

Figure 7.50 Retention plot for problem 7.1.

$$\tau = \eta_1 + \xi\left[1 + \frac{\phi a_m}{(1 + \chi_o + \beta\rho_1)^2}\right]; \rho_1 = \eta_1 - \tau_{inj}$$

and

$$\tau = \eta_2 + \left[1 + \frac{\phi a_m}{(1 + \chi_o + \beta\rho_2)^2}\right]; \rho_2 = \eta_2 - \tau_{inj}$$

a. Derive the equations that give the coordinates of the lines' point of intersection.

b. If the gradient is held constant (i.e., slope of gradient is zero), a shock wave will reach a constant strength of $\chi_f = KC_{M,f}$. What are the coordinates of this point in terms of dimensionless distance ($\xi = X/L$) and dimensionless time ($\tau = vt/L$)?

7.3 The velocity of three components are in Figure 7.51 at three different positions in the column, L_1, L_2, and L_3. The velocity of an unretained component is given by the dotted line. This representation is also given in terms of time and distance as shown in Figure 7.52. The material balance for ideal chromatography is given by

$$\frac{\partial c_m}{\partial t} + v\frac{\partial c}{\partial x} + \phi\frac{\partial q}{\partial t} = 0$$

where $q = K_D C_m$ (linear isotherm)

a. Show how these equations can be used to derive the expression below for the characteristic curves illustrated in Figures 7.51 and 7.52.

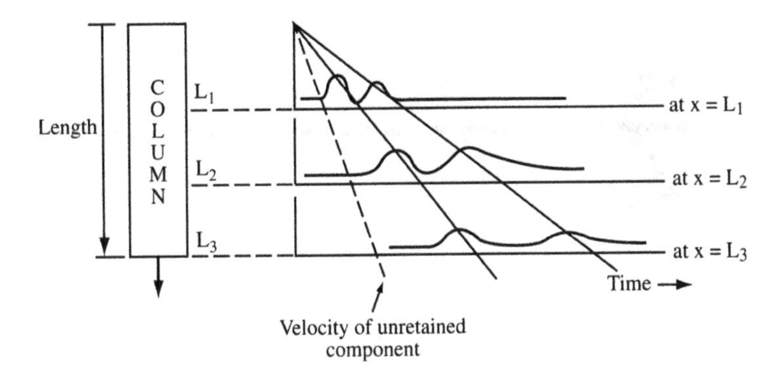

Figure 7.51 Representation of characteristic curves relative to peak retention for problem 7.3.

$$\left(\frac{dx}{dt}\right)_c = \frac{v}{1 + \phi K_D}$$

b. If the equilibrium relation for the salt (modulator) is given by

$$K_D(S) = \frac{\beta}{[S]^a}$$

where

$K_D(S)$ = salt dependent distribution coefficient for the protein

β = a constant for a given protein

S = salt concentration (ionic strength)

show the expression that results in representative characteristic curves.

c. What is the characteristic equation for a solute (protein) that has no interaction with the stationary phase and therefore defines the void volume?

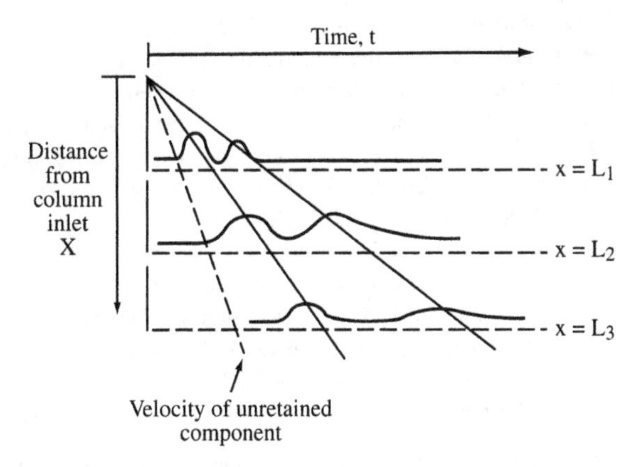

Figure 7.52 Representation of characteristic curves problem 7.3.

CHAPTER 8

PRINCIPLES OF BIOSEPARATIONS FOR BIOPHARMACEUTICALS AND RECOMBINANT PROTEIN PRODUCTS

INTRODUCTION

The large-scale purification of proteins and other bioproducts is the final step, prior to product packaging, in the manufacture of therapeutic proteins, specialty enzymes, diagnostic products, and value-added products from agriculture. The separations that purify biological molecules or compounds from biological sources are referred to as bioseparations. Large-scale bioseparations represent both art and a science that evolves from laboratory scale techniques. These are adapted and scaled up to satisfy the need for larger amounts of extremely pure product for analysis, characterization, testing of efficiency, clinical or field trials, and finally, full-scale commercialization. The rigorous quality control embodied in current good manufacturing practices, and the complexity and lability of the macromolecules being processed, provide other practical issues that must be addressed (Willson and Ladisch, 1990).

This chapter gives an overview of manufacturing approaches for selected recombinant proteins together with descriptions of their recovery and purification using various bioseparation techniques. Chromatography, is emphasized. Since the literature presents theory for analytical systems, and Chapters 5 to 7 of this book discuss modeling and scale-up, applications of process chromatographic separations are reviewed here. The goal of this chapter is to provide the reader with a framework of principles that can be used to initiate the scale-up of liquid chromatography, starting from a minimum amount of laboratory data. Once a process concept is defined, scale-up rules and correlations can be applied to estimate equipment size, from which costs can be calculated, as summarized in Chapter 6. Selected bioproducts are discussed to illustrate how recombinant biotechnology facilitates new products that could not be obtained by other means while providing unique tools

The author thanks Jeffrey C. Baker (Eli Lilly), Doug Cameron (while at University of Wisconsin), and Ayda Sarikaya and Craig Keim (Purdue University) for their review of this chapter and their many helpful suggestions. The materials in this chapter were first organized and presented by the author in the Kirk Othmer Encyclopedia of Technology (Ladisch, 1993).

that enable their production. The examples presented in this chapter are not intended to be viewed in the context of specific market information, sales data, investment trends, licensing arrangements, regulatory guidelines, etc. Rather the examples are intended to provide a frame of reference within which the vitality of this new industry and the impact of both technical and non-technical decisions can be cast and discussed. These examples do not represent economic regulatory or business metrics of the industry (particularly as it will be in 1999 and beyond). The industry will undergo rapid change in the coming decade. The examples were developed in the hope that they will provide insights on the factors that play a role in adaptation of new technology.

8.1 NEW BIOTECHNOLOGY PRODUCTS ARE THE FASTEST GROWING AREA IN BIOSEPARATIONS

Biotechnology was broadly defined in 1991 by the U.S. Office of Technology Assessment as "any technique that uses living organisms (or parts of organisms) to make or modify products, to improve plants or animals, or to develop microorganisms for specific uses." The *new* biotechnology, which was introduced in 1970, enables directed manipulation of the cell's genetic machinery through recombinant-DNA techniques and cell fusion. The new biotechnology was first applied on an industrial scale in 1979, and since then it has fundamentally expanded the utility of biological systems so that biological molecules, for which there is no other means of industrial production, can now be generated. Biotechnology, combined with existing industrial, government, and university intrastructure in biotechnology and the pervasive influence of biological substances in everyday life, has set the stage for unprecedented growth in products, markets, and expectations. Substantial manufacturing capability will be needed to bring about the full application of biotechnology for the benefit of society (Committee on Bioprocess Engineering, 1992). The recovery, purification, and packaging of these products for delivery to the consumer is making major contributions to the realization of this capability.

8.2 BIOSEPARATION PROCESSES HAVE A SIGNIFICANT IMPACT ON MANUFACTURING COSTS

The purification of the protein is a critical and expensive part of the process, and might account for 50% or more of the total production costs (Committee on Bioprocess Engineering, 1992). However, for small-volume very high value biotherapeutics (examples in Table 8.1), these costs may be considered secondary to the first to market principle unless the presence of a lower-cost competitor focuses attention on manufacturing costs or other factors limit prices. Annual sales of the products given in Table 8.1 were $700 million for human insulin (Thayer, 1996b), $290 million for tissue plasminogen activator, and $220 million for human growth hormone (Früst and Unterhuber, 1996; Thayer, 1996a). The most successful bioproduct in biotechnology history, recombinant erythropoetin (EPO), had worldwide sales estimated at $1.6 to 2.6 billion in 1995 (Rawls, 1996; Thayer, 1996b; Wells, 1996) for several kilograms of product. Epogen is a genetically engineered version of the protein erythropoetin, which is produced by the kidneys and stimulates blood stem cells to mature into red blood cells. Epogen can reverse the severe anemia often caused by kidney disease. Amgen's sales of this product, together with Neupogen (a recombinant protein that directs blood stem cells to become bacteria-fighting neutrophils), was about $1.8 billion, thus making it one of the most successful biotechnology companies in the mid 1990's (Roush, 1996). While Table 8.1 will quickly become dated in an era of a rapidly growing biotechnology industry, it nonetheless illustrates the value (and price) of bio-

TABLE 8.1 Unit Values and Relative Production Quantities for Selected Approved Biopharmaceuticals

Product	Year Approved	Approximate Selling Price $/g	Amount of Product for $200 Million in Sales, kg
Human insulin	1982	375[a]	530.
Tissue plasminogen activator	1987	23,000	8.7
Growth hormone	1985	35,000	5.7
Erythropoetein (Epogen)	1989	840,000	0.24
GM CSF	1991	384,000	0.52
G-CSF	1991	450,000	0.44

Source: Adapted from National Academy of Sciences, Committee on Bioprocess Engineering (1992), table 2-1, and data of Bisbee (1993).

[a]A 1996 estimate places the selling price at $130/g, thus indicating a significant decrease in price (Petrides et al., 1996).

pharmaceutical products relative to other, less expensive biotechnology products described earlier in this book.

8.3 BIOSEPARATION ECONOMICS ARE SECONDARY TO BEING FIRST TO MARKET . . . UNTIL AFTER THE NEW PRODUCT IS INTRODUCED

The development of biotechnology processes in the biopharmaceutical and bioproduct industries is driven by the precept of being the first to market, while achieving a defined product purity and developing a reliable process that will meet validation requirements. The cost of a lost opportunity in a tightly focused market where there is room for only a few manufacturers, with a single dominant player, can be devastating for products that take 5 to 10 years and 100 to 500 million dollars (U.S.) to develop. After the process and product are validated, the cost of change can be great, if only due to satisfying regulatory constraints. Hence, once the manufacturing process is in place, changes are likely to be considered only if very major improvements result (Wheelwright, 1991). However, the successful introduction of a new product to the market will inevitably attract competition. The cost of manufacture then becomes an issue because of downward pressure on price due to competing products with similar, if not superior, properties. This situation provides incentives for bioprocess scientists and engineers to "get it right the first time."

Genentech's Activase® (a recombinant tissue plasminogen activator or rt-PA introduced to the U.S. market in 1987) gave an example of how a product's success attracts competition—in this case it was Boehringer Mannheim's Retavase™ (being sold in Europe as Rapilysin). Activase is a glycosylated form of rt-PA (an alteplase), while Retavase is a smaller, nonglycosylated form of this protein (reteplase). Retavase was developed over a seven-year period, starting in 1989, at a cost of $140 million, before being approved for use in the European market in August 1996 by the European Medicines Evaluation Agency (EMEA) and by the FDA in October 1996 for use in the United States. The reteplase was available on the European market in October 1996 with 1997 sales expected to be $4 million in Germany and $30 to 70 million, worldwide. Boehringer Ingelheim (Ingelheim, Germany), which licensed Genentech's rt-PA for marketing, almost immediately filed a lawsuit against Boehringer Mannheim for infringing on Genentech's patent. The reteplase is claimed to be more effective than the alteplase based on results from ongoing clinical trials. Boehringer Mannheim says the new product is unique "because it represents the one and only designer molecule ever manufactured in Germany without any dependence on or licensing from foreign patents." Reteplase is the twenty-first recombinant drug to enter the German market (Fürst and Unterhuber, 1996).

8.4 BUT COST IS IMPORTANT WHEN SUCCESS OF A NEW PRODUCT GENERATES COMPETITION

Wheelwright (1991) presents a cogent analysis of the three main sources of competitive advantage in the manufacture of high-value protein products (page 1): "first to market, high quality (in terms of purity, activity, dependability, or flexibility) and low cost." The company that is the first to market a new protein biopharmaceutical, *and* is first with patent protection, enjoys a substantial advantage. The second company to enter the market may find itself with only one-tenth of the sales. In the absence of patent protection, product differentiation becomes very important. Differentiation will reflect a product that is more pure, more active, and/or has a greater lot-to-lot consistency.

The task of quickly specifying, designing and scaling-up a bioproduct separation can be daunting, particularly for biopharmaceuticals and protein products. These separations are carried out in a liquid phase, with macromolecules that are labile and whose conformation *and* heterogeneous chemical structure can undergo a subtle change during purification resulting in an unacceptable product. A process that was initially designed to result in a lower cost is particularly beneficial when a success of a new product generates competition. The lower cost initially provides a higher margin (and profit) and eventually prevails over a process that must be redesigned to reduce the cost of production, particularly since regulatory issues impose a high cost on process change.

8.5 MANUFACTURING PROCESSES FOR BIOLOGICS ARE PART OF THE PRODUCT'S REGULATION

Biologics are so named because it is generally not possible to define them as discrete chemical entities or demonstrate a unique composition. Biologics include blood fractionation products such as albumin and Factor VIII, and both live and killed viral vaccines. A drug (e.g., aspirin) is classified on the basis of being able to completely define its chemical nature and analytically demonstrate its purity, potency, and identity.

The process used to make a biologic is closely monitored and regulated by the regulatory agencies, since a significant change in the process may result in a product that is different from that previously reviewed and regulated, and hence may require a new license. Process changes made during the investigational new drug (IND) development stage, and before the license is approved, are more easily incorporated into a new product (from a regulatory point of view) than after the license is generated (Builder et al., 1989).

The definitions of biologics verses drugs continues to evolve, with assignment being made on a case by case basis (Sobel, 1991). The discussion presented here has the intent of introducing the student to nature of the definitions and regulations, rather than specific regulations, themselves, which change from one year to the next. Sobel (page 508) states that section 351 of the Public Health Service Act defines a biologic product as "any virus, therapeutic serum, toxin, antitoxin, vaccine, blood, blood component or derivative, allergenic product, or analogous product . . . applicable to the prevention, treatment, or cure of diseases or injuries in man." According to Solomon (Sobel, 1991), biologics are subject to licensing provisions that require both the manufacturing facility and the product be approved. All licensed products are subject to specific requirements for lot release by the FDA. In comparison, drugs are approved under Section 505 of the FD & C Act (21 USC 301–392), where there is not lot release by the FDA except for insulin products. Soloman gives the example that insulin, growth hormone, and many other hormones have been treated as drugs, while erythropoetin (abbreviated EPO), which also fulfills the criteria of a hormone, was reviewed in the biologic division of the FDA. (Note that insulin is derived from a bacterial fermentation while erythropoetin is obtained from mammalian cell culture.) Hormones, for the most part, will continue to be reviewed as drugs.

The analysis of regulations and their potential application to new bioproducts is not the intent of this chapter. Government regulations change, and the countries in which a product is sold influence the regulations that the product is subject to. Rather, this brief mention of government regulations is an attempt to reflect on other issues that could influence the design of bioseparation unit operations on a process scale, and the constraints on process development that grow as a recovery and purification scheme approach, and then pass, the licensing stages for commercial manufacture.

INSULIN CASE STUDY

Biosynthetic human insulin is generated biologically through fermentation of a transformed microorganism, and then modified using chemical and biochemical methods. The term "biosynthetic" can denote a combination of biological, recombinant, chemical, and/or biochemical steps followed by, or interspersed with, purification of the target molecule (i.e., product) (Baker, 1996).

8.6 BIOSYNTHETIC HUMAN INSULIN IS THE FIRST RECOMBINANT POLYPEPTIDE FROM *E. COLI* LICENSED FOR HUMAN USE

Insulin is a polypeptide hormone that stimulates anabolic reactions for carbohydrates, proteins, and fats and thereby produces a lowered blood glucose level. Porcine and bovine insulins were used to treat diabetes prior to the availability of human insulin. All three types are similar in amino acid sequence (Figure 8.1), although the sequence variation (of amino acid residues) could lead to immunogenic responses. The desire to not be restricted to animal tissue sources for insulin production reportedly led to interest in manufacture of human insulin by fermentation. Eli Lilly's human insulin was approved for testing in humans in 1980 by the FDA, and was placed on the market by 1982 (Bernon and Bodelle, 1991; Ladisch and Kohlmann, 1992). Since insulin is an endocrine hormone, it is regulated as a drug, whereas other types of hormones are likely to be viewed as biologics in the United States.

Human insulin was the first animal protein to be made in bacteria in a sequence identical to the human pancreatic peptide, and it resulted from the cooperation between Eli Lilly and Genentech. Expression of separate insulin A and B chains were achieved in *Escherichia coli* K-12 using genes for the insulin A and B chains synthesized at City of Hope National Medical Center, (Duarte, CA) and cloned in frame with the β-galactosidase gene of plasmid pBR322 (Chance et al., 1981a, 1981b). The product was a chimeric protein in an intracellular, granular form known as inclusion or refractile body. An example of a cell containing an inclusion body is shown in Figure 8.2. Insulin's small size (21 amino acids for the A chain, $M_r = 2300$; and 30 for B chain, $M_r = 3400$), and absence of methionine (Met) and tryptophan (Trp) residues in the A and B chains were critical elements in the decision to undertake cloning of this peptide hormone, as well as in the rapid development of the manufacturing process. The Met and Trp residues in the chimeric precursor, produced as a consequence of engineering and expression of the gene in *E. coli*, are hydrolyzed by the reagents used during the recovery process. The presence

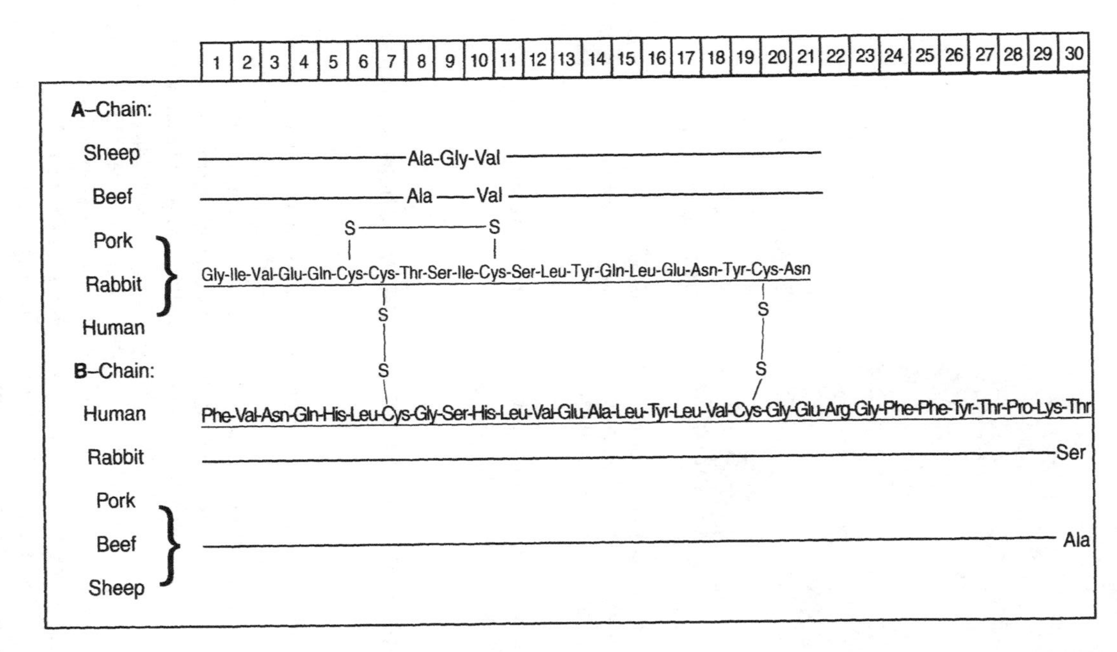

Figure 8.1 Amino acid sequence of A and B chains of insulin from different sources. (Reproduced with permission from Dolan-Heilinger, *Recombinant DNA and Biosynthetic Insulin: A Source Book*, Eli Lilly, Indianapolis, IN, 1982.)

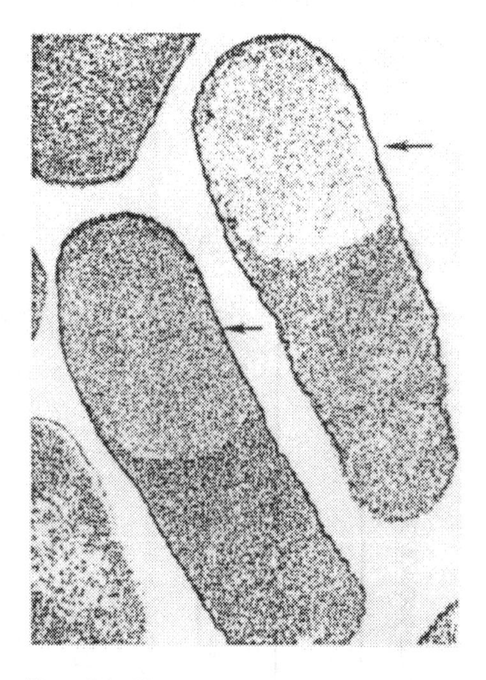

Transmission electron micrograph of *E. coli* (× 35,000) containing insulin A chain chimeric protein. Arrows indicate concentrations of this chimeric protein in the cells.

Figure 8.2 Electron micrograph of *E. coli* cell with inclusion body produced by accumulation of insulin protein (reproduced with permission from Dolan-Heitlinger, *Recombinant DNA and Biosynthetic Insulin: A Source Book*, Eli Lilly, Indianapolis, IN, 1982).

of these amino acids in insulin would have resulted in the hydrolysis and destruction of the product (Ladisch and Kohlmann, 1992).

8.7 RECOVERY AND PURIFICATION OF HUMAN INSULIN REQUIRES 27 STEPS

The production of human insulin manufactured by Eli Lilly requires 31 major processing steps of which 27 are associated with product recovery and purification (Prouty, 1991). This process is based on a fermentation that yields proinsulin. The literature indicates that the major steps in this purification sequence include centrifugation; ion exchange, reverse phase, and size exclusion chromatography; and precipitation as well as CNBr cleavage and enzyme transformation. A recent (1996) overview of a possible processing scheme is presented in Figure 8.3, taken from Petrides et al. (1996). Their evaluation of a process that manufactures 1500 kg of biosynthetic human insulin a year indicates that a total capital investment of $130 million would be needed for a plant producing 1500 kg of insulin a year. This assumes that an 18 hour fermentation is required and gives a concentration of *E. coli* cells (dry weight basis) of 37.5 g cells per liter of fermentor volume, with 20% of the cell weight corresponding to intracellular inclusion bodies containing the proinsulin fusion protein. Twelve hours are allotted for turnaround of the fermentor, resulting in a total time of 30 hours. On this basis the unit production cost was estimated by Petrides et al. (1996) to be $41/gram of insulin, with a selling price of $110/gram. This is significantly lower than the estimate of $375/g based on earlier data given in Table 8.1. This decrease likely reflects, in part, growth in use of recombinant forms of insulin, with its attendent economies of scale.

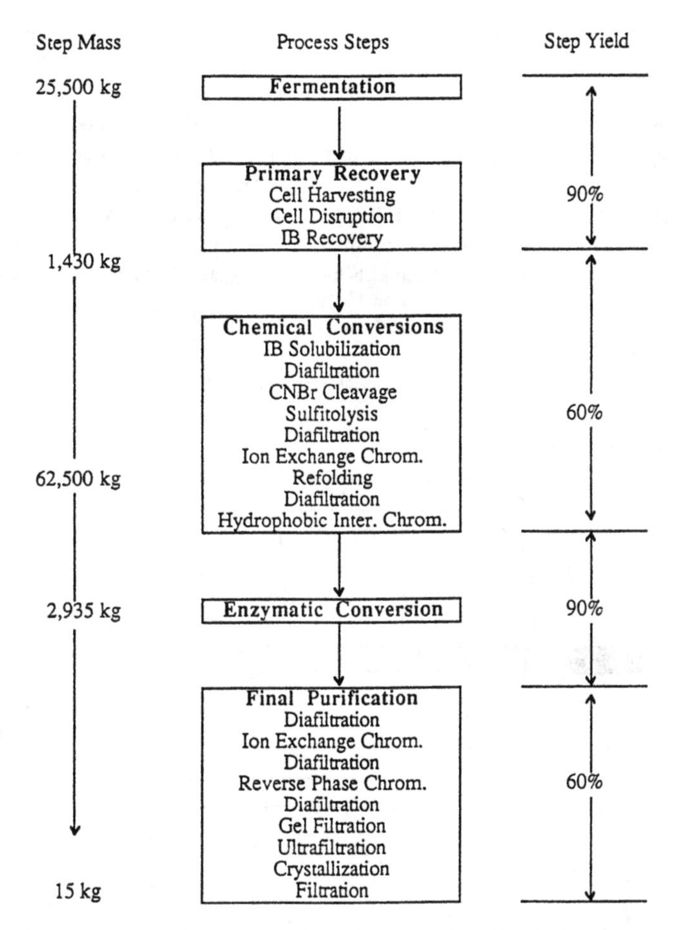

Figure 8.3 A proposed, general processing scheme for recombinant insulin. IB denotes inclusion body (from Petrides et al., 1996, with permission from Intelligen).

The production process for human insulin has a significant literature associated with it and provides an instructive case study on the range of unit operations that must be considered in the recovery and purification of a recombinant product from a bacterial fermentation. However, the exact sequence of the 27 steps has not been published, and hence it cannot be presented here in its entirety. Rather a synopsis of key steps is given.

The fermentation product is a fusion protein of a portion of the Trp enzyme connected to proinsulin through a Met residue. It is produced in *E. coli* that contains a plasmid with the proinsulin gene connected to the Trp promoter. The Trp operon is turned on when the fermentation media is allowed to become depleted of tryptophan.[1] This causes the production of a fusion protein of proinsulin to occur, which quickly accumulates intracellularly in the *E. coli*. An inclusion body (i.e., a large body of aggregated protein and nucleic acids) is formed that occupies about

[1]The formation of inclusion bodies causes cell growth to stop. Since the proinsulin fusion protein is an intracellular product, productivity is proportional to cell mass (i.e., the number of cells). It is important to maximize the number of cells before formation of inclusion bodies. Premature formation of inclusion bodies would result in a lower productivity. Hence the "Trp switch" is an important practical tool in maximizing productivity.

half of the cell volume. At this point the fermentation is complete, and protein recovery, dissolution, protein refolding, and purification are carried out (Figure 8.3) (Ladisch and Kohlmann, 1992, Petrides et al., 1996).

The fermentation broth is chilled, and the cells lysed. Following inclusion body recovery, CNBr (a hydrolytic agent that specifically attacks Met and Trp linkages) cleaves away the fusion protein from the proinsulin (see Figure 8.4). Since neither Met or Trp are among the amino acid residues that make up proinsulin, the proinsulin molecule is left intact. The proinsulin is then subjected to oxidative sulfitolysis, refolding to its proper conformation, purification, and enzyme treatment to remove the connecting (or C) peptide that connects the insulin A and B chains. These are sometimes

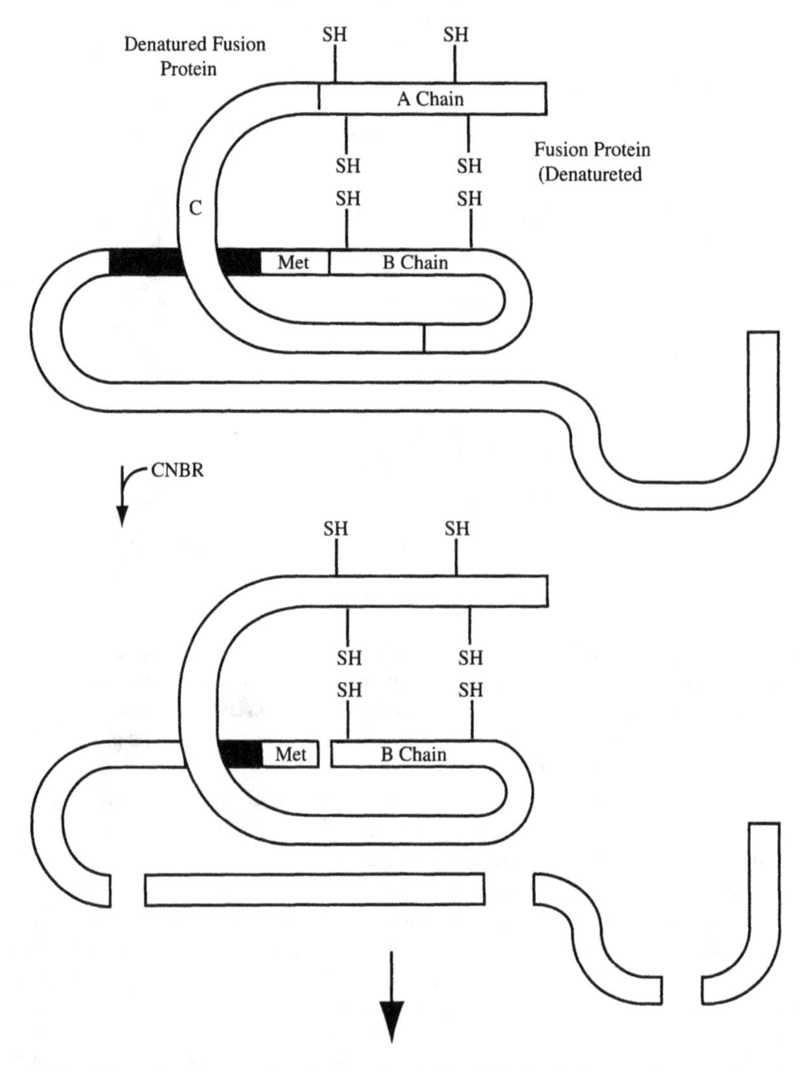

Figure 8.4 Schematic representation of major chemical and biochemical processing steps (continued).

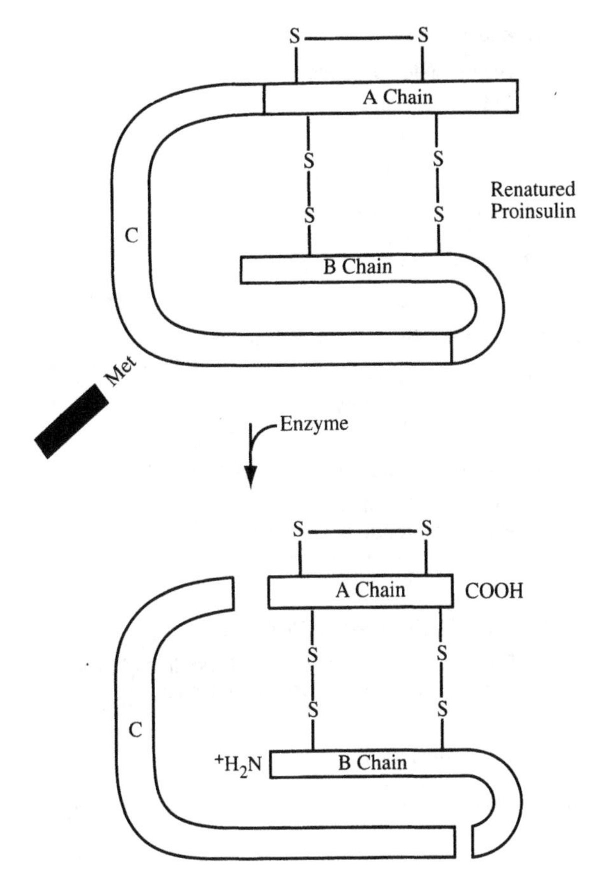

Figure 8.4 Continued.

referred to as front-end operations, since they are carried out early in the sequence of processing steps that leads to the final insulin product. Front-end operations generate, contain, and define process specific contaminants that must later be removed (Baker, 1996). The crude insulin consisting of A and B chains in their proper conformation are further purified using a sequence of ion exchange, reverse phase, and size exclusion chromatography steps as shown in Figure 8.5 (Kroeff et al., 1989; Wheelwright, 1991; Ladisch and Kohlmann, 1992). Another process summary is provided in Figure 8.6, which indicates locations of CNBr cleavage of the fusion tail of the chimeric protein, reduced disulfide bonds (proinsulin s-sulphonate), enzyme cleavage sites for proinsulin, and the refolded insulin as well as major purification steps.

Deamidation of asparagine or glutamine residues can occur readily in either acidic or neutral solutions, while disulfide exchange reactions causing the formation of isomeric monomers or aggregated forms (multimers) of the protein can occur at alkaline pH (Prouty, 1991). Deamidation products of insulin (also referred to as desamido insulin) can be formed during the processing of insulin. These variants of insulin require high-resolution chromatography techniques to remove. Therefore a multimodal sequence of chromatographic separations for the crude recombinant insulin is required and consists of the following:

Fermentation

1. Accumulate cell mass
2. Induce proinsulin formation

↓

Recovery of Inclusion Bodies

1. Cell recovery (centrifuge)
2. Cell lysis (release inclusion bodies)
3. Inclusion body recovery

↓

Recovery of Denatured Fusion Proteins

1. CNBr hydrolysis
2. Proinsulin recovery

↓

Recovery and Purification of Folded Proinsulin

1. Purify *s*-sulfonate proinsulin
2. Refold to proper conformation
3. Purify refolded proinsulin

↓

Recovery and Purification of Proinsulin

1. Enzyme hydrolysis to remove C peptide of Proinsulin
2. Chromatographic purification of insulin

↓

Formulation

Figure 8.5 Schematic representation of the major steps of product recovery and purification for recombinant insulin.

1. Ion exchange (removes most of the impurities)
2. Reversed phase (separates insulin from structurally similar insulinlike components)
3. Size exclusion (removes polymers, residual proinsulin, salts and small molecules from the insulin)

The best pH range of the acetonitrile mobile phase for step 2 was reported as 3.0 to 4.0, since it is well below the isoelectric pH of 5.4, gives excellent resolution, and minimizes the deamidation of insulin to monodesamido insulin if the residence time in the reversed phase column is less than several hours (Kroeff et al., 1989; Ladisch and Kohlmann, 1992). The sequence of steps given by Kroeff et al. (1989)—ion exchange, reversed phase, and gel permeation (size exclusion) chromatography—follows the principle of orthogonality of the separation sequence, where each step is based on a different property, namely charge, solubility, and size, respectively (Willson and Ladisch, 1990). Near the end of the chromatography sequence, the insulin may be concentrated by precipitation to form insulin zinc crystals. The procedure consists of adjusting the insulin concentration to 2 g/L with 0.25 M acetic acid, adjusting the pH to 5.9 with ammonium hydroxide, and adding a 1.8 moles zinc chloride per mole biosynthetic human insulin. The zinc insulin crystals are then recovered by filtration.

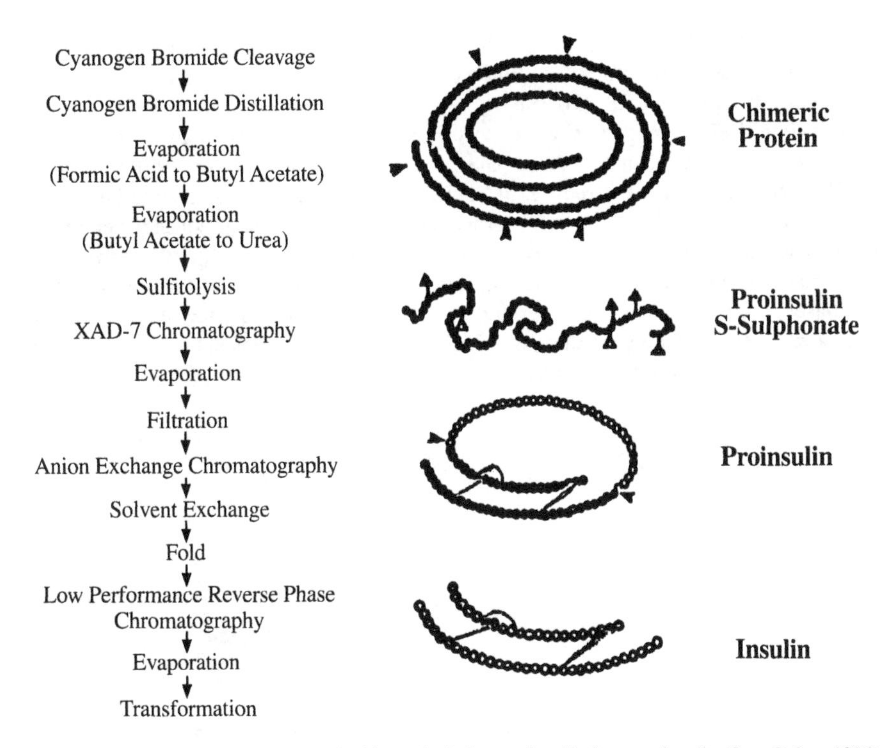

Figure 8.6 Recent process summary for biosynthetic human insulin from proinsulin (from Baker, 1996).

8.8 RECOVERY PROCESS EQUIPMENT VOLUMES ARE MODEST BY CHEMICAL INDUSTRY STANDARDS

The purification of insulin illustrates the many steps involved in product recovery and purification, and the unit operations required. These include cell lysis, centrifugation, refolding, buffer exchange, chromatography, precipitation, and filtration. Some of these steps are repeated. The volumes of the individual chromatography columns are estimated to range from 50 to 1000 L. While these volumes are small compared to other types of chemical recovery processes, they are large in the context of biotechnology manufacturing. An illustration of scale is provided by loadings reported for column sizes used in scale-up studies for the reversed phase insulin purification step illustrated in Figure 8.7, which is based on the data of Kroeff et al. (1989). The loading per unit column volume is about 15 g per liter of column volume, with the load increasing in direct proportion to the column volume. The mobile phase consumed was approximately 6 column volumes/run, or 536 L of mobile phase per kg of insulin loaded. This would correspond to 123 L acetonitrile if an average acetonitrile concentration of 23% is assumed. If the total amount of recombinant insulin produced annually on a worldwide basis is on the order of 7500 kg (Petrides et al., 1996), the chromatography column volume required would be on the order of 1500 L (assumes 1 run/day for 333 days/yr). The annual solvent usage would not be inconsequential with an estimated 804,000 L of mobile phase or 185,000 L of acetonitrile being consumed in a plant making 1500 kg/yr, assuming that the loadings per unit volume of stationary phase are the same at the commercial scale as was reported in Kroeff's scale-up studies.

Reversed phase liquid chromatography using small particle size stationary phases is also referred to as high-performance liquid chromatography or high-pressure liquid chromatography (abbreviated HPLC). The use of HPLC on a process scale probably evolved from analytical chromatography where insulin species differing by a single amino acid were readily separated with a reversed phase packing using an acetonitrile/dilute, aqueous 0.25% acetic acid gradient (Kroeff et al., 1989). The small particle size material used by this process (10 μm Zorbax C_8) posed a process engineering challange, since the complexities of efficiently and reproducibly packing large process columns with 10 μm particles are considerable. Consequently a self-packing, axial compression column was used for stationary phase (column volumes) of 8 L or higher. The principles of reversed phase media and axial compression columns are discussed in this chapter in Sections 8.4 to 8.5 on reversed phase chromatography and also in Section 5.8.

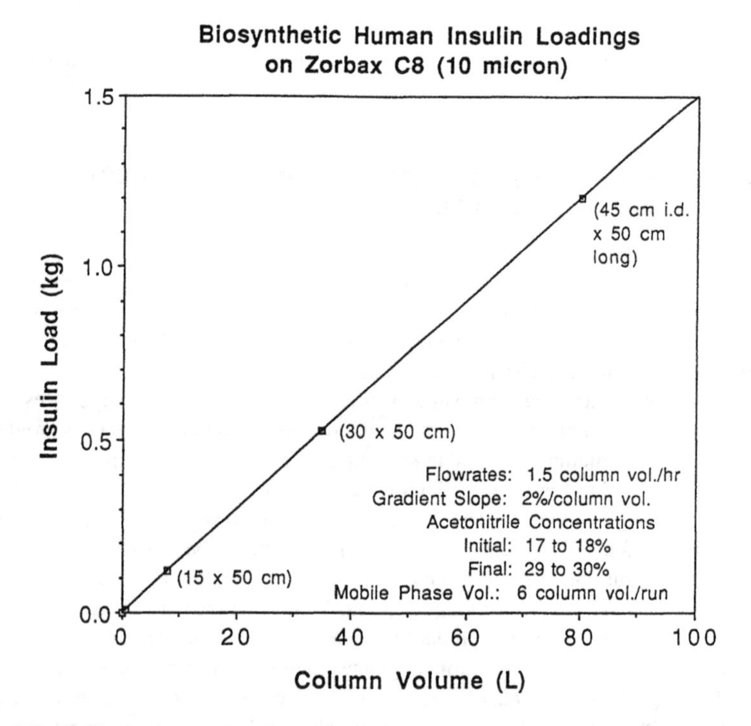

Figure 8.7 Insulin load as a function of column volume for reversed phase chromatography of biosynthetic human insulin over 10 micron diameter, Zorbax, process grade, C_8 reversed phase material. Column dimensions given in parenthesis (data from Kroeff et al., 1989).

8.9 YIELD LOSSES ARE AMPLIFIED BY THE NUMBER OF PURIFICATION STEPS

The numerous steps required for production of this recombinant protein incur a built-in yield loss. For example, if only 2% yield loss were to be associated with each step, the overall yield for a purification sequence of 10 steps would be

$$\eta = 100\left(1 - \frac{L}{100}\right)^n = 100\,(1 - 0.02)^{10} = 81.7\% \tag{8.1}$$

where η denotes percent yield, L the percent yield loss, and n the number of steps. If the yield loss at each step were 5%, the overall yield would only be 60%. Maximizing recovery at each step is important. The yields for each part of the process in Figure 8.3 range from 60 to 90%, with an overall yield of 29% being indicated.

During the recovery and purification of human insulin, its molecular structure, and hence, physicochemical properties, is altered five times during its recovery and purification. These forms are as follows:

1. Fusion protein
2. Denatured aggregate
3. Denatured monomers
4. Properly folded proinsulin
5. Insulin

Prouty (of the Lilly research laboratories) points out that "the changes in properties, such as pI, size, and hydrophobicity, present an opportunity to use various purification procedures repeatedly in the process, and expect them to purify away different contaminants at each stage. Although an apparent disadvantage because it introduces more steps in the process, the multiple forms of insulin generated in the process maximize the chance of removing contaminants that are not likely to change chemically in the same way as the insulin molecule. The final purification after transformation of proinsulin to insulin utilizes procedures that rely on multiple properties of the insulin, such as size, hydrophobicity, ionic charge, and crystallizability, further assuring the removal of contaminants, even below the parts per million range" (from Prouty, 1991, p. 237). The final purity level is reported to be > 99.99% (Builder et al., 1989).

8.10 LYS(B28), PRO(B29) BIOSYNTHETIC HUMAN INSULIN IS A HUMAN INSULIN ANALOG

Another recombinant protein has been developed and has recently entered the insulin market. This analogue of insulin is chemically identical to human insulin except that its sequence on the B chain has been altered so that the order of the amino acids in the B28 (i.e., insulin B chain, residue 28) and B29 positions are reversed to give a Lys–Pro instead of a Pro–Lys sequence. This insulin analogue has a more rapid onset of activity and rapid clearance than human insulin (Humulin R®), as shown in Figure 8.8, and it will enable insulin, taken 10 to 20 minutes before a meal, to be effective in controlling glucose. The lag time

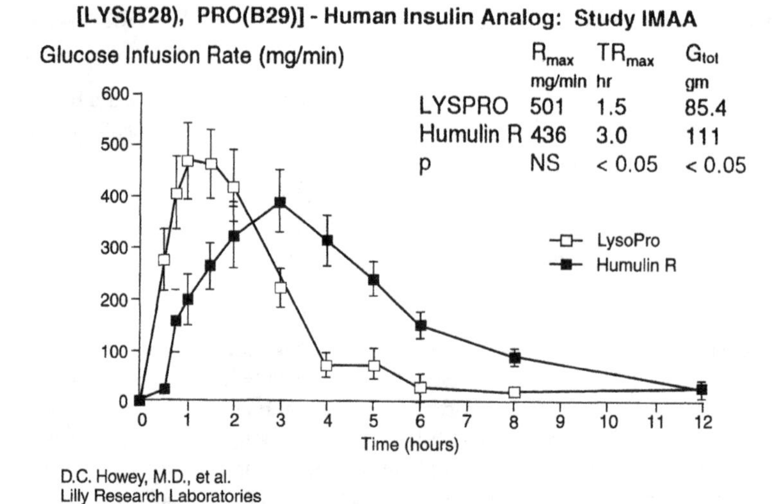

Figure 8.8 LysPro insulin's action is more rapid than human insulin (from Baker, 1996).

for LysPro insulin is less, and it translates into potentially better control of blood sugar levels in a diabetic (Baker, 1996).

8.11 THE FRONT-END STRATEGY FOR MANUFACTURE OF LYSPRO INSULIN IS DIFFERENT FROM BIOSYNTHETIC HUMAN INSULIN

LysPro insulin is also produced through fermentation of recombinant *E. coli.* However, its manufacture eliminates the need for CNBr. The removal of the *N*-terminal extension (abbreviated as MX in Figure 8.9) is achieved using a diamino peptidase from *Dyctiostelium discoidium* (Atkinson et al., 1995). The molecular weight of the fusion protein is lower, and the process of transforming the target molecule—from granule solubilization to enzymatic transformation—to an active insulin analog molecule is somewhat simplified (Figure 8.10). The purification steps for the insulin analogue that follow enzymatic transformation are likely to be similar to other insulin purification sequences.

TISSUE PLASMINOGEN ACTIVATOR

This example is also taken from a product developed during the early days of the industry. It illustrates how a biochemically complex molecule, much larger than insulin, was produced and tested. This case study also illustrates the effect of regulations on the product and process development. The intent is to introduce the student to effects of regulations, rather than the regulations themselves, which have changed over the last 15 years.

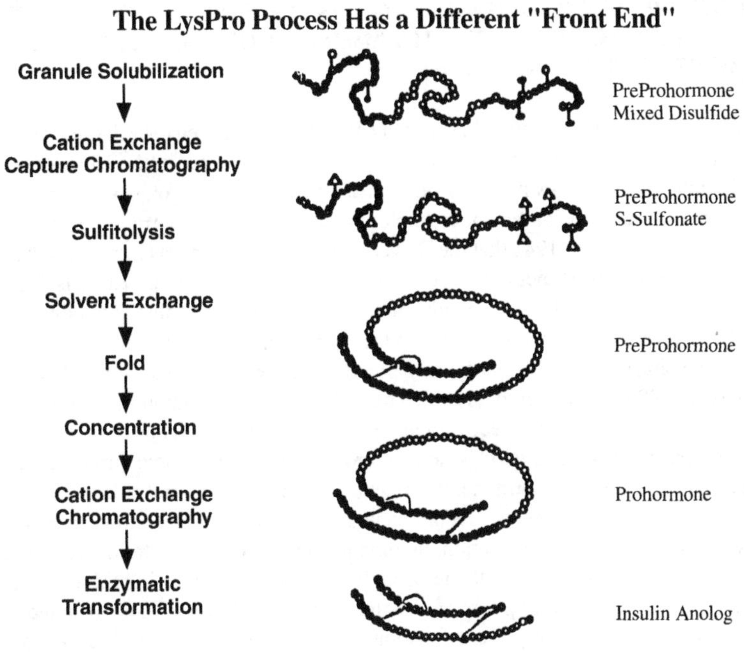

The LysPro Process Has a Different "Front End"

Granule Solubilization
↓
Cation Exchange
Capture Chromatography
↓
Sulfitolysis
↓
Solvent Exchange
↓
Fold
↓
Concentration
↓
Cation Exchange
Chromatography
↓
Enzymatic
Transformation

PreProhormone
Mixed Disulfide

PreProhormone
S-Sulfonate

PreProhormone

Prohormone

Insulin Anolog

Figure 8.9 Processing steps for LysPro insulin (from Baker, 1996).

The LysPro Process Uses A Different Front-End Strategy From That Used In the BHI Process

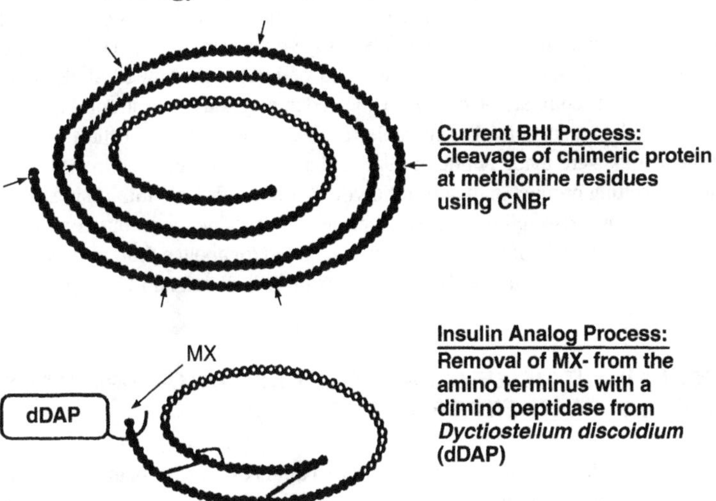

Figure 8.10 Comparison of LysPro Insulin analog to biosynthetic human insulin (from Baker, 1996).

8.12 TISSUE PLASMINOGEN ACTIVATOR IS THE FIRST RECOMBINANT PROTEIN PHARMACEUTICAL FROM MAMMALIAN CELL CULTURE FOR TREATMENT OF HEART ATTACKS

Tissue-type plasminogen activator (abbreviated t-PA and also referred to as tissue plasminogen activator in this chapter) was originally identified in tissue extracts in the late 1940s. Other known plasminogen activators include streptokinase from bacteria, urokinase from urine (but now manufactured using recombinant technology), and pro-urokinase from plasma (Harakas et al., 1988). It was not until 1981 that the Bowes melanoma cell line was found to secrete t-PA (abbreviated mt-PA) at 100 times higher concentrations than found in tissue extracts. This made it possible to isolate and purify this enzyme in sufficient quantities so that antibodies could be generated and assays developed, which led to the cloning of the gene for this enzyme and expression of the enzyme in both *E. coli* and a chinese hamster ovary (CHO) cell line (Builder and Grossbard, 1986; Builder et al., 1989). The CHO cells are an immortal cell line that has prompted special concerns in the recovery and purification of the t-PA product. Since recombinant tissue plasminogen activator is derived from a biological source (transformed CHO cells) and is not an endocrine hormone, it is regulated as a biologic—not a drug.

Comparison of the melanoma and recombinant forms of the enzymes showed the same molecular weights, amino acid composition, binding to fibrin, antigenic determinants, and pharmacokinetics (with a circulatory half-life of approximately 3 minutes in rabbits). A dose of 96,000 IU (approximately 1 mg) per body weight, administered over 4 hours, resulted in 65 to 75% fibrinolysis (fibrinolysis represents dissolution of the blood clot). A further series of tests with various types of primates, and comparison to urokinase (another thrombolytic agent) were positive and served as the basis for introducing recombinant t-PA into clinical trials in February 1984 (Builder and Grossbard, 1986). Two pilot studies with 2 patients with extensive venous thrombosis, and 7 patients with acute myocardial infarction (heart attack) demonstrated that mt-PA resulted in thrombolysis without significant fibrinogenolysis. (Fibrinogen is the precursor to fibrin and is important to the clotting of blood). Since mt-PA was only available in very limited quantities, recombinant t-PA was needed to carry out the first major trial with 50 patients. Doses of 0.375 to 0.75 mg rt-PA/kg body weight was found to be effective in achieving 70% recannalization. Another pilot study with 46 patients by the National Heart Lung and Blood Institute confirmed that a dose of 80 mg over 3 hours gave the same results (Builder and Grossbard, 1986). Later in 1984, two major trials were carried out in the United States and Europe. In the Unites States, the comparison of rt-PA (injected intravenously) to streptokinase (injected intracoronary) produced sufficiently favorable results for the rt-PA that the trial was ended early, and the results were made public in 1985 and a license was granted in 1987. The rt-PA became the best-selling product in the thrombolytic market, with $290 million in sales in 1995. Sold as Activase® (and also referred to as alteplase), this product contributed almost 30% of the $918 million in total sales of Genentech, and accounted for about a third of the $800 million in worldwide sales of drugs that dissolve blood clots in 1995 (Fürst and Unterhuber, 1996).

8.13 RECOMBINANT TISSUE PLASMINOGEN ACTIVATOR (t-PA) MAY BE EFFECTIVE FOR TREATMENT OF STROKES

Another trial showed that t-PA administered within 3 hours of a stroke (caused by a clot in the brain) facilitated full recovery of 31% of stroke patients. Hence another use of rt-PA is likely to develop, with demand for this product tied to the perception and treatment of stroke as a "brain attack" that requires immediate treatment within 3 hours. This application also provides an example of product differentiation against a potentially less expensive competitor (streptokinase) whose side effect—namely hemorrhage—prohibits its use, as well as a new use of an existing product. The rt-PA currently used in treating heart attacks could now be used for strokes.

Approximately 500,000 Americans suffer strokes each year, with 100,000 deaths. Many of the survivors suffer paralysis and impaired vision and speech. The combination of patient awareness (getting to the hospital right away), rapid diagnosis, medical response (treating strokes within 3 hours), and the availability of this recombinant biopharmaceutical (as well as other medicinals) could significantly improve the chance that stroke patients would be free of the need for rehabilitation and long-term care. Hence, while rt-PA is likely to be expensive (it costs $2,200/dose for treating heart attacks), its benefits could outweigh costs. Unlike heart attacks, where microbially derived streptokinase (which costs 10 times less) can be used, it was reported that rt-PA does not have a competing pharmaceutical (as of 1996), (Barinaga, 1996; Hamilton, 1996). Consequently the cost of manufacture of rt-PA may not be as dominant an issue as would be the case for other types of bioproducts, although this could change.

8.14 TISSUE PLASMINOGEN ACTIVATOR IS A PROTEOLYTIC ENZYME

This hydrophobic enzyme has a molecular weight of 66,000 and has 12 disulphide bonds, 4 possible glycosylation sites, and a bridge of 6 amino acids connecting the major protein structures (Figure 8.11) (Builder and Grossbard, 1986; Builder and Hancock, 1988; Watson et al., 1992). The hydrophobic enzyme is also referred to as an alteplase (Fürst and Unterhuber, 1996). Only three of the sites (Asn—117–184, –448) are actually glycosylated (Harakas et al., 1988). When administered to heart attack victims, it dissolves clots consisting of platelets in a fibrin protein matrix and acts by clipping plasminogen, an active precursor protein found in the blood, to form plasmin, which is a potent protease that degrades fibrin (Builder and Grossbard, 1986; Watson et al., 1992). Plasminogen activator is found in blood and tissues, but only at low concentrations (Builder and Grossbard, 1986). The concentration of t-PA in blood is 2 to 5 ng/mL (i.e., 2 to 5 ppb).

Plasminogen activation is accelerated in the presence of a clot, but the rate is still slow. The dissolution of a clot requires a week or more during normal repair of vasular damage (Builder and Grossbard, 1986). Prevention of irreversible tissue damage during a heart attack requires that a clot, formed by rupture of an atherosclerotic plaque, be dissolved in matter of hours to avoid extensive necrosis of the tissue that is deprived of blood flow since it is downstream of the blockage. At the same time, rapid thrombolysis (dissolution of the clot) must be achieved without significant fibrinogenolysis elsewhere in the patient. Hence injection of a thrombolytic is needed if timely dissolution of the clot is to occur.

Modification of the natural or recombinant t-PA to obtain a reteplase has been claimed (in 1996) to improve on rt-PA's clinical properties. Deletion of the "fibronectin-like" finger domain from the t-PA gave a molecule with reduced avidity for fibrin. It diffuses through the clot more rapidly, since it exhibits reduced binding at the surface of the clot. The engineering of this molecule so that it also lacks E- and kringle-1 domains (which promote clearance of active t-PA) and is not glycosylated, gives a more stable molecule with a half-life of 16 to 18 minutes. This change was reported to result in a plasminogen activator that restored blood flow in significantly more patients after 90 minutes than the alteplase (Fürst and Unterhuber, 1996). The modified, recombinant plasminogen activator (a reteplase) was introduced in 1996 as a competing product to the rt-PA (an alteplase), which had been available since 1987.

Figure 8.11 Schematic diagram of a possible structure of recombinant DNA-derived human tissue-type plasminogen activator (rt-PA). The one-letter code for each of the amino acids is given in the open circles. The cysteine residues are shaded. The solid black bars represent the potential disulphide bridges based on homology with other serine proteases. The arrow indicates the potential cleavage site between arginine and the isoleucine that results in the two-chain molecule from the one-chain form. The zigzag lines indicate potential glycosylation sites. The dark lines connect the six amino acids between the two kringles. (Reprinted with permission from Builder et al., 1988, Fig. 2, American Institute of Chemical Engineers, all rights reserved).

8.15 RECOMBINANT TECHNOLOGY PROVIDES THE ONLY PRACTICAL MEANS OF t-PA PRODUCTION

The amount of rt-PA required is on the order of 100 mg per dose. Up to 50,000 L of blood (containing 2 to 5 mg/L) would be required to produce a single dose. Cell lines of transformed chinese hamster ovary cells, selected for high levels of t-PA expression using methotrexate, are grown in large fermenters (Watson et al., 1992). The production of rt-PA illustrates the importance of recombinant technology as a means for producing biotherapeutic proteins that cannot be obtained in practical quantities by any other means (Builder et al., 1989).

8.16 PURIFICATION OF TISSUE PLASMINOGEN ACTIVATOR MUST REMOVE CELLS, VIRUS, AND DNA

The literature on the industrial practice of recovery and purification of rt-PA generated by suspension culture of CHO cells, while limited, is sufficient to provide insights into challanges and guidelines for recovering a protein derived from mammalian cells (Builder et al., 1989). The recombinant t-PA is purified from the culture medium by steps that appear in Figure 8.12:

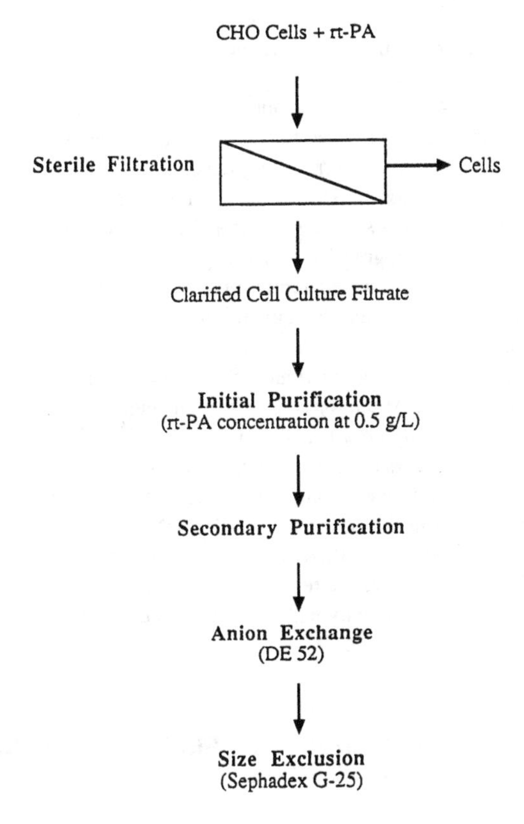

Figure 8.12 Diagram of steps for a possible recovery and purification sequence for recombinant tissue plasminogen activator derived from recombinant chinese hamster ovary cells.

1. Cell removal by sterile filtration
2. Protein purification accompanied by DNA and virus removal during initial and secondary purifications
3. Possible final purification using anion exchange and size exclusion chromatography

Sensitive, specific, and, if possible, rapid assays for product and potential contaminants are an essential part of the methods selected for the purification sequence. This sequence is consistent with a general scheme proposed for extracellular, soluble products derived from tissue culture. The steps of ultrafiltration followed by ion exchange chromatography and a final round of ultrafiltration have been proposed as a general method of concentrating a dilute protein while purifying it (Ho, 1990). Precipitation prior to chromatography could remove unwanted proteins before the sample is injected into the first liquid chromatography column.

The separation of cells from culture media or fermentation broth is the first step in a bioproduct recovery sequence. While centrifugation is probably common for recombinant bacterial cells, the final removal of CHO cells utilizes sterile filtration techniques familiar to the pharmaceutical industry for 30 years. Safety concerns with regard to contamination of the product with CHO cells were addressed by confirming the absence of cells in the product, and their relative noninfectivity with respect to immune competent rodents injected with a large number of CHO cells. While nodules formed initially at the site of injection, these growths disappeared as host-versus-graft response occurred.

The possibility that DNA from recombinant immortal cell lines (e.g., CHO cells) could cause oncogenic (gene-altering) events resulting in cancer (see Alberts et al., 1989, for an overview of oncogenes) was a major concern during development of the rt-PA purification sequence. While data from injection of DNA into rodents suggested that DNA, by itself, is inactive in vivo, the removal of DNA is still a concern. The goal for rt-PA purification was to reduce the DNA to below 10 picogram/dose (1 picogram = 10^{-12} g), with a level of 0.1 picograms being achieved. This represented approximately a 9 log reduction in the DNA and required special assays to be able to detect and quantify these very low levels of DNA in the final product (Builder et al., 1989).

The key steps in a possible purification sequence for mammalian cell culture derived protein are summarized in Figure 8.12. At the initial purification stage the rt-PA concentration is 0.5 g/L and the DNA is at 0.11 nanograms/mL (1 nanogram = 10^{-9}g). The use of anion exchange chromatography (DE step) appears to be particularly effective in removing DNA. Studies with a different protein, IgM, also derived from cell culture showed DNA clearance may be enhanced by predigestion (hydrolysis) of the DNA using DNAses prior to the anion exchange chromatography step (Dove et al., 1990). The use of hydrolytic enzymes for DNA clearance would require proof that the DNAses are removed in subsequent processing steps. Hence this approach would require additional analytical procedures and data as part of the validation of DNA clearance.

8.17 LIMITS ON ANALYTICAL DETECTION REQUIRE INDEPENDENT ASSAYS FOR PROVING VIRUS REMOVAL

Retroviruses and viruses can also be present in culture fluids of mammalian cell lines (Builder et al., 1989; Dove et al., 1990). The proof or demonstration of the *absence* of virus can prove difficult, since the possibility exists that the virus is present but at a concentration that is below the detection limit of the assays being used. Model viruses (NIH Rausher leukemia virus and NZB Xenotropic virus) were spiked into fluids being purified, and their removal subsequently validated when subjected to the same purification sequence as used for the product. Viral clearance can be achieved in a number of ways:

1. Chaotropes (urea, guanidine)
2. pH extremes
3. Detergents
4. Heat
5. Formaldehyde
6. Proteases or DNAases
7. Organic solvents, formaldehyde
8. Ion exchange or size exclusion chromatography

The protein product must be stable at the conditions used to deactivate or remove the virus. Only the inactivation or removal that can be measured is useful to prove validation. Hence a sequence of orthoganol removal/fractionation steps must be used with each step coinciding with the limits of detection of the assays used to confirm specific levels of viral clearance (Builder et al., 1989; Dove et al., 1990; Willson and Ladisch, 1990). Builder et al. (1989) give the example of a fluid spiked with 10^6 virus particles/mL. If the sensitivity of detection after each treatment is 10^2 particles, the analytical technique could only show a removal of 10^4 particles/mL. Hence, to achieve evidence of 12 logs of clearance (from 10^6 to 10^{-6} particles), three different mechanisms of analysis would need to be used if each gives 4 logs of clearance. Builder also notes that it is not valid to use the same approach (i.e., the same step repeated three times) to achieve the 12 log reduction. Total clearance, based on the product of three steps ($10^4 \times 10^4 \times 10^4 = 10^{12}$), requires that the three steps be totally independent.

Dove gives another example of virus clearance for IgM derived from human B lymphocyte cell lines where the steps are as follows:

1. Precipitation
2. Size exclusion preceded by DNA hydrolysis by DNAse (enzyme)
3. Anion exchange

A second sequence is presented consisting of

1. Cation exchange
2. Hydroxylapatite
3. Immunoaffinity chromatography

Each of these three-step sequences utilize steps that are based on different properties. The first sequence utilizes solubility, size, and anion exchange, while the second sequence consists of cation exchange, adsorption, and finally selective recognition based on an anti-u chain IgG (Dove et al., 1990).

8.18 THE PRESENCE OF u-PA COMPLICATES THE PURIFICATION OF t-PA

The sequence of steps that makes up the initial and final purifications of recombinant t-PA from CHO cells is not given. A detailed experimental protocol for t-PA derived from normal, nonrecombinant human cultured cells (ATCC CRL-1459, American Type Culture Collection, Rockville, MD), however, is published, and it provides insights into the types of chromatography steps that might be employed for purifying rt-PA (Harakas et al., 1988) if it were to occur with a similar enzyme, u-PA. Unlike t-PA, u-PA is glycosylated at only one site (at Asn-302). Otherwise, u-PA and t-PA are structurally similar and tend to co-purify. The sequence in Figure 8.13 shows the steps for fractionating the two PA's. However, as pointed out by Wheelwright (1991), "it is doubtful that this process could produce material of commercial quality in terms of both cost and purity," since the yield is 20 mg from 1400 L. Sufficient t-PA to treat 100 patients would require approximately 560,000 L of fermentation volume, and this is not practical. This case illustrates the critical role of a recombinant cell line in obtaining both high yields and higher selectivity in producing a specific type of protein.

The temperature of all purification steps in Figure 8.13 was 4°C, except for the size exclusion chromatography which was carried out at 22°C. Adsorption of this hydrophobic protein to

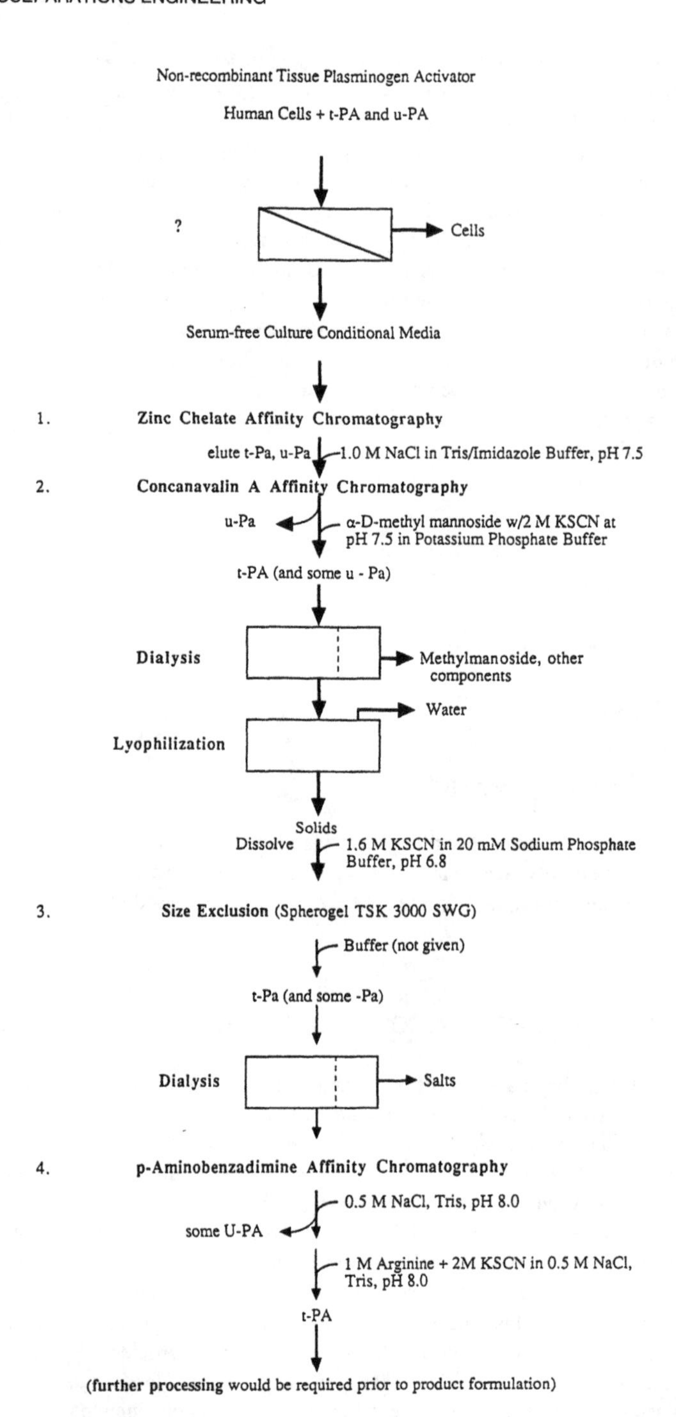

Figure 8.13 Overview of purification sequence for nonrecombinant tissue plasminogen activator (t-PA) containing urokinase-type plasminogen activator (u-PA). Serum-free culture conditioned media is from normal human cell line (generated from description in Harakas et al. 1988).

process equipment surfaces (consisting of either stainless steel or glass) was minimized by adding the detergent polyoxyethylene sorbitan monooleate (Tween 80) to the serum-free culture conditioned media at 0.01% (v/v). The equipment was also rinsed, before use, with phosphate-buffered saline (abbreviated PBS) containing 0.01% Tween 80. Hydrophilic plastic equipment was used whenever possible (Harakas et al., 1988). All buffers were sterile filtered (sterile filtration of liquids and gases is usually carried out using a 0.2 micron or 0.45 micron filters).

The culture media contained t-PA as well as urokinase type plasminogen activator (u-PA). Separation of these similar proteins required several extra affinity chromatography steps as well as dialysis/buffer exchange steps between the different chromatography steps in Figure 8.13. This particular sequence illustrates how salts and buffers, added during the purification sequence, must also be removed, hence adding significant complexity to the purification sequence. Desalting and buffer exchange constitute major separation steps in the production of almost all biotechnology protein products.

8.19 PROCESS CHANGES IN MANUFACTURE OF t-PA, A BIOLOGIC, ARE SUBJECT TO GOVERNMENT REGULATIONS

The difference between biologics and drugs is not only a matter of definition but also a process design issue. The examples of rt-PA derived from CHO cells, and t-PA/u-PA from a human cell line, illustrate how the manufacturing process, and, in particular, the cell line, give different results. Builder et al. (1989) effectively summarized this issue as follows: "To compensate for the incomplete analytical capability to define biologics, regulatory agencies have included, in the control and monitoring, parameters of a biologic the process used to make it. This situation exists because it is possible that if one changes the process significantly, it may yield a different product from that previously reviewed and approved. A new product will of course require a new license."

Because of this situation, according to Builder, substantial barriers exist in terms of effort, money, and time required to make significant changes in processes used to make licensed biologics. While process changes are expected during the IND (investigational new drug) phase and before the license is approved, significant changes after licensing are "accumulated into relatively large packages that are submitted infrequently and only when substantial benefit justifies the changes." The time that can elapse between conception of the idea for a process change and granting of new license may be two years and cost several million dollars. The events during this time include process development time, generation of material for clinical trials, carrying out clinical trials, filing the license, and receiving approval for the amendment.

CLASSES OF LIQUID CHROMATOGRAPHY

The equipment in which liquid chromatography is carried out consists of a solvent (eluent) reservoir(s), pump(s), injection valve, column, and detector (Bisbee, 1993) (as shown in Figure 8.14). The column is packed with an adsorbent (i.e., stationary phase). The mixture to be separated is pushed through the column by the eluent, or mobile phase. Isocratic chromatography is carried out at a constant flowrate, buffer composition, and temperature, and this is usually associated with size exclusion separations. Gradient chromatography typically uses a changing chemical composition at a constant flowrate and temperature. The composition is altered by mixing the contents of two or more buffer reservoirs in a prescribed manner to achieve a changing salt concentration or changing pH. The component used to form the gradient is also referred to as the modulator or modifier. The gradients that are formed, are reported in terms of concen-

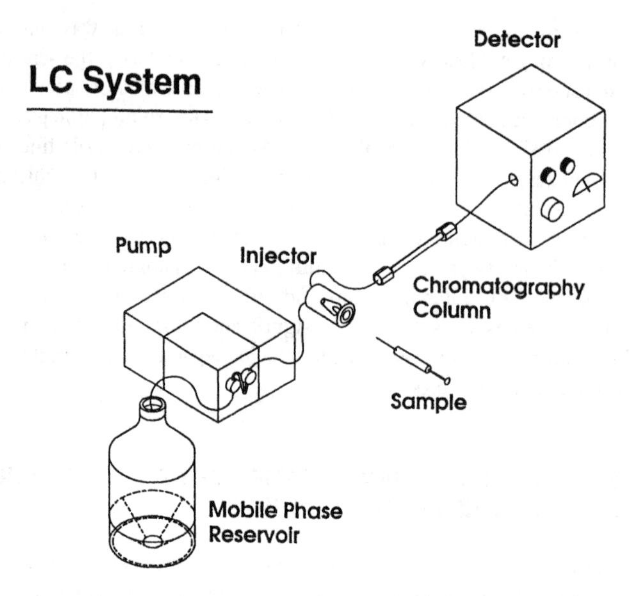

Figure 8.14 Generic representation of isocratic forming process liquid chromatography systems (courtesy of Kent Hamaker).

tration at the inlet of the chromatography column, while the protein peaks are detected at the column outlet. Hence a chromatogram of the type illustrated in Figure 8.15 needs to be interpreted accordingly. If the sample volume is small relative to the column volume, the gradient and start of the chromatogram are nearly the same. If the sample volume is large, there is a measurable, if not significant difference, between injection and start of the gradient. This will usually be shown as a region of zero or constant concentration of the modulator at the beginning of the chromatogram and is discussed in Chapter 7.

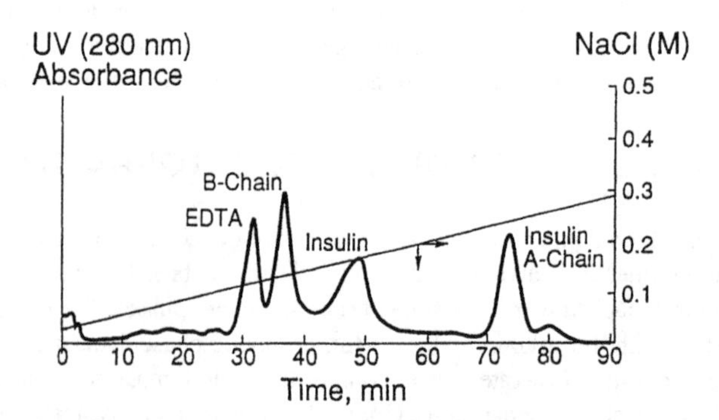

Figure 8.15 Separation of insulin, and insulin A and B chains. Sample volume of 0.5 mL with column volume at 18.7 mL. Protein concentration in sample of 1 mg/mL of each component. Eluent flow rate of 1.27 mL/min. Stationary phase is DEAE 650M (TosoHaas). (Reprinted with permission from Lin et al., 1988, fig. 4, copyright American Chemical Society.)

There are cases where the concentration profile of the gradient at the outlet of the column can be significantly different from the profile at the inlet (Velayudhan et al., 1995; Velayudhan and Ladisch, 1991, 1995). This occurs when the modulator in the eluting buffer adsorbs onto the stationary phase in a non-ideal manner and causes the gradient to deform as it passes through the column. This can give what appears to be anomolous peak behavior, including self-concentration of a peak, and appearance of shoulders or multiple peaks for single, sample components known to be homogeneous. Such an effect has been shown to be possible for gradients used in reversed phase, ion exchange, and affinity chromatography (Velayudhan, et al., 1995; Velayudhan and Ladisch, 1991, 1995). Analysis of this situation is offered in Chapter 7, Section 7.17.

8.20 PURIFICATION OF BIOLOGICS AND DRUGS ARE BASED ON FIVE CLASSES OF CHROMATOGRAPHY

Proteins and nucleotides are macromolecular biomolecules. Mixtures of biomolecules are fractionated based on differences in molecular weight, shape, size, charge, hydrophobic character, and types of active sites. The appropriate separation method(s) may be selected from five basic classes of chromatography. These classes fractionate molecules based on differences in properties (indicated in parenthesis):

1. Ion exchange (charge at a given pH)
2. Size exclusion, also known as gel permeation (size)
3. Reversed phase (solubility in aqueous organic solvents or hydrophobicity of the molecule)
4. Hydrophobic interaction (surface hydrophobicity as discussed in Chapter 4, Sections 4.9 to 4.12)
5. Affinity chromatography (stereoselective binding activity, discussed in Chapter 9)

TABLE 8.2 Qualitative Ranking by Percentages of Methods Used for Chromatography of Proteins 1987 to 2001

Method Type	Analytical[a] 1987–88	Preparativea 1987–88	Process Ranking[b] 1996	2001
Ion exchange	11	37	57	62
hydroxyapatite[c]	—	7	NS[d]	NS[d]
Gel filtration	11	25	10	3
Reversed phase	8	—	10	8
Affinity chromatography	—	19	0	21
Hydrophobic interaction	—	8	8	3
Other				
Chromatofocusing	—	3	NS[d]	NS[d]
Electrophoresis	70	1	NS[d]	NS[d]
Normal silica/alumina	—	—	8	8

[a]Based on a survey by Karlsson et al. (1989) of 100 articles published from 1987 to 1988.
[b]Based on a survey from 39 industrial respondents at the Purdue University Chromatography Workshop'96, W. Lafayette, IN, October 7, 1996. Respondents were asked to rank the top three process chromatography methods out of a list of 8 for 1996 and 2001.
[c]Considered as a form of ion exchange.
[d]NS denotes not surveyed.

An analysis of 100 publications of laboratory scale protein purification procedures showed that more than 30% of the purification steps used ion exchange and/or gel filtration, and at least 20% used affinity chromatography (Ho, 1990). The significant use of ion exchange and gel filtration is consistent with industrial practice, while affinity chromatography is less likely to be found in current manufacturing processes.

Reversed phase chromatography is widely used as an analytical tool for protein chromatography, but it is not as commonly found on a process scale for protein purification, since the solvents (acetonitrile, isopropanol, methanol, and ethanol) that make up the mobile phase reversibly or irreversibly denature proteins. Hydrophobic interaction chromatography appears to be the least common process chromatography tool, possibly due to the relatively high costs of the salts used to make up the mobile phases. Table 8.2 gives a qualitative ranking of chromatography methods likely to be used in industry.

ION EXCHANGE CHROMATOGRAPHY

8.21 ION EXCHANGE IS BASED ON COMPETITION OF CHARGED SPECIES FOR STATIONARY PHASE BINDING SITES

The basis for the ion exchange process is competition of the protein and a salt for the binding sites on the surface of the ion exchanger. As a first approximation, the energy ΔE gained by the formation of an ionic bond between charged group on the protein and a charge on the stationary phase is given by Coulomb's law:

$$\Delta E \propto \frac{Z_A Z_B e^2}{r_{AB}} \tag{8.2}$$

where ΔE is the change in energy as two charges A and B with Z_A and Z_B unit charges are brought within distance r_{AB} of each other represents the dielectric constant of the solution, assuming that there is a thick layer of solvent molecules between the two charges. The strength of binding is proportional to the charge on the protein for an oppositely charged ion exchanger. This principle is illustrated in Figure 8.16, with positively charged species adsorbing to the matrix while uncharged moleculés, or molecules with the same charge as the matrix are unretained. The ion exchange process for proteins is actually more complex than indicated by Eq. (8.2) and Figure 8.16, since the protein and support are not point charges (Kopaciewicz et al., 1983), and multiple binding sites on both the protein and stationary phase participate (Chicz and Regnier, 1988). Nonetheless, the proportionality is still valid.

Ion exchange matrices derivatized with negatively charged groups are cation exchangers, and they are given this name because they bind positively charged proteins, namely cations, when the mobile phase pH is below that of the pI of the protein. (The pI, or isoelectric point of the protein, is the pH at which a protein has an equal number of positive and negative charges and an amphoteric molecule has no net charge). Conversely, anion-exchangers are positively charged. Ion-exchangers utilized in the liquid chromatography of proteins are referred to as weak if their pK is close to 7, and strong for pK values far removed from 7. (The pK is the pH at which half of the ionizable groups are charged.) Hence a strong ion-exchanger does not denote strong protein binding, nor does a weak ion-exchanger denote weak binding. Rather, the pH of the mobile phase will determine whether the protein and stationary phase are oppositely charged and the extent to which the ionic groups dissociate, while the salt concentration in the mobile phase will determine if protein binding occurs, as both the salt and the protein compete for the charged sites on the stationary phase. The strong anion- and strong cation-exchangers

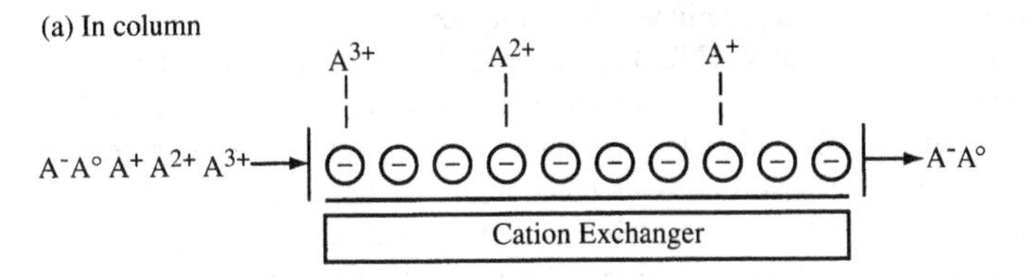

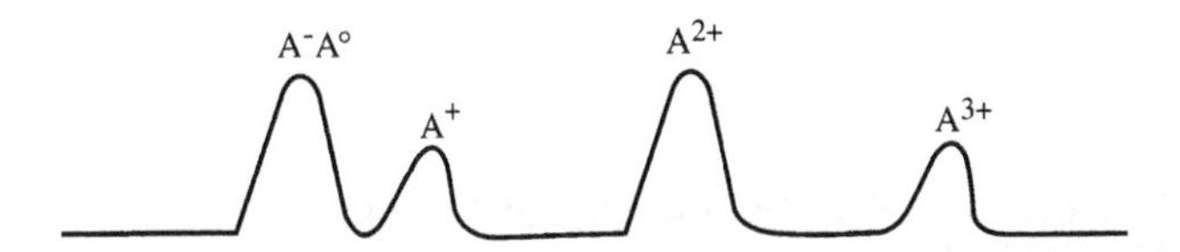

Figure 8.16 Principle of ion exchange chromatography. Species with several positive charges (A^{3+}) are adsorbed to the matrix; those with few charges move slowly, while those with no net charge or a net charge of the same sign as the adsorbent pass through the column unretained. (From Karlsson et al., 1989, fig. 4.1, VCH, New York, 1989.)

are almost fully ionized over pH's of 3 to 11, while the weak anion- and weak cation-exchangers have a narrower pH range over which they are ionized. At a sufficiently high salt concentration (typically between 0.2 to 2.5 M NaCl), the salt will occupy all of the charged sites, and protein will not bind. An example is shown in Figure 8.17 where the adsorption of the protein, bovine serum albumin (BSA), decreases as the increasing concentration of NaCl effectively competes for the anion exchange sites and inhibits protein adsorption.

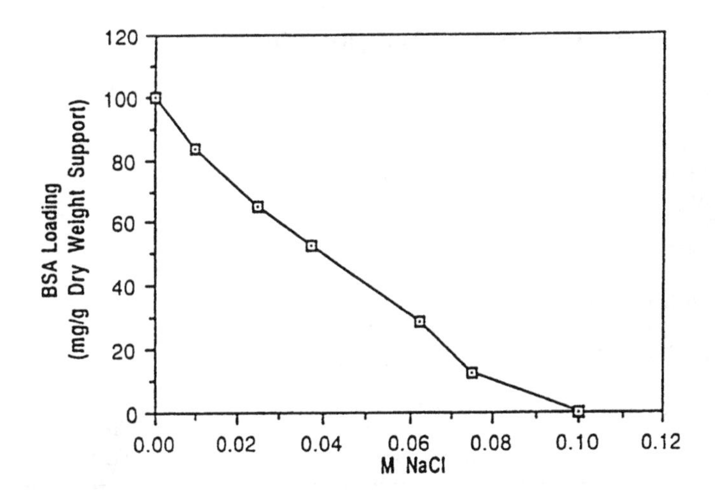

Figure 8.17 Equilibrium loading of bovine serum albumin on the anion exchanger DEAE 650 M at NaCl concentrations of 0 to 0.1 M at pH 7.2 in 16.7 mM Tris buffer at 30°C, for 30 min in a shaking incubator. (Reprinted from Ladisch et al., 1990, fig. 5, copyright, American Chemical Society, 1990.)

The binding of a charged protein will be strong for both strong and weak ion exchangers at low salt concentration and at a pH far from the pK of the ion exchanger. Proteins, unlike most ion exchange stationary phases, are amphoteric—their net charge can be either positive or negative, depending on pH. If the pH of the aqueous phase—in which the proteins are dissolved and the ion-exchanger is suspended—is much higher or lower than the pI of the protein, the protein binds so strongly to an oppositely charged ion-exchanger that it will not desorb at low ionic strength (i.e., low salt concentration). The pH and buffering capacity of the mobile phase, the isoelectric point of the protein (and therefore net charge at a specified pH), and the pK of the ion exchanger as well as the size of the protein, must all be considered in assessing strength of binding (Karlsson et al., 1989).

8.22 TITRATION CURVES GIVE pK AND GUIDE SELECTION OF STATIONARY PHASE

Either the anion- or cation-exchanger may be used as the stationary phase, depending on the pH of the solution in which the protein is dissolved and the eluting buffer. The pH of the buffer, as well as the type of ion exchange (anion or cation) stationary phase, are not only a function of the amphoteric nature of the protein but must also reflect protein stability as a function of pH. For example, if the protein is stable at a pH above 6.5, and has a pI of 5.5, an anion-exchanger is appropriate, with the separation run in a buffer at a pH of above 6.5. Alternately, if this protein were stable at pH 5 (below its isoelectric point at pH = 5.5), a cation-exchanger and a buffer pH of below 5.0 would be chosen.

There are a range of ion-exchangers available for protein chromatography. The ion exchange groups commonly associated with derivatized stationary phases are concisely summarized by Roe (1989) and are reproduced here in Table 8.3. Experience with these ion-exchangers makes it possible to select the most appropriate stationary phase, a priori, al-

TABLE 8.3 Ion-Exchangers Used in the Purification of Proteins

Type of Ion-Exchanger	Name	pK[a]	Abbreviation
Weak anion			
$-C_2H_4N^+H_3$	Aminoethyl	9.0–9.5	AE–
$-C_2H_4NH\ (C_2H_5)_2$	Diethylaminoethyl		DEAE–
Weak cation			
$-COO^-$	Carboxy	3.5–4	C–
$-CH_2\ COO^-$	Carboxymethyl		CM–
Strong anion			
$-CH_2N^+\ (CH_3)_3$	Trimethylaminoethyl	9.5	TAM
$-C_2H_4N^+\ (C_2H_5)_3$	Triethylaminoethyl		TEAE
$-C_2H_4N^+\ (C_2H_5)_2CH(OH)CH_3$	Diethyl-2-hydroxypropyl-aminoethyl		QAE
Strong cation			
$-SO_3^-$	Sulfonate	2	S–
$-CH_2SO_3^-$	Sulphomethyl	N/A	SM–
$-C_3H_6SO_3^-$	Sulphopropyl	2–2.5	SP–

Source: Adapted from Roe (1989), separation based on structure, in Protein Purification Methods: A Practical Approach, E. L. V. Harris and S. Angal (eds.), table 12, p. 207; reprinted by permission of Oxford University Press (Note: 1989 values of pK from Karlsson et al. (1989)).

[a]pK values at ionic strength of 1 from Karlsson et al. (1989).

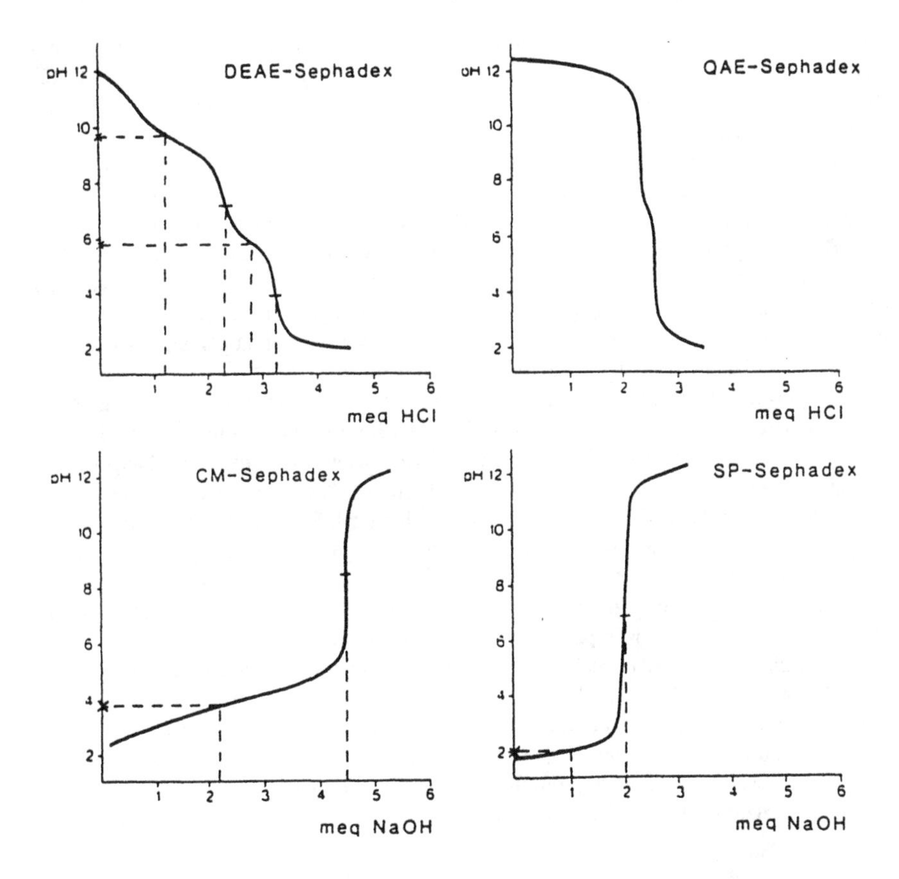

Figure 8.18 Titration curves for ion exchangers used in protein chromatography. The four curves show the titration of Pharmacia LKB DEAE-Sephadex A-50 (pK 9.5 and 5.8), CM-Sephadex C-50 (pK 4.0). QAE-Sephadex A-50 (the irregularity indicates that the ion-exchanger is not fully quaternized), and SP-Sephadex C-50 (pK 2.0). (From Karlsson et al., 1989, fig. 4-3, copyright VCH, New York, 1989.)

though titration of a new material is instructive, if not necessary, to confirm ion exchange properties.[2]

Titration curves for the weak and strong anion ion-exchangers, DEAE-Sephadex and QAE-Sephadex and weak and strong cation-exchangers CM-Sephadex and SP-Sephadex at 1.5 M NaCl in Figure 8.18 give examples of how titration curves result in useful information. Multiple pK values are possible for an ion-exchanger, since the derivatization chemistry can result in several different types of charged groups. Karlsson et al. (1989) give an example where DEAE-Sephadex (a crosslinked dextran gel derivatized with 2-chlorotriethylamine at alkaline pH) has two pK values (Figure 8.18). This material, which has a high degree of substitution, consists

[2]The extrapolation of pK values of an ion exchanger measured at one salt concentration to another is not recommended. Karlsson et al. (1989) give an example where the anion exchanger, DEAE cellulose, has pK values that are 1.5 units higher in 1 M KCl compared to the pK for the same material slurried in water. The titration curves for a given ion exchanger should be measured with the stationary phase suspended in the buffer and range of salt concentrations in which it will be used.

principly of DEAE groups, although some quaternary amines may also be present and will also contribute to higher pK values. These form when covalently bound DEAE groups are further derivatized with DEAE groups whose pK has been decreased due to interactions with nearby quaternary nitrogens.

The titration curve in Figure 8.18 for DEAE-Sephadex shows this material to have pK's of about 9.5 and 5.8. The multiple pK's indicate that different types of groups are present. The higher pK occurs at 1.1 milliequivalents (meq) HCl, which is halfway between no HCl and 2.2 meq. The lower pK of 5.8 is determined from the halfway point between 2.2 and 3.2 meq HCl. The determination of the pK is clearly not exact, since there is some interpretation involved in identifying starting and/or ending points in the titration. Nonetheless, the titration curves show there are two divergent pK's, which could give rise to several different pH dependent adsorption behaviors (i.e., binding strengths) for a given protein.

The strong anion-exchanger, QAE-Sephadex, has a much steeper titration curve Figure 8.18 with an apparent pK coinciding with pH $\cong$ 12 (equivalent to about 1.4 meq HCl). The slight shifting in the curve at between 2.3 and 2.6 meq HCl indicates that this ion-exchanger is "not fully quaternized," and it may contain some DEAE groups (Karlsson et al., 1989). Quaternary amines do not completely deprotonate, even at very high pH. They can be used at high pH because adsorbed species can still be desorbed by ion exchange with the appropriate salt. Indeed, if an extreme pH value (pH > 9) is to be used, the quaternary amine is needed. Both strong and weak anion exchangers will, in principle, give suitable binding over the pH range of 4 to 8, which is associated with most types of protein chromatography.

The titration curve for the weak cation-exchanger CM Sephadex shows that the pK occurs at about 4 (at 2.15 meq NaOH). The pK for the strong cation-exchanger SP-Sephadex (SP denotes sulfylpropyl) occurs at a pH value of about 2 (1 meq NaOH). A strong cation-exchanger is suitable for use at a pH of below 5, since the strong cation-exchanger is not fully deprotinized. In practice, however, the lower limit of the pH at which the stationary phases are employed is usually equivalent to the pK of the ion-exchanger.

8.23 pH, ION EXCHANGE, HYDROPHOBIC INTERACTIONS, AND VOLATILITY CHARACTERISTICS OF BUFFER COMPONENTS GUIDE SPECIFICATION OF BUFFER COMPOSITION

Buffer components are selected to control pH while keeping the buffer concentration relatively low (usually less than 0.05 M). A buffer is generally chosen so that its pK is within one pH unit at which ion exchange chromatography is to be carried out, since the buffering capacity is maximum when the pK and pH coincide and since a 90% reduction in buffering capacity occurs when the pK and pH differ by one unit. Another desirable characteristic of the buffer is that it not participate in ion exchange or binding with the stationary phase, since this may result in a change in pH. The pK_a of the buffer should be independent of temperature. The buffer should not absorb light in the ultraviolet range and should not chelate metals. Buffers that are commonly used for protein chromatography, together with their pK's and buffering ranges are summarized in Table 8.4.

Hydrophobic intereactions can sometimes cause loss of resolution and promote protein insolubility. Ethylene glycol, ethanol, area, and non-ionic detergents are added to the buffer in order to counteract such effects. The addition of organic modifiers can also decrease the dielectric constant from 80 (for pure water) to a lower value, thereby increasing the energy of binding as given in Eq. (8.2) (Karlsson et al., 1989). Volatility may also be considered if buffer salts must be removed from the product during lypholization of the protein (Table 8.5).

TABLE 8.4 Buffers for Ion Exchange Chromatography

Type of Ion-Exchanger	Buffer	pK	Buffering Range
Anion	L-Histidine	6.15	5.5 to 6.0
	Imidazole	7.0	6.6 to 7.1
	Triethanolamine	7.77	7.3 to 7.7
	Tris	8.16	7.5 to 8.0
	Diethanolamine	8.8	8.4 to 8.8
Cation	Acetic	4.76	4.8 to 5.2
	Citric	4.76	4.2 to 5.2
	Mes	6.15	5.5 to 6.7
	Phosphate	7.20	6.7 to 7.6
	Hepes	7.55	7.6 to 8.2

Source: Adapted from Roe (1989), Separation based on structure in protein purification methods: a practical approach, E. L. V. Harris and S. Angal (eds.), table 13, p. 209; reprinted with permission of Oxford University Press.

A widely used buffer that exhibits minimal ion exchange with an anion resin is the positively charged Tris (pK = 8.2) with a Cl^- counterion. Conversely, negatively charged phosphate, carbonate, acetate, or morpholino-ethane-sulfonate (MES) with K^+ or Na^+ counterions are employed for cation exchange chromatography. Exceptions have evolved from experience with chromatography of proteins where chaotropic properties of some types of salts, or complex ion exchange equilibria among buffer, protein, and stationary phase, have induced subtle effects that have enhanced or enabled difficult separations. Negatively charged acetate buffers have been employed with the positively charged anion-exchanger DEAE Sepharose CL-6B to purify human blood plasma proteins IgG, human serum albumin, and a glycoprotein using buffer step changes of pH 4.5, 5.0, and 5.2 at concentrations of 25, 150, and 20 mM sodium acetate, respectively. Cellulase enzymes have been fractionated over an anion-exchanger in an ammonium acetate gradient. Phosphate buffers have been employed in a similar manner (Karlsson et al., 1989).

Volatile buffer salts are particularly useful at the laboratory scale, when small amounts of proteins are being recovered, and lyophilization is used to concentrate them. Some of the volatile salts used for ion exchange chromatography, together with pK values of the buffering ions are summarized in Table 8.5. Ammonium acetate, at pH 7.0, is by itself a weak buffer, but it has a significant buffering capacity in conjunction with a strong cation exchanger that equilibrates with the soluble buffer component. Therefore the cation exchange stationary phase participates as one of the buffering salts.

Separations by ion exchange chromatography are driven by many factors, although in most cases ionic effects are likely to dominate other mechanisms. While buffer components may sig-

TABLE 8.5 Examples of Volatile Buffer Salts Used in Ion Exchange Chromatography

Buffer System		pK	
Base	Acid	Base	Acid
Ammonia	Formic acid	9.25	3.75
Ammonia	Acetic acid	9.25	4.76
Ammonium	Bicarbonate	9.25	6.5
Trimethylamine	Carbonate	9.25	6.50

Source: Adapted from Karlsson et al. (1989).

nificantly affect protein binding due to non-ionic effects, buffers are initially selected based on their capacity to maintain the desired pH. Other effects are difficult to anticipate. Hence, specification of a buffer composition that optimizes resolution and retention behavior ultimately requires experimental measurements.

8.24 AMINO ACIDS THAT FLANK CHARGED RESIDUES CAN ALTER PROTEIN BINDING

The retention behavior of genetically engineered variants of the enzyme subtilisin ($M_r = 27{,}500$ kD) from *Bacillus amyloliquefaciens* illustrates how a single substitution of an amino acid for glycine at position 166 of the protein results in large variations in protein retention over strong cation exchange resins. The average number of protein surface changes interacting with the stationary phase (sometimes called the *Z* number) varies from 2.60 to 3.77, as different amino acids are substituted for glycine at position 166. This results in an increase in retention time (the time required for the peak maximum to elute from the column) from 8.3 to 18.4 minutes when chromatography of the proteins are carried out over a strong cation exchange resin. The behavior of the two variants, which differs by only one methylene group, demonstrates that a subtle effect can lead to very large differences in retention behavior. Changes in retention behavior can also occur when neighboring amino acid residues affect the electrostatic charge (and therefore, pK) of the residues that bind with the ion exchange resin, and when steric perturbations are caused by hydrogen-bonded water molecules (Chicz and Regnier, 1989). Another example is given by LysPro (insulin, Section 8.10), which has the same amino acid composition but different chromatographic characteristics.

Binding characteristics are also likely to be affected by the buffer composition and/or the type of ion used to form the salt gradient to elute the protein. For example, the elution times are the same for the proteins α-lactalbumin and ovalbumin when NaI is used to form the salt gradient over an anion-exchanger (Mono Q). However, ovalbumin elutes ahead of α-lactalbumin with a sodium acetate gradient, and in the reverse order for NaCl and NaBr.

8.25 PROTEIN LOADING CAPACITY IS BASED ON EQUILIBRIUM MEASUREMENTS

The selection of an appropriate stationary phase and buffer sets the stage for determining the maximum amount of protein taken up by a given weight or volume of the ion-exchanger. This capacity is measured by incubating the ion exchange material in the buffer containing the selected protein at a series of initial concentrations using the general procedures described in Chapter 5 (Section 5.20) and Chapter 6 (Section 6.19). Equilibrium between protein on the ion-exchanger and protein dissolved in the buffer is typically achieved within a period of 6 to 24 hours at ambient temperature. The amount of protein adsorbed is calculated from the difference between the initial and final protein concentration measured in the supernate. A plot of the concentration of protein on the ion-exchanger (expressed either as μmole protein or mg protein, per mL or g of the ion-exchanger) as a function of the protein concentrations in the liquid phase at equilibrium gives the equilibrium isotherm. The leveled-off portion represents the maximum binding or saturation capacity (Figure 8.19). The saturation capacity is the protein loading where additional uptake due to direct interaction of the protein with stationary phase is negligible because all of the accessible surface area is covered with adsorbed protein.

The adsorption of lysozyme on the weak cation-exchanger IRC-50 gives an example of how the rate of adsorption and particle size of the stationary phase, as well as pH, affect the amount

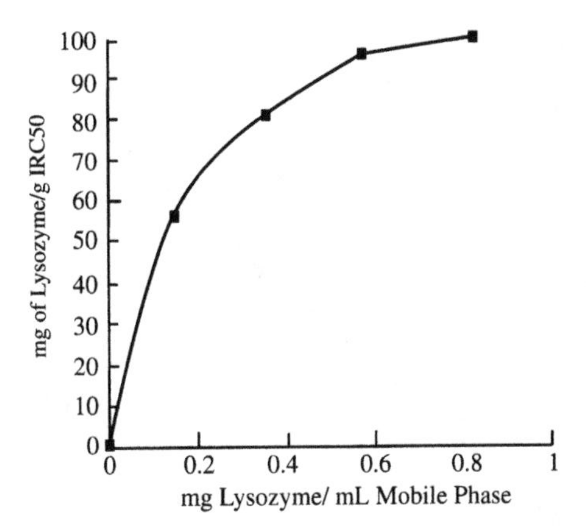

Figure 8.19 Equilibrium isotherm for lysozyme on IRC-50 at pH 7.3. (Data from Pollio and Kunin, 1971, *CEP*, 67, 108.) Reproduced with permission of the American Institute of Chemical Engineers. AIChE. All rights reserved.

of protein bound at equilibrium. IRC-50 is a methacrylic acid-divinylbenzene polymer with a carboxylic acid functionality. It has a pK of 6.1, ion exchange capacity of 3.5 meq/mL (or approximately 10 meq/g), porosity of 0.10 mL pore vol/mL stationary phase, an average pore size range of 800 to 4000 Ä, and a surface area of 1 to 2 m^2/g (Pollio and Kunin, 1971; Amberlite, 1974). Lysozyme uptake was measured at pH 7.3 with the protein dissolved in a MacIlvaine (phosphate/citrate) buffer. The rate of uptake of protein by the ion exchange resin is not instaneous—in fact more than 24 hours are required for the protein adsorbed on the resin to approach a steady-state value of about 140 mg of protein/g resin (Figure 8.20) (Pollio and Kunin, 1971).

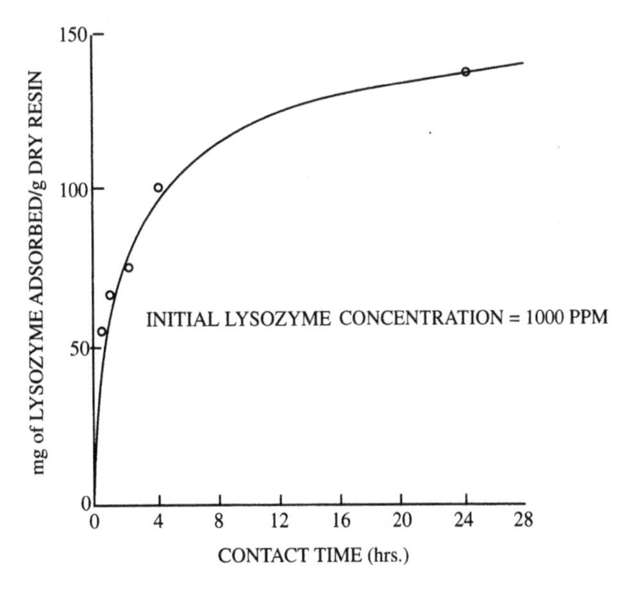

Figure 8.20 Rate of lysozyme adsorption at pH 7.3. (Reprinted from Pollio and Kunin, 1971, *CEP*, 67, 69, fig. 2.)

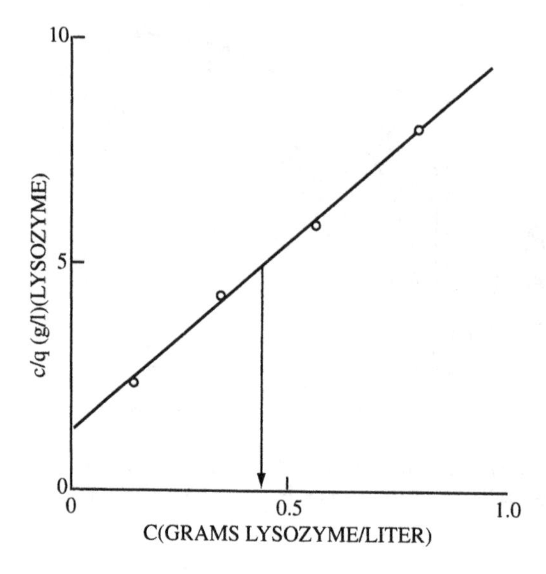

Figure 8.21 Replot of Langmuir equation. (Reprinted from Pollio and Kunin, 1971, *CEP*, 67, 71.)

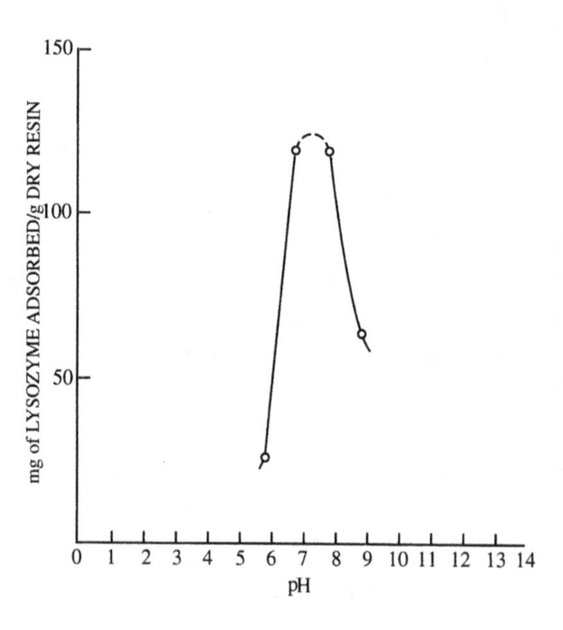

Figure 8.22 pH effect for binding of lysozyme to a carboxylic type ion exchanger. (Reprinted from Pollio and Kunin, 1971, *CEP*, 67, 70, fig. 4.)

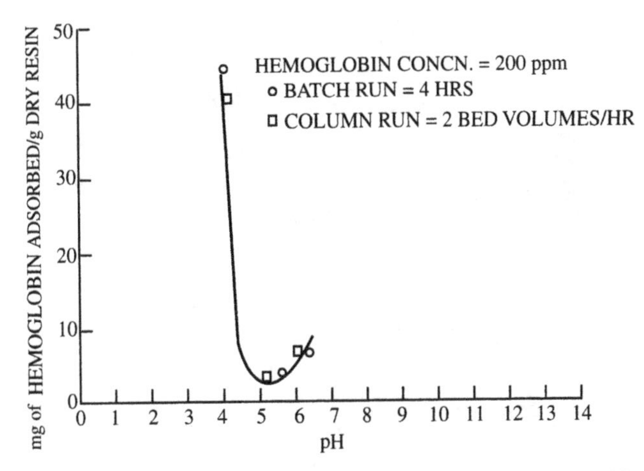

Figure 8.23 pH effect for hemoglobin sorption on IRC-50. (Reprinted from Pollio and Kunin, 1971, *CEP*, 67, 72, fig. 7.)

The amount of protein adsorbed as a function of contact time was measured at room temperature using a stirred, three-neck round bottom flask (1 liter size) under a nitrogen atmosphere. The resin was preconditioned in buffer, before 2 to 4 g (wet weight) were added to 200 to 400 mL buffer containing the dissolved protein. These equilibrium measurements showed that protein adsorption followed a Langmuir type curve (Figure 8.19) and that Langmuir constants could be obtained from the data (Figure 8.21). Further experiments used 30 to 50 mesh cut (average particle size of about 400 microns) for additional protein adsorption studies. The importance of pH is shown in Figures 8.22 and 8.23 for lysozyme and hemoglobin, respectively. When the pH is close to either the pK of the ion exchanger (pK = 6.1), or the isoelectric point of the two proteins (pI = 11.0 for lysozyme, and 6.7 for hemoglobin) the adsorption capacity is much lower than the maximum. This research, reported in 1971, is remarkable in that it anticipated the significance of large pores, accessible surface area, and methacrylate-based polymeric materials which were to become popular for protein chromatography 15 years later.

Bovine serum albumin, lysozyme, and hemoglobin are frequently used to determine binding capacities for both anion- and cation-exchangers, whose capacities will range from 0.6 to 350 mg protein/mL stationary phase at saturation, with capacities of 25 to 200 being typical (Karlsson et al., 1989). The equilibrium capacity is dependent on pH, and hence accurate reporting of loading capacities should include initial and final protein concentrations, the pI of the protein, the pH and composition of the buffer, and temperature. It is difficult to compare loading capacities of different ion-exchangers in the absence of knowledge of the conditions at which the measurements were made.

8.26 LARGE PROTEINS AND SMALL PORE SIZES DECREASE EQUILIBRIUM BINDING CAPACITY

The experimentally determined binding capacity may not coincide with complete utilization of all ion exchange sites, since the protein may be too large to penetrate some or all of the pores of the ion-exchanger. This is illustrated in Figure 8.24 for an anion-exchanger where the large

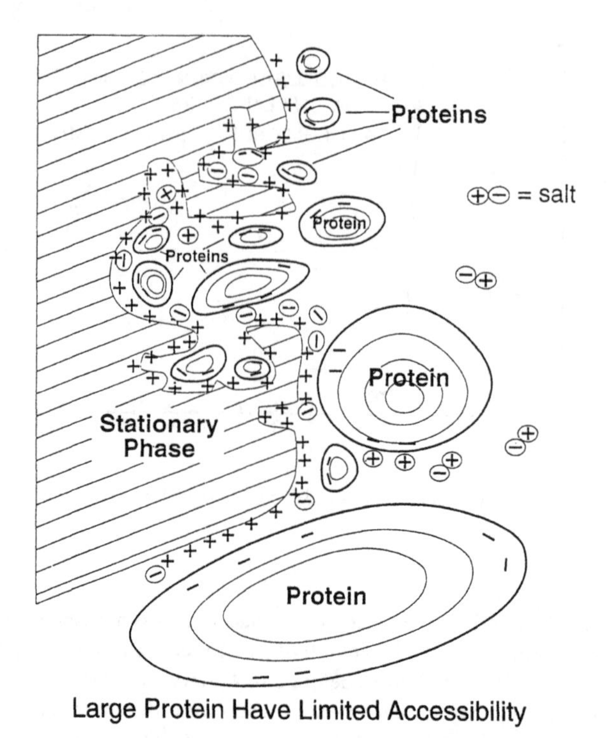

Figure 8.24 Schematic representation of effects of protein molecular weight and pore size on extent of adsorption of protein on an anion exchanger.

proteins are not able to enter the pores, while smaller proteins can explore the interior of the stationary phase.[3] The salt molecules, represented by the small circles, can explore the smallest pores, as well as being capable of diffusing between interfaces where proteins bind to the stationary phase. The proteins in this illustration are all negatively charged, thus indicating that their pI's are below the pH of the mobile phase.

An example of the effect of protein size is given by the capacity of a DEAE-cellulose (DEAE-Sephacel) that binds 160 mg/mL serum albumin (M_r = 66,000) but only 10 mg/mL of thyroglobulin (M_r = 670,000). The thyroglobulin cannot penetrate the pores of the stationary because it is too large (Karlsson et al., 1989). Another example is given by the decrease in the apparent binding capacity with increase in molecular weight of the small particle size, weak cation-exchanger, CM-cellulose at pH 6.0 shown in Figure 8.25 (Scopes, 1981). A similar consideration applies to reversed phase chromatography of insulin where pore sizes of 120 to 150 Ä permitted efficient mass transfer of insulin into and out of the pores but, at the same time, were small enough to maximize surface area for protein loading (Kroeff et al., 1989). These examples show why the pore size of stationary phases relative to protein size must be considered when interpreting protein binding capacities for purposes of selecting an appropriate stationary phase.

[3]Surface area and pore size are inversely related for porous materials. A larger number of smaller pores will give a larger surface area but decreased capacity when the protein molecule is too large to explore the pores, rendering the surface area inside the pores unavailable for binding.

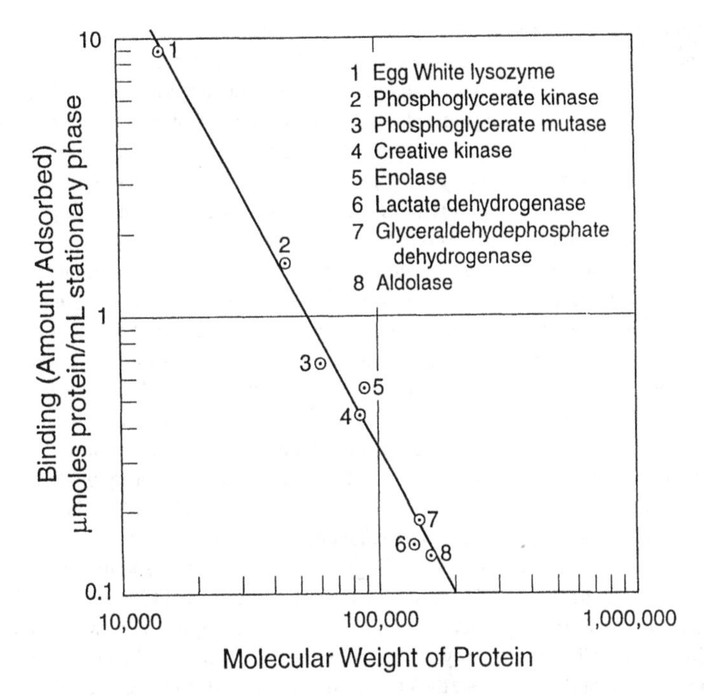

Figure 8.25 Effect of protein molecular weight on binding capacity of CM-cellulose for proteins of increasing molecular weight. Conditions: pH 6.0, potassium MES buffer, 0.01 M ionic strength, room temperature except for creative kinase (measured at 0°C). (Data from Scopes, 1981, with permission from Anal. Biochem.)

8.27 DYNAMIC (OPERATIONAL) CAPACITIES ARE SIGNIFICANTLY LOWER THAN EQUILIBRIUM CAPACITIES

A high surface area may not give correspondingly large protein capacities. Even if a high loading capacity is achieved, this will only represent a maximum, since at chromatographic conditions, where the sample is passed through the column in a matter of minutes, the slow rate of binding will give a lower extent of binding than that obtained for equilibrium reached after hours of contact between stationary phase and protein solution. The loading at chromatographic conditions is the dynamic capacity, and can be 5 to 10 times less than the equilibrium or maximum binding capacity. Hence the use of an ion-exchanger in column chromatography must factor in this sizable difference between static and dynamic capacities.

8.28 ION EXCHANGE CHROMATOGRAPHY SEPARATES PROTEINS BY DESORPTION USING INCREASING SALT GRADIENTS

Ion exchange chromatography is initiated by injecting the sample and eluting it through the column with a buffer that has no NaCl or other displacing salt in it. The protein binds by displacing anions or cations (usually from an inorganic salt) previously equilibrated on the stationary phase. The protein has charged sites spread over its surface, and these sites exchange with the salt counterions associated with the ion exchange stationary phase. A protein with a greater number and/or density of charged sites will displace or exchange more ions and hence will bind more strongly than a protein with a lower charge number or charge density.

Proteins deform and change their shape in response to their environment. Hence a protein left on the surface of an ion exchange resin for a day or longer can slowly start to unfold and expose an increasing number of charged sites to bind with the ion exchange resin. It is possible that this process can continue until the protein binds so strongly that it is impossible to desorb without dissolving it (e.g., in NaOH) and destroying it. This gives rise to the heuristic guideline ("rule of thumb") that ion exchange chromatography must be completed in a matter of hours.

After loading of the sample onto the column is completed, proteins of similar size and shape are separated by differential desorption from the ion-exchanger in an *increasing* salt gradient of the mobile phase. The weakly bound macromolecules elute first, and the most tightly bound elute last, at the highest salt concentration. An example of an anion exchange separation over diethylaminoethyl (DEAE) stationary phase is given in Figure (8.15). A mixture (volume of 0.5 mL) of insulin, EDTA, insulin A, and insulin B chain is injected into the column (10.9×200 mm long, volume of 18.7 mL). The protein is dissolved in a buffer of 16.7 mM Tris at pH 7.3. The eluent consists of the same buffer. The EDTA stabilized the solubility of the insulin. Prior to injection of the sample, the column was equilibrated with the 16.7 mM Tris buffer.

All of the proteins are initially retained on the anion exchange stationary phase during loading of the sample onto the column. The NaCl gradient (shown as a superimposed straight line on the chromatogram of Figure 8.15) elutes the proteins. The gradient is formed by the controlled mixing of buffers from two reservoirs of mobile phase (see Figure 8.26) where one reservoir contains the 16.7 mM Tris buffer only, while the second contains 0.5 M NaCl in the same buffer. Upon the elution of the last peak, the column is flushed with a buffer at a high salt concentration (2.5 M NaCl) to verify that all proteins are desorbed. In some cases a cleaning procedure is performed by passing methanolic NaOH through the column after the last protein has eluted (see Section 6.28). After the run the column is re-equilibrated with the starting, salt-free, buffer by pumping approximately 10 column volumes of the buffer through the stationary phase or until the pH of the effluent and influent are the same. The column is then ready for another injection.

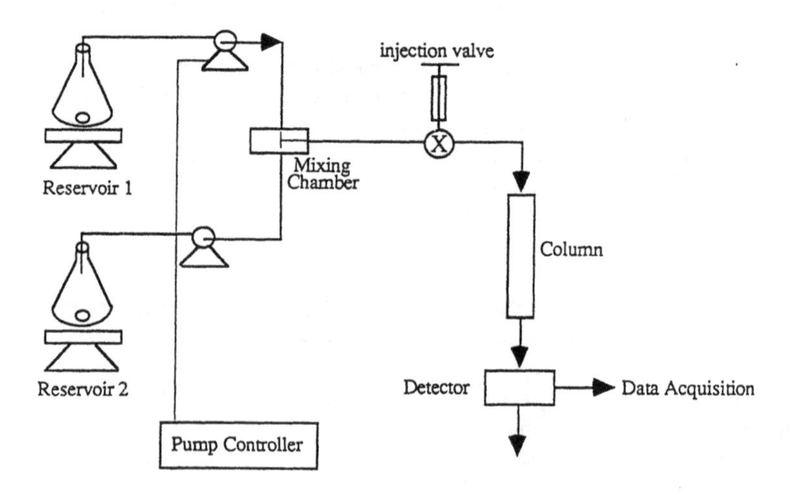

Figure 8.26 Generic representation of gradient forming process liquid chromatography systems.

8.29 AMPHOTERIC PROPERTIES OF PROTEINS DETERMINE CONDITIONS OF ION EXCHANGE CHROMATOGRAPHY

Proteins are amphoteric polymers of acidic, basic, and neutral amino acid residues and carry both negatively and positively charged groups on their surface. Structures of the amino acid side chains, illustrated in Figure 8.27, show groups that are ionized to impart charge to the protein (Alberts et al., 1989). Amino acids with basic and acidic side chains are nearly always found on the outer surface of proteins dissolved in water since these side groups are highly polar. The side chains of nonpolar amino acids (alanine, glycine, valine, leucine, isoleucine, proline, phenylalanine, methionine, tryptophan, and cysteine) are almost always found on the inside, since they tend to cluster together in a hydrophobic pocket. Amino acids with uncharged polar side chains are frequently exposed on the outer surface of the protein because of their hydrophilic nature (Alberts et al., 1989). Depending on pH, positive charges can be contributed by histidine, lysine, arginine, and N-terminal amino acid residues, while negative charges are from aspartic and glutamic acids, C-terminal carboxyl groups, and cysteine (Roe, 1989). A summary of charged groups in proteins, compiled by Karlsson et al. (1989) is given in Table 8.6. Other charged groups associated with proteins include sialic acid (carboxylate), α-carboxyglutamate in blood coagulation factors (carboxylate), and phosphoserine (phosphate) in phospho proteins.

Since the isoelectric point (or pI) is the pH at which a protein has a equal number of positive and negative charges, proteins in solution at a pH above the pI will have a net negative charge. Below the pI, the proteins will have a net positive charge (Figures 7.21 and 8.28). Since many proteins have a pI below 7 and are stable at pH > pI, they are processed using buffers with a pH of 7 to 8, using anion exchange (positively charged) stationary phases. Proteins with a low pI (e.g., bovine serum albumin and insulin) are referred to as acidic proteins, while proteins with

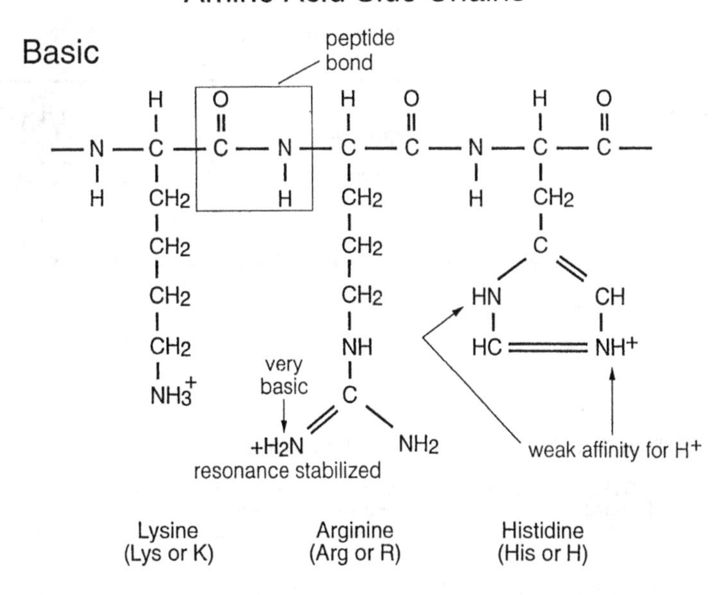

Figure 8.27 Amino acid side chains which ionize to give a protein its charge.

TABLE 8.6 Charged Amino Acid Side Chains and Other Groups in Proteins

Group	Structure	pK_a	Average Occurrence in Proteins, %
Basic			
Lysine	ε-amino	10.4	7.0
Arainine	Guanido	12	4.7
Histidine	Imidazole	6.2	2.1
Acidic			
Aspartic Acid	Carboxylate	4.5	5.5
Glutamic Acid	Carboxylate	4.6	6.2
Nonpolar			
Cysteine	Thiol	9.1 to 9.5	2.8
Uncharged polar			
Tyrosine	Phenol	9.7	3.5
Other			
α-amino	Amino	6.8 to 7.9	—
α-carboxyl	Carboxylate	3.5 to 4.3	—

Source: Adapted from Karlsson et al. (1989), table 4-2.

many basic amino acids have a high pI and are called basic proteins. If the pI is below 5, the protein is referred to as being strongly acidic, while a pI above 10 is associated with a strongly basic protein. Chicken egg white lysozyme and cytochrome C (pI = 10.5) are basic proteins (Karlsson et al., 1989). Separation of two proteins is possible when their net charge at a given

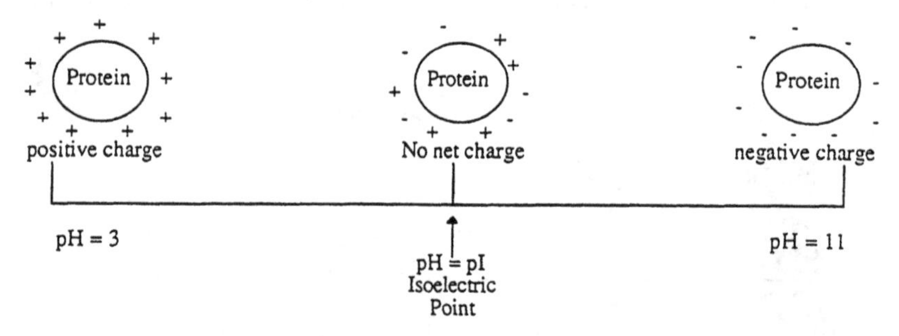

Stationary Phase Candidates:

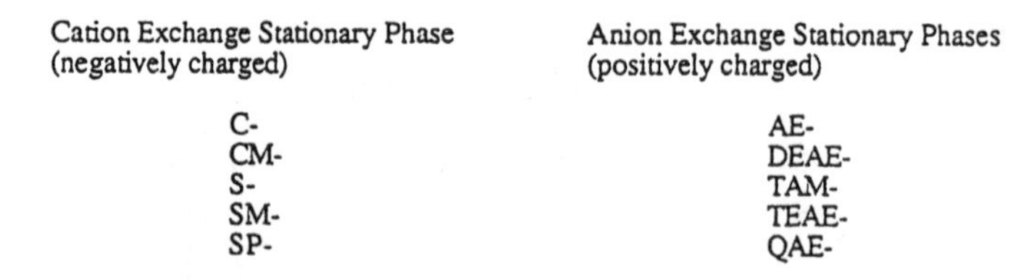

Stationary Phase Suppliers (Partial List - Alphabetical Order)

BTR, Bio-Rad, E. Merck, Mitsubishi, Pharmacia, Sartorius, Sepracor, Sepragen, Serva, Sterogene, Supelco, TosoHaas, Whatman

Figure 8.28 Schematic representation of amphoteric nature of protein. Stationary phase candidates indicate types of ion exchangers which may be used. See also Table 8.7.

TABLE 8.7 Net Charge of Human Serum Albumin and Hemoglobin as a Function of pH

	Average Net Charge[a]	
pH	Albumin	Hemoglobin
3.0	58	68.5
3.5	38.5	—
3.7	—	34
4.0	13	—
4.2	3	—
4.5	1.7	—
5.0	−1.5	20
6.5	−10.5	—
7.0	−14.0	0
9.0	—	−20.5
10.0	−32	—
10.5	—	−35

Source: From Mosher et al. (1989). Reprinted with permission from Anal. Chem., American Chemical Society.
[a]Average charge at a given pH, is $Z = Z^0 - \bar{\nu}$, where Z^0 is net charge of protein associated with j protons, and $\bar{\nu}$ is the number of protons removed per protein molecule at a given pH (hydrogen ion concentration).

pH is different. Two proteins with the same pI can be separated when their local charge densities are different and the proteins have different footprints.

An example of the effect of pH on charge for human serum albumin and hemoglobin is illustrated in Table 8.7, where the proteins have significantly different charges at the same pH (Mosher et al., 1989). Only a fraction of the charged residues participate in binding because the protein is a macromolecule with only a few of the charged residues being in close proximity to the surface of the stationary phase. This concept is illustrated in Figure 8.24.

8.30 HYDROXYAPETITE ($Ca_{10}(PO_4)_6OH_2$) IS A MIXED-MODE ION-EXCHANGER WITH BOTH WEAKLY ANIONIC AND WEAKLY CATIONIC FUNCTIONAL GROUPS

Hydroxylapetite is a form of calcium phosphate that can be prepared by slowly mixing calcium chloride and sodium phosphate to form a precipitate of brushite ($Ca_2HPO_4{\cdot}2H_2O$). The brushite, when boiled with 1% NaOH or ammonia, is converted to hydroxyapetite ($Ca_{10}(PO_4)_6OH_2$). Hydroxyapetite, which is also referred to as hydroxylapetite, is thought to adsorb proteins due to both Ca^{2+} and PO_4^{3-} groups on the crystal's surface, although the interaction between negatively charged protein surfaces and Ca^{2+} is thought to be stronger than the interaction between positively charged sites on the protein and the phosphate groups. Chemical longevity can be enhanced by boiling all buffers to remove dissolved CO_2 that will adsorb onto hydroxyapetite and form a hard crust. Tiselius and his coworkers are credited with showing that the conversion of brushite to hydroxyapetite gives enhanced protein adsorption properties for calcium phosphate. However, the flow properties of both materials are poor, and hydroxyapetite apparently forms fines more easily than brushite (Karlsson et al., 1989; Roe, 1989). The mechanical and chemical instability of hydroxyapetite has dampened its potential as a process separation tool until recently. The introduction of ceramic fluoroapetite may begin to address the need for mechanical stability while maintaining the chromatographic properties of hydroxyapetite. The availability of such a material may lead to wider use of this unique material particularly for separations and purification of monoclonal antibodies and removal of DNA (Karlsson et al., 1989).

Hydroxyapetite is particularly useful for the purification of medium and high molecular weight proteins and nucleic acids, since low molecular weight solutes generally have very little affinity. Hydroxyapetite can differentiate single- from double-stranded DNA, as well. The high selectivity of hydroxyapetite is thought to be principly due to the binding of proteins and other charged macromolecules with calcium ions. This is consistent with the observation that the number and distribution of carboxylates on the surface of the protein are key determinants of protein binding on this stationary phase. Acidic and neutral proteins probably bond to the hydroxyapetite's calcium while basic proteins adsorb to surface phosphate groups, although there are exceptions (Roe, 1989).

Roe (1989) gives an excellent synopsis of conditions associated with adsorption and elution of proteins. Eluent flowrates range from 10 to 25 mL/(cm^2hr), with porous spherical materials being stable at up to 100 mL/(cm^2hr). Adsorption capacities can range from 10 to 40 mg protein/mL of stationary phase, with the highest capacity being achieved at close to neutral pH. Salts, such as EDTA and citrate, have a stronger affinity for calcium than does phosphate. These salts decrease the protein binding capacity of the hydroxyapetite. While NaCl, KCl, or $CaCl_2$ in the mobile phase have little effect on adsorption of acidic proteins, these salts may reduce the adsorption capacity for basic proteins.

Chromatograms of a large protein, such as IgM, or DNA resemble those of an anion-exchanger. However, the two types of stationary phases do not have the same selectivity (Dove et al., 1990; Roe, 1989). An example given by Karlsson et al. (1989) is for isoforms of human ceruloplasmin. Ceruloplasmins are blue copper glycoproteins that differ only by their carbohydrate content. These fractionate over hydroxyapetite but are not separated by conventional anion exchange chromatography (Roe, 1989). The poor flow properties of hydroxyapetite, prompted batch adsorption experiments by Dove et. al. (1990). These trials showed that a high yield of IgM would be accompanied by high clearance of DNA from 0.05 to 0.1 M phosphate buffer. The DNA bound to a much greater extent than the IgM (Dove et al., 1990) at the low phosphate concentration, although the difference in binding between protein and DNA rapidly decreased as the buffer concentration increased.

Hydroxyapetite column chromatography is typically carried out by equilibrating the column in a low concentration of phosphate buffer at pH 6.8 and applying the sample, also dissolved in the same buffer. A linear gradient of phosphate of up to 0.5 M phosphate is then used to elute the proteins, although shallow gradients up to 0.1 M are often sufficient. Citrate is also a suitable salt. While the presence of other salts such as NaCl and $(NH_4)_2SO_4$ may decrease adsorption capacity, they do not affect the separation. Therefore halophilic enzymes, or proteins obtained directly from ammonium sulfate precipitation, can be fractionated over hydroxyapetite (Karlsson et al., 1989). Monoclonal antibodies are bound more strongly than contaminating proteins such as transferin and albumin at neutral pH and in a low ionic strength buffer (Perry and Kirby, 1990), hence enabling their separation.

SIZE EXCLUSION (GEL PERMEATION) CHROMATOGRAPHY

Size exclusion chromatography is often referred to as gel filtration or gel permeation chromatography, since the stationary phases are often soft spherical particles that resemble gels. Separation occurs by a molecular sieving effect where the larger molecule explores less of the intraparticle void fraction (i.e., pores) than a smaller molecule. (Early researchers envisioned this as a molecular filtration process, and hence gave rise to the term "gel filtration.") The larger molecule elutes first because it spends less time inside the stationary phase than the small molecules. Separation can be achieved if the porosity of the stationary phase is properly selected, *and* there is a significant difference in size of the molecules being separated (as measured by their hydrodynamic ratio). The type of apparatus utilized to carry out size exclusion of gel permeation chromatography is analogous to that shown in Figure 8.14, for isocratic operating conditions (constant flowrate, constant buffer composition) with the column packed with a gel filtration stationary phase.

8.31 SIZE EXCLUSION (GEL PERMEATION) SEPARATES PROTEINS BASED ON DIFFERENCES IN THEIR SIZE AND SHAPE

Selection of the stationary phase with the appropriate pore size requires that the size and shape of the proteins be known (Janson and Hedman, 1982). An example is given by the separation of IgM ($M_r = 800{,}000$) from albumin ($M_r = 69{,}000$) using FPLC. Superose 6, for a bench-scale separation (23.5 mL column) and Sepharose CL-6B for a large-scale run (86,000 mL column), as shown in Figure 8.29. Sepharose CL-6B and Superose 6 are crosslinked agaroses with a nominal molecular weight range of about 5000 to about 2 million. The chart given in Figure 8.30 compares Sepharose CL-6B to other commercially available stationary phases in the same size range and shows there are a number of materials available for separations of different size ranges of proteins (Preneta, 1989). While a variety of gel filtration matrices facilitate separations ranging from molecular weights of 50 to 10^8 are commercially available (Figure 8.30), a single gel having a porosity capable of sieving molecules over this entire size range does not currently exist. Rather, the gel must be selected according to the molecular weight and shape of the protein (Figures 8.30 and 8.31), as well as other considerations that include chemical and physical stability and cost (Pharmacia, 1983; Preneta, 1989). The development of Superdex® has resulted in significant improvements. Superdex® is a crosslinked dextron-agorose gel. (Superdex® is a registered trademark of Pharmacia Corporation).

The radius of gyraton, R_g, determines the size of pores that a molecule will be able to explore, and hence, the desired pore sizes in a size exclusion stationary phase. R_g is proportional to both molecular weight, M_r, and shape, a:

$$R_g \propto M_r^a \tag{8.2}$$

where $a = 1$ for rod shaped molecules, $\frac{1}{2}$ for flexible coils, and $\frac{1}{3}$ for spheres. For the same stationary phase the different shapes result in different distribution coefficients as shown at the bottom right of Figure 8.31. The radius of gyration is proportional to the hydrodynamic viscosity radius, R_h, for spherical proteins and flexible polymers (such as dextran) but not for rigid macromolecules (Hagel, 1989). The hydrodynamic viscosity radius, R_h, is obtained from the hydrodynamic volume, V_h, calculated from the intrinsic viscosity of the solute:

$$R_h = \left(\frac{3V_h}{4\pi}\right)^{1/3} \tag{8.3}$$

where

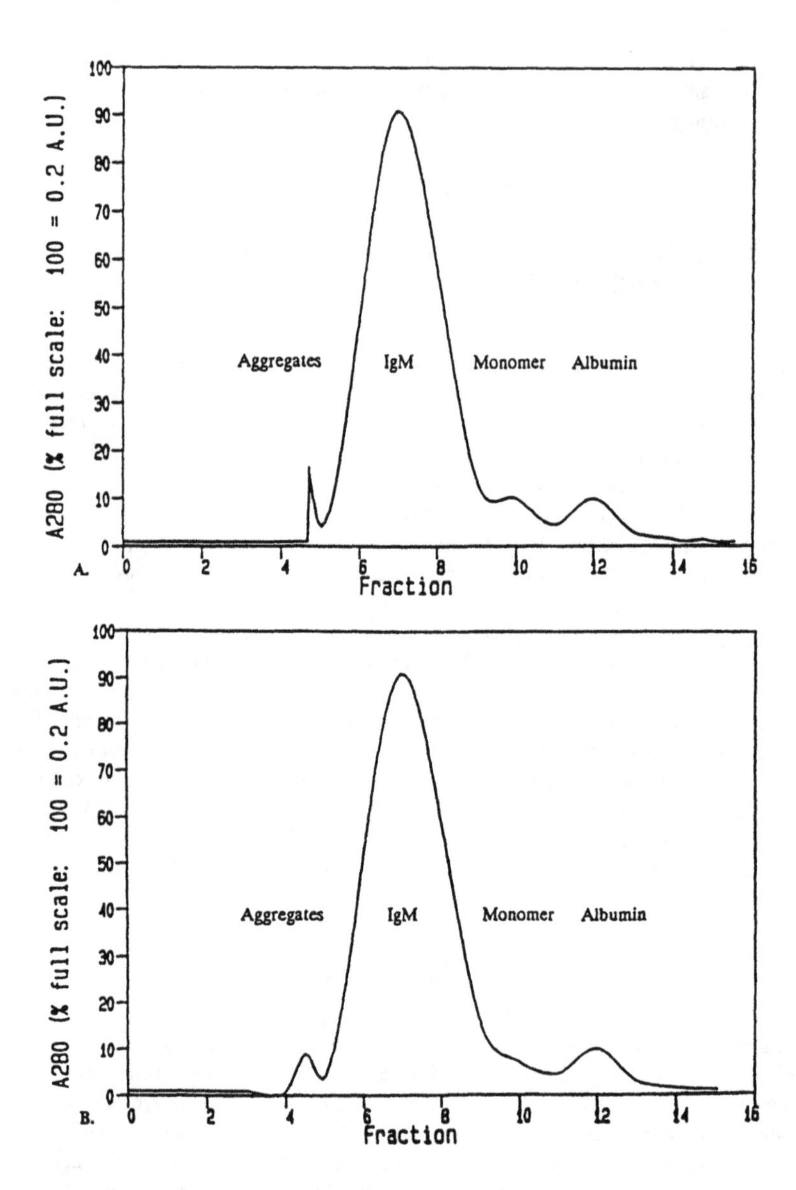

Figure 8.29 Chromatograms of size exclusion separation of IgM (M_r = 800 kD) from albumin (M_r = 69 kD) using (*a*) Superose 6 in a 1 cm × 30 cm long column, and (*b*) Sepharose CL-6B in a 37 column. (From Dove et al. 1990, fig. 1, reprinted with permission, American Chemical Society, copyright, 1990.)

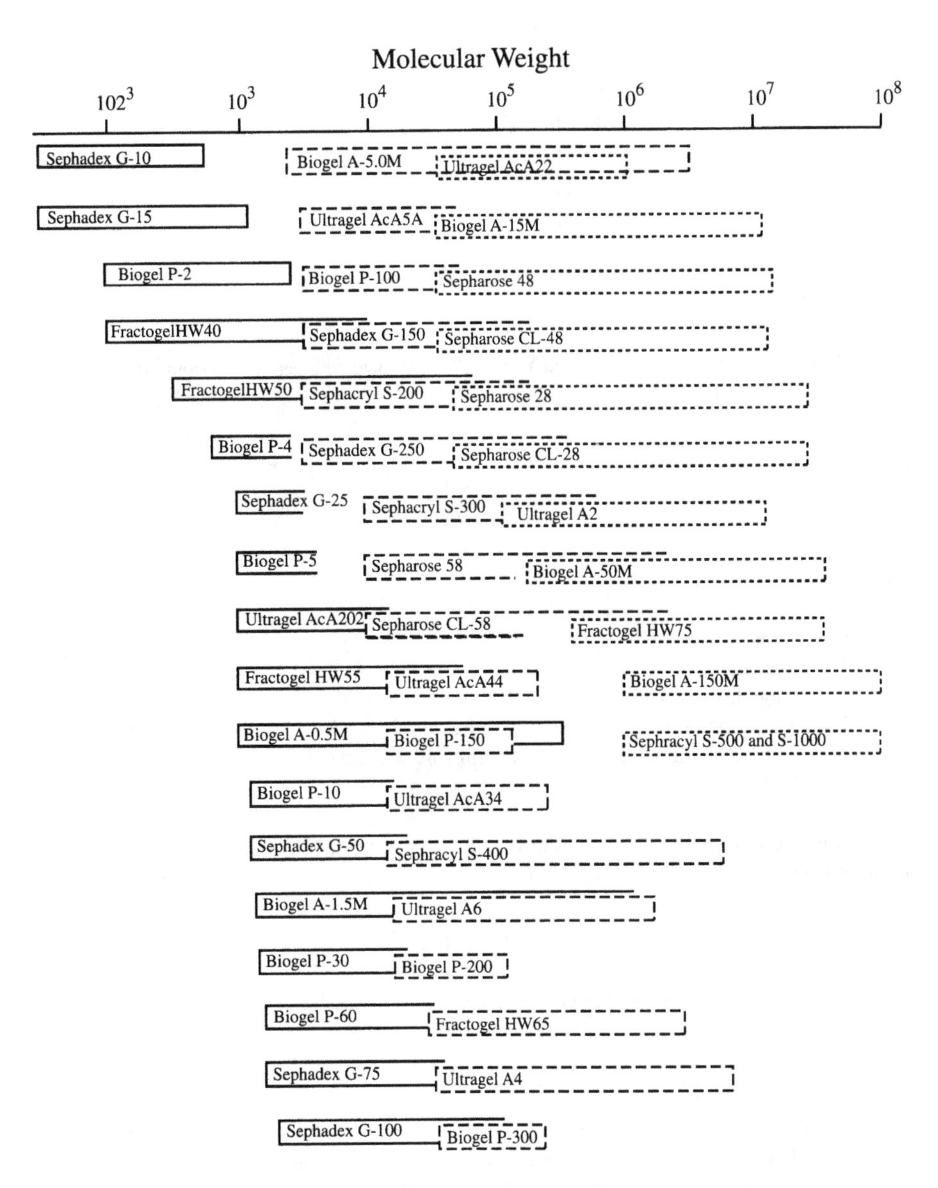

Figure 8.30 Fractionation ranges of commercially available gel filtration matrices with respect to solutes and biological molecules and particles. (Adapted from Preneta, 1989, fig. 2, reprinted with permission, Wiley, 1997.)

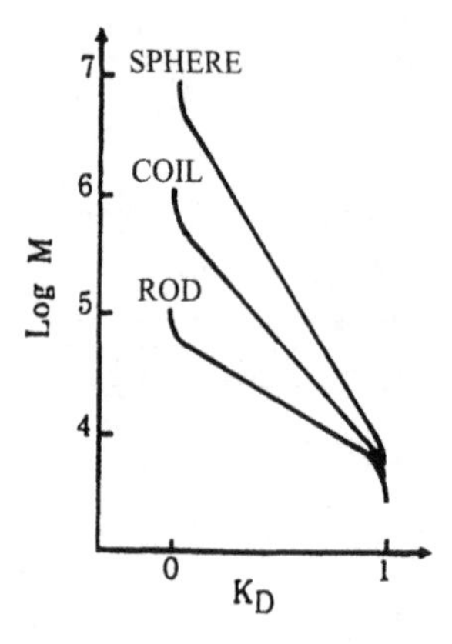

Figure 8.31 Effect of shape on K_D. (From Yau, 1980.) Reprinted with permission from Size Exclusion Chromatography, American Chemical Society.

$$V_h = \frac{[\eta] M_r}{V N_a} \tag{8.4}$$

with

$[\eta]$ = intrinsic viscosity
M_r = molecular weight
N_a = Avogadro's number = 6.022×10^{23}
V = a shape factor (referred to as the Simhas number) with a value of 2.5 for spheres and greater than 2.5 for ellipsoids. An ellipsoid with an axial ratio of 1:5 corresponds to $V = 5$.

The relation between the hydraulic viscosity radius and moleculàr weight for compact globular proteins is given by the Tanford equation:

$$R_h \approx 0.718 \times M_r^{1/3} \tag{8.5}$$

and by the Squire equation for globular solvated proteins:

$$R_h \approx 0.794 \times M_r^{1/3} \tag{8.6}$$

An alternate expression for compact globular proteins is based on the Stokes radius, R_{st}, where R_{st} is defined as the radius of a sphere that would have the same frictional coefficient as the protein. The Stokes radius is related to molecular weight by the equation

$$\log\left(\frac{R_{st}}{\bar{v}}\right) = (\log M) - (0.147 \pm 0.041) \tag{8.7}$$

according to Tanford et al. (1974). The partial specific volume $\bar{v}$, if unknown, is assigned a value of 0.74 (Hagel, 1989).

The general applicability (i.e., internal consistency) of the three measures of size are ranked by Hagel (1989) to be:

Hydrodynamic viscosity radius > Stokes radius >> molecular weight

Consequently well-defined, water-soluble polymers such as dextrans or polyethylene glycol are justified for use in calibrating a size exclusion column for characterization of the size of an unknown solute by gel permeation chromatography. Regardless of the definition that is chosen, there is likely to be some disparity between experimental results based on different methods, since differences in the shape, density, and conformation of the molecule relative to the standard is likely to give different results when different methods are employed. It is advisable, if not imperative, that measurements be done on an internally consistent basis. For example, the size of refolded proteins will differ from the same protein dissolved in a denaturing medium such as 6 to 8 M guanidine hydrochloride containing dithiotreitol (DTT) or mercaptoethanol, or a protein dissolved in sodium dodecyl sulphate (SDS). In the first case the disulfide bonds are broken by reductive cleavage with DTT or mercaptoethanol, and reoxidation is prevented by carboxymethylation. Protein dissolved in SDS results in a protein-detergent complex that is significantly larger than the native protein, and likely assumes a rod shape for larger proteins and a spherical shape as the proteins approaches a molecular weight of 15,000 (Hagel, 1989). The practical consequence is that the calibration curve for solutes of a given molecular weight show an increasing value of the distribution coefficient, K_D, as the shape changes from a rod, to a coil, and then to a sphere as in Figure 8.30 (Yau and Bly, 1980).

The distribution coefficient, K_D, represents the extent to which a solute can penetrate the stationary phase. A value of $K_D = 0$ corresponds to a solute whose size is so large that it cannot enter the gel particles. The distribution coefficient approaches a value of 1 for low molecular weight solutes that are small enough to be able to explore all of the pores. Equations used to fit data from analytical gel filtration columns have been summarized by Hagel (1989) and are shown in Table (8.8). These equations should be applicable to stationary phases used in industrial scale columns as well.

TABLE 8.8 Distribution Coefficients as a Function of Solute's Molecular Weight or Size

Equation	Investigator[a]
$K_D^{1/3} = a - b(M^{1/2})$	Porath
$K_D = a + b(\log M)$	Granath and Kvist
$K_D = a + b(\log([\eta]M))$	Benoit et al.
$\log(1 - K_D) = a + b(\log M)$	Baesdow et al.
$K_{av} = a + b(R_{st})$	Laurant and Killander
$\text{erfc}(1/K_D) = a + bR_{st}$	Ackers
$1/V_R = a + b(R_{st})$	Davis

Source: From Hagel (1989), Gel filtration, in Protein Purification: Principles, High Resolution Methods and Applications. Reprinted by permission of John Wiley & Sons, Inc.

[a]See (Hagel, 1989) for literature citations and description.

8.32 OPERATIONAL DEFINITIONS FOR ELUTION VOLUME IN GEL PERMEATION ARE BASIC TO ALL TYPES OF CHROMATOGRAPHY

Operational definitions in chromatography relate the observed retention of peaks to macroscopic and readily measureable parameters of stationary phases packed in chromatography columns. Differences in peak retention must occur if separation is to be achieved. The basis of principles for calculation of chromatographic elution (i.e., retention) volumes are briefly described in the following sections. Detailed development and discussions have been given previously in Chapters 5 and 6.

The equations are algebraically simple, and they introduce the concepts of mobile phase and stationary phase volumes, and how these can affect retention of a nonadsorbed solute. These concepts, and the associated terminology of void volume, elution volume, and capacity factor, are a starting point for equations used in the practice of all types of process chromatography. Extension of these concepts to ion exchange/gradient chromatography, as well as to calculation of elution profiles, was developed in Chapter 7.

The volume of the solvent between the point of injection and the peak maximum of the eluting protein is the elution volume, V_e (Figure 8.32) or retention volume, V_R (Section 5.13). The fluid volume between the particles of the stationary phase is the extraparticulate void volume or exclusion volume V_o, (Figure 8.33) as defined in Section 5.3. The discussion of void volumes as applies to gel permeation is further developed here. The porosity of the stationary phase, which is determined by the extent of crosslinking of the polymers that make up the particles of a gel permeation matrix, or the pore size distribution of rigid polymeric, silica, or alumina supports, determines the extent to which a protein or other solute can explore the intraparticulate void volume. The higher the crosslinking, the smaller is the effective pore size, and the lower is the molecular weight (or size) of the molecule excluded from the gel.

The apparent porosity of a gel permeation column is a function of the molecular probe used to measure it. The void volume, V_o, is equivalent to the extraparticulate void volume explored by a large molecule that is completely excluded. Molecules that are small enough to penetrate the gel have an elution volume that is greater than V_o. A small molecule, such as a salt, can potentially explore almost all of the bed volume, since

$$V_e = V_o + K_d V_s = V_R \tag{8.8}$$

where K_d represents the fraction of the volume of the mobile phase inside the particle, V_s, which can be explored by the molecular probe. For a probe that is small enough to explore all of the intraparticle void volume, K_d is 1, and the elution volume is $V_o + V_s$. Since the combined volume of the fluid between the particles (V_o) and inside the particles (V_s) cannot exceed the total volume of the column, V_t, V_e must be less than column volume V_t when a small sample volume is injected. (See Section 8.38 for report of industrial experience with large samples.)

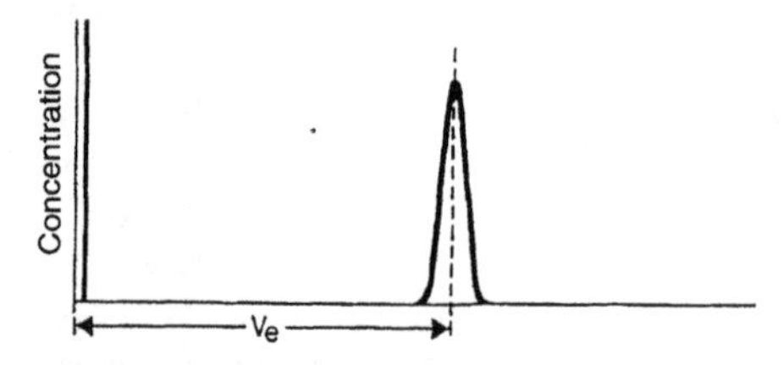

Figure 8.32 Representation of measurement of elution volume, V_e, as a function of sample volume is less than 2% of bed volume. (From Pharmacia, 1979.)

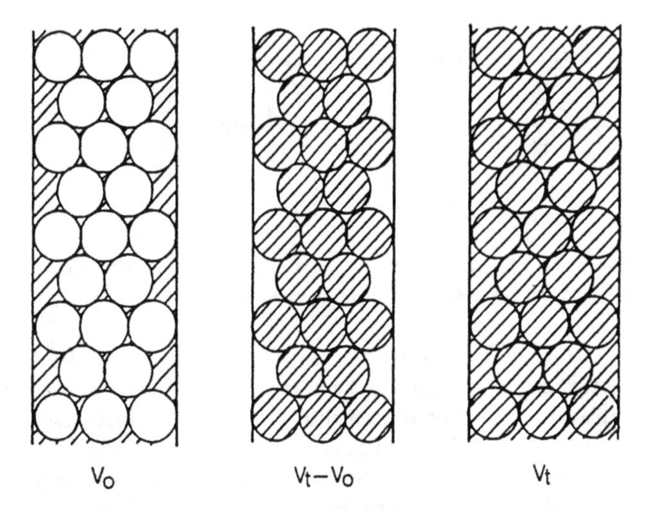

Figure 8.33 Diagrammatic representation of V_o and $V_t - V_o$. Note that V_t/V_o will include the volume of the solid material forming the matrix of each bead. The bead size is greatly exaggerated for purposes of explanation. (From Pharmacia, 1979, fig. 21; figure from Fischer, 1980, North-Holland, Amsterdam, copyright 1980.)

8.33 REPRESENTATION OF ELUTION VOLUME ON A CHROMATOGRAM DEPENDS ON INJECTION VOLUME OF SAMPLE

The elution volume is measured from the beginning of the injection to the center of the peak (coincides with the peak maximum for a Gaussian peak) if the sample volume is negligible relative to the elution volume Figure 8.32. A negligible injection volume is defined as being 2% of the elution volume. The elution volume for samples that are larger than this are measured from the half-way point of the volume injected to the center of the eluting peak. This is measured at the peak maximum for a Gaussian peak (Figure 8.34). The elution volume is measured from the start of the sample injection to the inflection point of rising part of the left side of the peak, for samples that are so large that a plateau region is obtained having the same concentration as the sample (Figure 8.35). The effect of sample volume on measurement of plate count or plate height, and equations for normalizing its effect are given in Section 5.16.

The dilution of small injected volumes can be 10-fold or more at the column outlet. Hence, proper representation of the inlet concentration in Figure 8.32, for example, would require that

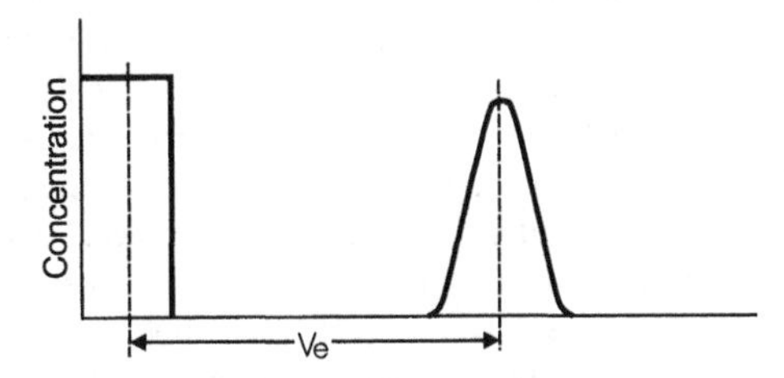

Figure 8.34 Sample volume is larger than 2%. (From Pharmacia, 1979, all rights reserved).

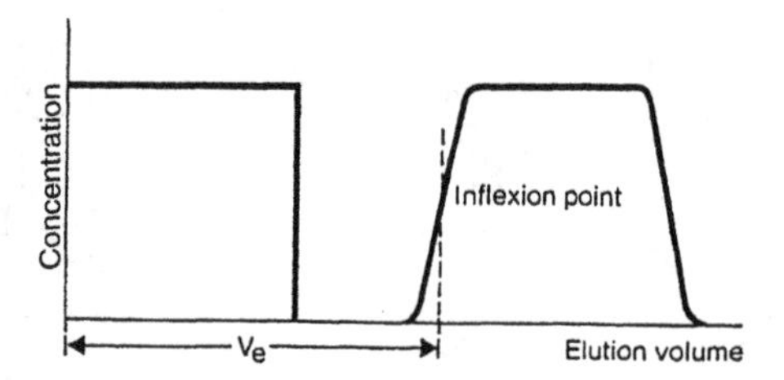

Figure 8.35 Sample volume is larger than 2% and gives a plateau region that has the same concentration as the injected sample. (From Pharmacia, 1979, all rights reserved).

the injection pulse be shown as a 5 to 10 × higher pulse since the area under the eluting peak and under the injected sample should be the same. The concentration of the injected pulse is not drawn to scale for the sake of this illustration. Similarly, dilution of the injected sample would occur in the case represented by Figure 8.34, although the difference in height between injected pulse and the maximum of the eluting peak would be larger since diffusion of the solute away from the center of mass occurs at the leading (left side) and tailing (right side) edges of the peak. If the peak is broad enough (i.e., the sample volume is large enough), the solute at the center of the peak is not diluted due to diffusion and the sample concentration at the peak maximum is therefore equal to that of the injected sample. Gel permeation stationary phases normally exhibit symmetrical (Gaussian) peaks since the partitioning of the solute between mobile and stationary phases is linear. According to Pharmacia, the pre-eminant manufacturer of such gels (Sephadex®, Sepharose®, and Sephacryl®), criteria more sophisticated than those represented in Figures 8.32–8.35 are seldom used (Anonymous, Pharmacia Fine Chemicals, 1979).

8.34 DISTRIBUTION COEFFICIENTS DEPEND ON MOLECULE SIZE AND GEL CROSSLINKING

The elution volume, V_e, and therefore the distribution coefficient, K_d, is a function of the size of solute molecule (i.e., hydrodynamic radius), and the porosity characteristic of the size exclusion media. A higher molecular weight, alone, is not sufficient to assure that a protein is larger than a protein with a lower molecular weight, since the hydrodynamic radii can be similar. An example is given by ovalbumin and lactalbumin whose molecular weights differ by 317% while their radii differ by only 121% Table 8.9. The baseline separation of α-lactalbumin from myoglobin by size exclusion chromatography is unlikely to be achieved given their similar molecular weights.

Some types of rigid size exclusion phases, based on silica or macroporous polymeric materials, have rigid pores and defined pore size distributions. But the dominant types of gels used in industry are still dextran crosslinked with epichlorophydran (Sephadex), agarose prepared from agar (Sepharose), or allyl dextran crosslinked with *N*, *N'*-methylene bisacrylamide (Sephacryl). The first stationary phases of this type were crosslinked dextrans as well as hydrophilic gels derived from starch and polyvinyl alcohol. The work of Lathe and Ruthven is credited by Porath and Flodin (1959) for giving a "good conception of the possibilities which are inherent in starch filtration." Restricted diffusion of different molecular weight species, passed through a bed of starch particles, resulted in retardation of the lower molecular weight components. Starch was also noted to have the undesirable properties of instability, poorly defined composition, and high resistance to flow. These disadvantages were perceived, in 1959, to be eliminated by "synthetic" gels such as dextran (i.e., Sephadex).

TABLE 8.9 Properties of Selected Standard Proteins Illustrating That Molecular Weight and Size Are Not Always Proportional

	M_r	pI	Radius, Å	Assymetry, f/f_o[a]
Bovine serum albumin	66,000	5.74	35.0	1.29
Ovalbumin	45,000	5.08	27.8	1.16
α-Lactalbumin	14,200	4.57	23.0	1.18
Myoglobin	16,900	7.1	24.0	1.18

Source: Basak and Ladisch (1995).
[a]Obtained from sedimentation: f = frictional coefficient of a solvated protein and f_o = frictional coefficient of a nonsolvated sphere of equal volume.

The hydrophilic character of the polymers that make up dextran and agarose gels require that they be crosslinked; otherwise, they would dissolve. Their hydrophilic character is compatible with the majority of industrially relevant proteins, since most of these can be denatured by hydrophobic surfaces but are preserved in active confirmation at hydrophilic conditions. However, the property of large effective pores can be offset by poor flow properties, particularly for lightly crosslinked gels, since these are soft and have a tendency to compress when flowrates exceed a threshhold. The maximum usable flowrate decreases with decreasing extents of crosslinking.

Sephadex G-10, for example, has the highest crosslinking and flow stability but the lowest specific bed volume and effective pore size or porosity. This limits its sieving capabilities to molecules that have a molecular weight of approximately 10 kD or less. Sephadex G-200, has the lowest crosslinking and flow stability, and the highest specific bed volume and effective pore size with a nominal molecular weight cutoff of about 200 kD. This makes Sephadex G-200 useful for separating high molecular weight proteins, but at relatively low flowrates (Table 8.10).

TABLE 8.10 Qualitative Comparison of Selected Sephadex Gel Permeation Stationary Phases

Type G	Specific Volume, mL water/g dry gel	Permeability, K_o	Maximum Operating Pressure, cm H_2O	Maximum Flowrate, mL water/$cm^2 \cdot$ hr	Nominal Molecular Weight Cutoff, kDal	Fractionation Range MW Peptides and Proteins
10	2–3	19	[a]	[a]	10	—
15	2.5–3.5	18	[a]	[a]	15	—
25	4–6	9–290[b]	[a]	[a]	25	1000–5000
50	9–11	13.5–400[c]	[a]	[a]	50	1500–30,000
75	12–15	—	160	18	75	3000–80,000
100	15–20	—	96	50	100	4000–150,000
150	20–30	—	36	23	150	5000–300,000
200	30–40	—	16	12	200	5000–600,000

Source: Adapted from Pharmacia Fine Chemicals (1979) and Janson and Hedman (1982).
[a]Use Darcy's law to calculate; maximum pressure of 310 cm H_2O (3 bar). $v = K_o(\Delta P/L)$ where v is linear flow as mL/($cm^2 \cdot$ hr), L is bed length in cm, ΔP is pressure drop over gel bed in cm H_2O, and K_o is specific permeability, which is a function of bead size and water uptake (regain).
[b]Depends particle size, d_p (as d_p increases, K_o also increases).
[c]Lists of size exclusion stationary phases are given in Anonymous (1979), Hagel (1989), and Janson and Hedman (1982). Reference to technical literature from the manufacturer that has appeared since then is strongly recommended for descriptions and applications of the most recent and improved materials.

Sephadex materials imbibe significant quantities of water with bed volumes ranging from 2 to 3 mL/g dry weight of stationary phase for Sephadex G-10 (nominal molecular weight cutoff of 10 kDal) to 20 to 25 mL/g dry weight of stationary phase Sephadex G-200 (nominal molecular weight cutoff of 200 kDal). Their structures resemble a crosslinked spiderweb, where the extent of crosslinking or association between hydrated polymer chains, rather than specific pore sizes, determine the apparent pore size distribution and the extent of penetration by proteins of a given size.

8.35 THE DISTRIBUTION COEFFICIENTS, K_d, AND K_{av}, ARE NOT EQUIVALENT

The partition coefficient, K_d, for size exclusion relates to the definition of k' by the relation $k' = \phi K$ (Section 5.20). Here K_d represents the fractional volume of a specific stationary phase explored by a given solute (the solute is sometimes also referred to as a molecular probe or target molecule). This is represented by a rearranged form of Eq. (8.8)

$$K_d = \frac{V_e - V_o}{V_s} \tag{8.9}$$

where V_o is the void volume, V_s is the volume of the solvent (usually aqueous buffer) inside the gel which is available to very small molecules, and V_e is the elution volume of a small volume of injected molecular probe as illustrated in Figure 8.31. The measurement of V_s is difficult because it requires use of an ion or small molecule that will freely diffuse into all of the fluid volume inside the gel particles, which does not adsorb or otherwise interact with the polymer backbone and which is readily detected at the outlet of the column. One manufacturer has indicated the use of radioactive ^{23}Na to satisfy these criteria, while our laboratory has used D_2O which is readily detected by a refractive index detector.

An indirect measurement of V_s is more convenient and is usually adequate. In this case, the column void volume (Figure 8.33) is measured using a soluble, high molecular weight target molecule that does not explore any of the internal fluid volume of the stationary phase and is only distributed in the mobile phase. We have used blue dextran (a water-soluble, sulfonated,

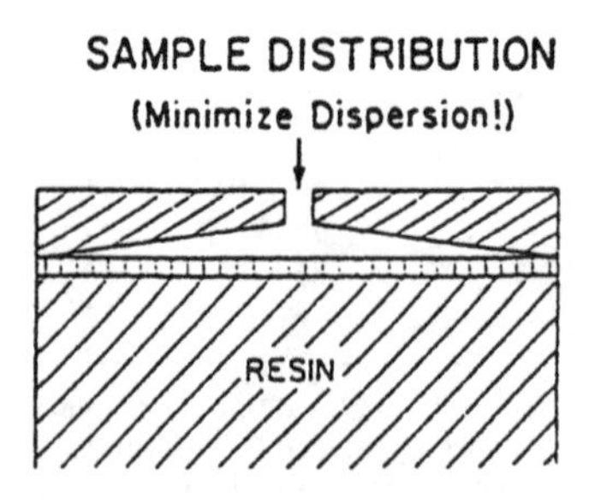

Figure 8.36 Column dead volume (i.e., volume not occupied by stationary phases) is minimized in chromatography columns in order to minimize backmixing and dispersion. (Reprinted with permission from Ladisch, 1987, fig. 9.2, of John Wiley & Sons, Inc.)

blue-colored dextran with $M_r > 669$ kD manufactured by Pharmacia) and DNA (Type III from salmon testes $M_r > 2{,}000$ kD) (Ladisch et al., 1990; Lin et al., 1988). The total column volume can be calculated from the dimensions of the bed, although we have found the direct measurement of column volume (before packing) to be more accurate. This is done by filling the column with water and draining it into a tared graduate cylinder or beaker and determining the volume (or weight) of water. The column volume (measured in triplicate) can differ from the calculated volume due to small differences in the actual and specified column diameters. The bed volume and column volume are synonymous in the practice of liquid chromatography, since an operational column has practically no dead volume between the ends of the bed and the inlet and outlet distribution screens (i.e., frits) or plates as illustrated in Figure 8.36 (Ladisch et al., 1984) and discussed in Section 5.5. The difference between the volume of an empty column, V_t, and the extraparticulate void volume, V_0, is then taken as an approximation of V_s (Figure 8.33). On this basis, K_{av}, the fraction of column volume displaced by the stationary phase that is available to a given solute species, is defined:

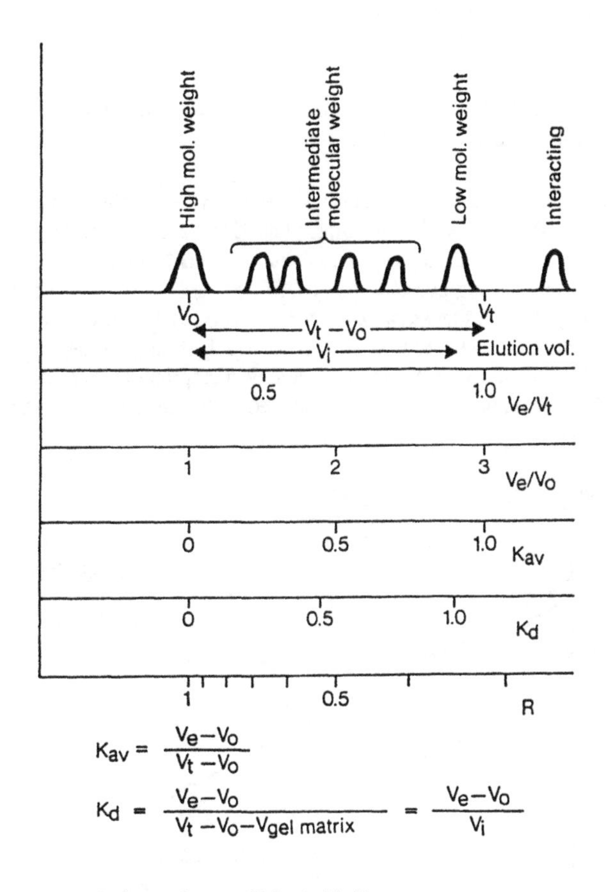

Figure 8.37 Illustration of relationships among K_{av}, K_d, V_e/V_t, V_e/V_0, and R. (From Pharmacia, 1979, fig. 21, copyright 1979.)

$$K_{av} = \frac{V_e - V_o}{V_t - V_o} \tag{8.10}$$

The constant K_{av}, *is not a true partition coefficient*, since the difference, $V_t - V_o$, includes the solids and the fluid associated with gel or stationary phase. By definition, V_s in Eqs. (8.8) and (8.9) represents only the fluid inside the stationary phase particles; it does not include the volume occupied by the solids that make up the gel. Although K_{av} is a property of the gel, it still defines solute behavior independently of the bed dimensions, as is the case for K_d in Eq. (8.39). The ratio of K_{av} to K_d should be a constant for a given component and gel packed in a specific column with $K_{av}/K_d > 1$ for nonexcluded components. Other definitions that have been used for quantitating elution behavior in gel permeation columns include V_e/V_t, V_e/V_o, and R (the retention coefficient = V_o/V_e) and are illustrated in Figure 8.37. Solutes that adsorb onto the stationary phase, and therefore interact with the stationary phase, emerge at an elution volume V_e (= V_R) which is greater than V_t (Figure 8.37). These definitions are mentioned here given the variety of conventions presented in the literature. The preferred definitions for the purposes of this book are discussed in Sections 5.1 to 5.6.

8.36 SELECTIVITY CURVES PLOT MEASURED VALUES OF K_{av} AGAINST LOGARITHMS OF MOLECULAR WEIGHT

Molecular weight determination of globular proteins having similar assymmetry factors (a sphere has an assymmetry factor of 1; an ellipsoid has a factor of greater than 1), using size exclusion chromatography, are based on selectivity curves. Such curves are linear, with the intercept increasing with increasing porosity (i.e., decreasing crosslinking) of the stationary phase (Figure 8.38). Extrapolation of these curves on the x-axis gives the molecular weights of probes that should be completely excluded from the gels, since these target molecules are larger than the largest pores. Theoretically the K_{av} for a given molecular probe should have a value between 0 and 1, where $K_{av} = 0$ denotes a completely excluded molecule and $K_{av} = 1$ for molecules that

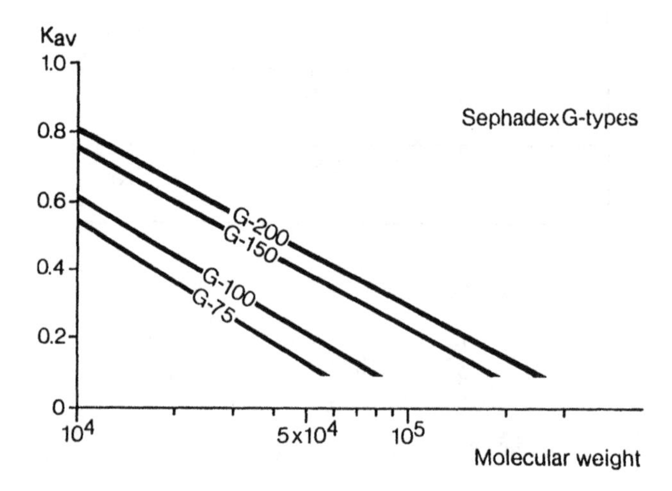

Figure 8.38 Example of selectivity curves for Sephadex G-75, G-100, G-150, and G-200 for globular proteins (From Pharmacia, 1979, fig. 2, copyright 1979.)

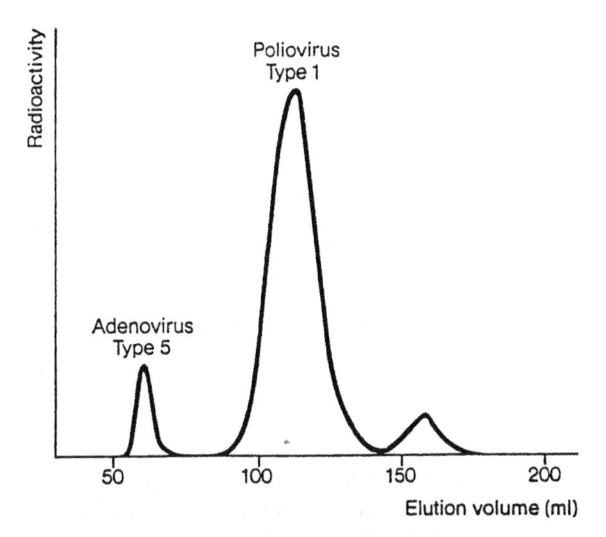

Figure 8.39 Separation of ^{32}P-labeled adenovirus and poliovirus on Sepharose 2B. Column: 2.1 × 56 cm; eluent: 0.002 M sodium phosphate, pH 7.2, containing 0.15 M NaCl; flowrate: 2 ml · cm^{-2} · h^{-1} (By courtesy of S. Bengtsson and L. Philipson, from Pharmacia, 1979, fig. 17, copyright 1979.)

are able to completely explore the fluid inside the stationary phase particle. If K_{av} is less than zero, channeling due to a poorly packed bed is a probable cause. If K_{av} is greater than 1, an interaction (adsorption) of the molecular probe with the stationary phase is a likely explanation.

The versatility of gel permeation media is indicated in Figure 8.30 and illustrated by Figure 8.39, where a separation of very high molecular weight adeno- and polio virus particles is illustrated. It is possible to carry out separations of proteins (and particles) whose size would be equivalent to a molecular weight of 40,000 kD using the agar-based Sepharose type gel. This particular gel has a limited temperature range for operation, however, since it melts upon heating to 40°C (Pharmacia Fine Chemicals, 1979).

The gels with a larger extent of crosslinking, particularly Sephadex G-10, G-15, and G-25, may retain (through weak adsorption) some types of aromatic molecules and consequently impart reversed phase properties to this type of gel. This may be due to the weakly hydrophobic character of the crosslinking agents used in the synthesis of these gels. Stationary phases for reversed phase chromatography methods are discussed in Sections 8.40 to 8.50.

8.37 BUFFER EXCHANGE AND DESALTING UTILIZE GEL PERMEATION

A major use of size exclusion chromatography has been in applications where salt or buffer is removed from the protein, namely in desalting and buffer exchange (Kurnik et al., 1995). Buffer exchange is used to remove denaturing agents in order to induce refolding of proteins, to remove buffers between purification steps, or to remove buffers and other reagents from the final product. Buffer exchange for these purposes is usually carried out at later steps in a recovery sequence (see Figures 8.1, 8.12, and 8.13). The difference in molecular weights is large, with salts generally having a molecular weight of below 200, while the proteins are between 10,000 and 60,000 Daltons.

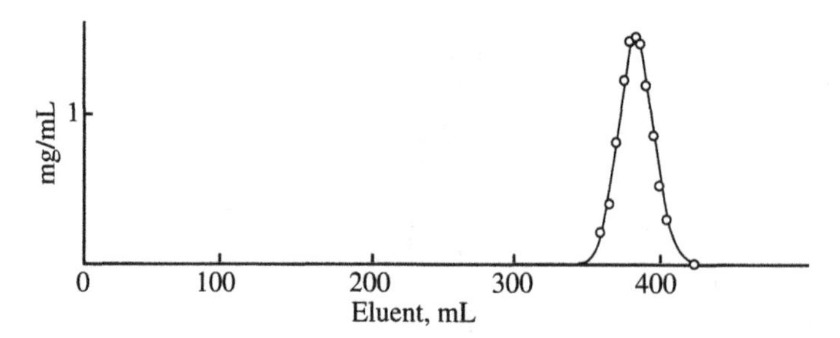

Figure 8.40 Elution profile for molecular sieving of solution containing 10 mg/mL of glucose. (From Porath and Flodin, 1959, fig. 1.)

The first work on gel filtration using crosslinked dextran gels addressed gel permeation as a method for desalting and group separations (Porath and Flodin, 1959). Separation of ammonium sulfate from serum proteins was carried out at an eluent flowrate of 25 mL/min (coincides to an interstitial velocity of 5.1 cm/hr) in a 25 mm i.d. × 175 mm bed packed with a dextran gel, while the principle of molecular sieving was illustrated using glucose and soluble dextran gels having molecular weights of 180, 1000 and 20,000 D, at a flowrate of 100 mL/hr or 20.4 cm/hr (Figures 8.40–8.43) for dextran gel packed in a 40 × 365 mm glass column (bed volume of 459 mL). The benefits of gel filtration over dialysis as a desalting method included the absence of the plugging or clogging associated with dialysis membranes, the ability to perform gel filtration with great rapidity (3 to 5 hr/run), relative to dialysis (10 to 30 hr), and the ability to readily scale up this technique. This new approach (in 1959) was also proposed as being particularly suited for buffer exchange and is illustrated in Figure 8.44 for separation of serum proteins (first peak) from ammonium sulfate (second peak) in a 25 × 175 mm column (volume of 86 mL). The main technologies currently available for buffer exchange are size exclusion chromatography (SEC), tangential flow filtration (TFF), and countercurrent dialysis (CCD). Debates on the merits of size exclusion relative to various types of dialysis/filtration as a process tool continuing to this day. While gel filtration has been a standard practice in the industry during the last 40 years, the search for methods with higher throughputs continues.

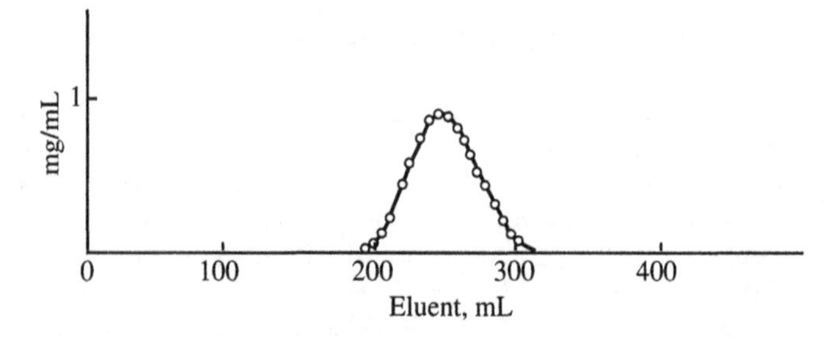

Figure 8.41 Elution profile for dextran of molecular weight = 1000. (From Porath and Flodin, 1959, fig. 1.)

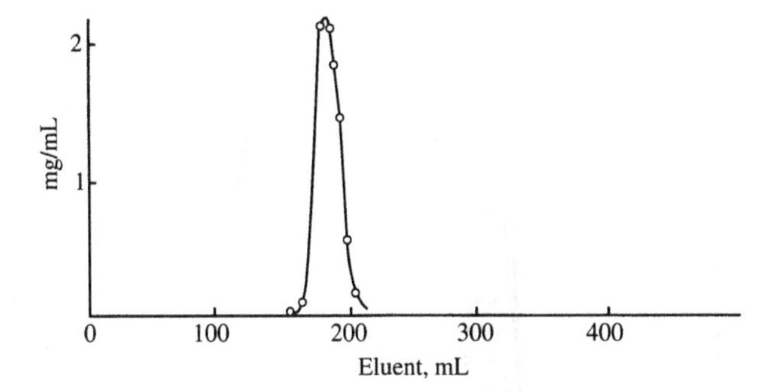

Figure 8.42 Elution profile for dextran of molecular weight = 20,000. (From Porath and Flodin, 1959, fig. 1.)

The recent testing of a novel size exclusion stationary phase that consists of a rolled woven textile (Section 5.11) and facilitates rapid preparative (process scale) separation of salts from protein in less than 7 minutes, shows that process size exclusion chromatography is capable of high throughput while at the same time reducing the volume needed to obtain proper refolding of the protein. This concept was demonstrated for the refolding of denatured recombinant secretory leukocyte protease inhibitor. The salts that cause the protein to be denatured are rapidly removed by size exclusion chromatography and decreased to a concentration where protein refolding occurs (Hamaker et al., 1996).

8.38 BUFFER EXCHANGE CAN ALSO BE ACHIEVED USING MEMBRANES

Alternate methods of desalting and buffer exchange include continuous diafiltration (a membrane separation technique) and crossflow filtration (also using membranes). These methods

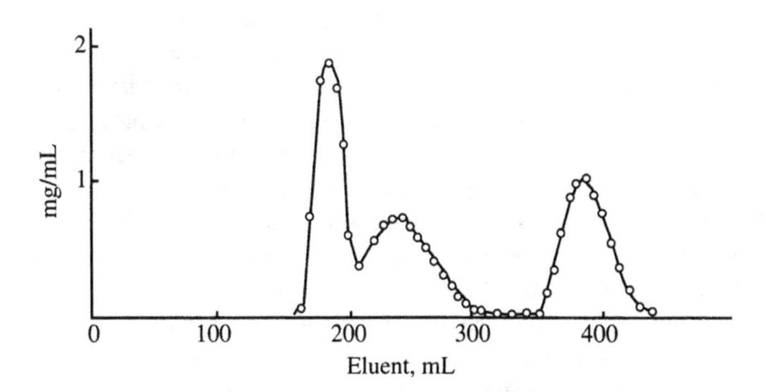

Figure 8.43 A mixture of all three components. A volume of 5 ml of 1% aqueous solutions of glucose and dextran fractions of Mn 1000 and 20,000, respectively, were filtered through a dextran gel column of the size 4.0 × 36.5 cm. (From Porath and Flodin, 1959, fig. 1.)

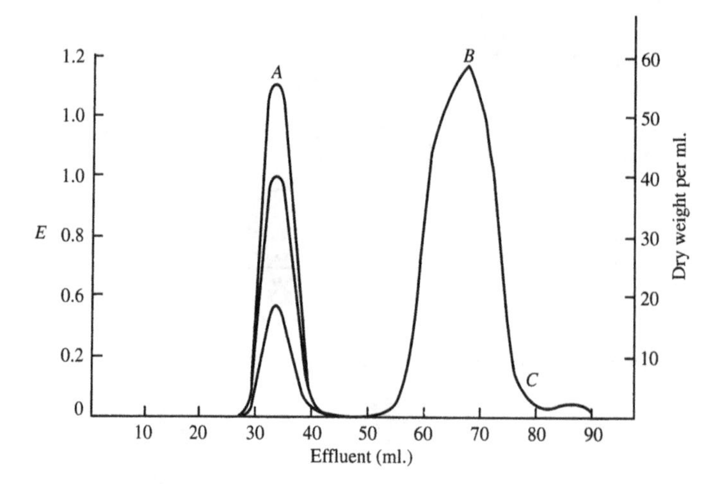

Figure 8.44 Diagram showing the separation of ammonium sulphate from serum proteins as obtained after filtration of the solution of the mixture through a dextran gel column (2.5 × 17.5 cm). Fractions (2 ml each) were collected and analysed by the method of Folin-Lowry (0.01 ml aliquots, – x –) and by ultraviolet light absorption at 280 mμ (–○–). The weight distribution of the substances in the effluent (–●–) was determined by weighing 1 ml aliquots after evaporation in vacuo. (From Porath and Flodin, 1959, fig. 1.)

rely on filtration at a molecular scale, using membranes with porosities that reject proteins but allow passage of salts. The membrane methods may be preferred for an unspecified protein because they are less costly and enable higher throughput than size exclusion chromatography for the particular protein being studied. Nonetheless, this is unlikely to be true for every case (Kurnik et al., 1995).

Equations for calculating separation efficiencies and recovery yields for all three methods, tested with an unnamed recombinant protein (molecular weight of 160 kDal), give insights into the benefits and disadvantages for each method. According to Kurnik et al. (1995) tangential flow filtration (TFF) requires multiple passes (typically 90 passes through a pump). In comparison, size exclusion chromatography and countercurrent dialysis are single pass. If there is protein aggregation or denaturation, SEC or CCD may be the appropriate choice for buffer exchange since all of the protein would be excluded. Another reason is that aggregates and denatured protein cause plugging of the TFF membrane. Other disadvantages of tangential flow filtration include the requirement of frequent change-out of membranes and the relatively large volumes of proteins needed for laboratory-scale tests (a disadvantage for new recombinant products). Nonetheless, this method had the benefits of 30% less floor space and double the throughput of gel permeation. Not all experimental details of the SEC system are given, although the calculated simulations of the column were done for a linear velocity of 60 cm/hr, for a column having 57 plates, and for a column height of 44 cm. While the elution volume is between 0.9 to 1.0 based on the definition in Figure 8.33, approximately 1.75 column volumes are needed to completely elute the salts because of the large sample volume equivalent to 35% of the column volume. The time for one run was on the order of 1.3 hours. For this particular case, Sephadex G-25M (gel filtration media), compared with tangential flow filtration or countercurrent dialysis, gave less complete ion removal, 130% or greater dilution, 66% higher (eluting) buffer requirement, and 30% higher cost (Kurnik et al., 1995).

Size exclusion chromatography is likely to continue to be a widely practiced technology in the industry. Rapid size exclusion columns for the purpose of buffer exchange have been developed, and they enable desalting to be achieved at linear velocities of 500 to 6000 cm/hr (Hamaker et al., 1996), thereby significantly increasing throughput and reducing operating time, plant floor space, and cost. Further, SEC using gel filtration media based on cellulosic stationary phases have a special niche for the *partial* and controlled separation of denaturing salts from recombinant proteins for purposes of refolding. The development of rapid desalting techniques are important because larger volumes of proteins will need to be processed in the industry. It is likely that membrane-based techniques and specialized chromatography media for use in industrial desalting will continue to grow and change the way in which gel permeation gels are viewed and used for biopharmaceutical applications.

8.39 GEL PERMEATION REQUIRES A LARGER COLUMN THAN ION EXCHANGE CHROMATOGRAPHY

Size exclusion chromatography columns are generally the largest columns on a process scale, since separation is strictly based on rates of diffusion of the molecules inside the gel particles. The proteins or other solutes are not adsorbed or otherwise retained due to an adsorption effect, so significant dilution of the sample volume can occur, particularly for small sample volumes. Hence the volumetric capacity (liters of feed/liter of column volume) of this type of chromatography is, in part, determined by the concentration of the proteins for a given volume of the feed placed on the column. All components injected into a size exclusion column should elute in one column volume (measured as shown in Figures 8.32 to 8.35), and the total stationary phase volume required to process a given feed stream will be proportional to the inlet concentration and volume of the feed. For example, for a typical inlet concentration of protein of 10 g/L, in a 100 L volume of feed, a column volume of at least 100 L is needed for size exclusion chromatography. In comparison, an ion exchange column whose adsorption capacity is 50 g/L would only require 20 L of column volume for the same feed, since the protein is adsorbed onto the stationary phase.

REVERSED PHASE CHROMATOGRAPHY

The term "reversed phase chromatography" is credited by Hearn to Boscott, Boldingh, and Howard, and Martin, who originally observed that polar components could be separated through weak, van der Waals type interactions between the solute and a nonpolar stationary phase (Hearn, 1989). Their work showed that elution of the solutes occurs in decreasing order of polarity, unlike normal or polar phase chromatography where elution order follows an increasing order of polarity. Since the solute retention is in the reverse order of normal phase chromatography, it was given the name "reversed phase chromatography." Hearn's historical account of the development of reversed phase chromatography stationary phases, summarized here, illustrates the importance of improvements in the mechanical and chemical properties of stationary phases on promoting their use.

Reversed phase bonded materials initially consisted of polymeric matrices that were coated with nonpolar, long-chain alkanes. Since these coatings were only immobilized through adsorption, their lifetimes were short. This was overcome through chemical bonding of the alkyl phase to agaroses and similar materials. Ultimately the development of small particle, silica-based bonded phases that had defined meso- and macroporous structures, narrow particle and pore size distributions, and average particle sizes in the 5 to 10 micron range facilitated both rapid and reliable reversed phase chromatography. These stationary phases, combined with the development of reliable high-pressure liquid chromatography systems and reproducible packing methods for high-pressure columns, made reversed phase chromatography a viable and high-resolution method for the separation of polypeptides and proteins.

8.40 REVERSED PHASE CHROMATOGRAPHY CAN SEPARATE PROTEINS BY SOLVOPHOBIC (HYDROPHOBIC) AND SILANOPHILIC (HYDROPHILIC) INTERACTIONS

The principles behind the retention of proteins and polypeptides are complex due to stationary phases that are intrinsically heterogeneous, to the opportunity of the macromolecules to interact with the stationary phase through ionic and hydrophilic mechanisms, and to hydrophobic effects. The most common reversed phase supports are probably alkylsilane-bonded materials. It has been estimated that for these materials only about half of the approximately 8 $\mu mol/m^2$ of surface groups on the silica particles are reacted with silane to form the bonded phase (Sedek et al., 1985). Consequently the characteristics of the silica itself affect retention. Trace amounts of adsorbed metal ions, the water content of the silica prior to bonding, and its relative Brönsted acidity all contribute to protein/stationary phase interactions. These interactions can be moderated through a uniform surface coverage of the silica using multiple-layer or branched-chain phases, and derivatization of the surfaces with polymeric layers. Alternately, the silica-based stationary phases can be replaced with a polymeric (organic) stationary phase altogether (Hearn, 1989). By definition, the dominating mechanism for both silica and nonsilica reversed phase supports should be the partitioning of the protein between stationary and mobile phases due to hydrophobic regions on the surface of the protein or polypeptide, or the hydrophobic character of lower molecular weight species.

8.41 DERIVATIZED SILICA PARTICLES ARE ONE TYPE OF REVERSED PHASE CHROMATOGRAPHIC MEDIA

The preparation of reversed phases consisting of silica derivatized with coatings consisting of 1 to 18 alkyl groups are reviewed by Snyder and Kirkland (1979) and are described by Karch et al. (1976). Stationary phases with covalently bound organic groups were apparently introduced in 1969, with the organic coatings bonded through reactions with alcohols to form silicalite esters:

$$-\mathrm{Si}-\mathrm{OH} + \mathrm{HOR} \longrightarrow \mathrm{Si}-\mathrm{OR} \tag{8.11}$$

or by chlorinating the silica surface with thionyl chloride, and then reacting chloride derivatives of the silanol groups with Grignard reagent or primary or secondary amines to yield Si–C (bonds) or amino silanes:

$$—Si—OH + SOCl_2 \longrightarrow Si—Cl \xrightarrow{C_6H_5MgBr} Si—C_6H_5 \quad (8.12)$$

$$Si—Cl \xrightarrow{H_2N(CH_2)_2NH_2} Si—NH\text{-}(CH_2)_2NH_2 \quad (8.13)$$

Grignard reactions sometimes result in a relatively low concentration of surface organic, and they can leave undesireable residues, Eq. (8.12). The Si–N bonded phases, Eq. (8.13), while more stable, are restricted to pH 4 to 8 when aqueous solvents are used, since the stationary phase surface degrades outside of this pH range. Consequently siloxanes, Eq. (8.14), are the most widely used bonded phases. These form by reaction of silanol groups with chlorosilane, which is stable over the pH range of 1 to 8.5 (silica starts to dissolve at above pH 8.5):

$$Si—OH + \begin{matrix} C\,or\,R_3 \\ ROSiR_3 \end{matrix} \longrightarrow —Si—O—SiR_3 \quad (8.14)$$

Stationary phases prepared by reaction (Ladisch and Kohlmann, 1992) are represented as materials for which the surface represents brushes, and where the monomeric, organic groups, stand on end. Although this type of coating results in good mass transfer efficiencies, it is also readily hydrolyzed off the stationary phase in the presence of water or alcohol (Snyder and Kirkland, 1979). Reaction (8.12) utilizes thionyl chloride to chlorinate the surface, and a Grignard reagent to form the silica–carbon bond. The hydrolytic stability of these materials are superior to silicate esters but the Grinard reaction produces a low organic content at the surface of the silica. The Si–N type bonded phases, while more stable than silicate esters, must still be run at pH 4 to 8 when aqueous solvants are used, in order to avoid hydrolysis and dissolution of the stationary phase.

Bonded phases prepared from siloxanes' reaction (8.14) are hydrolytically stable over the pH range of 2 to 8.5, and they are the basis of the popular reversed phases μ Bondapak C_{18} and Zorbax-CN (Snyder and Kirkland, 1979). The reaction (8.14) also lends itself to multiple covalent bonds, thereby further enhancing chemical stability as shown below:

```
                 O
                 |
Si — O — Si — O         R
                 |       \     /
                 O        Si          (8.15)
                 |       /     \
Si — O — Si — O         R
                 |
                 O
                 |
                 Si
```

TABLE 8.11 Reagents Used for Reversed Phase (Bonded) Materials

Type of Reversed Phase "Bristle"	Reagent
C_1	$ClSi(CH_3)_2$
C_4	$Cl_2Si(C_4H_9)_2$
C_{10}	$Cl_2Si(CH_3)C_{10}H_{21}$
C_{18}	$Cl_3SiC_{18}H_{37}$
$C_{18'}$	$Cl_2Si(CH_3)C_{18}H_{37}$

Source: Adapted from Karch et al. (1976).

These materials are stable to temperatures up to 220°C (Karch et al., 1976). Stationary phases, which are given the popular nomenclature of C_1, C_4, C_{18}, and so on, reflect the original naming systems where the surface coating was envisioned in terms of bristles have lengths of 1 to 18 methyl groups (summarized in Table 8.11). The ratio of the number of moles of silanol groups reacted to number of moles of derivatization reagent reacted is known as the F number, and it does not exceed 2. In the case of the trichlorosilanes, the silane is still attached to the silica surface phase by only two covalent bonds. According to Snyder and Kirkland values of $F = 3$ have never been found experimentally, probably due to steric hindrances.

The incomplete reaction of the surface with the silane reagent, or the formation of new Si–OH groups, results in a population of acidic Si–OH groups which can cause tailing of peaks, thereby reducing separation efficiency. This effect is suppressed by adding a small amount of a strong organic acid to the mobile phase. An excellent synopsis of the chemistry used to prepare these materials is given by Snyder and Kirkland (1979).

8.42 SMALL PARTICLE SILICA COLUMNS BENEFIT FROM AXIAL COMPRESSION

Process-scale reversed phase supports can have particle sizes as small as 10 to 25 microns. Unlike polymeric reversed phase sorbents, these small-particle, silica-based reversed phase supports require high-pressure equipment to be properly packed and operated. The introduction of axial compression columns (see Section 5.8) has helped promote the use of high-performance silica supports on a process scale, since its resolution, which approaches that of an analytical scale separation, can be achieved with columns that can also be quickly packed. These columns consist of a plunger fitted into a stainless steel column. The particles are placed into the column in a slurry. The plunger then squeezes or compacts the bed in an axial direction to give a stable, tightly packed bed. This type of column must be operated at pressures of up to 100 bar, but it also gives excellent resolution in run times of an hour or less. This technique, and its availability in an equipment package, illustrates how the availability of the appropriate equipment can encourage the scale-up of analytical separations to process-scale applications for high-value biochemicals (Colin et al., 1990). Further discussion on packing of chromatography columns is given in Sections 5.8 to 5.11.

8.43 POLYMERIC REVERSED PHASE CHROMATOGRAPHIC MEDIA ARE BASED ON POLYSTYRENE AND METHACRYLIC COPOLYMERS

The significance of adsorption of organic species by high-surface area styrene-divinylbenzene copolymers for the isolation, separation, and purification of water-soluble organic species was already recognized in 1968—a few years after synthesis of these macroporous beads had been achieved (Gustafson et al., 1968). These first polymeric adsorbents had porosities of 0.35 to 0.43, average pore diameters of 205 to 91 Å, and surface areas of 100 to 313 m^2/g, respectively. The polymeric adsorbents differed from gel permeation stationary phases, since the pores were discrete and had defined dimensions. This pioneering research showed that theophylline, caffeine, phenylpropanolamine, sodium cholate, phenylalanine, and glycine all adsorbed to some degree, and that heats of adsorption for butyric acid and sodium napthalenesulfate were on the order of several kcal/gmole. Other studies showed that elution of testosterone and p-chlorophenol could be achieved with the solvents (in order of decreasing effectiveness): acetone > dioxane > acetonitrile > *n*-propanol > ethanol > methanol, which is consistent with the solvent power for these mobile phases for reversed phase chromatography using bonded silica bases (Gustafson et al., 1968; Hearn, 1989). The binding of the solutes on the polymeric sorbent is sufficiently strong to remove various types of organic species from aqueous solution, but at the same time weak enough so that the adsorbed components can be eluted with alcohols and ketones. Unlike the carbon supports that they were designed to supplant, the polymeric adsorbents reversibly adsorb the organic species. The potential of underivatized polymeric stationary phases for polypeptide separations was also indicated by this early work.

The structures of various copolymers that make up polymeric phases are summarized in Table 8.12, which shows three types of adsorbents. The XAD-1, -2, and -4 are characterized by hydrophobic interactions between the solute and adsorbent, while XAD-7 and -8 exhibit both

TABLE 8.12 XAD Copolymers

Polymer	Structure	Average Surface Area, m^2/g	Average Pore Diameter, A	Porosity Volume %
XAD-1	Polystyrene/DVB[a]	100	200	37
XAD-2	Polystyrene/DVB	330	90	42
XAD-4	Polystyrene/DVB	780	50	51
XAD-7	Acrylic ester,[b] X = COOR′	450	80	55
XAD-8	Acrylic ester, X = crosslink	140	250	52
XAD-9	Acrylic ester, X = COOR′S(O)R′	70	370	NA
XAD-11	Acrylic ester, X = CONHR′C(O)R′	70	350	NA

Sources: Adapted from Pietrzyk and Chu (1977), table I. Amberlite (1974), summary chart.
[a]DVB = divinylbenzene.

```
            CH3     CH3     CH3
            |       |       |
b  B = - CH2- C- CH2- C-CH2-  C-CH2-
            |       |       |
            X      C=O      X
                    |
                    O
                    |
                    R
                    |
                    O
                    |
            X      C=O      X
            |       |       |
       -CH2- C-CH2- C-CH2-  C-CH2-
            |       |       |
           CH3     CH3     CH3
```

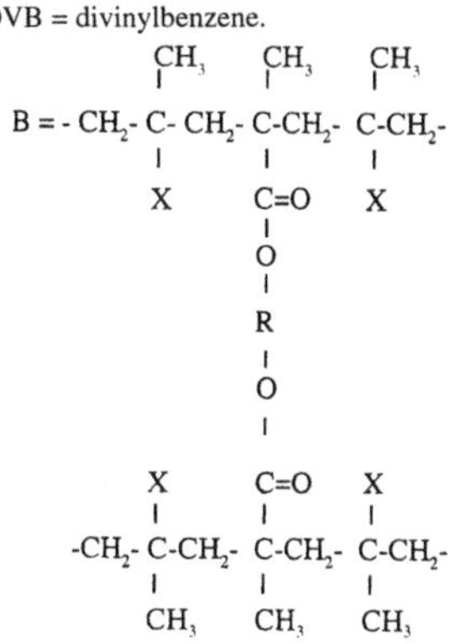

TABLE 8.13 Solvent Uptake for Several Copolymers

	Solvent/Dry Copolymer, g/g				
Solvent	P-8D[a,b]	XAD-1[a]	XAD-2[a]	XAD-4	XAD-7
Water	0.054	0.061	0.072	0.055	2.29
Methanol	0.050	0.488	0.699	0.981	1.90
Ethanol	0.062	0.507	0.719	1.02	1.90
Propanol	0.068	0.535	0.757	1.05	2.98
Hexane	0.015	0.540	0.515	0.857[c]	1.53[c]
Benzene	0.724	0.993	0.779	1.09	2.08
Chloroform	1.4	1.6	1.2	1.3	2.4

Source: Reproduced from Pietrzyk and Chu (1977), table II. Reprinted with permission from ACS.
[a]Solvent uptake (in g per g of dry copolymer) was determined by a centrifugation technique.
[b]Gel polystyrene-8% divinylbenzene copolymer.
[c]Heptane instead of hexane was used.

hydrophobic and hydrophilic behavior. The XAD-9 and -11 are principly hydrophilic (Pietrzyk and Chu, 1977).

The differences in the hydrophobic/hydrophilic character are indicated by the increasing extent of solvent uptake, although the XAD-7 can imbibe both water as well as nonpolar solvents as Table 8.13 shows (Pietrzyk and Chu, 1977). The chemical stability of this type of stationary phase at pH 12 is an attractive advantage over silica gel type reversed phases that leach away (dissolve) at alkaline pH (Pietrzyk and Stodola, 1981). The pH stability made it possible to demonstrate separation of alanine and isoleucine at pH 12, and alanine, isoleucine, and tyrosine in 0.02 M phosphoric acid with XAD-4 as the stationary phase. Also of interest is the ability of this type of adsorbent to separate the diastereomers D (L)-Ala-D(L)Ala from D(L)-Ala-L(D)Ala at acidic pH (Pietrzyk and Stodola, 1981).

The use of polymeric adsorbents for industrial reversed phase separations encompasses purification of various antibiotics (cephalosporin, thienamycin, actaplanin, vanomycin), vitamins (vitamin B_{12} and biotin), phenylalanine, and nucleosides (Pollio and Kunin, 1971; Takayanagi et al., 1996). The use of a polymethyacrylate resin for concentrating tissue plasminogen activator in a bioreactor is also mentioned (Takayanagi et al., 1996). The basic chemical structures of current versions of polymeric adsorbents seem to be similar to the XAD polymeric adsorbents. The backbone structure of the more hydrophobic stationary phase is given as

polystyrene/divinylbenzene

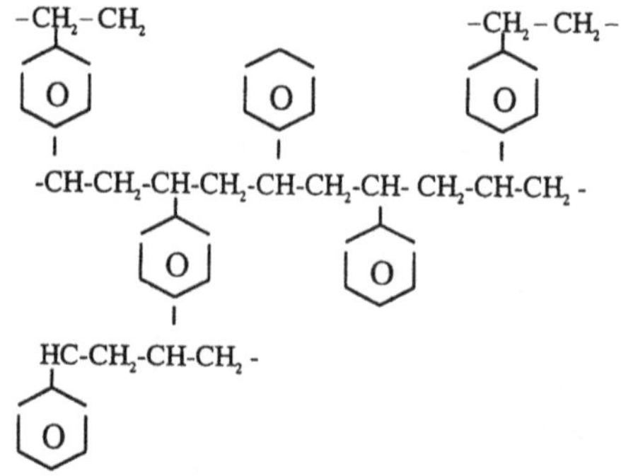

-CH_2-CH_2 -CH_2-CH_2-

-CH-CH_2-CH-CH_2-CH-CH_2-CH- CH_2-CH-CH_2 -

HC-CH_2-CH-CH_2 -

This is also prepared in a brominated form. The polymeric adsorbent based on polymethacrylate is represented as having the structure

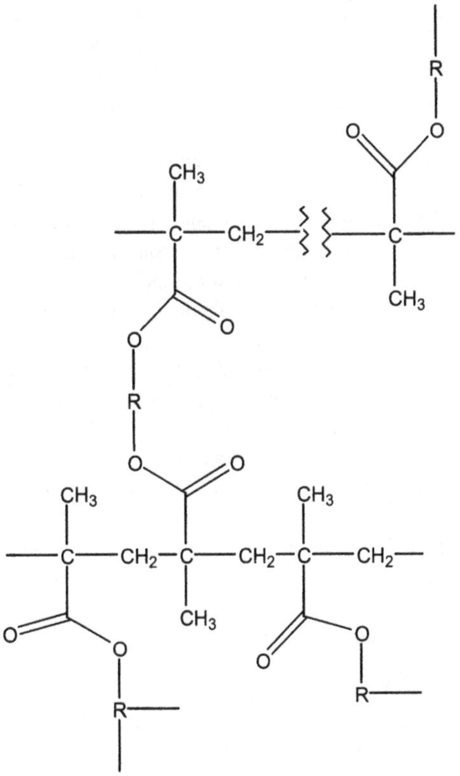

The particle sizes of these materials are 0.3 to 1 mm for industrial scale applications (Takayanagi et al., 1996).

All of these materials are characterized by large surface areas, and rigid, nearly permanent porous structures. Each bead is composed of a large number of microspheres agglomerated together into particles ranging from 0.010 to 1 mm in size. The microspheres, which themselves range from 3 to several hundred nanometers, provide a large internal surface area that is accessible by the large macropores found throughout the particles. Elution of adsorbed species is efficient due to the rapid rate of movement of solute and solvent through the pores of the resin (Gustafson et al., 1968; Pierzyk and Stodola, 1981).

8.44 EXPERIMENTALLY DETERMINED CHROMATOGRAMS ARE THE STARTING POINTS FOR SPECIFYING SOLVENT COMPOSITION FOR REVERSED PHASE CHROMATOGRAPHY

The choice of appropriate solvents for process chromatography on an a priori basis would be exceedingly complex were it not for the restrictions placed on solvent choice due to safety, viscosity, volatility, and toxicity considerations. The solvents that meet at least some of these criteria and/or are used on an industrial scale include water, acetonitrile, methanol, ethanol, isopropanol, tetrahydrofuran, acetone, ethyl acetate, and hexane. Consequently the range of solvents that would be considered for process chromatography is significantly less than the 74 candidates given by Snyder and Kirkland (1979). The selection of a solvent is nonetheless challanging, and requires the availability of experimentally obtained chromatograms with respect to a starting solvent composition. Once such data are available, the proportionalities of Eqs. (8.16) to (8.18), suggested by Snyder and Kirkland (1979), should prove useful in selecting starting solvent compositions for experimentally optimizing a reversed phase separations of bioproducts.

The systematic selection of mobile phase solvents for reversed phase chromatography is based on their strength and selectivity. The strength of a solvent is defined by its ability to dissolve a given solute, and this depends on the solvent's polarity relative to the solute. The polarity of the solvent is a function of the combined intermolecular interactions between the solute and mobile phase components based on dispersion, dipole, hydrogen bonding, and dielectric interactions. In reversed phase chromatography the solvent strength decreases with increasing polarity.

Solvent selectivity defines the ability of the solvent to preferentially dissolve one component from another. Selectivity can be adjusted by varying the composition of the mobile phase while keeping solvent strength constant. Snyder and Kirkland (1979) recommend that approximate compositions of mobile phases of similar strength can be estimated based on the ratio of polarities of the solvent. Their approach requires that an initial solvent composition consisting of two components (1 and 2) is first experimentally shown to give close to the desired chromatogram of the solutes to be separated. Substitution of component 1 by component 3 would then be carried out by making up a new mobile phase containing a volume fraction of component 3 in a mixture with component 2. The volume fraction of 3, is given by Eq. (8.16), with the new fraction of 2, $\varnothing_{2,\text{new}}$, being the balance as given by Eq. (8.17):

$$\varnothing_3 = \varnothing_2\left(\frac{P_2}{P_3}\right) = \varnothing_2\left(\frac{P_\text{w} - P_2}{P_\text{w} - P_3}\right) = \varnothing_2\left(\frac{S_2}{S_3}\right) \tag{8.16}$$

$$\varnothing_{2,\text{new}} = 1 - \varnothing_3 \tag{8.17}$$

where P_2 and P_3 are the polarities of solvents 2 and 3, and $\varnothing_{2,\text{new}}$ and $\varnothing_3$ are the respective volume fractions of the new mobile phase. Polarity usually refers to the mobile phase, while the solubility parameters (S_2 and S_3) represent the property of the mobile phase in the presence of a specific type of stationary phase. The use of polarity or solubility parameters in Eq. (8.16) are determined by the availability of data.

Solvents for reversed phase chromatography usually consist of an organic phase mixed with water. The second ratio in Eq. (8.16) is based on changes in polarity of the mixture with respect to water, where P_w is the solvent polarity parameter for water (= 10.2). For reversed phase chromatography, solvents 2 and 3 would be selected so that their polarity parameters are less than that of water. Values for polarity for commonly used solvents are given in Table 8.14.

The solvent strength parameters are a function of both solvent and stationary phase, and they will be different for different stationary phases. A comparison of solvent strengths for reversed phases and underivatized phases Si, Al, and carbon are given in Table 8.14. The solvent strength parameter, or alternately the polarity (which is a property of only the mobile phase), is particularly useful for nonaqueous mobile phases. The ratios, based on either solubility or polarity parameters, are approximate. Hence specification and validation of the final solvent composition

TABLE 8.14 Solvent Parameters of Selected Solvents for Reversed and Normal Phase Adsorbents

		Solvent Strength, S with respect to indicated stationary phase			
Component	Solvent Polarity Parameter, P^a	Reversed Phase	Silica	Underivatized Alumina	Carbon
Water	10.2	0	—	—	—
Methanol	5.1	3.0	0.7	0.95	0
Acetonitrile	5.8	3.1	0.5	0.65	0.04
Acetone	5.1	3.4	—	—	—
Dioxane	4.8	3.5	—	—	—
Ethanol	4.3	3.6	—	—	0.05
i-Propanol	3.9	4.2	—	—	—
Tetrahydrofuron	4.0	4.4	—	—	0.14

Source: Adapted from Pietryzk and Chu (1977) , Tables 9.4, 9.5, and p. 264, for comprehensive list of solvents.
[a]See Pietryzk and Chu (1977).

require that confirming chromatography runs be carried out. In this context the guidelines offered by Snyder and Kirkland, here in the form of Eq. (8.16), help to expedite specification of solvent compositions to initiate these experiments.

8.45 CHANGES IN RELATIVE PEAK RETENTION OCCUR WITH CHANGES IN MOBILE PHASE POLARITY

The capacity factor, k', is a measure of peak retention, and it represents the number of *additional* void volumes of mobile phase (not including the volume of the mobile phase initially present) required to elute the peak. The capacity factor for a given component changes as a function of solvent polarity.For solvents having polarities P_1 and P_2, this would be

$$\frac{k_2'}{k_1'} = 10^{\left(\frac{|P_2 - P_1|}{2}\right)} \tag{8.18}$$

A change of 2 in the relative polarity ($|P_2 - P_1|$) gives a 10-fold change in the capacity factor (since $10^{(|6-4|/2)} = 10'$). A detailed discussion of solvent effect correlations for retention times of organic molecules with respect to selected mobile and stationary phases, and extensive tables of parameters required for these calculations are given by Sedek et al. (1985).

8.46 REVERSED PHASE CHROMATOGRAPHY USES HYDROPHOBIC MEDIA AND INCREASING GRADIENTS OF AQUEOUS METHANOL, ACETONITRILE, OR ISOPROPANOL

Stationary phases for reversed phase chromatography consist principly of silica particles, silica supports with a hydrocarbon bonded phase, or polymeric materials based on either vinyl or styrene-divinylbenzene copolymers (Cantwell et al., 1984; Colin et al., 1990; Gustafson et al., 1968; Pietrzyk and Stodola, 1981; Pietrzyk and Chu, 1977; Morris and Fritz, 1993; Sedek et al., 1985).

Mobile phases commonly used in reversed phase chromatography are aqueous methanol, isopropanol, and acetonitrile with these organic modifiers being added to aqueous buffers to form gradients of increasing concentration. These solvents are often mixed with acidic buffers containing small amounts of acids such as trifluoroacetic acid or hexafluorobutyric acid. These acids reduce the pH of the mobile phase to 3 and give sharper peaks by surpressing ionization of silanol groups of silica based, reversed phase supports, and minimizing ionic effects (Sofer and Nystrom, 1989). The most prevalent type of stationary phase is made of silica or another type of inorganic support that has been derivatized and bonded with an octadecyl (C_{18}) or octyl (C_8) coating. Silicas that have not been derivatized are also sometimes utilized for process chromatography.

Much like ion exchange chromatography, the organic component (referred to as a modifier or modulator) is mixed with water or buffer to form an *increasing* gradient of the modifier. This gradient serves to elute the components, which are initially adsorbed onto the stationary phase in order of increasing hydrophobic character. Methanol is used as the modifier for eluting weakly adsorbed, hydrophilic peptides, while isopropanol is used to elute strongly adsorbed hydrophobic peptides. Methanol has a higher heat of mixing with water than the other solvents, and this can lead to solvent degassing and bubble formation in the column. The bubbles interfere both with the operation of the column (they cause peak dispersion) and with the detection of peaks at the column outlet (bubbles interfere with the optics of spectrophotometric and refractive index peak detection) (Sofer and Nystrom, 1989). Acetonitrile is probably the most widely used modifier for separations of proteins and many other types of molecules, since it exhibits favorable mass transfer properties, lower viscosity (and back pressure) than the other solvents, and good eluting strength.

8.47 INSULIN, A POLYPEPTIDE HORMONE, IS PURIFIED BY REVERSED PHASE PROCESS CHROMATOGRAPHY

An example of the purification of recombinant product by reversed phase chromatography is given by recombinant insulin, a polypeptide hormone. Insulin consists of 51 amino acid residues in two chains and is relatively small. Reversed phase chromatography is used after most of the other impurities have been removed by a prior ion exchange step (Ladisch and Kohlmann, 1992). The method, described by Kroeff et al. (1989), utilizes a process grade C_8 reversed phase support (Zorbex) with a particle size of 10 micron. Partially purified insulin crystals, dissolved in a water-rich mobile phase, were applied to the column and then eluted in a linear gradient generated by mixing 0.25 M acetic acid in water (no acetonitrile) with 60% acetonitrile in 0.25 M acetic acid. This example illustrates the probable interaction of the polypeptide with the silica backbone that causes the broadening of peaks—an effect that is moderated by the acidic mobile phase as discussed earlier in this chapter. The acidic mobile phase enabled insulin to be resolved from structurally similar insulinlike components (Figure 8.45). Under alkaline conditions the resolution is compromised (Figure 8.46). The ideal pH is from 3.0 to 4.0, which is below insulin's isoelectric pH of 5.4. Under mildly acidic conditions insulin may deamidate to monodesamido insulin, but if the reversed phase separation is done within a matter of hours, the deamidation can be minimized. The potential for scale-up of this method is illustrated by Figure 8.47 where a preparative scale column (volume of 10.4 mL) packed with 10 micron particle size reversed phase resolved the main product peak from a small amount of other contaminants. The significance of this separation is that it was achieved using a small particle size packing, and was later implemented on a production scale using a "self-packing," axial compression column (see Section 5.8).

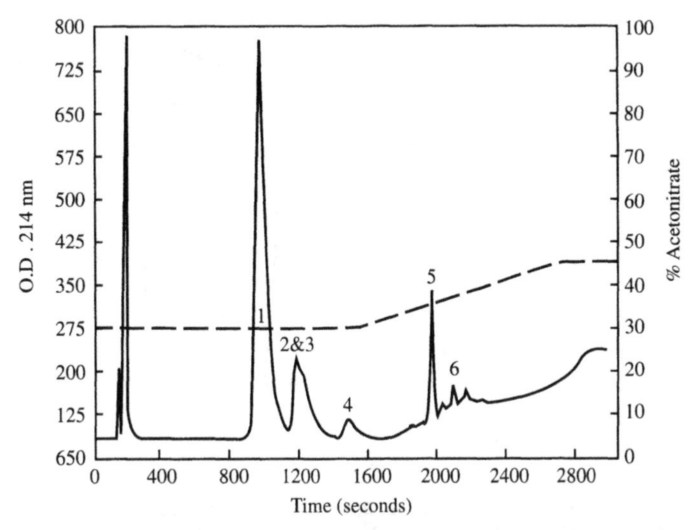

Figure 8.45 Reversed phase separations of biosynthetic human insulin and biosynthetic human insulin derivatives with acid. Flowrate at 1.0 mL/min. Total sample load of 8.8 μg insulin and insulin derivatives (7.5 μg BHI and 1.3 μg of BHI derivatives. Solutes: 1 = BHI, 2 = desamideo A-21 BHI, 3 = *N*-carbamoyl-Gly BHI, 4 = *N*-formyl-Gly BHI, 5 = *N*-carbamoyl-Phc BHI, 6 = BHI dimers. Mobile phases: Eluent A = 0.1 M phosphate (pH 2.1), eluent B = A-acetonitrile (1:1). BHI denotes biosynthetic human insulin. (Reprinted with permission from Kroeff et al., 1989. J. Chromatogr., with permission from Elsevier Science.)

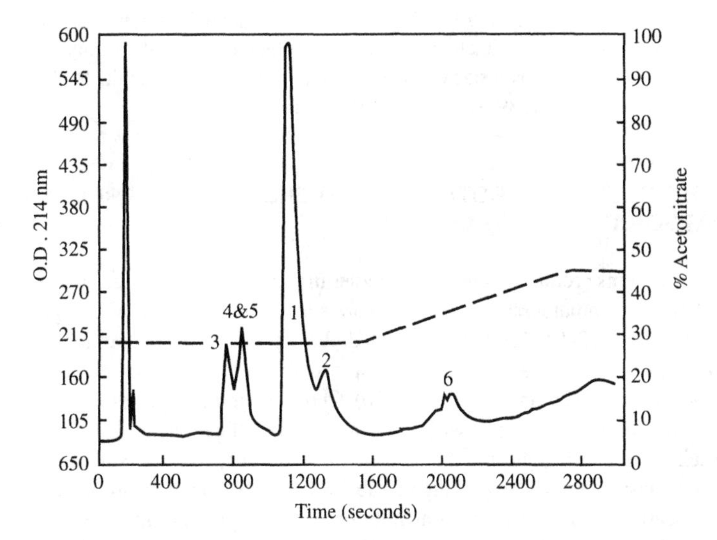

Figure 8.46 Reversed phase separations of biosynthetic human insulin and biosynthetic human insulin derivatives with alkaline mobile phases at 35°C with Zorbax C_8, 150 A, packed in 25 × 0.45 cm i.d. columns. Flowrate at 1.0 mL/min. Total sample load of 8.8 μg insulin and insulin derivatives (7.5 μg BHI and 1.3 μg of BHI derivatives). Solutes: 1 = BHI, 2 = desamideo A-21 BHI, 3 = *N*-carbamoyl-Gly BHI, 4 = *N*-formyl-Gly BHI, 5 = *N*-carbamoyl-Phc BHI, 6 = BHI dimers. Mobile phases: Eluent A = 0.1 M phosphate (pH 7.3), eluent B = eluent A-acetonitrile (1:1). (Reprinted with permission from Kroeff, et al., 1989. J. Chromatogr., with permission from Elsevier Science.)

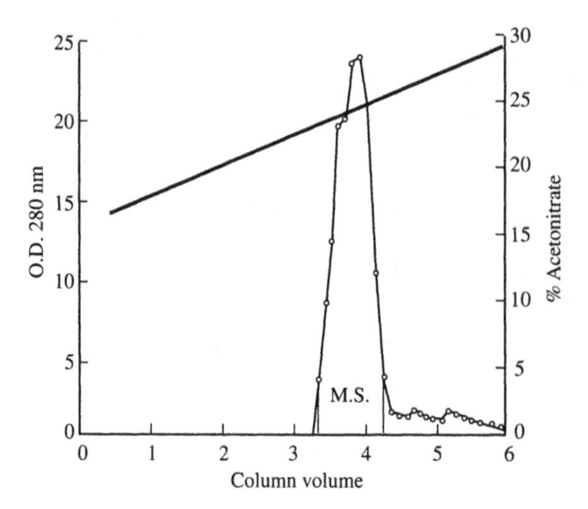

Figure 8.47 Preparative chromatogram obtained using 10 μm Zorbax C_8, in 15 × 0.94 cm i.d. column at flowrate 0.3 mL/min. Sample load of 153 mg (derived from proinsulin process): gradient, 17 to 29% acetonitrile in 0.25 M acetic acid in 6 column volumes (CV); flowrate 0.3 ml/min. Fractions 3.3 to 4.3 CV were pooled (mainstream). Fractions 3.2, 4.4–5.4, and the protein eluted in column regeneration (not shown) were combined (sidestream). (Reprinted with permission from Kroeff et al., 1989. J. Chromatogr., with permission from Elsevier Science.)

This reversed phase chromatography method was successfully used in a production-scale system to purify recombinant insulin. According to Kroeff et al. (1989) the insulin purified by reversed phase chromatography had a biological potency equal to that obtained from a conventional system that employs ion exchange and size exclusion chromatography. The reversed phase separation is followed by a size exclusion step to remove the acetonitrile eluent from the final product (Kroeff et al., 1989; Ladisch and Kohlmann, 1992).

8.48 PURIFICATION OF PROTEINS BY REVERSED PHASE PROCESS CHROMATOGRAPHY IS NOT COMMON

Recombinant proteins produced as inclusion bodies in bacterial fermentations may be amenable to reversed phase chromatography, since they are sometimes extracted with organic solvents (Sofer and Nystrom, 1989). The use of reversed phase process chromatography for other recombinant products does not appear to be widespread although its use for drug and small molecule purification is more common (Section 8.50). Chromatography options were examined and tabulated for an unspecified protein with pharmaceutical applications by Lawlis and Heinsohn (1993). Their industrial insight into the options currently available for commercial processes appears to be consistent with the ranking in Table 8.2. Ion exchange, hydrophobic interaction, and gel permeation, and perhaps even affinity chromatography, are more prevalent than reversed phase chromatography on a process scale.

8.49 REVERSED PHASE CHROMATOGRAPHY IS WIDELY USED FOR ANALYSIS OF PROTEINS AND PEPTIDE MAPPING

The major use of reversed phase chromatography has historically been in the analysis and process separations of peptides, amino acids, and organic compounds that are characterized by their lower molecular weight and solubility in acetonitrile or alcohol gradients. The mobile phases used for reversed phase separations are perceived to denature proteins and some polypeptides, since the mobile phases are aqueous organics at a low pH. Denaturation and/or less than complete protein recovery is not necessarily a general effect for chromatography with high coverage *n*-alkylsilicas eluted with aqueous organics—if appropriate operational conditions are defined and the protein is robust (Hearn, 1989). Nonetheless, reversed phase chromatography is not typical for *process-scale* purification of *proteins*. Rather the excellent protein resolving power of this type of chromatography is employed on analytical scale using columns packed with 2 to 10 mL of stationary phases having 1 to 5 micron particle sizes and for sample volumes that typically range from 1 to 10 microliters.

One purpose in monitoring a protein product is to detect the presence of a change in which as little as one amino acid has been chemically or biologically altered or replaced by another amino acid during the manufacturing process. These variant amino acid(s) in a protein may not affect protein retention during reverse phase chromatography because the three-dimensional structure of the protein can shield the variant residue from the surface of the reversed phase support (Builder and Hancock, 1988). Furthermore, since reversed phase chromatography discriminates between different molecules on the basis of hydrophobicity, and since large proteins may contain only small patches of hydrophobic residues, these patches may not correlate to the molecular modifications that a reversed phase analytical method seeks to detect. The reversed phase method must therefore be completely validated, and preferably combined with controlled chemical and/or proteolytic hydrolysis followed by chromatography or electrophoresis of the cleaved protein to give a detailed chromatogram (i.e., map) of the resulting peptide fragments (Builder and Hancock, 1988; Garnick et al., 1991).

A peptide map is generated by hydrolyzing a previously purified protein using chemicals or enzymes. These hydrolytic agents, which have known specificity, are used to perform limited proteolysis followed by resolution and identification of all the peptide fragments formed. Identification of changes, and reconstruction of the protein's primary structure, is then possible. Reagents and enzymes that cleave specific bonds include formic acid (cleaves bond between Asp-Pro at 37°C), cyanogen bromide (Met-Y, where cleavage is at the N-terminus for the residue Y), hydroxylamine (Asn-Gly), trypsin (Arg-Y, Lys-Y), pepsin (Phe-Y, Leu-Y, and pairs of nonpolar residues), and elastase (Ala-Y, Gly-Y) (Carrey, 1990).

An example of a tryptic (peptide) map generated by trypsin hydrolysis of recombinant tissue plasminogen activator is shown in Figure 8.48. The chromatogram shows the resolving power of reversed phase high-performance liquid chromatography in separating peptides obtained from t-PA in which the disulfide bonds have been reduced and alkylated prior to enzyme hydrolysis. The small peptides formed have little or no three-dimensional structure. Hence measurable shifts in elution profiles occur when there are variant amino acids, since a change of a single amino acid in the peptide has a larger effect on the solubility and retention of the peptides than the same change in a protein. The tryptic map in Figure 8.48 resulted from the formation of 57 peptides from the rt-PA that is made up of 527 amino acid residues. The replacement of arginine at position 275 in a normal rt-PA molecule, with glutamic acid in the mutant form of rt-PA, results in a significant peak shift of one of the peptide fragments obtained from cleavage of the rt-PA, as indicated by the arrows in Figure 8.48 (Garnick et al., 1991). This example shows how tryptic mapping can be a suitable method for monitoring lot-to-lot consistency of this particular recombinant product (Builder and Hancock, 1988).

The use of reversed phase high-performance liquid chromatography has come into use for estimating the purity of other proteins and peptides as well. However, a caveat is made that before it is used, a high-performance liquid chromatographic profile must be completely validated for its applicability to the analysis of a given protein (Garnick et al., 1991).

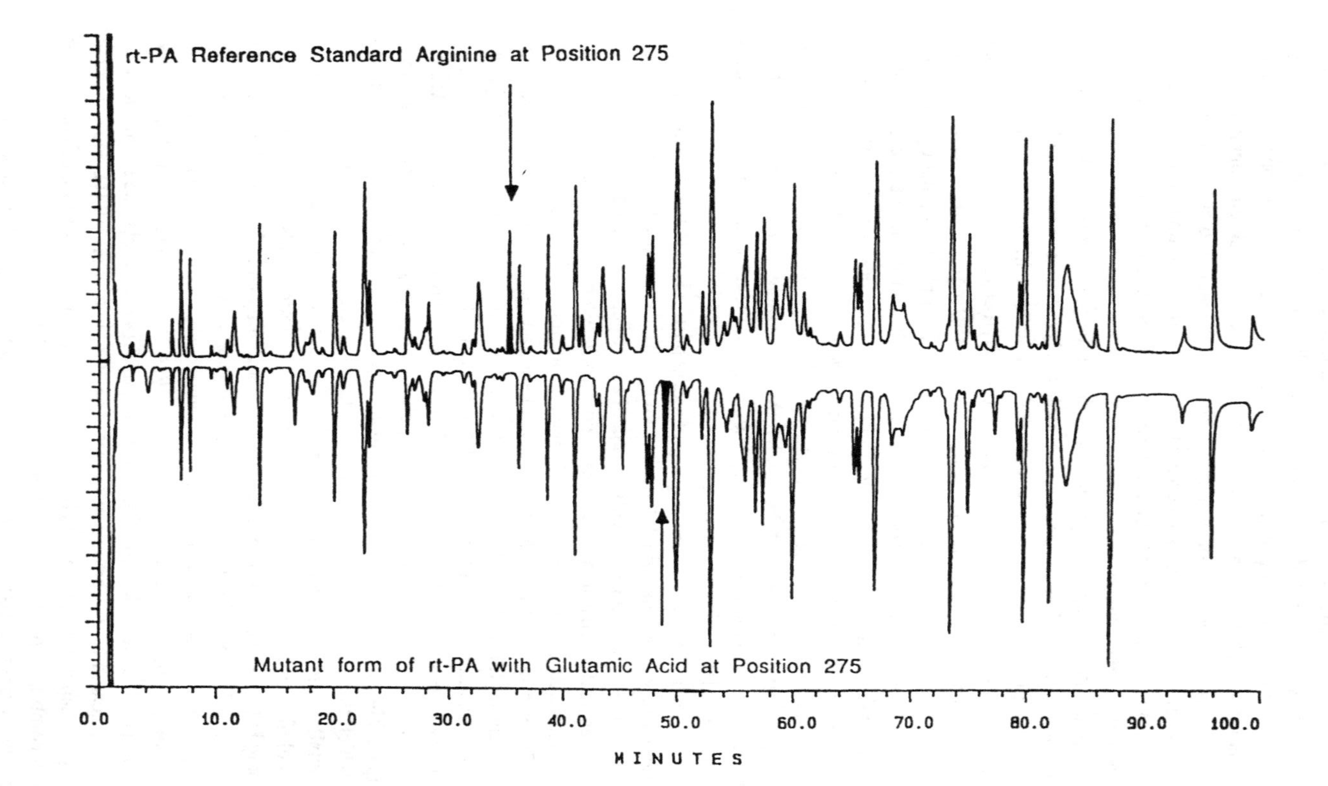

Figure 8.48 Tryptic map of rt-PA ($M_r = 66$ kD) showing peptides formed from hydrolysis of reduced, alkylated rt-PA. Separation by reversed phase octadecyl (C_{18}) column using aqueous acetonitrile with an added acidic agent the mobile phase. Arrows show the difference between a normal (top chromatogram) and mutant (bottom chromatogram) form of rt-PA where the glutamic acid residue has replaced the normal arginine residue at position 275, as indicated by the bottom arrow. (Reprinted with permission from Garnick et al., 1991, fig. 3, Marcel Dekker, copyright 1991.)

8.50 REVERSED PHASE PROCESS CHROMATOGRAPHY OF SMALL MOLECULES HAS A HISTORY OF USE

Polypeptides, peptides, antibiotics, alkaloids, and other low molecular weight compounds are amenable to process chromatography by reversed phase methods. There are numerous examples of bioproducts that have been purified using reversed phase chromatography. The manufacture of salmon calcitonin, a 32 residue peptide used for treatment of postmenopausal osteoporosis, hypercalcemia, and Paget's disease of the bone, includes reversed phase chromatography instead of more "tedious, extremely time-consuming, and expensive" solution (i.e., extraction and precipitation) methods. This peptide is commercially prepared on a kilogram scale by a solid phase synthesis, and then purified by a multimodal purification train. Reversed phase chromatography was the dominant technique used in the production and quality control of this peptide by Rhône Poulenc Rorer (Flanigan, 1991).

Another example is the purification of a β-lactam antibiotic, where process-scale reversed phase separations began to be used around 1983 when suitable, high-pressure process-scale equipment became available. A reversed phase, microparticulate (55 to 105 micron particle size) C_{18} silica column, with a mobile phase of aqueous methanol in 0.1 M ammonium phosphate at pH 5.3, was able to fractionate the product away from impurities not readily removed by liquid–liquid extraction (Cantwell et al., 1984). The principle impurity was tetrazole sulfonic acid disodium salt (II), shown as peak 2 in Figure 8.49; other inorganic and organic im-

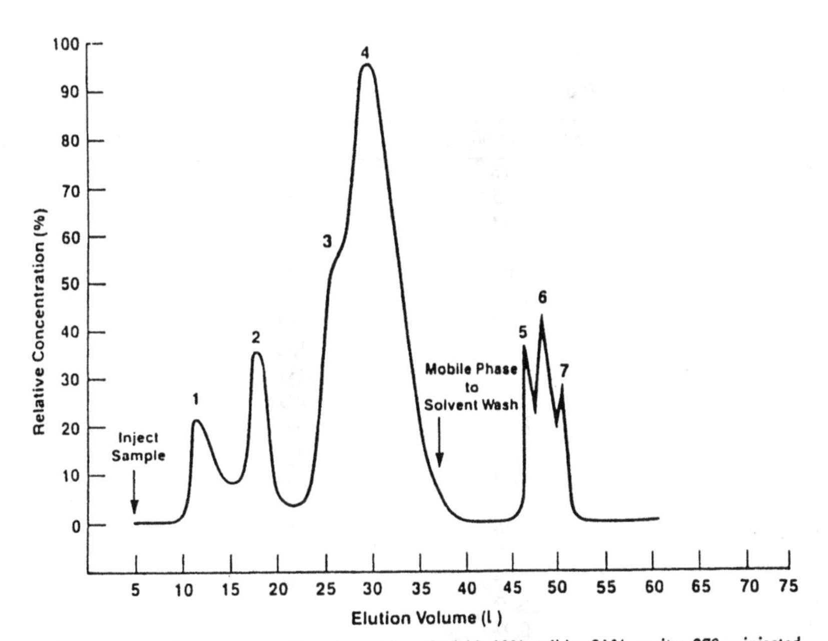

Process scale chromatogram. Sample: crude cefonicid, 10% solids, 81% purity, 270 g injected. Column: 60 × 20 cm I.D., 8.8 kg C_{18} silica, 55–105 μm. Mobile phase: water. Detection was by conductivity (peaks 1–4) and refractive index (peaks 5–7).

Figure 8.49 Process-scale chromatogram showing reversed phase separation of β-Lactam antibiotic (cefonicid). (Reprinted with permission from Cantwell et al., 1984, fig. 3, Elsevier Science Publishers, copyright 1984.)

purities (peaks 3, 5, 6, and 7) were also resolved. The product elutes in peak 4. Optimization of the separation resulted in recovery of product at 93% purity and 95% yield. This separation of an antibiotic or other small molecule differs markedly from protein purification in feed concentration (50 to 200 g/L for cefonicid compared to 1 to 10 g/L for a protein), molecular weight of impurities (M_r < 5 kD compared to 0.1 to 100 kD for proteins), and throughputs (1 to 2 mg product/(g stationary phase · min) compared to 0.01 to 0.1 mg/(g · min) for proteins).

Reversed phase separation was also found to purify diastereomer precursors used in the chemical synthesis of the insect sex phermone of *Limantria Dispar*, used to control this pest that attacks oak trees. The liquid chromatography columns tested had dimensions of up to 15 cm i.d. by 130 cm long, and they were able to purify up to 708 g of starting material dissolved in a sample volume of 4.1 L using a column packed with 23 L of stationary phase. The throughput is estimated to have been on the order of 0.2 to 0.4 mg/(g · min), with separation obtained using a gradient of hexane and diethyl ether (Pierri et al., 1983).

HYDROPHOBIC INTERACTION CHROMATOGRAPHY

Hydrophobic interaction chromatography (HIC) separates protein on the basis of differences in their surface hydrophobicity. While most of the hydrophobic amino acids are buried in the interior of globular proteins, some hydrophobic amino acids are at the protein's surface. A study of 46 monomeric globular proteins found that 50 to 68% of the proteins' surfaces consist of

TABLE 8.15 Hydrophobicity Scales for Amino Acids

	Scale I, kcal/mol[a]		Scale II, Fraction Buried[b]
Trp	3.77	Phe	0.87
Ile	3.15	Trp	0.86
Phe	2.87	Cys	0.83
Pro	2.77	Ile	0.79
Tyr	2.67	Leu	0.77
Leu	2.17	Met	0.76
Val	1.87	Val	0.72
Mct	1.67	His	0.70
Lys	1.64	Tyr	0.64
Cys	1.52	Ala	0.52
Ala	0.87	Ser	0.49
His	0.87	Arg	0.49
Arg	0.85	Asn	0.42
Glu	0.67	Gly	0.41
Asp	0.66	Thr	0.38
Gly	0.10	Glu	0.38
Asn	0.09	Asp	0.37
Ser	0.07	Pro	0.35
Thr	0.07	Gln	0.35
Gln	0.00	Lys	0.31

Source: Reproduced with permission from Eriksson (1989), table 7-1, John Wiley & Sons, Inc.

[a]Scale I is based on the solubility of the different amino acids, expressed as the free energy of transfer for the amino acids from ethanol or dioxane to water.

[b]Scale II is based on the fraction of the number of the different amino acids buried within proteins (average values obtained from 20 proteins with known structures).

amino acid residues containing side chains of carbon and sulfur (i.e., nonpolar side chains). An earlier study (in 1971) found the nonpolar surface area to be 41% for lysozyme, 48% for myoglobin, and 46% for ribonuclease (Eriksson, 1989). The amino acids that contribute to the hydrophobic character of the protein have been estimated based on their solubility in water, or the average values of the fraction of amino acids buried within the structure of 20 known proteins (Table 8.15).

Protein retention depends on the surface hydrophobicity of both the support and solute, as well as on the kosmotropic nature and concentration of the salt used in the mobile phase. The mobile phase in hydrophobic interaction chromatography is an aqueous solution of salts, unlike the mobile phases for reversed phase chromatography, which are aqueous organics. Hence hydrophobic interaction chromatography is classified as a mild method, since the salts used in HIC have a stabilizing influence on the protein structure. The strength of interaction between a HIC gel and protein depends on both the proportion of the nonpolar surface area and the distribution of the nonpolar areas. Preferential interactions of aqueous salts with the supports and proteins modify these interactions and result in different retention characteristics for different proteins (Eriksson, 1989; Roettger et al., 1990).

8.51 HYDROPHOBIC INTERACTION CHROMATOGRAPHY UTILIZES DECREASING AQUEOUS GRADIENTS OF KOSMOTROPES

Hydrophobic interactions of solutes with a stationary phase result in their adsorption on neutral or mildy hydrophobic stationary phases. The solutes are adsorbed at a high salt concentration, and then desorbed, in order of increasing surface hydrophobicity, in a decreasing kosmotrope gradient. This characteristic follows the order of the lyotropic series for the anions:

$$\text{Phosphates} > \text{sulfates} > \text{acetates} > \text{chlorides} > \text{nitrates} > \text{thiocyanates}$$

as well as cations:

$$Mg^{2+} > Li^{+} > Na^{+} > K^{+} > NH_4^{+}$$

Anions that precipitate proteins less effectively than chloride (nitrates and thiocyanates) are chaotropes or water structure breakers, and they have a randomizing effect on water's structure, thereby promoting solubility of proteins. The chloride ion is considered to be neutral with respect to water structure, while the anions preceding NaCl (phosphates, sulfates, acetates) are polar kosmotropes or water structure makers, and they promote precipitation of proteins (Eriksson, 1989; Roettger and Ladisch, 1989). Refer to Section 4.3 for a detailed discussion. Kosmotropes also promote adsorption of proteins and other solutes onto a hydrophobic stationary phase. Other beneficial characteristics include increasing the thermal stability of enzymes, decreasing enzyme inactivation, protecting against proteolysis, increasing the association of protein subunits, and increasing the refolding rate of denatured proteins. Hence hydrophobic interaction chromatography is an attractive method for purification of proteins where recovery of a purified protein in an active and stable conformation is desired (Roettger and Ladisch, 1989; Roettger et al., 1990). The selection of types of ions, and the thermodynamics of their interaction with proteins is developed in Sections 4.9 to 4.13 of this book.

8.52 THE TYPE AND CONCENTRATION OF SALT AFFECTS PEAK RETENTION IN HYDROPHOBIC INTERACTION CHROMATOGRAPHY

The definition of a capacity factor, k' in hydrophobic interaction chromatography (HIC) is different from the distribution coefficient K_{av} in gel permeation chromatography given previously in Eq. (8.3). The capacity factor is the number of additional bed void volumes of mobile phase (not including the interstitial volume initially present) required to elute the peak:

$$k' = \frac{V_e - V_o}{V_o} = \frac{V_e}{V_o} - 1 \tag{8.19}$$

where V_e is the elution volume and V_0 is the retention volume of a noninteracting (and excluded) solute. When protein retention due to adsorption occurs, the value of k' is greater than one (i.e., elution of the retained peak will not occur until after one column void volume of mobile phase has passed through the column).

The retention behavior of lysozyme on a polymeric hydrophobic interaction support follows the preferential interaction parameter of the lyotropic series of anions (Figure 8.50). The preferential interaction parameter is a measure of the net salt inclusion or exclusion in the hydration layer. The higher the value, the larger is the disrupting effect of the salt and the lower is the retention volume (compare lyotropic numbers and retention of lysozyme for NH_4Cl and $(NH_4)_2SO_4$ in Figures 8.50 and 8.51. This analysis has led to derivation and experimental validation of the capacity factor for lysozyme with respect to the

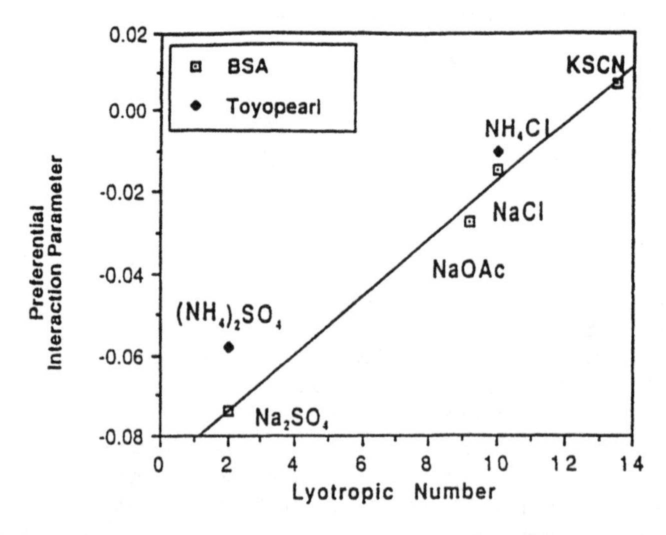

Figure 8.50 Preferential interaction parameter is a function of the lyotropic number. (Reprinted with permission from Roettger et al., 1990, American Chemical Society, Washington, DC, copyright 1990.)

lyotropic number of the anion. Chromatography was carried out using a hydrophilic vinyl polymer support (Toyopearl) with an average particle diameter of 30 microns and average pore size of 1000 Å. The capacity factor for HIC has the form

$$k' = a[C]^d[N_x - b] + h = -0.01224[C]^{6.603}[N_x - 10.57] + 0.03876 \tag{8.20}$$

where a, b, d, and h are protein specific parameters, N_x is the lyotropic number, and C is salt concentration in moles per liter. The values of the constants in Eq. (8.20) are for lysozyme with respect to Toyopearl HW-65S. This approach enables hydrophobic interaction parameters to be estimated from experimental peak retention data, and the change in retention time to be estimated with relation to a change in the type of salt which affects the lyotropic number, N_x or the salt concentration, C. An example of how salt type and concentration affect retention of a protein (lysozyme, in this case) is illustrated in Figure 8.52. A similar functional relation was found for myoglobin with respect to a hydrophilic vinyl polymer derivatized with butyl (C_4) groups (butyl-Toyopearl 650S (Roettger et al., 1989, 1990).

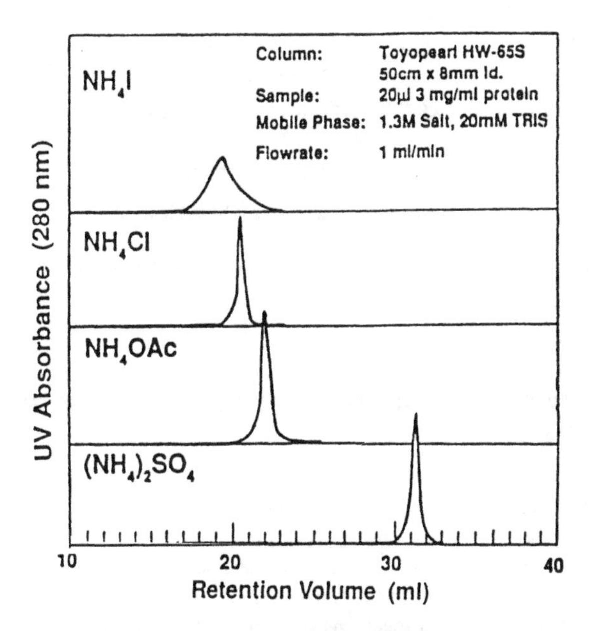

Figure 8.51 Retention of lysozyme increases with decreasing lyotropic number of salt used in the eluting buffer for hydrophobic interaction chromatography. (Reprinted with permission from Roettger et al., 1990, American Chemical Society, Washington, DC, copyright 1990.)

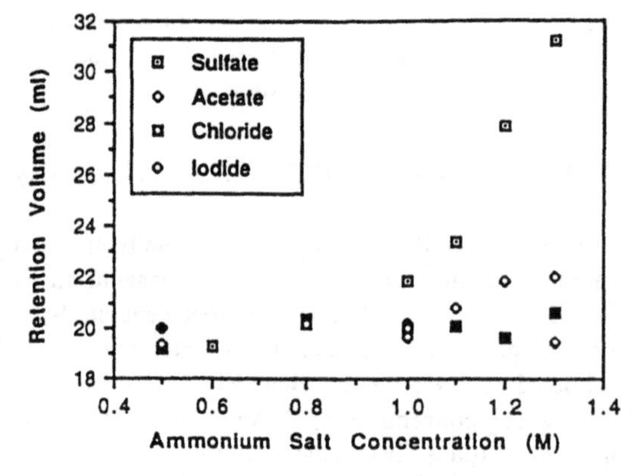

Lysozyme retention on Toyopearl HW-65S versus salt concentration.

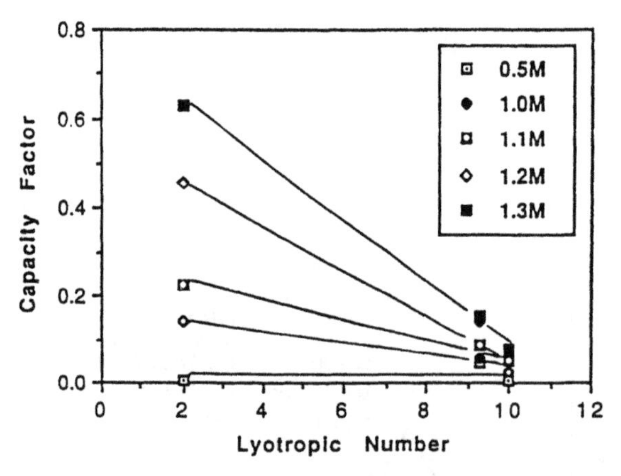

Lysozyme capacity factor versus lyotropic number.

Figure 8.52 Capacity factor for lysozyme which correlates with the lyotropic number and concentration of the salt in the eluting buffer. (Reprinted with permission from Roettger et al., 1990, American Chemical Society, Washington, DC, copyright 1990.)

8.53 HYDROPHOBIC INTERACTION CHROMATOGRAPHY IS SUITABLE FOR MANY PROTEINS

Various types of proteins have been purified using hydrophobic interaction chromatography. These include alkaline phophatase, estrogen receptors, isolectins, strepavidin, calmodulin, epoxide hydrolase, proteoglycans, hemoglobins, and snake venom toxins (Roettger and Ladisch, 1989). In the case of cobra venom toxins, the order of elution of the six cardiotoxins supported the hypothesis that their mechanism of action is related to hydrophobic interactions of the venom protein with the phospholipids in the membrane.

Chymosin is an enzyme that used to be obtained from the lining of the stomachs of calves and is used in cheese-making. The recovery of recombinant chymosin from a yeast fermentation broth showed that large-scale hydrophobic interaction chromatography could produce an acceptable product in one step. However, since the stationary phase capacity was limited, an alternative method was developed in which the enzyme was extracted into a two-phase polyethylene glycol–salt system. The partition coefficient for the chymosin into PEG was 100, and enabled efficient recovery of this enzyme in one step. When followed by an ion exchange step, this method gave purified and active chymosin at reasonable cost. The use of hydrophobic interaction chromatography gave a first indication that two-phase extraction is a viable approach (Lawlis and Heinsohn, 1993).

AFFINITY CHROMATOGRAPHY

The nature of the ligand differentiates affinity chromatography from ion exchange chromatography. Stationary phases used in both types of chromatography facilitate specific interactions between the protein and ligand, with the protein causing displacement of a competing, and usually lower molecular weight, component. The specificity of the affinity ligand, however, is different and sometimes greater than that of an ion-exchanger. The ion-exchanger functions best in adsorbing proteins on the basis of their charge when the salt concentration in the surrounding mobile phase is low (Sections 7.2 and 7.3). An affinity support has a ligand that can complex with the active site of the protein, or with a specific region on the protein's surface, and hence should be independent of salt concentration. In theory, affinity chromatography ligands will bind only one type of protein, while ion exchange ligands bind an entire class of protein (Asenjo and Leser, 1996). Applications of affinity chromatography are given in Sections 8.54 to 8.57 while the biology and biochemistry of binding is discussed in Chapter 9.

8.54 REVERSIBLE BINDING OF ENZYMES WITH IMMOBILIZED SUBSTRATES AND INHIBITORS LED TO AFFINITY CHROMATOGRAPHY

The concept of affinity chromatography is credited to Sarkenstein's discovery of biospecific adsorption in 1910, and its reintroduction as a means to purify enzymes in 1968 (Parikh and Cuatrecasas, 1985). Substrates and substrate inhibitors diffuse into the active sites of enzymes and then reversibly bind there. Conversely, if the substrate or substrate inhibitor is irreversibly attached (i.e., immobilized) through a covalent bond to a solid particle of stationary phase with large pores, the enzyme should be able to diffuse into and bind with the substrate or inhibitor (i.e., its binding partner or ligand). Since the substrate is small (with $M_r < 500$ D) and the enzyme is large ($M_r > 15$ kD), the diffusion of the enzyme to its binding partner at a solid surface can be sterically hindered. The placement of the substrate at the end of an alkyl or glycol chain tethered to the stationary phase's surface reduces hindrance. This concept has been applied to ion exchange chromatography, as well, under the terms of tentacle or fimbriated stationary phases and is not unique to affinity chromatography. However, not all types of affinity chromatography utilize long chains, such as in immobilized metal affinity chromatography which is discussed further in Chapter 9 (Johnson and Arnold 1995a,b).

The realization that enzymes could be selectively retained in a chromatography column packed with particles of immobilized substrates or substrate analogues led to experiments with other pairs of binding partners. Not surprisingly, numerous applications of affinity chromatography developed, given the specific and reversible, yet strong, affinity of biological macromolecules for numerous specific ligands or effectors. In vivo, these interactions form the basis of cell regulation by hormones, regulation of enzymes by cofactors and inhibitors, antibiotic action on bacteria (by attaching to cell wall proteins and preventing cell wall biosynthesis), base pairing of complimentary nucleic acids in polynucleotides, cell agglutination, and antibody/antigen interactions including antibody/virus binding. In vitro, these interactions have been exploited for purposes of highly selective, but sometimes expensive, protein purifications; recovery of mRNA in some recombinant DNA applications; and study of mechanisms of protein binding with effector molecules (Parikh and Cuatrecasas, 1985).

8.55 CAREFUL SELECTION OF AFFINITY STATIONARY PHASES MINIMIZES NONSPECIFIC BINDING

The purpose of affinity chromatography is the highly selective adsorption, and subsequent recovery of the target protein. Loss of specificity occurs when macromolecules, other than the proteins, also adsorb onto the stationary phase due to hydrophobic or ionic characteristics of the matrix that makes up the stationary phase. For example, a spacer arm that allows the ligand to be located away from the matrix surface can improve accessibility and reduce steric hindrance to the immobilized ligand, but decrease selectivity.

Hexamethylenediamine is a common spacer arm in affinity chromatography, but it produces strong, nonspecific binding because of its hydrophobic nature. The hydrophobic character is decreased by interposing an ether or secondary amine (e.g., 3, 3 diaminodipropylamine) in the middle of the spacer arm in order to impart a hydrophilic character.

The ideal matrices for anchoring the ligands are non-ionic, hydrophilic, chemically stable and physically robust. The most popular matrices are polysaccharide based, principally due to their hydrophilic character and history of use as size exclusion or gel permeation gels, although glass beads, polyacrylamide gels, crosslinked dextrans (Sephadex), and agarose synthesized into a bead form have all been used (Parikh and Cuatrecasas, 1985). In particular, agaroses such as Biogel A (Bio-Rad), Utragel A (IBF), and Sepharose (Pharmacia) are popular (Asenjo and Patrick, 1990; Parikh and Cuatrecasas, 1985). Crosslinked agaroses (Sepharose CL and Sepharose FF, by Pharmacia) are physically more stable than Sepharose and are suitable for attaching affinity ligands. Both forms of the agaroses have an open porosity that allows proteins to readily diffuse inside of them. Loading capacities are noted to be "reasonable," although a general range, in terms of mg protein/mL stationary phase, is not meaningful. The affinity chromatography results are usually reported in terms of specific activity of the final product (units of protein activity per mg of protein) and the amount of protein recovered (% yield).

8.56 ACTIVATION OF THE STATIONARY PHASE SURFACE PRECEDES COVALENT ATTACHMENT OF AFFINITY LIGANDS

Activation of polysaccharide, silica, or polyacrylamide stationary phases involve the formation of a reactive intermediate, covalently attached to the surface, to which a difunctional alkyl-, aryl-, or glycol spacer is subsequently reacted. One end of the spacer is attached, while the other end is subsequently reacted with the ligand. Cyanogen bromide (CNBr) has been widely used to activate agarose and dextran gels with formation of cyanate esters as the major product in agaroses, and imidocarbonates in dextrans (Parikh and Cuatrecasas, 1985). For agaroses a single hydroxyl will react with CNBr:

$$\text{(Agarose backbone)}\!-\!OH + CNBr \longrightarrow \text{(Activated, cyanate esters)}\!-\!O\!-\!C{\equiv}N \xrightarrow{H_2N\!\sim\!\sim\!NH_2} \quad (8.21)$$

$$\!-\!O\!-\!C(=NH)\!-\!NH\!\sim\!\sim\!NH_2 \quad \text{(Major product)}$$

Agarose backbone

Activated (cyanate esters)

Major product

where

$$H_2N\!\sim\!\sim\!NH_2$$

represents a difunctional spacer molecule. Dextran gels can form both carbamates (a minor and inert product) and imidocarbonates as the reactive intermediates:

$$\text{(Dextran backbone)}\begin{matrix}-OH\\-OH\end{matrix} \xrightarrow{CNBr} \text{(Cyanate esters)}\begin{matrix}-O-C{\equiv}N\\-OH\end{matrix} \longrightarrow \text{(Activated, imidocarbonate)}\begin{matrix}-O\\-O\end{matrix}\!\!>\!C{=}NH \quad (8.22)$$

$$\xrightarrow{H_2N\!\sim\!\sim\!NH_2} \begin{matrix}-O\\-O\end{matrix}\!\!>\!C{=}N\!\sim\!\sim\!NH_2$$

Dextran backbone

Cyanate esters

Activated (imidocarbonate)

The spacer is then reacted with the protein or other ligand to form the affinity support. In the case of a small molecule, attachment occurs with a single bond, while a protein (macromolecule) can have multiple covalent bonds. The occurrence of spontaneous release of the ligand attached through an imidocarbonate with two points of attachment to the stationary phase is 1000 times lower than release of a ligand attached to a cyanate ester through one point of attachment (Parikh and Cuatrecasas, 1985).

The cyanate esters of either the agarose or dextran gels can react with water to form small amounts of an unreactive carbamate:

$$\begin{matrix}-O-C{\equiv}N\\-OH\end{matrix} \xrightarrow{H_2O} \begin{matrix}-O-C(=O)-NH_2\\-OH\end{matrix} \quad (8.23)$$

Cyanate esters

Carbamates (minor, inert)

The attachment of ligands, and sometimes activation of supports, is still carried out in the laboratories of chromatography process developers, since fully prepared affinity stationary phases (with ligands already attached) are not as widely available as stationary phases for the other four major types of chromatography, namely, ion exchange, size exclusion, reversed phase, and hydrophobic interaction. Rather activated supports (with a reactive spacer arm attached) are often used to immobilize in-house generated affinity ligands.

8.57 AFFINITY CHROMATOGRAPHY WILL NOT BE A ONE-STEP PURIFICATION PROCESS

An excellent synopsis and industrial viewpoint of affinity chromatography is given by Prouty (1991). He points out that ligands for affinity chromatography range from the low molecular weight components: arginine and benzamidine (these bind "trypsinlike proteases"), triazine dyes, and metal chelates; to high molecular weight ligands ranging from protein A to immunoglobins and monoclonal antibodies. The blood factor VIII, purified by monoclonal affinity techniques, was approved by the FDA in 1988. While a very promising and potentially powerful technique dating back to at least 1969, affinity chromatography continues to exhibit some limitations for isolation of therapeutic agents on a production scale. These include the following:

1. Leaching at ppm levels of bound ligands from the column into the product
2. Nonspecific interactions resulting in contamination of the target molecule with impurities that are retained by the affinity matrix due to hydrophobic or ion exchange effects and then co-eluted with the affinity bound product
3. Failure of the affinity ligand to differentiate all variant forms of a protein or polypeptide

Examples that correspond to these limitations include an antibody for purifying interferon-$\alpha 2$, which leached into the product at 3 to 40 ppm, and copper ion, which leached from an immobilized metal chelate column at 30 to 50 ppm. Improvements in binding chemistry are expected to moderate ligand leakage, although research in this area has been a continuing theme for over 30 years (Parikh and Cuatrecasas, 1985; Prouty, 1991).

The issue of differentiation of variant forms of the target protein is less tractable. Small molecule ligands are recognized by just a few amino acid residues on the target protein (usually an antibody) while epitopes for the antibodies might also have surface residues that associate with the antigen. For example, polyclonal antibodies do not distinguish desamido-insulin, which contains a deamidated asparagine, from insulin. Structural, conformational, or chemical changes may not affect binding of the protein or polypeptide to the ligand, nor have an effect on chromatographic retention, since the affinity ligand probes only a small portion of the protein that might bind with it. Many antibody preparations cannot differentiate between minor structural changes in proteins. Affinity chromatography therefore must be followed by other separation steps, and it will not provide the one-step purification that is often conceptually associated with this otherwise elegant method. Nonetheless, affinity chromatography has the potential to simplify purification of recombinant proteins, since recombinant technology holds promise of enabling the design and production of specialized ligands for use in protein separations. In this context, affinity chromatography provides a bridge between molecular biology and separations science (as discussed in Chapter 9).

Protein Purification Strategies

This chapter has summarized key chromatography methods. However, a purification scheme requires that an appropriate sequence of steps be selected and developed. Major categories of bioseparations can be organized into solid/liquid separations, initial product concentration,

Typical Purification Scheme for Biopharmaceutical Proteins

- Harvest protein-containing material.
- Concentrate protein using ultrafiltration.
- Perform initial chromatography steps.
- Perform antiviral steps.
- Perform additional chromatographic steps.
- Concentrate protein using ultrafiltration.
- Perform formulation chromatography.

Chromatography Options*

Ion Exchange	Hydrophobic Interaction	Gel Permeation	Affinity	Adsorption	Reversed Phase
DEAE Sepharose Fast Flow LC	Toyopearl Butyl-650 M	Agarose, 16%	Blue Sepharose CL-6B	Silica	None
DEAE Sepharose Fast Flow HC	Butyl Spherodex M	Sephadex G25 Medium	Red Sepharose CL-6B		
DEAE Spherodex M	Toyopearl Ether-650 M	Sephadex G75	Blue Spherodex M		
DEAE Spherosil M	Octyl Spherodex M	Trisacryl Plus GF 03 M	Baseline Blue Trisacryl M		
DEAE Trisacryl Plus M	Phenyl Spherodex M	Trisacryl Plus GF 10 M	Heparin Sepharose CL-6B		
Toyopearl DEAE-650 (M)	Phenyl Sepharose Fast Flow	Trisacryl Plus GF 20 M	Heparin Spherodex M		
Fractogel EMD DMAE-650 (M)	Toyopearl Phenyl-650 M	Sephacryl S-100HR	Toyopearl AF-Chelate-650 M (Copper)		
Fractogel EMD DEAE-650 (M)	Bakerbond Hi-Propyl	Sephacryl S-200HR	Toyopearl AF-Chelate-650 M (Zinc)		
Q Sepharose Fast Flow		Toyopearl HW-50F			
QMA Spherodex M		Toyopearl HW-55F			
QMA Spherosil M					
QMA Trisacryl Plus M					
Toyopearl QAE-550 C					
SP Sepharose Fast Flow					
SP Sepharose High Performance					
SP Sepharose Big Bead					
SP Trisacryl Plus M					
Toyopearl SP-650 M					
Fractogel EMD SO_3-650 M					

Chromatography Parameters Investigated

- Buffer composition
- Load concentration
- Load volume
- Flow rate
- Pool size
- Linear gradient
- Stepped gradient
- Multiple cycles
- Regeneration buffers
- Backflushing
- Support and linkage stability
- Nonspecific adsorption
- Ligand density
- Resin particle size
- Resin poor size
- Column geometry
- Column design
- Temperature

*Sepharose, Sephadex, and Sephacryl resins are from Pharmicia. Spherodex, Spherosil, and Trisacryl resins are from Sepracor Inc. Toyopearl resins are from Tosol Iaas. Fractogel resins are from E. Merck. Bakerbond resins and silicas are from J. T. Baker.

Figure 8.53 Collage of chromatography options for protein pharmaceuticals, and criteria used to judge suitability of chromatography materials. (Adapted from Lawlis et al., 1993, pp. 724, 726, LC-GC, copyright 1993.)

purification, and final polishing steps as proposed by Belter et al. (1988). Further expansion of this approach is necessitated by the increased options and number of bioseparation methods that have become available since then. An organizational method for communicating process concepts using an indexing system that would assist communications among members of an interdisciplinary team of process developers is proposed here.

Bioseparation sequences follow a series of goals that meet the broad objective of producing a biologically active molecule of defined and reproducible purity, starting from a complex mixture of components found in a broth or culture fluid. The individual steps in a sequence are based on laboratory results obtained early in the product's development stage when product isolation and characterization, rather than process development, is the goal. While each of the individual steps of a laboratory procedure can appear to be straightforward, their assembly into a manufacturing process can result in sequences whose complexity only becomes apparent as the first flowsheets of the process are developed. Once the major types of chromatography steps are identified, there are numerous types of stationary phases from which to select an appropriate material, and at least 18 chromatography parameters to be considered. This is illustrated in Figure 8.53, which was assembled from a description of industrial perspectives on chromatography of biological molecules (Lawlis and Heinsohn, 1993). An attempt is made here to organize the many elements that go into the selection of separations methods and then to define an approach that enables a systematic development of separation strategies into a conceptual framework.

Development of a list showing a proposed sequence of bioseparation steps could prove useful to a research/process development team for communicating the flow of product as it moves through a plant. A conceptual flowsheet of this type would also help all members of such a team to gauge how separation steps, invented or selected and then developed in the laboratory, affect process complexity, robustness, and ultimately cost. Most important, scientists, engineers, and business managers would better understand how their efforts affect their colleague's options in attaining the common goal of being the first to market with a new product. An organizational method for generating a preliminary conceptual purification strategy early in the process development cycle could serve as a benchmark against which process development results can be compared. The SEP-OPS method organizes a separation sequence or process based on category (reflects separation goal), class (gives type of unit operation or method), and method code (gives specific equipment or technique that a method might use).

BIOSEPARATIONS PROCESS DEVELOPMENT

8.58 DOWNSTREAM PROCESSING REQUIRES MULTIPLE PURIFICATION STEPS BASED ON DIFFERENT MOLECULAR PROPERTIES

The five basic methods of chromatography—ion exchange, reversed phase, size exclusion, hydrophobic interaction, and affinity—are all capable of delivering a significant increase in purity in one step. These are referred to as downstream processing because they follow the biotransformation or product generation step (referred to as an upstream step). All of the chromatography methods, except for size exclusion, require that chemical reagents (buffers, salts, or displacing components) be added to the mobile phase to obtain the separation. Consequently multiple purification methods are needed, if only to remove the purifying agents, introduced in a previous purification step, from the target molecule. While the same type of separation method could be used in sequence (e.g., two anion exchange steps carried out one after the other), this is unlikely, for technical *and* regulatory guidelines recommend that preceding and succeeding steps be based on a different principle than the central purification step.

TABLE 8.16 Objective Codes for Major Categories of Bioseparations (Indexing System)

Separation Category Number	General Category	Objective Code: Two-Digit Number (Category · Separation Goal)	Separation Goal Number	Description of Separation Goals	Reason for this step
0	Disruption	0.1	1	Rupture cell wall	Release intracellular product from cells
1	Solid/Liquid	1.1	1	Cells from liquid broth	Recover and concentrate cells where cells contain intracellular product
		1.2	2	Cell debris from liquid broth	Recover broth or culture fluid which contains soluble protein or insoluble inclusion protein bodies
		1.3	3	Inclusion bodies from liquid broth	Recover inclusion bodies
		1.4	4	Colloidal materials from clarified broth or liquid	Remove particulates which would plug chromatography column
2	Wash	2.1	1	Broth components from cells	Minimize carryover of contaminating broth components or maximize recovery of product in liquid adhering to cells
		2.2	2	Broth components from inclusion bodies	Minimize carryover of contaminating reagents or broth components
		2.3	3	Broth from cell debris	Maximize recovery of protein product adhering to cells
		2.4	4	Dissolved product with liquid	Remove low molecular weight contaminants or replace one soluble component (such as a buffer) with another (i.e., buffer exchange)
3	Change in conformation or structure	3.1	1	Dilution of chaotropes (protein refolding)	Dilution (or partial removal) of chaotropic agent induces protein refolding, leading to an active or proactive form of protein
		3.2	2	Hydrolytic modification	Remove affinity tail or leader sequence from protein

(continues)

TABLE 8.16 Continued

Separation Category Number	General Category	Objective Code: Two-Digit Number (Category · Separation Goal)	Separation Goal Number	Description of Separation Goals	Reason for this Step
4	Separation of soluble macromolecules (proteins and DNA)	4.1	1	Protein from DNA	DNA clearance or DNA recovery
		4.2	2	Protein from low molecular weight (MW < 1000)	Desalts protein or recovers low molecular weight product
		4.3	3	Small protein from large protein	Separates product from other proteins based on difference in size
		4.4	4	Small protein from large protein	Separates product from similarly sized proteins with different molecular weights and amino acid composition, or from proteins with identical amino acid sequence but different conformation
		4.5	5	Large protein from large protein	Same as objective code 4.4 except for high molecular weight proteins
		4.6	6	Protein from variant of itself	Separates product protein from nearly identical macromolecules that differ by as little as one amino acid
5	Separation of soluble small (low molecular weight) molecules M_r < 1000)	5.1	1	Small molecule from cytoplasmic, broth, or culture fluid constituents	Recovers and concentrates the product while removing other broth components (DNA, cells, cell debris)
		5.2	2	Small molecule from other small molecules	Separates product from other molecules of disimilar molecular weight or chemical properties (charge, solubility, dipole moment)

TABLE 8.16 Continued

Separation Category Number	General Category	Objective Code: Two-Digit Number (Category · Separation Goal)	Separation Goal Number	Description of Separation Goals	Reason for this Step
		5.3	3	Small molecule from variants or chemical isomers of itself	Separates product from molecules of similar molecular weight or chemical properties
		5.4	4	Small molecule from stereoisomers of itself	Separates of *R*- and *L*- forms of the same molecule, which may have different reactivity or biological function (chiral compounds)
6	Change of phase (dissolution, vaporization)	6.1	1	Dissolution or protein inclusion body	Dissolves proteins (away from nonprotein constituent, if possible) in preparation for recovery and refolding
		6.2	2	Product precipitation	Concentrates and recovers solubilized partially purified product
		6.3	3	Product crystallization	Purifies and concentrates purified product from solution containing similar soluble constituents
		6.4	4	Two-phase extraction	Removes soluble protein from insoluble cellular components
		6.5	5	Distillation	Purifies volatile fermentation product from water and other constituents by difference in relative volatility
		6.6	6	Freeze concentration	Concentrates very dilute product into a readily recovered smaller volume if product is particularly sensitive to degradation at temperatures above ambient and needs to be kept in contact with solvent
		6.7	7	Evaporation	Concentrates dissolved nonvolatile product (usually dissolved in water) by evaporating the solvent
		6.8	8	Lyophilization	Concentrates and recovers heat-labile product in powdered form

8.59 SEPARATION GOALS DEFINE BIOSEPARATION SEQUENCES

Separation goals can be organized in terms of general separation categories as summarized in Table 8.16 together with objective codes that further sub-divide the categories into specific separation goals. An index or category number in column 1 is accompanied by a two-digit objective code. Analysis of many laboratory purification procedures, and a limited number of process case studies, indicate that individual bioseparation steps will fall into one of the categories: disruption (category 0), solid/liquid separation (1), washing (2), change in conformation (3), separation of soluble macromolecules (4), separation of small molecules (5), and separations based on a changes in phase (6). The categories assigned the lower numbers (0, 1, 2) would be expected to occur early in a bioseparation process, while those with higher values (3, 4) would generally come later. Separations based on a change in phase (category 6), however, could be found early in a process (e.g., aqueous 2 phase extraction of a fermentation broth; Forciniti et al., 1991, or distillation) or late in a purification where lypholization is carried out (Liu et al., 1991). There are a series of separation goals within each category which further delineate each category into subcategories at separation goals as indicated in Table 8.16. Table 8.17 indexes separation methods into 8 different classes and subdivides these classes by the methods or equipment associated with each method. The combination of separation categories (represented by objective codes) and separation methods (represented by method codes) results in a unique numerical identification of each separation step. The numerical scheme enables the tracking of a sequence of purification steps with the aid of a spreadsheet.

8.60 SPECIFICATION OF SEPARATION GOALS FACILITATES SELECTION OF SEPARATION METHODS

Separation classes and methods for bioproducts are listed in Table 8.17. Separation classes differ from separation categories since classes represent specific pieces of equipment or methods—unlike categories that state a separation goal that can be achieved by different methods. There are 8 major classes as indicated, which are built upon suggestions of Belter et al. (1988). Each class of separation is further divided into subclasses of the method. For example, the separation class for cell lysis is 0 while the specific method or equipment (a pressure press, in this case) is denoted by 2. The indexing system of Table 8.17 has been arranged so that the more complex or operationally more difficult separation methods have a higher separation class number. While there are some areas of discrepancy (membrane chromatography could be viewed as being more complex than size exclusion), these follow the sequence typically observed in laboratory and industrial practice. Hence, size exclusion chromatography (6.1) is rated as less complex than ion exchange (6.2), which requires salt gradients, but is more involved than a filtration system for large particles (2.1) where control of longitudinal dispersion is not an issue and pressure drop is less likely to be a factor than in a gel permeation chromatography column.

Another example is concentration (separation class 7 in Table 8.17) which is ranked here as being more complex than chromatography. Concentration usually follows partial or total purification of the product. Hence, extremely high yields are required, product lability must be considered, sterility maintained, and large volumes of salts and/or solvents, concentrated in a small area of the plant, must be handled safely and in an environmentally compatible manner. Water removal is ranked as a most difficult method since this unit operation requires relatively complex equipment and is subject to the same considerations as separation class 7.

TABLE 8.17 Summary of Separation Methods (Indexing System)

Method			Specific Method or Equipment	
Separation Class	Description	Method Code: Two-digit Number (Separation Class. Specific Method)	Specific Method Number	Description
0	Cell lysis	0.1	1	Grinding mill
		0.2	2	Pressure press
		0.3	3	Enzyme lysis
1	Centrifuge	1.1	1	Disk
		1.2	2	Nozzle
2	Filtration (solids/liquids)	2.1	1	Large particle
		2.2	2	Microfiltration
3	Dissolution/dilution	3.1	1	Dissolving protein precipitates by chaotrope addition (Guanidine Hcl, dithiothreital)
		3.2	2	Add water or buffer to protein solution
4	Extraction	4.1	1	Aqueous two phase
		4.2	2	Organic/aqueous
5	Membrane (molecular fractionation)	5.1	1	Ultrafiltration
		5.2	2	Reverse osmosis
		5.3	3	Chromatography
6	Chromatography	6.1	1	Size exclusion
		6.2	2	Ion exchange
		6.3	3	Reversed phase
		6.4	4	Hydrophobic interaction
		6.5	5	Affinity
7	Concentration	7.1	1	Precipitation by cooling
		7.2	2	Precipitation by salt addition
		7.3	3	Precipitation by organic addition
		7.4	4	Settling tank
8	Water removal	8.1	1	Lyophilizer
		8.2	2	Driers
		8.3	3	Triple effect evaporator
		8.4	4	Distillation columns
9	Hydrolytic modification	9.1	1	Proteolytic enzymes
		9.2	2	CNBr
		9.3	3	Hydroxylamine
		9.4	4	Other catalysts

The categories in Table 8.16 and classes in Table 8.17 are presented for bioproducts (and particularly proteins) which are heat sensitive, susceptible to a change in structure and function during their purification, and must be recovered in high yields in a single pass. Unlike some chemical products, reworking of off-specification product is unlikely to be an option in the manufacture of biotherapeutic proteins and many types of other bioproducts. Hence, purity and recovery must be achieved in a single pass. The ranking of the difficulty for the different classes and methods in Table 8.17 may be different for the purification, recovery, and packaging of chemicals, or some types of low molecular weight fermentation products when compared to protein products.

Adjustments and modifications of this scheme will be appropriate, if not necessary, as different processes are considered and new bioseparation technologies developed. The framework of presented in Tables 8.16 and 8.17 is intended to provide a starting point for a system that is amenable to being adapted to spreadsheet or project scheduling software. Capital costs of the individual unit operations may be estimated using scaling rules, (Sections 6.2, 6.4, 6.5, 6.18, and 7.14 for chromatography rules) if the volumes and product concentrations of the various streams are known. While most of the necessary cost correlations still need to be developed or published, the outline presented in Table 8.17 may prove to be a useful starting point for organizing the development of a first cost estimate.

8.61 SEPARATION STRATEGIES COMBINE CATEGORIES, OBJECTIVES, AND METHODS INTO A LOGICAL SEQUENCE OF PURIFICATION STEPS: EXAMPLE OF SEP-OPS METHOD

The sequence of objectives given in the example presented as Table 8.18 might be proposed, if the goal were to recover inclusion bodies from a recombinant bacterial fermentation broth and obtain the protein in a purified and active form. Instead of a written description, however, the separation goals are given as numbers taken from Table 8.16. Hence the first step, which might be the recovery of cells containing inclusion bodies from the broth, is given the objective code of 1.1 as shown in the first row, second column of Table 8.18. The other objectives are specified in the same manner. Examination of the objective codes obtained from Table 8.16 and listed in Table 8.18 gives a first indication whether an orthogonal sequence of separations has been proposed. If the value of the number to the left of the decimal point for two succeeding rows is different, the principle of the separation is different (e.g., the zero in 0.1 denotes a method of disruption while the one in 1.2 is a solid/liquid separation). If the first number in the index is the same, but the second number differs in adjacent rows, for example, the two in 4.2 and the one in 4.1 in rows 7 and 8, the nature of the sample is the same (i.e., macromolecules dissolved in a liquid) but the objectives are different in each step (i.e., separation by size in row 7 vs. separation by charge in row 8).

Once the separation goals are given, the list can be examined for obvious discontinuities in the order of the separation steps. Once method codes are selected from Table 8.17 and listed with the appropriate separation goals in Table 8.18, further comparison of separation steps is possible. For example, it is unlikely that dissolution of protein in an inclusion body (method code 3.1) will precede disruption of the cell wall (method code 0.1). This list should also indicate if the sequence of numbering is appropriate and intentional (e.g., centrifugation—method 1.1 which precedes disruption of cell wall—method 0.1). Alternately the need for an additional step might be indicated. For example, packed beds associated with liquid chromatography are subject to plugging if particulates or colloidal matter are present in the feed stream. Hence, if these must be removed, centrifugation or filtration would need to precede step 7. An objective

TABLE 8.18 Separation Objectives Planning and Strategy Method (SEP-OPS™): Example of a Separation Strategy. (The recovery of inclusiuon bodies from a recombinant bacterial broth and purification of protein from fermentation broth)

	From Table 8.16	From Table 8.17		From Table 8.17 (Method Code)		Resulting Streams Are:				Will Require Another Separation Step for Removal of Separating Agent	
Step	For Objective Code of:	Class of Method Is:	For Separation Based on Difference in:	Method Code - Method Is of Type:	For Input Stream of:	1	2	3	Separating Agent Added to Product Is:	Yes	No
1	1.1	1	Particle size, density	1.1	Cell broth	Cell paste	Supernatant	—	Mechanical force		•
2	0.1	0	Product partioning in biological material	0.2	Cell paste	Inclusion bodies + cell debris	—	—	Pressure (mechanical force)		•
3	1.2	1	Solubility	1.1	Inclusion bodies + cell debris	Inclusion bodies	Cell debris	—	Mechanical force		•
4	2.2	2	Solubility	2.2	Inclusion bodies	Inclusion bodies	Buffer + water	—	Buffer	•	
5	6.1	3	Solubility	3.1	Inclusion bodies	Dissolved protein + residues	—	—	Chaotrope	•	
6	3.1	3	Not applicable (change in conforma-tion)	3.2	Dissolved proteins + residues	Active (dissolved) proteins + residues	—	—	Water or buffer	•	

(continued)

TABLE 8.18 Continued

	From Table 8.16	From Table 8.17		From Table 8.17 (Method Code)		Resulting Streams Are:				Will Require Another Separation Step for Removal of Separating Agent	
Step	For Objective Code of:	Class of Method Is:	For Separation Based on Difference in:	Method Code - Method Is of Type:	For Input Stream of:	1	2	3	Separating Agent Added to Product Is:	Yes	No
6 (continued)	*1.4*	*2*	*Solubility*	*2.2*	*Colloidal particles + solution*	*Clarified solution*	*Colloidal particles*	—	*Mechanical force*		•
7	4.2	5 or 6	Molecular size	5.1 or 6.1	Active dissolved proteins + residues	Protein + some salts	Salt in buffer	—	Water or buffer	•	
8	4.1	6	Charge	6.2	Protein + salts	Protein + salts	DNA in buffer	Eluting buffer	Salt in buffer as a part of gradient	•	
9	4.4	6	Charge	6.2	Protein + salts	Purified protein + salts	Protein contaminant	Eluting buffer	Salt in buffer	•	
10	4.2	5 or 6	Size	6.1	Purified protein + salts	Purified protein in buffer	Salt contaminant	Eluting buffer	Final dissolving buffer		•

code of 1.4 would then be inserted between steps 6 and 7. Here Objective code 3.1 (step 6) in Table 8.18 is followed by 1.4 (shown in italics to indicate that it is inserted). Other appropriate methods and method codes might also be selected or inserted.

Table 8.18 is organized to identify resulting streams, the separating agents used, and the need for additional steps to remove the separating agent (right-hand side of Table 8.18). The step corresponding to the objective code 1.1 (step 1) results in two streams (cell paste and supernatant), and the separating agent or driving force is the mechanical force used in a centrifuge. The separating agent does not have to be subjected to another separation step for its removal, other than perhaps cooling. In comparison, objective code 4.1 (step 8 in Table 8.18) requires separation method 6 (liquid chromatography), which will generate 3 major streams (proteins and salts, DNA, and eluting buffer) with a separating agent, salt, being added (i.e., for the salt gradient) to achieve the desired fractionation in this step. The separating agent requires removal, as does the salt added in the next step (objective code 4.4 in step 9) and hence leads to a *subsequent* step (step 10 in Table 8.18). This is specified by objective code 4.2, from Table 8.16, using method 6.1 from Table 8.17, which involves buffer exchange/removal by size exclusion chromatography. The purified protein is assumed to be in the final buffer after this point, hence requiring no further separation. However, the resulting streams of the contaminating salt and eluting buffer must still be handled and processed into environmentally compatible waste streams. All steps are, in theory, orthoganol if succeeding method codes are different. Table 8.18 clearly lists the process steps, and shows that there will be at least 10 process steps that utilize six different methods utilized to obtain the final product. A distinguishing feature of chemical separations is that they are often continuous while bioseparations—particularly of proteins as represented here—are batch processes.

8.62 BIOTECHNOLOGY PROCESS DEVELOPMENT DIFFERS FROM CHEMICAL INDUSTRY PROCESS DEVELOPMENT

Upstream production (i.e., reactor) technology can often be integrated with downstream product recovery during process design for chemical manufacturing processes and carried out long after the basic laboratory research has defined the synthetic (upstream) chemistry. In the design of chemical processes, the impact of trace impurities or co-products on final product quality, and the difficulty and expense of obtaining the product, can be simulated by computer calculations. This approach reflects a strong heuristic knowledge base developed in the chemical industry during the course of a century of engineering practice in separations that often exploit a single thermodynamic property to achieve final purity. Thermodynamic correlations facilitate rapid and accurate estimation of physical properties that affect the separation. Computer software organizes this knowledge so that preliminary integrated designs for chemical processes can be quickly generated. Further, chemical processing, and particularly fractionation steps, are often continuous. Bioprocesses are typically batch.

A batch fermentation system can give products whose characteristics change with respect to fermentation time, while a continuous process, in principle, gives a consistent product. However, a continuous (flow type) bioreactor is also susceptible to contamination by undesirable organisms. Once a competing organism gains entry, the bioreactor must be shut-down, cleaned, sterilized, and then restarted. The choice of the upstream processing conditions reflects the characteristics of the organism that generates the product. The perceived benefits of a change to improve downstream processing by changing upstream (bioreactor) processing must consider productivity of the upstream step (Ogez et al., 1989). Regardless of the specific sequence that makes up the upstream process, separation methods are, *de facto*, specified early in biotech-

TABLE 8.19 Characteristics of the Biotechnology Toolbox

Characteristic	Motivation
Versatility	Purifications are multiple step
Variety	Each step should target different molecular property (size, shape, charge, solubility, hydrophobicity, affinity)
Compatibility	Each step should fit the next
Benchmarked	Initial selection based on prior experience with known proteins
Robustness	Process fluids are chemically and/or biologically complex

nology product development and often adapted as processing steps—even if not optimal—in the race to be first to market.

The downstream purification strategies for new biotechnology products should, ideally, be integrated with the development of upstream production technology. At this point in time (2000) the biotechnology industry is still developing fundamental knowledge of the characteristic properties of proteins that would facilitate the a priori calculations of separation sequences. A biotechnology toolbox with the characteristics summarized in Table 8.19 would assist early selection of appropriate biomolecule purification methods (Ladisch, 1996). The ability to manipulate biological expression systems that might tune product properties to facilitate facile recovery while leaving final biological functions unaltered have been suggested by laboratory results. However, many of these approaches still need to be scaled up. Software that quickly simulates and analyzes the economic consequences of hypothesized changes in overall manufacturing approaches are not as well developed for biotechnology as for the chemical industry although this is beginning to change.

A common denominator for both the chemical and biotechnology industries is the challange presented by the environmental impacts of manufacturing methods and the products that are generated. Both industries are seeking to develop environmentally friendly production methods for their products. This is sometimes referred to as "green chemistry."

8.63 UPSTREAM PRODUCTION METHODS AFFECT DOWNSTREAM PURIFICATION STRATEGIES

Controllable factors that occur upstream in a biotechnology production process provide opportunities to improve processing. At the same time the extent to which such factors can be selected are a function of the type of product. The upstream factors include the choice of host, expression systems, fermentation, or cell culture media, and type of fermentation or cell culture process. The host may be either a procaryote such as *E. coli* or eucaryotes such as yeast, or mammalian, plant, or insect cells. The eucaryotic systems would be used for proteins that are large, contain numerous disulfide bonds, and/or are glycosylated and secreted into the fermentation media or culture fluid as soluble protein. *E. coli* results in recombinant proteins that are initially recovered as inclusion bodies. In the case of *E. coli* the product may be concentrated in the form of insoluble protein found in the inclusion or refractile body. This form of intracellular protein is readily recovered from cell debris formed by disruption of the cells. Secreted products, such as those obtained from yeast or mammalian cells, need to be concentrated early in the downstream processing or purification sequence.

Molecular biology could enhance product recovery and concentration by adding "handles" or "tails" that strongly adsorb or otherwise interact with several classes of stationary phases.

The secretion of a protein with a "tail" is an upstream step, while its recovery is part of the downstream processing sequence. Ogez et al. (1989) review the examples of the fusion of a target protein (in this case insulinlike growth factor) to the C-terminus of a protein A leader, and a second example for the attachment of a polyarginine tail to the target protein. In the first case, the protein A leader (connected to the fusion protein) would strongly bind to an immunoaffinity column consisting of an immobilized antibody. The other proteins would pass through the column, since these did not have a "tail" which would strongly bind with the F_c region of the antibody. Elution of the fusion protein, followed by removal of its protein A tail using hydrolysis with hydroxylamine, and separation of the target protein from the hydrolyzed tail would complete this sequence. A possible sequence is summarized in Table 8.20 using the separations objective/planning and strategic indexing method that was proposed in Sections 8.59 to 8.61. In the case of the polyarganine tail/fusion protein, the polyarginine bound strongly to a cation exchange resin. The fusion protein was then eluted, and the protein A digested with the enzyme carboxypeptidase B, to yield basic amino acids and other compounds that could then be removed using anion exchange resin, leaving the partially purified protein product.

The goal of maximizing the ratio of product to contaminants might also be addressed by limiting or eliminating fermentation or culture media supplements that contain proteins. Supplements added to the media complicate product recovery and purification by introducing contaminants that must later be removed.

Summary and Perspectives

A major part of the manufacturing processes for biotechnology products are the sequences of separation steps that recover such products from complex mixtures of biologically derived fluids. They result in readily identifiable and purified molecules or macromolecules having known and predictable properties. The first steps in designing an appropriate bioseparation sequence are to identify the biological systems from which the products are to be derived, to determine properties (or "handles") that differentiate the product from other and often similar co-products, and to apply the principles of known bioseparation processes to select methods and conditions for initiating experiments on their recovery and purification. While the costs of bioseparations are important, being the first to market is critical for many new biotechnology products—particularly for those for which there is no other means of production. However, when such a product is successfully introduced, competition will develop. Bioseparation economics then gain more attention. However, by the time this occurs the plant has been built, the separations techniques established, and manufacturing processes validated through industrial and governmental regulatory systems. Changes can be made, but the existing manufacturing and regulatory infrastructure must be satisfied and some of the time-consuming and expensive steps to obtain regulatory approval must be repeated.

Two case studies—biosynthetic human insulin (Sections 8.6 to 8.10) and tissue plasminogen activator (Sections 8.11 to 8.19)—were presented to illustrate some of these issues. The case studies illustrated the impact of the biological system (procaryote vs. mammalian cells), regulatory approaches, protein characteristics, and types of co-products on the development of a process for the purification sequence. The principles of the five major liquid chromatographic techniques—ion exchange, size exclusion, reversed phase, hydrophobic interaction, and affinity—were outlined to provide insights on how these methods facilitate separation, and when they might be used. The separation concepts were finally presented in the context of a framework for organizing separation objectives and systematically specifying the sequence of purification methods for the purpose of developing a conceptual flowsheet. Several examples showed how the development of tables based on the separations objectives planning and strat-

TABLE 8.20 Separation Objectives Planning and Strategic Method: Example of a Separation Strategy Affinity Chromatography Separation Sequence

	From Table 8.16		From Table 8.17		Resulting Streams				Will Require Another Separation Step for Its Removal	
Step	Objective Code	Separation Goal	Method Code	Method	1	2	3	Separating Agent Added to Product Is:	Yes	No
1	1.1	Cells from broth	1.1	Disk centrifuge	Broth	Cells	—	Mechanical energy		•
2	4.2	Fermentation components from broth	5.1	Ultrafiltration	Proteins	Salts	—	Mechanical energy		•
3	4.3	Small protein from other (larger) proteins	6.5	Affinity chromatography	Target protein	Other proteins	Eluting salts	Eluting (low molecular weight) component	•	
4	3.2	Hydrolysis to remove protein A tail	9.3	Hydrolysis with hydroxylamine	Hydrolyzed protein A/insulinlike growth factor			Hydroxylamine	•	
5	4.4	Separate tail from insulinlike growth factor	6.2	Ion exchange chromatography	Insulinlike growth factor	Peptides	Salts, hydroxylamine, low molecular weight components	Salts	•	
6	4.2	Desalting	6.1	Size exclusion chromatography	Insulinlike growth factor	Peptides, salts, low molecular weight components		Final buffer		•

Note: The protein A tail is fused to an insulinlike growth factor.

egy method (SEP-OPS™) could provide an overview of the overall sequence of purification steps that would make up a bioseparations process.

The identification of individual separation steps that are likely to achieve the desired protein fractionation is essential to being able to define objectives and initiate strategic planning, much like a basketball coach must understand the capabilities of individual players when assembling a winning team. The key players in chromatography are the target molecule, stationary phase, and mobile phase. Their interactions, and therefore the ability of a stationary/mobile phase combination to purify the target molecule, is characterized by the distribution or partition coefficient in size exclusion chromatography, and capacity factors for reversed phase, ion exchange, hydrophobic interaction, and affinity chromatography.

The defining characteristic of the chromatography system is the target molecule's distribution coefficient or capacity factor that must be significantly different from the other molecules. In principle once the conditions are defined that result in such a difference, the separation is possible. Efforts to improving efficiency and throughput of each step, and the staging of one step with the next, are then justified. This chapter gave definitions of partition factors, and presented the rationale by which stationary phases are selected, and mobile phase compositions are defined, relative to the basic properties that describe the target molecule. Chapter 9 will present cases for molecular biology's impact on the future development of stationary phases involving very large capacity factors for target molecules and low capacity for all other impurities. A large difference in the capacity factors between product and co-products reduces the dependence of proper column operation on efficiency, and can lead to better scale-up characteristics. Particle size, mobile phase velocity, and column length affect resolution, as well, and these are related to column design through plate height concepts. The principles presented in this chapter were organized to enable a logical selection of separation methods. An appropriate selection of methods must proceed the synthesis of process sequences, bioseparations process development, and the engineering of individual chromatographic unit operations.

REFERENCES

Alberts, B., D. Bray, J. Lewis, M. Raff, K. Roberts, and J. D. Watson, 1989, *Molecular Biology of the Cell*, 2nd ed., Garland Publishing, New York.

Asenjo, J. A., and E. W. Leser, 1996, Process integration in biotechnology, in *Downstream Processing of Natural Products: A Practical Handbook*, M. Verrall, ed., Wiley, New York, 123–138.

Asenjo, J. A., and I. Patrick, 1990, Large-scale protein purification, in *Protein Purification Applications: A Practical Approach*, E. L. V. Harris and S. Angal, eds., IRL Press, Oxford, 1–28.

Atkinson, P. R., M. D. Hilton, and P. K. Lambooy, 1995, Purification and Preliminary Characterization of dDAP, a Novel Dipeptidylaminopeptidase from Dictyostelium discoideum, *Biochemistry*, 34, 10827–10834.

Baker, J. C., from Eli Lilly & Company, 1996, What is biosynthetic development? Guest Lecture in ABE 580 (intermediate level graduate course), Purdue University, April 17.

Barinaga, M., 1996, Finding new drugs to treat stroke, *Science*, 272(5262), 664–666.

Basak, S., and M. R. Ladisch, 1995, Correlation of electrophoretic mobilities of proteins and peptides with their physicochemical properties, *Anal. Biochem.*, 226, 51–58.

Belter, P. A., E. L. Cussler, and W.-S. Hu, 1988, *Bioseparations: Downstream Processing for Biotechnology*, Wiley-Interscience, New York.

Bernon, M., and J. Bodelle, 1991, Medical biotechnology: A brief historical survey, in *Drug Biotechnology Regulation*, Vol. 13, Y.-Y. H. Chin and J. L. Gueriguian, eds., Marcel Dekker, New York.

Bisbee, C. A., 1993, Current perspectives on manufacturing and scale-up of biopharmaceuticals, GEN, 13(14), 8–9.

Builder, S. E., and E. Grossbard, 1986, Laboratory and clinical experience with recombinant plasminogen activator, in *Transfusion Medicine, Recent Technological Advances*, 303–313.

Builder, S. E., and W. S. Hancock, 1988, Analytical and process chromatography in pharmaceutical protein production, *Chem. Eng. Progr.*, 84(8), 42–46.

Builder, S. E., R. van Reis, N. Paoni, and J. Ogez, 1989, Process Development and Regulatory Approval of Tissue-Type Plasminogen Activator, in *Proc. 8th Int. Biotechnology Symp.*, Paris (July 17–22).

Cantwell, A. M., R. Calderone, and M. Sienko, 1984, Process scale-up of (-lactam antibiotic purification by high performance liquid chromatography, *J. Chromatogr.*, 316, 133–149.

Carrey, E. A., 1990, Peptide mapping, in *Protein Structure: A Practical Approach*, T. E. Creighton, ed., IRL Press, Oxford, 117–144.

Chance, R. E., E. P. Kroeff, and J. A. Hoffmann, 1981a, Chemical, physical, and biological properties of recombinant human insulin, in *Insulin, Growth Hormone, and Recombinant Technology*, J. L. Gueriguian, ed., Raven Press, New York, 71–85.

Chance, R. E., E. P. Kroeff, J. A. Hoffmann, and B. H. Frank, 1981b, Chemical, physical, and biologic properties of biosynthetic human insulin, *Diabetes Care*, 4(2), 147–154.

Chicz, R. M., and F. E. Regnier, 1989, Single amino acid contributions to protein retention in cation-exchange chromatography: Resolution of genetically engineered subtilisin variants, *Anal. Chem.*, 61, 2059–2066.

Chicz, R. M., and F. E. Regnier, 1988, Surface-mediated retention effects of subtilisin type site-specific variants in cation exchange chromatography, *J. Chromatogr.*, 443, 193–203.

Colin, H., P. Hilaireau, and J. de Tournemire, 1990, Dynamic axial compression columns for preparative high performance liquid chromatography, *LC-GC*, 8(4), 302–312.

Committee on Bioprocess Engineering, National Research Council, 1992, *Putting Biotechnology to Work: Bioprocess Engineering*, Nat. Acad. Sci., Washington, DC, 2–22.

Dolan-Heitlinger, J., 1981, *Recombinant DNA and Biosynthetic Insulin*, Eli Lilly & Co., Indianapolis, IN.

Dove, G. B., G. Mitra, G. Roldan, M. A. Shearer, and M.-S. Cho, 1990, Purification alternatives for IgM (human) monoclonal antibodies, in *Protein Purification: from Molecular Mechanisms to Large-Scale Processes*, M. R. Ladisch, R. C. Willson, C.-D. C. Painton, and S. E. Builder, eds., Am. Chem. Soc., Washington, DC, 194–209.

Eriksson, K.-O., 1989, Hydrophobic interaction chromatography, in *Protein Purification: Principles, High Resolution Methods, and Application*, J.-C. Janson and L. Ryden, eds., VCH, New York, 207–226.

Fischer, L., 1980, Laboratory techniques in biochemistry and molecular biology, in *An Introduction to Gel Chromatography*, North-Holland, Amsterdam.

Flanigan, E., from Rhône Poulenc Rorer 1991, High performance liquid chromatography in the production and quality control of salmon calcitonin, in Purdue University Workshop on Chromatographic Separations and Scale-up, 207.

Forciniti, D., C. K. Hall, and M. R. Kula, 1991, Protein partitioning at the isoelectric point: Influence of polymer molecular weight and concentration and protein size, *Biotechnol. Bioeng.* 38(9), 986–994.

Fürst, I., and R. Unterhuber, 1996, First German recombinant drug ok'd by FDA and EMEA, *GEN*, 16(20), 1, 35.

Garnick, R. L., M. J. Ross, and R. A. Baffi, 1991, Characterization of proteins from recombinant DNA manufacture, in *Drug Biotechnology Regulation Scientific Basis and Practices*, Y.-Y. H. Chiu and J. L. Gueriguian, eds., Marcel Dekker, New York, 263–313.

Gustafson, R. L., R. L. Albright, I. Heisler, J. A. Lirio, and O. T. Reid, 1968, Adsorption of organic species by high surface area styrene-divinylbenzene copolymers, *Ind. Eng. Chem., Prod. R&D*, 7(2), 107–115.

Hagel, L. 1989, Gel filtration, in *Protein Purification: Principles, High Resolution Methods and Applications*, J.-C. Janson, and L. Ryder, eds., VCH, New York, 63–106.

Hamaker, K. H., J. Liu, R. J. Seely, C. M. Ladisch, and M. R. Ladisch, 1996, Chromatography for rapid buffer exchange and refolding of secretory leukocyte protease inhibitor, *Biotechnol. Progr.*, 12, 184–189.

Harakas, N. K., J. P. Schaumann, B. T. Connolly, A. J. Wittwer, J. V. Orlander, and J. Feder, 1988, Large-scale purification of tissue-type plasminogen activator from cultured human cells, *Biotechnol. Progr.*, 4(3), 149–158.

Hearn, M. T. W., 1989, High resolution reversed phase chromatography, in *Protein Purification: Principles, High Resolution Methods, and Applications*, J.-C. Janson and L. Ryden, eds., VCH, New York, 175–206.

Ho, S. V., 1990, *"Strategies for Large-Scale Protein Purification:" Protein Purifications From Molecular Mechanisms to Large-Scale Processes*, M. R. Ladisch, R. C. Willson, C.-D. C. Painton and S. E. Builder, eds., Am. Chem. Soc., Washington, DC, 14–34.

Janson, J.-C., and P. Hedman, 1982, Large-scale chromatography of proteins in advances in biochemical engineering, *Chromatogr.*, 25, 43–99.

Johnson, R. D., and F. H. Arnold, 1995a, Multipoint binding and heterogeneity in immobilized metal affinity chromatography, *Biotechnol. Bioeng.*, 48, 437–443.

Johnson, R. D., and F. H. Arnold, 1995b, The Tempkin model describes heterogeneous protein adsorption, *Biochem. Biophys. Aeta* 1247, 293–297.

Karch, K., I. Sebastian, and I. Halasez, 1976, Preparation and properties of reversed phases, *J. Chromatogr.*, 122, 3–16.

Karlsson, E., L. Rydén, and J. Brewer, 1989, Ion exchange chromatography, in *Protein Purification: Principles, High Resolution Methods, and Applications*, J.-C. Janson and L. Rydén, eds., VCH, New York, 107–148.

Kopaciewicz, W., M. A. Rounds, J. Fausnaugh, and F. E. Regnier, 1983, Retention model for high performance ion-exchange chromatography, *J. Chromatogr.*, 266, 3–21.

Kroeff, E. P., R. A. Owens, E. L. Campbell, R. D. Johnson, and H. I. Marks, 1989, Production scale purification of biosynthetic human insulin by reversed-phase high-performance liquid chromatography, *J. Chromatogr.*, 461, 45–61.

Kurnik, R. T., A. W. Yu, G. S. Blank, A. R. Burton, D. Smith, A. M. Athalye, and R. van Reis, 1995, Buffer exchange using size exclusion chromatography, countercurrent dialysis, and tongential flow filtration: Models development and industrial application, *Biotechnol. Bioeng.* 45, 149–157.

Ladisch, M. R., 1996, Building the biotechnology toolbox: Process chromatography of the future, *Purdue University Chromatography Workshop'96*, Purdue University, W. Lafayette, IN, p. 31.

Ladisch, M. R., 1987, Separation by sorption, in *Advanced Biochemical Engineering*, H. R. Bungay, G. Belfort, eds., Wiley-Interscience, New York, 219–237.

Ladisch, M. R., and K. L. Kohlmann, 1992, Recombinant human insulin, *Biotechnol. Prog.*, 8(6), 469–478.

Ladisch, M. R., R. L. Hendrickson, and K. L. Kohlmann, 1990, Anion exchange stationary phase for (-galactosidase, bovine serum albumin, and insulin, in *Protein Purification: From Molecular Mechanisms to Large-Scale Processes*, M. R. Ladisch, R. C. Willson, C.-D. C. Painton, and S. E. Builder, eds., Am. Chem. Soc., Washington, DC, 93–103.

Ladisch, M. R., M. Voloch, and B. Jacobson, 1984, Bioseparations: Column design factors in liquid chromatography, *Biotechnol. Bioeng. Symp.* 14, 525–541.

Lawlis, V. B., and H. Heinsohn, 1993, Industrial perspectives on the role of chromatography in the purification of biological molecules, *LC-GC* 11(10), 720–729.

Lin, J. K., B. J. Jacobsen, A. N. Pereira, and M. R. Ladisch, 1988, Liquid chromatography of carbohydrate monomers and oligomers, in *Methods in Enzymology*, 160, W. A. Wood and S. T. Kellogg, eds., Academic Press, San Diego, CA, 145–159.

Liu, W. R., R. Langer, and A. M. Klibanov, 1991, Moisture induced aggregation of lyphilized proteins in the solid state, *Biotechnol. Bioeng.*, 37(2), 177–184.

Morris, J., and J. S. Fritz, 1993, Separation of hydrophilic organic acids and small polar compounds using macroporous resin columns, *LC-GC*, 11(7), 513–517.

Mosher, R. A., D. Dewey, W. Thormann, D. A. Saville, and M. Bier, 1989, Computer simulation and experimental validation of the electrophoretic behavior of proteins, *Anal. Chem.*, 61, 362–366.

Ogez, J. R., J. C. Hodgdon, M. P. Beal, and S. E. Builder, 1989, Downstream processing of proteins: Recent advances, *Biotech. Adv.*, 7, 467–468.

Parikh, I., and P. Cuatrecasas, 1985, Affinity chromatography, *CEN*, 63, 17–32.

Perry, M., and H. Kirby, 1990, Example purifications: Monoclonal antibodies and their fragments, in *Protein Purification Applications: A Practical Approach*, E. L. V. Harris and S. Angal, eds., IRL Press, Oxford, 147–166.

Petrides, D. P., J. Calandranis, C. L. Cooney, 1996, Bioprocess optimization via CAPD and simulation for product commercialization, *GEN*, 16(16), 24, 28.

Pharmacia Fine Chemicals, 1979, *Gel Filtration—Theory and Practice*, Uppsala, Sweden.

Pharmacia Fine Chemicals, 1983, *Scale-up to Process Chromatography*, Uppsala, Sweden (May).

Pierri, G., P. Piccardi, G. Muratori, L. Cavalo, 1983, Scale-up for preparative liquid chromatography of fine chemicals, *La Chimica E L'Industria* 65(5), 331–336.

Pietrzyk, D. J., and C.-H. Chu, 1977, Amberlite XAD copolymers in reversed phase gravity flow and high pressure liquid chromatography, *Anal. Chem.*, 49(6), 757–764.

Pietrzyk, D. J., and J. D. Stodola, 1981, Characterization and applications of Amberlite XAD-4 in preparative liquid chromatography, *Anal. Chem.*, 53(12), 1822–1828.

Pollio, F. X., and R. Kunin, 1971, The use of macroeticular ion exchange resins of the fractionation and purification of enzymes and related proteins, *Chem. Eng. Symp. Ser.*, 67(108), 66–74.

Porath, J., and P. Flodin, 1959, Gel filtration: A method for desalting and group separation, *Nature*, 4676, 1657–1659.

Preneta, A. Z., 1989, Separation on the basis of size: Gel permeation chromatography, in *Protein Purification Methods: A Practical Approach*, E. L. V. Harris and S. Angal, eds., IRL Press, Oxford, 293–306.

Prouty, W. F., 1991, Production-scale purification processes, in *Drug Biotechnology Regulation*, 13, Y.-Y. H. Chien and J. L. Gueriguian, eds. Marcel Dekker, New York, 221–262.

Rawls, R., 1996, Small peptide can mimic erthyropoetin, *C&EN*, 74(32), 31.

Roe, S., 1989, Separation based on structure, in *Protein Purification Methods: A Practical Approach*, E. L. V. Harris and S. Angal, eds., IRL Press, Oxford, 200–216.

Roettger, B. F., and M. R. Ladisch, 1989, Hydrophobic interaction chromatography, *Biotechnol. Adv.*, 7, 15–29.

Roettger, B. F., J. A. Myers, M. R. Ladisch, and F. E. Regnier, 1990, Mechanisms of protein retention in hydrophobic interaction chromatography, in *Protein Purification: From Molecular Mechanisms to Large-Scale Processes*, M. R. Ladisch, R. C. Willson, C.-D. C. Painton, and S. E. Builder, eds., Am. Chem. Soc., Washington, DC, 80–92.

Roettger, B., J. Myers, M. Ladisch, and F. Regnier, 1989, Adsorption phenomena in hydrophobic interaction chromatography, *Biotechnol. Progr.*, 5, 79–88.

Roush, W., 1996, Biotech finds a growth industry, *Science* 273(5273), 300–301.

Scopes, R. K., 1981, Quantitative studies of ion exchange and affinity elution chromatography of proteins, *Anal. Biochem.*, 114, 8–18.

Sedek, P. C., P. W. Carr, R. M. Doherty, M. J. Kamlet, R. W. Taft, and M. H. Abraham, 1985, Study of retention processes in reversed-phase high-performance liquid chromatography by the use of the solvatochromic comparison method, *Anal. Chem.*, 57, 2971–2978.

Snyder, L. R., and J. J. Kirkland, 1979, *Introduction to Modern Liquid Chromatography*, 2nd ed., Wiley-Interscience, New York, 260–289.

Sobel, S., 1991, Regulatory evaluation of biotechnology drugs: Current trends in the US, in *Drug Biotechnology Regulation Scientific Basis and Practices*, Y.-Y. H. Chiu and J. L. Gueriguian, eds., Marcel Dekker, New York, 499–511.

Sofer, G. K., and L. E. Nystrom, 1989, *Process Chromatography: A Practical Guide*, Academic Press, London, 128–129.

Takayanagi, H., J. Fukuda, and E. Miyata, 1996, Non-ionic adsorbents in separation processes, in *Downstream Processing of Natural Products: A Practical Handbook*, M. Verrall, ed., Wiley, Chichester, 159–178.

Thayer, A. M., 1996a, Biotech stocks strong in 1995 despite lackluster earnings, *CEN*, 74(1), 22–23.

Thayer, A. M., 1996b, Market investor attitudes challenge developers of biopharmaceuticals, *CEN*, 74(33), 13–21.

Velayudhan, A., and M. R. Ladisch, 1995, Effect of modulator sorption on gradient shape in ion exchange chromatography, *Ind. Eng. Chem. Res.*, 34(8), 2805–2810.

Velayudhan, A., and M. R. Ladisch, 1991, Role of modulator in gradient elution chromatography, *Anal. Chem.*, 63(18), 2028–2032.

Velayudhan, A., R. L. Hendrickson, and M. R. Ladisch, 1995, Simultaneous concentration and purification through gradient deformation in gradient elution chromatography, *AIChE J.*, 41(5), 1184–1193.

Watson, J. D., M. Gilman, J. Witkowski, and M. Zoller, 1992, *Recombinant DNA*, 2nd ed., Freeman, New York, 458–460.

Wells, J. A., 1996, Hormone mimicry, *Science*, 273(5274), 449–450.

Wheelwright, S. M., 1991, *Protein Purification: Design and Scale up of Downstream Processing*, Hanser Publishers, Munich.

Willson, R. C., and M. R. Ladisch, 1990, Large-scale protein purification: Introduction, in *Protein Purification: From Molecular Mechanisms to Large-Scale Processes*, ACS Symp. Ser. 427, M. R. Ladisch, R. C. Willson, C.-C. Painton, and S. E. Builder, eds., Am. Chem. Soc., Washington, DC, 1–13.

Yau, W. W., and D. D. Bly, 1980, in *Size Exclusion Chromatography (GPC)*, T. Provder, ed., Am. Chem. Soc. Washington, DC, 197.

SUGGESTED READINGS

Anonymous, 1996, Huge biotech harvest is a boon for farmers—and for Monsanto, *Wall Street J.*, A1, A8 (October 24).

Anonymous, 1994, Bovine hormone sales booming despite attacks, *CEN*, 72(17), 8.

Arnold, F. H., and B. L. Haymore, 1991, Engineering metal-binding proteins: Purification to protein folding, *Science*, 252, 1796–1797.

Bailon, P., and D. V. Weber, 1988, Receptor affinity chromatography, *Nature*, 335(6193), 839–840.

Bailon, P., D. V. Weber, M. Gately, J. E. Smart, H. Lorberboun-Galski, D. Fitzgerald, and I. Pastan, 1988, Purification and partial characterization of an interleukin-2-pseudomanas exotoxan fusion protein, *Biotechnol.*, 7, 1326–1329.

Bailon, P., D. V. Weber, R. F. Keeny, J. E. Fredericks, C. Smith, P. C. Familletti, and J. E. Smart, 1987, Receptor-affinity chromatography: A one-step purification for recombinant interleukin-2, *Bio/Technol.*, 5, 1195–1198.

Carter, P., 1990, Site specific proteolysis of fusion proteins, in *Protein Purification: From Molecular Mechanisms to Large-Scale Processes*, M. R. Ladisch, R. C. Willson, C.-D. C. Painton, and S. Builder, eds., Am. Chem. Soc., Washington, DC, 181–193.

Cavuco-Paulo, A., and L. Almeida, 1996, Cellulase activities and finishing effects, *Textile Chemist Colorist*, 28(6), 28–32.

Chamov, S. M., and A. Ashkenazi, 1996, Immunoadhesions: Principles and applications, *TIBTECH*, 14, 52–60.

Din, N., H. G. Damude, N. R. Gilkes, R. C. Miller, R. A. J. Warren, and D. G. Kilburn, 1994, C_1–C_X revisited: Intramolecular synergism in a cellulase, *Proc. Natl. Acad. Sci.*, 91, 11383–11387.

Gilkes, N. R., E. Jervis, B. Henrissat, B. Tekant, R. C. Miller, R. A. J. Warren, and D. G. Kilburn, 1992, The adsorption of a bacterial cellulases and its two isolated domains to crystalline cellulose, *J. Biological Chem.*, 267(10), 6743–6749.

Gulati, M., P. J. Westgate, M. Brewer, R. Hendrickson, and M. R. Ladisch, 1996, Sorptive recovery of dilute ethanol for distillation column bottoms stream, *Appl. Biochem. Biotechnol.*, 57158, 103–119.

Hamilton, J. O'C. 1996, A miracle drug's second coming, *Business Week*, 3478, 118, 122.

Haymore, B. L., G. S. Bild, W. J. Salsgiver, N. R. Staten, and G. G. Krivi, 1992, Introducing strong metal-binding sites onto surfaces of proteins for facile and efficient metal-affinity purifications, in *Methods: A Companion to Methods in Enzymology*, 4, 25–40.

Ladisch, M. R., 1998, Bioseparations, in Kirk Othmer *Encyclopedia of Chemical Technology*, Wiley, New York.

Ladisch, M. R., K. W. Lin, M. Voloch, and G. T. Tsao, 1983, Process considerations in the enzymatic hydrolysis of biomass, *Enz. Microb. Technol.*, 5, 82–102.

Ladisch, C. M., Y. Yang, A. Velayudhan, and M. R. Ladisch, 1992, A new approach to the study of textile properties with liquid chromatography, *Textile Res. J.*, 62(6), 361–369.

Le, K. D., N. R. Gilkes, D. G. Kilburn, R. C. Miller, J. N. Saddler, and R. A. J. Warren, 1994, A streptavidin–cellulose-binding domain fusion protein the binds biotinylated proteins to cellulose, *Enzy. Microb. Technol.*, 16, 496–500.

Long, J., 1993, First bioengineered animal drug approval, *CEN*, 71(46), 9.

Luong, C. B. H., M. F. Browner, and R. J. Fletterick, 1992, Purification of glycogen phosphorylase isozymes by metal affinity chromatography, *J. Chromatogr.*, 584, 77–84.

Morgan, D. A., F. W. Ruscetti, and R. C. Gallo, 1976, Selective in vitro growth of T-lymphocytes and normal human bone marrows, *Science*, 220, 1007–1008.

Ong, E., J. B. Alimonti, J. M. Greenwood, R. C. Miller, R. A. J. Warren, and D. G. Kilburn, 1995, Purification of human interleukin-2 using the cellulose-binding domain of a prokaryotic cellulase, *Bioseparations*, 5, 95–105.

Rohm and Haas Company, 1974, *Amberlite Ion Exchange Resins Laboratory Guide*, Philadelphia, PA.

Romicez, C., J. Fung, R. C. Miller, R. Antony, J. Warren, and D. G. Kilburn, 1993, A bifunctional affinity linker to couple antibiotics to cellulose, *Bio/Technol.*, 11, 1570–1573.

Smith, M. C., J. A. Cook, T. C. Furman, P. D. Gesellchen, D. P. Smith, and H. Hsiung, 1990, Chelating peptide-immobilized metal-ion affinity chromatography, in *Protein Purification: From Molecular Mechanisms to Large-Scale Processes*, M. R. Ladisch, R. C. Willson, C.-D. C. Painton, and S. E. Builder, eds., Am. Chem. Soc., Washington, DC, 169–180.

Sutton, C., 1992, Lectin affinity chromatography, in *Protein Purification Methods: A Practical Approach*, E. L. V. Harris and S. Angal, eds., IRL Press, Oxford, 268–282.

Tanford, C., Y. Nozaki, J. A. Reynolds, and S. Mikano, 1974, Molecular Characterization of Proteins in Detergent Solutions, *Biochem*, 13, 2369–2376.

Tomme, P., R. A. J. Warren, and N. R. Gilkes, 1995a, Cellulose hydrolysis by bacteria and fungi, *Adv. Microbial Physiol.*, 37, 1–81.

Tomme, P., D. P. Driver, E. A. Amandoron, R. C. Miller, R. Antony, J. Warren, and D. G. Kilburn, 1995b, Comparison of a fungal (family I) and bacterial (family II) cellulose binding domain, *Bacteriol.*, 177(15), 4356–4363.

Tomme, P., N. R. Gilkes, R. C. Miller, A. J. Warren, and D. G. Kilburn, 1994, An internal cellulose-binding domain mediates adsorption of an engineered bifunctional xylanase/cellulose, *Protein Eng.*, 7(1), 117–123.

Wallis, M., S. L. Howell, and K. W. Taylor, 1985, *The Biochemistry of Polypeptide Hormones*, Wiley, Chichester, 184–221.

Weber, D. V., R. F. Keeny, P. C. Familletti, and P. Bailon, 1988, Medium-scale ligand-affinity purification of two soluble forms of human interleukin-2 receptor, *J. Chromatogr.*, 431, 55–63.

PROBLEMS

8.1 Recombinant insulin case study. The development of a process to manufacture insulin using *E. coli* was the result of a major research effort and clever insights into the unique properties of the insulin molecule that made it amenable to manufacture via recombinant technology. Both the upstream (fermentation) and downstream (recovery and purification) steps have characteristics that are useful as case studies for other types of recombinant proteins. Describe some of these characteristics, as well as the importance of the engineering interface between upstream and downstream processing, by answering the questions below.

a. Why did the absence of Met and Trp residues in insulin make its recovery from the fusion protein possible?

b. Could CNBr be used in a downstream processing step for a protein with Met or Trp residues in its structure?
If you answered YES, briefly explain why. If you answered NO, briefly explain why not.

c. What is the Trp LE operon?
How did it contribute to a major improvement in productivity of rDNA insulin? (5 points)
How was this operon used to decrease the number of manufacturing steps by about 50%?
Why did this operon improve the translational yield of the fusion protein compared to the B-Gal operon?

8.2 Match the correct items on the right with the terms on the left.

____ A chain
____ B chain

____ Proinsulin

____ Strong promoter
____ Number of insulin refold products
____ This component elutes first in SEC of insulin components
____ This peptide would be hydrolyzed by CNBr

a. 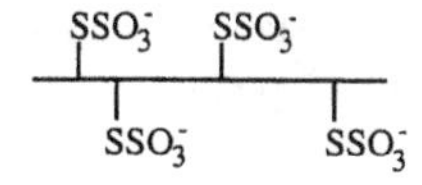 SSO_3^- SSO_3^- SSO_3^- SSO_3^-

b. 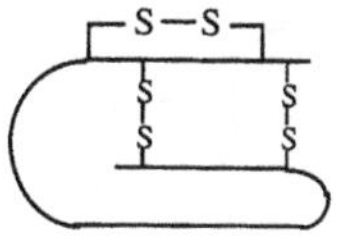 S—S S S S S

c. Trp LE

d. 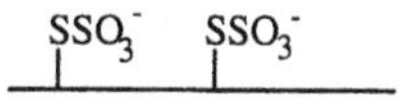 SSO_3^- SSO_3^-

e. β-gal

f. Trp

g. Leu

h. Met

i. β-gal

____Proinsulin formation induced when this amino acid is absent from the fermentation medium

j. Gly-Ile-Val-Glu-Gln-Cys-Cys-Thr-Ser-Ile-Cys-Ser-Leu-Tyr-Gln-Leu-Glu-Asn-Tyr-Cys-Asn
k. Phe-Phe-Tyr-Tyr-Thu-Pro-Lys-Lys-Met-Gly-Thr-Thr-Asn
l. 1
m. 5
n. more than 20
o. insulin
p. insulin agregate

8.3 After the fermentation has been completed, the inclusion bodies must be recovered, concentrated, and provided to the subsequent processing step as a slurry of inclusion bodies in buffer. The slurry must be free of cells, cell debris, fermentation broth components, and processed in five steps or less. Starting with Figure 8.54, draw a process schematic diagram, and briefly describe the key steps required to obtain this slurry, and construct a SEO-OPS table. Are all of the steps orthogonal?

8.4 The development of a bioseparations process requires the specification of process steps and their sequence. These steps could include filtration, centrifugation, precipitation, and chromatography. In the case of the development of a chromatographic separation for a protein of known properties (size, stability, and isoelectric point), the parameters to be specified include the stationary phase type and particle size, column dimensions, and eluent velocity, pH, and ionic strength.

In the case study presented in this problem, IgG (immunoglobin G) is to be separated from other blood serum proteins (principally human serum albumin) by ion exchange chromatography. One of the uses of IgG is in improving the resistance of burn patients to infection.

The objective of the chromatography step is to fractionate the IgG in one peak and the other proteins in a second peak. The isoelectric point of the IgG is about 7.4, while the other serum protein peak has an apparent isoelectric point of 4.9. The proteins are to be separated by adsorbing both onto a stationary phase and then sequentially eluting them as two peaks using a step change in the ionic strength of the eluting buffer.

a. Type of stationary phase. The buffer to be tried first for this separation is at pH 8.0. Which type of stationary phase would you recommend first—assuming that the

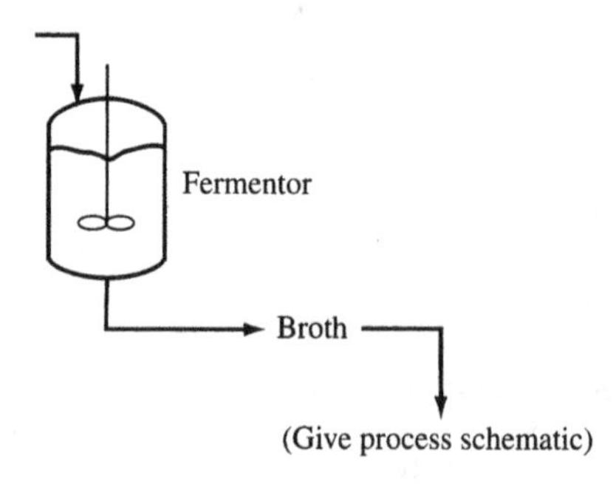

Figure 8.54 Schematic diagram for starting problem 8.3.

proteins will dissolve and are stable at the pH and ionic strength of this buffer? Briefly explain the reason for your choice.

b. Particle size. The initial separations development is to be carried out on a column having dimensions of 4 mm i.d. × 20 cm long. Which particle size would you select: 8 micron, or 15 micron, or 40 micron? Briefly explain the reason for your choice.

c. When the protein solution is injected at pH 8.0 in a buffer containing 10 mM salt followed by elution with the same buffer, no peaks are detected even after 60 minutes at an interstitial buffer velocity of 2 cm/min. The eluent is then changed to a buffer of the same pH but with an ionic strength of 50 mM NaCl. A peak with a shoulder elutes within 15 minutes. When the ionic strength is further increased to 500 mM NaCl, a second, single peak elutes again within 15 minutes. A third increase of the salt concentration to 2 M gives no further peaks. Give a possible explanation for the observed behavior.

d. A second approach is taken. This time the protein solution is loaded using a buffer at pH of 6.8 and an ionic strength equivalent to 10 mM salt. The buffer flowrate is again equivalent to an interstitial velocity of 2 cm/min, but this time a protein peak elutes starting after about 5 minutes. When the eluting buffer is changed to obtain an ionic strength of 500 mM NaCl, a second peak elutes. A further increase of the eluting buffer to an ionic strength of 2 M NaCl gives no further peaks. Explain. What is the major protein component likely to be found under the first peak? What is the major protein component likely to be associated with the second peak? What analytical technique might you use to confirm the identity of the first and second peaks? Give a brief explanation.

8.5 Human insulin produced by recombinant technology involves fermentation, isolation, purification, refolding, and further purification. Develop a conceptual process flowsheet and give the logic behind the sequence of steps and conditions for this flowsheet by reading the material provided and answering the questions below.

a. An existing process uses the lac operon (i.e., beta-Gal) promoter on the plasmid in *E. coli* K-12 cells. The fusion protein that results is intracellular and insoluble. You have the choice to use a new *E. coli* that has a plasmid with a trp LE promoter, or the existing *E. coli* recombinant strains in doubling the capacity of the process. Which process would you consider first? Explain your choice in terms of potential productivity gains, and choice of equipment. Be specific (but concise) in your answers.

b. Draw a conceptual process diagram based on your choice in part a. Show the sequence of all key steps starting with fermentation, proceeding through product recovery, and including refolding.

c. How many different refolding products are possible?

d. Show the 5 possible refolding products if the insulin A and insulin B chain fermentation approach were used.

e. Name a solvent used in the insulin recovery refolding process. Briefly explain any special conditions required for refolding.

f. If the objective is to produce 50 kg per day of final insulin product, and 50% of the *E. coli* mass is assumed to be chimeric insulin protein at the completion of the fermentation, what magnitude of order of fermentation volume would be required? Assume that the fermentation requires 24 hours (including turnaround) with a final cell count of 100 million cells/mL. State your assumptions and show your work.

g. Draw the structure of the active insulin product. Show the amino acid sequence using the appropriate abbreviations.

h. Draw the structure of proinsulin.

i. Why is the absence of Met residues a major factor in the development and commercializaton of the insulin process which you chose above? Be specific in your answer.

j. Construct a SEP-OPS table. Are all of the steps orthogonal? Use the table to identify steps that require additional purification methods to remove reagents added to achieve the separation in the effected steps.

CHAPTER 9

AFFINITY CHROMATOGRAPHY: BRIDGE BETWEEN MOLECULAR BIOLOGY, COMBINATORIAL METHODS, AND SEPARATIONS SCIENCE

INTRODUCTION

Affinity chromatography will form a bridge between separations science and applications of molecular biology and combinatorial methods by utilizing protein receptors generated by these techniques. While molecular biology enables probing of organisms and discovery of existing receptors, combinatorial methods are resulting in synthesis of new receptors based on randomly assembled chemical molecules. Both molecular biology and combinatorial chemical techniques have the potential to generate templates for affinity ligands, suitable for immobilization on stationary phases, in a cost-effective manner. The bioseparation techniques addressed in this chapter have a strong basis in the biology and molecular biology of viruses, or the molecular biology of single-celled organisms. The discussion of HIV illustrates how fundamental knowledge of receptors may lead to understanding of why therapies work or don't work, while an example of bacterial virus shows how a virus can be used to propagate affinity ligands capable of highly specific binding to a target molecule.

Molecular biology not only enables identification of protein structure/function and generation of protein ligands, but it also facilitates construction of chimeric proteins specifically de-

The author thanks Professor Douglas Kilburn, Director of the Biotechnology Laboratory, University of British Columbia, Vancouver, for review and comments on the cellulose binding domain case study, Dr. Michael Thien (Merck) for his review and comments on protease inhibitors, Dr. Pascal Bailon (Roche) for his review and comments on receptor affinity chromatography, and Dr. Gary Forrest (Wyeth-Ayerst Research), Dr. Harold Staack (Abbott Laboratories), Nathan Mosier (while a graduate student at Purdue University), and Dr. Rae Record (Purdue) for their reading and suggestions for this chapter.

signed to bind with these ligands. The chimera[1] consists of a peptide attached to the *N*-terminus of the target protein. The peptide, and therefore the protein to which it is fused, binds to affinity ligands with high specificity, while unbound impurities are washed away. A change in mobile phase conditions then causes the chimera to elute. Finally, the peptide is hydrolyzed from the purified chimera resulting in peptide fragments and the target protein. The protein/peptide mixture is further fractionated using methods described in Chapter 8 to give the final, highly purified product.

The possibilities of affinity chromatography for protein separations have excited researchers for more than 30 years. While this type of bioseparation meets three of the five biotechnology toolbox criteria given in Table 8.19 (versatility, variety, and compatibility), affinity chromatography is perceived to be more expensive and less robust than other types of liquid chromatography. As a consequence its benchmarking as a process chromatography tool is less extensive than for other types of chromatography that are considered when process scale-up becomes an imminent prospect. This situation is compounded by unrealistic expectations for affinity chromatography. Its large selectivity is often misinterpreted to be synonymous with single step recovery and purification of the product directly from biological fluids or fermentation media. Cells, cell debris, waste metabolites, residual (growth) media components, and nonprotein, cytoplasmic constituents in these fluids may foul the solid stationary phases to which the ligands are attached. Nonspecific adsorption of the contaminating components compromises the benefits gained from the specificity of affinity ligands, and this reduces the operational lifetime of affinity supports because of the relatively harsh regeneration conditions needed to remove the adsorbed solutes. Furthermore repeated regeneration can gradually destroy the attached ligands. Consequently affinity chromatography must be preceded by cell removal and preliminary adsorption or chromatography steps.

Ligands must be available and affordable if process scientists and engineers are to be encouraged to experiment to identify operational conditions and immobilization methods where the ligands retain specificity and binding capacity, while avoiding the leaching of the ligand into the product for hundreds to thousands of cycles. The application of molecular biology and combinatorial chemistry for identifying large numbers of affinity ligands, accompanied by the use of recombinant organisms (for proteins) or chemical synthesis (for small molecules and peptides) to produce large quantities of new ligands, will promote their availability and facilitate benchmarking. This chapter describes how application of molecular biology and combinatorial methods to separation science may lead to large-scale purification of biomolecules by affinity chromatography.

AFFINITY LIGANDS FROM COMBINATORIAL LIBRARIES

Ligands from combinatorial methods are likely to be small molecules—carbohydrates, peptides and nucleotides. These are randomly synthesized, selected for their ability to bind a specified target protein using rapid screening techniques, and then characterized with respect to chemical structure. The chemical structure serves as the template or prototype from which a large-scale synthesis method is developed. The ligand is then made in a large enough quantity so that it can be studied as a separations tool and ultimately scaled up for use in process-scale affinity chromatography.

[1]Composite protein (and other polymeric) molecules have been named in a loose analogy to the creature of Greek mythology having the head of a lion, body of a goat, and tail of a serpent—that is, the chimera. The term "chimera" was originally associated with composite DNA molecules into which foreign DNA had been inserted (Old and Primrose, 1981).

The template provides the structural information needed to devise a chemical synthesis procedure for obtaining molecules of the same composition, size, and shape. This type of template is analogous to a pattern used to fabricate a cookie cutter. The cutter enables the rapid reproduction of numerous cookies having the same shape and size from dough of a given composition. A second type of template is based on use of pure receptor molecules to individually define or imprint the molecular structure of every binding site on a stationary phase. This would be analogous to using a large number of identical cookie cutters and discarding each one after it had been used to make one cookie. This second approach is unlikely to be practical, since the cost of the purified receptor molecules would be large relative to their one time use.

9.1 COMBINATORIAL CHEMISTRY CREATES LIBRARIES OF NEW PEPTIDE SEQUENCES

Combinatorial methods have the objective of assembling a large number of permutations of a group of molecular building blocks. An example of a popular combinatorial approach is the split and mix technique in which a single molecular building block with an attached chemical tag is covalently bound to a large pool of small, insoluble polymer beads. The derivatized beads are recovered, washed, mixed, and then subdivided into a preset number of batches as illustrated in Figure 9.1. Different molecular building blocks and tags are then added to each of the batches. Derivatization is carried out and the beads are once again recovered, washed, mixed together, and then redivided into the same number of batches (three batches are used for this example). Each batch now contains at least some beads having all of the pairs of molecular building blocks (in this case dipeptides), although not all of the beads react during each cycle, as illustrated by beads 3, 7, 9, 10, 13, 15, 17, 21, 22, 24, 25, 26, and 30 in step 2(a) of Figure 9.1. A different set of building blocks would again react with each of the three batches, and the dipeptide would be derivatized to give a tripeptide. The beads would then be recovered, mixed, and subdivided once more so that each batch has beads with all of the different variations of the tripeptides. This sequence of steps is known as the couple, mix, divide, and couple synthesis approach. Three batches with a total of 36 beads are used for the purpose of this explanation although a typical combinatorial procedure is based on a much larger number of batches with 10^7 to 10^9 beads per batch and many different building blocks.

Table 9.1 shows how the two cycles of split and mix illustrated in Figure 9.1 result in 13 different combinations. The process is repeated numerous times to generate a large number of unique peptide sequences, which are referred to as "libraries." The resulting beads are then screened for their ability to bind a specific protein. The best candidates are isolated, chemically characterized, and then synthesized in larger quantity to confirm binding specificity (Buettner et al., 1996; Service, 1996).

The generation of millions of unique peptide sequences was demonstrated by Furka in 1991 according to Buettner et al. (1996), while the concept itself is credited to Geysen who introduced a method for simultaneously synthesizing a large number of peptides on solid supports (Schultz and Schultz, 1996). The conceptual simplicity of this approach made it attractive for synthesizing polypeptides and oligonucleotides, since polymerization of amino acids and nucleotides can be achieved by repeating a well understood chemical reaction with defined reagents (Service, 1996). This approach would not have been useful without the development of rapid screening techniques for identifying beads with peptide sequences exhibiting the desired binding properties. Affinity (protein) target molecules, and protein-specific dyes or substrates, screen combinatorial libraries by differentiating beads capable of selectively binding the target

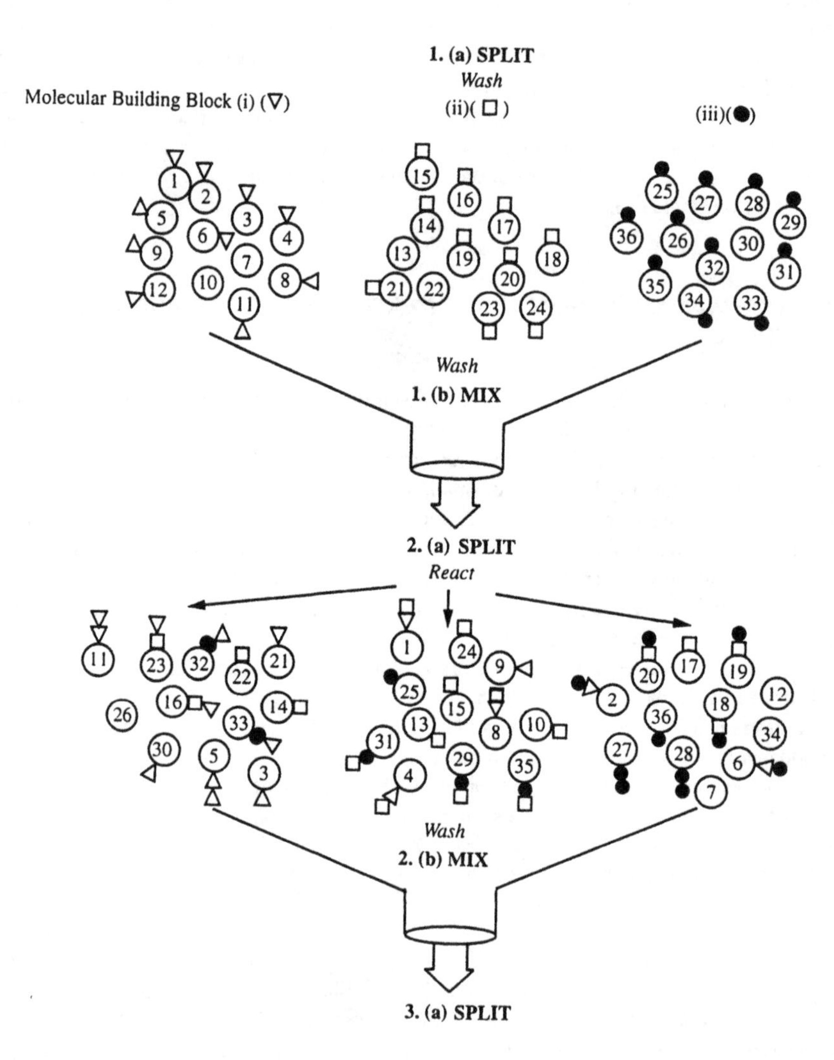

Figure 9.1 Concept of combinatorial reaction scheme. Molecular building blocks (*i*) (▽); (*ii*) (□); (*iii*) (●) represent different molecules (e.g., three different amino acids). These are reacted with particles 1 to 12; 13 to 24; and 25 to 36 in three different reaction vessels (not shown). After washing, the particles are mixed (step 1*b*), divided into three equal parts (step 2a), and again reacted with the three molecular entities (*i*) (▽); (*ii*) (□); (*iii*) (●) in separate reactors. These are again washed and mixed (step 2*b*). This gives particles with different (random chains) as indicated by the symbols, with a few particles *not* reacting at each step as indicated by beads that have no ligand or only one ligand attached after the second cycle. (Although only 36 particles are shown, each lot will consist of millions of particles.)

TABLE 9.1 Number of Different Beads in Steps 1 and 2 of Figure 9.1

Step	Total Number of Different Beads	Ligands (Types of Combined Building Blocks)	Symbol (in Figure 9.1)
1	4	Unreacted (step 1)	
		i	
		ii	
		iii	
2	13	Unreacted (step 1) + ligands i, ii, iii	
		Unreacted (steps 1 and 2)	
		i + i	
		i + ii	
		i + iii	
		ii + i	
		ii + ii	
		ii + iii	
		iii + i	
		iii + ii	
		iii + iii	

Note: Assume that no more than 1 ligand reacts per cycle.

protein from beads giving a false positive due to nonspecific adsorption or interaction with the detecting reagents. The beads themselves must give good synthesis capacity and high loading of the ligand, exhibit chemical resistance to synthesis reagents, and be sufficiently rigid for use in column chromatography (Buettner et al., 1996).

9.2 VISUAL IDENTIFICATION FINDS A NEEDLE IN THE HAYSTACK: ONE AFFINITY LIGAND AMONG THOUSANDS OF PEPTIDES: FACTOR IX EXAMPLE

Peptide affinity columns have substantial advantages over less specific forms of liquid chromatography, but the number of known naturally occurring peptide sequences that can serve as ligands are limited. The construction of combinatorial libraries will expand the pool of affinity peptides. When combined with rapid and accurate screening techniques, these will help to promote the development of affinity chromatography columns specifically designed for a particular separation. An example is given by Buettner et al. (1996) who synthesized, screened, and optimized peptide ligands for the purification of trace amounts of the serine protease coagulation Factor IX from blood plasma.[2]

A polyhydroxylated methacrylate polymer with 1000 Å pores (currently marketed for metal chelate affinity chromatography) was modified to generate a free amino group suitable for initiating peptide synthesis. Peptide sequences were synthesized on the solid phase using a peptide synthesizer with α-fluorenylmethoxycarbonyl (abbreviated Fmoc) as the α-amino protection agent. The beads (65 μm particle size) were added in 0.5 g amounts to each of 18 reaction vessels and 18 of the 20 naturally occurring L-amino acids (cysteine and methionine were not used) were coupled to the beads. After reaction, the beads from each vessel were pooled, washed with dimethylformamide, and equally redistributed into the 18 reaction vessels in an approach that is analogous to that presented in Figure 9.1. The next amino acid was then coupled. The procedure was repeated until a library of hexamers was generated (Buettner et al., 1996). The weight of beads (9 g) was equivalent to approximately 128 million beads. Many of the beads did not have properly synthesized sequences attached. Other peptide sequences, while properly synthesized, would not bind the target protein. A rapid, semi-automated method for probing these libraries is of obvious importance.

A stain system was developed in which beads stain blue for nonspecifically reacted immuno reagents used to probe the library. Beads, which specifically bound the target protein, would stain red. The red-stained beads are then handpicked under a dissecting microscope using a pipettor. Six beads (out of a possible 128 million!) were isolated in this manner. Two of the beads each contained about 10 pmol of peptide (0.1 mequiv/g). This was sufficient to obtain the peptide sequence. The peptide consisted of the amino acid sequence *YANKGY*.[3] This example illustrates the importance of a visual method for rapidly screening a large number of particles for identifying peptides exhibiting a potentially desirable binding characteristic.

9.3 LIQUID CHROMATOGRAPHY USING 300 MG OF BEADS CONFIRMS SPECIFICITY OF FACTOR IX BINDING

The sequence YANKGY was synthesized on a larger quantity of beads (approximately 0.3 g). These beds were packed into a 2.1 mm × 150 mm chromatography column having a volume of 0.52 mL. The stationary phase gave selective binding of Factor IX over human serum albumin (HSA). The Factor IX bound in the presence of 1 M NaCl but eluted when a weak acid was introduced to this column. Analysis of the data suggest that binding interactions between the immobilized YANKGY and Factor IX occurred at the surface of the protein, with the sequence KGY preferentially binding Factor IX over human serum albumin. While human serum albumin adsorbed on the bead, thus indicating nonspecific binding of the human serum albumin, it was eluted by 1 M NaCl (Figure 9.2).

[2] Factor IX is a protein involved in blood clotting. A deficiency of Factor IX causes hemophilia B, which is a rare inherited disease affecting 3300 patients in the United States and 10,000 worldwide. Control of the disease has been achieved by infusions of Factor IX products made from human blood that is treated to reduce risks of infection by virus or pathogens from the donor. In 1997 the FDA approved sale of Factor IX made using recombinant technology. Since it is not derived from human blood, the risk of infection by human viruses or pathogens is eliminated. Factor IX treatments currently cost $20,000 to $25,000/year per patient in the United States (Staff reporter, 1997).

[3] The abbreviations for the amino acids are Y = tyrosine, A = alanine, N = asparagine, K = lysine, and G = glycine (rules for remembering these abbreviations are given by Brandon and Tooze, 1991).

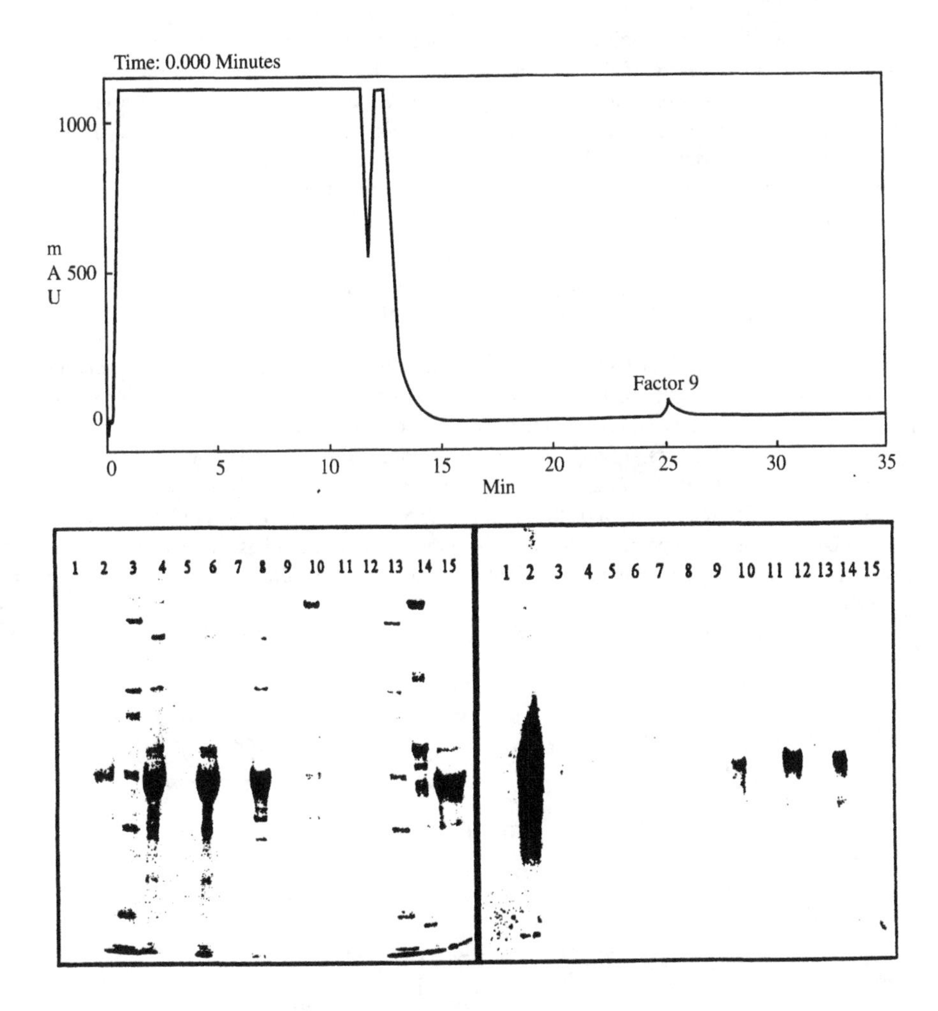

Figure 9.2 Chromatogram, SDS-PAGE and Western blot of 1.0 mL citrated human plasma on the acetyl-YANKGY-TSK resin. Approximately 10 μg were loaded in each lane and the electroblotted gel was loaded with the same amount of sample as the Coomassie stained gel. Lanes 1, 5, 7, 9, and 11 were blank. Lane 2, 1.0 μg of Factor IX; and 12, 0.0625 μg of Factor IX; lanes 3 and 13, molecular weight standards; lanes 4 and 15, human plasma; lane 6, flow through (t = 0–12 min); land 8, flow through (t = 12–15 min); lane 10: the acid eluate; lane 14, a partially purified Factor IX intermediate. The Western blot on the lower right shows no Factor IX in the starting material, flow throughs, or salt peak A strong immuno-detected band corresponding to the Factor IX zymogen is in the acid eluate. (Reproduced with permission from Buettner et al., 1996, fig. 10, p. 81, Munksgaard International Publishers Ltd. Copenhagen, Denmark.)

9.4 COST-EFFECTIVE SYNTHESIS OF KILOGRAM QUANTITIES OF AN AFFINITY PEPTIDE REMAINS A CHALLENGE: ESTIMATING PEPTIDE COST

The amount of Factor IX bound by the YANKGY column was equivalent to 0.44 mg/mL of stationary phase. While the conditions were not optimized to attain maximum loading, the challenges inherent to scale-up of the preparation of an affinity ligand are illustrated by the YANKGY example. Assume that a chromatogram is run every 30 minutes. Then recovery of 0.44 mg of Factor IX per mL column volume per run corresponding to 0.88 mg Factor IX/(mL · hr) or 0.88 g/(L · hr) is possible. If production of 100 kg/yr of Factor IX is the goal, and if the column's specificity and loading last indefinitely, then the chromatography column volume, V_{col}, is given by

$$V_{col} = \frac{P_{an}/4200}{P_{col}/1000} = \frac{100/4200}{0.88/1000} = 27.1 \text{ L} \tag{9.1}$$

where V_{col} is in liters, P_{an} is the annual production of purified target molecule in kg; P_{col} denotes the hourly throughput in g/(L · hr), while 1000 is the number of g per kg and 4200 is the maximum number of hours that a column would be operated on an annual basis. An optimistic value of 350 days/year at 12 hours per day (= 4200 hours, total) was assumed, leaving 15 days for downtime. The bed volume of 27 L (or 15.6 kg of beads) may appear to be modest. However, the synthesis of 1.56 equivalents of peptide required to obtain 27 L of stationary phase, to which 0.1 milliequivalents of the peptide is attached per gram of stationary phase, may be expensive. The cost is estimated by calculating the weight of peptide corresponding to 1.56 equivalents of YANKGY.

Peptides are condensation polymers of amino acids joined by amide linkages (peptide bonds):

$$H_2N\text{—}CH(R_1)\text{—}C(=O)OH + H_2N\text{—}CH(R_2)\text{—}C(=O)OH \xrightarrow{-H_2O} H_2N\text{—}CH(R_1)\text{—}C(=O)\text{—}NH\text{—}CH(R_2)\text{—}C(=O)OH \tag{9.2}$$

where R_1 and R_2 represent the side groups on the amino acids. The molecular weight of the peptide YANKGY (denoted by $M_{r,YANKGY}$) is given by

$$M_{r,YANKGY} = \left[\sum_{i=1}^{j} (n_i(M_{r,AA,i} - 18)) \right] + 18 \tag{9.3}$$

where n_i is the number of times that the amino acid residue i of molecular weight $M_{r,AA,j}$, appears in the peptide, and j represents the number of different types of amino acid residues that appear in the peptide. The value 18 is the molecular weight of water.

TABLE 9.2 Summary of Molecular Weight Calculation for YANKGY

Amino Acid Residue	i	n_i	$M_{r,AA,i} - 18$	$n_i(M_{r,AA,i} - 18)$	Charge (at pH = 7)
Y (tyrosine)	1	2	163	326	0
A (alanine)	2	1	71	71	0
N (asparaginine)	3	1	114	114	0
K (lysine)	5	1	128	128	+1
G (glycine)	4	1	57	57	0
Total: $\sum_{i=i}^{5} n_i(M_{r,AA,i} - 18)$				697	1

YANKGY has five different residues ($j = 5$), and one of them (residue Y) appears twice ($n_1 = 2$). The calculation, which is readily carried out using a hand calculator or spreadsheet,[4] is illustrated in Table 9.2. Hence Eq. (9.3) gives a molecular weight of 715 (= 697 + 18). If a net charge of 1 is assumed, 1.56 equivalents of the peptide YANKGY corresponds to 1115 g of pure peptide (= $M_{r,YANKGY} \times 1.56$). The cost of the peptide, $C_{peptide}$, is given by

$$C_{peptide} = \left(\frac{C_{residue} N_{residue}}{W_{unit}}\right)(W_{total})(1000) \tag{9.4}$$

where $C_{residue}$ is the synthesis cost in \$/residue, $N_{residue}$ is the number of residues in the peptide, W_{unit} is the unit weight, in mg, of peptide upon which $C_{residue}$ is based, W_{total} is the total weight, in g, of the peptide to be synthesized, and 1000 is the number of mg per g. Several 1997 advertisements show that custom peptides are synthesized at \$15 to \$18 per amino acid residue at 10 to 40 mg quantities (Advertisement, 1997). Assume a price of \$18/residue for 40 mg of the six-residue peptide YANKGY. The corresponding price for 1.11 kg would be unrealistically large:

$$\frac{\$18/\text{residue} \times 6 \text{ residues}}{40 \text{ mg}} \Big| 1115 \text{ g peptide} \Big| \frac{1000 \text{ mg}}{\text{g}} \Big| = \$3{,}010{,}500 \tag{9.5}$$

The synthesis of a peptide on a kg scale, however, will benefit from economies of scale, with the expectation that the cost would be decreased by a factor of 100 to 1000. An example of manufacture of a peptide on a kg scale is given by the synthesis of the 32 residue peptide calcitonin, which had a global market estimated to be \$800 million/year and is used for treatment of post-menopausal osteoporosis (Anonymous, 1997; Flanigan, 1991). An alternate approach would be to synthesize the polynucleotide corresponding to this peptide, and then clone the polynucleotide into a microorganism that would express the peptide in a fermentation process.

9.5 COMBINATORIAL SYNTHESIS OF CARBOHYDRATES AND SCREENING WITH BIOTIN LABELED LECTIN GIVES NEW AFFINITY LIGANDS FOR LECTINS

Carbohydrates in the form of oligo- and polysaccharide chains, covalently bound to membrane proteins, or oligosaccharides bound to lipids, are found on the surface of all encaryotic cells (Al-

[4] A generalized spreadsheet that lists the molecular weights of all of the individual amino acids is recommended. These can then be summed according to Eq. 9.3, with $n_i = 0$ if the amino acid i is not present.

berts et al., 1989, pp. 298–299). Cell surface carbohydrates play central roles in normal and pathological recognition processes, and they have been implicated in chronic inflammation, viral and bacterial infection, tumorigenesis, and metastasis. Synthesized carbohydrates that bind their protein targets more strongly than natural carbohydrates at the same binding site on the protein could prevent or treat receptor mediated diseases (Liang et al., 1996). Lectins are proteins that have sites that can recognize and bind specific sequences of sugar residues on the surfaces of cells. These proteins were first isolated from seeds, where they occur in large quantities and are sometimes highly toxic to deter animals from eating the seeds. Lectins also occur in other organisms and are thought to participate in cell–cell recognition (Alberts et al., 1989). Development of chemistries to generate combinatorial libraries of carbohydrates that bind lectins has been stimulated by successes with peptide and oligonucleotide libraries, and the desire to learn about the role of carbohydrates at cell surfaces.

The synthesis of carbohydrate molecules requires special catalysts and a broader range of reactions than the polymerization of amino acids or nucleotides into polypeptides or oligonucleotides. Sulfoxide glycosylation is key to synthesizing a solid phase carbohydrate library. Using this approach, Kahne, Still, and co-workers synthesized two carbohydrate ligands that bind lectin from *Bauhinia purpurea* more strongly than the natural ligand (Liang et al., 1996). A split and mix strategy was used to obtain different forms of glycosyl molecules on a TentaGel resin (bead), starting with six different carbohydrate monomers. Twelve different glycosyl sulfoxide donors were coupled separately to the subdivided mixture of beads containing all six starting carbohydrates. Excess glycosyl donor avoided incomplete reactions by driving the reaction to completion and thereby preventing mixtures of products from forming on individual beads. After each reaction the beads were mixed, and the sugar azides reduced to amines. The beads were split again. The separate fractions of the beads were then N-acylated with 18 different reagents. Finally the beads were recombined, deprotected, and screened for binding with lectin (Liang et al., 1996).

The importance of a visible and rapid screen is illustrated in Figure 9.3 which schematically illustrates a dark bead indicating strong binding of the lectin. The colorometric assay of the

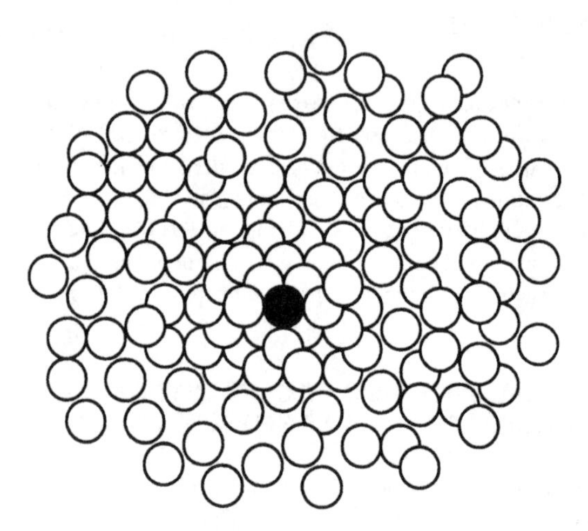

Figure 9.3 Schematic representation of a portion of the beads in a carbohydrate affinity ligand library after 5 minutes of staining. The dark bead in the center was identified as a hit (Liang et al., 1996).

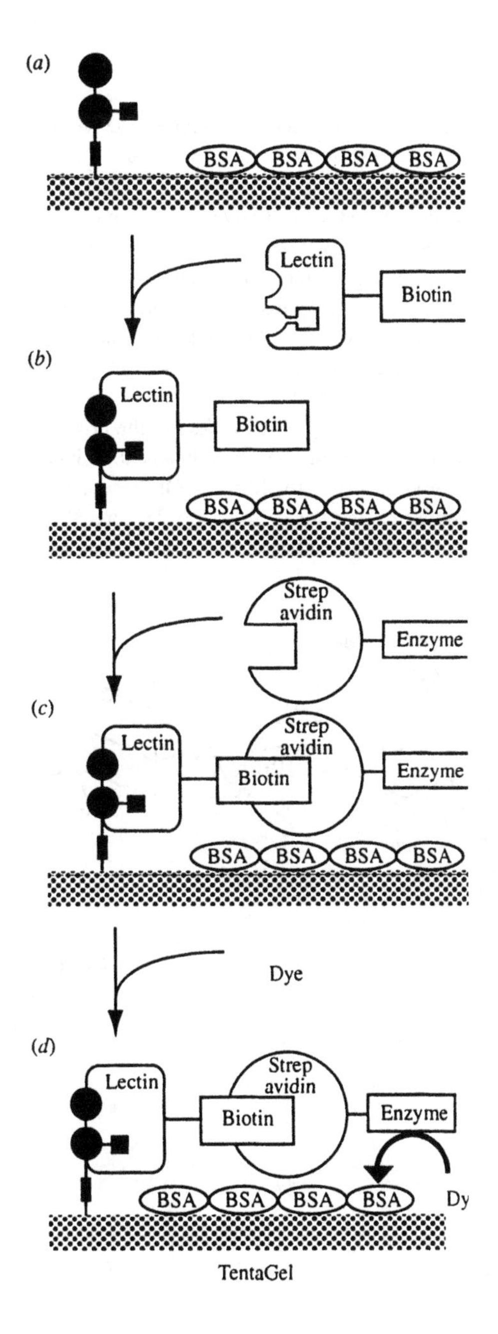

Figure 9.4 Enzyme-based assay for carbohydrate ligands synthesized on beads. (*a*) Derivatized TentaGel beads (10 mg) are washed three times with 1 mL of PBST buffer consisting of 10 mM sodium phosphate, pH 7.2; 150 mM sodium chloride (NaCl); 0.05% Tween-20, and then suspended in 1 mL of PBST containing 3% bovine serum albumin (BSA). After being shaken for 30 minutes at room temperature, the beads are washed three times with 1 mL of PBST containing 1% BSA. (*b*) The beads are incubated at room temperature for 3 hours on a rotary shaker in 1 mL of a biotin-labeled lectin solution (10 μg/mL in PBST containing 1% BSA) and then washed three times with 1 mL of TBST buffer [20 mM tri(hydroxymethyl) aminomethane hydrochloride (tris·HCl), pH 7.5; 500 mM NaCl; 0.05% Tween-20) containing 1% BSA. (*c*) The beads are incubated on a rotary shaker for 20 min at room temperature in 1 mL of alkaline phosphatase-coupled streptavidin (10 μg/mL in TBST containing 1% BSA). The beads are washed three times with 1 mL of alkaline phosphatase buffer (100 mM tris·HCl, pH 9.2; 100 mM NaCl and 5 mM $mgCl_2$) and kept in the alkaline phosphatase buffer prior to staining. (*d*) A portion of the beads was transferred to a petri dish and the alkaline phosphatase buffer was replaced with 200 μL of a solution containing 5-bromo-4-chloro-3-indolyl phosphate (BCIP) and nitro blue-tetrazolium (NBT). Color development was observed under a low-power microscope. The staining was terminated by washing the beads twice with 200 μL of sodium EDTA solution (20 mM, pH 7.4). The colored beads were picked out manually under the microscope for decoding. (Adapted with permission from Liang et al. (1996), fig. 3.)

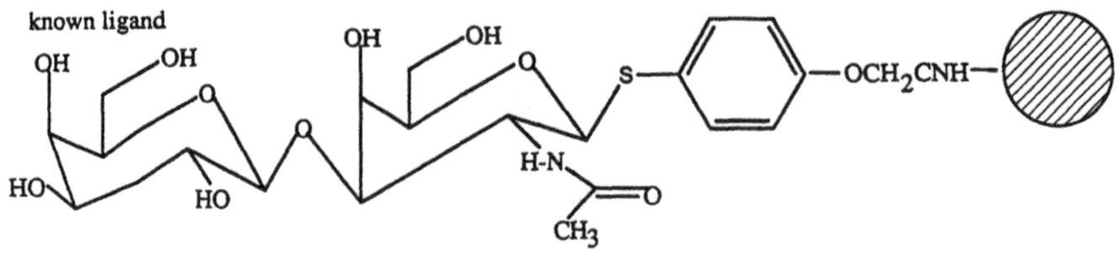

Figure 9.5 Known ligand for lectin: Galactose-α-1, 3-glucose NR-α-thiophenyl glycoside. Shaded circle represents solid bead.

beads entailed incubation of 10 mg of derivatized TentaGel (equivalent to approximately 9000 beads) with 1 to 3% bovine serum albumin, at pH 7.5, followed by biotin-labeled lectin in a solution of 1% bovine serum albumin (BSA). A chimeric protein of streptavidin-alkaline phosphate was added next with streptavidin part of the protein binding with the biotin of the biotin-labeled lectin. After a final washing and dilution, substrate and dye were added, incubated, and then removed by washing Figure 9.4. The colored beads are picked out manually (Liang et al., 1996).

The enormity of the task, and the critical role of the screen, becomes apparent when it is considered that the 9000 beads contained about 6 copies of each of the 1300 different compounds (i.e., the library). Of these, 25 stained beads were obtained, of which 13 beads contained the same disaccharide core (1-3-GlcNR-α-thiophenyl glycoside). The hit ligands (so named because they give a heavy stain after 5 minutes, indicating binding of the

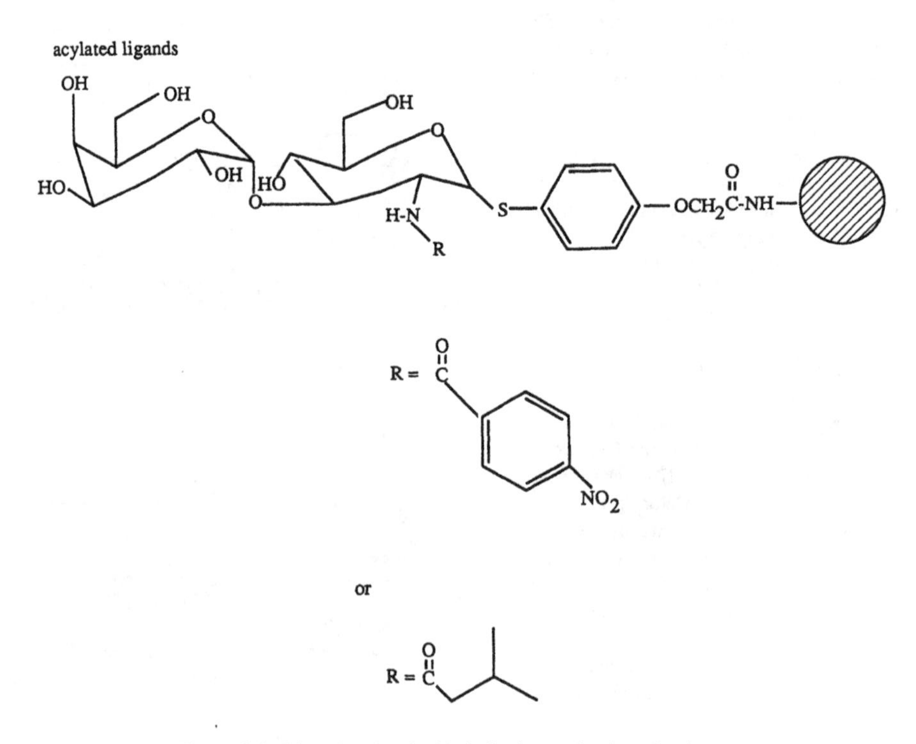

Figure 9.6 Ligand acylated with 4-nitrobenzoyl or isovaleryl groups.

lectin) were identified as being acylated with either 4-nitrobenzoyl or isovaleroyl groups. The acetylated disacharides and the known disaccharide were independently resynthesized on the TentaGel beads, which were then mixed and stained. The beads derivatized with the known ligand (structure in Figure 9.5) stained preferentially to most other beads derivatized with other carbohydrates. Only those disaccharides that were acylated (Figure 9.6) gave a rapid staining, while the known disacharide ligand of Figure 9.5 stained much more slowly. Hence the synthesized carbohydrates (Figure 9.6) bind lectin better than the previously known carbohydrate.

The importance of this method transcends affinity chromatography since it could be used to study and probe binding interactions that normally occur at surfaces. A paradox of carbohydrate binding is that carbohydrate binding proteins may have similar affinities for a range of saccharides in solution, yet exhibit remarkable specificity for saccharides on the surfaces of cells as encountered in cell to cell recognition. It is believed that polyvalent presentation of carbohydrates on cell surfaces may amplify the affinity and specificity of carbohydrate binding proteins containing multiple carbohydrate binding sites. Hence on-bead screens are proposed to be biologically more relevant (Liang et al., 1996).

9.6 BOVINE SERUM ALBUMIN (BSA) SUPPRESSES NONSPECIFIC BINDING OF THE LECTIN ON AN AFFINITY SUPPORT

Bovine serum albumin adsorbs on hydrophobic, ionic, or other nonselective protein adsorption surfaces on the bead that would otherwise bind the lectin in the absence of a large background of BSA. The biotin-labeled lectin will bind to both ligands and the surfaces of the bead. The surface bound portion of the lectin will not be desorbed by alkaline phosphatase buffer (pH 9.2) used to wash the beads. Hence the streptavidin alkaline phosphatase chimera, when subsequently added as represented in Figure 9.4, could bind with the nonselectively retained lectin, as well as biotin-labeled lectin that is complexed with the ligand. This would ultimately lead to false positives, and to selection of beads to which the lectin does not bind the ligand but only adsorbs nonspecifically to the surface of the stationary phase. When preceded by incubation of the beads with BSA, binding of the biotin-labeled lectin is limited to the ligands.

9.7 THE COMBINATION OF COMBINATORIAL CHEMISTRY AND COMPUTERS WILL ACCELERATE DRUG DISCOVERY AND CATALYST DEVELOPMENT

Methods for producing small organic molecules of molecular weight of 500 by combinatorial techniques were reported and *then* rapidly assimilated into automated synthesizers starting in 1991 and 1992. This type of equipment is now readily available. Computer software to optimize combinatorial reactions and substituents to give compounds most likely to exhibit structure-activity relationships of a particular type of drug or chemical began to become available in 1996 (Borman, 1997). Future developments will adapt fast chemical reactions to combinatorial synthesis, use rational drug design[5] to guide selection of molecular building blocks, and automate

[5]Rational drug design is a process where techniques including X-ray crystallography, structure-function studies, and computer simulation of 3-dimensional structures generate hypotheses of how small molecules can be designed to bind to protein targets. The protease inhibitor Viracept, used to treat AIDS, is reported to be a product of rational drug design, with the time from synthesis to market being a record—only 36 months (Armstrong, 1997).

the combinatorial split and mix technique. These developments are projected to accelerate drug discovery, since most medicinal compounds are in the low molecular weight range, and also result in new types or organic and inorganic catalysts (Borman, 1997; Hill and Gall, 1996).

Rapid screens for target molecules and analysis of the diversity or similarity of compounds in large libraries is a major part of automating combinatorial reactions. As summarized by Glaser (p. 19, 1996), "The power of combinatorial chemistry is in the numbers—the ability to synthesize thousands of new compounds in a week, rather than hundreds in a year. The more compounds you test against a target, the greater the chances of identifying a lead. Pfizer is a good example: during the past 40 to 50 years, its medicinal chemists have synthesized approximately 300,000 compounds. Within a 4 to 6 week period, the company used combinatorial chemistry to produce two libraries containing more than 500,000 compounds each. The marriage of combinatorial libraries and high throughput screening technology, both of which rely on advanced automation and robotics systems is dramatically shortening the time to identify and optimize lead drug candidates." While rational drug design[5] imparts a sense of direction, the randomness of combinatorial chemistry contributes a sense of speed (Glaser, 1996).

9.8 COMBINATORIAL LIBRARIES OF SMALL MOLECULES CAN BE GENERATED BY SOLID PHASE SYNTHESIS OR SOLUTION-BASED CHEMISTRY

The benefit of solid phase synthesis lies in the ease with which excess reagents can be added to drive the reaction to completion, and then washed away before the next step. Split synthesis (Figure 9.1) makes it possible to generate large libraries in which each bead holds only one compound. A soluble form of the compound can be obtained by cleaving it from the support after a sequence of reactions is completed. Further, only a relatively small number of reaction vessels is needed to generate in a large number of compounds by the split, mix, react, and split sequence illustrated in Figure 9.1. An estimate of the minimum number of reaction vessels, $N_{vessels}$, is given by

$$N_{vessels} \cong \sqrt[n]{N_{compounds}} \tag{9.6}$$

where n is the number of times each vessel is used (i.e., the number of steps) and $N_{compounds}$ is the number of different compounds to be synthesized. Three steps ($n = 3$) would be required to generate 10,000 compounds from 22 vessels each containing one of 22 different substituents and 100 mg amounts (millions) of solid beads. If 10,000 compounds were to be generated using a liquid reaction, 10,000 reaction vessels would be needed at each step, or 30,000 for the three steps. Despite the logistical challenge of solution phase, combinatorial synthesis, the development of solution phase combinatorial synthesis is a high priority, since a much wider range of organic reactions is available for solution phase synthesis than for solid phase reactions. An approach that combines the benefits of both solution and solid phase methods has been proposed in which the reaction is carried out in a solution phase, and unused reactants are subsequently removed via covalent bond formation to solid phase particles added upon the completion of the reaction. Other hybrid approaches are likely (Borman, 1997).

RECEPTORS, AFFINITY LIGANDS, AND ACQUIRED IMMUNE DEFICIENCY SYNDROME (AIDS)

The acquired immune deficiency syndrome (AIDS) was first recognized in 1981 when symptoms of unusual infections and cancer in healthy, young individuals, principly men, were reported. These symptoms were quickly tracked to compromised immune systems caused by the human immunodeficiency virus (HIV) transmitted through sexual contact, parenteral (e.g., needle sticks) contact with blood or blood products containing infectious virions, or perinatal exposure. HIV was first identified in 1983 and given the name lymphadenopathy-associated virus or human T-cell lymphotropic virus type III. The virus was named HIV in 1985 by an international committee. A second strain of HIV was discovered in West Africa in 1986, and found to be a cause of AIDS in Southern Europe and given the name HIV-2. The strain of HIV identified in the United States and France was denoted HIV-1 (Yang et al., 1996). By the beginning of 1998 there were 33 million people infected with HIV, with two-thirds in sub-Saharan Africa (News Item, 1998).

Receptors on the surface of T-cells serve as affinity ligands for the binding of human immunodeficiency virus (HIV) particles. The blocking of this function by surrogate molecules is an objective of AIDS research. Conversely, the human antibody for the p24 core protein of an HIV particle (see Figure 9.7), serves to detect the presence of the p24 protein, and therefore AIDS, in the blood serum of HIV infected people. The antibody is immobilized on beads and binds the antigen to the HIV core protein in a manner that leads to visible color formation. Both the detection of AIDS, and perhaps, someday, the control of HIV replication in infected individuals by blocking T-cell receptor sites, utilizes affinity ligands. The field of HIV research continues to evolve, and some of the biological mechanisms described here are likely to be simplistic, or even incorrect, when compared to the state of the knowledge in the field when this chapter appears in print. Nonetheless, this case study serves to illustrate the effect of advances in the life sciences on the business of biotechnology, and advances in treatment of a viral disease. The next sections describe how affinity ligands are used in AIDS diagnostic methods and give an overview of the mechanisms of anti-AIDS drugs, the role of receptors in

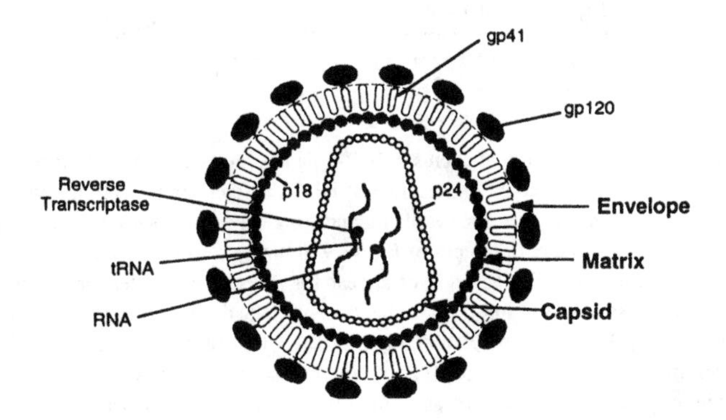

Figure 9.7 Schematic representation of an HIV-1 particle. The envelope lipid bilayer is shown with gp41 embedded in the membrane. The gp120 protein is attached to the membrane by binding the gp41 protein. If we cut away the envelope, the HIV-1 matrix is exposed. The matrix itself contains predominantly one kind of protein, p18. If we cut away the matrix shell, the cone-shaped HIV-1 capsid is exposed. Inside the capsid shell is the core of HIV-1. (From Yang et al. (1996), fig. 7.3, p. 163.)

the infection process, and the impacts of viral mutation and clearance on drug development. Since much has been written, it was possible to develop this case study in a manner that illustrates how the understanding of receptors contributes rationale for medical treatments, as well as drug development.

9.9 HIV-1 INITIALLY INCREASES ANTIGENS TO HIV CORE PROTEIN p24 IN THE SERUM OF INFECTED PERSONS

An excellent synopsis by Pidgeon et al. (Yang et al., 1996) describes the initial events associated with HIV. A spike in p24 protein in the blood of a newly infected person occurs within the first 3 to 6 weeks of infection as shown on the left side of Figure 9.8. The p24 protein induces antigens to develop to the HIV-1 circulating in the blood. These antigens bind predominantly with the HIV core protein p24, and coincide with the flu-like symptoms that occur shortly after infection. The immune system then generates antibodies against the HIV-1 antigens, and the antibodies bind the circulating p24 antigens making them unavailable to a human antibody based assay; that is, the antigens disappear from the blood. The presence of the p24 antigen is not detectable in 90% of HIV-1 seropositive persons once antibodies to the HIV core protein

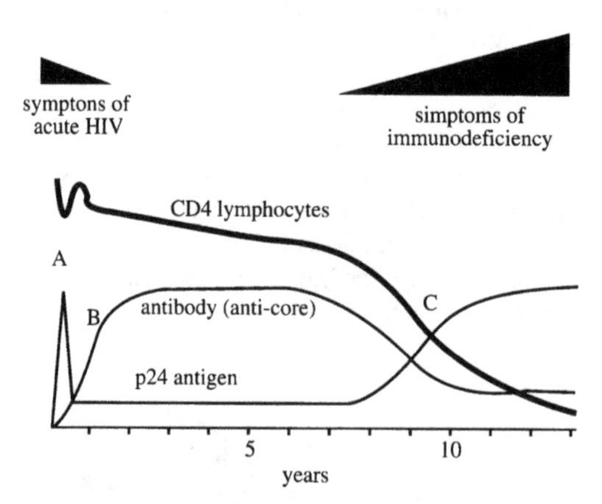

Figure 9.8 Schematic representation of events in HIV-1 pathogenesis. At the early stage of the infection, there is slight change in the CD4 count and a surge of HIV-1 antigens in the serum. Symptoms of acute infection are also observed such as fevers, swollen glands, night sweats, rashes, fatigue, and weight loss. However, antibodies against the core protein of HIV-1 will not be developed until several months later. Symptoms of clinical immunodeficiency do not appear for several years after the infection. An initial effective immune response to HIV-1 infection is able to control the disease for a prolonged period, usually lasting years. Eventually the virus, through a highly inaccurate replication mechanism, is able to generate mutants that can escape the immune response, and the disease then progresses. The number of CD4 + T-cells present in the patient's blood is used as an indication of the patient's disease status. As the relative number of CD4 + T-cell decreases, the patient becomes susceptible to opportunistic infections and certain cancers that define AIDS, the advanced stage of HIV-1 infection. Thus CD4 + T-cell counts are routinely used to monitor disease progression. However, the patient's CD4 counts only reflect immunologic competency and serve as an indirect or "surrogate" marker for viral replication. Equally important for evaluation of the patient's clinical status is the number of HIV-1 particles in the patient's blood. This is known as the "viral load."

p24 have developed. After this point the p24 protein level is low until the onset of the symptoms of immunodeficiency.

9.10 AN IMMOBILIZED HUMAN ANTIBODY AGAINST THE ANTIGEN FOR THE HIV CORE PROTEIN, p24, DETECTS AIDS: ELISA EXAMPLE

An accurate test for HIV is based on detection of p24 during the initial phase of HIV infection by an enzyme-linked immuno sorbent assay (abbreviated ELISA) and is an example of a cost-effective use of an affinity ligand. Pidgeon et al. (Yang et al., 1996) discuss Abbott Laboratories' HIVAG-1™ assay which is based on an immobilized human antibody. Enzyme-linked immuno-adsorbent assays are attractive because detection is accomplished using an enzyme rather than a radioisotope. This eliminates problems associated with the safety, expense, and short shelf life of radioimmuno assay reagents (Yang et al., 1996).

Antibodies[6] consist of high molecular weight and low molecular protein regions and a variable region which selectively binds a single antigen as illustrated in Figures 9.9 to 9.12. The HIV assay is based on the binding of an immobilized human antibody with the antigen to HIV-1 core protein p24. The assay is started by adding the serum of the patient to a well containing a single bead to which the human antibody is immobilized, as shown in Figure 9.13, step *i*. HIV-1 p24 antigen, if present, will bind with the antibody and be retained on the bead when the bead is washed to remove the serum (Figure 9.13, steps *ii* and *iii*). Next a rabbit anti-p24 polyclonal antibody is added to and incubated with the bead (Figure 9.13, step *iv*). The antibody will complex with the antigen to protein p24, if it is present. Unbound rabbit anti p24 polyclonal antibody is removed by washing. A third antibody, IgG (immunoglobulin G), derived from a goat and altered so that it has the enzyme, horseradish peroxidase, attached to it. It binds with the F_c region of the rabbit anti-p24 polyclonal antibody (Figure 9.13, step *v*). Finally, a substrate that is hydrolyzed by the horseradish peroxidase (consists of a solution of *o*-phenylenediamine (OPD) containing H_2O_2), is added and incubated in the well. If the enzyme, horseradish peroxidase, is present, the *o*-phenylenediamine will be hydrolyzed at a rate that is proportional to the concentration of the enzyme, which in turn is proportional to the amount of goat to rabbit IgG antibody bound to the rabbit antibody against the p24 antigen. The hydrolysis of *o*-phenylenediamine gives a yellow color that can be quantitated by measuring spectral absorbance at 420 nm (Figure 9.13, step *vi*). The details of this assay, data analysis, preparation of a standard curve, and follow-up assays performed to prevent false positives are given by Pidgeon et al. (see Cai et al., 1996). This type of assay represents a type of affinity chromatography since the affinity ligand—an antibody—is fixed to a bead, and is used to capture a protein from the surrounding solution.

[6]Antibodies, which are also known as immunoglobulins, are found both on the surface of B-cells and the bloodstream. They act as antigen receptors on the surface of B-cells. The B-cells are stimulated to secrete antibodies into the bloodstream when the antibodies on the surface of these cells bind antigens. The human body generates about 50 million new antibody producing cells every day.

A different type of cell, the T-cell, also has receptors for antigens, but they only exist on the surface of the cell and are not secreted into the bloodstream. Binding of antigen onto the T-cell receptor contributes to the activation of the T-cell whose principal function is to recognize and destroy virus-infected cells. For further description of recognition of foreign molecules by the immune system, see Brandon and Tooze (1991) and the computer based tutorial by Pidgeon et al. (Cai et al., 1996; Yang et al., 1996).

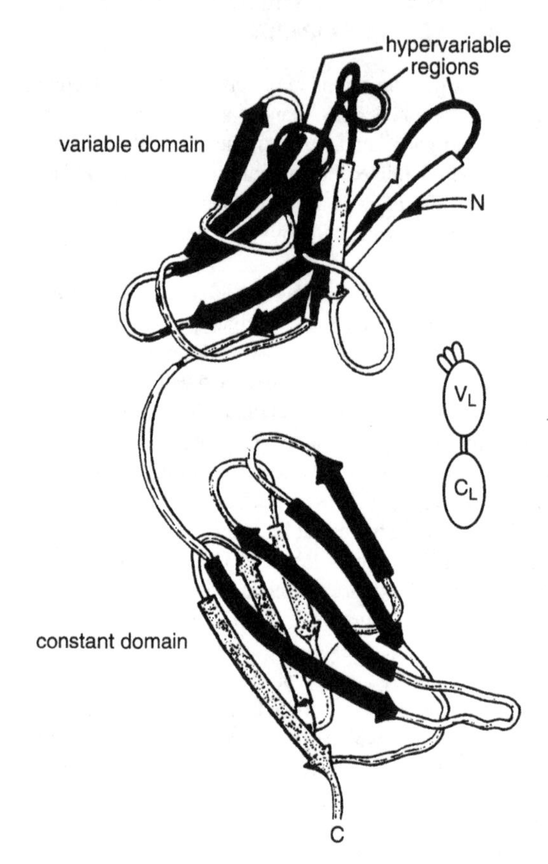

Figure 9.9 Conceptual representation of antibodies. Illustration of how variable regions (V_H and V_L in Figure 9.10) interact with antigen; the variable and constant domains in the light chain of immunoglobulins are folded into two separate globular units. In both domains the four-stranded β sheet is indicated by the dark color. The hypervariable CDR regions are at one end of this elongated molecule. (From Branden and Tooze, Figure 12.9(a), p. 187, copyright Garland Publishing, 1991).

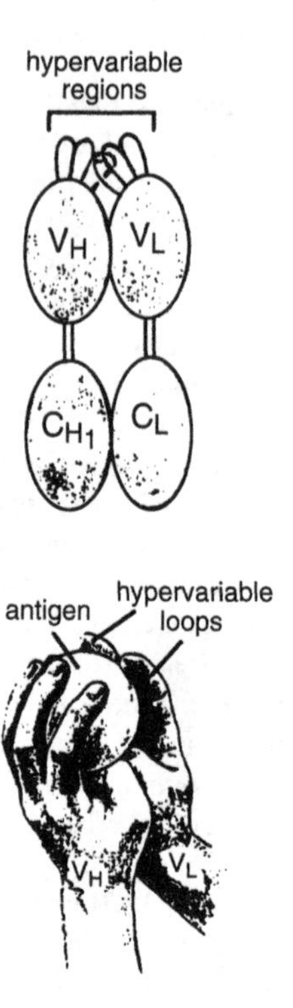

Figure 9.10 In the Fab fragment as well as in the intact immunoglobulin molecules the domains associate pairwise so that V_H interacts with V_L and C_{H1} with C_L. By this interaction the CDR regions of both variable domains are brought close to each other and together form the antigen binding site. (From Branden and Tooze, Figure 12.9(b), p. 187, copyright Garland Publishing, 1991).

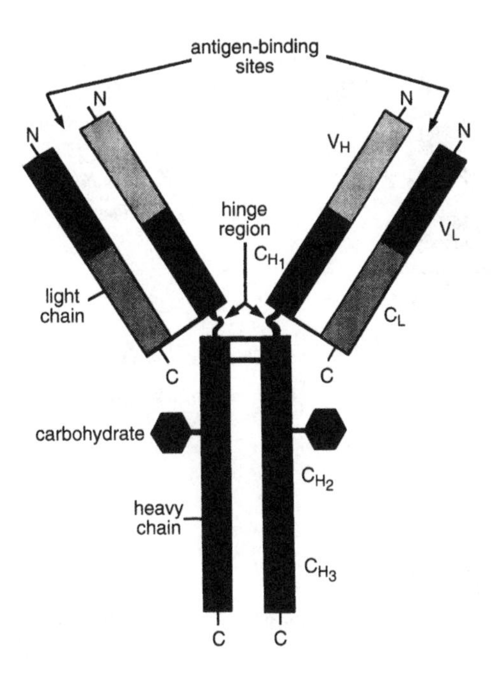

Figure 9.11 Diagram showing the heavy and light chains (i.e., high and low molecular weight fragments), variable regions (denoted V_H and V_L) and hinge region. (From Branden and Tooze, fig. 12.2, p. 180, copyright Garland Publishing, 1991).

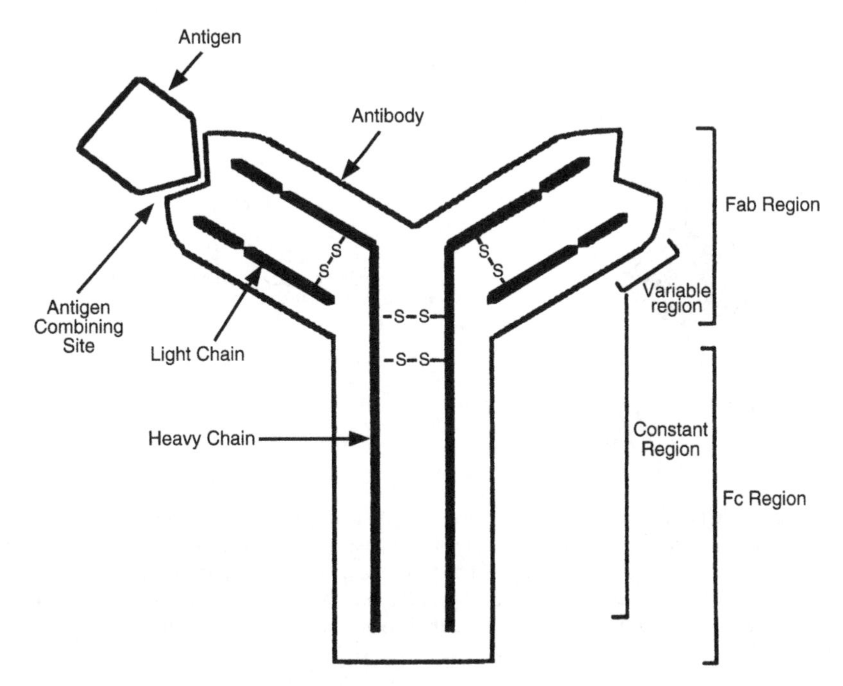

Figure 9.12 Schematic summary of interactions between antigen and antibody. (From Cai et al., 1996, fig. 8.1, p. 177.)

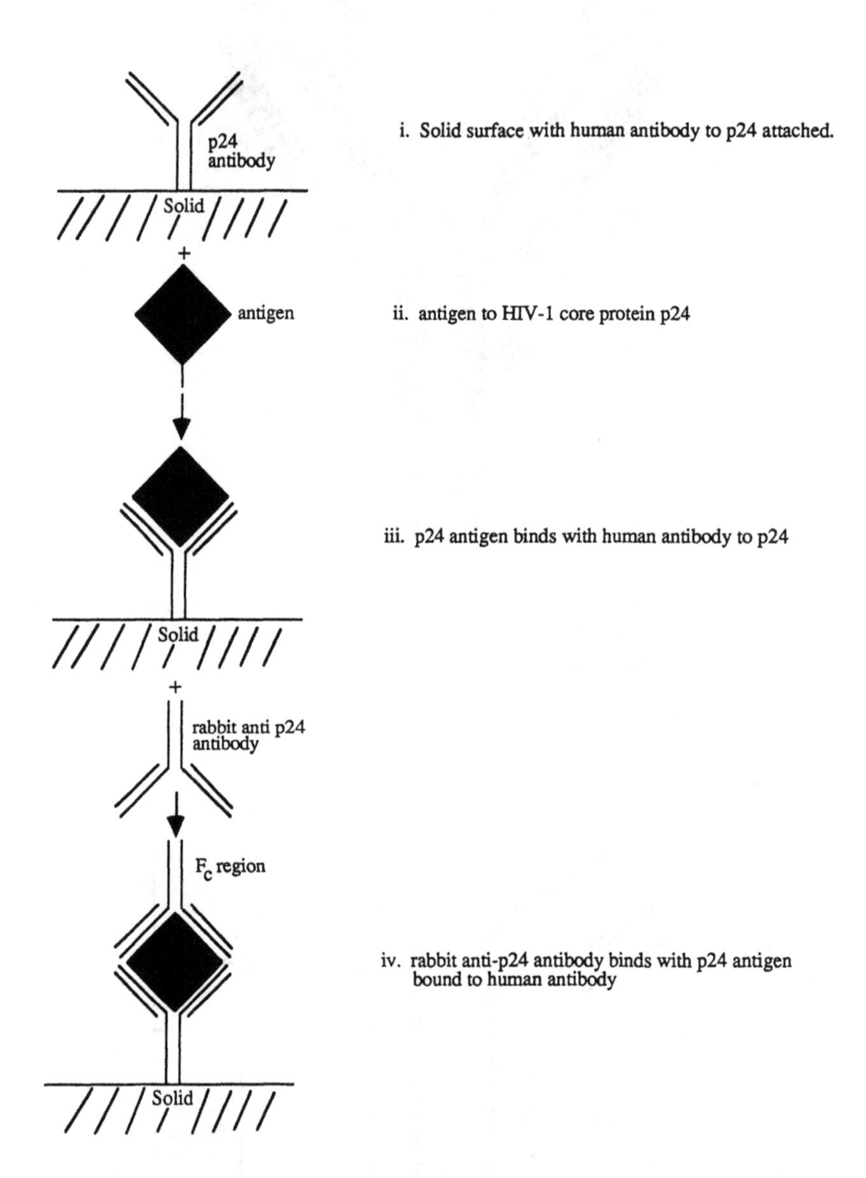

Figure 9.13 Schematic representation of format of HIVAG-1 assay for HIV-1 antigen for p24. The HIV-1 p24 antigen is bound between two antibodies (human Ab to HIV-1 p24 on the bead and rabbit Ab). This format is called a "sandwich." The colored reaction product is directly proportional to the concentration of HIV-1 p24 Ag in the specimen. The intensity of the color provides the basis for quantitation of p24. The affinity between the antigen and antibodies determines the sensitivity of the assay. (Adapted with permission, Cai et al., 1996, fig. 8.3, p. 188.)

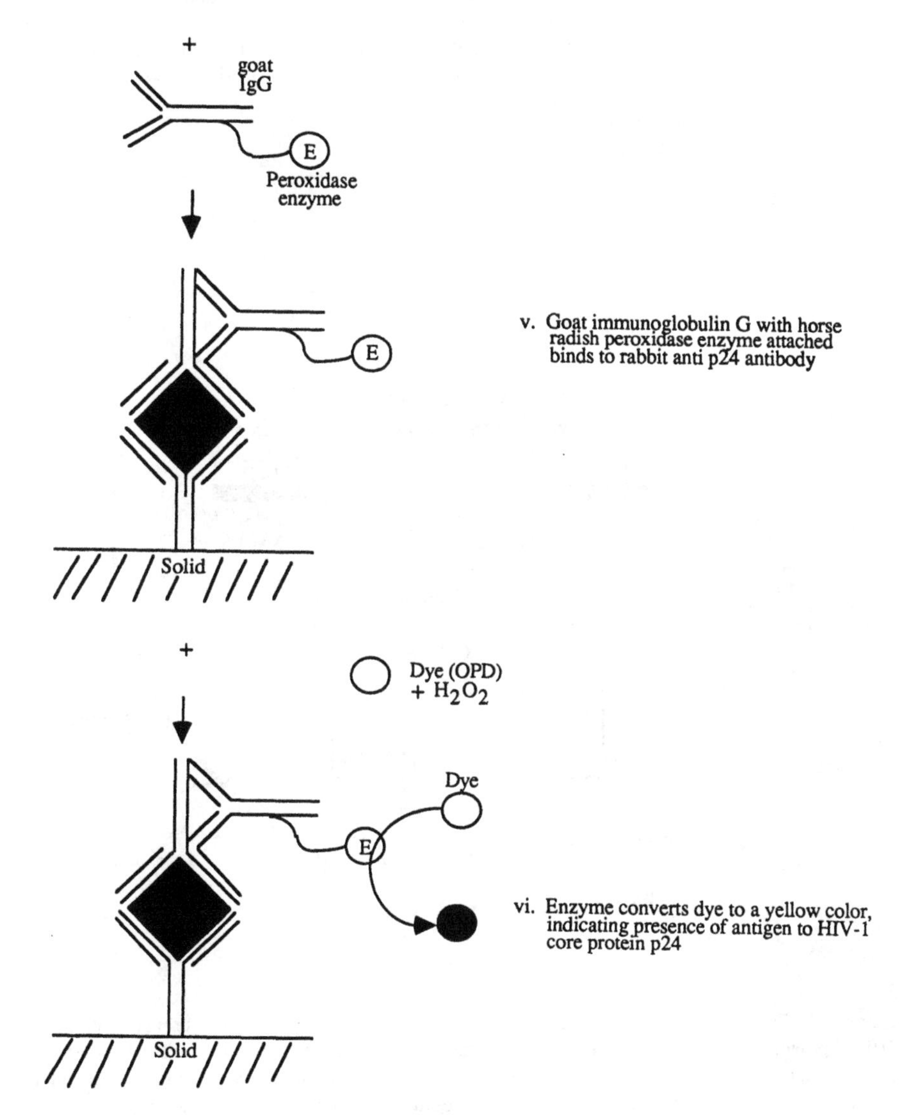

Figure 9.13 Continued

TABLE 9.3 Genes of HIV-1

Gene	Function
env	Encodes virus coat proteins
gag	Encodes proteins of the inner core
net	Function not known but acts in vivo to main high level of infection
pol	Encodes viral enzymes, including reverse transcriptase
rev	Encodes Rev protein, which regulates transfer of unspliced RNA to autoplasm
tat	Encodes Tat protein, which induces high-level expression of HIV genes
vif	Increases infectivity of virus particles
vpr	Encodes transcriptional activator
vpu	Participates in viral assembly and budding

Source: From Watson et al. (1992), table 25-1.

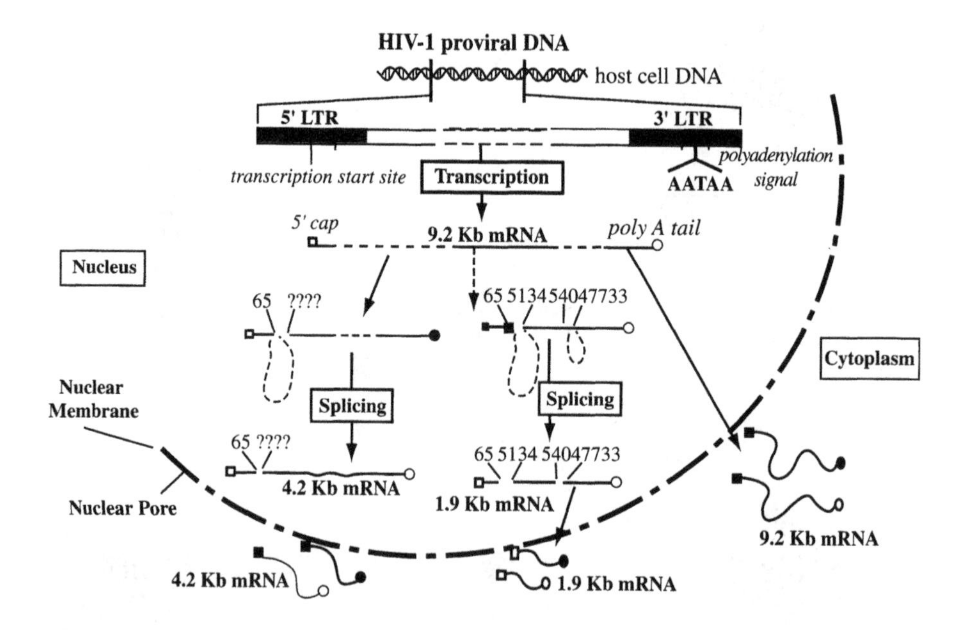

Figure 9.14 Events that occur within the nucleus and cytoplasm of an infected cell during transcription of HIV-1 proviral DNA (black) incorporated into the host cell genome. The entire proviral DNA is initially transcribed into a 9.2 kb mRNA. This large mRNA transcript contains a mGppp cap at the 5′ end and a poly-A tail at the 3′ end. The 3′ end poly-A tail facilitates mRNA transport out of the nuclear pores. From the 9.2 kb mRNA, one splicing event generates a 4.2 kb HIV-1 mRNA, and two splicing events generate the 1.9 kb mRNA. All three HIV-1 transcripts enter the cytoplasm. However, the three transcripts do not enter the cytoplasm at the same time. The temporal dependence of the number of mRNA copies in the cytoplasm depends on the HIV-1 proteins in the cytoplasm. The first HIV-1 mRNA to enter the cytoplasm in high copy number is the 1.9 kb mRNA, which codes for Tat, Rev, and Nef. The nonspliced 9.2 kb mRNA is part of the mature HIV-1 virion, and copious amounts of this 9.2 kb are present. (From Yang et al., 1996, fig. 7.5, p. 165).

9.11 HIV IS A LENTIVIRUS, A TYPE OF RETROVIRUS WITH CYTOPATHIC ACTIVITY

The genome of the lentivirus HIV has 9.2 kilobase pairs and is more complex than other retroviruses. HIV has at least four other genes that regulate viral expression in addition to the *gag*, *env*, and *pol* genes common to other retroviruses, as shown in Table 9.3 (Watson et al., 1992).

Usually when a retrovirus infects a cell, its RNA genome is converted into DNA by the virial enzyme, reverse transcriptase, and is then permanently integrated into the host cell's genome so that the viral DNA is replicated together with the host DNA each time that the cell divides. The integrated viral DNA generates messenger RNA for the translation and expression of viral proteins, as well as the RNA that becomes part of the genome of new virus particles (Figure 9.14). The virus is assembled in the cytoplasm, and then buds from the cell membrane to form new virus particles, while leaving the cell intact (Figure 9.15). This differs from lytic viruses, which enter cells, take over, replicate massively, and then release in large numbers by rupturing the cell wall and killing the cell. Unlike most retrovirus, however, HIV kills the host cells that it infects after it reproduces hundreds of copies of itself.

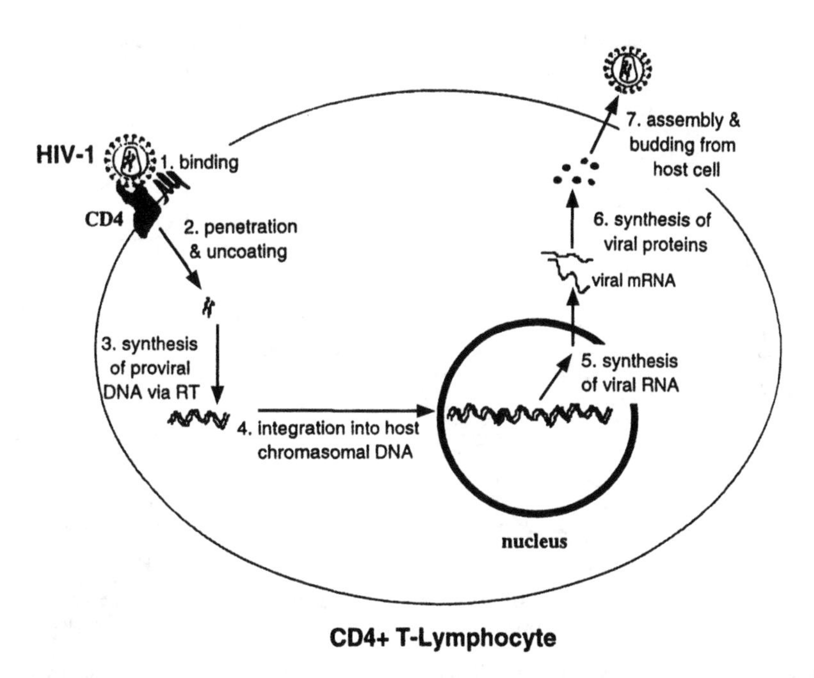

Figure 9.15 Schematic representation of the HIV-1 life cycle leading to formation of new virus particles. Step 1 depicts HIV-1 binding to T-cells via the gp120-CD4 receptor interaction. In step 2, the virus fuses with the host cell membrane and uncoats itself in the cytoplasm. In step 3, RT generates proviral DNA from HIV-1 RNA. In step 4, the proviral DNA forms a duplex that integrates into the host cell genome using the viral enzyme integrase. After a latent period, the HIV-1 gene is expressed and viral mRNA is synthesized (step 5). Step 6 shows the production of HIV-1 proteins from the mRNA and the subsequent budding of the virus particles. (Reprinted with permission, Yang et al., 1996, fig. 7.7, p. 169.)

9.12 AZIDOTHYMIDINE (AZT) STOPS HIV REPLICATION BY INHIBITING THE ENZYME, REVERSE TRANSCRIPTASE

Among the first drugs demonstrated to inhibit HIV was azidothymidine (abbreviated AZT). In AZT, the 3′ hydroxyl on deoxy-thymidine is replaced by the azido ($-N_3$) group as shown in Figure 9.16 (Watson et al., 1992). The DNA polymerases of the cell prefer thymidine triphos-

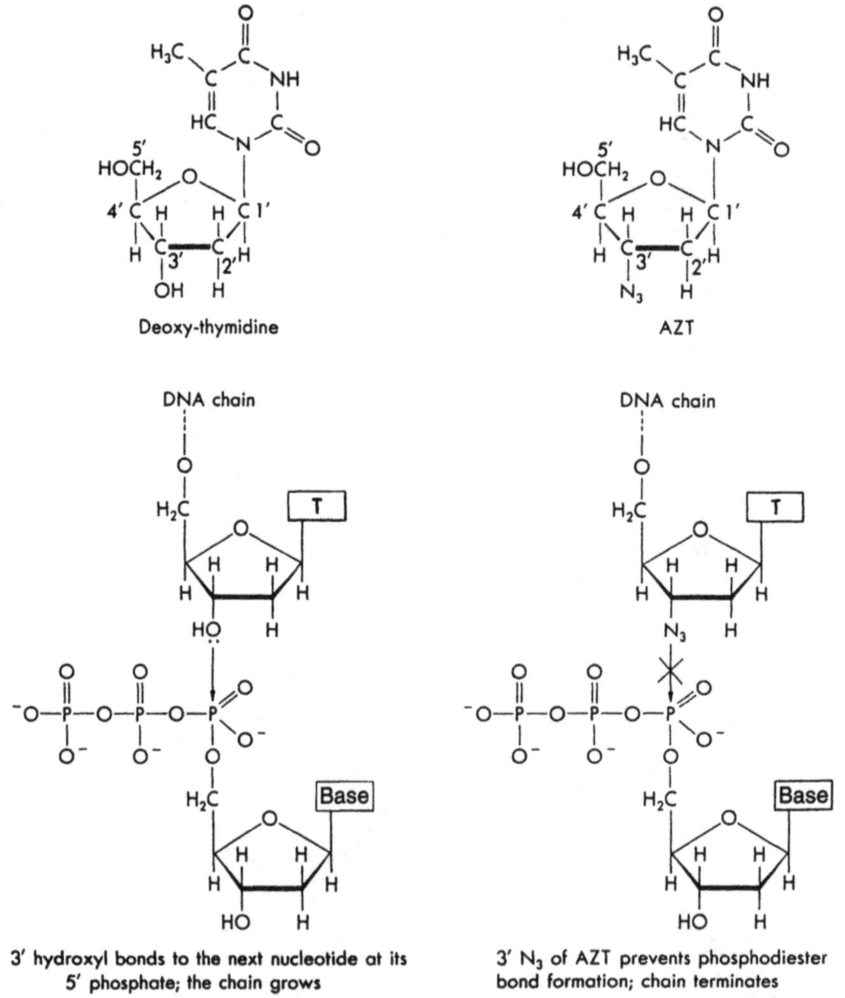

Deoxy-thymidine

AZT

3′ hydroxyl bonds to the next nucleotide at its 5′ phosphate; the chain grows

3′ N_3 of AZT prevents phosphodiester bond formation; chain terminates

Figure 9.16 AZT works as a chain terminator during the synthesis of the proviral DNA by reverse transcriptase. AZT is an analogue of thymidine in which the hydroxyl group on carbon 3 of the deoxyribose ring has been replaced by an azido group (N_3). Nucleosides are phosphorylated before they are added to the growing DNA chain. A phosphodiester bond is made between the hydroxyl group on the last nucleotide incorporated, and the incoming deoxyribonucleoside triphosphate. If the last nucleotide incorporated was AZT, the hydroxyl group is missing, the next nucleotide cannot be added, and synthesis stops. AZT is a dideoxypyrimidine analog. There are also dideoxypurine analogues that have similar activities. One of these, dideoxyinosine, has been useful in treating childhood AIDS. (From Watson et al., 1992, fig. 25-8, p. 495.)

TABLE 9.4 Sales of Reverse Transcriptase Inhibitors in 1995

Abbreviation	Chemical Name	Supplier	U.S. Sales ($ millions)
AZT (zidovudine)	3′-Azido-3′-deoxythymidine	Glaxo-Welcome	141
3TC (lamivudine)	-2′, 3′-Dideoxy-3′-thiacytidine	Glaxo-Welcome (licensed from Biochem Pharma)	NA[a]
d4T		Bristol-Myers Squibb	44.5
ddC (zalcitabine)	Dideoxycytosine	Roche	33.0
ddI (didanosine)	Dideoxyinosine	Bristol-Myers Squibb	19.6

[a]Revenues of $39.9 million from January to March 1996.
Source: From Cohen (1996b).

phate while HIV's reverse transcriptase preferentially incorporates AZT, taken up by an infected cell, instead of thymidine into the growing DNA strand. DNA synthesis is stopped because the 3′ hydroxyl on the thymidine required for making the sugar phosphate backbone of the DNA strand is missing and the reverse transcriptase enzyme cannot join the AZT to the next nucleotide (Figure 9.16). AZT is an effective therapeutic agent because it inhibits HIV replication at a concentration that is 10 to 20 times less than the concentration that is toxic to other cells. The complexity of HIV treatments is reflected in their costs.

AZT was licensed in 1987 and was still extensively used in 1995, as indicated by its sales Table 9.4. It grossed $141 million in the United States and $317 million worldwide in 1995, according to its manufacturer, Glaxo-Welcome, with cumulative worldwide sales of $2.5 billion by June 1996 (Cohen, 1996b). The U.S. sales of all types of reverse transcriptase inhibitors totaled approximately $379 million in 1995. Since then, a second reverse transcriptase inhibitor (3TC), known as EPIVIR, as well as a different type of drug known as a protease inhibitor (discussed in Section 9.16) have been added to drugs used to treat HIV (Carey, 1998, 1999). The total number of anti-HIV drugs were thirteen in 1999 (Waldholz, 1999). Drug cocktails introduced in 1996 combined reverse transcriptase inhibitors with protease inhibitors into "drug cocktails" that proved to be more effective than individual drugs, taken alone (Waldholz, 1998; Carey, 1999). Sales of protease inhibitors, alone, exceeded $1 billion in 1997–98. In 1997, sales were $582 million worldwide (of which $299 million was in the U.S.) for Merck's Crixivan, and $467 million for Agouron's Viracept in the year ending June 30, 1998 (Rundle and Waldholz, 1999). (Note, in 1999, Warner Lamber planned to acquire Agouron Pharmaceuticals). By 1999, there were about a dozen AIDS treatments on the market. State-run AIDS Drug Assistance Programs (or ADAPs) in the US who purchase therapies for HIV infected people that lack insurance and are ineligible for state Medicaid programs had outlays that increased from $52 million in fiscal 1996 to an estimated $700 million by 2000 (Waldholz, 1998). While an overall estimate of annual sales of anti-HIV drugs is not known to this author, it may be between $1 to 2 billion.

Mutations in the *pol* gene lead to expression of reverse transcriptase that is resistant to AZT but not necessarily to other deoxynucleotides. This initially led to a therapy of alternating treatments between different reverse transcriptase inhibitors (Watson et al., 1992). Subsequent research showed that strategies for combining antiretroviral drugs must be carefully chosen, since adding a single new drug to a failing or inadequate regimen can select for drug-resistant variants of the virus (Lange, 1997). An increased understanding of the process of rapid mutation of the virus has since shown that larger doses of reverse transcriptase inhibitors in combination with other drugs are needed (News Item, 1996a; Wilson, 1996), and that combinations of nucleoside reverse transcriptase inhibitors are superior to AZT, alone. Combinations of reverse transcriptase inhibitors are able to maintain plasma HIV levels to one-tenth of their initial value for about two years (Richman, 1996). The combination of the reverse transcriptase inhibitor, 3TC, with AZT and a protease inhibitor appears to have a synergistic benefit. However, by 1999 newly infected individuals began to carry viruses that were resistant to at least one of thirteen drugs used in combination drug therapy (Waldholz, 1999). At about the same time, Dupont introduced a non-nucleoside reverse transcriptase inhibitor (named Sustiva) that could be taken orally, once a day, compared to other therapies that required multiple pills (Waldholz, June 30, 1998).

9.13 PROTEASE INHIBITORS PREVENT CLEAVAGE OF *gag*- AND *pol*-ENCODED VIRAL PROTEINS REQUIRED FOR VIRAL REPLICATION

Virus-encoded aspartyl protease cleaves and proteolytically modifies the long polyprotein produced from the *gag*- and *pol*- genes. The *gag* Pr 55 protein precursor gives rise to four structural proteins of the virus coat, while the *gag-pol* Pr 160 polyprotein is cut into three viral enzymes including reverse transcriptase and the protease itself (Watson et al., 1992). The protease is needed to cut these polyproteins into the mature proteins found in infectious virions (virus particles). HIV protease inhibitors are substrate-based inhibitors that bind at the active site of the protease and thereby block the enzyme from acting on the polyprotein substrate as illustrated in Figure 9.17 (Ridky and Leis, 1995). However, unlike AZT, where the increase in resistance

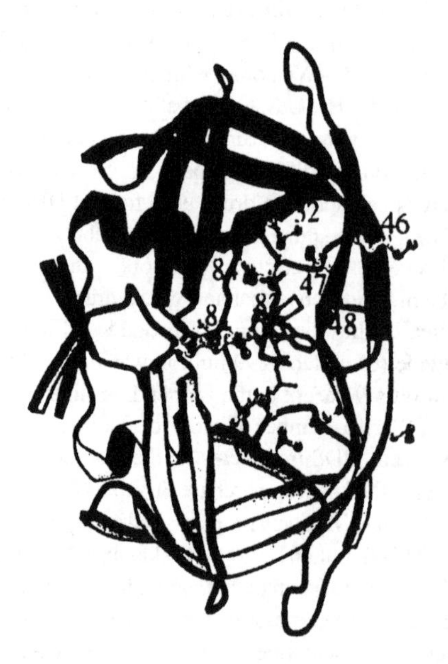

Figure 9.17 The structure of the HIV-1 protease complexed with the inhibitor Ro-8959 (Sequinavir; Structure in Figure 9.20). The inhibitor, which is in a β-sheet conformation, is shown by the center. Enzyme amino acid residues that change in developing drug-resistant protease phenotypes are depicted by the *ball* and *sticks*. Individual enzyme subunits are shown in light and dark green. The residue numbers for only one of the two identical subunits are shown. This figure was prepared by Dr. Alex Wlodawer, NCI-Frederick Cancer Research Center. (From The American Society for Biochemistry & Molecular Biology, Ridkey and Leis, 1995, fig. 1, p. 29622.)

to the drug is measured in years, protease inhibitors, when used alone, can lose their effectiveness "within months, sometimes only weeks" (Wilson, 1996). Furthermore mutation of the virus and slight changes in the protease's structure at the active site requires that protease inhibitors be given in high enough doses to completely suppress viral replication. Otherwise, the virus mutates to a drug-resistant form. Hence, protease inhibitors are used in combination with other types of anti-HIV drugs.

Combinatorial synthesis has resulted in new protease inhibitors. For example, in 1997 Novartis reported identification of an oligomer of a peptoid (i.e., a peptide analogue) and a peptide, which was screened from a library of 3 million combinatorial compounds. This oligomer, named CPG 64222, is resistant to proteolytic activity and inhibits HIV replication in human lymphocytes at nanomolar concentrations. Simplified analogues are being studied to identify molecules with even greater in vivo effectiveness (Borman, 1997).

9.14 MUTATION OF HIV TO A DRUG-RESISTANT FORM IS PROMOTED BY RAPID TURNOVER OF A LARGE POOL OF INFECTED T-CELLS

HIV's primary target is helper T lymphocytes (i.e., T-cells, Figure 9.18) that display CD4 and chemokine receptors on their surfaces (Figure 9.19). The virus also infects macrophages and dendritic cells (Watson et al., 1992; Dagani, 1996). The viruses in macrophages dominate during the symptom-free stage of HIV and play a role in transmission of HIV from person to person (Dagani, 1996). The half-life of the virus is approximately 6 hours, and an estimated 10^{10} particles are generated daily (Richman, 1996). While HIV and most single-stranded RNA viruses undergo 3×10^{-5} mutations/nucleotide per replication cycle, the high levels of 10^{10} virions generated each day and a genome size of 10^4 nucleotides means that virtually all possible mutations are generated daily. The number of new cells infected each day is estimated to be 10^9 during

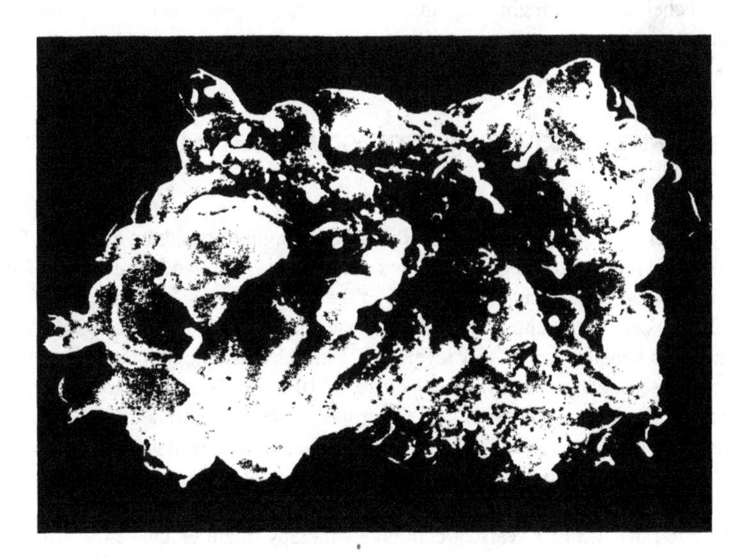

Figure 9.18 Helper T-cell (large body) under attack. Like piranhas attacking a baby whale, HIV particles (tiny dots) gang up on a T-cell (an immune system cell). Each virus particle measures 10 nanometers across, whereas the diameter of a T-cell is roughly 1000 times as large; AIDS virus (small particles). (Copyright Boehringer Ingelheim International GmbH. Reproduced with permission, Roberts, 1996 from FASEB J.)

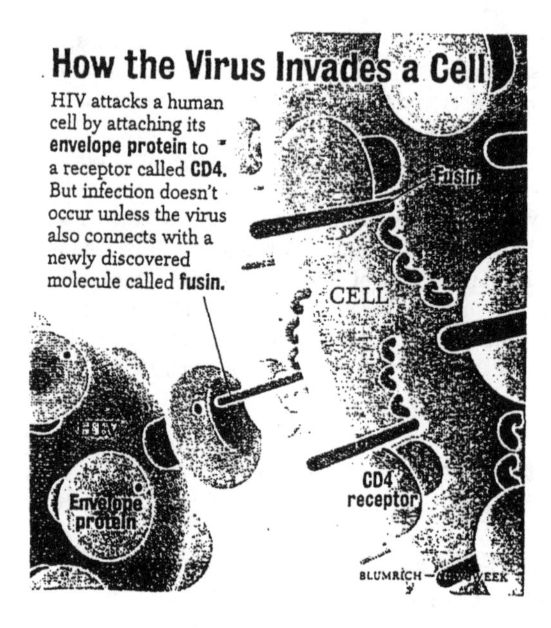

Figure 9.19 Conceptual representation of virus particle docking believed to occur in later stages of HIV infection. (Reproduced with permission from Newsweek, Cowley, 1996, p. 62.)

latent or steady-state stages. High turnover combined with the low fidelity of the reverse transcriptase, genetic recombination between the HIV genome and endogenous viral sequences, and other mechanisms can lead to rapid emergence of drug resistance (Ridky and Leis, 1995). Hence potent chemotherapy regimes against a broad range of potential mutations are needed to prevent formation of drug-resistant mutants.

9.15 NEW DRUG DISCOVERY IS NECESSITATED BY RAPIDLY MUTATING HIV VIRUS

Combination therapies, such as the use of the protease inhibitor, indinavir (Figure 9.20), together with the reverse transcriptase inhibitors zidovudine (AZT) and lamivudine (Table 9.4) can reduce viral load to less than 1% with 85% of the patients having undetectable plasma viremia after 24 weeks of treatment. Precise benefits await documentation (Ho, 1996). Although a few patients have kept the virus at low levels for 21 months after stopping therapy, others have seen the virus rise to dangerous levels in as little as 10 days (Waldholz and Tanouye, 1999). Another combination therapy consisting of the protease inhibitor nelfinavir (structure in Figure 9.20), zalcitabine (ddC) and 3TC, resulted in a rapid decrease in viral load during the first two weeks followed by a second slower decay. A proposed explanation is that active viruses are quickly killed, while the latent viruses (in the cells) are killed off more slowly. A two-stage model suggested that 1.5 to 3 years of continuous therapy might eradicate the virus (Wilson, 1996), but this turned out not to be the case. Further analysis also indicated that some forms of DNA are not detectable by the immune system or accessible to current drugs.

The first results were obtained with 6 patients and showed the combination of Norvir, AZT, and 3TC after 6 months resulted in no detectable virus in the blood, nor in the tonsils. The tonsils

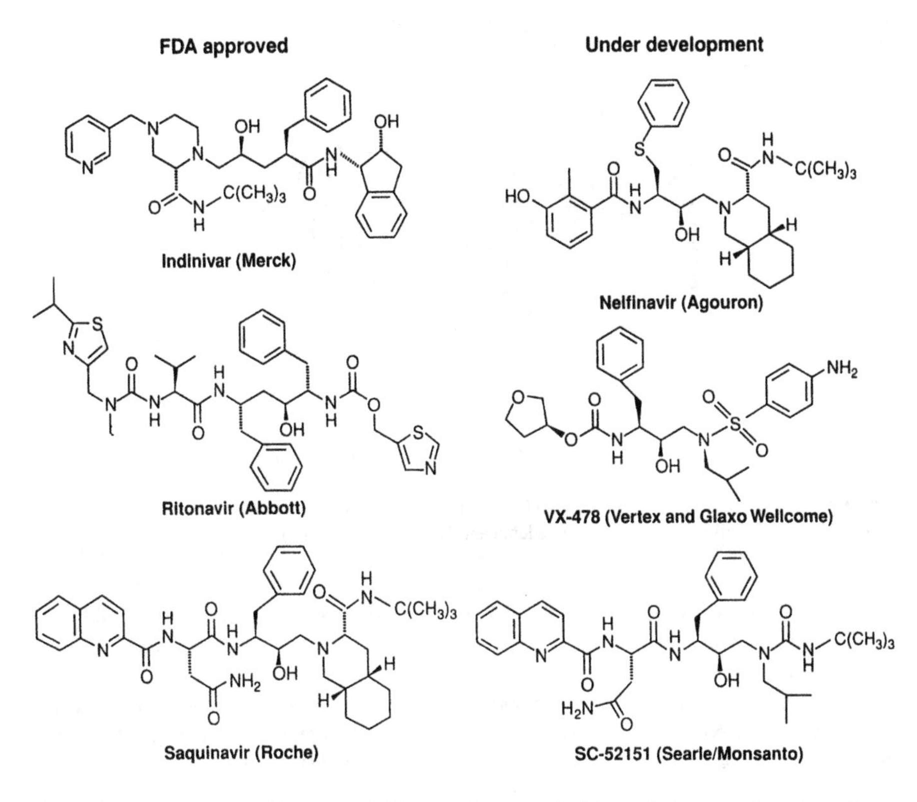

Figure 9.20 Structures of HIV protease inhibitors. (Reproduced with permission from Chem. Eng. News, Wilson, 1996, p. 43.)

are made of lymphatic cells and lymphatic cells are believed to harbor much of the HIV that is not circulating in the blood stream (Waldholz, 1996b). However, there were exceptions. The viral load returned from undetectable levels to "27,000 particles/sample" after 72 weeks in one case. This was of obvious concern as 100,000 people are taking drug combinations and the longevity of beneficial effects were not known in 1996 (Waldholz, 1996a). HIV integrates its genes into the host cell's DNA so that, when the cell divides, the viral DNA produces new virus particles. But if the HIV host cell does not divide, the viral DNA or provirus is not detectable by the immune system and is invulnerable to current (1997) AIDS drugs. Even if there is no detectable virus in lymph node cells, the infection can still persist in the form of HIV DNA, as was shown to be the case with at least one patient (Cohen, 1997). The near-term prospects were perhaps best summarized by Richman (June 1996, p. 1887). "In less than a year, the prospects for treatment of HIV-infected patients, at least those socioeconomically privileged, have improved dramatically. Important new questions about HIV pathogenesis can now be asked and investigated. Nevertheless, the prospects of drug resistance, the toxicity of current drugs, and the need for even greater antiretroviral activity will require the discovery and development of still more and better drugs" (Richman, 1996). This analysis did not include the subsequent development of gene-based detection methods that can now help to identify which anti-AIDS drugs a virus may be resistant to. Since there were 335,000 people in the U.S. who take anti-HIV

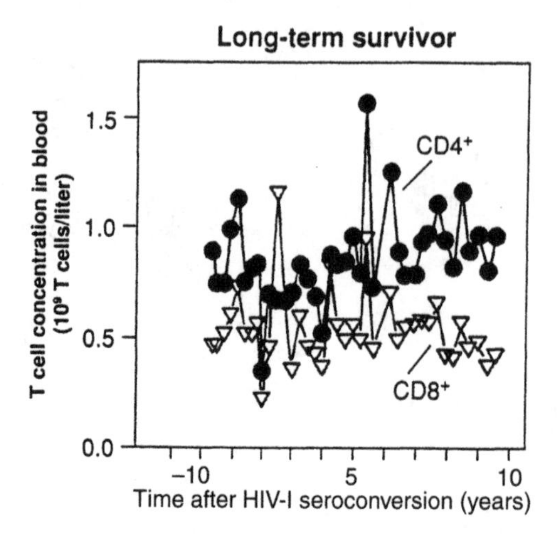

Figure 9.21 Natural history of HIV-1 infection. Long-term survivors of HIV infection maintain normal $CD4^+$ and $CD8^+$ T-cell numbers. (From Miedema and Klein, 1996, p. 505.)

drugs, tests for mutations of the virus that would make it resistant to a specific drug could yield benefits by finding the best drugs to use. The annual cost, however, is estimated at $270 million (Carey, 1998).

Virus replication and reinfection of blood cells occurs during the time period between when the patient is infected and symptoms of immunodeficiency become evident (Figure 9.8). HIV clearance may be balanced by its production during this period (Ho, 1996):

$$k_c V = k_\alpha N T^* \tag{9.7}$$

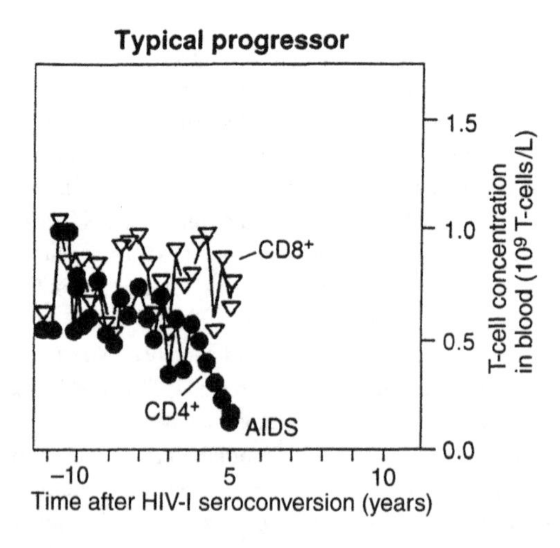

Figure 9.22 Natural history of HIV-1 infection. Long-term survivors of HIV infection maintain persistent Gag-specific CTL responses, and low numbers of $CD4^+$ T-cells infected with HIV-1. Individuals who eventually show symptoms of disease (typical progressors). (From Miedema and Klein, 1996, p. 505.)

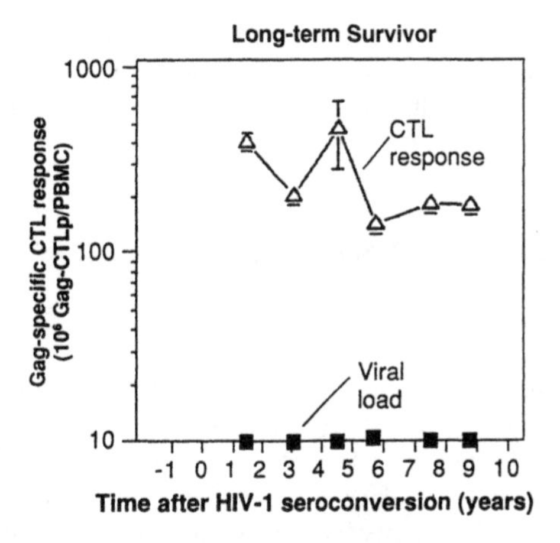

Figure 9.23 Natural history of HIV-1 infection. Long-term survivors of HIV infection maintain relatively stable numbers of $CD4^+$ T-cells during a variable latency period. (From Miedema and Klein, 1996, p. 505).

where V is the virion concentration, T^* is the number of virus producing cells, N is the number of virions produced per infected cell (burst size), k_c is the rate constant for virus clearance, and k_α is the rate constant for the loss of virus producing cells T^*. V is proportional to NT^* across a range of infected persons, since the constants k_c and k_α do not vary significantly among patients. The measurement of virial loads gives a useful marker for clinical outcomes of HIV infections. Data from Wolinsky et al. (1996a), as presented by Miedema and Klein (1996) in Figures 9.21 to 9.24, appear to support this hypothesis, although there continues to be debate on whether viral diversity (Nowak et al., 1996; Wolinsky et al., 1996b), viral load, and the onset of AIDS are correlated (Figures 9.23 and 9.24) with long-term survivors having low viral loads (Figures

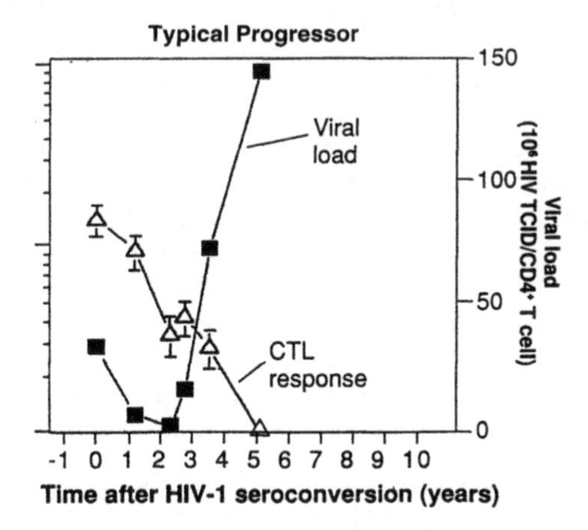

Figure 9.24 Decline of HIV-1-specific CTL responses coinciding with increasing viral load, leading to total collapse of the immune system and inevitable development of AIDS. TCID, tissue culture infectious dose; PBMC, peripheral blood mononuclear cell; CTLp, CTL precursors. (From Miedema and Klein, 1996, p. 505.)

9.21 and 9.23). These observations have also led to the hypothesis that lowering viral loads with antiviral drugs should result in improved prognosis. However, the ability of the virus to mutate rapidly, and its ability to integrate its genes into the host cell's DNA, makes it difficult to completely eradicate it.

9.16 PRODUCTION SCALE-UP IS PART OF PRODUCT DISCOVERY: CASE STUDY OF PROTEASE INHIBITORS

The accelerated discovery of new drugs and affinity ligands does not ensure an accelerated pace of development of new products. The introduction of new drugs must consider how the transition from the laboratory to a manufacturing facility is to be achieved where being among the first to market is critical. The story behind introduction of the protease inhibitor Crixivan® (indinavir sulfate), which is used to treat HIV infection in combination with reverse transcriptase inhibitors, yields insights on the impact of competitive pressures on the time line of introducing a new life-saving product. Reporters' accounts on protease inhibitors also give insights into the magnitudes of investments, market sizes, and costs associated with development of new drugs.

Protease inhibitors were quickly approved by the FDA in late 1995 and early 1996 for three of the dozen companies that are manufacturing them, with a fourth licensed in March 1997 (Cohen, 1996a; Ingersoll, 1997).[7] Three others were in human trials (Cohen, 1996a). By 1999, there were 13 HIV medicines on the market that were either protease inhibitors or reverse transcriptase inhibitors. Key developments in protease inhibitors (as of March 1996 through March 1997), summarized by Waldholz (1996a,b,c), Cohen (1996a), Tanouye (1996), and Rundle (1997), give insights into the rapidity with which a new class of drug can be adopted.

The development of the protease inhibitors indivinar and ritonavir may have benefited from preexisting efforts (prior to 1987) to develop inhibitors of renin, another protease that regulates blood pressure. It was already known that the inhibitor would be hydrophobic, and to be effective as a drug it would need to be water soluble if it is to be taken orally. Merck scientists derivatized renin inhibitors to make them soluble while at the same time minimizing loss in potency. While the first candidate was shelved because it caused liver damage in laboratory animals, the knowledge gained on synthesis techniques eventually led to the synthesis of Indinavir. Abbott also developed a first-generation protease inhibitor whose solubility was limited, and therefore of limited utility. This led to ritonavir in 1992, which entered trials in 1993, at about the same time as Merck's product (Cohen, 1996a,b; Tanouye, 1996).

Hoffmann-LaRoche won the race to market by licensing saquinavir (tradename of Invirase®; structure in Figure 9.20) and putting it on the market by December 1995 (Cohen, 1996a; Wilson, 1996). However, this inhibitor had limited bioavailability on a systemic basis, possibly due to its rapid metabolism—"and thus less effective" (Cohen, 1996a). Indinavir (Crixivan®) and ritonavir (Norvir®) were introduced by Abbott Laboratories and Merck, respectively, by mid-1996 (structures also given in Figure 9.20) (Cohen, 1996a; Wilson, 1996). Abbott's trials with 1090 patients showed that Norvir®, used in combination with reverse transcriptase inhibitors decreased AIDS-related disease and death in half after 7 months, while Merck's studies showed that Crixivan® together with reverse transcriptase inhibitors AZT plus 3TC gave no de-

[7]Hoffman-LaRoche's saquinavir came to market December 1995. Abbott Laboratories' ritonavir was approved on March 1, 1996—72 days after the request was filed. Merck's indinavir was approved March 13, 1996—42 days after filing (Cohen, 1996a,b; Waldholz, 1996a). According to a news report, top researchers at Merck made an effort to have FDA consider indivinar along side of ritonavir (Tanouye, 1996). Agouron Pharmaceuticals' Viracept, and Abbott's Norvir were approved in March 1997 to treat pediatric AIDS patients (Ingersoll, 1997). In April 1997 market shares of the four protease inhibitors were 52% for Crixivan (Merck), 23% for Invirase (Roche), 14% for Norvir (Abbott), and 11% for Viracept (Agouran) (Armstrong, 1997).

tectable HIV after 6 months in 20 of the 22 patients treated with this combination. By March 1997 Abbott's Norvir®, and Agouran Pharmaceutical's Viracept® were approved for children infected with HIV, of which 7200 have been diagnosed with AIDS since 1981 compared to 550,000 adults in the United States (Ingersoll, 1997).[8]

A smaller previous trial had shown that patients treated with lower doses of indinavir, immunotherapy developed resistant strains of HIV (Cohen, 1996a). Resistance to all of these protease inhibitors can occur, since HIV replicates and acquires resistance-associated mutations. These drugs can lose potency within months, particularly if there is poor patient compliance with the dosing schedule. There are other side effects including nausea, kidney stones, and metabolic interference with other drugs (Wilson, 1996). Nonetheless, the benefits of combination therapies of protease and reverse transcriptase inhibitors resulted in a rapid increase in the growth of demand for saquinavir, which grossed $35 million in revenues between December 1995 and March 1996, when it was first introduced. A case history of indinavir by Tanouye (1996) gives insights into the challenges inherent to the rapid scale-up and marketing of a new drug.

Some projections placed the need for protease inhibitors at up to 1 million kg/yr (Tanouye, 1996), since a dose of 2.4 g/day gave encouraging results with indinivar. This corresponds to a kilogram of drug per year per patient. The amount of indivinar for 90,000 HIV patients was reported to be equivalent to producing enough Vasotec® (Merck's hypertension drug) for 21 million people (News Item, 1996a; Tanouye, 1996). Indinavir is chemically more complex, since it has 5 chiral centers compared to Vasotec's 3. Large-scale chemical synthesis of such a complex molecule required a synthesis process that lasts 6 weeks and involves 15 steps, while other pharmaceuticals required about one-third of the number of steps and one-third of the time (Tanouye, 1996).

Competitive pressures, namely the race to be first in the market, pushed Merck's submission of the application to the FDA and full-scale production to be done sooner than had been planned for (Tanouye, 1996). During its application for a license, Merck assured the FDA that it would make enough indinavir sulfate to initially supply up to 30,000 patients. This was a major commitment, since a shortfall would place patients, who were taking indinavir sulfate, at risk of developing resistance to the drug even if they stopped taking it for a short period of time. Abbott Laboratories faced similar challenges in scaling up the production of Norvir®, and addressed it by contracting out 80% of its production to suppliers in Japan, Italy, and France.

Merck decided to handle much of the production in-house in order to minimize the chance of missed deadlines by an outside supplier. The company was willing to commit to building production capacity before the drug was approved by the FDA and to abandon its investments if the drug began to fail in clinical trials. Seven of the early steps were handled by two suppliers, Mitsui (Japan) and DSM NV (the Netherlands), while the rest were handled in-house, at Elkton, VA, and Albany, GA. Construction of the bulk manufacturing plant had begun in March 1995, with initiation of the first bulk drug batch on May 27, 1996. The first batch of bulk drug was completed at the Elkton site on July 6. Demand had increased during this period and keeping the production ahead of demand continued to be a challenge. Sales of Crixivan® approached

[8]Agouran Pharmaceuticals, Inc. raised $500 million from investors since 1984 and spent over $100 million since 1994 to develop and market Viracept® (Armstrong, 1997; Rundle, 1997). Its wholesale price was set at $5650/year, which was 29% higher then the $4380 or Merck's Crixivan® for adults (Rundle, 1997; Walsholz, 1996a), but lower than Hoffman-LaRoche's Invirase® at $5800, and Abbott Laboratories Norvit® at $6500 (Rundle, 1997). Another type of anti-HIV drug is Dupont's Sustiva® with an annual wholesale price of $3920 (Waldholz, 1998). Sustiva inhibits the reverse transcriptase enzyme. The high costs of these and other drugs have limited access of people living in poor nations to treatment and has prompted large companies to reduce prices by up to 90% for treatment programs administered by the UN (Waldholz, 2000).

$500 million and was being taken by 130,000 people within 8 months of its introduction (Tanouye, 1996).

9.17 CORECEPTORS (AFFINITY LIGANDS) FACILITATE HIV BINDING AND INFECTION: CXCR4, CCR5, AND GP 120

The 1996 discoveries of the role of the cell membrane receptor protein, fusin, on T-cells and CCR5 on the surface of microphages helped to explain why CD4 animal cells and human cells containing only the CD4 receptor[9] are not infected (Cohen, 1996c; Dagani, 1996). Fusin is so named because it is a cofactor for HIV fusion with T-cells (it is now renamed CXCR4, and has also been called LESTR and HUMSTR) (Cohen, 1996d). Since two types of receptors are involved in docking of the virus on the surface of the cell (Cohen, 1996d; Feng et al., 1996; Balter, 1998; Wyatt and Sodroski, 1998), a chemokine or small molecule that binds the chemokine receptor (i.e., fusin protein site at the cell's surface) could block binding by the HIV virus (Figure 9.25). Subsequent research showed that the CD4 and CXCR4 receptors function together to remove and bind gp120 (Figure 9.7) protein in a manner that allows the virus' gp41 protein to breach the membrane of the T-cell, as shown in Figure 9.26 (Cohen, 1996d).

Experiments with natural and modified forms of the chemokine, RANTES, indicate that this protein binds to the fusin receptor and prevents HIV from infecting some types of cells that display both CD4 and chemokine receptors. The deletion of the first eight amino acids from the N-terminus of RANTES apparently avoids the normal response of the cell to RANTES binding where the cell is activated and produces significant amounts of other chemokines that cause inflammation (News Item, 1996b). The cloning and expression of fusin (Feng et al., 1996) provides a new affinity ligand for in vitro studies. This ligand could also find use in the capture and purification of a chemokine (not yet identified; see Dagani, 1996; Feng et al., 1996) which is believed to bind to the fusin receptor in vivo.

Another coreceptor, CCR5, is thought to be the main coreceptor in the sexual transmission of HIV, while CXCR4 is likely to be important in the more virulent forms of an established HIV. There are likely to be other coreceptors as well, although they have not yet been identified. While these coreceptors are attractive targets for antiviral therapy, the selection of the appropriate targets and antagonists is complicated by the high rate of mutation of the virus. For example, the blockade of CCR5 in an established infection might drive the virus to mutate to CXCR4 or other coreceptors resulting in a more virulent form of the disease (Moore, 1997).

According to a review by Wyatt and Sodroski, CCR5 is a coreceptor on the surface of CD cells that is used by most primary, clinical isolates of primate immunodeficiency viruses to enter CD4 cells during the early phases of infection. The protein CCR5 is thought to be a required coreceptor, since people who are genetically different and whose CD4 cells are unable to express CCR5 are also relatively resistant to HIV-1 infection. As the disease progresses, a portion of the structure of the gp 120 protein on the surface of the HIV-1 virus particles changes so that this protein, and therefore the virus, binds through another chemokine receptor, CXCR4, in the later stages of infection. The region of gp 120 that changes is referred to as the V3 (third variable loop) region. When antibodies bind to this region, or when the V3 region is deleted from the gp 120 protein, the modified gp 120 protein does not bind to the CCR5 receptor on the surface of the CD4 cell (Wyatt and Sodroski). The CCR5 and CD4 binding sites on the T-cell are believed to be essentially

[9]CD4 is a receptor on the surface of helper T lymphocytes to which the virus particles will bind. The name of CD4 containing T-cells is abbreviated to CD4 cells.

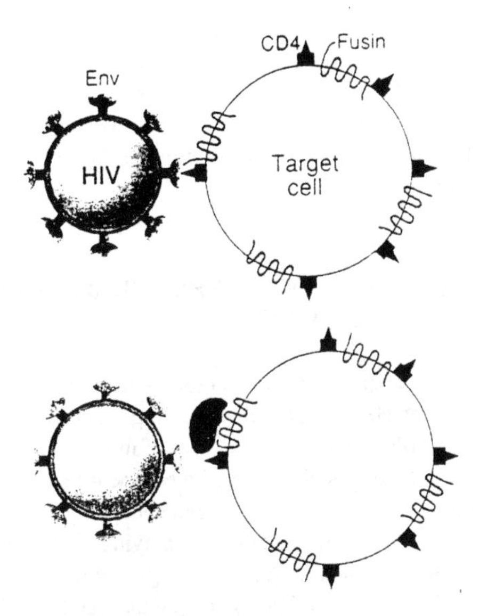

Figure 9.25 Hypothesized approach utilizing chemokine proteins to block infection. (Reproduced with permission from Cohen, 1996c, p. 809, Science.)

invisible to the immune system, and hence HIV infected people produce few antibodies that are effective in blocking these receptors. The increasing knowledge of the receptors will help to guide vaccine development. There is consensus that vaccines will be required to control the epidemic (Balter, 1998; Jasney and Clery, 1998).

Preliminary tests of AIDS vaccines were initiated in 1999 by companies including Merck, Rhone-Poulenc, and Vaxgen. Research on developing a vaccine dates' back to 1987 (Barrett et al., 1999). A vaccine based on protein from the virus particle's envelope was tried in the early 1990s and was unsuccessful. The cause for failure is now beginning to be understood as the role of receptors in virus docking and entry becomes better defined. Current vaccines, under development, are based on either generating antibodies against the virus, or promoting formation of cells in the human immune system that would attack the

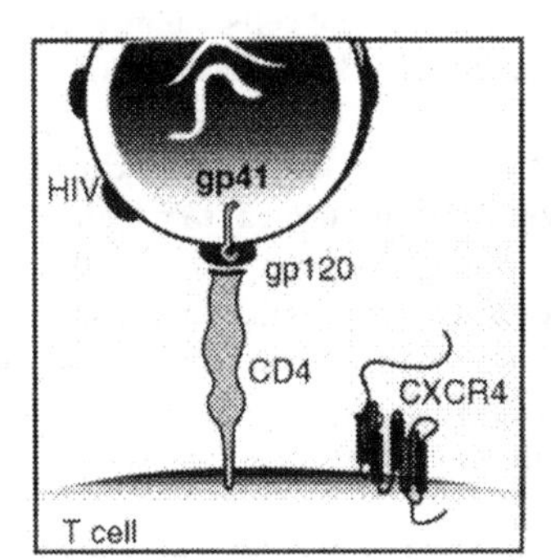

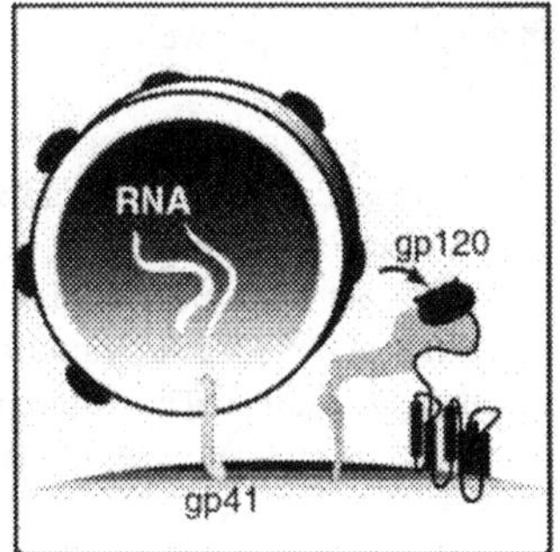

Figure 9.26 After docking onto a T-cell's CD4 receptor, HIV's gp 120 binds to chemokine receptors like CXCR4, allowing gp41 to breach the cell membrane. (Reproduced with permission from Cohen, 1996d, p. 502, Science.)

virus (Waldholz and King, 1999; Carey and Barrett, 1999; Barrett et al., 1999). Some of the vaccines that are likely to enter trials in 2000 are DNA-based. These are designed to deliver selected HIV genes to human cells, in order to trigger production of sufficient HIV associated proteins (which do not cause AIDS) to induce HIV-fighting T-cells to be generated (Barrett et al., 1999). Once induced, these cells would be ready to attack HIV, if the virus later appears.

9.18 RECEPTORS ARE IMPORTANT TO THE DYNAMICS OF HIV TRANSMISSION AND PATHOGENESIS

Mathematical modeling of the interactions of receptors with chemokines could add to the fundamental understanding of how HIV infects cells. The importance of receptors is indicated not only through advances in cell biology but also by cases of infected individuals who do not progress to AIDS and of repeatedly exposed people who escape infection. It is proposed that elevated levels of β-chemokine expression are involved in controlling HIV load (Weiss, 1996). As stated by Weiss (p. 1886), "The realization that HIV is playing on the keyboard of 7-transmembrane (7tm) G protein-coupled chemokine receptors is giving us new insights into virus-host cell interactions," with "cell biology illuminating the population dynamics of pathogenesis." A recent monograph on receptors suggests ways in which mathematical modeling of receptor/ligand interactions might provide a bridge between cell biology and engineering (Lauffenberger and Linderman, 1993).

9.19 A SENSITIVE AND RAPID SCREEN HELPS FIND THE CXCR4 (FUSIN) PROTEIN RECEPTOR IN CD4 CELLS

Progress in identifying the first of the HIV docking cofactors, known to exist since 1986, is attributed to the sensitivity and versatility of the assay used to detect fusion of an HIV surrogate (fibroblast cells) to Murine NIH 3T3 (mouse) cells modified to express human CD4 (protein receptors) on the cell's surface. Vaccinia virus[10] was used to convert fibroblast cells into Env-expressing cells by cloning the *env* gene which encodes for HIV virus coat proteins into the DNA of the vaccinia virus. Fusion between the transformed cells and CD4-expressing cells leads to activation of a reporter gene (*E. coli* lacZ) that expresses β-galactosidase. Murine cells mixed with Env-expressing fibroblast cells (abbreviated Env-cells) in microtiter plates (10^5 cells per well), and incubated for 3.5 hours at 37°C, can be readily examined for syncytia.[11]

The expression of β-galactosidase (which occurs if the two types of cells bind) can be detected by color (measured spectroscopically at 540 nm) formed from hydrolysis of chlorophenol red-β-D-galactopyranoside (Feng et al., 1996). Two, 96-well flat-bottom microtiter plates with 1×10^5 cells per well for each cell type with a total volume of 0.2 mL were used in this assay. After 3.5 hour incubation periods, samples on one plate were fixed, stained with crystal violet, examined microscopically for syncytia, and photographed. Aliquots, 50 microliter in volume, of replicate samples on the second plate were transferred to a third 96-

[10]This is a large DNA-containing virus that replicates entirely in the cytoplasm of the infected cell. See Watson et al. (1992) for description of general techniques that employ this virus for high-level protein production.

[11]Syncytia are very large multinucleated cells that form when infected cells (in this case, Env-cells) fuse with uninfected CD4 cells (Watson et al., 1992).

well plate where they were mixed with 50 μL of the substrate for β-galactosidase and then incubated at ambient temperature. High absorbance at 570 nm indicated hydrolysis of the chromophores from the substrate and therefore the presence of β-galactosidase. Since β-galactosidase is expressed when the two types of cells bind or fuse to each other, color formation correlates with fusion of cells and therefore, the presence of the receptors. The Env-cell is attractive because experiments with the HIV virus itself are avoided in these initial, rapid screens.

The rapid throughput screens were used to assay mouse cells engineered to express human CD4 receptors, into which a library of cDNAs from human cells, known to be susceptible to HIV, were transfected.[12] Repeated examination of the cDNA library, combined with isolation of individual colonies on agar plates, led to identification of a cDNA insert of about 1.7 kb (kb is the abbreviation for kilobase) that was capable of allowing CD4 NIH 3T3 murine cells to fuse with other NIH 3T3 cells engineered to express vaccinia encoded Env (HIV coat protein). DNA sequence analysis, cloning into a plasmid (pSC59), expression of the protein in BS-C-1 cells, and analysis showed the protein to have a molecular weight of about 46,000. Other tests confirmed that NIH 3T3 cells expressing both CD4 and fusin bound Env-cells would bind while cells expressing only one or the other did not bind. Mink, monkey, and human cells that are typically not infected with HIV will fuse with the Env-cells when fusin cDNA is cloned and the protein expressed by these mammalian cells. Further confirming evidence on the role of fusin was provided by the observation that antibodies against fusin blocked both fusion and infection of susceptible cells by HIV itself (Feng et al., 1996).

The case study of Sections 9.9 to 9.19 not only illustrates the importance of receptors but also of the ability of viruses to achieve expression of foreign proteins. An example of how a virus can be used to generate proteins that have utility as affinity ligands is given by phage display technology, which employs a bacterial virus and enables generation of microproteins that are potentially useful in purification of biotherapeutics.

PHAGE AND RIBOSOME DISPLAY

Viruses that attack bacteria and multiply inside the microorganisms are called bacteriophages. A single-parent phage particle can multiply to hundreds of progeny particles in less than an hour (Watson et al., 1992). The cycle of infection, replication, and production of new phage particles is illustrated in Figure 9.27. One of the genes in the bacteriophage codes for a protein positioned on the external surface of the phage (Figure 9.28). This gene, referred to as gene III, can be modified to incorporate the DNA that codes for a specific microprotein. The fusion protein of the gene III protein and the microprotein is then displayed on the external surface of the bacteriophage, hence giving rise to the term "phage display" (MacLennan, 1995). A large number of variants of a small protein can be quickly generated and used to discover new ligands for use in affinity chromatography.

[12]"Transfection" is a term that arises from early gene transfer experiments carried out with DNA tumor viruses. DNA isolated from the purified viruses was introduced into uninfected cells, some of which eventually produced fully infectious viruses. This DNA-mediated transfer was called transfection to distinguish it from natural entry of such viruses by infection. Watson et al. (1992) give a thorough description of techniques used to transfer genes into mammalian cells.

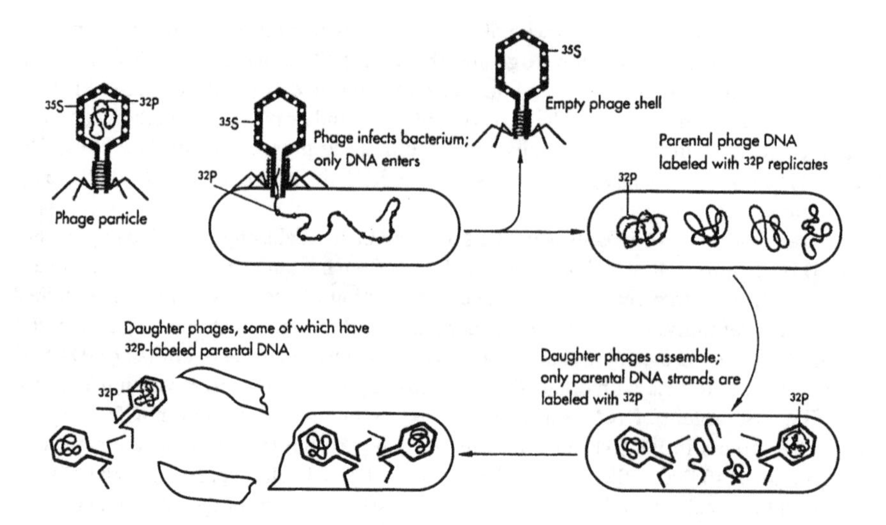

Figure 9.27 Radiactively labeled bacteriophage T2 (labeled with ^{35}S and ^{35}P) was used to demonstrate that its genetic material is DNA, not protein. (From Watson et al., 1992, fig. 2-3, p. 17.)

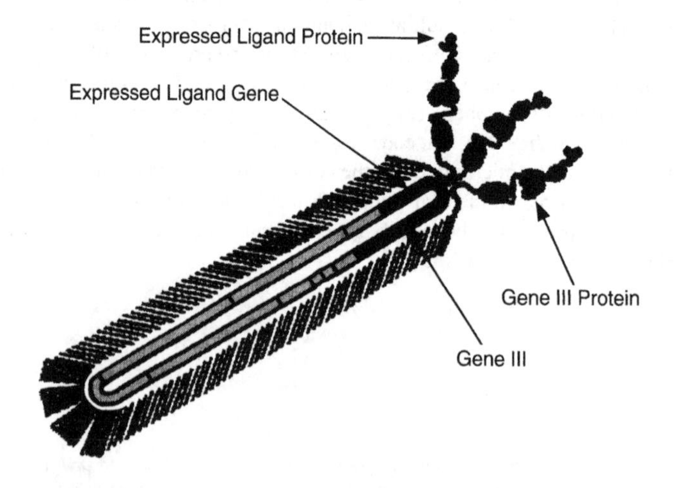

Figure 9.28 Schematic diagram showing display of expressed protein ligand from gene III. (From MacLennan, 1995, fig. 1, p. 1180.)

9.20 BIOCOMBINATORIAL GENERATION OF SMALL PROTEIN LIGANDS UTILIZES BACTERIAL VIRUSES: PHAGES

DNA for microproteins was chosen for generating small protein ligands because the microproteins consist of less than 60 amino acids and form one or more disulfide bonds. The disulfide bonds hold the protein in a conformation in which the amino acids at the outer surfaces of the protein can be deduced and their binding characteristics delineated. Gene III (also referred to

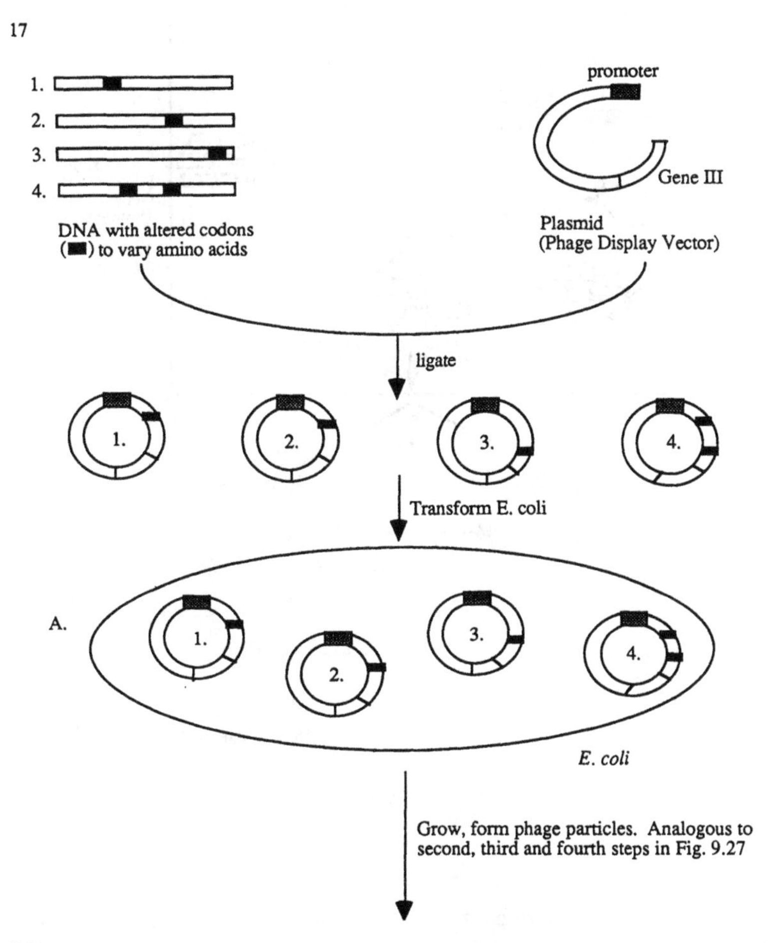

Figure 9.29 Schematic representation of phage display technique, shown here for 4 genes having altered codons corresponding to varied amino acids. The phage display vector, once ligated with the altered genes, is used to transform the *E. coli*, with the phage display vector (a plasmid) which also containing the information needed to regenerate the phage particles that display the protein corresponding to the altered gene. These are then purified as illustrated in steps *B* and *C*. Steps *A*, *B*, and *C* are repeated to enrich the phage population in the desired component, which is finally cloned, generated in larger quantities, and recovered so that the gene corresponding to the displayed protein can be sequenced, and the sequence used to generate larger quantities of the protein using molecular biology techniques. (Adapted from Watson et al., 1992, fig. 23-10, p. 467.) (continued)

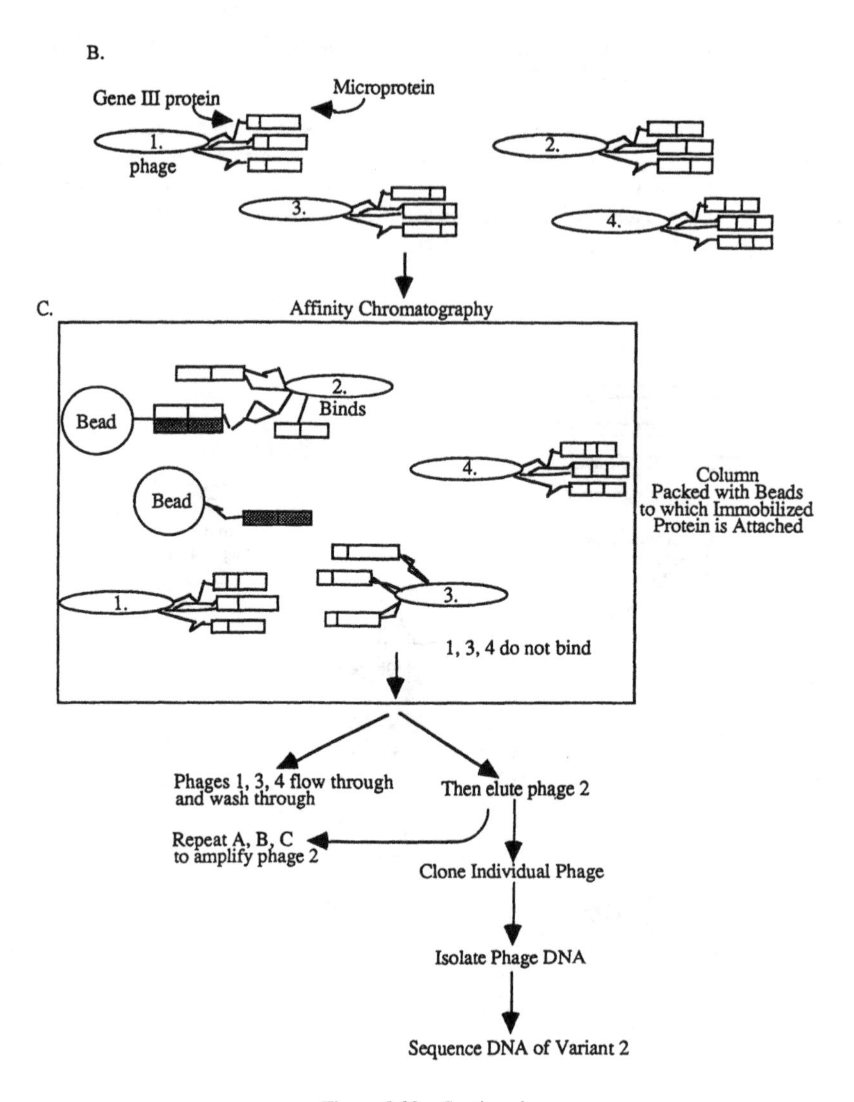

Figure 9.29 Continued.

as the pIII gene) of the bacteriophage is responsible for expressing protein on the surface of the phage. The combination of gene III with a gene for a protein that would serve as an affinity ligand results in a recombinant gene that will cause the protein for the affinity ligand to be coupled to the gene III protein, when the recombinant gene III is expressed (Figure 9.28). Since the protein for the ligand is small, it is sometimes referred to as a microprotein. The microprotein gene can be varied through DNA mutagenesis techniques so that the amino acid residues that make up the micro protein are changed at predetermined locations. Expressed microprotein ligands having the desired binding characteristics are identified by screening technique.

Variations in single amino acids or combinations of amino acids in the microprotein quickly generates numerous protein structures whose amino acid sequences differ by at least one amino acid as illustrated by steps *A* and *B* in Figure 9.29. The number of possible variants is very large. Changes in 3 amino acids result in 8000 different combinations, while changes in 10 different amino acids give 10^{13} different proteins (Watson et al., 1992).

9.21 BINDING CHARACTERISTICS OF SMALL PROTEIN LIGANDS ARE DETERMINED BY ADDING THE PHAGES TO TARGET MOLECULES IMMOBILIZED ON BEADS

Affinity chromatography is used to screen the libraries of microprotein ligands, displayed by the phages, for their binding characteristics (step *C* in Figure 9.29). The pH, buffer strength, and temperature are selected to conform to conditions at which the ligands are used. After incubation of the phages with the target molecules, the nonbinding phages are washed off. The phages that display microproteins that bind to the target protein or molecule immobilized on the stationary phase are then desorbed by changing the pH or salt concentration. Alternately, a component can be introduced into the mobile phase that competes for the binding sites and displaces the bound phage particles. The desorbed phages are collected, analyzed, and used to infect bacteria. The bacteria are then grown to generate a larger number of phages that are cycled through another sequence of capture and recovery. Once the target phage with the desired microprotein is amplified, the cell-free growth medium itself is used to dissociate phages from the affinity support. This method selects for the phages that display high affinity for the target molecule in the presence of other components and proteins in the media used to grow the bacteria and generate the phages (MacLennan, 1995).

Once the phage population that binds with the immobilized protein has been narrowed from 20 million to about 100 candidates, the remaining phages are plated and single-clone populations are generated using an appropriate microbial expression system (e.g., yeast). The amino acid sequence of the microprotein for each of the remaining phages is deduced from sequencing the DNA of the microprotein gene contained in gene III. Positions in the protein ligand, displayed by the phage, that differ by only 1 or 2 amino acids mark the amino acid sequences likely to be important to binding. If the properties of these amino acid sequences are confirmed by another round of binding studies, the role of the phage selection/identification is complete. Otherwise, the screening procedure must be repeated. Once the DNA sequence that corresponds to the desired microprotein is identified, the amino acid sequence is known. The sequence can then be directly synthesized, using a peptide synthesizer, or alternately, the gene can be incorporated into a microbial expression system to generate large quantities of the ligand. The entire process may be completed in as little as five weeks to generate a customized affinity ligand (MacLennan, 1995).

[13]Kunitz domains are small (58 amino acid residues or less), very stable, and have a three-dimensional structure that tolerates changes in amino acid sequence. These amino acid sequences can have new binding domains engineered into them, can be produced efficiently in yeast, and have a history of use in humans (Markland et al., 1996a). The bovine pancreatic trypsin inhibitor (abbreviated BPTI and also known as aprotinin or Trasylol) has a Kunitz domain and inhibits trypsin by formation of a tight complex in which 14 residues of BPTI contact 25 residues of trypsin in the catalytic site region (Markland et al., 1996b).

9.22 PHAGE DISPLAY TECHNOLOGY HAS GENERATED A PROTEASE INHIBITOR WITH HIGH AFFINITY AND SPECIFICITY FOR PLASMIN

Plasmin (PLA, EC 3.4.21.7) is a serine protease that has a central role in hemostasis, and whose role can be compromised in some types of surgical procedures, thereby causing excessive bleeding. Plasma-derived products or, in extreme cases, bovine pancreatic trypsin inhibitor have been used to stop bleeding in some patients, although these agents have disadvantages (Markland et al., 1996a).

A phage display technique was used to identify an amino sequence that is particularly effective in binding human plasmin. The amino acid sequences displayed by the phage, and strongly binding the plasmin, were identified by isolating, amplifying, and recovering the phage. The pIII gene in the phage was then sequenced, and the resulting gene cloned into yeast. The transformed yeast was grown, resulting in the expression of 100 to 500 μg of the protein in 500 mL of fermentation broth. Several proteins generated in this manner were used to study inhibition of plasmin activity. An inhibitor having a dissociation constant of 87 picomolar (10^{-12} moles/L) with potentially interesting therapeutic properties was isolated (Markland et al., 1996a).

A biocombinatorial approach also enabled 13 amino acids to be varied in human lipoprotein-associated coagulation inhibitor (LACI-Dl, also known as tissue pathway inhibitor-I). These amino acids belonged to a kunitz domain[13] with the positions selected so that the amino acid residues that are changed would influence binding, but not structure, of the LACI-Dl protein. These amino acid residues are at the surface of the protein (Figure 9.30).

Each round of protein generation and screening dealt with libraries where gene III of the bacteriophage was altered by the insertion of designed DNA sequences (Figure 9.31). A region of the protein was first varied in a defined manner by substitution of codons to obtain different amino acid residues. The resulting phages displayed the corresponding proteins. Phages that displayed proteins that bound poorly to plasmin immobilized to agarose beads were eliminated.

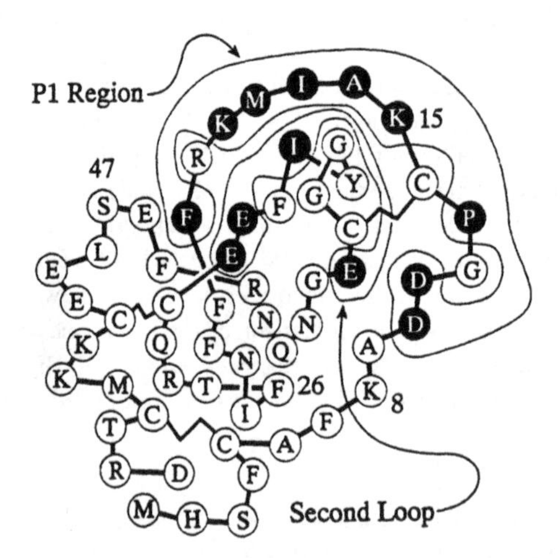

Figure 9.30 Schematic of LACI-D1 showing positions (in black) varied in present study. The P1 (positions 10–21) and second loop (positions 31–39) varied regions are enclosed. Backbone taken from Brookhaven model of BPTI 2TPA with modifications for clarity. (From Markland et al., 1996, fig. 2, p. 8047.)

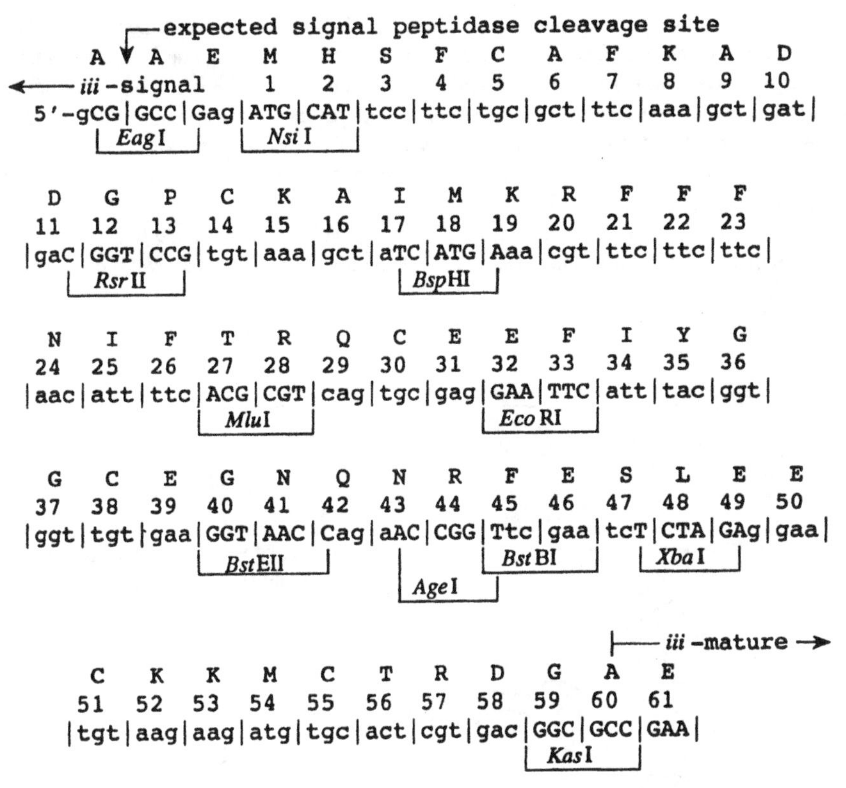

Figure 9.31 Designed *laci-d1* portion of a *iii-signal::laci-d1::iii-mature* gene showing DNA and deduced amino acid sequence. The DNA sequence maximizes use of codons contained in *E. coli* genes known to be highly transcribed and expressed except as needed for placement of restriction sites. The restriction sites are unique to the gene and the display vector used in these studies. (From Markland et al., 1996, fig. 1, p. 8047.)

Those that bound strongly were retained and recovered. A second region was varied, and the procedure repeated with strongly interacting pairs of amino acids being identified in the second domain. This process led to definition of small stable Kunitz domains having amino acid sequences that are close to sequences in human proteins that inhibit plasmin but with higher affinity and specificity for plasmin than the human lipoprotein-associated coagulation inhibitor. These molecules have potential as candidate therapeutics, imaging agents, or lead compounds for drug development.

9.23 GENERATION AND SCREENING OF PROTEINS BY IN VITRO RIBOSOME DISPLAY SEPARATES AND AMPLIFIES mRNA FOR A SPECIFIC PROTEIN

Phage display involves a sequence of steps in which in vitro changes in the gene are followed by in vivo generation and screening of the resulting protein. Ribosome display carries out these steps completely by in vitro techniques. It uses a cell-free, enzyme-based expression system to transcribe a DNA library into messenger RNA (mRNA). An in vitro ribosome system then translates the mRNA into proteins Figure 9.32. The proteins are screened against immobilized biological target molecules, and those proteins that bind also have the ribosome and associated messenger RNA attached. The unbound ribosomes are removed by washing, and the bound ribosome/protein complexes are desorbed and dissociated (step 4*a* in Figure 9.32) or dissociated (step 4*b* in Figure 9.32). Finally, the mRNA undergoes reverse transcription to cDNA and the cDNA is amplified by the polymerase chain reaction, which are also in vitro techniques. The resulting DNA then enters the next cycle of enrichment as illustrated in Figure 9.32 (Borman, 1997; Hanes and Plückthun, 1997).

In vitro translation methods have previously been developed and applied to peptides using immunoprecipitation of polysomes (Hanes and Plückthun, 1997). Recent developments added

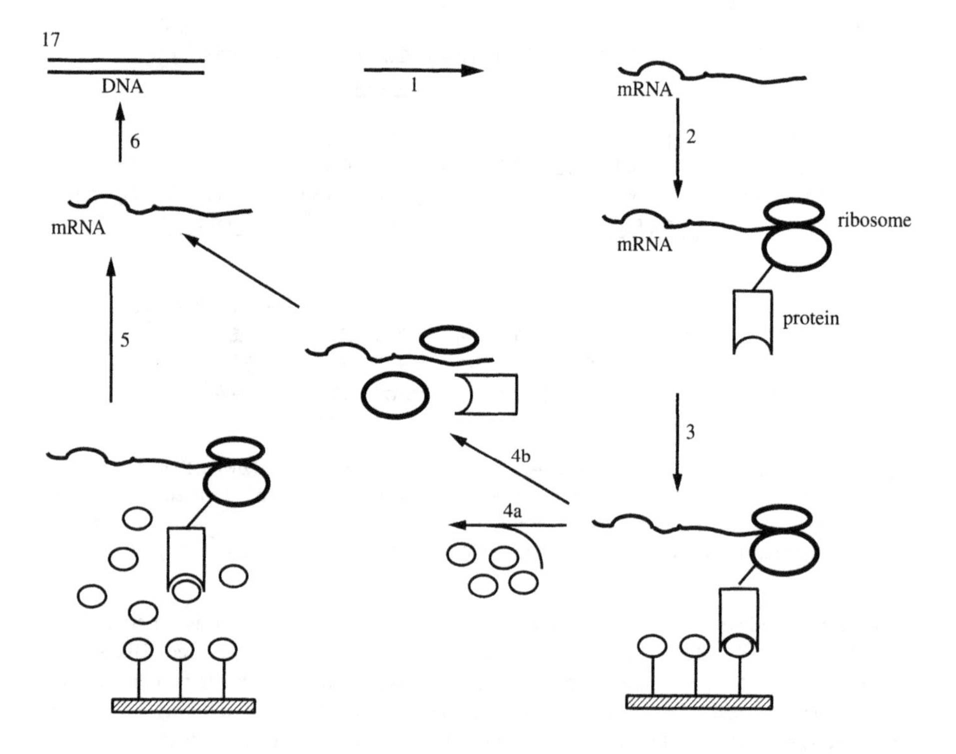

Figure 9.32 Principle of in vitro ribosome display for screening native protein (scFv) libraries for ligand (antigen) binding. Step 1: A DNA scFv library is first amplified by PCR, whereby a T7 promotor, ribosome-binding site, and stem-loops are introduced, and then transcribed to RNA. Step 2: After purification, mRNA is translated in vitro in an *E. coli* S-30 system in the presence of different factors enhancing the stability of ribosomal complexes and improving the folding of the scFv antibodies on the ribosomes. Translation is stopped by cooling on ice, and the ribosome complexes are stabilized by increasing the magnesium concentration. Step 3: The desired ribosome complexes are affinity selected from the translation mixture by binding of the native scFv to the immobilized antigen. Unspecific ribosome complexes are removed by intensive washing. Step 4: The bound ribosome complexes can then be dissociated by EDTA (step 4*b*), or whole complexes can be specifically eluted with antigen (step 4*a*). Step 5: RNA is isolated from the complexes. Step 6: Isolated mRNA is reverse transcribed to cDNA, and cDNA is then amplified by PCR. This DNA is then used for the next cycle of enrichment, and a portion can be analyzed by cloning and sequencing and/or by ELISA or RIA (RIA is an abbreviation that denotes radioimmunoassay).

the capability to prevent the ribosome complex from dissociating by removing the stop codon from the mRNA and adding structural changes to the mRNA, inhibiting of release and degradation of protein through the use of an antisense oligonucleotide to prevent attachment of an RNA tag that marks the protein molecule for degradation when a stop codon on the mRNA is missing, and inhibiting ribonuclease from degrading mRNA by using a transition state analogue. These methods increased the yield of mRNA from 0.001% to 0.2% per round of generation and selection. This yield reflects the combined efficiency of covalent attachment of mRNA to the ribosome, protein folding, ligand binding to the immobilized target molecule, ribosome capture, RNA release, and amplication by PCR (Hanes and Plückthun, 1997). This technique and the potential of *in vitro* amplification was demonstrated for a mixture of mRNA's for single, light chains of anti-hemagglutinin and the anti-β-lactan antibodies. These were mixed in the ratio of 1 to 10^8. After five cycles of the type illustrated in Figure 9.31, with an immobilized hemagglatinin peptide used as the affinity ligand, 90% of the ribosome complexes contained the anti-hemagluttin antibody.

RECEPTOR AFFINITY CHROMATOGRAPHY

Receptor affinity chromatography is a form of immunoaffinity chromatography that is more selective. Immunoaffinity chromatography uses antibodies immobilized to a stationary phase (i.e., the immunoadsorbent) to selectively capture antigens dissolved in an aqueous buffer, and thereby purify the antigen. The selection of antibodies for use as ligands includes consideration of desorption conditions and the need to recover the protein under nondenaturing conditions (Nachman-Clewner et al., 1993; Sado and Katoh, 1993). Such antibodies can be obtained from the antisera against the target molecule (a hapten, peptide, or protein) raised in rabbits. Once it is purified from the antisera using a variety of chromatographic techniques, the antibody (i.e., ligand) is immobilized to the stationary phase to obtain the immunoadsorbent. Alternately, the ligand can be a monoclonal antibody generated by B-cell clones grown in vitro. The growth of immunoaffinity chromatography as a separations tool was promoted by the development of monoclonal antibody producing hybrid cell lines by Kohler and Milstein in 1975.

The high affinity of the antibody for the antigen may also necessitate extreme pH's or denaturing agents to be used to elute the antigen, thereby denaturing it. While an immobilized antibody protein can exhibit excellent selectivity for its antigen, it can also bind inactive and improperly folded forms of the target molecule. The conflicting requirements of immunoaffinity chromatography—high affinity during adsorption and low affinity during desorption at mild conditions—and its lack of specificity promoted the development of receptor affinity chromatography.

9.24 RECEPTOR AFFINITY CHROMATOGRAPHY SELECTIVELY BINDS ACTIVE FORMS OF PROTEIN

An immobilized receptor should strongly bind only the fully active biomolecule in its native conformation while releasing the biomolecule when a mild defoaming reagent (e.g., low pH buffer) is used as the displacer. Receptor affinity chromatography is differentiated from immunoaffinity chromatography principly by the manner in which the ligands are generated. By definition, receptor affinity chromatography utilizes specific receptors for a target protein generated by recombinant DNA techniques, while immunoaffinity chromatography is based on antibodies that bind a range of target molecules and hence are less selective.

9.25 RECEPTOR AFFINITY CHROMATOGRAPHY COMBINES UPSTREAM AND DOWNSTREAM PROCESSING STEPS

The protein or polypeptide ligands used in preparing receptor affinity supports are themselves products of fermentation of recombinant microorganisms, and they are subjected to a separate sequence of purification steps prior to being reacted with a functionalized stationary phase to form the affinity support. Initially the development of this technique required that sufficient quantities of receptors be available for research purposes. Ultimately its scale-up as a manufacturing tool will require that the receptors be prepared and purified in a cost-effective manner (Bailon et al., 1987; Bailon and Weber, 1988). The upstream process (and overall economics) for producing the receptor depends on the successful packaging of the receptor into an affinity chromatography configuration that is cost effective and amenable to scale-up. While the resulting affinity chromatography columns are expensive when viewed on the basis of cost of support per unit volume of stationary phase, the cost–benefit ratio could still be attractive because process-scale columns have small volumes on the order of 1 to 10L. For this reason they enable recovery of dilute but highly active proteins that are difficult to purify by other means.

Receptor affinity chromatography also illustrates the synergies between fermentation or cell culture (upstream processing) and recovery (downstream processing) to purify bioproducts by integrating upstream with downstream processing. The downstream purification, in the case of receptor-affinity separations, depends on an upstream culturing step that generates the protein (i.e., ligand) to which the receptor binds. The receptors generated by a separate fermentation can then be recovered in sufficient quantity to prepare an immobilized stationary phase.

9.26 INTERLEUKIN-2 IS A CASE STUDY IN THE DEVELOPMENT OF RECEPTOR AFFINITY CHROMATOGRAPHY

Interleukin-2 (abbreviated IL-2) was found to sustain the long term growth of human T-cells, and it was originally called the T-cell growth factor. This observation was made by Morgan et al. in 1976, and subsequently it received attention as a potential therapeutic agent for cancer and for treatment of immunodeficiency syndromes (Bailon et al., 1987, 1988; Danheiser, 1997; Waldholz, 1995; Morgan et al., 1976). Possible therapeutic uses (in 1988) included autoimmune thyroid disease, organ transplant rejection, rheumatoid arthritis, allograft rejection, and Crohn's disease (a digestive disorder) (Bailon et al., 1988). The only interleukin drug currently marketed in the United States (in 1997) is Proleukin® (IL-2 produced by Chiron) for treatment of metastatic renal cell carcinoma (Danheiser, 1997). The laboratory purification of this protein involved gel filtration in combination with isoelectric focusing, use of dye matrices (another form of affinity chromatography), or reversed phase chromatography. Yields were low, and the number of steps required, as well as the need to use organic solvents, prompted examination of immunoaffinity chromatography to produce homogeneous interleukin-2 for further studies.

The bench-scale production of monomeric, water-soluble, low-affinity interleukin-2 receptor (abbreviated IL-2R) from conditioned mammalian cell culture supernatants made development of receptor affinity chromatography practical (Bailon et al., 1987; Nachman-Clewner et al., 1993). While the extent to which this technique is currently used as a manufacturing tool is not clear, the potential for its use in the future is likely to grow as production of receptors becomes cost effective *and* leakage of affinity ligands into the product is minimized. The history of the development of this technique for interleukin-2 illustrates how recombinant technology can drive the development of an advanced separation technique, and how the separation technique, in turn, has the potential to facilitate an increase in the availability of a recombinant product such as interleukin-2.

9.27 EITHER INTERLEUKIN-2 OR ITS RECEPTOR IS AN AFFINITY LIGAND FOR CHROMATOGRAPHY OF THE OTHER

Initially there was no facile means to generate sufficient receptors to make receptor-affinity chromatography feasible. While the goal of affinity chromatography is a one-step purification procedure, both the preparation of the receptor, and the subsequent purification of the interleukin product using the resulting receptor affinity column, require other separation steps: diafiltration, dialysis, DEAE silica chromatography, and gel filtration (Bailon et al., 1987; Bailon and Weber, 1988). A modified gene for the IL-2 receptor was prepared in which the nucleotide sequence, corresponding to the 28 amino acids sequence (ΔNae-IL-2R) at the carboxyl terminus, was deleted. Protein expressed from this gene, which lacks N- and O-linked glycosylation sites that are otherwise naturally occurring and are believed to prevent the secretion of the receptor protein by the transfected rodent cells, is obtained with 72 amino acids deleted. The resulting IL-2R is denoted ΔMST-IL-2R. The amino acid deletions resulted in the removal of the presumptive transmembrane and cytoplasmic domains. This enabled the transformed CHO cells, containing the gene for the interleukin-2 receptor, to secrete the protein that was otherwise properly glycosylated. This particular protein was given the name IL-2R-Δ-Nae (Weber et al., 1988).

Interleukin-2 receptor proteins were secreted by chinese hamster ovary (CHO) cells, grown in medium-scale continuous fermentation in a 3L airlift reactor. The conditioned medium from the CHO cells was collected and filtered through a 0.8/0.2 micron filter. The filtrate was subjected to further purification using an affinity support to which interleukin-2 was immobilized. It is derived from CHO cells, which are capable of generating properly glycosylated protein, rather than from fermentation of transformed *E. coli*, which does not give glycosylated proteins. The first purification of IL-2R did not utilize affinity chromatography over IL-2, since sufficient quantities of IL-2 were not available (Bailon et al., 1987).

The interleukin-2 was generated through the Bioprocess Development Department at Hoffmann-LaRoche either from a recombinant *E. coli* cell line or from the conditioned medium of a CHO cell line transfected with the human IL-2 gene (Bailon et al., 1987; Weber et al., 1988). *E. coli*, does not glycosylate the protein, thus indicating that glycosylation of IL-2 is not needed to obtain the desired binding activity. The recovery sequence for recombinant IL-2 derived from transformed *E. coli* is summarized in Figure 9.33.

The recombinant interleukin-2 from the *E. coli* cells was immobilized to polyaldehyde NuGel-AF poly-aldehyde support. A volume of 140 mL of the support material (equivalent to 100 g) was washed with cold water, and transferred to 140 mL of 0.1 M phosphate buffer containing 0.1 M NaCl, pH 5.0, and 5 mg/mL of the recombinant interleukin-2. The mixture was gently swirled at 4°C in the presence of 0.05% sodium cyanoborohydride for 16 hours. The supernatant was removed and assayed for the amount of remaining (uncoupled) IL-2. The support was then resuspended in 0.5 M ethanolamine hydrochloride at pH 7.0, which contained 0.05% sodium cyanoborohydride and further agitated for 4 to 6 hours, in order to neutralize the remaining unreacted aldehyde groups in the affinity support. The support (a gel) was washed with phosphate buffered saline (PBS) and stored in 0.02% sodium azide at 4°C until use. The sodium azide is a potent inhibitor of microbial growth and is used as a preservative.

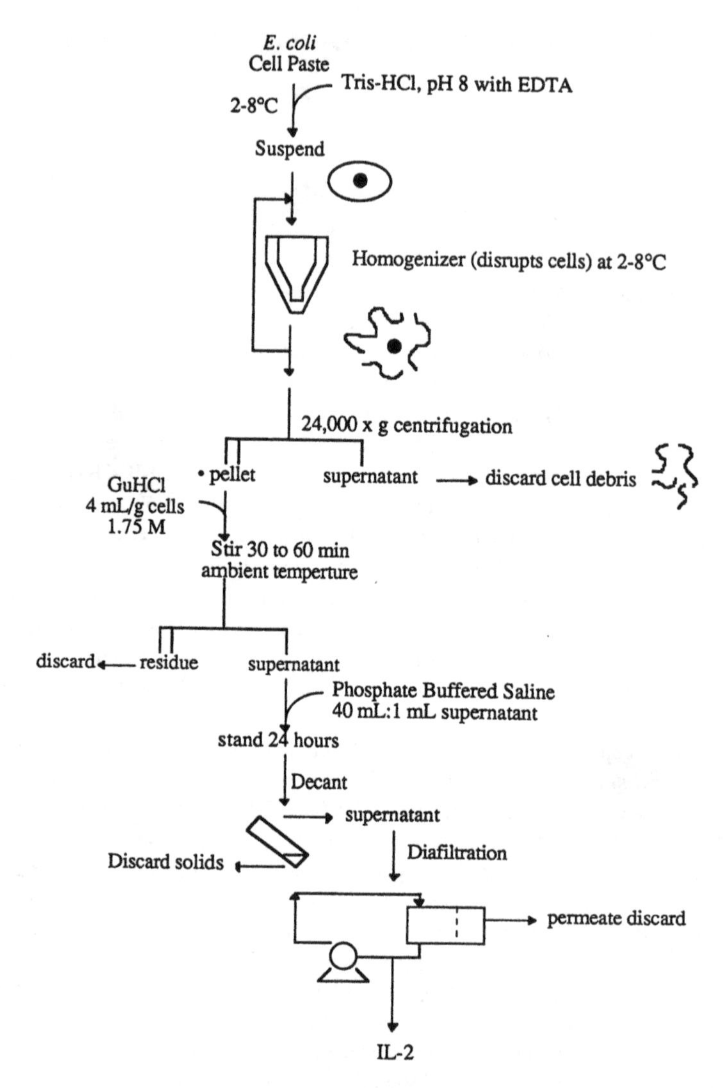

Figure 9.33 Recovery sequence of recombinant interleukin-2 (IL-2) from *E. coli* inclusion bodies. (Reconstructed from description of Bailon et al., 1987.)

9.28 MULTIPLE DISCRETE PROCESSING STEPS ARE ASSOCIATED WITH AFFINITY CHROMATOGRAPHY OF THE IL-2 RECEPTOR

The IL-2 affinity adsorbent is packed into an Amicon 6.6 × 4.4 cm column, fitted with two end caps and equilibrated with the phosphate buffered saline pumped through the column. A volume of 18 L of CHO-conditioned and filtered medium, containing the interleukin-2 receptor (abbreviated IL-2R-Δ-Nae), is passed over the 100 mL affinity column at 20 mL/min, followed by washing with PBS until the UV absorbance returns to the baseline. The IL-2R activity, which has reversibly bound to the immobilized interleukin, is then eluted with less than 200 mL of 0.2

M NaCl dissolved in 0.2 M acetic acid. The IL-2R is concentrated by over 90-fold as well as being purified. The IL-2 receptor, which was initially contained in 18 L, is recovered in less than 0.2 L, as shown in Figure 9.34 (Weber et al., 1988). Subsequent fractionation by either size exclusion (using Sephacryl) or anion exchange chromatography (using DEAE silica) shows that monomeric and dimeric forms of the IL-2 receptor are present together with other impurities. Both forms are glycosylated with molecular weights of 33 and 67 kD, as determined by independent assays (Weber et al., 1988).

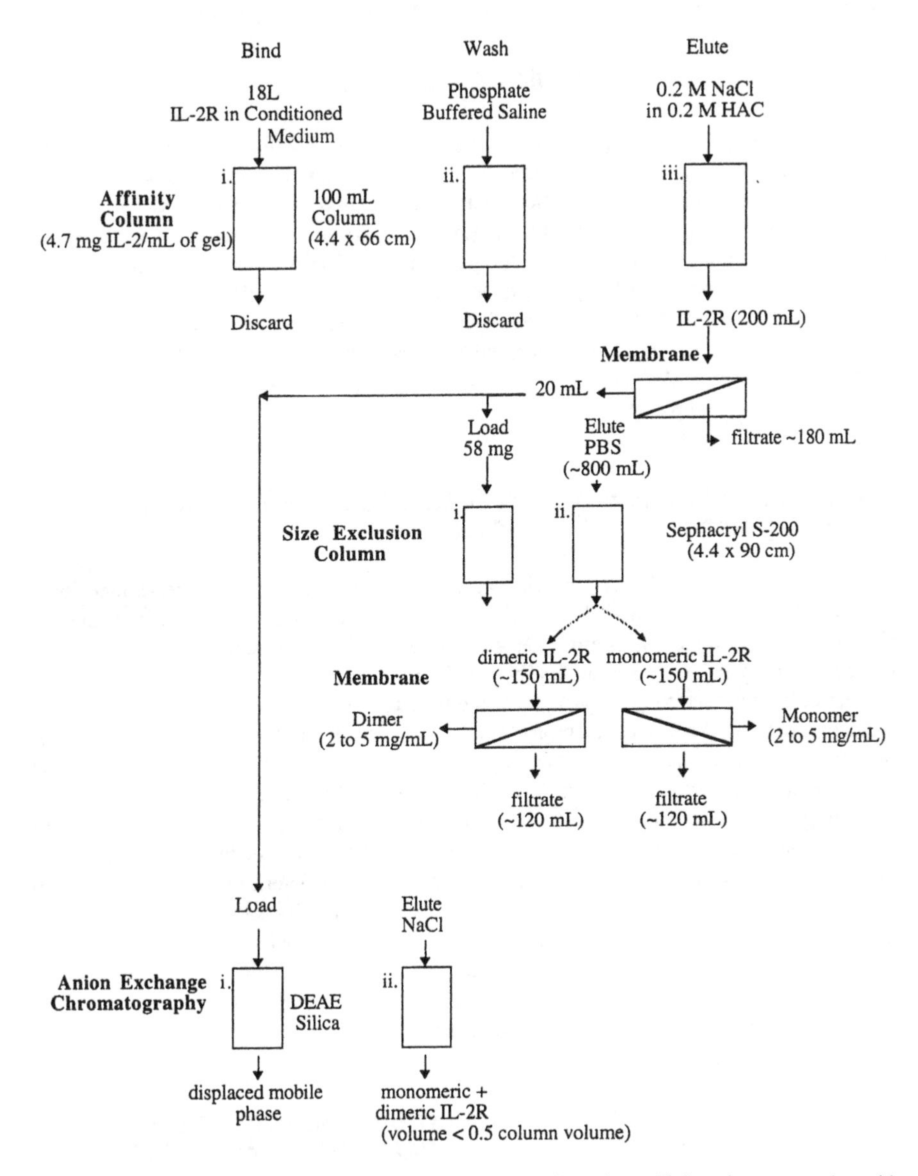

Figure 9.34 Overview of fractionation sequence for IL-2R using affinity chromatography with immobilized IL-2 column, followed by SEC or DEAE chromatography. (Adapted from description in Weber et al., 1988.)

The sequence of purification steps and the approximate volumes are summarized in Figure 9.34. This figure also illustrates how different types of fractionation steps might be combined, and how the volume containing the target molecule (IL-2R) changes as it moves through the purification sequence. Significant volumes of buffer are needed at each step, and more than one unit operation is required for purifying this molecule. This illustrates why affinity chromatography is unlikely to be a one-step purification procedure. Receptor affinity chromatography gives an example of the synergy between upstream and downstream processing, and the impact of recombinant methods in developing a new affinity chromatography method.

LECTINS AS LIGANDS

Lectins are carbohydrate binding proteins that were originally recognized for their ability to agglutinate human type A, B, and O erythrocytes by binding to specific receptors on the surface of these cells. This class of lectins is often referred to as agglutinins. Other types of lectins that initiate proliferation or morphological response of lymphocytes are referred to as mitogens. Commonly agglutinins are concanavilin A (abbreviated as Con-A), lentil lectin, and soybean lectin, all of which exist in multimeric form. Con-A is a dimer below pH 5.6 (M_r = 55 kD) and a tetramer at pH 5.6 to 7 (M_r = 110 kD), and it occurs in an aggregated form above pH 7.0. The lentil and soybean lectins exist as a dimer and tetramer with molecular weights of 49 and 110 to 120 kD, respectively. An excellent synopsis of their properties, immobilization procedures, and use is given by Sutton (1992).

9.29 LECTINS ARE AN IMPORTANT LABORATORY TOOL FOR AFFINITY CHROMATOGRAPHY OF GLYCOPROTEINS

Divalent metal ions (usually Ca^{2+} or Mn^{2+}) are required to transform the three-dimensional structure of the lectin protein so that it will form a saccharide binding site. Hence lectins are often immobilized to the spacer arm of an affinity support in the presence of a divalent metal ion and a carbohydrate in the solution. The metal ion ensures that the lectin is immobilized in the correct form to bind to glycoproteins, while the sugar (e.g., methyl mannoside, mannose, or glucose for Con-A) occupies and protects the binding site by preventing it from reacting with the activated group on the affinity stationary phase. These precautions help to ensure that the glycoprotein binding sites will be preserved during the crosslinking reaction.

Affinity chromatography is carried out by passing the sample through a column packed with the lectin containing stationary phase. In the absence of other sugars the lectin immobilized on the stationary phase will bind glycosylated proteins through its interactions with the oligosaccharides on the protein. Proteins that do not have the appropriate oligosaccharides will not bind and flow through the column. The adsorbed protein is then desorbed using displacing components such as glucose, mannose, or methyl mannoside. These sugars bind with the lectin at the same site as oligosaccharides associated with the glycosylated protein, thereby causing the less strongly bound glycosylated protein to be released.

9.30 CONCAVALIN-A CHROMATOGRAPHY OF α-GALACTOSIDASE DEMONSTRATES AFFINITY CHROMATOGRAPHY OF A PLANT ENZYME

Soybean seeds are a source of the plant enzyme, α-galactosidase. This protein comprises a small fraction of the protein of the seed but has potential value in the hydrolysis of the α-1,

TABLE 9.5 Modified Purification of Soybean α-Galactosidase

Fraction	Volume, mL	Activity, units/mL	Protein, mg/mL	Specific Activity, units/mg	Purification (fold)
Ammonium sulfate[a] (20–50%)	87.0	895	2.64	339	2.5[b]
QAE550C	84	503	0.674	746	5.5
Carboxy-methyl	20	1332	0.093	14,330	106
Concanavalin-A	10	839	0.015	55,933	414

Source: Reproduced with permission from Porter et al. (1991). The material is reproduced with permission by Wiley-Liss, Inc., a subsidiary of John Wiley, Inc.

[a]Ammonium sulfate fraction after storage and dialysis at 4°C.

[b]Calculated with respect to original large-scale cryoprecipitation.

6 linkages in stachyose and raffinose to give sucrose and galactose. This enzyme was studied and purified from soybean given its potential as a processing aid in the crystallization of sucrose, and the in vitro hydrolysis of difficult-to-digest oligosaccharides to readily digestible monosaccharides in soybean meal used in the diets of monogastric animals and humans. An aliquat of this enzyme, when extracted from 45.5 kg soybean containing 17.1 kg protein, initially has a specific activity of 90 to 100 units per mg of protein. One unit of activity is defined as the amount of enzyme that hydrolyzes 1 nmol of *p*-nitrophenyl-α-D-galactoside (abbreviated PNPG) at pH 4.8 and 37°C in 20 min.

A sequence of ammonium sulfate precipitation, strong anion exchange chromatography, weak cation exchange chromatography, and finally Con-A affinity chromatography resulted in an enzyme preparation with a specific activity of 56,000 (Table 9.5). The partially purified protein from the carboxy-methyl cellulose column was dialyzed overnight against 0.1 M sodium acetate, pH 5.2, containing 1 mM each of $MnCl_2$, $MgCl_2$, $CaCl_2$, and 0.09% NaCl, before loading onto the Con-A column (Porter et al., 1991).

A chromatogram showing elution of the α-galactosidase peak using a methyl mannoside gradient is given in Figure 9.35. The chromatogram and corresponding SDS-page (silver stain) gel show that the Con-A column selectively removed the α-galactosidase enzyme from the rest of the protein (compare lanes 4 to 6 to lanes 7 to 8 in Figure 9.36).

The affinity chromatography of soybean α-galactosidase provided a bridge to molecular biology by facilitating preparation of a significant quantity of protein suitable for generation of a polyclonal antibody. In this case the purified α-galactosidase, injected into a rabbit, would provide a rabbit antibody against soybean α-galactosidase. The polyclonal antibody, once recovered and purified from the rabbit's blood, would be appropriate for use as a tool to probe a soybean cDNA library for the gene for the α-galactosidase.[14]

Despite the widespread use of immobilized lectins in laboratory-scale procedures, and purification of proteins for research purposes, this author is not aware of published examples of chromatography using Con-A immobilized ligands for manufacture of biotherapeutic proteins.

[14]The cDNA is cloned into an appropriate host microorganism, which in turn expresses the protein corresponding to the cDNA. The host containing the cDNA for the target protein is identified when the antibody, added to the culture broth causes a precipitate to form. The presence of small amounts of protein impurities in the target protein preparation can cause the polyclonal antibody to contain activity against the impurities as well as the target protein. Hence an extremely pure protein is required to generate the polyclonal antibody if a meaningful positive test (precipitate) for a particular protein in the presence of background contaminant is to be achieved.

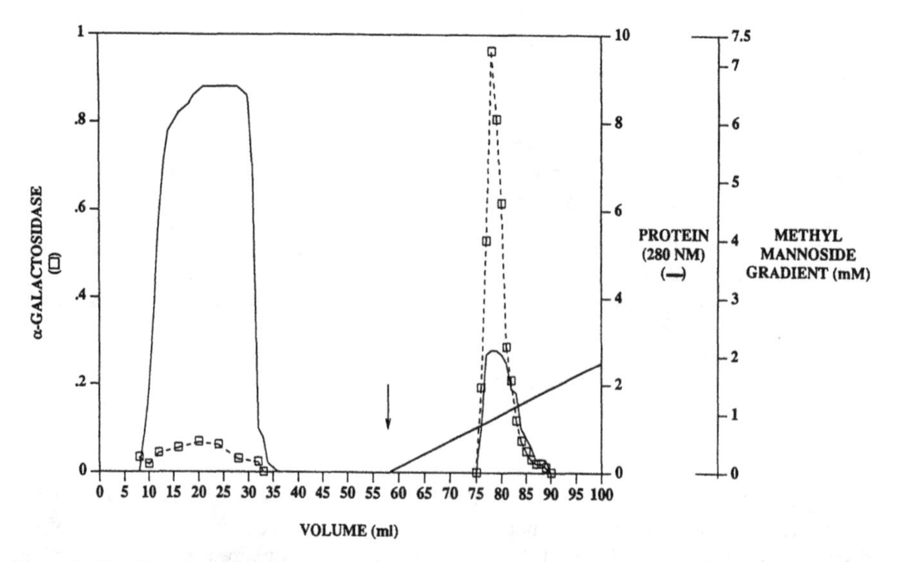

Figure 9.35 Chromatogram of α-galactosidase purified using lectin affinity chromatography. (Reprinted with permission from Porter et al., 1991, fig. 8, Biotechnol. Bioeng. Reprinted by permission of Wiley-Liss, Inc., a subsidiary of John Wiley & Sons, Inc.)

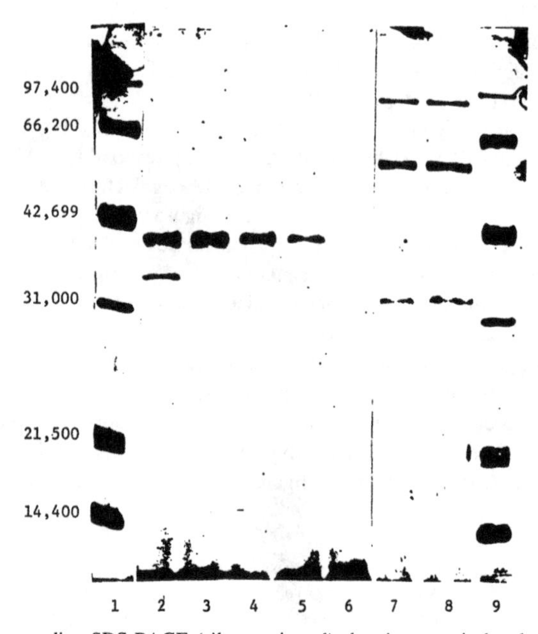

Figure 9.36 Corresponding SDS-PAGE (silver stain gel) showing protein bands corresponding to the elution profile in Figure 9.35. (Reprinted with permission from Porter et al., 1991, fig. 9, Biotechnol. Bioeng. Reprinted by permission of Wiley-Liss, Inc., a subsidiary of John Wiley & Sons, Inc.)

PROTEIN A

Protein A is a monomeric protein derived from the cell wall of *Staphylococcus aureus* with M_r = 41 kD, which binds antibodies in their F_c region (see Figure 9.12 for a schematic representation of an antibody). Immunoprecipitation of proteins was originally carried out by adding preparations of dry cells of *S. aureus* to the protein, and identifying mixtures in which a precipitate formed. Purification of the cell wall proteins led to the identification of protein A as the protein that binds the immunoglobins. Properties of protein A as an affinity ligand include a small dissociation constant ($K_d = 10^{-9}$M) with five potential binding domains. Affinity chromatography with immobilized protein A derived from *S. aureus* has been widely used for recovery of immunoglobulins for purposes of both research and therapy, and the purification of monoclonal antibodies from ascitic fluid and cell culture (Boyle et al., 1993; Kenney and Chase, 1987). Protein A can be obtained in commercial quantities from both native *S. aureus* as well as recombinant sources (Boyle et al., 1993; Chamov and Ashkenazi, 1996).

9.31 PROTEIN A IS ONE OF MANY BACTERIAL IMMUNOGLOBULIN-BINDING PROTEINS USED IN AFFINITY CHROMATOGRAPHY

Protein A is part of a family of bacterial surface molecules that react with immunoglobulin molecules in a nonimmune manner. Six types of immunoglobulin G (IgG) binding proteins have been reported for gram-positive bacteria. These are classified based on their ability to bind with the F_c region of immunoglobulin of different subclasses. The IgG binding proteins have been classified based on study of receptors found in *Staphylococcus aureus* isolated from humans. These are (Boyle et al., 1993):

Protein A	Type I from *Staphylococcus*
	Type II from streptococci
Protein G	Type III from streptococci, which binds human serum albumin, and has a wide range of species and subclass activities. Other variants are IV, V, and VI.
Protein B	Reacts with IgA subclasses
Protein L	Reacts selectively with human immunoglobulin light chains. Protein L is derived from *Peptococcus magnur*.

The order of increasing reactivity (or decreasing selectivity) for proteins A and G with respect to IgG subclasses is

$$\text{I} < \text{II} < \text{III}$$

Proteins that bind IgG have also been found on the surfaces of *Haemophilis somnus*, *clostridium perfringens*, and *Brucella abortus*. While bacterial proteins that react with constant regions in IgA and IgD from different species have been identified, bacterial proteins with absolute specificity for the F_c region of IgM from all different types of animal species have not been reported. A toxin protein from *Clostridium difficile* and a protein from *Brucella abortus* displays reactivity with respect to murine and bovine IgG, respectively (Boyle et al., 1993). The affinity chromatography stationary phase is prepared by immobilizing the desired protein on beads using the approaches described in Chapter 8.

The separation process using a protein A affinity stationary phase is succinctly and clearly described by Boyle et al. and is illustrated in Figure 9.37 (Boyle et al., 1993). The F_c-binding protein is immobilized. The immobilized protein is used to selectively bind certain types or subclasses of antibodies at a given pH or ionic strength of the mobile phase, thereby adsorbing or binding the immunoglobulin. The nonbinding proteins (impurities) flow through, and those that remain in the void volume are washed out with buffer. The selectively bound immunoglobulins are then eluted by changing the pH and/or ionic strength of the mobile phase to reverse the binding of the immunoglobulin to the immobilized bacterial immunoglobulin-binding protein.

The selection of protein A or protein G as the immobilized affinity ligand depends on the source of the immunoglobulin species to be purified. Guidelines, offered by Boyle et al. (1993), are summarized in Table 9.6. The efficiency of immunoglobulin binding to immobilized proteins is enhanced when an 8 to 10 carbon chain length spacer arm is used. The spacer arm moderates steric hindrance, which occurs when the high molecular weight immunoglobulin approaches the surface of the stationary phase (see Chapter 8).

Another use of bacterial IgG-binding proteins is in the immobilization of specific antibodies for detecting or purifying antigens to these antibodies. This approach is viable as long as the pH and ionic strength at which the immunoglobulin binds to the immobilized protein (i.e., the "adsorb immunoglobulin step" in Figure 9.37) is different from the pH and ionic strength required to bind or desorb the antigen. Otherwise, both the antibody and antigen would elute, thereby defeating the purpose of immobilizing the antibody. Examples, given by Boyle et al. (1993) are summarized in Table 9.7.

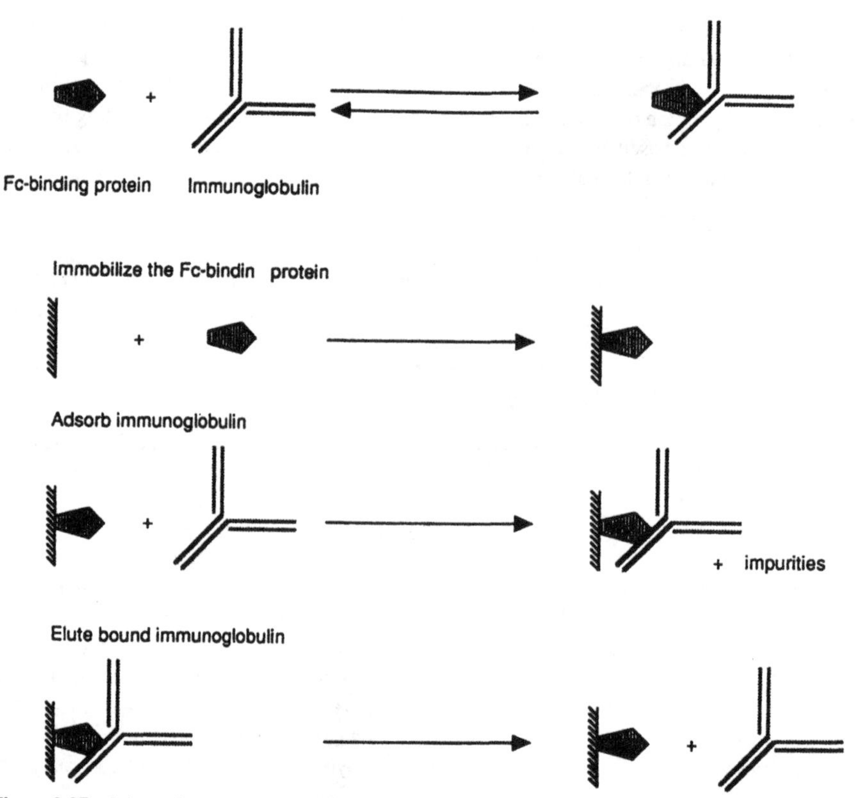

Figure 9.37 Schematic representation of immobilized bacterial immunoglobulin-binding proteins for the affinity purification of proteins. (Reprinted with permission from Boyle et al., 1993, fig. 1, in Molecular Interactions in Bioseparations, reprinted with permission from Kluwer Academic/Plenum Press, New York.)

TABLE 9.6 Comparison of Immunoglobulin Species and Subclass Reactivities of Proteins A and G

		Reactivity with:				Reactivity with:	
Species	Subclass	Protein A	Protein G	Species	Subclass	Protein A	Protein G
Human	IgG1	+++	+++	Rabbit	IgG	+++	+++
	IgG2	+++	+++	Cow	IgG1	–	+
	IgG3	–[a]	+++		IgG2	+	+++
	IgG4	+++	+++	Sheep	IgG1	–	+
	IgA	+/–[b]	–		IgG2	+	+++
	IgM	+/–[b]	–	Goat	IgG1	+/–	+
Mouse	IgG1	+/–	+		IgG2	+	+++
	IgG2a	+	+++	Horse	IgG(ab)	+	+++
	IgG2b	+	+++		IgG(c)	+	+++
	IgG3	+	+++	Dog	IgG	+	NT[c]
Rat	IgG1	–	+/–		IgM	+	NT
	IgG2a	–	+		IgA	+	NT
	IgG2b	–	+				
	IgG2c	+/–	+				

Source: Modified from Boyle et al. (1993), table 1, p. 94, in Molecular Interactions in Bioseparations, reprinted with permission, Kluwer Academic/Plenum Press, New York.

Note: Symbols +++, strong reactivity; ++, intermediate reactivity; +, weak reactivity; +/–, very low reactivity; –, no reactivity.

Derived from studies by Langone (1982a), Reis et al. (1984b), Boyle et al. (1985), Wallner et al. (1987), Reis et al. (1986), Nilsson et al. (1982), and Nilsson et al. (1986). See Boyle et al. (1993), for literature citations.

The majority of these studies were carried out with myeloma proteins, and many instances are noted in which the expected reactivity with a given mammalian subclass was not observed, as has been particularly evident in studies of mouse monoclonal antibodies. Consequently this table should only be used as a guide for probable reactivities. Each antibody should be checked directly for reactivity with protein A or protein G prior to use for any immunochemical procedure.

[a]Certain IgG3 allotypes have been reported to react with the type I Fc-binding protein (Haake et al., 1982). See Boyle et al. (1993), for literature citations.

TABLE 9.7 Use of Bacterial IgG-Binding Proteins to Immobilize Specific Antibodies for Affinity Chromatography

Bacterial Binding Protein Support	Specific Antibody Source	Antigen Purified or Detected	Reference
Protein A	Polyclonal rabbit	Human IgA	Russell et al., 1984
	Polyclonal rabbit	Human IgM	Russell et al., 1984
	Mouse monoclonal	Common acute leukemia-associated antigen	Schneider et al., 1982
	Mouse monoclonal	HLA-AB antigen	Schneider et al., 1982
	Transferrin rabbit polyclonal antigen-antibody complex	Transferrin receptor	Schneider et al., 1982
Protein G	Mouse monoclonal	Transferrin receptor	Schneider et al., 1982
	Mouse monoclonal	Hen egg lysozyme	Webb-Walker, 1990

Source: From Boyle et al. (1993), table 4, p. 106, in Molecular Interactions in Bioseparations, reprinted with permission from Kluwer Academic/Plenum Press, New York.

9.32 PROTEIN A IS A LIGAND FOR PROCESS AFFINITY CHROMATOGRAPHY OF IMMUNOADHESIONS

Affinity chromatography with protein A (or protein G) can purify antibody-like molecules known as immunoadhesions (Figure 9.38). Immunoadhesions are fusion proteins which combine immune and adhesion functions and are obtained by combining framework (gene) sequences from a human monoclonal antibody with sequences from a human protein that carries a target recognition function. These are useful for study of biological functions of novel receptors and ligands.

Preclinical studies suggest immunoadhesions have potential to help combat viral infection, inflammatory and autoimmune disease, and tumor metastasis. These proteins share many of the properties of monoclonal antibodies, and they may provide an alternate in cases where monoclonal antibodies are difficult to generate (Chamov and Ashkenazi, 1996). Efficient mammalian cell expression, combined with ease of purification over protein A (99% in one step), has fostered development and characterization of immunoadhesions. Both recombinant and native forms of protein A affinity stationary phases are available from commercial suppliers, with immunoadhesion purification being carried out in the same manner as for immunoglobulins.

A typical laboratory purification using a relatively small bed (1 × 1.3 cm) of a commercial protein A stationary phase demonstrated the facile nature of the purification. The material was initially equilibrated with a starting buffer of 0.2 M Tris buffer at pH 7.4 containing 0.15 M NaCl. The sample was then applied to the column and a loading of 30 mg/mL bed volume obtained. The column was again washed with starting buffer to remove unadsorbed components. The protein was then eluted using 0.05 M citric acid at pH 3.0 containing 20 (weight/vol) % glycerol. The product, while 99% pure, would likely require further processing to remove the

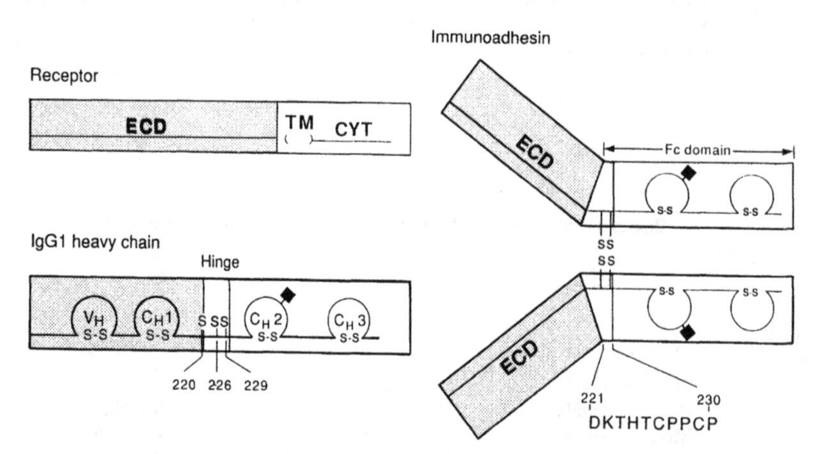

Figure 9.38 Schematic structure of a prototypic immunoadhesin. An immunoadhesin is usually derived from the parental ligand-binding protein, in this case a type I transmembrane receptor and an IgG1 heavy chain molecule. ECD, TM, and CYT refer to the extracellular, transmembrane and cytoplasmic domains, respectively, of the receptor. The variable (V_H) and constant (C_H1, hinge, C_H2 and C_H3) regions of IgG1 heavy chain are shown. Shading has been used to distinguish regions of the two parent molecules that are used to form the immunoadhesin. The *N*-linked carbohydrate chain in C_H2 is indicated by a closed square. The hinge-region sequence indicates the location of proteolytic cleavage to generate monovalent reactor fragments. (Redrawn, with permission, from Chamov and Ashkenazi, 1996, fig. 1; redrawn from Chamov and Ashkenazi, 1995. Reprinted from Trends in Biotechnology, Vol. 14, Immunoadhesions: Principles and Applications. With permission from Elsevier Science.)

glycerol, other buffer components, and the remaining 1% impurities. Based on these results, production of 11.3 kg of the immunoadhesion, expressed at a concentration of 750 mg/L in 15,000 L of culture filtrate would require a 56.5 L column of the protein A affinity stationary phase if the 15,000 L of filtrate were to be processed over 10 cycles in 1500 L batches. Feasibility of large-scale production still needs to be confirmed.

9.33 THE LEACHING OF LIGANDS FROM PROTEIN A AFFINITY COLUMNS IS DIFFICULT TO DETECT

The extracorporeal treatment of plasma from cancer and autoimmune patients over protein A columns has shown that the protein A ligand can leach from the column and cause toxic reactions in the patients (Boyle et al., 1993). Proteins used in therapeutic applications and purified over protein A must be free of the protein A ligand, namely contaminants of bacterial binding protein. The leaching of the bacterial protein ligand from the stationary phase into the product must be avoided during its purification over protein A. It is difficult to detect low levels of bacterial binding proteins complexed with immunoglobulins, since there is a low concentration of protein A present in a high concentration of immunoglobulins and since detecting reagents for protein A, the antigen, react with the same F_c regions of the antibodies that complex with the antigen. Specialized antibody reagents and procedures are needed to detect the very small quantities of potentially toxic antigens.

The paradox of high ligand specificity and contamination with low levels of ligands leached from a column can be addressed by following the affinity chromatography step with other purification methods. These would remove the contaminant at a pH and/or ionic strength that will dissociate the antigen from the antibody, and use a different mechanism of separation to selectively adsorb one of the two components on another type of stationary phase. This concept, and the need for orthogonal purification steps, is discussed as part of the SEP-OPS™ method in Sections 8.59 to 8.61. It is unlikely that protein A affinity chromatography will be the final purification step in the production of monoclonal antibodies or other immunoproteins intended for therapeutic applications.

IMMOBILIZED METAL IONS

Ion exchange resins exchange cations and anions in aqueous streams, and some types of these resins are widely used in water treatment. A different application of ion exchange resins immobilizes metal ions by chelating the ions in a manner that facilitates formation of coordination bonds with sugars and amino acids. Chelating ligands fixed to the surface of chromatographic stationary phases enable protein purification through interactions of the metal ion with amino acids in peptide sequences associated with the protein, or hydroxyl groups of some types of saccharides.

The significance of the partially filled set of d-orbitals of the copper (II) ion with respect to chromatographic separations was apparently realized as early as 1971 (Bourne et al., 1971). Bourne et al. demonstrated complete separations of D-glucose from D-glucitol and D-mannose from D-mannitol over a cation exchange resin (Amberlite IR 120) treated with 2.5% aqueous copper (II) acetate monohydrate. They found that reduced mono- and disaccharides (i.e., sugar alcohols) form complexes with the copper (II) ion but that the reducing sugars does not (except for D-ribose, D-xylose, and D-gulose) (Bourne et al., 1971). Later work showed that the silver form (Ag^+) of the cation exchange resin Aminex A-5 give separation of oligosaccharides (Scobell and Brobst, 1981) and that immobilized Ag^+ is likely to exhibit linear coordination and form strong monodentate complexes with polyols, while Cu^{2+} exhibits square planar coordination with bidentate complex formation. Chromatographic elution data for a series of sugars and sugar alcohols indicate that Ca^{2+}, Sr^{2+}, and Ba^{2+} immobilized on a cation exchange resin (Aminex A-5) form relatively strong tidentate chelates when three adjacent hydroxyl groups in pyranose rings are in an axial-equitorial-axial conformation, or adjacent hydroxyls are in a furanose ring are cis-cis (Goulding, 1975).

Iminodiacetic acid, tris (carboxymethyl) ethylenediamine and nitrilotriacetate fixed to stationary phase surfaces are able to retain Cu^{2+}, Zn^{2+}, or Ni^{2+} metal ions. Immobilized Cu^{2+} and Zn^{2+} are particularly effective in complexing with histidine residues in proteins or peptides. This property has been applied to recombinant fusion proteins of the target molecule and His polypeptides. The His residues facilitate capture of the fusion protein by the affinity matrix from contaminating proteins which do not have the histidine tail. As succinctly described by Hochuli (1988, pp. 293–294), "attractive genetic approaches to large-scale purification of recombinant proteins have been developed. Hybrid proteins were prepared by fusing the coding sequence of the protein of interest with the coding sequence of a peptide with high affinity for the affinity resin together with the sequence of a specific cleavage site between the peptide and the protein. Such fusion proteins could efficiently be purified, taking advantage of the specific binding of the affinity peptide (affinity tail) to the affinity resin. After purification of the fusion protein, the affinity tail could be split off at the designed cleavage site." Examples of this approach include staphyloccal protein A as an affinity tail, which binds with IgG, and histidine tails, which complex an immobilized metal ion affinity support.

9.34 COUNTERIONS BOUND TO SULFONIC ACID CATION EXCHANGE RESINS COMPLEX SUGARS THROUGH THEIR HYDROXYL GROUPS

Sulfonic acid-based cation exchange resins with the immobilized counterions Ca^{2+} and Li^{+} have been used for the liquid chromatographic separations of sugars with water as the eluent since 1960 (Jacobson, 1982; Jones et al., 1960; Jones and Wall, 1960; Lin et al., 1988), while fructose, galactose, and glucose were separated over a strong anion exchanger in 1951 (Khym and Zill, 1951). Although the term "ion exchange" is used to describe the stationary phase, the mechanism of separation does not involve ion exchange. Rather metal ions (or counterions) bound to the resin form reversible complexes with the hydroxyl groups on monosaccharides, with the size of the counterion and the orientation of the sugars' hydroxyl groups resulting in different strengths on binding, and therefore different retention behaviors for different sugars. Water-soluble oligomers of glucose, xylose and arabinose (Brobst et al., 1973; Ladisch et al., 1978; Ladisch and Tsao, 1978; Scobell and Brobst, 1981; Scrobell et al., 1977), and numerous monosaccharides including glucose, fructose, xylose, arabinose, and mannose are readily separated by liquid chromatography by this mechanism. The separation of the oligosaccharides occurs with the retention time increasing as the molecular weights of the oligosaccharides decrease, and hence this is attributed to a size exclusion effect for some types of resins (Ladisch et al., 1978). The complexing of monosaccharides with an ion sequestered with an ion exchange group on the resin is referred to as the counterion effect and separates monosaccharides based on differences in their strength of binding with the counterions (Jacobson, 1982). Other counterions effective for separating sugars have included Na^{+}, Rb^{+}, Mg^{2+}, Ag^{2+} and Cu^{2+} (Angyal et al., 1979). Cation exchange resins with Ca^{2+} as the counterion were probably instrumental in the growth of the high-fructose corn syrup industry both for the purposes of analytical monitoring of the separation of glucose from fructose (Brobst et al., 1973) and process-scale separations (Ching, 1983).

Goulding (1975) and Angyal et al. (1979) proposed that strong tridentate complexes are formed with adjacent hydroxyl groups that have an axial-equatorial-axial sequence in pyranose rings or a cis-cis sequence in furanose rings as shown in Eq. (9.8):

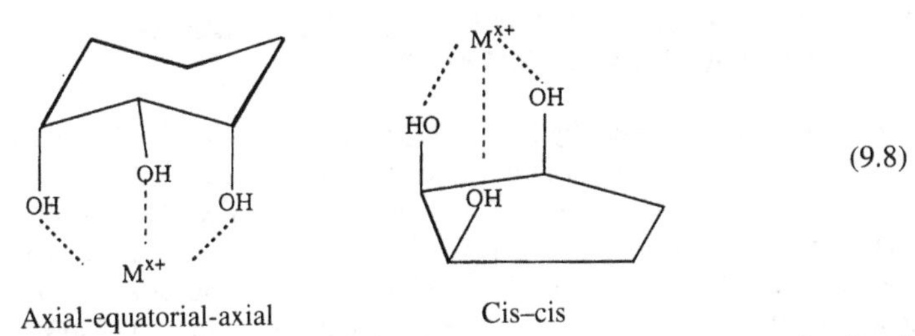

(9.8)

Examples of sugars that form tridentate complexes are allose and talose (pyranose rings) and ribose (furanose ring). A bidentate complex formed with two adjacent hydroxyl groups in an axial-equatorial sequence is weaker:

(9.9)

HO OH M^{x+}

Axial-equatorial

with mannose, glucose, arabinose, and xylose exhibiting this behavior. The size of the cation affects the type of complex formed, with small counterions and large counterions forming the weaker bidentate complexes, with moderately sized counterions forming the stronger tridentate complexes (Goulding, 1975). The larger metal ions form a different type of bidentate complex, however, where equatorial-equatorial and 1, 3 axial-axial diol sequences complex with the ion to form relatively weak bidentate complexes. The separation achieved with the Ca^{2+} form of a cation exchanger is illustrated in Figure 9.39.

COMPOSITE CHROMATOGRAM OF CARBOHYDRATE SEPARATED
ON AMINEX 50W-X4 USING WATER AS ELUENT

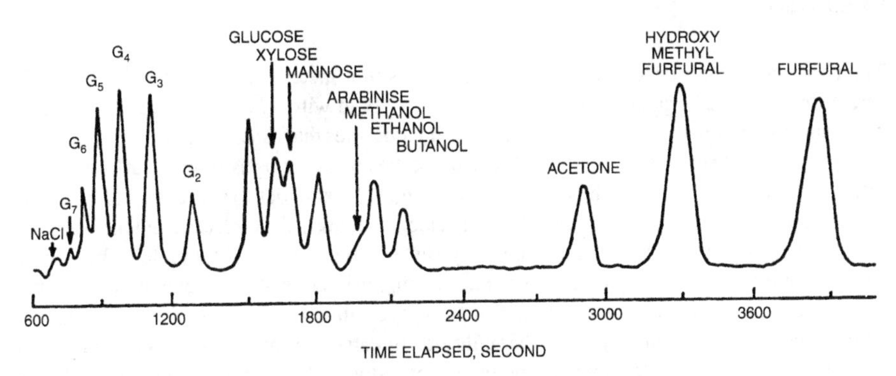

Figure 9.39 Chromatogram of sugars showing resolution of oligosaccharides on the basis of decreasing size, monosaccharides by differences in complexing with the Ca^{2+} counterion bound to the ion exchange resins and other molecules due to hydrophobic interaction with the stationary phase. Conditions: 6 mm i.d. × 600 mm column packed with 4% crosslinked styrene-divinyl benzene cation exchanger (Aminex 50 W × 4 from Bio-Rad) in Ca^{2+} form. Eluent was water at 0.5 mL/min. Column temperature at 85°C, pressure drop at 100 psig. Sample size of 20 μL. (Reprinted with permission from Ladisch and Tsao, 1978, fig. 9, p. 99, J. Chromatogr.)

9.35 Ca^{2+} BOUND TO SULFONIC ACID RESIN SEPARATES STEREOISOMERS OF 2,3-BUTANEDIOL

The fermentation of xylose by *Klebsiella pneumoniae* (ATCC 8724) produces *meso* and *non-meso* 2, 3 butanediol. These stereo-isomers were resolved using Aminex 50 W-X4 (Ca^{2+} form) at 85°C with water as the eluent (Voloch et al., 1981). A combination of ^{13}C-NMR, gas chromatography combined with mass spectroscopy of trimethylsilyl ethers of the meso and non-meso butanediols, and melting points of nitrobenzene/ethanol esters of the two forms were used to identify the two peaks obtained. The meso and non-meso butanediols

Meso D- or L- D- or L- (eclipsed)

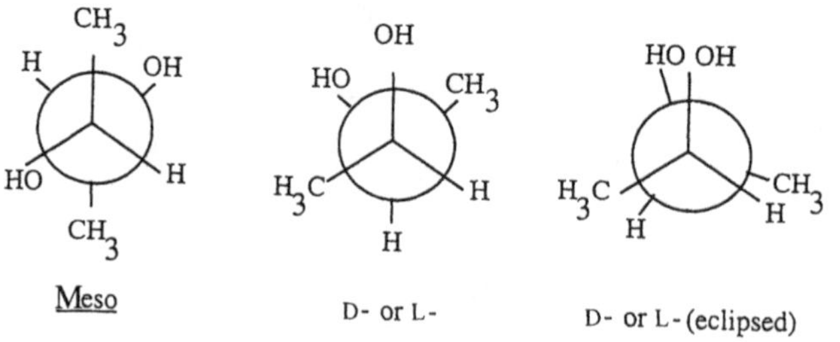

complex with different affinities to the Ca^{2+} counterion, resulting in their separation. This further illustrates the utility of an immobilized or bound counterion in discriminating between, and therefore resolving, similar molecules. This property is even more useful for fractionating proteins, where complexing of immobilized transition metal ions with histidine residues provide a way to selectively adsorb proteins.

9.36 COPPER (Cu^{2+}) AND OTHER TRANSITION METAL IONS, BOUND TO IMINODIACETATE, COMPLEX PROTEINS THROUGH THEIR HISTIDINE RESIDUES

The first row of the periodic table includes the transition metal ions: Ni^{2+}, Cu^{2+}, and Zn^{2+}. These ions have fast ligand exchange properties, will bind with an immobilized chelating ligand, and retain from one to three available coordination sites that can bind to histidine residues in peptides and at surfaces of proteins. These properties form the basis of metal affinity separation techniques where strong binding of partially ligated metal ions to side chains of histidine and cysteine residues of proteins result in their selective uptake. This selective uptake was first described for separations of small molecules by Helferrich in 1962, and for proteins by Porath and co-workers in 1975. Homologous proteins having different histidine contents were originally used to delineate the molecular basis of protein retention by chelated metal ions. The advent of recombinant technology and the ability to selectively replace individual amino acid residues in a protein enabled study of the impact of individual amino acids, and their relative spacing in a peptide sequence, on binding of protein with chelated metals (Haymore et al., 1992; Johnson and Arnold, 1995a).

Metal chelating resins have been prepared from Trisacryl (an acrylic based stationary phase), Sephadex (crosslinked dextran gel) and Sepharose (agarose) (Johnson and Arnold, 1995a; Luong et al., 1992; Smith et al., 1990). These stationary phases are derivatized with iminodiacetates, which in turn complex some types of metal ions with vacant coordination sites.

These immobilized metal ions complex with His or Cys residues on proteins. Copper appears to be a preferred chelating metal ion, since it is neutral over the pH range of approximately 3 to 8. With copper, nonspecific binding effects of ion exchange are minimized, allowing coordination bonds to form with the proteins at a useful pH range. The complexes formed by copper at high and low pH are illustrated in Figure 9.40. While iminodiacetate complexes of copper, nickel, and zinc can bind histidine residues of proteins, only Cu^{2+} and Ni^{2+} gave good separations; Zn^{2+}, Co^{2+}, Fe^{2+}, and Fe^{3+} gave poor results due to leaching of the metal ion from the column and/or weak protein binding (Haymore et al., 1992; Luong et al., 1992). Copper-iminodiacetate matrices gave the best results. Hence studies on the copper (II) complex domi-

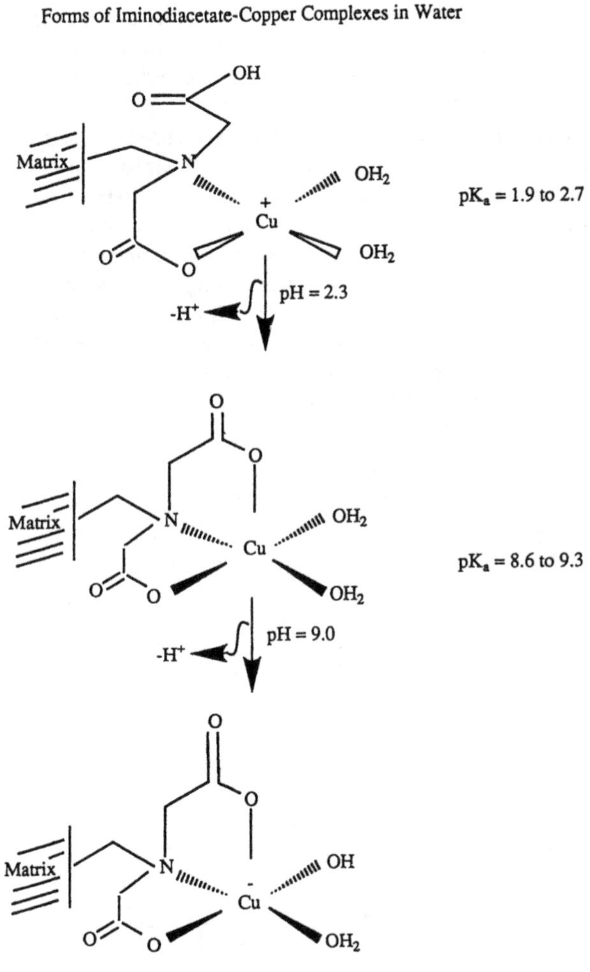

Forms of Iminodiacetate-Copper Complexes in Water

Figure 9.40 Iminodiacetate-copper complex (abbreviated (IDA) Cu $(OH_2)_2$) exists in a neutral form over pH 3 to 8. Matrices used for derivatization with iminodiacetate include Sephadex, Sepharose, Trisacryl, and Styrene-divinylbenzene (modified from Luong et al., fig. 3, p. 83, 1992).

nate the literature (Arnold and Haymore, 1991; Haymore et al., 1992; Johnson and Arnold, 1995a; Luong et al., 1992; Smith et al., 1990).

9.37 ELUTION BUFFER AND pH DISSOCIATE METAL-PROTEIN COMPLEXES TO ACHIEVE SEPARATION

The manner in which a protein forms reversible coordination bonds is illustrated in Figure 9.41. The metal ion is initially loaded and the column washed (Figure 9.42). Open coordination sites are occupied by water. The His residues on a protein displace the water at a higher pH, while other proteins that do not have exposed His will not bind. The bound protein is eluted by lowering the pH of the buffer and/or using a buffer (e.g., imidazole) which competes for the coordination sites. At a lower pH the donor atoms on the His residues are protonated so that they can no longer coordinate with the immobilized metal ions. Similar binding characteristics have been noted for cysteine residues and predicted for tryptophan residues, although, according to Smith et al. (1990), there are no known examples of tryptophan coordinating metal ions through the indole group, either as amino acid complexes, peptide complexes, or in metal co-proteins.

The choice of eluting buffers used to displace the protein must consider the strength of their binding to the metal ion. A thorough and elegant study by Haymore et al. (1992) recommended release buffers (so named since they cause complexed proteins to be released or desorbed) to include imidazole and *N*-acetyl histidine for strongly bound proteins, and glycinamide and 4- and 2-picoline for weakly bound proteins. Glycine, histidine, iminodiacetic acid, and EDTA were poor releasing agents because of their "overly strong" metal-protein binding, and hence caused difficulties in regenerating the metal ion complex. Metal-binding pH buffers, such as Tris, were avoided—presumably for the same reason. Buffers containing 0.1 to 2 M NaCl were used to suppress ion exchange properties of the matrix, and hence minimize nonspecific binding of other proteins to the affinity matrix.

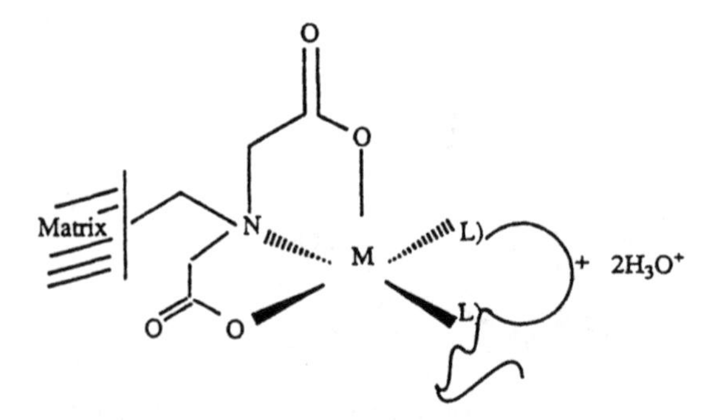

Figure 9.41 Schematic illustration of forming the coordinating metal ion complex at pH 7.0 in immobilized metal affinity chromatography (adapted from Smith et al., fig. 1, p. 173, 1990).

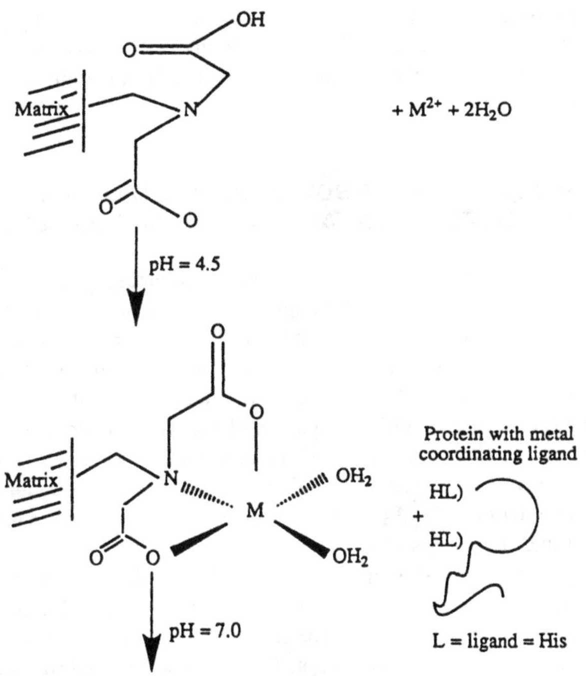

Figure 9.42 Schematic illustration of immobilized metal affinity chromatography: addition (or release) of protein: coordination complex of protein with metal ion. (Adapted from Smith et al., 1990, with description taken from Haymore et al., 1992. With permission from Academic Press.)

9.38 A CHELATING PEPTIDE MUST EXHIBIT RAPID KINETICS FOR COMPLEX FORMATION TO BE USEFUL AS A PURIFICATION HANDLE

As clearly stated by Smith et al. (1990, p. 171), "A slow rate of complex formation would defeat the purpose of using a chelating peptide as a purification handle, since the complex must form during the time of the chromatographic separation." This has led to the choice of transition metals in the first row of the periodic table: Fe^{2+}, Co^{2+}, Ni^{2+}, Cu^{2+}, and Zn^{2+}. These metal ions are known to have rapid water and ligand exchange rates. The amino acids that bind metals in metalloproteins (histidine, cysteine, aspartate, glutamate, methionine, lysine, and tyrosine) do not necessarily make good purification handles. A careful analysis by Haymore et al. (1992), showed that Ni^{2+}, Cu^{2+}, and Zn^{2+} can bind to imidazole ($pK_a \approx 6.7$) and thiolate groups ($pK_a \approx 8.5$) from His and Cys residues, respectively. The reason that His, rather than Cys, dominates the literature (and research) on this type of separation is that "exposed sulfhydryl groups on protein surfaces are relatively rare and when present are often susceptible to oxidation" (Arnold and Haymore, 1991, p. 1796). Binding interactions between copper and imidazole are on the order of 5 kcal/gmole. Under some conditions the amino terminus ($pK_a \approx 7.7$) and aspartate and glutamate residues ($pK_a \approx 3.9$) can contribute to metal binding, but lysine and arginine residues at physiological pH (6 to 8) contribute little to metal binding unless their pK_a's are abnormally low. The presence of aromatic residues (Trp, Tyr, and Phe) can enhance the binding strength of other metal-binding sites, but these residues *do not* directly bind to immobilized metals in aqueous solutions (Haymore et al., 1992).

9.39 RECOMBINANT FORMS OF BOVINE SOMATOTROPIN FACILITATED STUDIES OF IMMOBILIZED METAL AFFINITY CHROMATOGRAPHY

While it is not clear that immobilized metal affinity chromatography is used to purify recombinant bovine somatotropin (Wheelwright, 1991), the availability of recombinant forms of the hormone made it possible to strategically design bovine somatotropin molecules for studying complexes of histidine residues with copper chelates. Recombinant methodologies were used to introduce strong metal-binding recognition sites onto surfaces of recombinant proteins and to facilitate rapid, single-step purifications from crude protein mixtures. A key parameter was found to be the accessibility of the His residues, and the separation of the residues by three, intervening amino acid residues in an α helix (represented by the abbreviation –His X_3 His) or a single amino acid residue between two His residues in a β strand (His X His).

Bovine somatotropin, which has an accessible histine residue at position 19 (His^{19}), was altered and expressed in *E. coli* to add a single His residue at position 15, to give a His^{15}–X_3 His^{19} variant (represented as His^{15}–bST). The His^{15}–bST variant strongly bound to the Cu^{2+} iminodiacetate complex, while the bST did not. The His^{15}–bST variant was recovered in 95% yield and 97% purity from a crude cell lysate in a biologically active form (Arnold and Haymore, 1991; Haymore et al., 1992). Binding constants for genetically engineered His X_3 His sites in exposed α-helices range from 2×10^4 to 2×10^6 M^{-1}. A good chelating dihistidine site should have a binding constant that is 100 times larger than that for the same copper complex binding to a similar protein with two exposed nonchelating histidine residues (Arnold and Haymore, 1991).

9.40 PURIFICATION OF RECOMBINANT BOVINE SOMATOTROPIN BY IMMOBILIZED METAL AFFINITY CHROMATOGRAPHY REQUIRES MANY STEPS

Bovine somatotropin, also referred to as bovine growth hormone, is a naturally occurring protein hormone that is produced in the cow's pituitary gland. This hormone has a molecular weight of 22,000 and consists of either 190 or 191 amino acid residues in a single polypeptide chain with two disulfide links (Wallis et al., 1985). The recombinant form of this hormone, injected into cows on a biweekly basis, boosts milk production by 10 to 15% (Kirschner, 1994). It is produced through fermentation of recombinant *E. coli*.

Recombinant bovine somatotropin (abbreviated BST) was approved for use on November 5, 1993 (after 10 years of review by the FDA), entered the market as Posilac® in February 1994, and "booked its first profit" in September 1996. Recombinant bovine somatotropin is the first bioengineered animal drug approved by the FDA, and it is being supplied to Monsanto on a contract manufacturing basis by Biochemie (Kundl, Austria). About 17,000 U.S. dairy farmers (15% of all U.S. producers) have purchased bovine somatotropin making it "the largest selling dairy animal health product in the US in just two years" (Anonymous, 1996). Design of a manufacturing facility to be built in Augusta, GA, was started in June 1996. The breakeven point in sales was estimated at $100 million, with the eventual U.S. market estimated at $500 million/yr.

However, in 1998 Posilac generated about $200 million annual sales for Monsanto with about 13,000 U.S. dairy farmers using it a $5.00/dose (Heylin, 1998; Kirschner, 1994). The potential utility of the immobilized affinity chromatography method is illustrated by the analytical chromatogram of somatropin variants (Figure 9.43) and somatotropin (His^{15}–bST) purified from a crude cell lysate (Figure 9.44). The analytical column was a 1.0 × 10.2 cm glass column (8 mL volume) packed with washed iminodiacetate-Trisacryl loaded with 22 μmol Cu/mL of stationary phase. The matrix was then washed with 25 mL of 50 mM $Cu(C10_4)_2$ at pH 4.5 with the IDA functionalized matrix adsorbing nearly all of the copper ions. After a sequence of buffer washes similar to that given in Table 9.8, a 0.5 mL sample of the bST was injected onto the column and developed in a linear gradient (i.e., linear at the inlet) of 1 to 66 mM imidazole buffer at pH 7.0 over 6 hours. Figure 9.43 shows a typical chromatogram. Metal ion affinity chromatography is conceptually simple, but operationally complex, as illustrated by the elution protocol given in Table 9.8 for purification of bovine somatotropin (Haymore et al., 1992).

The preparative column consisted of a 2.2 × 21 cm glass column (80 mL volume). In this case cell-free supernatant consisting of a complex and crude mixture of proteins (10 g) containing 2 to 20 mg bST was directly placed on the column. Elution was with a 1 to 44 mM imidazole gradient, to give the chromatogram of Figure 9.44. The purification of His^{15}–bST was carried out at loadings as high as 11 mg His^{15}–bST/mL IDA-Trisacryl with greater than 95% recovery, and 96 to 98% purity in a single step. The IDA-Trisacryl was exposed to 300 cycles of loading of clarified crude cell lysate containing the bST without a change in recovery or purity, and only a 5% decrease in copper binding capacity (Haymore et al., 1992). At a loading of 11 mg/mL (11 g/L) per day, the volume of stationary phase required to produce 10 tons product/yr (about 30 kg/run) would be about 2800 L, assuming that there are 300 runs/yr.

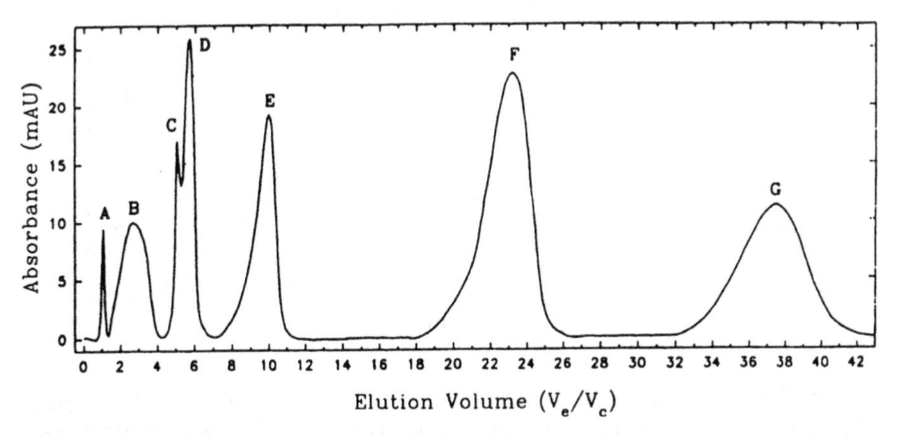

Figure 9.43 Chromatogram of somatotropin variants. Elution was carried out with a 1-66 mM imidazole gradient over 45 column volumes using the 8 mL Cu-IDA-Trisacryl column. Absorbance was monitored at 280 nm. (*A*) Injection peak; (*B*) 120 μg Ser_{169}-pST; (*C*) chromatographic artifact; (*D*) 150 μg native pST; (*E*) 200 μg His_{149}-bST; (*F*) 550 μg His_{15}-bST; (*G*) 380 μg $His_{26}His_{30}$-bST. (From Haymore et al., fig. 1, p. 29, 1992, reprinted with permission, Academic Press.)

TABLE 9.8 Typical Buffer Schedule for Developing a bST Elution Profile

Column Volumes	Flowrate (mL/min)	Solution
2	1.0	1 mM imidazole (loading buffer)
25	1.0	1-45 mM imidazole linear gradient
1	1.0	45 mM imidazole (release buffer)
3	2.0	100 mM NaCl
3	2.0	50 mM EDTA, pH 7.0
2[a]	2.0	100 mM NaCl
5[a]	2.0	100 mM NaOH
2[a]	2.0	100 mM NaCl
3[a]	2.0	50 mM NaH_2PO_4, pH 7.0
2	2.0	100 mM NaCl
3	2.0	50 mM $Cu(ClO_4)_2$, pH 4.5
3	2.0	100 mM NaCl
4	2.0	45 mM imidazole (release buffer)
4	2.0	1 mM imidazole (loading buffer)

Source: The material is reproduced with permission of Academic Press from Haymore et al., 1992.
[a]Used during wash cycle.

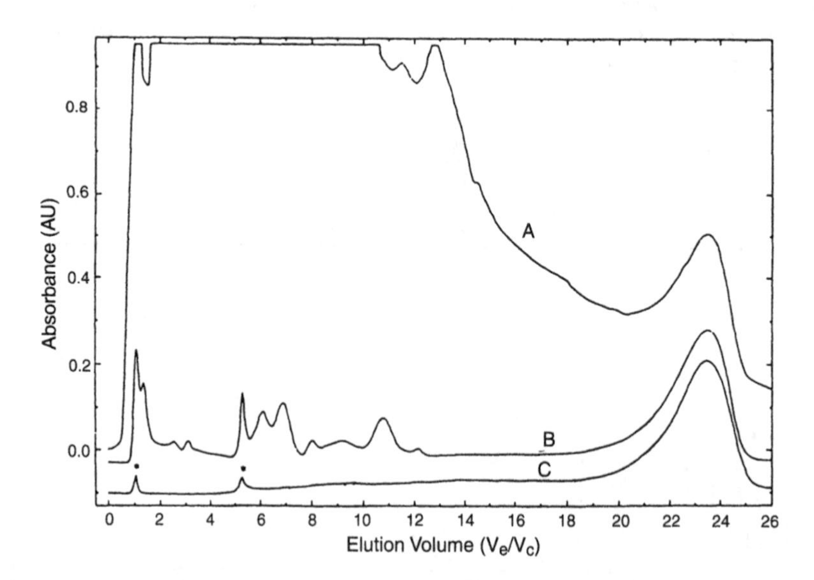

Figure 9.44 Purification of His_{15}-bST from a crude cell lysate. Sample contained over 10 g total protein and 20 mg somatotropin variant. Elution was carried out with a 1-44 mM imidazole gradient over 30 column volumes using the 80 mL Cu-IDA-Trisacryl column. Absorbance was monitored at 280 nm. Asterisks (*) indicate injection and artifact peaks. (*A*) Chromatogram of crude cell lysate; (*B*) Chromatogram of late eluting peak from (*A*) after buffer exchange to reduce imidazole content; (*C*) third pass through column does not improve impurity and shows no impurities. (Reproduced with permission from Haymore et al., (1992), fig. 1, p. 29, and fig. 2, p. 30, 1992, copyright Academic Press.)

9.41 HETEROGENEITY OF PROTEIN BINDING IN IMMOBILIZED METAL AFFINITY CHROMATOGRAPHY FOLLOWS THE TEMKIN ISOTHERM

The ability to manipulate the number of interactions between stationary phase and protein through the density of surface metal sites or by using site-directed mutagenesis to change the number of exposed metal coordinating histidine groups on the protein enables binding heterogeneity to be studied (Johnson and Arnold, 1995a). The Langmuir adsorption model, derived for gas adsorption on a solid surface, does not adequately describe heterogeneity of protein adsorption to an affinity matrix where the interaction between a protein and metal complex on the surface of an iminodiacetate stationary phase affects binding of additional protein molecules. The Langmuir equation reflects a single maximum binding capacity, and it does not account for the increase in the maximum amount of binding that occurs with increasing surface histidine content of the protein, or the change in binding property due to a significant local change in amino acid conformation and/or composition (Haymore et al., 1992; Johnson and Arnold, 1995a).

The Temkin isotherm model, like the Langmuir model, was developed for gas phase adsorption at a solid surface (Johnson and Arnold, 1995a; Johnson and Arnold, 1995b), but it gives a good fit of data for immobilized metal affinity chromatography of 10 yeast cytochrom C variants. The model is

$$Q(C) = q_T \ln(1 + K_T C) \tag{9.10}$$

where q_T is the differential increase in the limiting capacity for protein adsorption with increasing protein binding affinity (moles/mL stationary phase), K_T is the maximum binding affinity (L/mole), and C is the concentration of the protein at equilibrium (moles/L). A plot of $Q(C)$ as a logarithmic function of $K_T C$ should give a single characteristic curve, even for a wide range in the values of the binding affinity of the protein, K_T. This was shown to be the case for 10 cytochrome C variants ranging from zero to three surface histidines, and native horse cytochrome C at 9 different copper densities on an iminodiacetate stationary phase (Figure 9.45). While the values of K_T and C obtained from fitting of the data are not given, the form of the Tempkin isotherm and the uniformity of the data plotted in this manner (Figure 9.45) show the applicability of this model.

9.42 PROTEINS BIND LESS RANDOMLY IN BIOLOGICAL RECEPTOR SYSTEMS THAN AT SURFACES OF CHROMATOGRAPHIC STATIONARY PHASES

The binding of proteins with receptors and antigens with antibodies at cell surfaces may be better described by the distribution curves of the Hill or Langmuir isotherms (Figure 9.46) rather than the uniform distribution of the Tempkin isotherm (Figure 9.47). This comparison illustrates the difference between the binding energy distributions between surfaces on chromatographic or solid supports that are more random (e.g., the Tempkin isotherm) compared to biological systems with more specific binding (Langmuir isotherms) particularly at low protein loadings (Johnson and Arnold, 1995b). This divergent behavior offers a starting hypothesis for the observed differences in binding of biocombinatorially derived molecules in free solution compared to binding at solid surfaces, as discussed earlier in this chapter.

The significance of these observations is summarized by Johnson and Arnold (1995a, p. 441): "A protein will show the highest affinity for surface arrangements which best match its

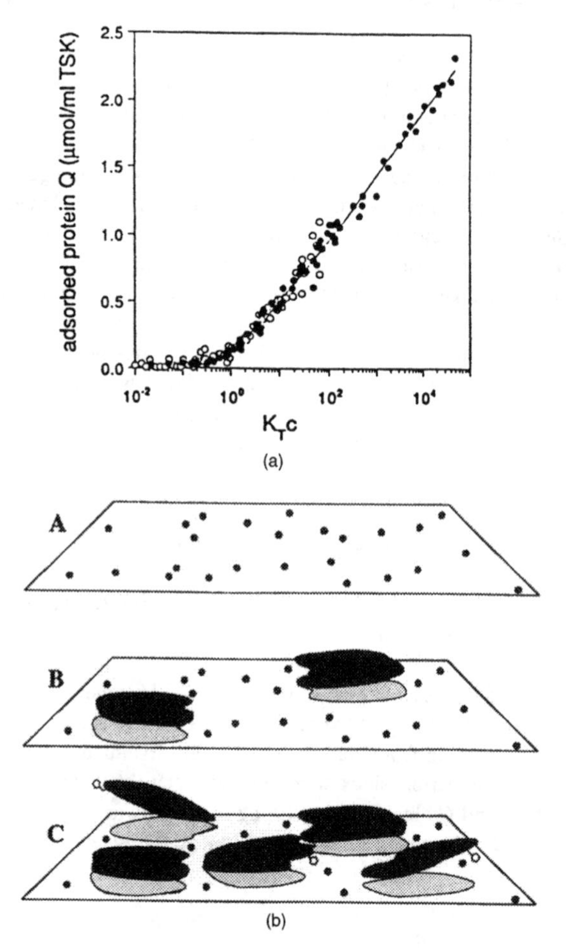

Figure 9.45 Temkin model for heterogeneous adsorption in IMAC. (*a*) Free protein concentration is scaled by the maximum binding constant (K_T) determined by nonlinear regression of adsorption data to the Temkin isotherm (allowing both K_T and q_T to vary freely). Adsorption data are presented for cytochromes *c* to CuIDA-TSK; (●) 10 yeast cytochrome *c* variants (zero to three surface histidines) and tuna cytochrome *c* (one surface histidine) at maximum copper loading (18.6 μmol Cu/mL TSK); and (○) horse cytochrome *c* at 9 different copper loadings (6.1 to 18.6 μmol Cu/mL TSK). Curve represents the Temkin isotherm calculated using $q_T = 0.21$ μmol/mL TSK. (Reprinted with permission, Johnson and Arnold (1995).) (*b*) Heterogeneous protein adsorption to a random arrangement of surface binding sites. (*A*) Surface has a nonuniform arrangement of ligands for protein binding (copper sites in IMAC). (*B*) Protein binds with highest affinity to surface "binding sites" in which the ligand pattern complements arrangement of protein functional groups (surface histidines). (*C*) Protein adsorbs with lower affinity to surface "binding sites" with less optimal ligand patterns, resulting in a range of binding energies for protein adsorption to the surface. (From Johnson and Arnold (1995), fig. 4, p. 440, and fig. 5, p. 441, Biotechnol. Bioeng., reprinted by permission of Wiley-Liss, Inc., a subsidiary of John Wiley & Sons, Inc.)

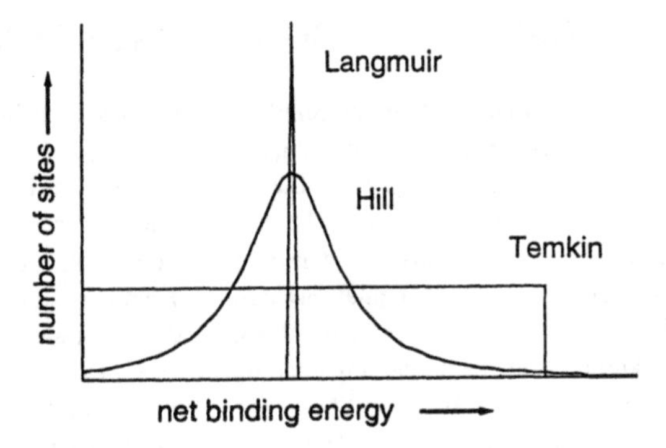

Figure 9.46 Form of adsorption isotherm expression depends on distribution on in binding energies. Langmuir isotherm is described by a spike at a particular binding energy. Hill isotherm (exponent less than one) is described by a bell-shaped distribution in binding energies. Temkin isotherm is represented by a uniform distribution, up to some maximum binding energy. (Reproduced with permission from Johnson and Arnold (1995), p. 296, Biotechnol. Bioeng.)

own distribution of functional sites. Protein adsorption particularly at the relatively low coverage typical of chromatographic applications, will therefore be dominated by those surface sites which optimize the interaction between protein and chromatographic support." Adsorption can occur over a wide range of binding energies, and protein adsorption equations must be capable of reflecting a heterogeneous population of binding sites, particularly at higher protein loadings. Examples of chromatography in which there are multiple binding sites include charge–charge interactions in ion exchange chromatography, a metal-histidine coordination interaction in metal affinity chromatography, and ligand-binding interactions in other types of affinity chromatography.

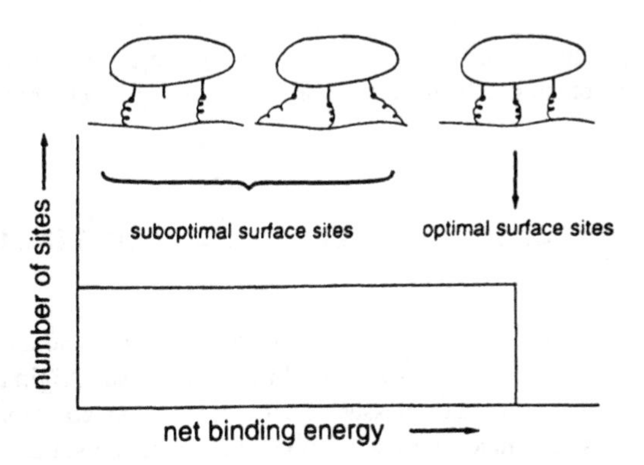

Figure 9.47 Heterogeneous protein described as a distribution of binding energies over the population of surface sites. Distribution in binding energies results from multiple contacts between protein and random arrangement of surface ligands (connected to the surface via a flexible spacer arm). Protein binds with highest affinity to surface sites in which ligand pattern complements arrangement of protein functional groups. Less optimal surface sites must incur additional energy penalties to support the same degree of protein binding, resulting in a range of net binding energies for protein adsorption.

CELLULOSE BINDING DOMAIN CASE STUDY

The complex interactions between systems of cellulase enzymes and an insoluble, cellulose substrate have been studied since 1950 with the objective of developing rapid, cost-effective, and environmentally friendly means to convert renewable cellulosic materials into glucose, and other fermentable sugars (Ladisch et al., 1983). These sugars can be fermented to alcohols, organic acids, and amino acids that are currently derived from petroleum sources or, in the case of ethanol, from corn. Significant fundamental research has been carried out on the structure and function of cellulase enzymes since 1950, and this has led to a number of beneficial, practical developments, ranging from the processing of cotton textiles[15] to development of affinity-bifunctional linkers to couple recombinant antibodies to cellulosic stationary phases (Cavuco-Paulo and Almeida, 1996; Romicez et al., 1993) while minimizing use of chemical reagents of the type described in Chapter 8.

Fundamental studies have identified noncatalytic, protein segments of cellulases that serve to anchor the catalytic domain of the enzyme either to a cellulosome or to cellulose (Gilkes et al., 1992; Tokatlidis et al., 1991). The latter is referred to as the cellulose-binding domain (abbreviated CBD), and it has a number of potential applications. A hybrid protein consisting of CBD and a target molecule, connected through a removable amino acid residue, would allow the hybrid protein to be adsorbed on cellulose particles, the cellulose filtered and washed, and the target molecule now free of contaminants, to be recovered by hydrolyzing it away from the cellulose-binding protein attached to the cellulose particle. This approach has been used to recover a hybrid protein consisting of a cellulose-binding domain coupled to IL-2 (Section 9.26). Another application is the construction of a hybrid protein consisting of CBD and an affinity ligand. Once expressed, the hybrid protein is captured and immobilized on a cellulose scaffold, without the need to use extraneous chemicals to activate the cellulose or covalently bind the ligand. A hybrid immobilized ligand of this type, for example, protein A for the purification of IgG, is then ready for use as an affinity ligand once it is adsorbed by the cellulose.

This chapter ends with a case study on the cellulose binding domain, since it illustrates the impact of long-range fundamental research in an unrelated area (enzyme hydrolysis of cellulose) on the development of a new separations approaches in affinity chromatography, and shows how affinity chromatography provides a bridge between molecular biology and separations science.

9.43 CELLULASES CONSIST OF A CELLULOSE-BINDING DOMAIN AND A CATALYTIC DOMAIN

Cellulases are modular proteins that can hydrolyze cellulose principally to cellobiose, while β-glucosidase further hydrolyzes the cellobiose to glucose. Originally Reese and his co-workers proposed that two major classes of cellulase enzyme components are required for converting cellulose to glucose: a C_1 component to make the solid substrate more ac-

[15] A soon to be introduced cellulase formulation from an unspecified bacteria was isolated from an alkaline soda lake. This enzyme is named cellulase 103, and it is reported to be able to break down microscopic fuzz of cellulose fibers that traps dirt on surfaces of cotton textiles while minimizing hydrolysis to the main part of the cotton fibers. According to Genencor, the market for detergent enzyme additives is $600 million/year, in which cellulases could find significant use (Pennisi, 1997).

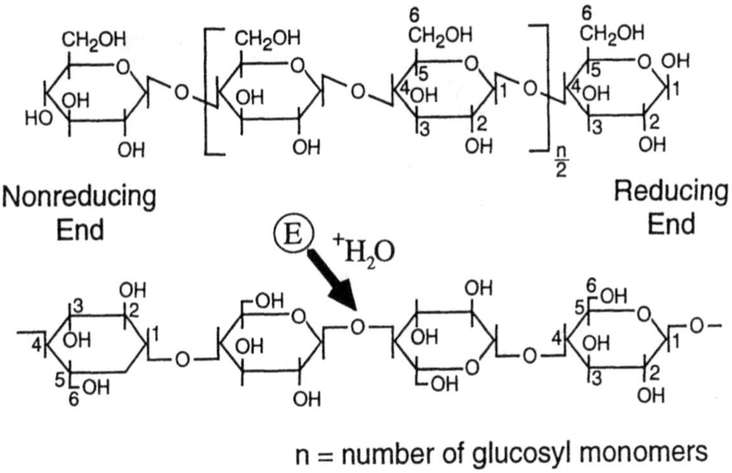

Nonreducing End

Reducing End

n = number of glucosyl monomers

Figure 9.48 Schematic representations of cellulose molecule and the glycosidic bonds that are cleaved by enzymes denoted E.

cessible and C_x to carry out the hydrolysis. It is now known that many cellulases have discrete independently functioning cellulose-binding domains that are devoid of hydrolytic activity to which a catalytic domain is attached. The catalytic domain depolymerizes the cellulose to soluble oligomers and glucose by adding water to the β-1,4 bonds (Figure 9.48) (Ladisch et al., 1983; Din et al., 1994).

Studies with a cellulase derived from the bacterium *Cellulomonas fimi*, and compared to a cellulase from the fungus *Trichoderma reesei*, showed that the cellulose-binding domain for an exocellulase from *C. fimi* consists of 100 amino acid residues while that from *T. reesei* is 36 residues long.[16] The cellulose-binding domains of both *C. fimi* and *T. reesei* cellulases contain a large number of aromatic residues. The three tryptophan residues on the surface of the *C. fimi* cellulose-binding domain, and the three tryosines on the surface of the *T. reesei* cellulose-binding domain, are thought to interact with the cellulose surface and promote binding. If this proves so, it would be similar to the adsorption interface of maltose-binding protein from *E. coli*, for which aromatic residues have been shown to play an important role (Gilkes et al., 1992).

The part of the protein responsible for hydrolysis (i.e., the catalytic domain) is attached to the binding domain by a 10 to 20 residue polythreonyl linker[17] (Tomme et al., 1995a), or linker sequences rich in proline, serene, or threonine amino acid residues (Tomme et al., 1994). This concept is illustrated in Figure 9.49. The dimensions of the cellulose-binding

[16]There are 12 different families of cellulase catalytic domains and 10 families of cellulose binding domains, known at this time (Tomme et al., 1995a,b). Enzymes are grouped into families based on similarities of their amino acid sequence (Tomme et al., 1995a). Cellulose-binding domains may vary from 36 to 240 amino acid residues (Romicez et al., 1993), and have been identified in over 100 β-1,4-glucanases (Creagh et al., 1996).

[17]This linker was engineered in the construct. Linkers typically vary from 3 to 40 residues, but much larger linkers have been observed for cellobiohydrolase from *T. reesei* (Tomme et al., 1995a; Gilkes et al., 1992a, 1992b).

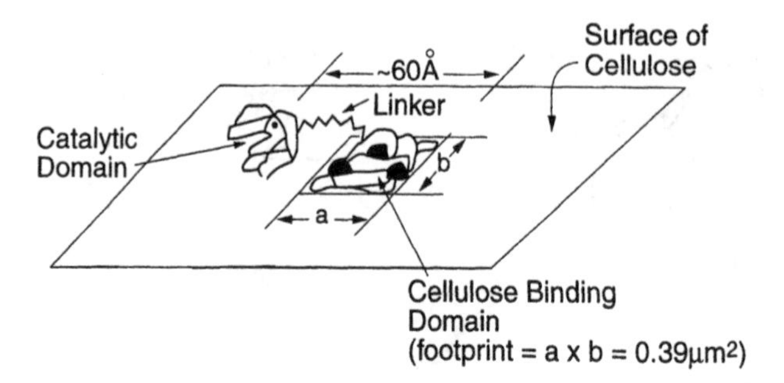

Figure 9.49 Schematic representations of role of cellulase catalytic and cellulose binding domains: Cellulose binding domains anchors catalytic domain to cellulose.

domain are estimated to be on the order of 45 × 25 × 25A (corresponds to the surface area of 39 cellobiose units) for protein from the family II cellulose-binding domain from *C. fimi* compared to 30 × 18 × 10A for the cellulose-binding domain. This is represented in Figure 9.49.

The mode of action of the two cellulose-binding domains differs in that the cellulose-binding domain of the exocellulase from *C. fimi* which has a larger surface packing density than the cellulose-binding domain protein from *T. reesei*. This is inconsistent with the measured size and number of amino acid residues for the *C. fimi* protein, which are greater than those for the cellulose-binding domain protein from *T. reesei*. This anomaly may be due to the action of the cellulose-binding domain of the exoglucanase from *C. fimi*, which disrupts the cellulose fiber structure of crystalline cellulose and creates extra surface area (not accounted for in calculating surface packing density) compared to the cellulose-binding domain from *T. reesei*, which does not exhibit such activity.

9.44 CELLULOSE-BINDING DOMAINS MODIFY COTTON CELLULOSE BUT NOT BACTERIAL CELLULOSE NOR AVICEL

The family II cellulose binding domain from *C. fimi*, with or without the catalytic domain attached, releases small particles from cotton and ramie fibers (Din et al., 1994; Tomme et al., 1995a), but not from bacterial microcrystalline cellulose (Din et al., 1994) or Avicel® (Le et al., 1994; Romicez et al., 1993; Tomme et al., 1995a). The family II (bacterial) cellulose-binding domain also prevents flocculation of the bacterial cellulose (Tomme et al., 1995a). The cellulose-binding domain associated with family I cellobiohydrolase I from *T. reesei* does not disrupt the fiber structure and does not create surface area for cotton, bacterial, or microcrystalline cellulose. The *C. fimi* derived cellulose-binding domain is thought to roughen the surfaces of cotton fibers at a molecular level and release particles, presumably because it binds to and penetrates surface discontinuities, causing noncovalently bound cellulose fragments to be sloughed off. This is postulated to uncover the ends of cellulose chains (Din et al., 1994).

9.45 CELLULOSIC AFFINITY CHROMATOGRAPHY STATIONARY PHASE MUST MATCH BINDING PREFERENCES OF THE BINDING DOMAIN PROTEIN AND GIVE GOOD FLOW PROPERTIES

The binding and particle-releasing characteristics of the two proteins are an important consideration when designing a scaffold upon which a cellulose binding domain/fusion protein might be immobilized to form an affinity support. Studies of this property with respect to bacterial microcrystalline cellulose, cotton or ramie cellulose, and a microcrystalline cellulose derived from wood pulp (known as Avicel®)[18] have been carried out for proteins from both bacterial (*C. fimi*) and fungal (*T. reesei*) sources. The cellulose-binding domains from these two sources have different properties.

If cotton cellulose were to be used, the material would need to be pretreated in some manner to remove adsorbed particles first. Otherwise, the cellulose-binding domain protein could cause release of cellulose particles, which in turn would migrate through the column to the retaining frit at the outlet and cause it, and the column, to plug. The crystallinity of the cellulosic material will affect binding capacity. Avicel, a microcrystalline cellulose, does not exhibit particle release and hence represents a good starting point for testing immobilization of a CBD fusion protein, although flow properties of packed beds of this particulate cellulose are poor. Bacterial microcrystalline cellulose could also be chosen but it is not as readily available, and is more expensive than Avicel and cotton. A suitable matrix is suggested by the work of Yang, Ladisch et al. (1992) in which packed beds of woven cotton and ramie textiles—rolled in a cylindrical configuration—have the desired porosity characteristics and mechanical flow stability when inserted in a chromatography column.

9.46 COMBINATIONS OF PROTEINS WITH CELLULOSE BINDING DOMAINS CAN BE OBTAINED THROUGH GENETIC ENGINEERING

The advent of recombinant biotechnology makes it possible to genetically engineer combinations of protein domains (i.e., chimeric proteins) having special functions and to express (i.e., produce) the chimeras in useful quantities using *E. coli* or other organisms. This approach was proposed and tested, for example, to "engineer" proteins which combine catalytic domains for hydrolysis of either xylan (a polymer of xylose) or cellulose with a polypeptide linker for attachment to a cellulose binding domain protein. The resulting protein could conceptually have both hemicellulose (xylan) and cellulose (glucan) hydrolysis activity, although the objective of combining the cellulose binding domain with two catalytic domains was to demonstrate that two different catalytic domains separated by a cellulose binding domain protein could give a cellulose binding domain which is still fully active. This type of protein was also intended to give further insights into the nature of multidomain architectures of bacterial and fungal cellulases which can hydrolyze more than one type of polysaccharide (Tomme et al., 1994). This is of fundamental and practical importance since hemicellulose and cellulose are found in similar proportions and are closely associated with each other in biomass.

[18]Bacterial microcrystalline cellulose is a uniform suspension of cellulose microfibrils from the cellulose pellicle obtained from cultures of *A. xylinum*. This cellulose consists of bundles of highly crystalline microfibrils that are 20 to 50 nm wide and several thousand nm long, and are extended β 1,4-glucose chains arranged in a parallel orientation. Avicel, in comparison, is a solid material obtained from acid hydrolysis, washing, and spray drying of wood fiber. It has a crystallinity of approximately 50% and is physically more heterogeneous than the bacterially derived microcrystalline cellulose (Gilkes et al., 1992). Cotton fibers are made of a naturally occurring cellulose, with many naturally occurring surface irregularities.

This fusion protein bound strongly to Avicel®, suggesting that chromatography over Avicel® (due to CBD's strong affinity for cellulose) was a possible method for purifying the cellulase. However, the affinity of the cellulose binding domain protein was so strong that it could only be eluted with 1% SDS (sodium dodecyl sulfate–a detergent) or 8 M guanidine HCl. These conditions also denatured the attached catalytic domains, and hence were not practical for purifying cellulase-using cellulose (Tomme et al., 1994). However, purification of a cellulose-binding domain linked to a different protein (protein A) was later shown to be feasible (Romicez et al., 1993). Binding domains with lower avidity for cellulose have been proposed although strong binding is nonetheless desirable for immobilizing an affinity ligand. An example of this is given by a fusion protein of Staphylococcal A protein linked to the cellulose-binding domain of the exoglucanose from *C. fimi.* The protein A retained its capacity to bind immunoglobin G molecules while the cellulose binding domain, to which it was linked, was firmly associated (bound) to Avicel (Romicez et al., 1993).

9.47 NONCOVALENT BONDING OF CELLULOSE BINDING DOMAIN PROTEINS COULD COUPLE RECOMBINANT AFFINITY LIGANDS TO CELLULOSE

The analysis of cellulase activity in the context of the catalytic and binding domains, and the impact of these protein domains on cellulase activity on cellulose are not yet completely understood[19] nor is their potential impact on separations science fully appreciated. It is conceivable that chimeric proteins consisting of a cellulose binding domain and a noncatalytic domain, could result in a new family of cellulose based, affinity chromatography stationary phases. The cellulose binding domain would tightly anchor the noncatalytic domain (i.e., the ligand) to a cellulose stationary phase while avoiding the use of CNBr or other organic reagents to activate the stationary phase and immobilizing ligands through covalent bonds (see Section 8.56). This could significantly enhance the prospects of simple recovery of valuable proteins via a detachable affinity-binding domain. The potential impact of a cellulosic binding domain on the practice of affinity chromatography illustrates how fundamental research can affect both upstream and downstream process technology—thereby shaping future developments in protein biosepa-rations.

9.48 PROTEIN A/CELLULOSE-BINDING DOMAIN PROTEIN FROM *E. COLI* IS AN EASILY PURIFIED AND IMMOBILIZED AFFINITY LIGAND

Protein A is a component of the cell wall of *Staphylococcus aureus* and has specific affinity for the F_c portion of immunoglobulin G (IgG) as described earlier in this chapter. In *S. aureus*, protein A consists of a single polypeptide chain with five homologous IgG binding domains (E, D, A, B, and C). Protein A is used to detect antibodies and to carry out immunoprecipitation. Protein A, added to an antibody solution, causes precipitation of IgG. Immobilized protein A is used to purify polyclonal and monoclonal antibodies, and to remove immunocomplexes from plasma as means to reduce the risk, prior to transplantation, of organ rejection.

[19]Mutiple component mixtures of cellulases from genetically transformed *Trichoderma reesei* remove pills from cotton fabric due to "a careful balance between cellulase activity and mechanical action (Cavuco-Paulo and Ameida, 1996)." The individual effects of the catalytic and binding domains have yet to be delineated for textile processing, but once understood could result in new enzyme formulations for non-invasive textile processing which fractionate malformed cellulose fibers (i.e., pills) from the bulk of the fabric.

Chemical methods of attaching this affinity ligand to a stationary phase require the protein to be coupled to a chemically activated support via a nucleophilic group. An alternate approach would be to engineer and express a protein A–cellulose-binding domain fusion protein, with the strong binding of the cellulose binding domain to cellulose enabling facile recovery of the protein A (Romicez et al., 1993). This has been done through gene fusion of a plasmid, that contains the sequence for the leader peptide and IgG binding domains of *S. aureus* protein A, with a polynucleotide fragment encoding for a Pro-Thr linker and the cellulose-binding domain of the exoglucanase, Cex, from *C. fimi*. The resulting fusion protein, expressed by *E. coli*, consists of the 271 amino acid sequence of protein A followed by the Pro-Thr linker and then the cellulose-binding domain. This fusion protein has a molecular weight of 42 kD (Romicez et al., 1993).

The protein A–cellulose-binding domain (abbreviated A-CBD) is produced by fermentation of the transformed *E. coli* cells. The intracellular protein is constitutively expressed so that an inclusion body is not formed, and protein recovery is possible by disrupting the cells by sonification, centrifuging down the cell debris, and processing the supernatant. Avicel is added to the supernatant containing the cellulose-binding domain–protein A. After incubation, the supernatant is filtered away from the Avicel over the glass fiber filter (Whatman GF-A). The Avicel is then removed, washed, and eluted with 7 M guanidine-HCl. Buffer exchange (using a membrane) removes the guanidine-HCl and replaces it with 50 mM potassium phosphate at pH 7.0, with the protein refolding correctly during the course of this buffer exchange step. This procedure results in recovery of a protein preparation from the supernatant that is 95% pure, and is suitable for immediate use as a protein A affinity stationary phase.

9.49 PROTEIN SIZE DETERMINES OPTIMUM IMMOBILIZED DENSITY OF PROTEIN A LIGANDS

The maximum amount of the recombinant protein A–cellulose-binding domain bound to Avicel is 7 mg/mL (where 1 mL of column volume = 500 mg of Avicel on a dry weight basis). This is equivalent to 159 nmoles of cellulose binding domain—protein A/mL of Avicel (1 nmole = 10^{-9} mole). Hence the loading of CBD–protein A on Avicel is similar to commercial protein A stationary phases consisting of protein A chemically immobilized to crosslinked agarose, polystyrene/divinyl-benzene, or porous glass particles, at protein A concentrations of 3 to 9 mg/mL. This preparation of CBD–protein A was able to bind a maximum of 159 nmoles of IgG (Romicez et al., 1993).

Native protein A, which is not immobilized, would normally bind 2 moles of IgG/mole protein A. The maximum loading of IgG bound to CBD–protein A on Avicel corresponds to a ratio of only 0.3 nmoles IgG/nmole immobilized CBD–protein A. Steric hindrance between IgG molecules near the surface of the solid cellulose particles to which the protein A is immobilized may help to explain the lower ratio of IgG to Protein A given the large size and high molecular weight of IgG ($M_r \cong 160$ kD). If the amount of protein A is controlled to be only 3 to 5 nmoles per mL Avicel, 2 molecules of IgG bind per molecule of protein A, which further supports the hypothesis of a steric effect. An optimum level of protein A–cellulose-binding domain protein is reported to be 30 nmoles/mL Avicel®, which binds 30 nmoles of IgG/mL Avicel® (equivalent to 6.5 mg/mL) with 95% recovery when the IgG is eluted using 0.2 M potassium phosphate at pH 4.0. This is comparable to commercial gels that bind 5 to 10 mg of IgG/mL at the stationary phase (Romicez et al., 1993). At higher levels of protein A, complete uptake and recovery of IgG from solution is lost. Higher ligand densities result in decreased recovery due to steric effects of protein molecules that are too close together and thus interfere with each other's adsorption and desorption.

One objective of affinity chromatography is to bind, selectively, all of the target protein in the mobile phase and then to recover all of the bound protein in a purified form. Losses of the target molecule can be costly, since affinity chromatography usually recovers expensive proteins from fermentation or cell culture broths, or is the last step in a purification sequence, after the protein has been significantly purified and is in a relatively concentrated (and very valuable) form. Consequently a gel with a higher "loading" (i.e., higher density of protein A affinity ligands) is not necessarily better if the extent of recovery of the product is compromised. Rather, the concentration of the ligand on the surface of the cellulose needs to be carefully selected and optimized with respect to the size of the protein to be purified.

9.50 ADDITIONAL PURIFICATION STEPS MUST FOLLOW AFFINITY CHROMATOGRAPHY OF IgG OVER PROTEIN A

As is the case for other affinity chromatography techniques, one-step purification of a protein must be followed by additional cleanup steps. Elution of the target molecule, then buffer exchange using a membrane or size exclusion chromatography will likely follow elution of the protein from the column. Leaching of trace amounts of the immobilized ligand from the Avicel into the final product could also be a concern, although cleanup could be achieved by passing the eluting protein over a second cellulose column. Alternately, the molecular weight difference between the protein A–cellulose-binding domain or related proteins ($M_r \approx 40$ kD) and IgG ($M_r \approx 160{,}000$ kD) is sufficiently large that such a mixture could be separated by size exclusion chromatography, discussed in Sections 6.2 and 6.6.

9.51 STREPTAVIDIN–CELLULOSE-BINDING DOMAIN PROTEIN ON AVICEL IMMOBILIZES CATALYTICALLY ACTIVE ENZYMES

Affinity ligands associated with affinity chromatography can bind many types of antigens, including some types of proteins that have enzymatic activity. An immobilized enzyme is used much like a plug flow reactor containing a fixed bed of catalyst over which the reactants are passed. An example is the streptavidin–cellulose-binding domain fusion protein immobilized onto Avicel, which in turn can bind catalytically active biotinylated proteins onto the streptavidin.

Streptavidin from *Streptomyces avidinii* is a protein with $M_r = 60$ kD, and a high affinity for the water soluble vitamin D (biotin). Since streptavidin is not glycosylated, and since only a small portion of the protein is needed to maintain avidity for biotin, it is amenable to production using transformed *E. coli* into which the streptavidin-biotin gene has been cloned. Production of fusion proteins of streptavidin with protein A, metallothionein, or a polycysteine peptide at the C terminus further extends the utility of this system and provides another example of how upstream developments (fusion proteins from *E. coli*) affect downstream applications (binding domains for affinity chromatography).

Streptavidin covalently linked to Sepharose has proved to be an excellent immobilization matrix because it is stable and can immobilize a second affinity ligand that has biotin attached, without the need to use other chemical steps to achieve immobilization. This concept has been extended to streptavidin–cellulose-binding domain for the noncovalent attachment of streptavidin to cellulose matrices (Le et al., 1994). Immobilization by chemically crosslinking the protein with stationary phase can result in loss of protein activity due to reaction of a crosslinker with the active site or region of the protein.

The gene was constructed of a 0.7 kb DNA fragment containing the sequence coding for the cellulose-binding domain ligated with a plasmid containing a 2.7 kb fragment for the sequence coding for streptavidin. The resulting plasmid was used to transform *E. coli* cells, which were screened and selected for further growth. The *E. coli* cells were grown in 500 mL volumes in a shaker. Once the cells were grown, expression of the fusion protein was induced using isopropylthiogalactoside. After four to five hours, the cells were harvested, washed, and lysed. The inclusion bodies containing the fusion protein were separated from the cell debris and then dissolved in 6 M guanidine HCl at pH 7.0. Further processing, including buffer exchange, gave an overall yield of about 4% renatured protein. The protein was contacted with Avicel to allow the streptavidin–cellulose-binding domain protein (M_r = 25 kD) to adsorb onto the Avicel.

The enzyme, β-glucosidase, was immobilized onto this matrix. The β-glucosidase had previously been biotinylated with biotin *N*-hydroxysaccinimide ester. The biotin residue on the enzyme bound with the strepavidin, thereby immobilizing the enzyme. The entire sequence is represented in Figure 9.50.

The solid particles, to which enzyme had been immobilized, were then packed and operated in a column over which 10 mM cellobiose in 0.1 M sodium acetate at pH 4.8 was passed at 50°C for two weeks. There was no loss in enzyme activity as measured by the rate of hydrolysis of cellobiose to glucose (Le et al., 1994). While the yield of streptavidin fusion protein from the *E. coli* is too low to be practical, this approach nonetheless provides an example of how affinity chromatography concepts are amenable to development of an immobilized enzyme reactor.

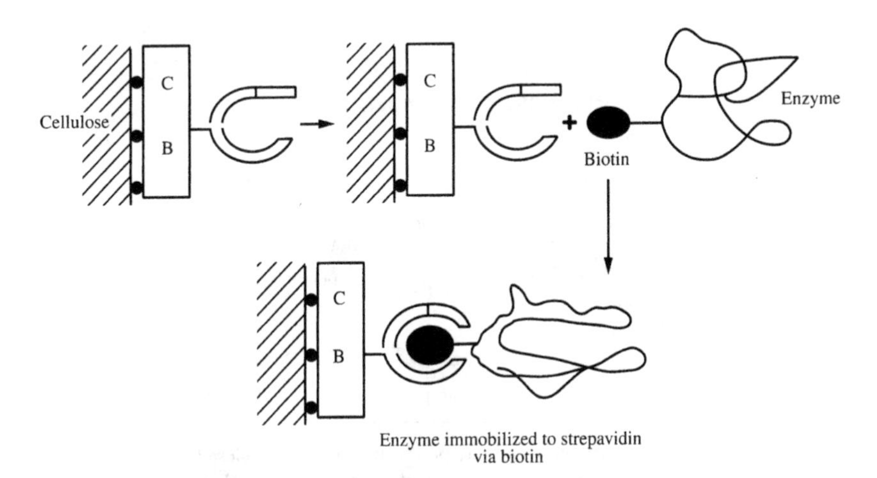

Figure 9.50 Schematic representation of enzyme immobilization via biotin.

9.52 HUMAN INTERLEUKIN-2 CAN BE COUPLED TO, PRODUCED WITH, AND RECOVERED FROM CELLULOSE BINDING DOMAIN PROTEIN

Interleukin-2 is a 15 kD protein produced by T lymphocytes upon antigen stimulation, and it is a key component in the activation of the immune system. This protein is involved in the proliferation and differentiation of lymphoid cells (natural killer cells, and antigen-stimulated B- and T-cells). Potential therapeutic uses include treatment of cancer and autoimmune disorders, as well as an adjuvent to vaccination and the immune response in intracellular infection (Ong et al., 1995).

This protein, and others, can be produced with an additional protein domain that serves as an affinity tag fused to it. A specific peptide sequence is inserted between the product molecule and the tag, allowing for subsequent removal of the protein by hydrolysis. Site-specific cleavage can be achieved chemically using, for example, CNBr in 70% formic acid to cleave Met residues. However, CNBr must be handled carefully because it is toxic, and it can introduce heterogeneity in the target protein due to undesirable side reactions. Thus a protease, such as blood coagulation Factor X_a, can be attractive for site-specific proteolysis of fusion proteins—although it too has the disadvantage of having broad specifity. Nonetheless, Factor X_a has been widely used with a number of fusion proteins including protein A, β-galactosidase, the maltose-binding protein (Carter, 1990), and interleukin-2 fused with a cellulose-binding domain.

An affinity tag for interleukin-2 was constructed based on the cellulose-binding domain protein of *C. fimi* fused to interleukin-2 via a proline-threonine-rich linker with a factor X_a cleavage sequence (IleGluGlyArg) on the N-terminus of IL-2. The concept is to generate an IL-2/cellulose binding domain fusion protein through fermentation of transformed *E. coli*, or through *Streptomyces lividans* cells, or by culture of transformed mammalian COS cells. The fermentation and culture yields were low (3 to 10 μg/L for bacterial fermentation and 600 μg/L for the mammalian cells; i.e., 3 to 600 ppb). However, the capture of these low levels of proteins on Avicel (cellulose) was successful, and release of 65% of the interleukin-2 from the cellulose-binding domain immobilized on the Avicel was achieved upon hydrolysis with Factor X_a (Ong et al., 1995). This concept, illustrated in Figure 9.51, further illustrates the versatility of an affinity system, and the potential synergy between upstream and downstream processing.

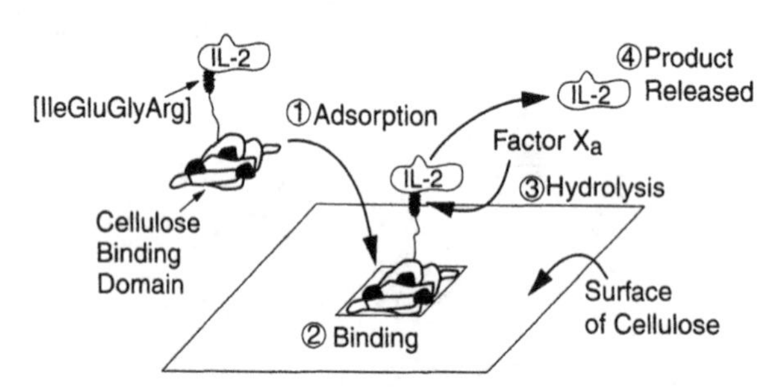

Figure 9.51 Schematic representations of affinity purification of IL-2 cellulose binding domain hybrid protein. IL-2-cellulose binding domain is recovered by adsorption on cellulose. Hydrolysis releases the IL-2 molecule from the cellulose binding domain.

Summary and Perspectives

Affinity chromatography has the potential of becoming a cost-effective, custom-designed method for process-scale purification of proteins based on specific interactions between the protein product and the affinity ligand. Sequencing, chemical synthesis, and recombinant techniques have been adapted to the generation of new affinity ligands, molecules to probe receptors at cell surfaces and lead compounds for drugs. Molecular biology techniques to construct organisms for the production of small proteins or polypeptides on which these ligands are based enables research on their structure and function by providing means by which milligram quantities of the ligands can be generated.

The rapidity with which new affinity ligands are found will depend on how well apparent hits are screened so that the number of final candidates is kept at a manageable level. Those that are selected should have a high probability of success in satisfying all five of the separations toolbox criteria outlined in Table 8.19: versatility, variety, compatability, benchmarking, and robustness. An engineering challenge remains, however. Once drug candidates or affinity ligands are identified, their synthesis will need to be scaled up, and their generation, recovery, and purification assembled into cost-effective manufacturing technologies in a timely manner. Combinatorial or other approaches, when combined with high throughput screens, will result in a large number of small molecule drug candidates. The potential exists that viable drug candidates will become available at such an accelerated pace that production scale-up (for producing test amounts) will become a bottleneck, unless addressed in parallel with the discovery process.

Case studies and examples for the production of Factor IX, protease and reverse transcriptase inhibitors for HIV, IL-2, immunoadhesions, bovine somatotropin, and protein A illustrate the potential of affinity ligands for purifying proteins, or probing the nature of receptors responsible for some types of disease states. In the case of Factor IX, biocombinatorial techniques were able to identify the affinity ligand. In the case of HIV, an understanding of the interactions between substrate and enzyme, combined with role of the enzyme in propagating the virus, led to affinity methods for drug discovery, purification, and treatment of HIV. IL-2 purified through receptor affinity chromatography showed how molecular biology was the only practical means to generate the receptor (i.e., affinity ligand) as well as the product, IL-2. The purification of immunoadhesions is yet another example of how an understanding of the specific nature of affinity between antibody and antigen facilitates facile recovery and purification of a potentially valuable therapeutic product, which itself exists because of molecular biology techniques. The case of bovine somatotropin illustrates a different type of affinity ligand, an immobilized metal ion capable of coordination bonding, particularly with His. Hence engineering of hybrid proteins to which specific sequences of His residues are attached enables the protein's purification. Molecular biology again provided a bridge between product and purification technique. The potential of immobilizing a ligand fused to a protein domain with special ability to adsorb to cellulose shows how physical binding may someday supplant the need to use chemical reagents to achieve the same effect.

All of these examples have the common element that molecular biology makes possible the existence of industrially produced forms and quantities of these molecules. All of the proteins discussed here can be purified by affinity chromatography, with some providing affinity ligands for purification of other molecules. Affinity chromatography provides the bridge between separations science and molecular biology and combinatorial techniques for producing new molecules for therapeutic applications, animal health products, and for purifying products from biological sources.

REFERENCES

Advertisement, 1997, *Science*, 275(5299), 586.

Alberts, B., D. Bray, J. Lewis, M. Raff, K. Roberts, and J. D. Watson, 1989, Molecular biology of the cell, 2nd ed., Garland Publishing, New York, 298–299.

Angyal, S. J., G. S. Bethell, and R. J. Beveridge, 1979, The separations of sugars and of polyols on cation exchange resins in the calcium form, *Carbohydrate Res.*, *73*, 9–18.

Anonymous, 1996, Monsanto to build it first U.S. bovine somatotropin plant, *C&EN*, *74*(26), 13.

Anonymous, 1997, . . . as PPL turns to rabbits for protein research, *Chem. Eng. News*, *75*(13), 12.

Armstrong, L., 1997, Besting AIDS and the drug giants, *Business Week*, *3530*, 79–80.

Arnold, F. H., and B. L. Haymore, 1991, Engineering metal-binding proteins: Purification to protein folding, *Science*, *252*, 1796–1797.

Bailon, P., and D. V. Weber, 1988, Receptor affinity chromatography, *Nature*, *335*(6193), 839–840.

Bailon, P., D. V. Weber, R. F. Keeny, J. E. Fredericks, C. Smith, P. C. Familletti, and J. E. Smart, 1987, Receptor-affinity chromatography: A one-step purification for recombinant interleukin-2, *Bio/Technology*, *5*, 1195–1198.

Bailon, P., D. V. Weber, M. Gately, J. E. Smart, H. Lorberboun-Galski, D. Fitzgerald, and I. Pastan, 1988, Purification and partial characterization of an interleukin-2-pseudomonas exotoxan fusion protein, *Biotechnology*, *7*, 1326–1329.

Balter, M., 1998, Revealing HIV's t-cell passkey, *280*(5371), 1833–1834.

Barrett, A., J. Carrry, and D. Dawley, 1999. Closing in on an HIV vaccine, *Business Week*, *3641*, 41.

Borman, S., 1997a, Combinatorial Chemistry, *Chem. Eng. News*, *75*(8), 43–62.

Borman, S., 1997b, Ribosome Display, *Chem. Eng. News*, *75*(20), 11–12.

Bourne, E. J., F. Searle, and H. Weigel, 1971, Complexes between polyhydroxy compounds and copper (II) ions, *Carbohydrate Res.*, *16*, 185–187.

Boyle, M. D. P., E. L. Faulmann, and D. W. Metzger, 1993, Applications of bacterial immunoglobulin-binding proteins to the purification of immunoglobulins, in *Molecular Interactions in Bioseparations*, T. T. Ngo, ed., Plenum Press, New York, 91–112.

Branden, C., and J. Tooze, 1991, Recognition of foreign molecules by the immune system, in *Introduction to Protein Structure*, Garland, New York, 179–199.

Brobst, K. M., H. D. Scobell, and E. M. Steele, 1973, Analysis of carbohydrate mixtures by liquid chromatography, *Am. Soc. Beer Brewing Chemists Proc.*, 43–46.

Buettner, J. A., C. A. Dadd, G. A. Baumbach, B. L. Masecar, and D. J. Hammond, 1996, Chemically derived peptide libraries: A new resin and methodology for lead identification, *Int. J. Peptide Protein Res.*, 47, 870–893.

Cai, S.-J., T. J. Holzer, and C. Pidgeon, 1996, Principles of enzyme-linked immunosorbent assays (ELISAs), in *Tutorials for the Biomedical Sciences: Animations, Simulations, and Calculations Using Mathematica®*, C. Pidgeon, ed., VCH, New York, 177–192.

Carey, J., and A. Barnett, 1999, An aids vaccine is no longer a dream, *Business Week*, *3645*, 94–98.

Carey, J. 1999, Aids drugs: Giving it a rest, *Business Week*, *3633*, 94, 98.

Carey, J. 1998, Outsmarting the virus, *Business Week*, *3617*, 142–143.

Carter, P., 1990, Site-Specific Proteolysis of Fusion Proteins, in Protein Purification: from Molecular Mechanisms to Large-scale Processes, M. R. Ladisch, R. C. Willson, C-D. C. Painton, and S. E. Builder, Am. Chem. Soc., Washington, D.C., 181–193.

Cavuco-Paulo, A., and L. Almelda, 1996, Cellulase activities and finshing effects, *Textile Chemist and colorists*, *28*(6), 28–32.

Chamov, S. M., and A. Ashkenazi, 1996, Immunoadhesions: Principles and applications, *TIBTECH*, *14*, 52–60.

Ching, C. B., 1983, A theoretical model for the simulation of the operation of the semi-continuous chromatographic refiner for separating glucose and fructose, *J. of Chem. Eng., Japan*, 15(1), 49–53.

Cohen, J., 1997, Stubborn HIV reservoirs vulnerable to new treatments, *Science*, *276*, 898–899.

Cohen, J. 1996a, Protease inhibitors: A tale of two companies, *Science*, *272*(5270), 1882–1883.

Cohen, J., 1996b, Money matters: The marketplace of HIV/AIDS, *Science*, *272*(5270), 1880–1881.

Cohen, J., 1996c, Likely HIV cofactor found, *Science*, *272*(5263), 809–810.

Cohen, J., 1996d, Investigators detail HIV's fatal handshake, *Science*, *274*(5287), 502.

Cowley, G., 1996, Another key to HIV: Scientists identify a long-sought molecule that helps AIDS virus penetrate human cells, *Newsweek*, 62 (May 20).

Creagh, A. L., E. Ong, E. Jervis, D. G. Kilburn, and C. A. Haynes, 1996, Binding of the cellulose-binding domain of exoglucanase cex from *Cellulomonas fimi* to insoluble microcrystalline cellulose is entropically driven, *Proc. Natl. Acad. Sci.*, *93*, 12229–12234.

Dagani, R., 1996, HIV-infection of cells: Second cofactor is chemokine receptor, *Chem. and Eng. News*, *74*(26), 8 (June 24).

Danheiser, S. L., 1997, Novel interleukins demonstrate promise in cancer therapy and management, *GEN*, 17(8), 17, 28.

Din, N., H. G. Damude, N. R. Gilkes, R. C. Miller, R. A. J. Warren, and D. G. Kilburn, 1994, C_1-C_x revisited: Intramolecular synergism in a cellulase, *Proc. Natl. Acad. Sci.*, *91*, 11383–11387.

Feng, Y., C. C. Broder, P. E. Kennedy, and E. A. Burger, 1996, HIV-1 entry cofactor: Functional cDNA cloning of a seven-transmembrane, G protein-coupled receptor, *Science*, *272*(5263), 872–877. (New development, *Science*, *274*, pg. 602, 1996).

Flanigan, E., 1991, (Rhône Poulenc Rorer) High performance liquid chromatography in the production and quality control of Salmon Calcitonin, in Purdue University Workshop on Chromatographic Separations and Scale-up, 207.

Gilkes, N. R., B. Henrissat, D. G. Kilburn, R. C. Miller Jr., and R. A. J. Warren, 1992a, Domains in microbial β-1, 4-glycanases, sequence, conservation and enzyme families, *Microbiol. Rev.*, 55, 303–315.

Gilkes, N. R., E. Jervis, B. Henrissat, B. Tekant, R. C. Miller, R. A. J. Warren, and D. G. Kilburn, 1992b, The adsorption of a bacterial cellulases and its two isolated demains to crystalline cellulose, *J. Biological Chem.*, *267*(10), 6743–6749.

Glaser, V., 1996, Companies develop novel techniques for maximizing combinatorial library potential, *GEN*, 16(10), 1, 19.

Hanes, J., and A. Plückthun, 1997, *In vitro* selection and evolution of functional proteins by using ribosome display, *Proc. Natl. Acad. Sci.*, 94, 4937–4942.

Haymore, B. L., G. S. Bild, W. J. Salsgiver, N. R. Straten, and G. G. Krivi, 1992, Introducing strong metal-binding sites onto surfaces of proteins for facile and efficient metal-affinity purifications, in *Methods, A Companion to Methods in Enzymology*, *4*, 25–40.

Heylin, M., 1998, BGH fracas erupts again in U.S., Canada, *C&EN*, *76*(51), 7–8.

Hill, C. L., and R. D. Gall, 1996, The first combinatorially prepared and evaluated inorganic catalysts, Polyoxometalates for the aerobic oxidation of the mustard analogue tetrahydrothiophene (THT), *J. Mol. Catalysis A: Chem.*, 114, 103–111.

Ho, D. D., 1996, Viral counts count in HIV infection, *Science*, *272*(5265), 1124–1125.

Hochuli, E., 1988, Large scale chromatography of recombinant proteins, *J. Chromatogr.*, 444, 293–302.

Ingersoll, B., 1997, FDA approves AIDS drugs for treating children, *Wall Street J.*, B4 (March 17).

Jacobson, B. J., 1982, A fundamental study of cation exchange columns in aqueous liquid chromatography of carbohydrates, M.S. thesis, Purdue University, December.

Jasney, B. K., and D. Clery, 1998, AIDS Research-1998, *Science*, 280(5371), 1855.

Johnson, R. D., and F. H. Arnold, 1995a, Multipoint binding and heterogeneity in immobilized metal affinity chromatography, *Biotechnol. Bioeng.*, *48*, 437–443.

Johnson, R D. and F. H. Arnold, 1995b, The Tempkin model describes heterogeneous protein adsorption, *Biochem. Biophys. Acta*, *1247*, 293–297.

Jones, J. K. N., R. A. Wall, and A. O. Pittet, 1960, The separation of sugars on ion-exchange resins, Part I; Separation of oligosaccharides, *Can. J. Chem.*, 38(12), 2285–2289.

Jones, J. K. N., and R. A. Wall, 1960, The separation of sugars on ion exchange resins, Part II, Separation of monosaccharides, *Can. J. Chem.*, 38(12), 2290–2294.

Kenney, A. C., and H. A. Chase, 1987, Automated production scale affinity purification of monoclonal antibodies, *J. Chem. Tech. Biotechnol.*, 39, 173–182.

Khym, J. X., and L. P. Zill, 1951, The separation of monosaccharides by ion exchange, *J. Am. Chem. Soc.*, 73, 2399–2400.

Kirschner, E., 1994, Bovine Sales Booming Despite Attacks, *CEN*, *72*(17), 8.

Ladisch, M. R., A. L. Huebner, and G. T. Tsao, 1978, High speed liquid chromatography of cellodextrins and other saccharide mixtures using water as the eluent, *J. Chromatogr.*, 147, 185–193.

Ladisch, M. R., and G. T. Tsao, 1978, Theory and practice of rapid liquid chromatography at moderate pressures using water as the eluent, 166, *J. Chromatogr.* 85–100.

Ladisch, M. R., K. W. Lin, M. Voloch, and G. T. Tsao, 1983, Process considerations in the enzymatic hydrolysis of biomass, *Enz. Microb. Technol.*, *5*, 82–102.

Ladisch, C. M., Y. Yang, A. Velayudhan, and M. R. Ladisch, 1992, A New Approach to the Study of Textile Properties by Liquid Chromatography, Comparison of Void Volume and Surface Area of Cotton and Ramie Using a Rolled Fabric Stationary Phase, *Textile Res. J.* *62*(6), 361–369.

Lange, J. M. A., 1997, Current problems and the future of antiretroviral drug trials, *Science*, *276*(5312), 548–550.

Lauffenberger, D. A., and J. J. Linderman, 1993, Receptors: Models for binding, trafficking and signaling, Oxford University Press, New York.

Le, K. D., N. R. Gilkes, D. G. Kilburn, R. C. Miller, J. N. Saddler, and R. A. J. Warren, 1994, A streptavidin-cellulose-binding domain fusion protein the binds biotinylated protein to cellulose, *Enzyme Microb. Technol.*, *16*, 496–500.

Liang, R., L. Yan, J. Loebach, M. Ge, Y. Uozumi, K. Sekanina, N. Horan, J. Gildersleeve, C. Thompson, A. Smith, K. Biswas, W. C. Still, and D. Kahne, 1996, Parallel synthesis and screening of a solid phase carbohydrate library, *Science*, 274(5292), 1520–1522.

Lin, J. K., B. J. Jacobson, A. N. Pereira, and M. R. Ladisch, 1988, Liquid chromatography of carbohydrate monomers and oligomers, in *Methods in Enzymology*, 160, W. A. Wood and S. T. Kellogg, eds., 145–159.

Luong, C. B. H., M. F. Browner, and R. J. Fletterick, 1992, Purification of glycogen phosphorylase isozymes by metal affinity chromatography, *J. Chromatogr.*, *584*, 77–84.

MacLennan, J., 1995, Engineering microprotein ligands for large-scale affinity purification, *Biotechnol.*, 13, 1180–1183.

Markland, W., A. C. Rey, S. W. Lee, and R. C. Ladner, 1996a, Iterative optimization of high affinity protease inhibitors using phage display. 1 Plasmin, *Biochem.*, 35(24), 8045–8057.

Markland, W., A. C. Ley, and R. C. Ladner, 1996b, Iterative optimization of high-affinity protease inhibitors using phage display. 2. Plasma kallikrein and thrombin, *Biochem.*, 35(24), 8058–8067.

Miedema, F., and M. R. Klein, 1996, AIDS pathogenesis: A finite immune response to blame?, *Science*, 272(5261), 505–506.

Morgan, D. A., F. W. Ruscetti, and R. C. Gallo, 1976, Selective *in vitro* growth of T-lymphocytes and normal human bone marrows, *Science*, *220*, 1007–1008.

Moore, J. P., Coreceptors: Implications for HIV pathogenesis and therapy, *Science*, 1997, 276(5309), 51–52.

Nachman-Clewner, M., C. Spence, and P. Bailon, 1993, Receptor-affinity chromatography (RAC), in *Molecular Interactions in Bioseparations*, T. T. Ngo, ed., Plenum Press, New York, 139–167.

News Item, 1996a, Protease inhibitors: Studies probe HIV resistance to drugs, *Chem. Eng. News*, 74(11), 6.

News Item, 1996b, Modified chemokine locks out AIDS virus, *Chem. Eng. News*, 74(41), 30.

News Item, November 25, 1998, *Wall Street Journal*, A1.

Nowak, M. A., R. M. Anderson, M. C. Boerlijst, S. Bonhoeffer, R. M. May, and A. J. McMichael, 1996, Technical comments: HIV-1 and disease progression, *Science*, 274(5289) 1008–1010.

Old, R. W., and S. B. Primrose, 1981, *Principles of Gene Manipulation: An Introduction to Genetic Engineering*, 2nd ed., University of California Press, Berkeley, Blackwell Scientific, Oxford, 1–11.

Ong, E., J. B. Alimonti, J. M. Greenwood, R. C. Miller, R. A. J. Warren, and D. G. Kilburn, 1995, Purification of human interleukin-2 using the cellulose-binding domain of a prokaryotic cellulase, *Bioseparations*, *5*, 95–105.

Pennisi, E., 1997, Industry, extremophiles begin to make their mark, *Science*, 276(5313), 705–706.

Porter, J. E., M. R. Ladisch, and K. M. Herrmann, 1991, Ion exchange and affinity chromatography in the scale-up of the purification of α-galactosidase from soybean seeds, *Biotechnol. Bioeng*., 37, 356–363.

Richman, D. D., 1996, HIV therapeutics, *Science*, 272(5270), 1886–1888.

Ridky, T., and J. Leis, 1995, Development of drug resistance to HIV-1 protease inhibitors, *J. Biol. Chem*., 270(50), 29261–29263.

Roberts, S. L., 1996, Blood safety in the age of AIDS, *FASEB J.*, 10.

Romicez, C., J. Fung, R. C. Miller, R. Antony, J. Warren, and D. G. Kilburn, 1993, A bifunctional affinity linker to couple antibiotics to cellulose, *Bio/Technology*, *11*, 1570–1573.

Rundle, R. L., 1997, Agouron gets go-ahead to join the AIDS-drug battle, *Wall Street J.*, B4 (March 17).

Rundle, R. L., and M. Waldholz, Jan. 27, 1999, Warner-Lambert agrees to buy Augoron, *Wall Street Journal*, A3.

Sado, E., and S. Katoh, 1993, Suitable antibodies as ligands in affinity chromatography of biomolecules, in *Molecular Interactions in Bioseparations*, T. T. Ngo, ed., Plenum Press, New York, 205–211.

Schultz, J. S., and J. S. Schultz, 1996, The combinatorial library: A multifunctional resource, *Biotechnol. Prog*., 12(6), 729–743.

Scobell, H. D., and K. N. Brobst, 1981, Rapid high-resolution separation of oligosaccharides on silver form cation-exchange resins, *J. Chromatogr*., 212, 51–54.

Scobell, H. D., K. M. Brobst, and E. M. Steele, 1977, Automated liquid chromatographic system for analysis of carbohydrate mixtures, *Cereal Chem*., 54(4), 905–?.

Scopes, R. J., 1981, Quantative Studies . . . , *Anal. Biochem*., *114*, 8–18.

Service, R. F., 1996, Combinatorial chemistry hits the drug market, *Science*, 272(5266), 1266–1268.

Smith, M. C., J. A. Cook, T. C. Furman, P. D. Gesellchen, D. P. Smith, and H. Hsiung, 1990, Chelating Peptide-immobilized metal-ion affinity chromatography, in *Protein Purification: From Molecular Mechanisms to Large-Scale Processes*, M. R. Ladisch, R. C. Willson, C-D. C. Painton, and S. E. Builder, *Am. Chem. Soc*., Washington, DC, 169–180.

Staff Reporter, 1997, American home unit gets approval to sell clot drug, *Wall Street J*., B14 (February 13).

Sutton, C., 1992, Lectin affinity chromatography, in *Protein purification methods: A practical approach*, E. L. V. Harris and S. Angal, eds., IRL Press, Oxford, 268–282.

Tanouye, E., 1996, Success of AIDS drug has Merck fighting to keep up the pace, *Wall Street J*., A1, A6 (November 5).

Tokatlidis, K., S. Salamitou, P. Beguin, P. Dhurjati, and J.-P. Aubert, 1991, Interaction of the duplicated segment by *Clostridium thermocellum* cellulases with cellulosome components, *FEBS*, 291(2), 185–188.

Tomme, P., N. R. Gilkes, R. C. Miller, A. J. Warren, and D. G. Kilburn, 1994, An internal cellulose-binding domain mediates adsorption of an engineered bifunctional xylanase/cellulose, *Protein Engineering*, 7(1), 117–123.

Tomme, P., R. A. J. Warren, R. C. Miller Jr., D. G. Kilburn, and N. R. Gilkes, 1995a, in *Enzymatic Degradation of Insoluble Polysaccharides*, J. N. Saddler and M. Penner, eds., Am. Chem. Soc., Washington, DC, 142–143.

Tomme, F., D. P. Driver, E. A. Amandoron, R. C. Miller, R. Antony, J. Warren, and D. G. Kilburn, 1995b, Comparison of a fungal (Family I) and bacterial (Family II) cellulose binding domain, Bacteriology, 177(15), 4356–4363.

Voloch, M., M. R. Ladisch, V. W. Rodwell, and G. T. Tsao, 1981, Separation of meso and racemic 2, 3 butanediol by aqueous liquid chromatography, *Biotechnol. Bioeng.*, 23, 1289–1296.

Waldholz, M., May 11, 2000, Makers of AIDS Drugs Agree to Slash Prices, *Wall Street Journal*, A1, A12.

Waldholz, M., Feb. 5, 1999, Drug-resistan HIV becomes more widespread, *Wall Street Journal*, B5.

Waldholz, M., and R. T. King, Feb. 9, 1999, Rhone-Poulenc Testing Aids Vaccine in Africa; VaxGen plans Thai trial, *Wall Street Journal*, B2.

Waldholz, M., and E. Tanouye, Jan. 25, 1999, HIV patients who quit drug combo and thrive give researchers clues, *Wall Street Journal*, A1, A8.

Waldholz, M., Nov. 19, 1998, Some states refuse to pay for HIV drug unless maker Dupont agrees to discounts, *Wall Street Journal*, A8.

Waldholz, M., June 30, 1998, New drug mix would simplify HIV therapy, *Wall Street Journal*, B1, B7.

Waldholz, M., March 26, 1998, Aids-drug cocktails in use since 1996 cause steep drop in deaths, study finds, *Wall Street Journal*, A10.

Waldholz, M., 1996a, Merck's AIDS drug is cleared, priced 30% less, *Wall Street J.*, B5 (March 15).

Waldholz, M., 1996b, Some AIDS cases defy new drug cocktails, *Wall Street J.*, B1, B9 (October 10).

Waldholz, M., 1996c, For first time, drug cocktail seems to eliminate HIV in its hiding place, *Wall Street J.*, B6 (November 7).

Waldholz, M., 1995, Chiron's IL-2 boosts cells hurt by AIDS, new study reports, *Wall Street J.*,(March 2).

Wallis, M., S. L. Howell, and K. W. Taylor, 1985, The biochemistry of polypeptide hormones, John Wiley & Sons, Chichester, 184–221.

Watson, J. D., M. Gilman, J. Witowski, and M. Zoller, 1992, *Recombinant DNA*, 2nd ed., Scientific American Books, Freeman, New York.

Weber, D. V., R. F. Keeny, P. C. Familletti, and P. Bailon, 1988, Medium-scale ligand-affinity purification of two soluble forms of human interleukin-2 receptor, *J. Chromatogr.*, *431*, 55–63.

Weiss, R. A., 1996, HIV receptors and the pathogenesis of AIDS, *Science*, *272*, 1885–1886.

Wheelwright, S. M., 1991, Protein Purification: Design and Scale-up of Downstream Processing, Hanser Publishers, Munich, 205–207.

Wilson, E. K., 1996, AIDS conference highlights hope of drug cocktails, chemokine research, *Chem. Eng. News*, 74(31), 42–46.

Wolinsky, S. M., B. T. M. Korber, A. U. Neumann, M. Daniels, K. J. Kuntsman, A. J. Whetsell, M. R. Furtado, Y. Cao, D. D. Ho, J. T. Safrit, and R. A. Koup, 1996a, Adaptive evolution of human immunodeficiency virus—Type I during the natural course of infection, *Science*, 272(5261), 537–542.

Wolinsky, S. M., K. J. Kunstman, J. T. Safrit, R. A. Koup, A. U. Neumann, and B. T. M. Korber, 1996b, Response to technical comments, *Science*, 274(5289), 1010–1011.

Wyatt, R., and J. Sodroski, 1998, The HIV-1 envelope glycoproteins: Fusogens, antigens, and immunogens, *Science*, 280(5371), 1884–1888.

Yang, Y., A. Velayudhan, C. M. Ladisch, and M. R. Ladisch, 1992, Protein Chromatography Using a Continuous Stationary Phase, *J. Chromatogr.*, *598*, 169–180.

Yang, C. Y., S.-J. Cai, T. J. Holzer, R. M. Novak, and C. Pidgeon, 1996, Introduction to human immunodeficiency viruses, in *Tutorials for the Biomedical Sciences: Animations, Simulations, and Calculations Using Mathematica®*, C. Pidgeon, ed., VCH, New York, 161–175.

PROBLEMS

9.1 Why are on-bead screens more relevant than a screen based on a liquid solution method, for screening the potential of carbohydrates to bind with some types of proteins?

9.2 What is the role of BSA (in solution with the target protein) in a binding assay where the ligand is attached to a bead?

9.3 What is p24 antigen, and why does it spike and then level off after a person is newly infected with HIV? What is an ELISA and how does it work in detecting this protein?

9.4 Consider the hypothetical case where a contaminating protein that binds goat IgG adsorbs on the surface of the solid to which the human antibody to p24 is attached. A reagent blank that does *not* contain p24 antigen still gives a yellow color at the completion of the assay. Explain the result. How could this false positive be avoided?

9.5 How do coreceptors (affinity ligands) facilitate HIV binding and infection?

9.6 What is a protease inhibitor, and how does it prevent cleavage of gag- and pol-encoded viral proteins?

9.7 A new type of affinity chromatography stationary phase has been proposed. It would consist of a protease inhibitor immobilized onto the surface of a Sephadex gel. The purpose of this affinity support would be to capture and concentrate the proteases that may be present in samples of HIV patients. Would you expect for this method to be successful? Why or why not?

AUTHOR INDEX

SUBJECT INDEX